WASTEWATER TREATMENT FUNDAMENTALS III

ADVANCED TREATMENT

2022

Water Environment Federation
601Wythe Street
Alexandria, VA 22314–1994 USA
http://www.wef.org

Water Professionals International
9400 Plum Drive, Suite 160
Urbandale, Iowa 50322
http://www.gowpi.org

ISBN: 978-1-57278-411-6

About WEF

The Water Environment Federation (WEF) is a not-for-profit technical and educational organization of 30,000 individual members and 75 affiliated Member Associations representing water quality professionals around the world. Since 1928, WEF and its members have protected public health and the environment. As a global water sector leader, our mission is to connect water professionals; enrich the expertise of water professionals; increase the awareness of the impact and value of water; and provide a platform for water sector innovation. To learn more, visit www.wef.org.

About WPI

Water and wastewater treatment professionals are on the front lines of protecting public health and the environment. Since 1972, Water Professionals International (formerly known as the Association of Boards of Certification) has been the central water industry authority that ensures these women and men are prepared to meet the standards that their communities can trust in through our testing and certification services.

WPI includes 100+ certifying authorities representing more than 40 states, 10 Canadian provinces and territories, and several international and tribal programs. WPI has assisted in the certification of more than 500,000 water and wastewater treatment professionals, including those who have earned a WPI Certification or Professional Operator (PO) Certification.

Prepared by

Chapter 1 Industrial Wastewater and Pretreatment

Alison Ling

Sara Arabi, PhD, PE, BCEE

Carlos D. Claros

Tyson Schlect, PE

Tat Ebihara

Hemanth Haft

Chapter 2 Advanced Safety Considerations

Nicole Zerbel Ivers

Tim Page-Bottorff

Megan Yoo Schnieder, PE

Chapter 3 Physical and Chemical Treatment

Raj Chavan

Kamrun Ahmadi

Kristin O'Neill, Brown and Caldwell

Chapter 4 Advanced Activated Sludge

John C. Kabouris, PhD, PE, ENV SP, Stantec

Thomas Nogaj, PhD, PE, Stantec

Parsa Pezeshk

Chapter 5 Nontraditional Disinfection

Gary Hunter, PE, BCEE, ENVSP

Leonard W. Casson, PhD, PE, BCEE, ENVSP

Bernadette Drouhard

Isaiah Spencer-Williams

Chapter 6 Membranes

Kerry J. Howe, PhD, PE, BCEE, Howe Water Science LLC

Elise Chen, PE, Trussell Technologies, Inc.

Rodrigo Tackaert, PE, Trussell Technologies, Inc.

Aleksey N. Pisarenko, PhD, Trussell Technologies, Inc.

Chapter 7 Water Reuse

Anya Kaufmann, PE, Trussell Technologies, Inc.

Rodrigo Tackaert, PE, Trussell Technologies, Inc.

Eileen Idica, PhD, PE, Trussell Technologies, Inc.

Aleksey N. Pisarenko, PhD, Trussell Technologies, Inc.

Chapter 8 Characterization and Sampling of Sludge

Mahmudul Hasan

Adrian Romero

Arifur Rahman, PhD, PE

Chapter 9 Management of Solids

Kari Brisolara

Rachel Watson

Keerthisaranya Palanisamy

Chapter 10 Additional Stabilization Methods

Anthony Tartaglione

James Tiberius Schiera, PE

Jay S. Kemp

Jorj Long

Chapter 11 Odor Control

Chriso Petropoulou

Mike Hickey

Jesse Varsho

Chapter 12 Instrumentation

Sidney Innerebner, PhD, PE, CWP, PO, WEF Fellow, Indigo Water Group

Stacy J. Passaro, PE, WEF Fellow, Passaro Engineering, LLC

EURING Oliver Grievson, CEng, CEnv, CSci, CWEM, FCIWEM, FIEnvSc, FIET, FInstMC, Z-Tech Control Systems

Melody White

Chapter 13 Supervisory Control and Data Acquisition (SCADA) Systems

Bob Daly

Gerald Di Noia

Chapter 14 Leadership and Management

Tricia H. Kilgore, PE, Beaufort-Jasper Water & Sewer Authority

Melissa Darr, PE

Dan O'Sullivan

Ryan Lacharity, MASc, PEng, Region of Waterloo

Christel Dyer

Aaron J. Rivard, Genesee County Drain Commissioner: Water & Waste Services

With assistance and review provided by

Chandra Baker
Vanessa Borkowski, PE
Chein-Chi Chang, PhD, PE, Chang Tech International
James P. Cooper, PE
Frank DeOrio, PO
Houssam B. Eljerdi, Chemical Engineer, PE
Richard Finger
Val S. Frenkel, PhD, PE, DWRE, GREELEY and HANSEN
Ronald Gehr
Georgine Grissop, PE, BCEE
Vaughan Harshman, PE
Marialena Hatzigeorgiou, PE
Sidney Innerebner, PhD, PE, CWP, PO, WEF Fellow, Indigo Water Group
Matthew Jalbert
Jeannette Klamm
Dale Kocarek
Anna Lacourt, MEng, PEng
(Emy) Wenxin Liu, PhD, PE
Yanjin Liu, PhD, PE
Jennifer Loudon
Indra Maharjan, Ontario Clean Water Agency
Karthik Manchala
Andy O'Neil, PO
Stacy J. Passaro, PE, WEF Fellow, Passaro Engineering, LLC
Natalia Perez, PE, New York City Department of Environmental Protection (NYCDEP)
Lance Salerno
Frederick Tack, PE, D.WRE
Steve Walker, CWP

Under the direction of the **Technical Practice Committee**

Andrew R. Shaw, PhD, PE, *Chair*
Jon Davis, *Vice-Chair*
Dan Medina, PhD, PE, *Past Chair*

H. Azam, PhD, PE
G. Baldwin, PE, BCEE
S. Basu, PhD, PE, BCEE, MBA
P. Block, PhD
C.-C. Chang, PhD, PE
R. Chavan, PhD, PE, PMP
A. Deines
M. DeVuono, PE, CPESC, LEED AP BD+C
N. Dons, PE
T. Dupuis, PE
T. Gellner, PE
C. Gish
V. Harshman, PE
G. Heath, PE
M. Hines
M. Johnson
N.J.R. Kraakman, Ir., CEng.
J. Loudon
C. Maher
M. Mulcare
C. Muller, PhD, PE
T. Page-Bottorff
A. Rahman, PhD, PE
J. Reina
V. Sundaram, PhD, PE
M. Tam, PE
E. Toot-Levy

Contents

Water Professionals International Formulas

Wastewater Treatment, Collection, Industrial Waste, & Wastewater Laboratory Exams

*Pie wheel format for this equation is shown at the end of the formulas.

$$\text{Alkalinity, mg/L as CaCO}_3 = \frac{\text{(Titrant Volume, mL)(Acid Normality)(50 000)}}{\text{Sample Volume, mL}}$$

$$\text{Amps} = \frac{\text{Volts}}{\text{Ohms}}$$

$$\text{Area of Circle*} = (0.785)(\text{Diameter}^2)$$

$$\text{Area of Circle} = (3.14)(\text{Radius}^2)$$

$$\text{Area of Cone (lateral area)} = (3.14)(\text{Radius})\sqrt{\text{Radius}^2 + \text{Height}^2}$$

$$\text{Area of Cone (total surface area)} = (3.14)(\text{Radius})(\text{Radius} + \sqrt{\text{Radius}^2 + \text{Height}^2})$$

$$\text{Area of Cylinder (total exterior surface area)} = [\text{End \#1 SA}] + [\text{End \#2 SA}] + [(3.14)(\text{Diameter})(\text{Height or Depth})]$$

Where SA = surface area

$$\text{Area of Rectangle*} = (\text{Length})(\text{Width})$$

$$\text{Area of Right Triangle*} = \frac{(\text{Base})(\text{Height})}{2}$$

$$\text{Average (arithmetic mean)} = \frac{\text{Sum of All Terms}}{\text{Number of Terms}}$$

$$\text{Average (geometric mean)} = [(X_1)(X_2)(X_3)(X_4)(X_n)]^{1/n}$$

The *nth* root of the product of *n* numbers

$$\text{Biochemical Oxygen Demand (seeded), mg/L} = \frac{[(\text{Initial DO, mg/L}) - (\text{Final DO, mg/L}) - (\text{Seed Correction, mg/L})][300\text{ mL}]}{\text{Sample Volume, mL}}$$

$$\text{Biochemical Oxygen Demand (unseeded), mg/L} = \frac{[(\text{Initial DO, mg/L}) - (\text{Final DO, mg/L})][300\text{ mL}]}{\text{Sample Volume, mL}}$$

$$\text{Blending or Three Normal Equation} = (C_1 \times V_1) + (C_2 \times V_2) = (C_3 \times V_3)$$

Where $V_1 + V_2 = V_3$; C = concentration, V = volume or flow; Concentration units must match; Volume units must match

$$\text{\# CFU/100 mL} = \frac{[(\text{\# of Colonies on Plate})(100)]}{\text{Sample Volume, mL}}$$

$$\text{Chemical Feed Pump Setting, \% Stroke} = \frac{\text{Desired Flow}}{\text{Maximum Flow}} \times 100\%$$

$$\text{Chemical Feed Pump Setting, mL/min} = \frac{(\text{Flow, mgd})(\text{Dose, mg/L})(3.785\text{ L/gal})(1\,000\,000\text{ gal/mil. gal})}{(\text{Feed Chemical Density, mg/mL})(\text{Active Chemical, \% express as a decimal})(1440\text{ min/d})}$$

$$\text{Chemical Feed Pump Setting, mL/min} = \frac{(\text{Flow, m}^3\text{/d})(\text{Dose, mg/L})}{(\text{Feed Chemical Density, g/cm}^3)(\text{Active Chemical, \% express as a decimal})(1440\text{ min/d})}$$

Circumference of Circle = (3.14)(Diameter)

$$\text{Composite Sample Single Portion} = \frac{(\text{Instantaneous Flow})(\text{Total Sample Volume})}{(\text{Number of Portions})(\text{Average Flow})}$$

$$\text{Cycle Time, min} = \frac{\text{Storage Volume, gal}}{(\text{Pump Capacity, gpm}) - (\text{Wet Well Inflow, gpm})}$$

$$\text{Cycle Time, min} = \frac{\text{Storage Volume, m}^3}{(\text{Pump Capacity, m}^3/\text{min}) - (\text{Wet Well Inflow, m}^3/\text{min})}$$

$$\text{Degrees Celsius} = \frac{(°\text{F} - 32)}{1.8}$$

Degrees Fahrenheit = (°C)(1.8) + 32

$$\text{Detention Time} = \frac{\text{Volume}}{\text{Flow}}$$ *Units must be compatible*

Dilution or Two Normal Equation = $(C_1 \times V_1) = (C_2 \times V_2)$ *Where* C = Concentration, V = volume or flow; Concentration units must match; Volume units must match

Electromotive Force, V* = (Current, A)(Resistance, ohm - Ω)

$$\text{Feed Rate, lb/d*} = \frac{(\text{Dosage, mg/L})(\text{Flow, mgd})(8.34\ \text{lb/gal})}{\text{Purity, \% expressed as a decimal}}$$

$$\text{Feed Rate, kg/d*} = \frac{(\text{Dosage, mg/L})(\text{Flow rate, m}^3/\text{d})}{(\text{Purity, \% expressed as a decimal})(1000)}$$

$$\text{Filter Backwash Rate, gpm/sq ft} = \frac{\text{Flow, gpm}}{\text{Filter Area, sq ft}}$$

$$\text{Filter Backwash Rate, L/(m}^2\text{·s)} = \frac{\text{Flow, L/s}}{\text{Filter Area, m}^2}$$

$$\text{Filter Backwash Rise Rate, in./min} = \frac{(\text{Backwash Rate, gpm/sq ft})(12\ \text{in./ft})}{7.48\ \text{gal/cu ft}}$$

$$\text{Filter Backwash Rise Rate, cm/min} = \frac{\text{Water Rise, cm}}{\text{Time, min}}$$

$$\text{Filter Yield, lb/sq ft/hr} = \frac{(\text{Solids Loading, lb/d})(\text{Recovery, \% expressed as a decimal})}{(\text{Filter Operation, hr/d})(\text{Area, sq ft})}$$

$$\text{Filter Yield, kg/m}^2\text{·h} = \frac{(\text{Solids Concentration, \% expressed as a decimal})(\text{Sludge Feed Rate, L/h})(10)}{(\text{Surface Area of Filter, m}^2)}$$

Flowrate, cu ft/sec* = (Area, sq ft)(Velocity, ft/sec)

Flowrate, m³/sec* = (Area, m²)(Velocity, m/s)

$$\text{Food-to-Microorganism Ratio} = \frac{\text{BOD}_5\text{, lb/d}}{\text{MLVSS, lb}}$$

$$\text{Food-to-Microorganism Ratio} = \frac{\text{BOD}_5\text{, kg/d}}{\text{MLVSS, kg}}$$

Force, lb* = (Pressure, psi)(Area, sq in.)

Force, newtons* = (Pressure, Pa)(Area, m^2)

$$\text{Hardness, as mg CaCO}_3\text{/L} = \frac{(\text{Titrant Volume, mL})(1000)}{\text{Sample Volume, mL}}$$

Only when the titration factor is 1.00 of ethylenediaminetetraacetic acid (EDTA)

$$\text{Horsepower, Brake, hp} = \frac{(\text{Flow, gpm})(\text{Head, ft})}{(3960)(\text{Pump Efficiency, \% expressed as a decimal})}$$

$$\text{Horsepower, Brake, kW} = \frac{(9.8)(\text{Flow, m}^3\text{/s})(\text{Head, m})}{(\text{Pump Efficiency, \% expressed as a decimal})}$$

$$\text{Horsepower, Motor, hp} = \frac{(\text{Flow, gpm})(\text{Head, ft})}{(3960)(\text{Pump Efficiency, \% expressed as a decimal})(\text{Motor Efficiency, \% expressed as a decimal})}$$

$$\text{Horsepower, Motor, kW} = \frac{(9.8)(\text{Flow, m}^3\text{/s})(\text{Head, m})}{(\text{Pump Efficiency, \% expressed as a decimal})(\text{Motor Efficiency, \% expressed as a decimal})}$$

$$\text{Horsepower, Water, hp} = \frac{(\text{Flow, gpm})(\text{Head, ft})}{3960}$$

Horsepower, Water, kW = (9.8)(Flow, m^3/s)(Head, m)

$$\text{Hydraulic Loading Rate, gpd/sq ft} = \frac{\text{Total Flow Applied, gpd}}{\text{Area, sq ft}}$$

$$\text{Hydraulic Loading Rate, m}^3\text{/(m}^2\cdot\text{d)} = \frac{\text{Total Flow Applied, m}^3\text{/d}}{\text{Area, m}^2}$$

Loading Rate, lb/d* = (Flow, mgd)(Concentration, mg/L)(8.34 lb/gal)

$$\text{Loading Rate, kg/d*} = \frac{(\text{Flow, m}^3\text{/d})(\text{Concentration, mg/L})}{1000}$$

Mass, lb* = (Volume, mil. gal)(Concentration, mg/L)(8.34 lb/gal)

$$\text{Mass, kg*} = \frac{(\text{Volume, m}^3)(\text{Concentration, mg/L})}{1000}$$

$$\text{Mean Cell Residence Time or Solids Retention Time, days} = \frac{(\text{Aeration Tank TSS, lb}) + (\text{Clarifier TSS, lb})}{(\text{TSS Wasted, lb/d}) + (\text{Effluent TSS, lb/d})}$$

Milliequivalent = (mL)(Normality)

$$\text{Molarity} = \frac{\text{Moles of Solute}}{\text{Liters of Solution}}$$

$$\text{Motor Efficiency, \%} = \frac{\text{Brake hp}}{\text{Motor hp}} \times 100\%$$

$$\text{Normality} = \frac{\text{Number of Equivalent Weights of Solute}}{\text{Liters of Solution}}$$

$$\text{Number of Equivalent Weights} = \frac{\text{Total Weight}}{\text{Equivalent Weight}}$$

$$\text{Number of Moles} = \frac{\text{Total Weight}}{\text{Molecular Weight}}$$

$$\text{Organic Loading Rate-RBC, lb SBOD}_5\text{/1000 sq ft/d} = \frac{\text{Organic Load, lb SBOD}_5\text{/d}}{\text{Surface Area of Media, 1000 sq ft}}$$

$$\text{Organic Loading Rate-RBC, kg SBOD}_5\text{/m}^2\cdot\text{d} = \frac{\text{Organic Load, kg SBOD}_5\text{/d}}{\text{Surface Area of Media, m}^2}$$

$$\text{Organic Loading Rate-Trickling Filter, lb BOD}_5\text{/1000 cu ft/d} = \frac{\text{Organic Load, lb BOD}_5\text{/d}}{\text{Volume, 1000 cu ft}}$$

$$\text{Organic Loading Rate-Trickling Filter, kg/m}^3\cdot\text{d} = \frac{\text{Organic Load, kg BOD}_5\text{/d}}{\text{Volume, m}^3}$$

$$\text{Oxygen Uptake Rate or Oxygen Consumption Rate, mg/L}\cdot\text{min} = \frac{\text{Oxygen Usage, mg/L}}{\text{Time, min}}$$

$$\text{Population Equivalent, Organic} = \frac{\text{(Flow, mgd)(BOD, mg/L)(8.34 lb/gal)}}{\text{0.17 lb BOD/d/person}}$$

$$\text{Population Equivalent, Organic} = \frac{\text{(Flow, m}^3\text{/d)(BOD, mg/L)}}{\text{(1000)(0.077 kg BOD/d}\cdot\text{person)}}$$

$$\text{Power, kW} = \frac{\text{(Flow, L/s)(Head, m)(9.8)}}{1000}$$

$$\text{Recirculation Ratio-Trickling Filter} = \frac{\text{Recirculated Flow}}{\text{Primary Effluent Flow}}$$

$$\text{Reduction of Volatile Solids, \%} = \left(\frac{\text{VS in} - \text{VS out}}{\text{VS in} - (\text{VS in} \times \text{VS out})}\right) \times 100\%$$

All information (In and Out) must be in decimal form

$$\text{Removal, \%} = \left(\frac{\text{In} - \text{Out}}{\text{In}}\right) \times 100\%$$

$$\text{Return Rate, \%} = \frac{\text{Return Flowrate}}{\text{Influent Flowrate}} \times 100\%$$

$$\text{Return Sludge Rate-Solids Balance, mgd} = \frac{\text{(MLSS, mg/L)(Flowrate, mgd)}}{\text{(RAS Suspended Solids, mg/L)} - \text{(MLSS, mg/L)}}$$

$$\text{Slope, \%} = \frac{\text{Drop or Rise}}{\text{Distance}} \times 100\%$$

$$\text{Sludge Density Index} = \frac{100}{\text{SVI}}$$

$$\text{Sludge Volume Index, mL/g} = \frac{\text{(SSV}_{30}\text{, mL/L)(1000 mg/g)}}{\text{MLSS, mg/L}}$$

$$\text{Solids, mg/L} = \frac{\text{(Dry Solids, g)(1 000 000)}}{\text{Sample Volume, mL}}$$

$$\text{Solids Capture, \% (Centrifuges)} = \left[\frac{\text{Cake TS, \%}}{\text{Feed Sludge TS, \%}}\right] \times \left[\frac{\text{(Feed Sludge TS, \%)} - \text{(Centrate TSS, \%)}}{\text{(Cake TS, \%)} - \text{(Centrate TSS, \%)}}\right] \times 100\%$$

$$\text{Solids Concentration, mg/L} = \frac{\text{Weight, mg}}{\text{Volume, L}}$$

$$\text{Solids Loading Rate, lb/sq ft/d} = \frac{\text{Solids Applied, lb/d}}{\text{Surface Area, sq ft}}$$

$$\text{Solids Loading Rate, kg/m}^2\text{·d} = \frac{\text{Solids Applied, kg/d}}{\text{Surface Area, m}^2}$$

Solids Retention Time: *see Mean Cell Residence Time*

$$\text{Specific Gravity} = \frac{\text{Specific Weight of Substance, lb/gal}}{\text{8.34 lb/gal}}$$

$$\text{Specific Gravity} = \frac{\text{Specific Weight of Substance, kg/L}}{\text{1.0 kg/L}}$$

$$\text{Specific Oxygen Uptake Rate or Respiration Rate, (mg/g)/h} = \frac{\text{(OUR, mg/L}\cdot\text{min)(60 min)}}{\text{(MLVSS, g/L)(1 h)}}$$

$$\text{Surface Loading Rate or Surface Overflow Rate, gpd/sq ft} = \frac{\text{Flow, gpd}}{\text{Area, sq ft}}$$

$$\text{Surface Loading Rate or Surface Overflow Rate, L/m}^2\text{·d} = \frac{\text{Flow, L/d}}{\text{Area, m}^2}$$

$$\text{Total Solids, \%} = \frac{\text{(Dried Weight, g)} - \text{(Tare Weight, g)}}{\text{(Wet Weight, g)} - \text{(Tare Weight, g)}} \times 100\%$$

$$\text{Velocity, ft/sec} = \frac{\text{Flowrate, cu ft/s}}{\text{Area, sq ft}}$$

$$\text{Velocity, ft/sec} = \frac{\text{Distance, ft}}{\text{Time, sec}}$$

$$\text{Velocity, m/s} = \frac{\text{Flowrate, m}^3\text{/s}}{\text{Area, m}^2}$$

$$\text{Velocity, m/s} = \frac{\text{Distance, m}}{\text{Time, s}}$$

$$\text{Volatile Solids, \%} = \left[\frac{\text{(Dry Solids, g)} - \text{(Fixed Solids, g)}}{\text{(Dry Solids, g)}}\right] \times 100\%$$

Volume of Cone* = (1/3)(0.785)(Diameter^2)(Height)

Volume of Cylinder* = (0.785)(Diameter^2)(Height)

Volume of Rectangular Tank* = (Length)(Width)(Height)

$$\text{Water Use, gpcd} = \frac{\text{Volume of Water Produced, gpd}}{\text{Population}}$$

$$\text{Water Use, L/cap·d} = \frac{\text{Volume of Water Produced, L/d}}{\text{Population}}$$

$$\text{Watts (AC circuit)} = (\text{Volts})(\text{Amps})(\text{Power Factor})$$

$$\text{Watts (DC circuit)} = (\text{Volts})(\text{Amps})$$

$$\text{Weir Overflow Rate, gpd/ft} = \frac{\text{Flow, gpd}}{\text{Weir Length, ft}}$$

$$\text{Weir Overflow Rate, L/m·d} = \frac{\text{Flow, L/d}}{\text{Weir Length, m}}$$

$$\text{Wire-to-Water Efficiency, \%} = \frac{\text{Water hp}}{\text{Motor hp}} \times 100\%$$

$$\text{Wire-to-Water Efficiency, \%} = \frac{(\text{Flow, gpm})(\text{Total Dynamic Head, ft})(0.746\ \text{kW/hp})(100\%)}{(3960)(\text{Electrical Demand, kW})}$$

Pie Wheels

- To find the quantity above the horizontal line: multiply the pie wedges below the line together.
- To solve for one of the pie wedges below the horizontal line: cover that pie wedge, then divide the remaining pie wedge(s) into the quantity above the horizontal line.
- Given units must match the units shown in the pie wheel.
- When US and metric units or values differ, the metric is shown in parentheses, e.g. (m^2).

Area of Circle

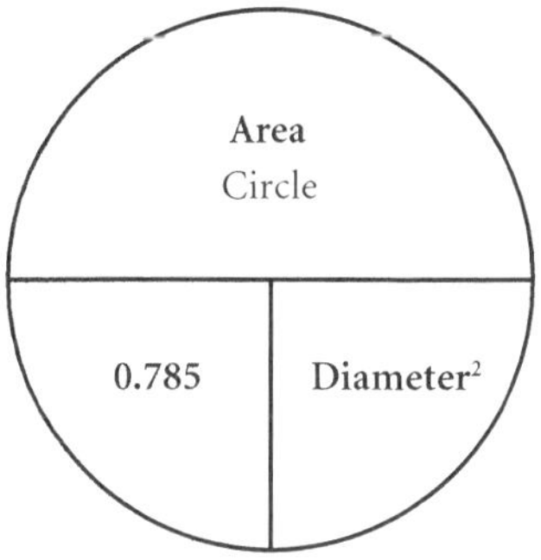

Area of Rectangle

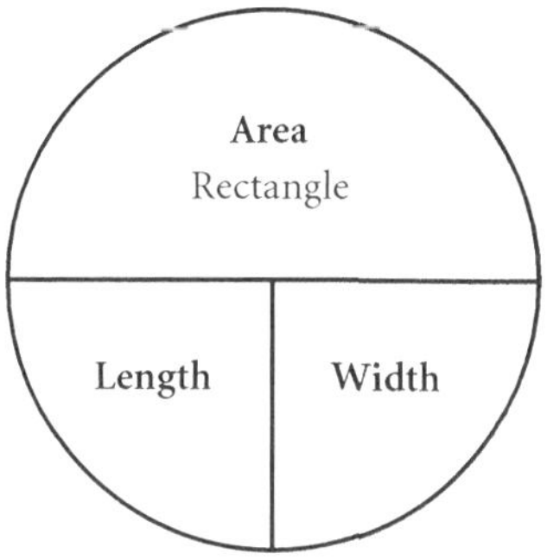

Area of Right Triangle

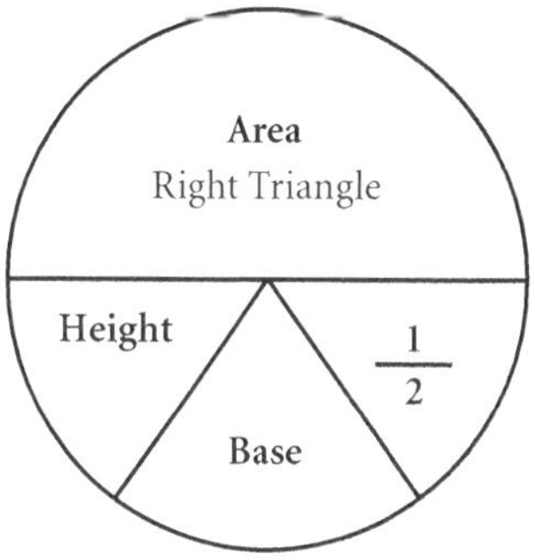

Electromotive Force (EMF), Volts

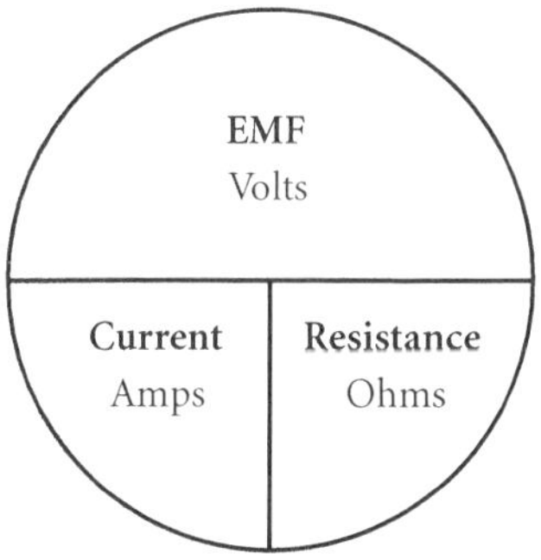

Feed Rate, lbs/day (kg/day)

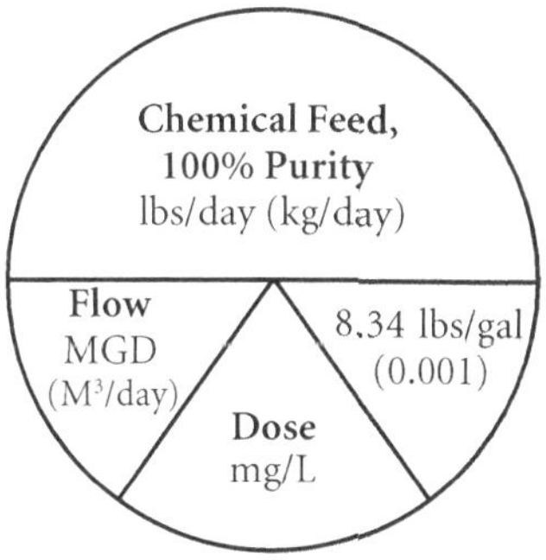

Flow Rate, ft³/sec (m³/sec)

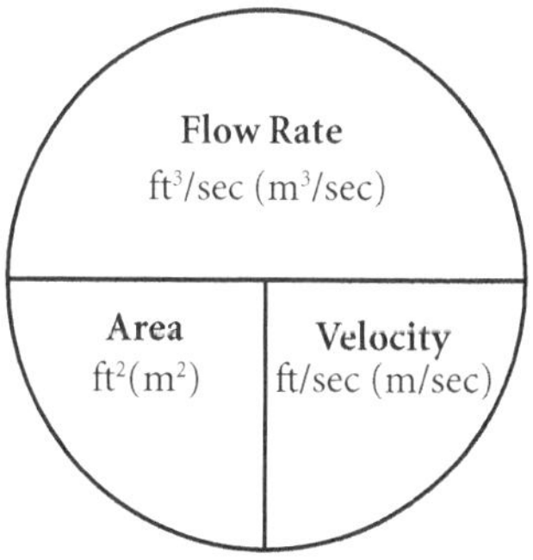

Force, lbs (Newtons)

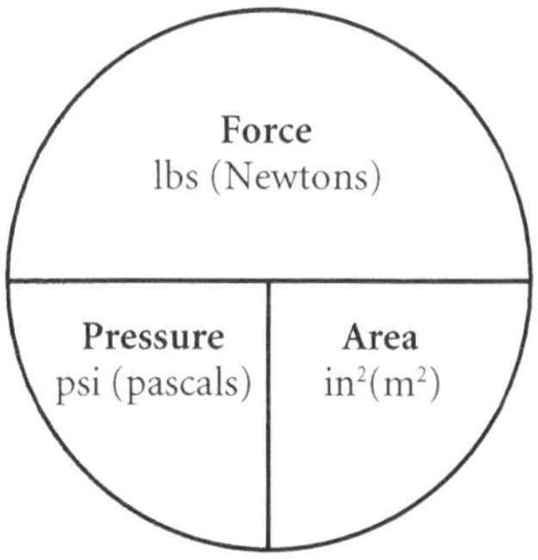

Loading Rate, lbs/day (kg/day)

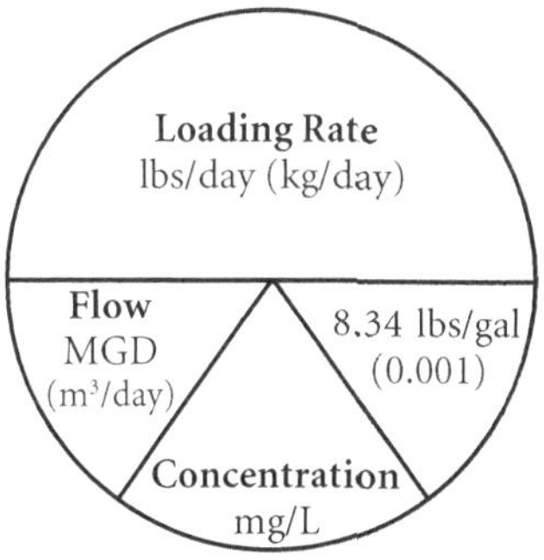

Mass, lbs (kg)

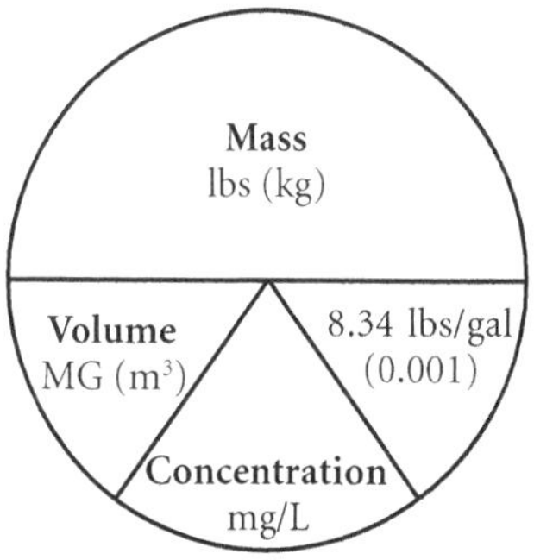

Volume of Cone

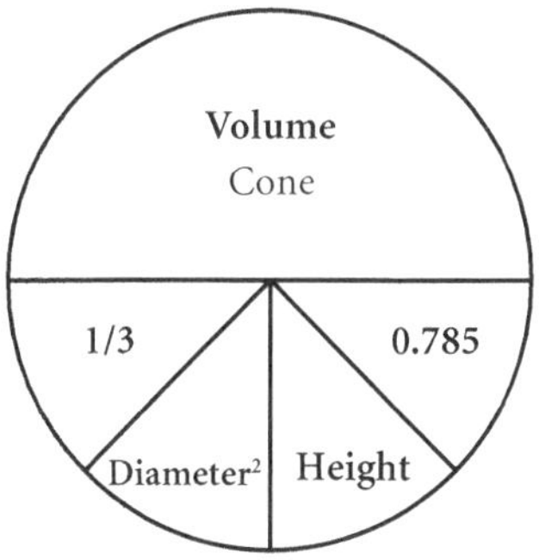

Volume of Cylinder

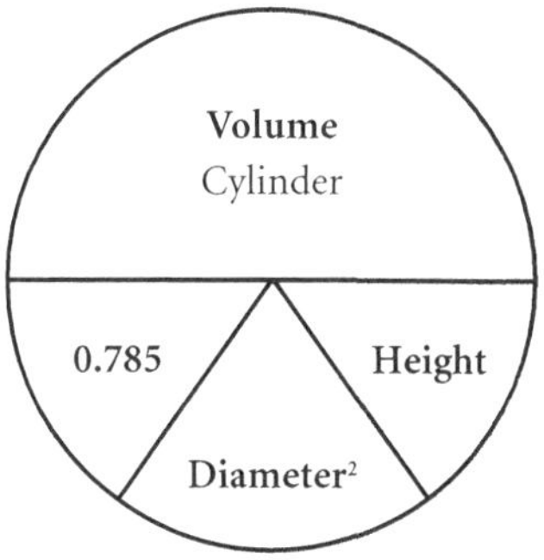

Volume of Rectangular Tank

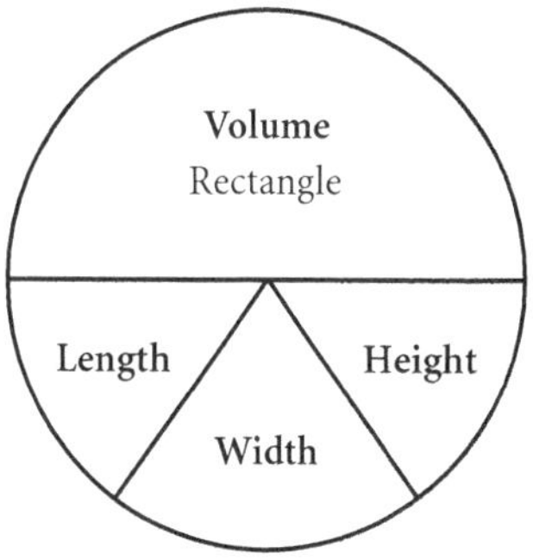

Conversion Factors

1 ac = 4046.9 m^2 or 43 560 sq ft

1 ac ft of water = 326 000 gal

1 atm = 33.9 ft of water
= 10.3 m of water
= 14.7 psi
= 101.3 kPa

1 cfs = 0.646 mgd
= 448.8 gpm

1 cu ft of water = 7.48 gal
= 62.4 lb

1 ft = 0.305 m

1 ft H_2O = 0.433 psi

1 gal (US) = 3.79 L
= 8.34 lb of water

1 gr/gal (US) = 17.1 mg/L

1 ha = 10 000 m^2

1 hp = 0.746 kW
= 746 W
= 33 000 ft lb/min

1 in. = 25.4 mm or 2.54 cm

1 L/s = 0.0864 ML/d

1 lb = 0.454 kg

1 m of water = 9.8 kPa

1 m^2 = 1.19 sq yd

1 m^3 = 1000 kg
= 1000 L
= 264 gal

1 metric ton = 2205 lb

1 mile = 5280 ft

1 mgd = 694 gpm
= 1.55 cfs
= 3.785 ML/d

Population equivalent (PE), hydraulic = 378.5 L/cap·d
= 100 gpd/cap

PE, organic = 0.077 kg BOD/cap·d
= 0.17 lb BOD/cap/d

1 psi = 2.31 ft of water
= 6.89 kPa

1 ton = 2000 lb

1% = 10 000 mg/L

π or pi = 3.14

CHAPTER 1

Industrial Wastewater and Pretreatment

Introduction

Treatment of industrial wastewater is as varied as industry itself. From food and beverage production to metalworking and chemical production, industries produce wastewater streams with a wide variety of flows, contaminant concentrations, and production schedules. Industrial wastewater can be discharged, either with or without pretreatment, to a water resource recovery facility, discharged directly to surface waters or aquifers, or managed off-site by an external company. Historically, both industrial and municipal wastewater treatment systems were referred to as *wastewater treatment plants*, with the term *publicly owned treatment works* (POTW) specifically referring to public facilities. The updated term *water resource recovery facility* (WRRF) has since been adopted by the industry and the Water Environment Federation to reflect efforts to reclaim wastewater flow and constituents for beneficial use. In this chapter, we use the term WRRF to refer to public treatment facilities that may receive industrial wastewater flows. However, the term POTW is still in common use among industrial operators. When referring to industrial treatment facilities, we use the terms *wastewater treatment system* or *wastewater pretreatment system*. Source control measures, including changes in operations or product formulations as well as implementation of water reuse, can also be used to manage industrial wastewater. Figure 1.1 outlines management options for industrial wastewaters. Industrial discharges may be direct or indirect dischargers. Direct dischargers own and operate their own treatment plants. They discharge directly to a stream, river, lake, or groundwater. Indirect discharges may or may not fully or partially treat their wastewater. They discharge to a WRRF via a wastewater collection system.

Source control measures minimize wastewater strength and volume using operational adjustments, like changing manufacturing processes, substituting ingredients, and installing low water use systems.

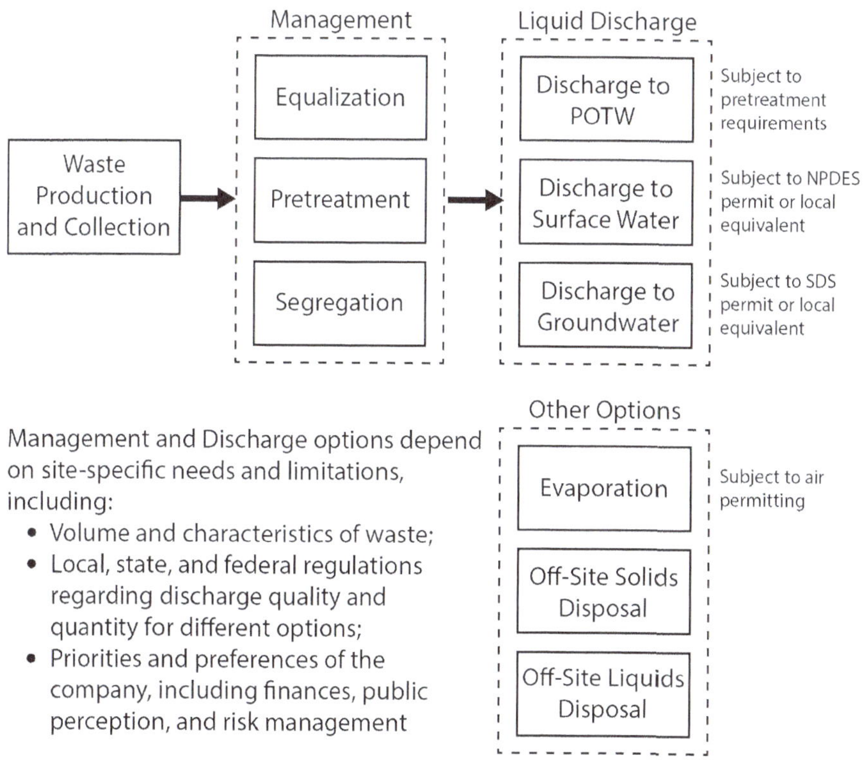

Figure 1.1 Industrial Pretreatment Flowchart

A *slug load* refers to a significant short-term increase in wastewater flow or loading of a specific contaminant reporting to a WRRF, often due to the timing of industrial discharges.

Water resource recovery facilities manage upstream industrial wastewater through pretreatment programs. Industrial wastewater flowrates can vary widely from less than 6 m^3/d (1 gpm) to greater than 20 000 m^3/d (5 mgd) and may be released to the sewer at an irregular frequency. In addition, the potential for industrial wastewater to upset WRRF activated sludge processes can be significant, especially with slug loads. This can be especially concerning for facilities whose discharge flowrates or the amount of contaminant routed to the WRRF changes significantly over time. This chapter is intended to address regulatory considerations for industrial pretreatment and discharge as well as treatment considerations across a wide range of industries. We use the term *industrial wastewater* to refer to liquid industrial wastes.

The content of this chapter is divided into two main categories that consist of the following sections:

- Regulatory Considerations
  - Pretreatment requirements
  - Direct discharge of industrial wastewater
  - Evolving regulations and contaminants of emerging concern
- Treatment Considerations
  - Industrial wastewater sources and characteristics
  - Industrial pretreatment processes

LEARNING OBJECTIVES

Learning objectives for this chapter are outlined below.

- Pretreatment and discharge requirements:
  - Understand the requirements of National Pretreatment Program standards and contaminants of concern.
  - Know typical requirements for sampling and monitoring for pretreatment programs.
  - Understand the requirements for reporting and recordkeeping.
- Evolving regulations and emerging contaminants:
  - Be able to describe characteristics of an emerging contaminant.
  - Know some key emerging organic and inorganic contaminants of concern.
  - Be familiar with regulation and control technologies for industrial per- and poly-fluoroalkyl substances (PFAS).
- Industrial wastewater sources and characteristics:
  - Understand different types of industrial wastewater.
  - Compare and contrast industrial versus municipal waste characteristics.
  - Understand how wastewater is generated for some common sources of industrial wastewater.
  - Know typical biochemical oxygen demand (BOD), total suspended solids (TSS), total Kjeldahl nitrogen (TKN), and total phosphorus (TP) concentrations from key industrial wastewater sources.
  - Understand uniquely industrial pollutants such as metals, polymers, acids, caustics, and inorganic chemical species.
  - Understand the range of effects of industrial waste on municipal treatment performance.
- Industrial pretreatment processes:
  - Identify typical sources and characteristics of industrial wastewater that require pretreatment control technologies.
  - Know several key pretreatment technologies used to remove different parameters.

Purpose and Function

The goal of this chapter is to provide municipal operators with an understanding of industrial wastewater streams that may be discharging to their WRRF and industrial pretreatment programs to manage them. The focus of this chapter is to provide a broad overview as well as specific operational and management strategies to address industrial wastewater streams, with the goal of minimizing effects on WRRFs and receiving waters.

Pretreatment Requirements

NATIONAL PRETREATMENT PROGRAM

The National Pretreatment Program provides the overall framework for joint regulation by local, state, and federal authorities to control pollutants that may pass through or interfere with WRRF treatment processes or deteriorate the wastewater sludge quality. Wastewater operators and company management share the common goal of effectively managing industrial waste to minimize effects to human health and the environment. This section helps you gain familiarity with important terminology and regulations affecting industrial wastewater control technologies and management approaches.

"Pass-through" describes the ability of a pollutant to go through WRRF treatment processes without being removed. These may be described as "pass-through" pollutants or as "passing through" the WRRF. Pass-through pollutants are discharged to the water body receiving WRRF effluent and may affect the WRRF's ability to meet its National Pollutant Discharge Elimination System (NPDES) permit limit. Pass-through can occur when pollutants are discharged to a WRRF that is not designed to treat them or in excess of a WRRF's designed capacity to treat them.

The National Pretreatment Program is a component of the National Pollutant Discharge Elimination System (NPDES) program. It is a cooperative effort of federal, state, and local environmental regulatory agencies established to protect water quality. Similar to how the United States Environmental Protection Agency (U.S. EPA) authorizes the NPDES permit program to state, tribal, and territorial governments to perform permitting, administrative, and enforcement tasks for discharges to surface waters (NPDES program), U.S. EPA and authorized NPDES state pretreatment programs approve local municipalities to perform permitting, administrative, and enforcement tasks for discharges into the WRRFs.

The National Pretreatment Program requires non-domestic dischargers that introduce pollutants to WRRFs to comply with pretreatment standards to ensure the goals of the Clean Water Act (CWA) are attained.

Pretreatment standards and requirements include:

- general and specific prohibitions,
- categorical pretreatment standards, and
- local limits.

The General Pretreatment Regulations for Existing and New Sources of Pollution, presented in 40 CFR Part 403 (U.S. EPA, 1978), establish responsibilities of industry, the public, and various levels of governments to implement Pretreatment Standards. Pretreatment Standards are intended to control pollutants that may pass through or interfere with WRRF processes or that may contaminate wastewater sludge. Control of pollutants before the discharge of wastewater to WRRFs minimizes their adverse effects to WRRFs and downstream waters (U.S. EPA, 2011). The General Pretreatment Regulations require specific WRRFs to establish local pretreatment programs. This includes all larger WRRFs designed to treat flows of more than 19 000 m^3/d (5 mgd) that also receive discharges subject to the pretreatment standards and any smaller WRRFs with significant industrial discharges. These local programs must enforce all national pretreatment standards and requirements in addition to any more stringent local requirements necessary to protect site-specific conditions at the WRRF.

The goal of the WRRF Pretreatment Program is to protect the WRRF by ensuring that industrial users (often referred to as "IUs") comply with applicable pretreatment standards and requirements. Regulatory enforcement generally becomes the responsibility of local control authorities. The General Pretreatment Regulations define the term *control authority* as a WRRF that is authorized to control discharges to its system using an approved pretreatment program. The *approval authority* is responsible for reviewing and approving the pretreatment program and overseeing implementation of the program by the control authorities. Generally, the WRRFs act as the pretreatment control authorities, and either the state or U.S. EPA acts as the pretreatment approval authority. In the absence of an approved WRRF pretreatment program, the state or U.S. EPA approval authority also serves as the control authority. This is common for small WRRFs with few industrial discharges that may struggle to implement a full pretreatment program.

The General Pretreatment Regulations require control authorities to adopt and implement an Enforcement Response Plan (ERP). The ERP regulations establish a framework for WRRFs to formalize procedures to investigate and respond to industrial user noncompliance. With an approved ERP in place, WRRFs can use objective criteria in the regulatory framework to protect their infrastructure against negative effects from industrial discharges. Review of monitoring data and supporting documentation can inform the compliance status of an industrial user's discharge, the quality and quantity of its contribution to the WRRF, and the costs of treating the waste contribution.

Volatilization is the process by which chemicals in the water phase transfer into the gas phase. So, volatilization potential from a safety standpoint means whether harmful gases can come out of the discharged water in sewer pipes or at the WRRF.

Health and safety concerns are considered in the development of local limits. Pollutants that could cause fire and explosion hazards or other worker health and safety concerns are evaluated for volatilization potential and their expected concentration in air. Chapter 2 presents more details on health and safety considerations.

Categorical industrial users are defined as an industrial user subject to national categorical pretreatment standards based on the types of industrial processes used. Four criteria used to define an SIU under National Pretreatment Regulations are

1. An industrial user that discharges an average of 95 m^3/d (25 000 gpd) or more of process wastewater to the WRRF,
2. An industrial user that contributes process wastewater making up 5% or more of the average dry weather hydraulic or organic capacity of the WRRF,
3. An industrial user designated by the control authority as such because of its reasonable potential to adversely affect the WRRF's operation or violate any pretreatment standard or requirement, or
4. An industrial user subject to federal categorical pretreatment standards.

The General Pretreatment Regulations apply to all non-domestic or industrial users that introduce pollutants to a WRRF. Because not all industrial users create problems for WRRFs, U.S. EPA developed designations for significant industrial users (SIUs) and categorical industrial users (CIUs) based on Standard Industrial Classification. The National Pretreatment Program identifies specific requirements that apply to all industrial users, additional requirements that apply to all SIUs, and certain requirements that only apply to CIUs. The objectives of the National Pretreatment Program are achieved by applying and enforcing three types of discharge standards, as outlined in Table 1.1:

- Prohibited discharge standards, which are designed to protect against pass-through and interference generally;
- Categorical pretreatment standards, which are designed to ensure that industrial users implement technology-based controls to limit the discharge of pollutants; and
- Local limits, which address the specific needs and concerns of a WRRF and its receiving waters. A local limit is determined by the WRRF. It limits the mass and concentrations of specific pollutants that may be discharged to the WRRF. A local limit may be set to ensure that a WRRF does not exceed its own discharge permit limits or to protect the use of biosolids. A local limit may result in permit limits for a CIU that are lower than those specified in 40 CFR, Parts 405–471.

Table 1.1 Summary of Pretreatment Standards

	Prohibited Discharge Standards	Categorical Pretreatment Standards	Local Limits
Development	Established at the federal level	Established at the federal level	Developed by the WRRFs
Reference	40 CFR 403.5 (a) & (b)	40 CFR parts 405–471	Requirements for development found in 40 CFR 403.5 (c) & 403.8(f). Local limits are often found in the local sewer use ordinance.
Applicability	All industrial users	CIUs only	Commonly for all industrial users or all SIUs, but depends on the allocation method used when developing limits.
Purpose	Provide general protection of WRRFs. Categorical requirement standards or local limits may be more stringent.	Minimum standards based on available pretreatment technology and pollution prevention measures for controlling non-conventional and toxic pollutants that could cause pass-through, interference, and such at the WRRF. Local limits may be more stringent.	Provides site-specific protection for a WRRF and its receiving waters. Categorical standards may be more stringent.

Prohibited Discharge Standards

Prohibited discharge standards are general, national standards applicable to all industrial users, regardless of whether or not the WRRF has an approved pretreatment program or the industrial user has been issued a discharge permit. These standards are designed to prevent contaminants from passing through the WRRF without treatment, interfering with treatment operations, interfering with biosolids or sludge disposal, damaging infrastructure, or harming operators.

Specifically, the discharge standards include the following prohibitions:

- Pollutants that will create a fire or explosion hazard in the WRRF with a closed-cup flashpoint of less than 60 °C (140 °F) using the test methods specified in 40 CFR Part 261.21;
- Pollutants that will cause corrosive structural damage to the WRRF, but never discharges with pH lower than 5.0, unless the facility is specifically designed for such discharges, per 40 CFR, Part 403 (U.S. EPA, 1978);
- Solid or viscous pollutants in amounts that will cause obstruction of the flow in the WRRF;
- Any pollutant, including oxygen-demanding pollutants (BOD) and suspended solids, discharged at a flowrate or pollutant concentration that interferes with WRRF operation;
- Heat in amounts that will inhibit biological activity in the WRRF, resulting in interference, but never heat in such quantities that the temperature at the WRRF is higher than 40 °C (104 °F), unless specifically authorized;
- Petroleum oil, non-biodegradable cutting oil, or products of mineral oil origin in amounts that will cause interference or pass-through;
- Pollutants that result in the presence of toxic gases, vapors, or fumes within the WRRF, in a quantity that may cause acute worker health and safety problems; and
- Any trucked or hauled pollutants, except at discharge points designated by the WRRF.

Closed-cup flashpoint is the lowest temperature at which a pollutant in the gas phase can ignite in a closed container.

These prohibited discharge standards are intended to provide general protection for WRRFs. However, their lack of specific pollutant limitations creates the need for additional controls, namely categorical pretreatment standards and local limits.

Categorical Discharge Standards

For CIUs, U.S. EPA has established categorical pretreatment standards (for indirect dischargers) for 35 industrial categories including aluminum forming, coil coating, electroplating, ink manufacturing, and iron and steel manufacturing and other similar industries, as listed in 40 CFR.

Local Limits and Specific Utility Considerations

Local limits may be used to translate general prohibitions in the General Pretreatment Regulations into site-specific needs and to control industrial discharges of pollutants not regulated by the categorical standards. A local limit supersedes the categorical pretreatment standard for a given pollutant if it is more stringent than the categorical standard. However, a local limit typically applies at the point of discharge to the sewer system, whereas a categorical pretreatment standard generally applies immediately after treatment. Typically, local limits are developed to regulate the discharge from all SIUs, not just CIUs, and they are usually imposed at the end-of-pipe discharge from an industrial user.

Among the factors that should be considered in developing local limits are the WRRF's treatment efficiency; its compliance with its NPDES permit limits; the condition of the water body that receives its treated effluent; any water quality standards that are applicable to the water body receiving its effluent; the WRRF's retention, use, and disposal of wastewater sludge; and worker health and safety concerns. U.S. EPA has specific guidance for developing local limits (U.S. EPA, 2004).

The following are recommended guidelines for establishing local limits by the control authorities:

- Conduct an industrial waste survey to identify all industrial users that might be subject to the pretreatment program;
- Determine the character and volume of pollutants contributed to the WRRF by these industries;

- Determine which pollutants have a reasonable potential for pass-through, interference, or sludge contamination;
- Conduct a quantitative evaluation to determine the maximum allowable WRRF headworks (influent) loading for at least arsenic, cadmium, chromium, copper, cyanide, lead, mercury, molybdenum, nickel, silver, and zinc;
- Identify additional pollutants of concern;
- Determine contributions from unpermitted sources to determine the maximum allowable WRRF headworks loading from "controllable" industrial sources; and
- Implement a system to ensure that these loadings will not be exceeded.

The methods commonly used to derive local limits are the Maximum Allowable Headworks Loading (MAHL) and Maximum Allowable Industrial Loading (MAIL) methods. The MAIL is the total daily mass that a WRRF can accept from all permitted industrial users and still ensure that the WRRF is protecting against pass-through and interference. Water resource recovery facility data are used to calculate removal efficiencies for each pollutant to back-calculate the MAHLs before applying the most stringent criteria (i.e., water quality, sludge quality, NPDES permit, or pollutant inhibition levels). Subtracting contributions from unpermitted sources, the available industrial loading is then either evenly distributed among the industrial users or allocated on an as-needed basis to those industrial users discharging the pollutant above background levels.

Pretreatment Surcharge Programs

Water resource recovery facilities often charge industrial users a wastewater disposal fee for treating their industrial wastewater. The disposal fee typically consists of three components: a fixed fee, a volumetric fee, and waste surcharge fee, which is based on wastewater quality and quantity. Typically, WRRFs establish surcharges for conventional pollutants that the WRRF can remove, such as BOD and TSS. Surcharges are additional charges to recover the cost to treat wastewater. These are typically assessed when industrial discharge concentrations are above typical WRRF influent concentrations. Ordinances should clearly distinguish between concentrations that require surcharges and local limits.

Both domestic and industrial users should pay for their share of wastewater treatment costs. Surcharges should aim to ensure industrial discharges are paying the proportional cost of treating their wastewater relative to domestic users. A utility should not push operating costs onto industrial users in place of raising domestics rates.

MONITORING AND REPORTING

Identification and monitoring of industrial users are key to any pretreatment program. Monitoring and enforcement activities are initiated by the municipalities based on each industry's potential for affecting the WRRF. Wastewater sampling is conducted at both industrial facilities and WRRFs. Sampling and analysis at an industrial facility is performed for compliance monitoring, process control, and troubleshooting. Water resource recovery facilities specify permit limits that industrial discharges must comply and apply surcharge rates when certain parameters exceed specified concentrations. To ensure that these requirements are met, WRRFs sample wastewater from industrial users to verify compliance and monitor for constituents that may adversely affect the WRRF. For example, bulking can be caused by industrial wastewater with high BOD, low nutrients, and high influent sulfide concentrations that cause filamentous microorganisms to grow. Sampling can also be conducted as part of design steps, including wastewater characterization surveys and treatability testing.

Oxidation–reduction potential is a measure of how oxidized or reduced conditions in a water are. Higher values indicate more oxidized conditions that would reflect more oxygen present and more particulate metals, whereas lower values reflect more reduced conditions with low oxygen and more dissolved metals. Whole effluent toxicity testing is a type of test intended to reflect overall toxicity of a water. These tests involve growing aquatic animals like minnows and water fleas in various dilutions of the discharge water and evaluating its effect on growth and young production.

In addition to the pH, BOD, and TSS requirements, industrial discharge permits may include parameters such as total dissolved solids (TDS), oxidation–reduction potential (ORP), heavy metals, total petroleum hydrocarbons, total organic carbon (TOC), whole effluent toxicity (WET) testing, and a toxicity characteristic leaching procedure (for sludge).

Knowledge of the types of industries contributing to the collection system and the color of their discharges can help the operator identify the source of an upset condition. A few troubleshooting techniques and waste identification methods outlined in *Operation of Water Resource Recovery Facilities* (Manual of Practice 11; Water Environment Federation [WEF], 2017), and relevant to industrial discharges, are presented below.

1. The color of wastewater depends on the amounts and types of dissolved, suspended, and colloidal matter present. Normal raw wastewater is gray. Other colors could indicate the presence of industrial discharges. For example, green, blue, or orange discharges may come from plating operations; red, blue, or yellow discharges are often dyes from textile industry effluents.
2. Fresh wastewater typically produces a musty odor. Other wastewater odors, such as petroleum, solvents, or other abnormal scents, indicate an industrial spill.
3. Grab sampling is done to determine a "slug" discharge's composition and to identify its source and potential effects on WRRF treatment processes.
4. Portable analyzers such as ORP and conductivity sensors are used to monitor for potential toxic slug discharges. Such measurements offer reliable, early warnings of industrial slug discharges as they enter the WRRF. The warnings can sometimes allow the facility operator to take the steps necessary to protect the WRRF from an industrial discharge.
5. Because some of the compounds present in wastewater may be toxic, caution must be used when smelling wastewater, especially when smelling bottled samples. Detection of unusual odors in a facility, particularly in confined areas, requires exercising caution and strict adherence to safety procedures.

Sampling for Pretreatment Programs

An industrial survey is a study conducted by a WRRF. It begins with compiling an inventory of industries and their discharge streams in the WRRF's service area. To complete the survey, this preliminary inventory is updated for each facility using a questionnaire, in some cases followed by a physical inspection. The questionnaire usually seeks specific information about the type of processes and discharges, water usage, chemical usage, and any pretreatment processes.

Sampling and analyses of industrial discharges provide water quality data to support an effective enforcement program. These data serve as the basis for determining compliance with local, state, and national discharge requirements. Before doing a sampling study, facility inspections are needed to select the type and duration of sampling. Inspections can be conducted as scheduled, unscheduled, or random, depending on the situation.

The type of sampling selected for a facility should be based on the type of discharge and the nature of the operations. In general, flow proportional sampling is recommended wherever possible, except for those parameters that cannot be stored for the sampling duration and must be collected as grab samples. However, physical constraints at facilities may make flow proportional sampling difficult. In these cases, properly executed time proportional sampling could be an acceptable substitute.

Flow proportional sampling is conducted by an automated composite device that collects volumes of water in proportion to how much water is being discharged. This method requires a flow meter that dictates when and how the sampler collects water.

The type and complexity of data management systems required to operate an industrial pretreatment program is governed by the number of industrial facilities to be monitored and the amount of information and data to be stored for future use. Information that should be stored includes the following:

- Industry survey data;
- Self-monitoring report data;
- Company enforcement history;
- Reports of WRRF inspections;
- Analytical data from WRRF sampling studies; and
- WRRF influent, effluent, and residuals.

Minimum pretreatment program reporting requirements for industrial users are specified in 40 CFR 403.12. Because WRRFs are responsible for communicating applicable standards and requirements to industrial users and for receiving and analyzing reports, they should understand and implement the industrial user reporting requirements in the General Pretreatment Regulations. Those reporting requirements for SIUs and CIUs are outlined in Table 1.2. In cases where the WRRF itself collects all the information required for a specific report, including flow data, the industrial user is not required to submit that report.

U.S. EPA has published two useful guidance manuals to support WRRF efforts to monitor and audit industrial users. *Industrial User Inspection and Sampling Manual for POTWs* (U.S. EPA, 2017) recommends procedures to be used by WRRF personnel when inspecting or sampling an industrial user's facility. *Control Authority Pretreatment Audit Checklist and Instructions* (U.S. EPA, 2010) provides a framework to

Table 1.2 Summary of Reporting Requirements for Pretreatment Requirements

Reporting for CIUs	Reporting for SIUs
Baseline Monitoring Report [40 CFR 403.12(b)]	Periodic Compliance Report [40 CFR 403.12 (h)]
Compliance Schedule Progress Report [40 CFR 403.12(c)(3)]	Slug Control Plan [40 CFR 403.8(f)(2)(vi)]
90-Day Compliance Report [40 CFR 403.12(d)]	Bypass Report [40 CFR 403.17]
Periodic Compliance Report [40 CFR 403.12 (e)]	
Upset Report [40 CFR 403.16]	

be used by WRRF personnel when auditing an industrial user's facility for compliance. Also, CFR Title 40, Part 136 (U.S. EPA, 2007) presents testing guidelines for the analysis of pollutants, including those pertinent to wastewater sampling. Both may be downloaded from U.S. EPA's National Environmental Publications Internet Site.

Industrial users must notify the WRRF when specific events or changes occur. Notification requirements for industrial users and CIUs are applicable to all industrial users, except for notification of upset and Notification of Middle-Tier Categorical Industrial User, which are applicable to CIUs. These notification requirements for industrial users and CIUs are listed below:

- Notification of Production Level Change in the Equivalent Limit Calculation [40 CFR 403.6(c)(9)]
- Notification of Material/Significant Change in the Alternative Limit Calculation [40 CFR 403.6(e)]
- Notification of Waived Pollutant Present [40 CFR 403.12(e)(2)(vi)]
- Notification of Middle-Tier Categorical Industrial User [40 CFR 403.12(e)(3)(iv)]
- Notification of Potential Problems Including Slug Loading [40 CFR 403.12(f)]
- Noncompliance Notification and Repeat Sampling Report [40 CFR 403.12(g)(2)]
- Notification of Changed Discharge [40 CFR 403.12(j) and 40 CFR 122.42(b)]
- Notification of Discharge of Hazardous Wastes [40 CFR 403.12(p)]
- Notification of Upset [40 CFR 403.16]
- Notification of Bypass [40 CFR 403.17]

Inspections

Water resource recovery facilities are required to inspect all SIUs at least once a year pursuant to 40 CFR 403.8(f)(2)(v). Although U.S. EPA regulations establish a minimum frequency for inspections, additional inspections by the WRRF might be necessary depending on issues such as the variability of an SIU's effluent, the effect of the SIU's discharge on the WRRF, and the facility's compliance history. Routine WRRF inspections of SIUs typically consist of three activities—preparation, on-site assessment, and follow-up. Scheduled inspections are useful when the POTW wants to gather specific information from the facility that necessitates meeting with specific SIU contacts. However, on-demand inspections or unscheduled inspections might more accurately reflect SIU compliance status because scheduled inspections sometimes interrupt normal operations.

Noncompliance/Violations

In addition to requirements for permitting, sampling, and inspecting industrial users, the General Pretreatment Regulations require WRRFs to review industrial user reports and plans and to respond to instances of industrial user noncompliance. The ERP regulations at 40 CFR 403.8(f)(5) establish a framework for WRRFs to formalize procedures for investigating and responding to instances of industrial user noncompliance. U.S. EPA strongly recommends that the WRRF include an enforcement response guide (ERG) as part of the approved ERP. The ERG is based on type and duration of violation, frequency of violation, effect of violation, economic benefit gained by the violator, and the attitude of the violator. The range of enforcement mechanisms include informal notice to the industrial user, informal meetings, notice of violation, administrative orders and compliance schedules, administrative fines, civil suits, criminal prosecution, and termination of service.

TEST YOUR KNOWLEDGE

1. The Federal Pretreatment Act defines industrial and commercial users differently from one another.
   - ☐ True
   - ☐ False
2. What two parameters are most often considered when calculating pretreatment surcharges for industrial users?
   a. COD and BOD
   b. BOD and TSS
   c. TSS and TDS
   d. TOC and BOD
3. A pass-through pollutant reporting to a WRRF from an industrial source can affect its ability to meet permit limits for that pollutant.
   - ☐ True
   - ☐ False
4. Pretreatment of industrial wastes is regulated under the NPDES program.
   - ☐ True
   - ☐ False
5. Pretreatment permitting can only be done by the local WRRF receiving industrial wastes.
   - ☐ True
   - ☐ False
6. Which of the following is not included in pretreatment standards and requirements?
   a. general and specific prohibitions
   b. source control measures
   c. categorical pretreatment standards
   d. local limits
7. All WRRFs are required to establish local pretreatment programs.
   - ☐ True
   - ☐ False
8. Generally, a WRRF acts as the pretreatment "control authority" and the state or U.S. EPA acts as the pretreatment "approval authority."
   - ☐ True
   - ☐ False
9. Pretreatment control authorities must establish an Enforcement Response Plan (ERP), which includes a framework for which of the following:
   a. Procedures to investigate and respond to industrial user noncompliance
   b. Procedures to develop local limits
   c. Procedures to report industrial user noncompliance to the U.S. EPA
10. Which of the following criteria is NOT used to define a significant industrial user (SIU) under National Pretreatment Regulations?
    a. An industrial user that discharges an average of 95 m^3/d (25 000 gpd) or more of process wastewater to the WRRF
    b. An industrial user that contributes process wastewater making up 5% or more of the average dry weather hydraulic or organic capacity of the WRRF
    c. An industrial user subject to federal categorical pretreatment standards.
    d. An industrial user that produces more than 10 000 tonnes (11 000 tons) of hazardous waste per month.
11. A local limit supersedes the categorical pretreatment standard for a given pollutant if it is more stringent.
    - ☐ True
    - ☐ False
12. Maximum Allowable Headworks Loading (MAHL) is a method used to establish local limits based on how much of a specific pollutant the WRRF can accept from all permitted industrial users and still ensure protection against pass-through and interference.
    - ☐ True
    - ☐ False
13. The most common parameters measured in industrial discharges are pH, BOD, and TSS.
    - ☐ True
    - ☐ False
14. Time proportional sampling is recommended for industrial discharges where possible.
    - ☐ True
    - ☐ False
15. 40 CFR 403 specifies different reporting requirements for categorical industrial users (CIUs) versus SIUs.
    - ☐ True
    - ☐ False
16. The difference between the local limit and categorical pretreatment standard is where the criteria are applicable (point of discharge or after treatment).
    - ☐ True
    - ☐ False
17. Water resource recovery facilities are required to inspect all industrial dischargers at least once a year.
    - ☐ True
    - ☐ False

Evolving Regulations and Contaminants of Emerging Concern

When industrial NPDES, State Pollutant Discharge Elimination System, or pretreatment permits are renewed, new pollutants and/or limitations may be added as new water quality criteria are identified. Industrial wastewaters often contain higher concentrations of contaminants of emerging concern (CECs) than municipal wastewater. As such, evolving regulations associated with CECs in waters are of special concern in industrial settings. Changes in regulations and increased awareness for a given contaminant usually follow new knowledge regarding either health risks or improved ability to measure them. In some cases, emerging contaminants can affect downstream WRRFs. They can affect WRRF treatment effectiveness through inhibition of biological processes or overloading treatment processes. They can also affect discharge quality if the contaminants are not removed through WRRF treatment (i.e., a pass-through contaminant).

Contaminants of emerging concern describe pollutants with increasing regulatory scrutiny due to any combination of increasing use, evolving understanding of public and environmental health risks, and improving analytical detection capabilities.

ORGANIC CATION EXCHANGE CAPACITIES

Organic CECs are typically human-made compounds produced for some specific industrial or commercial purpose. These specific purposes are sometimes at odds with environmental outcomes, resulting in toxicity or persistence in environmental settings. Treatment for these contaminants is most often through oxidation or sorption but depends on the specific chemistry of each contaminant. Organic CECs include (Glassmeyer et al., 2017):

- Pharmaceuticals and personal care products, including antibiotics and endocrine-disrupting compounds—variable in chemical structure and characteristics, with potential biological effects to humans and aquatic animals
- Emerging herbicides and pesticides, including neonicotinoids—variable in chemical structure and characteristics, but designed to be toxic to targeted plants or animals
- Per- and polyfluoroalkyl substances—designed to be long-lasting, also have long-term health effects
- Cyanotoxins—produced by algae, largely in surface waters affected by excess nutrients

Endocrine-disrupting compounds can negatively affect reproduction of downstream aquatic life.

INORGANIC CATION EXCHANGE CAPACITIES

Inorganic CECs are typically less varied but may still be difficult to treat. These include:

- Metals not previously included in water quality or biosolids standards, including tungsten and manganese—can be from industrial source materials or additives, treatment options similar to other metals
- Treatment byproducts, including perchlorate and bromate—produced during industrial or treatment processes. Treatment options depend on the type of chemical
- Nanomaterials and microplastics—designed for a specific purpose, often long lasting in the environment and difficult to treat due to small size

This group of human-made chemicals includes perfluorooctanoic acid (PFOA) and perfluorosulfanoic acid (PFOS) as well as other more recent derivatives and relations. These compounds are present in a wide range of consumer goods, from nonstick cookware to clothing with fire retardants. They have demonstrated human health effects at part per trillion (ppt) concentrations. These compounds are most prevalent in industrial wastewaters from facilities where PFAS are used in the industrial manufacturing process.

PFAS—AN EXAMPLE OF CATION EXCHANGE CAPACITY REGULATION AND TREATMENT

One example of a class of CECs with evolving regulation is PFAS. While these compounds are currently treated in some industrial wastewater applications, they are not yet widely treated in municipal wastewater applications.

U.S. EPA issued "provisional health advisory" values for PFOA and PFOS in 2009, reflecting potential health-based hazards. Six PFAS compounds were then added to the Unregulated Contaminant Monitoring Rule (UCMR) UCMR3 list by 2015 to support additional sampling and analysis for future regulatory actions. The PFOA and PFOS "drinking-water health advisory" levels were then updated to a total of 70 ppt (the current limit) in 2016 and issued a "PFAS Action Plan" in 2019. The action plan recommends future actions, including steps to establish maximum contaminant levels, hazardous substance designations, and groundwater cleanup recommendations for PFOA and PFOS (U.S. EPA, 2019, 2020). Numerous states continue to publish drinking water advisories for PFAS compounds. In response to these actions, industrial wastewater dischargers have come under additional scrutiny in multiple states, and some have implemented pretreatment systems to remove PFAS before discharge. Sampling and treatment of PFAS is complicated by the ubiquity of these compounds in consumer products.

The Unregulated Contaminant Monitoring Rule includes specific lists of CECs that may be present in drinking water but are currently not regulated there.

TEST YOUR KNOWLEDGE

1. Per- and polyfluoroalkyl substances (PFAS) have established regulatory limits at both the state and federal level.
   - ☐ True
   - ☐ False
2. Industrial facilities cannot directly discharge wastewater to a surface water via an NPDES permit; they must discharge to a WRRF.
   - ☐ True
   - ☐ False
3. Contaminants of emerging concern (CECs) are called that instead of "emerging contaminants" because many have been present for decades but are only now being subjected to increased regulatory and public health scrutiny as measurement methods improve and we understand more about their toxicity.
   - ☐ True
   - ☐ False
4. Treating organic CECs is similar, regardless of their specific chemistry.
   - ☐ True
   - ☐ False

Industrial Wastewater Sources and Characteristics

As mentioned earlier in this chapter, industrial wastewater presents operational challenges for treatment facilities due to its complexity and variability. This is true of both flows and loads and in a few different ways: by industry, by time, and by downstream effects.

First, industrial wastewater varies by industry. Potato wastewater, for example, is high in starches and easily biodegradable organics (e.g., BOD), whereas mining wastewater is high in metals and TDS, but low in organics. Industrial wastewater varies even within the same industry; some dairy product producers have different wastewater characteristics than others that need to be considered.

Second, industrial wastewater can be highly variable over time. Similar to how municipal wastewater often has daily cycles of flowrate, BOD, TSS, and nitrogen loading, industrial wastewater flows and loads fluctuates as well—sometimes by a factor of 10 or more—throughout the day. Similar variation occurs throughout the year as different industrial products have higher demand during certain times of the year. For example, some dairy producers have higher BOD loads during the holiday season because of production of eggnog, and sugar beet factories have much higher loads during the winter when they actively "slice" beets during and after beet harvest. Many industrial processes, especially food processors, clean or disinfect their factory floors at the end of every shift. This can generate significant volumes of high-strength wastewater. In addition, unintentional loss of product due to spills on the factory floor can create sudden increases in pollutant loading. Swings in pH are also common for industries that use acids and bases for clean-in-place processes. Industrial wastewaters, especially from the pulp and paper industry, can also be nutrient deficient compared to municipal wastewaters. This means that they may not contain enough nitrogen, phosphorus, and other nutrients for WRRF biological processes to degrade the BOD they contain.

> Clean-in-place processes or rinses using acids, bases, or other chemicals are often used to clean industrial manufacturing equipment. Clean-in-place processes are typically used when equipment needs to be cleaner than can be achieved with only water rinses.

Third, industrial wastewater may be considered variable in that its effects on WRRF treatment processes may be variable and difficult to identify. Constantly evolving WRRF operational setpoints and strategies, combined with daily variable industrial wastewater sources and production, make the effects of industrial wastewater at the WRRF highly variable.

Wastewater generation comes from many different portions of each industrial facility. To understand how industrial processes produce wastewater and the general characteristics of each, refer to Figure 1.2, which outlines where wastewater comes from, in general, for several types of industrial processes.

WASTEWATER CHARACTERIZATION

It is important for both existing and new industrial facilities to understand their wastewater characteristics to apply for and maintain compliance with a permit. Typical wastewater characteristics for specific

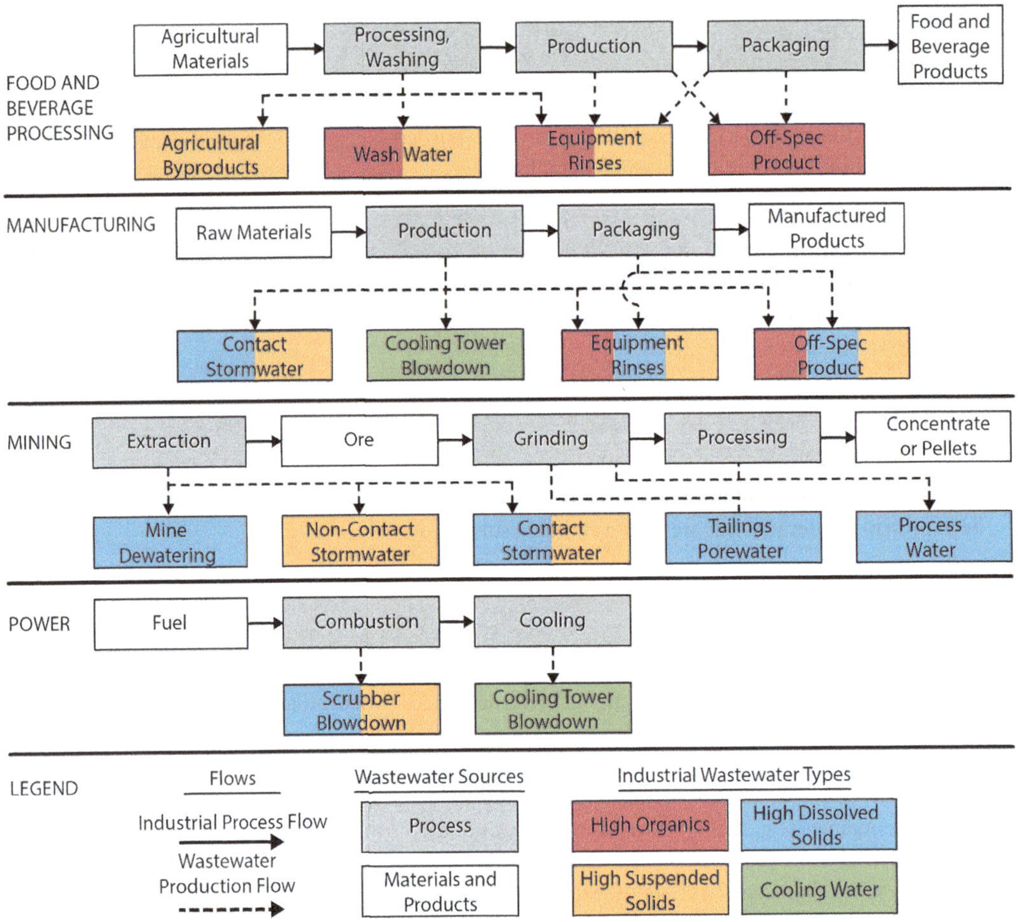

Figure 1.2 Example Wastewater Sources and Characteristics by Industry

industries are covered in further detail in the next few sections. For industrial facilities, it is important to understand which chemical products are used at the facility and which pollutants may be present in wastewater discharges. These products can influence the quality of treated effluent wastewater and compliance with the facility's NPDES or pretreatment permits. It should be noted that under 40 CFR 117.12(A)(2), certain discharges of hazardous substances may be exempted from the requirements of CWA.

Equalization of waste streams at the industrial facility can balance variability of flow and influent quality and minimize effects on downstream wastewater treatment operations. Equalization can temper peak concentrations of slug loads and pH and provide acidification of complex organic substrates for easier downstream treatment.

Despite the high degree of variability, ranges of typical characteristics for industrial wastewater discharges have been identified; these are presented in Table 1.3 and discussed in following sections. This information can be used to determine the potential effect of a certain type of industrial wastewater on a municipal WRRF facility's operation and loading. When possible, obtain actual operating data from the industrial user or a similar facility producing the same product.

Food and Beverage Sector

The food and beverage sector is characterized by heavy water consumption and processes to prepare raw food into consumable product. Wastewater and its components are sourced from either raw agricultural materials or additives included during processing steps and include high organic loads. Four examples of food and beverage sector wastewaters are outlined below:

Table 1.3 Typical Wastewater Characteristics by Industry

Industry	Sources	Biochemical Oxygen Demand (BOD) or Chemical Oxygen Demand (COD)	Total Suspended Solids	Total Kjeldahl Nitrogen	Total Phosphorus	Total Dissolved Solids (TDS)	Other Notable Compounds
Units		mg/L	mg/L	mg/L as N	mg/L	mg/L	
Food and Beverage							
Potato Processing	3,5,6	1150–3000 (BOD)	400–1210	20–230	66–80	1488	
Milk and Dairy	2	104–5000 (BOD)	390–5070	6–60	5–41		
Vegetable Processing	5	2737 (BOD)	1394	77	24	1625	
Sugar Beet Processing	4,5	1100–3300 (COD)	367–459	35–133	0.8–1.9		Nonvolatile dissolved solids 782–2797 mg/L
Pulp and Paper	1	474 (BOD)	342	8	2		Adsorbable organic halogen 32 mg/L
Computer Chip	9	100–200 (COD)	1000–1500	250–350		900–1000	Chloride ~100 mg/L Fluoride ~600–700 mg/L
High Strength Nylon and Synthetic Rubber Manufacturing	8	112 600–350 000 (COD)					
Mining	7		66–3834			267–2895	

[1]Adapted from Cabrera et al. (2017)
[2]Carvalho et al. (2013)
[3]U.S. EPA (1969)
[4]USDA (1980)
[5]HDR (2020)
[6]Goodman and Rife (2015)
[7]Estimated from U.S. EPA (2008). Assuming TDS = 0.67x [conductivity, μmhos/cm]
[8]Eckenfelder & Musterman (1995)
[9]HDR (2020), Courtesy of Anonymous Computer Chip Manufacturer

- Potato processing wastewater is high in starch and other organic molecules.
- Sugar beet processing wastewater is high in salts, sugars, and complex organics.
- Dairy wastewater is high in solids, dissolved proteins, and FOG.
- Brewery wastewater is high in solids, sugars, and alcohols.

Industrial pretreatment of high-strength wastewater typically uses biological processes similar to those at a typical WRRF. Biological treatment can consist of aerobic, anoxic, and anaerobic treatment as well as combinations of these. Many industrial users also use anaerobic digestion to treat high-strength wastewater.

Food industries often use strong disinfectants such as quaternary ammonium compounds (QACs) that can affect biological treatment processes. Many food and beverage wastewaters are warmer than domestic wastewater, as high as 38 °C. This can affect WRRF biological treatment processes and temperature limits for final WRRF effluents. Management considerations for high solids and high-organic industrial wastewaters are further described in Table 1.4.

Manufacturing Sector

Manufacturing wastewater is primarily generated from rinses of production equipment, and includes components that reflect the final product produced as well as any additives used during the production process. Because a given facility may produce different products at different times, the composition of

Fats, oils, and grease (FOG) refer to oily substances in wastewater. Primary sources of FOG in industrial wastewaters are from plant and animal materials, as with wastewater from dairies and tanneries, and petroleum products, as with refinery wastewater.

"Aerobic," "anoxic," and "anaerobic" refer to what microbes in biological treatment processes breathe. *Aerobic* means breathing with oxygen, *anoxic* means breathing with something else, like nitrate or sulfate, and *anaerobic* means using fermentation, similar to microbes in anaerobic digesters or alcohol production.

Quaternary ammonium compounds are an effective disinfectant that kills many types of pathogens and biofilm-causing bacteria; they are widely used as a cleaning agent in industrial facilities.

these wastewaters can vary widely with time. This industry sector is varied and includes manufacturing of pharmaceutical, agricultural, industrial, construction, and consumer products. Specifically under consideration for this chapter are pulp and paper discharges and computer chip discharges; refer to Table 1.3 for characteristics. Table 1.4 further describes management considerations for high-solids, high-organics, and salty industrial wastewaters. Some manufacturing wastes contain recalcitrant organic material that can be difficult for WRRF biological treatment processes to remove.

Recalcitrant organic compounds are those that are not readily biodegraded, and thus cannot be effectively removed using biological treatment.

Mining Sector

Mining wastewater is typically generated from stormwater runoff or groundwater that comes in contact with mine surfaces and pits (i.e., contact stormwater or mine dewatering). It contains salts and metals leached from ore and rock surfaces during mining and material processing and can also contain dissolved solids sourced from stormwater runoff. Table 1.4 further describes management considerations for salty and high-metals industrial wastewaters.

Power Sector

Wastewater from the power sector is primarily sourced from cooling towers, but also includes some heavy metals and potentially inhibitory compounds depending on the kind of power generation in question. Cooling tower blowdown is the discharge stream from an open recirculating cooling water system. Recirculating water cooling "loops" take water from the cooling tower basin and pass cooled water through process equipment, then pass the heated water through the cooling tower where evaporation cools the bulk water stream. Evaporation concentrates dissolved impurities in the recirculating water stream. Fresh source water makeup, added to maintain a constant level in the cooling tower basin, maintains the recirculating water concentration at steady-state concentrations. Chemical additives are regularly used in the cooling tower basin to help reduce biofouling, scaling, and corrosion within the recirculation loop and process equipment.

The quality of the cooling tower blowdown is defined by source water impurities, chemicals added to prevent scaling and fouling, atmospheric dust accumulation, and evaporative concentration of these components. Chemicals added to reduce biofouling, scaling, and corrosion can contribute metals and nutrients to the water, which may necessitate additional treatment (NALCO Water, 2018).

Flowrates of cooling tower blowdown, or once-through cooling water discharge, can be significant at some facilities and may be regulated as separate outfalls with permit limitations specific to cooling water. At other industrial facilities, cooling tower blowdown may be combined with process wastewater and regulated as a combined discharge. Thermal discharges are also regulated by the Clean Water Act, Part 316, and are required to meet temperature requirements within the receiving waters. Some of the key concerns for regulators include: heat and thermal load in the discharge; residual chlorine; chemical additives used in the cooling system; WET (if multiple chemical additives are used); and concentrated metals, dissolved solids, and other parameters in the cooling tower blowdown.

Non-contact cooling water discharges generally have a lower risk of containing pollutants and may be regulated separately from cooling waters that have direct contact with industrial materials or process equipment. Although the risks of oil or chemical leakage for non-contact cooling water discharges are considered to be low, these risks are taken into account when determining how a regulatory authority identifies specific limitations for non-contact cooling waters as meeting the overall requirements of CWA.

EFFECTS OF INDUSTRIAL WASTEWATER SOURCES ON MUNICIPAL TREATMENT WORKS

Industrial wastewater discharges can negatively affect the WRRF treatment process. Process effects at municipal WRRFs due to industrial influents can take many forms. The most common three effects are (1) toxicity and/or biological inhibition that kills or inhibits activated sludge organisms, (2) chemistry incompatibilities that cause settleability issues in clarifiers, and (3) fluctuating nutrient loadings that adversely affect biological nutrient removal (BNR).

A prominent example of toxicity and biological inhibition is the discharge of cleaning agents from industrial plants into the WRRF influent. Their strength as biocides can also adversely affect bacteria in WRRF activated sludge systems (Condidi et al., 2018). A slug load of QACs can rapidly affect the WRRF's nutrient removal performance. Sometimes, the ability to nitrify can be knocked out completely by QAC toxicity. Long-term discharge of QACs to a WRRF can consistently inhibit performance of activated sludge and anaerobic digestion processes. In that case, the WRRF should design the activated sludge system for reduced biological kinetic rates so that permit requirements can still be met. Toxicity issues can also be caused by heavy metals, organic solvents, pharmaceuticals, and pesticides. Some WRRF biological treatment processes incorporate activated carbon to adsorb toxic compounds or chemical oxidation to break down recalcitrant organic molecules to be more biodegradable.

An example of chemical incompatibility is inhibition of WRRF coagulants or flocculants by similar chemicals in industrial discharges. A WRRF may use these chemicals to enhance settling in secondary clarifiers or the chemical phosphorus removal process. If an industrial user discharges an oppositely charged polymer used for pretreatment or within facility processes, it may interfere in coagulant or flocculant performance at the WRRF.

Coagulants are chemicals added to convert dissolved contaminants into particles or smaller particles into larger particles for settling. Some coagulants produce a targeted charge to promote particle agglomeration. Flocculants are chemicals, often polymers. Polymers produce a targeted change to link particles into larger flocs to improve settleability of solids and can be either positively charged (cationic) or negatively charged (anionic).

An example of fluctuating nutrient loadings that affect WRRF performance is the treatment of landfill leachate. One recent study found that a WRRF accepting landfill leachate flows equal to 2% of the total facility flow realized a significant drop in nitrification due to high levels of ammonia in the leachate (Brennan et al., 2017). Fluctuating influent loads can be a difficult challenge for WRRF operators because it requires working closely with the industrial discharger to implement or improve their own pretreatment system or to coordinate the timing of nutrient discharge loads to the WRRF. Depending on local political history and the nature of industries within the community, the WRRF may need to upgrade treatment to accommodate high degrees of fluctuation in nitrogen and phosphorus loadings to the WRRF. Large organic slugs can also exceed available oxygen supply or biomass capacity of WRRF biological treatment systems.

Landfill leachate is liquid collected from landfills; it commonly contains high levels of organic material and nutrients. Depending on the type of material in the landfill, these flows can also contain heavy metals and CECs. Many landfills discharge collected leachate to their local WRRF.

Chapter Problems

Problem: A municipal WRRF experienced an overnight event that killed the majority of activated sludge and caused effluent noncompliance.

Issue: It was determined that a food and beverage industry in the collection system had a major spill of quaternary ammonia during a cleaning regimen the prior evening and had failed to notify the operator on call.

Solution: The WRRF worked with the industry through their local limits program to better manage cleaning chemical use and established a notification protocol for future spill events.

Problem: A municipal WRRF experienced high effluent turbidity and a greenish hue in the secondary clarifiers, leading to noncompliance with turbidity limits.

Issue: It was determined that a dairy industry that discharged to the system had high TDS, resulting in fine particles at the WRRF that remained in suspension and boosted effluent turbidity. The greenish hue was from sunlight reflecting off the small particles.

Solution: The WRRF bench tested several different coagulants to determine a backup plan for compliance in the future and worked with the industry through their local limits program to better manage cleaning chemical usage and established a notification protocol for future spill events.

Refer to Chapter 9 "Nutrient Removal" in *Wastewater Treatment Fundamentals I—Liquid Treatment* (WEF & ABC, 2018).

Problem: A municipal WRRF saw influent ortho-phosphate concentrations that ranged from 8 mg/L to 42 mg/L as PO_4-P (note that the "typical" range for domestic wastewater is 4 to 12 mg/L as PO_4-P). The biological phosphorus removal process, therefore, struggled to remove phosphorus reliably. The biological phosphorus removal process depends on the BOD-to-phosphorus ratio.

Issue: It was determined that an onion processor and soda bottling facility both periodically discharged high levels of phosphorus as part of their operations due to off-spec product and/or extra cleaning requirements.

Solution: The WRRF bench tested alternative polyaluminum chloride compounds to better remove phosphorus and retain settleability in the clarifiers. They also discussed the issue with the industries to gain advanced notice on high phosphorus discharges.

TEST YOUR KNOWLEDGE

1. The effects of polymers used in industrial wastewater processes are neutralized by the activated sludge at the WRRF and do not affect settleability.
   - ☐ True
   - ☐ False
2. All food processing wastewater is high in BOD and TSS but low in nutrients.
   - ☐ True
   - ☐ False
3. A high-volume cooling water discharge to a WRRF may be a concern if:
   a. Thermal load could affect WRRF biological treatment performance.
   b. Water volume could increase WRRF receiving water temperature.
   c. Either a or b
4. Quaternary ammonium compounds (QACs) discharged by industrial users typically affect downstream WRRF processes by inhibiting bacteria in the activated sludge process.
   - ☐ True
   - ☐ False
5. Polymer flocculants typically affect downstream WRRF processes by inhibiting bacteria in the activated sludge process.
   - ☐ True
   - ☐ False
6. Landfill leachate discharged to WRRFs often results in nutrient overloading, which can cause the WRRF to exceed nutrient permit limits.
   - ☐ True
   - ☐ False
7. Landfill leachate discharges can have a negative effect on WRRF processes, even if it makes up less than 5% of the total WRRF influent flow.
   - ☐ True
   - ☐ False
8. Equalization at the industrial site refers to upstream industrial processes meant to control the amount of wastewater produced.
   - ☐ True
   - ☐ False
9. Clean-in-place (CIP) processes at industrial sites typically use acids, bases, or organic chemicals to clean equipment.
   - ☐ True
   - ☐ False
10. Wastewater discharges from the food and beverage sector do not exhibit this characteristic:
    a. High salt concentrations
    b. High temperatures
    c. High concentrations of organic compounds
    d. High heavy metals concentrations
11. Bench testing can be a good option for WRRFs to evaluate alternative chemical options and doses to counter interferences from industrial discharges.
    - ☐ True
    - ☐ False

Overview of Industrial Pretreatment Processes

Table 1.4 provides a summary of pretreatment processes that industrial facilities may use to remove contaminants from water before discharge to a WRRF.

Table 1.4 Summary of Industrial Wastewater Pretreatment Technologies

Pretreatment Process/ Unit Operation	Brief Description	Common Types (or) Classification	Example Applications
Suspended Solids and FOG-Laden Industrial Wastewater			
Screening	Devices of various size openings physically separate coarse solids and FOG from the water.	Based on Opening Size ■ Coarse screens like bar screens and trash racks ■ Fine screens like rotary drums and vibratory screens ■ Micro screens Based on cleaning method ■ Manual (or) mechanically cleaned	■ Coarse screening ahead of gravity separators ■ Fine screens to remove small particles, or sometimes plastic pellets ■ Trash racks in stormwater management applications
Gravity Separation	Settleable material consisting of grit, sand, and heavy organics settle out while FOG and free oil globules float to the surface of clarifiers.	■ Grit chambers, diffused air grit tanks ■ Conventional sedimentation tanks (clarifiers) ■ Oil-water separators – American Petroleum Institute (API), corrugated plate interceptor (CPI) separators	■ Clarifiers with solids handling to remove settled solids and surface skimming to remove oil and scum ■ Oil-water separators such as API separators in refineries and chemical plants
Flotation	Fine air (or) gas bubbles attach to oil and solids to bring them to the surface of the treatment tank to be skimmed off for disposal.	■ Dissolved air flotation ■ Dissolved gas flotation ■ Induced air flotation) ■ Induced gas flotation	■ Dissolved gas flotation process at oil refineries to protect from risk of fire hazard due to the presence of volatile organic compounds (VOCs)
Coagulation and Flocculation	Coagulation changes particle electrical charge, and flocculation forms larger particles that will either settle or float in gravity separation.	■ Static mixer ■ Rapid mix ■ Flocculation tanks	■ Heavy metals removal, emulsified oil, and grease separation ■ Removal of dissolved organics, as with surface water treatment for drinking water
Membrane Filtration	Sheets or hollow-fiber filter media with pore openings to keep out fine solids from treated water.	■ Microfiltration ■ Ultrafiltration	■ Ultrafiltration used to remove fine particulate metals from air pollution control scrubber water
High-Organics Industrial Wastewater			
Aerobic Biological Treatment	Bacteria remove soluble organics by using it as a food source while breathing oxygen.	■ Activated sludge with or without BNR ■ Aerobic lagoons ■ Sequencing batch reactors (SBRs) ■ Membrane bioreactors (MBRs)	■ Conventional activated sludge process. ■ The SBR for small to medium flow applications to limit number of tanks ■ MBR systems to limit footprint (higher biomass concentration achievable)
Powdered Activated Carbon (PAC) Biological Treatment	Soluble toxic pollutants stick to surfaces of activated carbon and improve activated sludge treatment by lowering toxicity to microorganisms.	■ PACT® ■ PAC-MBR	■ Used to capture and degrade recalcitrant organic chemicals, and organic micropollutants
Anaerobic Biological Treatment	Microbes degrade organic matter in absence of oxygen, which decreases sludge production and aeration costs relative to aerobic biological treatment.	■ Anaerobic contact reactors ■ Anaerobic lagoons or ponds ■ Upflow anaerobic sludge blanket (UASB) reactors	■ Anaerobic treatment in combination with aerobic treatment to treat high-BOD wastewater before discharge to WRRF ■ Anaerobic lagoons or ponds to pretreat high-strength organic wastewaters from industries such as meat-packing, slaughterhouse, and dairy
Chemical Oxidation	The oxidizing chemical reacts with organic compounds in the water to change them to smaller, less toxic forms.	■ Hydrogen peroxide/ Fenton's Reagent ■ Chlorine ■ Ozone ■ Permanganate	■ Pretreatment of industrial wastewater with recalcitrant organics can be pretreated using ozone before downstream biological treatment. ■ Oxidant demand reflects how much of a chemical oxidant will be used up by the water.

Table 1.4 Summary of Industrial Wastewater Pretreatment Technologies (*Continued*)

Dissolved Solids and Heavy Metals Industrial Wastewater			
Chemical Precipitation	Changes dissolved metals into settleable solids.	■ Metal hydroxide precipitation ■ Metal sulfide precipitation ■ Anionic metal co-precipitation	■ Lime addition to precipitate cationic metals as metal hydroxides in electroplating wastes ■ Lime addition to precipitate sulfate as gypsum in mining applications ■ Sulfide scavenger addition to remove trace metals from steel mill wastewaters
Chemical Oxidation/Reduction	Converts metals to less toxic or more treatable forms.	■ Hexavalent chrome removal ■ Cyanide removal	■ Destruction of cyanide ■ Reduction of hexavalent chromium to trivalent chromium ■ Oxidation of arsenite to arsenate
Membrane Separation	Sheets or hollow-fiber filter media with pore openings to keep out fine solids, molecules, and ions from treated water	■ Reverse osmosis ■ Nanofiltration ■ Electrodialysis reversal	■ Water reuse applications in which the high quality treated water is needed ■ Treatment of mine water for direct discharge in locations with strict regulations
Adsorption	Heavy metals and trace organics stick to granular media as water passes through the filter bed.	■ Granular activated carbon (GAC) ■ Activated alumina	■ Used as a polishing step following other metals removal processes ■ GAC adsorption can also be applied to remove trace organic pollutants that are not readily biodegradable, such as PFAS.
Ion Exchange	Metals and molecular ions stick onto the surfaces of synthetic media as water passes through the filter bed.	■ Anionic exchange resin ■ Cationic exchange resin ■ Natural materials like zeolite	■ Conventional in-home water softeners ■ Removal of trace ■ Selenium removal
Biological Treatment	Microbes or live plants absorb or chemically change the metals to less-toxic forms.	■ Anoxic fixed-film reactors ■ Constructed wetlands	■ Biological reduction of selenium at mine sites ■ Passive wetland treatment of cationic metals at mine sites to sequester metal sulfides

TREATMENT OF SUSPENDED SOLIDS AND FATS, OILS, AND GREASE

Pretreatment of FOG may be needed in some cases to further reduce organic loading to the WRRF or to prevent fouling of sewers and primary treatment facilities at WRRFs. Fats, oils, and grease typically float to the surface of clarifiers. Suspended solids and FOG are often removed concurrently in gravity-phase separators, with FOG scraped from the surface using skimming arms. Oil-water separators such as API separators and CPIs are used for oil removal in refineries and chemical plants. In an API separator, free (non-emulsified) FOG and free oil globules rise to the surface of a long rectangular tank due to the difference in specific gravity between the oil and wastewater. The accumulated oil layer at the top of the tank, entrained with water, is removed using surface skimmers. A CPI separator consists of a more compact design with submerged plates in the separator chamber to increase its surface area. Heavy solids and sediment collect at the bottom of a gravity-phase separator tank. Flotation systems use fine gas bubbles to more quickly bring solids and/or FOG to the surface of the treatment tank. Flotation systems include dissolved air flotation, dissolved gas flotation, induced air flotation, and induced gas flotation.

Fats, oils, and grease refer to oily substances in wastewater. Some can be from plant and animal materials, as with wastewater from dairies and tanneries, whereas others are sourced from petroleum products.

TREATMENT OF HIGH-ORGANICS INDUSTRIAL WASTEWATER

High-strength or high-organics industrial wastewater may contain elevated concentrations of BOD and COD, as indicated in Table 1.3, which includes examples of industries that contain high-strength organics waste. Wastewater characterization was covered in further detail in the previous Wastewater Characterization section of this chapter. Some of these industries generate wastewater in which a large fraction of high organic content is present as suspended solids and/or FOG. Anaerobic biological treatment is an example of a pretreatment process for high-strength wastewater to reduce BOD surcharge costs and organic mass loading to WRRFs. In addition, BOD pretreatment yields the benefit of lower aeration requirements and waste sludge production at WRRFs.

TREATMENT OF TOTAL DISSOLVED SOLIDS AND HEAVY METALS

The presence of heavy metals can adversely affect biological processes at the WRRF due to their toxic characteristics and accumulation in biosolids. Pass-through of contaminants could result in WRRF discharge violations of their NPDES permit. Therefore, WRRFs typically specify maximum allowable concentrations for specific metals in the discharge permits issued to industrial users. Heavy metals also tend to accumulate in the solids generated in activated sludge systems, and aerobic and anaerobic digesters. As the organic material is biologically degraded, total sludge volume reduces and metals can accumulate significantly. Metals can then remain with solids as they are settled, thickened, and dewatered. This could lead to a WRRF's failure to comply with biosolids regulations.

Heavy metal-laden wastewater can be treated to make wastewater streams suitable for discharge or reuse. The initial treatment step is often to convert dissolved metals to a solid form, or chemical precipitate. This is typically followed by a single or combination of solids removal "polishing" processes, including chemical precipitation and sedimentation, filtration, membrane separation, adsorption, ion exchange, and, occasionally, biological treatment. Selection of treatment processes depends on what metals are present, whether they are dissolved or particulate, and other water quality characteristics. Because of the wide range of inorganic constituents found in industrial wastewater, dissolved heavy metals are best treated in segregated streams at the source rather than combined diluted streams.

EXAMPLE APPLICATION: TREATMENT TRAIN FOR HEAVY METALS

Figure 1.3 presents a simplified process scheme for treatment of heavy metals. The system consists of pH adjustment, chemical precipitation and co-precipitation, coagulation, and sedimentation in an inclined plate clarifier. The chemical dosage is monitored and adjusted daily via jar testing to account for wastewater quality variability stemming from production changes in the manufacturing processes. Ultrafiltration followed by reverse osmosis is included as a polishing step for producing high-quality effluent for reuse. Sludge drawn off from the clarifier tank is thickened and then dewatered in a filter press.

TREATMENT OF COOLING WATER BLOWDOWN

The treatment processes required for cooling tower blowdown treatment are highly dependent on the use or final discharge requirements. For cost-effective cooling water blowdown reuse, industrial facilities seek to match blowdown quality with reuse quality objectives so that minimal treatment is required. Typically, the main components of cooling tower blowdown requirement treatment are removal of hardness components, such as calcium or magnesium, TSS, and TDS. These are accomplished by lime softening, clarification, reverse osmosis membrane separation, or electrodialysis reversal, and ion exchange treatment. Chapters 3, 4, and 6 provide additional details of these unit operations.

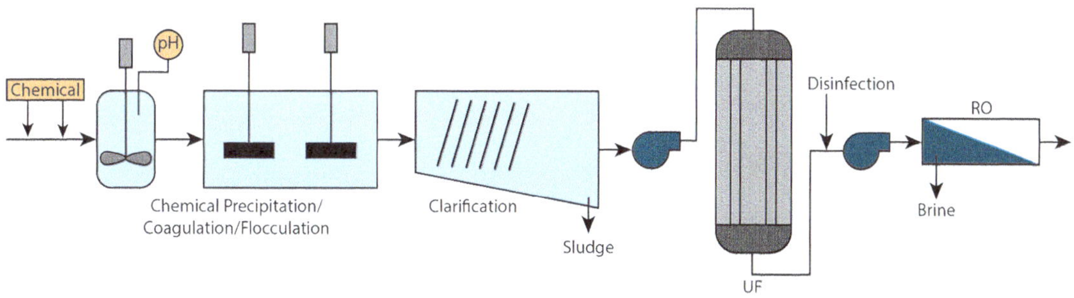

Figure 1.3 Example Process Train for Treatment of Heavy Metals Wastewater for Water Reuse

TEST YOUR KNOWLEDGE

1. Flotation treatment systems such as dissolved air flotation (DAF) work by adding fine gas bubbles to float solids and FOG to the surface.
   - ☐ True
   - ☐ False
2. High-strength wastewater is wastewater that contains high amounts of dissolved salts or TDS.
   - ☐ True
   - ☐ False
3. Pretreatment of wastewater with heavy metals is effective because the heavy metal concentration is diluted before treatment.
   - ☐ True
   - ☐ False
4. Heavy metals can accumulate in WRRF solid residuals like waste activated sludge and aerobic and anaerobic digesters.
   - ☐ True
   - ☐ False
5. Why might a dairy plant wastewater discharger to a WRRF pretreat high-strength BOD wastewater?
   a. Reduce toxicity of BOD to WRRF activated sludge treatment
   b. Reduce BOD surcharge costs
   c. Reduce TSS surcharge costs
   d. Both b and c
6. Why would heavy metal-laden wastewaters be pretreated before discharge to a WRRF?
   a. Reduce toxicity
   b. Minimize effect on biosolids quality
   c. Both a and b
7. Ion exchange is typically the initial pretreatment process used to remove metals.
   - ☐ True
   - ☐ False

CHAPTER SUMMARY

	Industrial Wastewater Treatment
PURPOSE AND FUNCTION	■ WRRFs have limited hydraulic, organic, and solids capacity. ■ Industrial discharges to the collection system can cause a WRRF to exceed one or all capacity types. ■ WRRFs may issue discharge permits to upstream industries to limit their effects on the WRRF and collection system. ■ Two primary discharge options for industrial wastewater are discharge to a WRRF under a pretreatment permit or discharge to a surface water under an NPDES permit. Some states also permit land application, deep-well injection, or other alternate disposal methods.
PRETREATMENT REQUIREMENTS	■ Pretreatment requirements and permits are typically issued by the local authority operating the POTW to be receiving the pretreated water, but in some case, pretreatment permits can be issued by state or federal agencies. ■ Pretreatment requirements are governed by the National Pretreatment Program and the associated General Pretreatment Regulations for Existing and New Sources of Pollution are presented in 40 CFR Part 403 (U.S. EPA, 1978). Within them, three different types of standards are included: prohibited discharge standards, categorical requirements, and local limits. ■ Industries may fall into categories for SIUs or CIUs, which may bring additional requirements. ■ Significant industrial users and CIUs trigger the need for a formal pretreatment program to discharge to a WRRF.
EVOLVING REGULATIONS AND CECs	■ The evolving nature of industry, and industrial production, means that industrial wastewater management often overlaps with emerging contaminants, toxic substances, and high production volume chemical regulation. ■ Per- and polyfluoroalkyl substances reflect a recent example of emerging regulations and treatment for a specific class of CECs.
INDUSTRIAL WASTEWATER SOURCES AND CHARACTERISTICS	■ Wastewater flows and loads can vary more significantly in industrial settings than municipal settings. These characteristics vary by industry type and individual facility.
MANAGEMENT OF SUSPENDED SOLIDS AND FOG-LADEN INDUSTRIAL WASTEWATER	■ High-solids wastewaters are those that contain significant suspended solids and can be produced by a wide variety of industries, including breweries, fruit and vegetable processing facilities, and any facility managing stormwater runoff. ■ Total suspended solids removal can be done through filtration/screening, gravity settling, or flotation, any of which can be improved by adding specific chemicals. All solids-removal processes produce a solid waste that requires disposal. ■ Removal of oils or FOG can be achieved through oil-water separators.

MANAGEMENT OF HIGH-ORGANICS INDUSTRIAL WASTEWATER	■ Industries that typically produce high-organic water include dairies, breweries, pulp/paper manufacturers, refineries, and other manufacturers. ■ Water with high organics typically requires treatment to lower organics surcharge costs (as measured by BOD or COD) and sometimes to remove nutrients or specific organic compounds regulated by a facility's permit. Treatment typically involves aerobic and/or anaerobic biological treatment.
MANAGEMENT OF HEAVY METALS WASTEWATER	■ Salty or high-metals wastewaters are produced in many industrial facilities. Example industries that produce water with high metals concentrations include mining, electrical manufacturing, and metal plating. ■ Removal of specific salts or metals can be done with chemical precipitation or ion exchange. Some metals can also be removed using biological processes or oxidation. ■ Bulk saltiness reduction is achieved through membrane separation processes, such as reverse osmosis and electrodialysis reversal. ■ Brines produced as byproducts of membrane or ion exchange processes require special management, and can add additional cost and operational complexity.
MANAGEMENT OF COOLING TOWER BLOWDOWN	■ Cooling water also classifies as a salty wastewater due to the concentration of original source water. ■ Cooling water management should be conducted with consideration for water quality as well as temperature, chemical additives, and concentration ratios.

References

Brennan, R. B., Clifford, E., Devroedt, C., Morrison, L., & Healy, M. G. (2017). Treatment of landfill leachate in municipal wastewater treatment plants and impacts on effluent ammonium concentrations. *Journal of Environmental Management, 188*, 64–72.

Cabrera, M. N. (2017). Pulp mill wastewater—Characteristics and treatment. In R. Farooq (Ed.), *Biological wastewater treatment and resource recovery*. IntechOpen. http://doi.org/10.5772/62795

Carvalho, F., Prazeres, A. R., & Rivas, J. (2013). Cheese whey wastewater: Characterization and treatment. *Science of the Total Environment, 445–446*(15), 385–396.

Condidi, D., Andalib, M., Andres, C., Bye, C., Umble, A., & Dold, P. (2018). Quaternary ammonium compound inhibition of biological nutrient removal activated sludge: The significance of good modeling practice on process design. *Proceedings of the 91st Annual Water Environment Federation Technical Exhibition and Conference*, New Orleans, LA, United States, Sept 29–Oct 3. Water Environment Federation.

Eckenfelder, W. W., & Musterman, J. L. (1995). Activated sludge: Treatment of industrial wastewater (pp. 5–26). Technomic Publishing Company, Inc.

Glassmeyer, S., Furlong, E., Kolpin, D., Batt, A., Benson, B., Boone, S., Conerly, O., Donohue, M., King, D., Kostich, M., Mash, H., Pfaller, S., Schenck, K., Simmons, J. E., Varughese, E., Vesper, S., Villegas, E., & Wilson, V. (2017). Nationwide reconnaissance of contaminants of emerging concern in source and treated drinking waters of the United States. *Science of the Total Environment, 581–582*, 909–922.

Goodman, A., & Rife, J. (2015). Redefining MBR nutrient removal for drinking water equivalent reuse in a large potato industry. *Proceedings of the 88th Annual Water Environment Federation Technical Exhibition and Conference* [CD-ROM], Chicago, IL, United States, Sept 24–30. Water Environment Federation.

HDR Engineering, Inc. (2020). Reference database. Data courtesy of HDR Engineering, Inc., Omaha, NE.

NALCO Water. (2018). Cooling Water Treatment Programs, Chapter 19. In *The NALCO water handbook* (4th ed.). McGraw-Hill Education. https://www.accessengineeringlibrary.com/content/book/9781259860973/toc-chapter/chapter19/section/section32

Smith, J. H., & Hayden, C. W. (1980). *Treatment and disposal of sugarbeet processing waste water by irrigation.* Conservation research report number 25. U.S. Department of Agriculture.

U.S. Environmental Protection Agency. (1969). Current practice in potato processing waste treatment. Water Pollution Control Research Series DAST-14.

U.S. Environmental Protection Agency. (1978). Title 40 CFR, Part 403, General pretreatment regulations for existing and new sources of pollution. *Federal Register*, Vol. 43, No. 123.

U.S. Environmental Protection Agency. (2004). *Local limits development guidance* (EPA 833/R-04-002A). https://www3.epa.gov/npdes/pubs/final_local_limits_guidance.pdf

U.S. Environmental Protection Agency. (2007). Title 40 CFR Part 136, Guidelines establishing test procedures for the analysis of pollutants. *Federal Register*, Vol. 72, No. 47.

U.S. Environmental Protection Agency. (2008). *Coal mining detailed study* (EPA 821/R-08-012).

U.S. Environmental Protection Agency. (2010). *Control authority pretreatment audit checklist and instructions* (EPA 833/B-10-001). https://www3.epa.gov/npdes/pubs/final_pca_checklist_and_instructions_%20feb2010.pdf

U.S. Environmental Protection Agency. (2011). *Introduction to the National Pretreatment Program* (EPA 833/B-11-001). U.S. Environmental Protection Agency, Office of Wastewater Management. https://www.epa.gov/sites/production/files/2015-10/documents/pretreatment_program_intro_2011.pdf

U.S. Environmental Protection Agency. (2017). *Industrial user inspection and sampling manual for POTWs* (EPA 831/B-17-001). U.S. Environmental Protection Agency, Office of Compliance. https://www.epa.gov/sites/production/files/2017-01/documents/iuinspect.pdf

U.S. Environmental Protection Agency. (2019). Per- and polyfluoroalkyl substances (PFAS) action plan. https://www.epa.gov/sites/production/files/2019-02/documents/pfas_action_plan_021319_508 compliant_1.pdf

U.S. Environmental Protection Agency. (2020). PFAS action plan: Program update February 2020. https://www.epa.gov/sites/production/files/2020-01/documents/pfas_action_plan_feb2020.pdf

Water Environment Federation. (2017). *Operation of water resource recovery facilities* (7th ed., WEF Manual of Practice No. 11).

Water Environment Federation & Association of Boards of Certification. (2018). *Wastewater treatment fundamentals I—Liquid treatment.* Water Environment Federation.

Suggested Readings

Claros, C. D. (2011). Anaerobic treatment of high-strength wastewater in the food and beverage and agricultural industries. *Proceedings of the 2011 AIChE Spring National Meeting,* Chicago, IL, United States, March 13–17. AIChE.

International Petroleum Industrial Environmental Conservation Association. (2010). Petroleum refining water/wastewater use and management. Operations Good Practice Series. https://www.ipieca.org/resources/good-practice/petroleum-refining-water-wastewater-use-and-management/

Metcalf & Eddy. (2003). *Wastewater engineering treatment and reuse* (4th ed., pp. 310–325). McGraw-Hill Higher Education.

U.S. Environmental Protection Agency. (1971). State of art sugarbeet process waste treatment. Water Pollution Control Research Series 12060 DSI 07/71.

U.S. Environmental Protection Agency. (2011). Introduction to the National Pretreatment Program (EPA 833/R-11-001). https://www.epa.gov/sites/default/files/2015-10/documents/pretreatment_program_intro_2011.pdf

U.S. Environmental Protection Agency. (2020a). Learn about effluent guidelines. https://www.epa.gov/eg/learn-about-effluent-guidelines

U.S. Environmental Protection Agency. (2020b). Overview of identifying and restoring impaired waters under Section 303(d) of the CWA. https://www.epa.gov/tmdl/overview-identifying-and-restoring-impaired-waters-under-section-303d-cwa

Water Environment Federation. (2008). *Industrial wastewater management, treatment and disposal* (3rd ed., WEF Manual of Practice No. FD-3, pp. 258–264).

CHAPTER 2

Advanced Safety Considerations

Introduction

As advancements in wastewater treatment at water resource recovery facilities (WRRFs) have been made, so have approaches to occupational health and safety performance and development of safety culture. Because there have been many Water Environment Federation publications focusing on safety and health in traditional wastewater treatment, this chapter will focus on hazards and risks associated with advanced wastewater treatment methods as well as advanced safety management approaches in the safety world today.

Developing a high-performance safety culture requires much more than policies and regulatory compliance. In this chapter, you will gain a deeper understanding of how to eliminate or reduce safety and health hazards using the hierarchy of safety controls (e.g., elimination, substitution, engineering controls, administrative controls, and personal protective equipment (PPE), as well as an organization and system approach.

LEARNING OBJECTIVES

Upon completing this chapter, you will be able to:

- Recognize the general risks and hazards associated with wastewater treatment and advanced treatment methods.
- Identify ways to eliminate or engineer out hazards using advanced safety management approaches.
- Identify ways to protect the worker.
- Understand safety guidelines when working with chemical and mechanical equipment.
- Describe established safety procedures (e.g., PPE, lockout/tagout, confined spaces, hazard communication, fall protection).
- Describe spill requirements.
- Describe approaches to safety and how to develop a program.
- Identify training needs.

Safety Programs/Incident Prevention

The main goal of any safety and health program is to prevent workplace injuries, illnesses, and deaths as well as the suffering and financial hardship these events can cause for workers, their families, and employers. However, a high-performing system brings a number of other benefits as well. They can help the facility to:

- Prevent workplace injuries and illnesses
- Improve compliance with laws and regulations
- Reduce costs, including significant reductions in workers' compensation premiums
- Engage workers
- Enhance total worker well-being
- Improve quality
- Increase positivity and enhance overall business operations

A health and safety program is more than a large binder of rules on a shelf. It should be developed as your facility-specific guide to effectively implement the program. According to U.S. Occupational Safety and Health Administration (OSHA) guidance, the core elements of any program should contain the following elements: management leadership, worker participation, hazard identification and risk assessment, education and training, and program evaluation and improvement. These are discussed in the following sections.

MANAGEMENT LEADERSHIP

This section of your facility-specific plan should describe the leadership, vision/policy, and resources needed to implement the program. It should describe the roles and responsibilities and expectations of each worker's role. It should also clearly state the facility's program goals and how they will be measured.

WORKER PARTICIPATION

The key to an effective health and safety program is worker participation. Workers benefit the most from a successful program and have the most to lose from a failing program. This element should define the ways that information will be communicated to them; how to provide feedback and raise safety concerns; and how to report injuries, illnesses, incidents. It should also contain a system for reporting non-injury-related events that have the potential to cause injury. These are often referred to as "near misses."

Establishing a facility health and safety committee is one way to have workers actively participate in the implementation of the program. The committee should be made up of workers from all levels of the organization, regardless of their skill level, education, or language. The committee meetings provide a venue to assess the performance of the program, provide feedback, discuss any injury or non-injury events reported, and what lessons can be learned.

Finally, a reward and recognition program can also be part of the program to encourage worker participation. A reward and recognition program should focus on encouraging positive safety behaviors or proactive actions in an effort to improve the overall culture. Note that a reward and recognition program should not be mistaken for an incentive program. Prizes or monetary rewards should be used with caution so as to not discourage injury and illness reporting; they should also never be linked to incident counts, labor hours, or rates.

HAZARD IDENTIFICATION AND RISK ASSESSMENT

A hazard is a situation or condition that creates a threat to life, safety, health, property, the environment, company reputation, and so on.

A risk is the probability or likelihood of getting injured by that hazard or damage occurring.

Hazard identification and risk assessment is another key part of the health and safety program.

In WRRFs, hazards can be numerous and can include treatment chemicals and laboratory chemicals, gases such as methane and hydrogen sulfide (H_2S), low oxygen environments as in confined spaces, and tanks containing water and/or wastewater.

Risks can include falling, drowning, being caught in an excavation collapse, being struck by equipment or caught in machinery, and inhaling chlorine gas or other toxic chemicals. The purpose of hazard identification and risk assessment is to identify potential hazards and controls, or actions to eliminate the hazards, based on the severity and probability of an incident. This process includes the following:

- **Hazard assessment:** The Identification of the hazards and potential hazards, and determination of the risks and the risk rating based on severity and probability (or likelihood).
- **Hazard control:** Elimination or control of the hazards and the risks associated with the hazard. This may include training and supervision on the hazards, risks and controls for employees affected by the hazards.
- **Job hazard analysis:** A job hazard analysis (JHA), also known as a *job safety analysis*, is a technique that focuses on jobs tasks as a way to identify hazards before they occur. It focuses on the relationship between the worker, the task, the tools, and the work environment.

EDUCATION AND TRAINING

An education and training program is essential to provide operators and other WRRF employees the skills necessary to do their work safely and recognize hazards along the way. Per OSHA, the employer must ensure their employees have received training for the hazards that exist in their workplace, and the equipment, policies, and procedures used to address those hazards. The training must be documented with the employee's name, date of training, and a proof of training comprehension. This is typically a short quiz, test, or assessment that confirms the employee retained the knowledge.

Through an effective training program, workers will understand the unique hazards present in the facility and throughout the various processes, and how to control them. The program should also consist of specific trainings to help workers fulfill each of their expectations per the health and safety program. Training may include a variety of topics such as safety and health awareness to safety leadership.

Basic Health and Safety Awareness Training. Training must be provided to employees on the safety and health policies, procedures, and laws and regulations surrounding their job hazards (see Table 2.1).

Note that many government WRRFs may be exempt from the OSHA requirements; however, providing proper training is still an essential part of maintaining an injury- and illness-free workplace.

Role Responsibilities. Each employee must understand their expectations and be trained to fulfil them. For example, if an operator is responsible for being part of a learning event team (sometimes referred to as an *accident investigation*), they must be trained on how to perform such an analysis and may include root cause analysis training.

Hazard Identification. Workers must be trained on the specific hazards of their job responsibilities and how to recognize hazards, implement protective measures, and complete a JHA. They must also be trained to wear, inspect, and maintain certain PPE. Very specialized training may be required for emergency response, facility startup and shutdowns, maintenance, and repair because these are nonroutine activities.

PROGRAM EVALUATION AND IMPROVEMENT

Once a program is in place, it must be treated as a "living" document, meaning that it must be continuously evaluated and improved upon based on shortcomings or opportunities for improvement. Your plan must identify the process for measuring and evaluating goals and metrics. These will include both lagging and leading indicators:

- **Lagging indicators:** Lagging indicators measure events that occurred in the past. These include the number of injuries, illnesses, or other unintended events and incident and severity rates. Although it's necessary to track these metrics, they are not indicators or predictors of future performance. They give little insight to the conditions that led to the incident.
- **Leading indicators:** Leading indicators are a proactive metric used to gauge the effectiveness of a safety program or preventative efforts. They can include the number of trainings offered and completed, audits or inspections, completion of permit-to-work, and reporting of incidents—both injury and non-injury

Table 2.1 Basic Regulatory Requirements per OSHA's 29 CFR 1910: OSHA Standards for General Industry

Bloodborne Pathogens	Fall Prevention	Heavy Equipment Safety	Respiratory Protection
Caught in or between hazards	Fire protection/use of fire extinguishers	Ergonomics	Stairway and ladder safety
Confined space entry	Forklift operation	Machine guarding	Struck by hazards
Chemical safety/ laboratory safety	Hazard communication	Industrial hygiene	Trench and excavation safety
Electrocution hazards/ lockout/tagout	Hearing conservation	Process safety	Working over/near water
Emergency action/ response plans	Heat and cold stress	PPE	

events (near misses), area observations and walkarounds, participation in safety committees, safe observations, number of corrective actions and time to action, number of recognitions for safe behavior, percent of maintenance time spent on planned versus unplanned maintenance, and so on.

Leading and lagging indicators should be tracked, analyzed, and reported regularly. These can be discussed at least monthly during management meetings and safety committee meetings. The scope and frequency of a full program evaluation will depend on the scope, complexity, and maturity of the program and on the types of hazards it must control. At a minimum, they should be completed annually, but might also be triggered by a change in process or equipment or an incident such as a serious injury, significant property damage, or an increase in safety-related complaints. Any changes made to the program should be documented, updated in all copies (both soft and hard) in circulation or replaced with the current version, and communicated to all employees.

TEST YOUR KNOWLEDGE

1. **The main goal of a safety and health program is to avoid citations from OSHA.**
   - ☐ True
   - ☐ False
2. **A facility health and safety committee can be a way to actively participate in a safety program, provide feedback, and discuss opportunities for improvement.**
   - ☐ True
   - ☐ False
3. **The following is true about reward and recognition programs:**
   a. They encourage worker participation and positive safety behaviors.
   b. They are important to achieving low injury rates.
   c. They should always be in the form of money or large prizes.
   d. They are the same as an "incentive."
4. **Hazard identification is ___________.**
   a. A permit to perform work where the permit is signed by the most senior operator
   b. Not important in a health and safety program because risks are commonly known and expected
   c. A process used to evaluate if any particular situation, item, thing, and so on may have the potential to cause harm
   d. The safety professional's responsibility to identify and then notify the other workers in the facility

The Hierarchy of Controls is the idea that control methods at the top of graphic are typically more effective and protective than those at the bottom—like PPE.

Hierarchy of Controls

Following a hierarchy of controls leads to the implementation of systems in which the hazard is removed, eliminated, or separated from the worker. The last two controls, administrative and PPE, don't remove the hazard, they just lesson the consequence. The Hierarchy of Controls should be the first stage in any PPE program development. See Figure 2.1 for the hierarchy of controls as represented by the National Institute for Occupational Safety and Health (NIOSH).

ELIMINATION AND SUBSTITUTION

Elimination and/or substitution tend to be the most effective at reducing hazards because they remove the hazard entirely or replace the hazard with a less harmful condition. An example of this would be replacing traditional chlorine gas disinfection treatment options for another method, or using a less toxic chemical like sodium hypochlorite.

ENGINEERING CONTROLS

Engineering controls may not control the hazard as much as elimination or substitution, but they are more preferable than administrative controls or PPE. They are typically put in place to separate the hazard and the worker. Some of the most common engineering controls in a facility include guardrails, catwalks, lockout/tagout (LOTO) devices, and equipment guards.

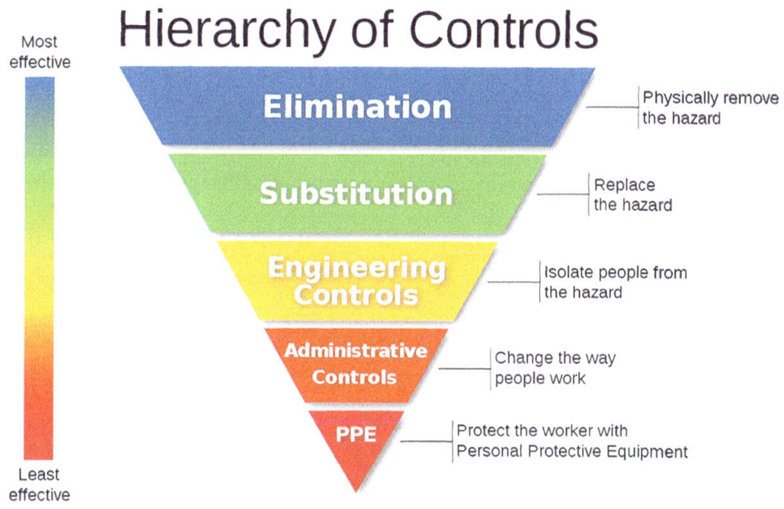

Figure 2.1 The Hierarchy of Controls Defined by NIOSH (NIOSH, 2015)

ADMINISTRATIVE CONTROLS AND PERSONAL PROTECTIVE EQUIPMENT

Administrative controls and PPE are the least effective safety measures. These should be the last to be implemented. Administrative controls typically consist of spreading an exposure over multiple workers over a period of time. For example, instead of using engineering controls to reduce the noise being emitted by a pump, workers would be limited to how long they can each work near the pump.

Although basic PPE such as head, eye, and hand protection is considered standard in most facilities, it should not be the only practice—or safety precaution—an employer or employee implements. Personal protective equipment cannot protect against all hazards and is very dependent on the worker to wear and to use it properly.

TEST YOUR KNOWLEDGE

1. The last line of defense is PPE. Barriers and procedures should come first.
   - ☐ True
   - ☐ False
2. An example of substitution is
   a. Installing guardrails around a tank
   b. Using a water-based paint instead of solvent-based
   c. Wearing a hard hat and safety goggles
   d. Putting a guard on a pump
3. There are five levels in the hierarchy of controls.
   - ☐ True
   - ☐ False

Personal Protective Equipment

As described previously, PPE should be the last level of protection used to reduce exposure to a hazard. However, even when other controls are implemented, it may be infeasible to remove all hazardous conditions and, therefore, PPE may still be required.

A safety data sheet (SDS) is integral to deciding what level of PPE might be appropriate for specific hazards. Eliminating the hazard during the construction phase, engineering controls, and administrative work practices should always be the first steps before deciding to use PPE.

PERSONAL PROTECTIVE EQUIPMENT—BASIC

Most facilities require basic or standard PPE. As shown in Figure 2.2, these include protective clothing, head protection, eye/face protection, and hand and foot protection.

SPILL CLEANUP PERSONAL PROTECTIVE EQUIPMENT

Additional or different PPE may be required when handling or cleaning up spilled chemicals. Refer to the SDS to determine the appropriate or required PPE. Figure 2.3 shows examples of common PPE for this activity. The type and/or level of PPE chosen is based on the level of protection required. For example, gloves made from one type of rubber can be protective against certain chemicals, whereas another glove made from a different type of rubber may be permeated or broken down by a chemical.

Some examples of chemical-resistant gloves include:

- **Butyl gloves**—made of synthetic rubber; protect against a wide variety of chemicals, such as peroxide, highly corrosive acids (nitric acid, sulfuric acid, hydrofluoric acid, and red-fuming nitric acid), strong bases, alcohols, aldehydes, ketones, esters, and nitrocompounds. Butyl rubber does not perform well with aliphatic and aromatic hydrocarbons and halogenated solvents.
- **Natural (latex) rubber gloves**—good general purpose gloves that protect against water solutions of acids, alkalis, salts, and ketones; may cause allergic reactions in some people.
- **Neoprene gloves**—made of synthetic rubber and protect against hydraulic fluids, gasoline, alcohols, organic acids, and alkalis.
- **Nitrile gloves**—made of a copolymer and provide protection from chlorinated solvents such as trichloroethylene and perchloroethylene, oils, greases, acids, caustics, and alcohols, but are generally not recommended for use with strong oxidizing agents, aromatic solvents, ketones, and acetates.

Table 2.2 from the OSHA Personal Protective Equipment Guide (OSHA, 2004) rates various gloves as being protective against specific chemicals and will help you select the most appropriate gloves.

Figure 2.2 Typical PPE for Operators

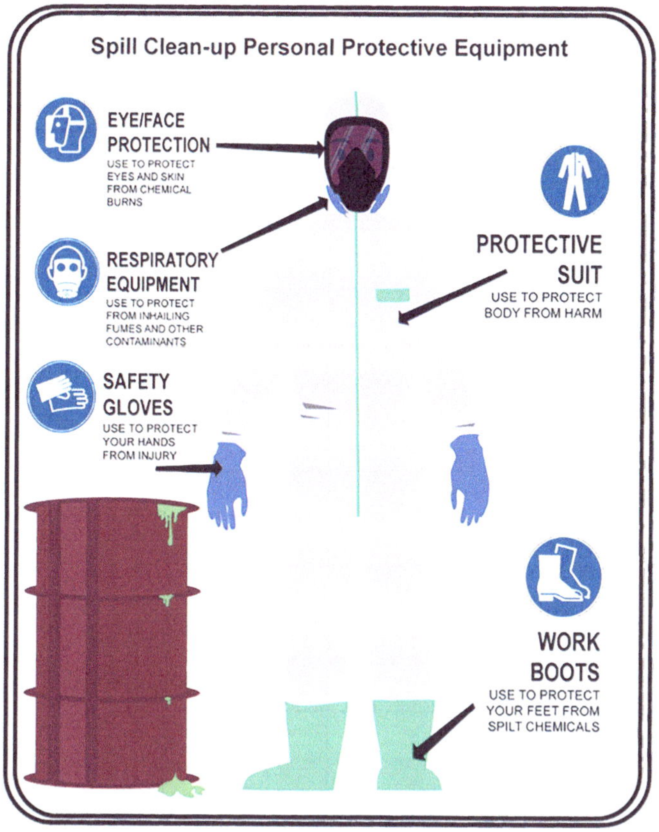

Figure 2.3 Spill Response Personal Protective Equipment

PERSONAL PROTECTIVE EQUIPMENT IN ELECTRICAL WORK

The U.S. Occupational Safety and Health Administration requires that workers be provided with PPE appropriate for the work to be performed and for the parts of the body that need protection. These include, but are not limited to, safety glasses, proper clothing, gloves, foot protection, hard hats, and hearing protection. The American National Standards Institute (ANSI) has established special designations for many items for personal protection when performing electrical work. Management should search out and use equipment approved for electrical work designed to protect workers from electrical shocks and other injuries. Refer to the section on PPE in this book for more information.

Arc-rated PPE is typically expressed in small calories of heat energy per square centimeter (cal/cm^2). Tests for determining arc rating are defined in American Society for Testing and Materials (ASTM) F1506, Standard Performance Specification for Flame Resistant Textile Materials for Wearing Apparel for Use by Electrical Workers Exposed to Momentary Electric Arc and Related Thermal Hazards (ASTM, http://www.astm.org/standards/F106.htm).

Selection of appropriate PPE, given a certain task to be performed, is typically handled in one of two possible ways. The first method is to consult a hazard category classification table like that found in the National Fire Protection Association's (NFPA) *Handbook for Electrical Safety in the Workplace* 70(E) (Table 130.7[C][9][a]) (NFPA, 2021). It lists a number of typical electrical tasks performed at various voltage levels and recommends the category of PPE that should be worn. The minimum rating of PPE necessary for any category is the maximum available energy for that category.

The second method of selecting PPE is to perform an arc flash hazard calculation to determine the available incident arc energy. The Institute of Electrical and Electronics Engineers Standard 1584, Guide for Performing Arc-Flash Hazard Calculations (IEEE-SA, 2002), provides a guide to perform these calculations if the bolted fault current, duration of the fault, and other general equipment information are known. Once the incident energy is calculated, the appropriate ensemble of PPE that offers protection greater than the energy available can be selected.

Table 2.2 U.S. Occupational Safety and Health Administration Personal Protective Equipment Guide (OSHA, 2004)

Chemical	Neoprene	Latex/Rubber	Butyl	Nitrile
Acetaldehyde*	VG	G	VG	G
Acetic acid	VG	VG	VG	VG
Acetone*	G	VG	VG	P
Ammonium hydroxide	VG	VG	VG	VG
Amy acetate*	F	P	F	P
Aniline	G	F	F	P
Benzaldehyde*	F	F	G	G
Benzene*	P	P	P	F
Butyl acetate	G	F	F	P
Butyl alcohol	VG	VG	VG	VG
Carbon disulfide	F	F	F	F
Carbon tetrachloride*	F	P	P	G
Castor oil	F	P	F	VG
Chlorobenzene*	F	P	F	P
Chloroform*	G	P	P	F
Chloronaphthalene	F	P	F	F
Chromic acid (50%)	F	P	F	F
Citric acid (10%)	VG	VG	VG	VG
Cyclohexanol	G	F	G	VG
Dibutyl phthalate*	G	P	G	G
Diesel fuel	G	P	P	VG
Diisobutyl ketone	P	F	G	P
Dimethylformamide	F	F	G	G
Dioctyl phthalate	G	P	F	VG
Dioxane	VG	G	G	G
Epoxy resins, dry	VG	VG	VG	VG
Ethyl acetate*	G	F	G	F
Ethyl alcohol	VG	VG	VG	VG
Ethyl ether*	VG	G	VG	G
Ethylene dichloride*	F	P	F	P
Ethylene glycol	VG	VG	VG	VG
Formaldehyde	VG	VG	VG	VG
Formic acid	VG	VG	VG	VG
Freon 11	G	P	F	G

Table 2.2 U.S. Occupational Safety and Health Administration Personal Protective Equipment Guide (OSHA, 2004) (*Continued*)

Chemical	Neoprene	Latex/Rubber	Butyl	Nitrile
Freon 12	G	P	F	G
Freon 21	G	P	F	G
Freon 22	G	P	F	G
Furfural*	G	G	G	G
Gasoline, leaded	G	P	F	VG
Gasoline, unleaded	G	P	F	VG
Glycerin	VG	VG	VG	VG
Hexane	F	P	P	G
Hydrazine (65%)	F	G	G	G
Hydrochloric acid	VG	G	G	G
Hydrofluoric acid (48%)	VG	G	G	G
Hydrogen peroxide (30%)	G	G	G	G
Hydroquinone	G	G	G	F
Isooctane	F	P	P	VG
Kerosene	VG	F	F	VG
Ketones	G	VG	VG	P
Lacquer thinners	G	F	F	P
Lactic acid (85%)	VG	VG	VG	VG
Lauric acid (36%)	VG	F	VG	VG
Lineolic acid	VG	P	F	G
Linseed oil	VG	P	F	VG
Maleic acid	VG	VG	VG	VG
Methyl alcohol	VG	VG	VG	VG
Methylamine	F	F	G	G
Methyl bromide	G	F	G	F
Methyl chloride*	P	P	P	P
Methyl ethyl ketone*	G	G	VG	P
Methyl isobutyl ketone*	F	F	VG	P
Methyl metharcrylate	G	G	VG	F
Monoethanolamine	VG	G	VG	VG
Morpholine	VG	VG	VG	G
Naphthalene	G	F	F	G
Napthas, aliphatic	VG	F	F	VG
Napthas, aromatic	G	P	P	G

Table 2.2 U.S. Occupational Safety and Health Administration Personal Protective Equipment Guide (OSHA, 2004) (*Continued*)

Chemical	Neoprene	Latex/Rubber	Butyl	Nitrile
Nitric acid*	G	F	F	F
Nitric acid, red and white fuming	P	P	P	P
Nitromethane (95.5%)*	F	P	F	F
Nitropropane (95.5%)	F	P	F	F
Octyl alcohol	VG	VG	VG	VG
Oleic acid	VG	F	G	VG
Oxalic acid	VG	VG	VG	VG
Palmitic acid	VG	VG	VG	VG
Perchloric acid (60%)	VG	F	G	G
Perchloroethylene	F	P	P	G
Petroleum distillates (naphtha)	G	P	P	VG
Phenol	VG	F	G	F
Phosphoric acid	VG	G	VG	VG
Potassium hydroxide	VG	VG	VG	VG
Propyl acetate	G	F	G	F
Propyl alcohol	VG	VG	VG	VG
Propyl alcohol (iso)	VG	VG	VG	VG
Sodium hydroxide	VG	VG	VG	VG
Styrene	P	P	P	F
Styrene (100%)	P	P	P	F
Sulfuric acid	G	G	G	G
Tannic acid (65)	VG	VG	VG	VG
Tetrahydrofuran	P	F	F	F
Toluene*	F	P	P	F
Toluene diisocyanate (TDI)	F	G	G	F
Trichloroethylene*	F	F	P	G
Triethanolamine (85%)	VG	G	G	VG
Tung oil	VG	P	F	VG
Turpentine	G	F	F	VG
Xylene*	P	P	P	F

Note: VG = very good; G = good; F = fair; P = poor (not recommended). Chemicals marked with an asterisk (*) are for limited service.

TEST YOUR KNOWLEDGE

1. The basic forms of PPE recommended for most WRRF operators may include the following:
   a. Long pants, long sleeves, gloves, and respirator
   b. Goggles, gloves, and protective clothing
   c. Hard hat, protective clothing, and eye protection
   d. Hard hat, protective clothing, and eye and hearing protection

2. Choose the best glove for handling diesel fuel:
   a. Neoprene
   b. Latex
   c. Butyl
   d. Nitrile

Housekeeping

Housekeeping is more than just cleaning up. Effective housekeeping is a good way to reduce or eliminate hazards in the facility. Poor housekeeping contributes to injuries, incidents, poor safety culture, and inefficiency. Risks from poor housekeeping include:

- Slips, trips, and falls on wet, greasy, or slippery floors, or over loose objects on walking surfaces
- Being hit by dropped objects
- Being struck by materials or poorly organized items
- Being cut or punctured by materials
- Exposure to chemicals or biological hazards spilled or uncleaned
- Ergonomic injuries due to working or moving around cluttered or disorganized workspaces
- Fire

In addition to reducing the risk of these injuries or incidents, there are other benefits that exist from a clean, organized plant. They include:

- Increased worker morale, and thereby reduced stress, from a neat, organized, and safe work environment
- Improved productivity. When the workspace is tidy and organized, it allows for more effective use
- Reduced time spent accessing equipment, materials, tools, and so on
- Reduced money spent on lost or damaged equipment or inefficient movement around the facility

Plan for housekeeping like other tasks in the facility. Create a system for where materials and tools go with labeled shelves or hooks. Create a schedule to ensure cleaning and organizing tasks are assigned and performed. Ensure cleanup is complete at the end of each shift and not left for the next shift of workers. Finally, perform regular inspections of housekeeping along with the safety inspections to identify and correct hazards as soon as possible.

Fall Protection and Prevention

Falls are one of the leading causes of injuries and deaths on the job. They can be caused by slips and trips, impacts from moving equipment, environmental conditions caused by cold or moisture, medical reasons, or simply a loss of balance. Occupational falls can be complex events; they often involve faulty or missing equipment and almost always are associated with improper preventive work procedures. Many injuries and deaths could be prevented with the appropriate equipment and procedures in place.

Fall protection is a combination of methods and devices used to protect workers from falling off, onto, or through working levels. Fall protection methods and devices are typically divided into two categories: those that prevent falls (fall restraint) and those that arrest falls. Fall protection also includes methods and devices that protect workers from being struck by tools, debris, or other objects falling from repair activities overhead. Examples of fall protection methods and devices include rails, guards, guardrails (see Figure 2.4), barriers, fall-arrest systems, positioning-device or fall-restraint systems, safety nets, hole covers, and various work practices and procedures.

Figure 2.4 Catwalk with Guardrail System Over Clarifier

A "passive" system is like a guardrail. It stays in place and doesn't rely on the worker as much as an active system. An "active" system includes fall restraint or fall arrest and requires the worker to wear the protection. In the hierarchy of controls (refer to Figure 2.1), passive systems should always be used before relying on active systems.

According to OSHA 1910.28, fall protection is required where a walking or working surface has an unprotected side or edge that is 1.2 m (4 ft) or more above a lower level. The worker must be protected from falling by one or more of the following: guardrail systems; safety net systems; or a personal fall protection system, such as personal fall arrest, travel restraint, or positioning systems, as discussed above. Note that safety net systems are rarely used anymore. The most common methods of fall prevention are guardrails (preferred) and personal fall protection systems (typically used when guardrails are not present or feasible). The exception to this 1.2-m (4-ft) rule are ladders. Safe ladder use is discussed further below.

Fall-restraint equipment (see Figure 2.5) consists of a safety belt or harness and a rope that is tethered to a substantial anchor point. These body belt or body harness systems are intended to be set up so that a worker can free fall no farther than 0.6 m [2 ft]; they should be secured to an anchorage capable of supporting at least twice the potential impact load of an employee's fall or 13.3 kN [3000 lb], whichever is greater. It is important to recognize that fall restraint is different from fall arrest in that the worker is restrained from falling more than 0.6 m [2 ft], which greatly reduces the fall-arresting force. Requirements for snap hooks, D-rings, and other connectors used with positioning device systems must meet the same criteria as those for personal fall-arrest systems. Whenever practical, guiderails or handrails should be installed as added protection for workers.

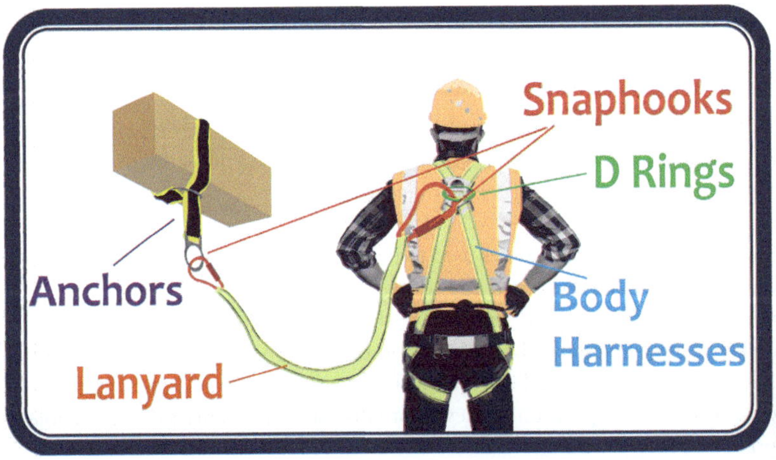

Figure 2.5 Personal Fall Arrest System Components

For fixed ladders (installed before November 19, 2018) that extend more than 7.3 m (24 ft) above a lower level, 1910.28(b)(9) of the OSHA regulation requires that the ladder be equipped with a personal fall arrest system, ladder safety system, cage, or well. All new fixed ladders installed after November 19, 2018, must be equipped with a personal fall arrest system or a ladder safety system. Where a personal fall arrest system or ladder safety system is used, there must be rest platforms every 45.7 m (150 ft). Those with a cage or well must be offset from adjacent sections and have a landing platform every 15.2 m (50 ft). Side clearances must also be in accordance with Figure 2.6 from the OSHA 1910.23 (OSHA, 2019) standard for ladders.

For non-fixed ladders—or portable ladders—including stepladders and extension, rungs and steps are to be slip resistant; portable ladders used on slippery surfaces are to be secured and stabilized; portable ladders are not to be moved, shifted, or extended while a worker is on them; top steps and caps of stepladders are not to be used as steps; ladders are not to be fastened together to provide added length unless designed for such use; and ladders are not to be placed on boxes, barrels, or other unstable bases to obtain added height (OSHA, 2016). Extension ladders should be placed at a 4:1 angle, meaning, for every 1.2 m (4 ft) of height you have to climb, move the base 0.3 m (1 ft) away from the wall that the ladder is resting against (see Figure 2.7).

Step ladders, or A-frame ladders, are one of the most common types of ladders used. General guidelines for safe use of these ladders include the following:

- Use the right type of ladder for the job. For example, do not use an A-frame stepladder against a wall where an extension ladder would be more appropriate.
- Always use the stepladder with spreaders unlocked.
- Get the correct height ladder. Never use the top step or cap as a step. This is more likely to happen when you don't have the right height ladder it the first place. Also, platform or podium ladders provide a more stable platform to work from.
- Don't move or shift the ladder while you are on it or with another person on it.
- Don't use cross bracing on the rear of stepladders for climbing.
- Always inspect the ladder before use. If the ladder is damaged, do not use it and remove it from service.
- Don't leave tools/materials/equipment on the stepladder. Use a proper toolbelt to hold your tools and equipment.
- Never use a metal stepladder near power lines or electrical equipment. Fiberglass ladders are safer to use in general so that a metal one doesn't inadvertently get used near electrical equipment.

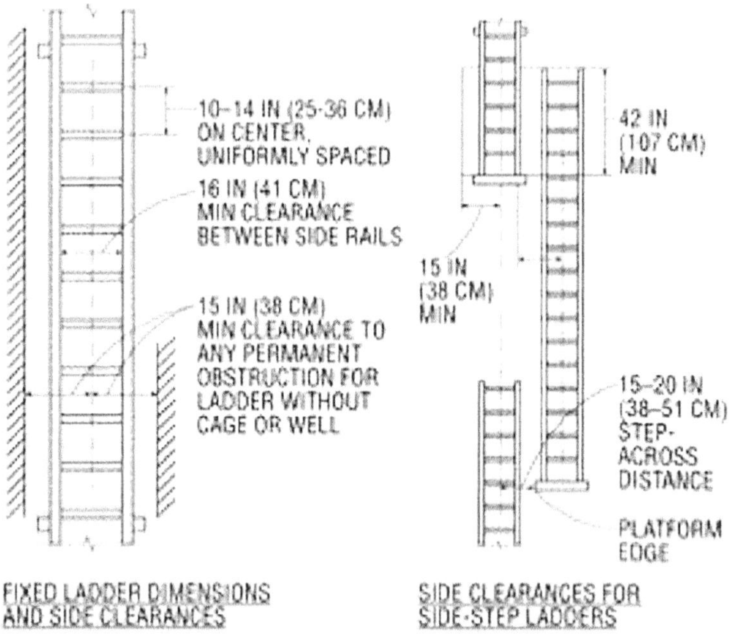

Figure 2.6 Side-Step Fixed Ladder Sections (OSHA, 2019)

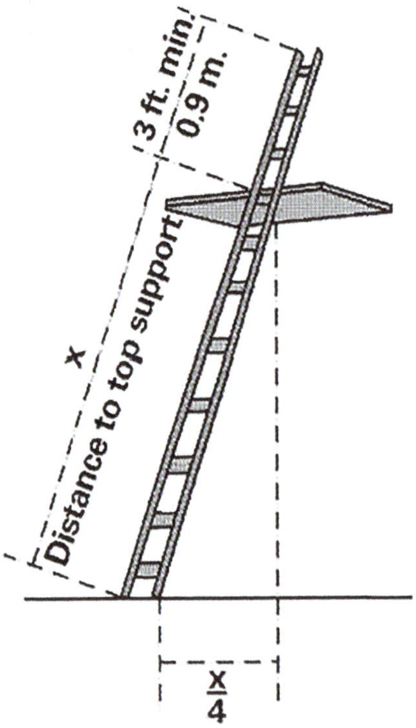

Figure 2.7 Portable Ladder Setup (OSHA, 2019)

- Use three points of contact when climbing the ladder. Obviously, this is not possible when doing work from the ladder so this is where platform or podium ladders can provide a safer and more stable work platform when you cannot maintain three points of contact.

When employees discuss fall protection, they mostly discuss harnesses, anchorage devices, and equipment. The discussion should be more robust to include passive systems such as guardrails, hole covers, and other activities. Fall protection means so much more than just wearing a harness.

Finally, ladders are a common cause of falls and injuries. Eliminating the use of ladders wherever possible is the best way to eliminate risk of falling. Stairs, scaffolds, and lifts are all safer options to ladders wherever possible.

OSHA requires employers to provide training for each employee in recognizing and minimizing the hazards of falls. The training program needs to include hazards and the procedures that need to be followed to minimize those hazards. Each worker exposed to a fall hazard should be trained by a competent person in identifying fall hazards and safe work practices to avoid or eliminate fall hazards. Awareness about falls and the surrounding area comes with human factors management and practicing better habits (Page-Bottorff, 2018).

TEST YOUR KNOWLEDGE

1. **OSHA requires training for fall protection hazards.**
   - ☐ True
   - ☐ False

2. **Fall protection is required where a walking or working surface has an unprotected side or edge that is how many meters (feet) or more above a lower level?**
   a. 1.8 m (6 ft)
   b. 3 m (10 ft)
   c. 7.6 m (25 ft)
   d. 1.2 m (4 ft)

3. **Falls are one of the leading causes of injuries and deaths on the job.**
   - ☐ True
   - ☐ False

4. **A passive fall protection system includes which of the following?**
   a. Fall harness
   b. Lanyard
   c. Guardrail
   d. Anchor point

5. Fall-restraint equipment consists of a harness and a lanyard that is tethered to a substantial anchor point.
   - ☐ True
   - ☐ False

6. Fixed ladders must be equipped with a personal fall arrest system, ladder safety system, cage, or well when it exceeds how many meters (feet)?
   a. 1.2 m (4 ft)
   b. 1.8 m (6 ft)
   c. 7.3 m (24 ft)
   d. 9.1 m (30 ft)

Lockout/Tagout (Control of Hazardous Energy)

In simple terms, a lockout/tagout (LOTO) program is a formalized program that has routine steps, typically to place an energy isolation device on a machine or piece of equipment so that it become inoperable. This is most common where machines or equipment are being repaired or maintained. "Hazardous energy" includes not only electricity but other forms of energy such as pneumatic pressure, hydraulic pressure, water, and gas. The purpose of LOTO procedures is both to prevent direct exposure to this energy and to prevent harm caused by any machinery or objects that could be moved by that energy. For example, to turn off power to an electrical system, machinery, or equipment, the main panel shutoff will typically have a lock placed on it by the person doing the work. In a place like a large WRRF, there could be multiple employees working on the same item at once and the LOTO system of locks isolates the equipment so it can be safely repaired and maintained without being turned on.

Isolation may also include placing valve locks or plugging pipes to prevent water from being turned on or entering a certain location. These may include ball valve locks, gate valve locks (see Figure 2.8), and plug valve locks.

In some cases, one lock is adequate. But in some cases, where multiple crews are working on, for example, a clarifier, the main shutoff for the clarifier would have a lockout hasp—a device that would fit through the hole in the main power shutoff and allowing for multiple locks to be used (see Figure 2.9). Then, locks from all three crews and the supervisor would fit into the holes of the hasp. There would be a tag stating "do not use" if this is not printed on the device. When each crew has completed their individual projects, their lock is removed from the device. When the job is complete, the supervisor would take the lock off of the device, remove the tag, and allow the device to be turned on.

Locks should not be cut off of a lockout device or a piece of equipment except in cases where the key is lost, the worker who applied it were to quit without returning keys, or the worker is unable to return to work. Then, the decision to cut the lock becomes an administrative decision—only cut by an authorized person after obtaining permission from the person in charge of the LOTO program. Locks with a master key should not be used.

Figure 2.8 Example of Lockout Device on Valve

Figure 2.9 Multi-Lock HASP at a Clarifier

To ensure the safety of employees, employers should have an energy control program in place. The energy control program ensures that any equipment is fully isolated from its energy source and rendered inoperative before servicing whenever the possibility of an unexpected machine or equipment startup could be present or when the unexpected release of stored energy could occur and cause injury. In summary, a LOTO program that meets OSHA specifications covers employee training in the three following aspects:

An energy control program consists of energy control procedures, employee training, and periodic inspections to ensure that before any employee performs any servicing or maintenance on a machine or equipment where the unexpected energizing, startup, or release of stored energy could occur and cause injury, the machine or equipment shall be isolated from the energy source and rendered inoperative.

- The employer's energy control program,
- Elements of energy control programs that are relevant to the employee's duties, and
- Various requirements of the OSHA LOTO standard.

The following are mandates that an employer must follow to protect employees:

- Develop, implement, and enforce an energy control program.
- Use lockout devices for equipment that can be locked out. Tagout devices can be substituted in lieu of lockout devices only if tagout devices can provide the same degree of protection.
- Ensure that new or overhauled devices are capable of being locked out.
- If existing equipment cannot be locked out, develop a tagout program.
- Develop, document, implement, and enforce energy control procedures.
- Use only LOTO devices authorized for the particular application and ensure that they are durable, standardized, and substantial.
- Ensure that LOTO devices identify the individual users.
- Establish a policy that permits only the employee who applied the device to remove it (see 29 CFR 1910.147 [e][3] for exceptions).
- Inspect and update energy control procedures annually.
- Provide effective training for all employees covered by the standard.
- Comply with the additional energy control procedures in OSHA standards when machines are tested or repositioned, when outside contractors work at the site, in lockout situations, and during shift and personnel changes.

TEST YOUR KNOWLEDGE

1. **Lockout/tagout devices can be removed by another employee for the following reasons:**
   a. If they have permission from the employee who applied the device.
   b. If they are authorized to do so by the person in charge of the LOTO program.
   c. If the employee who put on the lock is on lunch.
   d. If the work is completed.

2. **Which of the following is true about energy control procedures?**
   a. They cover maintenance schedules.
   b. They include periodic inspections.
   c. They do not need to include OSHA requirements.
   d. Energy control procedures are optional.

3. **Lockout/tagout guidance is only a general recommendation and is not legally required by OSHA.**
   ☐ True
   ☐ False

Confined Space Entry

Conducting work in a WRRF may involve activities in which the operator or mechanic will be required to work in a space where access is restricted or difficult, where breathing air may be harmful because of toxic contaminants or atmosphere, or where there is a greater risk for an explosion because of residual vapors. Additionally, there may be other hazards from mechanical motion, flooding of the space from varied sources, and physical entrapment that could lead to crushing or death by asphyxiation. If any of these conditions exist, the work environment is likely a confined space.

Examples of confined spaces found in WRRFs and collection systems include, but are not limited to, manholes, meter vaults, wet wells, valve chambers, and pumping stations (see Figure 2.10).

Figure 2.10 Example of a Confined Space (Lift Station)

The U.S. Occupational Safety and Health Administration has defined a *confined space* using three conditional criteria. All three criteria must be met before the space is considered a confined space. First, the space must be large enough and have a configuration where a worker can enter the space. That seems simple enough. The space must be big enough to get into and there must be a way in. Second, the space has a limited means of access or egress (i.e., the way to, through, and away from an exit). Although this seems relatively straightforward, there are conditions or circumstances that may warrant further consideration. For example, a 0.75-m [30-in.] manhole located near the bottom of a tank may appear to provide an adequate access point; however, could five workers exit through that opening in a timely manner under emergency conditions? Workers must be able get in or out of the space unassisted, self-rescue must be a realistic option, and there must be sufficient access points so that all workers can exit the space in a timely manner.

The third criterion for a confined space is that the space is not designed for continuous human occupancy. This criterion needs some interpretation. For instance, some people might think that only a space such as a living room would qualify as being designed for continuous occupancy. However, if the design of the space took into consideration the work to be performed and provided adequate space, lighting, and clean air for breathing, then it was intended for continuous occupancy. For more information, the reader is referred to OSHA's *Letter of Interpretation—Permit Required Confined Space Standard as it Applies to Certain Operations* (OSHA, 1993).

A confined space may be classified as either a non-permit-required confined space or a permit-required confined space. Under certain conditions, a permit-required confined space may allow for alternate entry procedures. It is never appropriate to assume the classification of a confined space based solely on past practice or to underclassify a space for reasons influenced by available staffing, rescue personnel, or time. When classifying a confined space, it is important to take into consideration design and physical configuration of the space and known hazards that must be considered that are supported by actual testing or the past history of working within the space. If for any reason (e.g., the purpose of the entry is not routine) the conditions are not as expected or the work to be performed will cause or create a health or safety hazard, then the space must be evaluated and classified appropriately as the actual conditions dictate.

A non-permit-required confined space is a confined space that does not contain or have the potential to contain any recognized hazard capable of causing death or serious physical harm. This also includes any hazard or potential hazard associated with breathing air within the space.

Permit-required confined space entries are expected to be relatively infrequent and involve little to no risk because hazards have been eliminated or controlled. Most entries will be scheduled and well planned for the purpose of inspections or routine maintenance. Typically, all work in a permit-required confined space involving significant maintenance, construction, or repair work, or lasting for extended periods of time, will be performed by qualified contractors. However, in rare instances (i.e., emergency repairs), trained personnel may be required to perform work in a permit-required confined space while a hazard may still exist.

A permit-required confined space entry will include, at a minimum, the following three individuals: the attendant, the entrant, and the entry supervisor. An emergency rescue plan, appropriate for hazards associated with the entry, must be considered. Entrant self-rescue or rapid retrieval are typical rescue plans. In rare instances, on-site rescue services may be necessary. Figure 2.11 presents a typical confined entry permit; the permit must contain the items shown in the example. The permit is be used to establish appropriate conditions before entry. A permit is typically only valid for one shift.

Alternate entry procedures may be used for entering a permit-required confined space provided that the only hazard posed by the permit space is an actual or potential hazardous atmosphere and continuous forced air ventilation alone is sufficient to maintain that permit space safe for entry. Water resource recovery facilities must review this procedure in great detail before an entrant is given permission to enter.

CONFINED SPACE ENTRY PERMIT

LOCATION AND DESCRIPTION OF CONFINED SPACE ______________________

ENTRY SUPERVISOR ______________________ DATE __________

ATTENDANT ______________________ TIME __________

PURPOSE OF ENTRY ______________________ EXPIRATION __________

____

HAZARDS ASSOCIATED WITH CONFINED SPACE

Yes No N/A		Yes No N/A	
☐ ☐ ☐	Chemicals ______________	☐ ☐ ☐	Slip/Trip/Falls
☐ ☐ ☐	Strike Against	☐ ☐ ☐	Thermal Burn
☐ ☐ ☐	Inhalation	☐ ☐ ☐	Engulfment
☐ ☐ ☐	Explosive Atmoshpere	☐ ☐ ☐	Electric Shock
☐ ☐ ☐	Special Hazards	☐ ☐ ☐	Other Permitted Activities

SPECIAL REQUIREMENTS AND SUPPORT EQUIPMENT

Yes No N/A		Yes No N/A	
☐ ☐ ☐	Lockout/De-energization	☐ ☐ ☐	Fire Extinguisher
☐ ☐ ☐	Line breaking/Capped/Blinded	☐ ☐ ☐	Fall Protection/Retrieval Equipment
☐ ☐ ☐	Purge/Flush	☐ ☐ ☐	Ventilation
☐ ☐ ☐	Respiratory Protection	☐ ☐ ☐	Lighting
☐ ☐ ☐	Secure Area	☐ ☐ ☐	Personal Protective Equipment

Communication means ☐Voice ☐ Radio ☐ Other __________

ATMOSPHERIC TESTING

Is continuous monitoring required? Yes☐ No☐ If yes, for what. __________

Atomospheric Testing	Acceptable Range	Time	Result	Time	Result	Time	Result	Time	Result
% of Oxygen	>19.5% or <23.5%								
% of Lower Explosive Limit (LEL)	<10% of LEL								
Other:									

Tester Signature ______________________

ATMOSPHERIC TESTING EQUIPMENT USED

☐Oxygen Meter ☐ Multi Pump ☐Manufacturer Name __________

☐Explosion Meter ☐ Multi Color Pac ☐Other __________

ENTRY PERSONNEL

NAME	DESIGNATION	IN	OUT	IN	OUT	IN	OUT

PRE-ENTRY BRIEFING

Have all entrants been briefed in the hazards associated with the confined space? Yes☐

Has the SDS of any chemicals in or being used in the confined space been discussed? Yes☐ N/A☐

All Emergencies Call: 911

I hereby certify that the above mentioned area has been deemed safe for confined space entry as long as the above restrictions are observed and monitoring is completed.

____

Entry Supervisor Signature

POST-ENTRY BRIEFING

Did any problems or unexpected hazards occur during the entry Yes☐ No☐

If yes explain ______________________

34

Figure 2.11 A Typical Confined Entry Permit

HAZARDOUS ATMOSPHERES

Hazardous atmospheres encountered in confined spaces can be divided into four distinct categories: flammable, toxic, irritant and/or corrosive, and asphyxiating. A flammable atmosphere generally arises from enriched oxygen atmospheres, vaporization of flammable liquids, byproducts of work, chemical reactions, concentrations of combustible dusts, and desorption of chemical from inner surfaces of the confined space. An atmosphere becomes flammable when the ratio of oxygen to combustible material in the air is neither too rich nor too lean for combustion to occur. Combustible gases or vapors will accumulate when there is inadequate ventilation in areas such as a confined space. Flammable gases such as acetylene, butane, propane, hydrogen, methane, and natural (hydrogen sulfide) or manufactured gases or vapors from liquid hydrocarbons can be trapped in confined spaces; because many gases are heavier than air, they will seek lower levels as in pits, sewers, and various types of storage tanks and vessels. In a closed-top tank, it should also be noted that lighter-than-air gases may rise and develop a flammable concentration if trapped above the opening. The byproducts of work procedures can generate flammable or explosive conditions within a confined space. Specific kinds of work such as spray painting can result in the release of explosive gases or vapors. Welding in a confined space is a major cause of explosions in areas that contain combustible gas. Chemical reactions forming flammable atmospheres occur when surfaces are initially exposed to the atmosphere, or when chemicals combine to form flammable gases. In a dry permit-required confined space, compounds have the potential to explode upon percussion or exposure to increased temperature. Another class of chemical reactions that form flammable atmospheres arises from deposits of pyrophoric substances (ferric chloride, ferrous oxide, ferrous sulfate, iron, etc.). These tanks containing flammable deposits will spontaneously ignite upon exposure to air.

Substances to be regarded as toxic in a confined space can cover the entire spectrum of gases, vapors, and finely divided airborne dust in industry. The sources of toxic atmospheres encountered may arise from the following: (a) the treatment process; (b) the product stored (removing decomposed organic material from a tank can liberate toxic substances, such as hydrogen sulfide); or (c) the operation performed in the confined space (e.g., welding or brazing with metals capable of producing toxic fumes). During loading, unloading, formulation, and production, mechanical and/or human error may also produce toxic gases that are not part of the planned operation.

Carbon monoxide is a hazardous gas that may build up in a confined space. This odorless, colorless gas that has approximately the same density as air is formed from incomplete combustion of organic materials such as wood, coal, gas, oil, and gasoline; it can be formed from microbial decomposition of organic matter in sewers, silos, and fermentation tanks. Carbon monoxide is an insidious toxic gas because of its poor warning properties. Early stages of carbon monoxide intoxication are nausea and headache. Carbon monoxide may be fatal at 1000 ppm in air, and is considered dangerous at 200 ppm because it forms carboxyhemoglobin in the blood, which prevents the distribution of oxygen in the body. Carbon monoxide is a relatively abundant colorless, odorless gas; therefore, any untested atmosphere must be suspect. It must also be noted that a safe reading on a combustible gas indicator does not ensure that carbon monoxide is not present. Carbon monoxide must specifically be tested for. The formation of carbon monoxide may result from chemical reactions or work activities; therefore, fatalities due to carbon monoxide poisoning are not confined to any particular industry. There have been fatal accidents in WRRFs due to decomposition products and lack of ventilation in confined spaces.

In welding operations, oxides of nitrogen and ozone are gases of major toxicologic importance, and incomplete oxidation may occur and carbon monoxide can form as a byproduct.

Another poor work practice, which has led to fatalities, is the recirculation of diesel exhaust emissions. Increased carbon monoxide levels can be prevented by strict control of the ventilation and the use of catalytic convertors.

Irritant or corrosive atmospheres can be divided into primary and secondary groups. The primary irritants exert no systemic toxic effects (i.e., effects on the entire body). Examples of primary irritants are chlorine, ozone, hydrochloric acid, hydrofluoric acid, sulfuric acid, nitrogen dioxide, ammonia, and sulfur dioxide. A secondary irritant is one that may produce systemic toxic effects in addition to surface irritation. Examples of secondary irritants include benzene, carbon tetrachloride, ethyl chloride, trichloroethane, trichloroethylene, and chloropropene. Irritant gases vary widely among all areas of industrial activity. They can be found in plastics plants, chemical plants, the petroleum industry, tanneries, refrigeration industries, and paint manufacturing and mining operations. Prolonged exposure at irritant or corrosive concentrations in a confined space may produce little or no evidence of irritation. This may result in a general weakening of the defense reflexes from changes in sensitivity. The danger in this situation is that the worker is typically not aware of any increase in their exposure to toxic substances.

Asphyxiating atmospheres are those where oxygen is deficient. The normal atmosphere is composed approximately of 20.9% oxygen and 78.1% nitrogen, and 1% argon with small amounts of various other gases. The reduction of oxygen in a confined space may be the result of either consumption or displacement. The consumption of oxygen takes place during combustion of flammable substances, as in welding, heating, cutting, and brazing. A more subtle consumption of oxygen occurs during bacterial action, as in the fermentation process. Oxygen may also be consumed during chemical reactions as in the formation of rust on the exposed surface of the confined space (iron oxide). The number of people working in a confined space and the amount of their physical activity will also influence the oxygen consumption rate.

A second factor in oxygen deficiency is displacement by another gas. Examples of gases that are used to displace air, and, therefore, reduce the oxygen level, are helium, argon, and nitrogen. Carbon dioxide may also be used to displace air and can occur naturally in sewers, storage bins, wells, tunnels, wine vats, and grain elevators. Aside from the natural development of these gases, or their use in the chemical process, certain gases are also used as inerting agents to displace flammable substances and retard pyrophoric reactions. Gases such as nitrogen, argon, helium, and carbon dioxide are frequently referred to as *non-toxic inert gases*, but have claimed many lives. The use of nitrogen to inert a confined space has claimed more lives than carbon dioxide. The total displacement of oxygen by nitrogen will cause immediate collapse and death. Carbon dioxide and argon, with specific gravities greater than air, may lie in a tank or manhole for hours or days after opening. Because these gases are colorless and odorless, they pose an immediate hazard to health unless appropriate oxygen measurements and ventilation are adequately carried out. Oxygen deprivation is one form of asphyxiation.

While it is desirable to maintain the atmospheric oxygen level at 20.9% by volume, the body can tolerate deviation from this ideal. When the oxygen level falls to 17%, the first sign of hypoxia is a deterioration to night vision, which is not noticeable until a normal oxygen concentration is restored. Gas meters are set to alarm for oxygen concentrations at 19.5% and at 23.5%.

ENTRY PROCEDURES

There are three major roles to be assigned in a permit required confined space (PRCS) entry. The first, and the one at highest risk, is the entrant. Before going in, the entrant needs to have direct authorization from their employer to enter the space. The OSHA confined space standard states that the entrant must:

- Know the hazards associated with confined space entry, and, in particular, the hazards associated with the PRCS being entered.
- Know how to use all required equipment.
- Know the procedures for communication with the attendant.
- Know how to alert the attendant of hazardous or prohibited conditions.
- Know how to exit the space if necessary (i.e., self-rescue).

The second party of the confined space team is the attendant. There must be at least one attendant on each PRCS team. The attendant, arguably, has the most amount of responsibility on the confined space team because they have the highest number of duties required by OSHA. These duties are to:

- Know the hazards. In the case of the attendant, this can often include using air-monitoring equipment to keep a close watch on the atmospheric conditions inside the confined space and communicate any changes observed.
- Know the signs and symptoms of potential exposure to hazards.
- Be able to identify the authorized entrants.
- Remain outside until relieved.
- Communicate with entrants throughout the work period.
- Monitor and evacuate entrants if necessary.
- Summon rescue, if needed.
- Warn away unauthorized persons.

The third and last required party on a PRCS team is the entry supervisor. In most cases, the entry supervisor is the employer or directly represents the employer. It is a good idea for this person to be trained and ready to serve as an entrant or attendant if necessary as well. The entry supervisor is responsible for determining whether acceptable entry conditions exist, authorizing the entry, overseeing entry operations, terminating the entry, and canceling the entry permit. Per the OSHA regulation, the entry supervisor must:

- Know the hazards.
- Verify safe entry conditions.
- Terminate entry and cancel permit.
- Verify availability and effectiveness of rescue services.
- Remove unauthorized persons.
- Ensure acceptable entry conditions are maintained.

While OSHA has established an exposure limit of 50 ppm for carbon monoxide, other organizations who study and recommend exposure limits have recommended a lower exposure limit for carbon monoxide of 25 ppm. Further, the OSHA PEL for hydrogen sulfide is 50 ppm, whereas NIOSH recommends a limit of 10 ppm.

ATMOSPHERIC MONITORING

Before anyone enters a confined space, the air must be tested (from a safe location outside of the confined space). The equipment must be capable of detecting flammability, oxygen levels, carbon monoxide, and toxicity (typically hydrogen sulfide). Typically, these are referred to as *4-gas meters*. Table 2.3 shows the atmospheric limits for entry based on the permissible exposure limit (PEL) issued by OSHA.

Depending on the conditions, additional toxic gases may need to be tested for and may require additional or different equipment. The meter should also have alarms, preferably both visual and audio.

TEST YOUR KNOWLEDGE

1. The U.S. Occupational Safety and Health Administration has defined a *confined space* using three conditional criteria: (1) large enough and employee can enter, (2) limited means of access or egress, and (3) not designed for continuous human occupancy.
   - ☐ True
   - ☐ False
2. Permit required confined spaces can be declassified as non-permitted confined spaces.
   - ☐ True
   - ☐ False
3. An entrant needs to know the hazards before entering a confined space.
   - ☐ True
   - ☐ False
4. Which of the following is a responsibility of the attendant?
   a. Communicate with the human resources department.
   b. Know the weather forecast.
   c. Know the signs or symptoms of potential exposure by the entrant.
   d. Terminate and cancel entry permit.
5. Hazardous atmospheres include:
   a. Nitrogen deficiency
   b. Combustible atmospheres
   c. Nontoxic chemicals or gases
   d. Dirt and dust

Table 2.3 Limits for Atmospheric Monitoring

Oxygen	19.5–23.5%
Carbon monoxide	0–50 ppm (OSHA PEL)
Hydrogen sulfide	0–50 ppm (OSHA PEL)
Combustible gases	Below 10% of lower explosive limit

Hazard Communication

A hazard communication program (HAZCOM) is typically recognized as the standard for employees' right to know. There are three broad components to a written HAZCOM program. Each has detailed requirements listed in the OSHA standard that outline specific rules for employers and employees to follow: (1) labels and other forms of warning for chemicals in the workplace, (2) SDSs, and (3) employee information and training (29 CFR 1910.1200). See Figure 2.12 for an example cover of an SDS.

Before 2012, SDSs were referred to as *material safety data sheets* (MSDSs).

Chemical manufacturers, importers, and distributors of hazardous chemicals are all required to provide the appropriate labels and MSDSs to the employers to which they ship the chemicals. The information is to be provided automatically. Every container of hazardous chemicals must be labeled, tagged, or marked with the required information.

Suppliers must provide end users a properly completed SDS at the time of the first shipment of the chemical, and with the next shipment after the SDS is updated with new and significant information about the hazards.

Employers that use or store hazardous chemicals must have a program to ensure the information is provided to exposed employees.

HAZARD COMMUNICATION PROGRAM

The U.S. Occupational Safety and Health Administration requires employers to have a HAZCOM that includes:

- A written hazard communication program
- Labels and other forms of warning
- Safety data sheets
- Employee information and training

IDENTIFY HAZARDOUS CHEMICALS

The OSHA standard requires a list of hazardous chemicals in the workplace as part of the written hazard communication program. The list will eventually serve as an inventory of everything for which an SDS must be maintained.

- Survey the workplace.
- Check purchase records.
- Establish receiving procedure.
- Forward all SDSs to the proper person.

Identify chemicals in containers, including pipes, but also think about chemicals generated in the work operations. For example, welding fumes, dusts, and exhaust fumes are all sources of chemical exposures. Read labels provided by suppliers for hazard information. Make a list of all chemicals in the workplace that are potentially hazardous. Note on the list the locations of the products within the workplace.

VERIFY THE SAFETY DATA SHEET

Check SDS files against the chemical inventory. If any are missing, contact your supplier and request one. Document these requests, either by copy of a letter or a note regarding telephone conversations. Do not allow employees to use any chemicals for which you have not received an SDS.

WRITTEN PROGRAM

All workplaces where employees are exposed to hazardous chemicals must have a written plan that describes how the OSHA standard will be implemented. The written program has to reflect what you are doing in your workplace. If you use a generic program, it must be adapted to address the facility it covers. The written plan must list the chemicals present at the site, indicate who is to be responsible for the various aspects of the program, and tell where written materials will be made available to employees.

Chlorine

Safety Data Sheet E-4580

according to the Hazardous Products Regulation (February 11, 2015)

Date of issue: 10-15-1979 Revision date: 01-01-2021 Supersedes: 09-27-2018

SECTION 1: Identification

1.1. Product identifier

Product form	: Substance
Substance name	: Chlorine
CAS No	: 7782-50-5
Formula	: Cl2
Other means of identification	: Dichlorine, Molecular chlorine, Betholite
Product group	: Core Products

1.2. Recommended use and restrictions on use

Recommended uses and restrictions	: Industrial use Use as directed.

1.3. Supplier

Linde Canada inc.
1200 – 1 City Centre Drive
Mississauga - Canada L5B 1M2
T 1-905-803-1600 - F 1-905-803-1682
www.lindecanada.ca

1.4. Emergency telephone number

Emergency number	: 1-800-363-0042 Call emergency number 24 hours a day only for spills, leaks, fire, exposure, or accidents involving this product. For routine information, contact your supplier or Linde sales representative.

SECTION 2: Hazard identification

2.1. Classification of the substance or mixture

GHS-CA classification

Ox. Gas 1	H270
Liquefied gas	H280
Acute Tox. 2 (Inhalation: gas)	H330
Skin Corr. 1A	H314
Eye Dam. 1	H318
STOT SE 3	H335

2.2. GHS Label elements, including precautionary statements

GHS-CA labelling

Hazard pictograms :

Signal word	: DANGER
Hazard statements	: MAY CAUSE OR INTENSIFY FIRE; OXIDIZER CONTAINS GAS UNDER PRESSURE; MAY EXPLODE IF HEATED CAUSES SEVERE SKIN BURNS AND EYE DAMAGE FATAL IF INHALED CORROSIVE TO THE RESPIRATORY TRACT (This statement supercedes H335)
Precautionary statements	: Do not handle until all safety precautions have been read and understood Keep away from clothing and other combustible materials Keep valves and fittings free from oil and grease

This document is only controlled at the time received electronically directly from Linde or while on the Linde Canada inc. website where a copy of this controlled version is available for download. Linde cannot assure the integrity or accuracy of any version of this document after it has been electronically downloaded or removed from our website.

EN (English - CA) SDS ID : E-4580 1/10

Figure 2.12 Example SDS for Chlorine (always check the manufacturer's website for the most current SDS) (Reprinted with permission from Linde)

LABELING SYSTEM

The written program must describe how the requirements for labels and other forms of warning, MSDSs, and employee information and training are going to be met in your facility. In-plant containers of hazardous chemicals must be labeled, tagged, or marked with the identity of the material and appropriate hazard warnings. If chemicals are transferred from a labeled container to another container, the new container must be properly labeled. There are no specific requirements for size or color, or any specified text.

TEST YOUR KNOWLEDGE

1. **Hazard communication is also known as the:**
   a. The employers right to refuse
   b. The employers right to know
   c. The employees right to know standard
   d. None of the above

2. **What are the three broad elements to a written HAZCOM program?**
   a. Labeling, SDSs, and employee information and training
   b. Science, math, and technology
   c. U.S. Occupational Safety and Health Administration needs, employer needs, and chemical manufacturer needs
   d. There are no requirements for a written program

3. **SDS is an acronym for:**
   a. Safety down the street
   b. Safely describe sewers
   c. Safety data sheet
   d. Sewers destroy safety

Chemical Handling, Storage, and Response

Water resource recovery facility workers are at risk of serious chemical-related injuries and illnesses because of the concentrations of chemicals being used. When working with chemicals, health effects can also be chronic and may not be apparent for many years. Health effects can be acute and cause immediate and serious damages to the body upon exposure. There are various ways to be exposed and adversely affected by chemicals. These include:

- Chemical receiving
- Chemical transfer
- Changing containers
- Repairing leaks
- Connecting tanks, cylinders, feed systems, and containers
- Responding to incidental releases

HANDLING

When handling chemicals, even in small quantities, it is vital to know the chemical's hazardous properties and the safest means of handling the chemical. This includes understanding and using proper engineering controls and administrative controls and wearing the required PPE. Employees also need to be aware of incompatibility issues, proper spill response measures, and which chemicals may react and emit off-gases (i.e., fumes) when heated or mixed.

To become more familiar with chemicals, employees should consult with their supervisor or coworkers or review the chemical's SDS. The SDS provides detailed information on the chemical's hazardous properties and provides preventive measures such as first aid, engineering controls, and PPE that should be worn when working with the chemical.

Engineering controls are methods implemented to protect workers by removing hazardous conditions or by placing a barrier between the worker and the hazard. Examples include local exhaust ventilation to capture and remove airborne emissions or machine guards to shield the worker. Administrative controls may include, but are not limited to, safety rules, policies, and training (NIOSH, https://www.cdc.gov/niosh/engcontrols/).

STORAGE

Proper chemical storage is important not only for safety, but for quality control. Chemicals should be dated, properly labeled, and inventoried regularly. See Figure 2.13 for proper and improper labeling. Care must be taken to not store chemicals with other incompatible chemicals because they might react with each other and cause toxic off-gasing (i.e., fuming) and/or fires and explosions. Table 2.4 shows the storage compatibility for various chemical groups.

Therefore, it is important to be sure that only containers marked for that chemical be used. For example, potassium permanganate placed into a bucket that contains organic substances can explode. Chemical storage must also provide secondary containment to keep spills, no matter how small, from contaminating surrounding surfaces, which could cause an injury to an unsuspecting worker. If secondary containment is not designed into chemical storage cabinets, a rubber tub or polypropylene tray can be used as long as it is compatible with the materials being stored.

SPILL RESPONSE

Chemical responses (spills) at WRRFs could involve anything from chlorine rail cars to small bottles of laboratory reagents. Regardless of the chemical type, facility operators will need to be prepared for whatever spill might come their way. The first step is to determine the type of spills a facility might expect. This means your chemical inventory required by HAZCOM must be up to date. This includes updating chemical inventory, ensuring all SDSs have been collected and placed in a designated location, and updating a written program.

OSHA's HAZCOM standard requires all employers with hazardous chemicals in their workplaces prepare and implement a written hazard communication program, ensure that all containers are labeled, provide employees access to SDSs, and have an effective training program for all potentially exposed employees.

The U.S. Occupational Safety and Health Administration defines three types of spills. The first type of spill is an incidental release. This type of release can typically be handled by one individual and does not require additional help other than what's immediately available to the first responder. The second type of spill is an operational type of spill to which an operational spill team will respond (see Figure 2.14). This type of spill exceeds a single individual department's capabilities and, therefore, other facility participation might be needed. (If an operator needs to call for help, then it is classified as an operational spill. The help could be internal to the facility). The third and most dangerous is an emergency response (technician) level spill. This type of spill requires the use of an outside elevated team. An elevated team could include a local third-party hazardous materials (HAZMAT) team, the fire department, or other city/municipal responders. This elevated team is commonly referred to as the *HAZMAT team*.

A HAZMAT response team is an organized group of individuals trained to respond to spills or leaks of hazardous materials. A HAZMAT team can be internal, third-party, or the local fire department that has specific first responder technician level training.

Figure 2.13 Example of Proper and improper labeling. Containers with and without U.S. Department of Transportation labeling

Table 2.4 Storage Compatibility for Various Chemical Groups

	Flammable Liquids	Oxidizers	Organic Acids	Inorganic Acids	Bases	Water Reactives	Aqueous Solutions	Cyanides
Flammable Liquids		X		X				
Oxidizers	X		X					
Organic Acids		X		X	X	X		X
Inorganic Acids	X		X		X	X		X
Bases			X	X				
Water Reactives			X	X			X	
Aqueous Solutions						X		X
Cyanides			X	X			X	

Water resource recovery facility operators need to train their employees to respond to each of the respective levels of spills. If it is expected that employees will be part of a larger HAZMAT team, the minimum level of training is 24 hours (American Chemical Society, 2018). If the facility will only respond to operational level spills, then 8 hours of training is all that is needed. For specific training requirements, you will need to refer to the OSHA regulation titled "HAZWOPER" (Hazardous Waste Operations and Emergency Response) at 29 CFR (Code of Federal Regulations) 1910.120 paragraph (q), which provides the training requirements for emergency responders.

Remember, if it is expected that WRRF employees will respond to chemical spills, training is required, but it is also important to make sure employees have the tools to perform the response. It is standard practice to have PPE, plug and patch kits, containment equipment, cleanup equipment, and decontamination equipment. For those under the jurisdiction of OSHA, OSHA requires the employer to pay for all equipment, PPE, and training.

Figure 2.14 Example Operational Spill Team

TEST YOUR KNOWLEDGE

1. Which chemical spill can be handled by a single individual?
   a. Incidental
   b. Operational
   c. Emergency response
   d. None of the above

2. How many hours of training are needed at the operational level?
   a. 2 hours
   b. 4 hours
   c. 8 hours
   d. 24 hours

3. Who should pay for required equipment, PPE, and training?
   a. The employee
   b. The employer
   c. The community
   d. The federal government

4. What does the acronym HAZWOPER mean?
   a. Hazardous Wastes Operations Team
   b. Hazardous Waste Operations and Emergency Response
   c. Hazardous Worker Personnel
   d. Hazardous Work Operations

Heavy Equipment and Machinery

Water resource recovery facilities often handle large and heavy materials. This section covers various pieces of equipment and machinery (as well as guarding for that machinery) often found in WRRFs. These include forklifts, cranes, hoists, machinery, and lifts.

CRANES AND HOISTS

There is a significant amount of heavy lifting using hoists and cranes in WRRFs. Injuries involving cranes, hoists, and the lifting of heavy loads can result in a significant injury or even death due to the weight and force involved. Always follow the manufacturer's directions to minimize risk. Never lift more weight than a hoist or crane is rated for. Always lift straight up. Lifting at an angle may cause a hoist or crane to tip or drop its load.

The types of cranes or hoists typically used at WRRFs include the following:

- Bridge cranes (see Figure 2.15): These are often used for changing and lifting plant filters (American Crane, 2021). They are considered an "overhead" crane, can be "single-girder" or "double-girder," and typically have the largest capacity of 100 tons (9 0744 kg) or more (Crane Repair Company, n.d.; American Crane, 2021).
- Gantry Cranes: These cranes have two legs that support a girder, which spans over the area between the legs (Crane Repair Company, n.d.). They look and function similar to a bridge crane except that they have legs on wheels that extend to the ground. They are often used for lifting, moving, and placing pumps, motors, valves, and blowers, as well as for digester cover removal and rehabilitation (American Crane, 2021). Gantry cranes allow for more mobility and height adjustments than bridge cranes, especially when moving loads outdoors.

Figure 2.15 Bridge Crane

The U.S. Occupational Safety and Health Administration's 29 CFR 1910.179 applies to overhead and gantry cranes, including semi-gantry, cantilever gantry, wall cranes, storage bridge cranes, and others having the same fundamental characteristics that were installed after August 31, 1971. Per OSHA, these cranes are grouped because they all have trolleys and similar travel characteristics. A unique aspect of this OSHA regulation is that it refers to the requirements of the ANSI Standard for Overhead and Gantry Cranes, ANSI B30.2.0-1967, and is "incorporated by reference."

Trolley: The "trolley" is a wheeled mechanism part of the crane from which a hoist is suspended. The trolley consists of a cart that moves along the jib horizontally along a beam/bridge rail.

Cranes can be modified and load capacity rerated as long as the modifications and associated structure are thoroughly checked for the new rated load by a qualified engineer or the equipment manufacturer (Grainger, 2018).

The rated load of the crane should be plainly marked on each side of the crane; if more than one hoist is present, each hoist must have its rating shown (Grainger, 2018). If the crane is not marked, or if you are unsure of its capacity, do not use it.

Incorporation by reference is defined in 29 CFR 1910.6(a)(1) of the OSHA General Industry Standards. The U.S. Occupational Safety and Health Administration lists government standards as well as consensus organization standards (e.g., ANSI, American Society of Mechanical Engineers, American Petroleum Institute, American Society for Testing and Materials, Compressed Gas Association, National Fire Protection Association, and Society of Automotive Engineers), which are incorporated by reference. These publications have the same force and effect as other OSHA standards. Not all standards are incorporated by reference in the OSHA standards.

Clearance must be maintained above and to the side of cranes (Grainger, 2018). Do not store boxes, equipment, or chemicals near them.

Walkways should not be placed in a crane operating zone that would compromise employee safety when the crane is in operation (Grainger, 2018). If this is observed in your WRRF, discuss with the facility engineers or managers about redirecting or adding additional protection for employees.

Parallel cranes must have adequate clearance between the two bridges if no walls or structures are between them (Grainger, 2018).

Only designated personnel are permitted to operate a crane (Grainger, 2018).

Crane inspections must be conducted per OSHA 1910.179(j)(1). These include "frequent" and " periodic" inspections. Table 2.5 lists the items to be inspected and their inspection frequency.

Frequent is defined as daily to monthly intervals. *Periodic* is defined as 1- to 12-month intervals.

Table 2.5 Frequent Inspections List

Item to Be Inspected	Frequency
■ Functional operating mechanisms for maladjustment ■ Deterioration or leakage in lines, tanks, valves, drain pumps, and other parts of air or hydraulic systems ■ Hooks with deformation or cracks (visual) ■ Hoist chains and end connections for excessive wear, twist, or distortion interfering with proper function, or stretch beyond manufacturer's recommendations (visual)	Daily
■ Hooks with deformation or cracks (written record with signature of inspector and date) ■ Hoist chains and end connections for excessive wear, twist, or distortion interfering with proper function, or stretch beyond manufacturer's recommendations (written record with signature of inspector and date) ■ Running rope and end connections for wear, broken strands, and so on (written record with signature of inspector, rope identity, and date)	Monthly
Moving pieces and critical parts for excessive wear	Daily to monthly
Rope reeving according to manufacturers' recommendations *(Note: "Rope reeving" is a mechanical device with several pulleys. It allows the crane to lift heavy loads while limiting the force applied on the hoisting ropes.)*	As recommended by manufacturer

Complete inspections of the crane are to be performed periodically and include the following items:

- Deformed, cracked, or corroded members
- Loose bolts or rivets
- Cracked or worn sheaves and drums
- Worn, cracked, or distorted parts such as pins, bearings, shafts, gears, rollers, and locking and clamping devices
- Excessive wear on brake system parts, linings, pawls, and ratchets
- Load, wind, and other indicators over their full range, for any significant inaccuracies
- Gasoline, diesel, electric, or other power plants for improper performance or noncompliance with applicable safety requirements
- Excessive wear of chain-drive sprockets and excessive chain stretch
- Electrical apparatus, for signs of pitting or any deterioration of controller contactors, limit switches, and pushbutton stations

Always repair or replace worn parts promptly.

A preventive maintenance program—based on the manufacturer's recommendations—is required to be established in the United States per the OSHA regulation. Creating an effective program can prevent unplanned breakdowns and significant safety incidents. The maintenance program should include defined, safe procedures to ensure those performing maintenance are not injured. Refer to 29 CFR 1910.147 to ensure requirements for the control of hazardous energy or LOTO are incorporated to the program.

Operation of the crane is covered under OSHA 1010.179(n). Only designated personnel shall be permitted to operate a crane covered by this section of the OSHA regulation. Operators of the crane must be aware of the size of the load, how to attach the load, and how to move the load, and be able to test the hoist switch limit before putting a load on. Loads should never be moved over the heads of people.

POWERED INDUSTRIAL VEHICLES—FORKLIFTS

The OSHA Powered Industrial Vehicle (PIV) Standard, 29 CFR 1910.178, covers a range of PIVs/forklifts (see Table 2.6). Specifically, the standard covers fire protection, design, inspection, maintenance, use, operation, and training for forklifts powered by electric motors or internal combustion engines. Although this OSHA standard does not apply to compressed air or nonflammable compressed-gas-operated forklifts, following the requirements is recommended. And while the standard refers to these as "powered industrial vehicles" or "PIVs", they will be referred to as forklifts for this section.

TYPES OF FORKLIFTS

The U.S. Occupational Safety and Health Administration classifies forklifts into the following seven classifications (see also Table 2.7):

- Class I: Electric Motor Rider Trucks
- Class II: Electric Motor Narrow Isle Trucks
- Class III: Electric Motor Hand Trucks or Hand/Rider Trucks (i.e., pallet jack)
- Class IV: Internal Combustion Engine Trucks on Solid or Cushion Tires (i.e., common forklift)
- Class V: Internal Combustion Engine Trucks on Pneumatic Tires
- Class VI: Electric and Internal Combustion Engine Tractors
- Class VII: Rough Terrain Forklift Trucks

Important factors to consider when operating a forklift include the following.

Center of Gravity: Most forklifts are used for lifting and/or moving materials or equipment throughout the facility. It is especially important to remember that many forklifts operate on a center of gravity that allows the forklift to stay stable, especially when carrying loads. To prevent the forklift from tipping, weight must be distributed evenly, especially when carrying irregular-sized loads. Loads should also be kept as close to the front wheels of the forklift as possible and in the center of the forks, when applicable. Always follow the manufacturer's directions and review operations and maintenance manuals before use.

Table 2.6 Types of Forklifts

Diesel	Electrical	Gasoline	Liquified Petroleum/ Propane Gas
D = diesel (minimal to no safeguards against fire hazards)	E = electrical (minimal to no safeguards against fire hazards)	G = gasoline (minimal to no safeguards against fire hazards)	LP = liquified petroleum/ propane (minimal to no safeguards against fire hazards)
DS = diesel with additional safeguards against inherent fire hazards	ES = electrical with additional safeguards to electrical system	GS = gasoline with additional exhaust, fuel, and electrical system safeguards	LPS = LP with additional exhaust, fuel, and electrical system safeguards
DY = safeguards of DS, but with enclosed electrical and other safeguards	EE = safeguards of E and ES PIVs, but have enclosed electrical motors and equipment		
	EX = safeguards of E, ES, and EE models, but specifically constructed and assembled to be used in atmospheres containing flammable vapors or dusts		

Load Center: According to OSHA (OSHA, 2008), the load center is the distance from the face of the forks to the center of gravity of the load. Many forklifts are rated using a 61-cm [24-in.] load center, which means that the center of gravity of the load must be 61 cm [24 in.] or less from the face of the forks (OSHA, 1998). Before handling a load, the load center must be determined for the forklift, and the load positioned accordingly.

U.S. Occupational Safety and Health Administration eTools for load handling can be found at https://www.osha.gov/SLTC/etools/pit/operations/loadcomposition.html.

Hazards and Hazard Controls: The most common risk to forklift drivers is the forklift tipping over, especially when the driver is not wearing a seatbelt. Drivers also need to be aware of pedestrians or other workers in the area that they are working in. Besides being hit by the forklift itself, an unstable load can also fall on a person. To reduce the likelihood of a tip-over or dropping a load, drivers should be aware and familiar with workplace conditions and pedestrian traffic and be able to properly load and carry materials. Drivers should remember that pedestrians always have the right of way.

In addition to physical dangers, there are atmospheric hazards to be aware of when operating certain types of forklifts in potentially hazardous or poorly ventilated areas. The OSHA standard (based on Table N-1 located in 1910.178[c]) presents information on the various types, classes, and divisions of forklifts in addition to recommendations for which forklifts can be used in certain locations. Many areas in WRRFs

Table 2.7 Forklift Classes

Classes	Description of Classes
Class I locations	Flammable gases or vapors are or are potentially present in quantities (in the air) sufficient to produce explosive or ignitable mixtures
Class II locations	Locations which are hazardous because of the presence of combustible dust
Class III locations	Easily ignitable fibers or flying present, but not likely to be in suspension in quantities sufficient to produce ignitable mixtures
Unclassified	Locations not possessing atmospheres as described in rows above

are considered unclassified locations. An unclassified location has no hazardous atmosphere, that is, an atmosphere where no flammable or explosive gas is present. Diesel, electric, gasoline, and liquid propane gas-operated forklifts can all be used freely in areas with no flammable or explosive atmospheres. If the area contains flammable gases that are contained and not expected to be released except in instances of malfunction or accident, then diesel, electric, gasoline, and liquid propane gas-operated forklifts may all be used. However, Table N-1 should be consulted for operating in specific hazardous atmospheres. Finally, diesel and gasoline-operated forklifts always need proper ventilation to prevent buildup of carbon monoxide.

Training Requirements: The OSHA standard requires initial training that includes procedures for the safe operation of a forklift, types of forklifts, safety hazards associated with operating and working near forklifts, and the requirements of the OSHA PIV (forklift) standard. The following are key points of an effective training program:

- Use knowledgeable and experienced instructors.
- Use a combination of instruction including lecture, discussion, interactive computer learning, video, written material, and practical training.
- Evaluate the operator's competency.
- Recertify drivers' competency at least every 3 years and provide refresher training whenever a driver shows deficiency in their forklift driving skills (based on operations requirements listed in the following section) or is assigned to a new type of forklift.
- Keep training records for each driver who has been trained and evaluated in accordance with the OSHA standard. Each training record must include the operator's name, the date of training, date of evaluation, and the name of the person(s) performing the training or evaluation. The OSHA standard for training is located in 1910.178(1)(3).

PREQUALIFICATION OF CANDIDATES

Because forklifts can pose a hazard to drivers and passersby, potential operators should be determined to be competent before operating a forklift. This can be achieved through completion of training and a successful evaluation. In addition, a potential forklift driver must have or meet the following basic requirements:

- A valid driver's license
- Good driving record
- Good vision (adequately corrected by glasses or contacts)
- Good hearing (adequately enhanced by hearing aid)
- Physical condition that allows for full and safe operation of the vehicle
- No impairments that affect the balance and/or consciousness of the operator
- No medications or other drugs that affect vision, perception, or other physical capabilities

Once a PIV operator has fulfilled minimum competency requirements, they may participate in initial training. As listed in OSHA standard 1910.178(1)(3)(i)(A) through (M), training should include the following topics:

- Operating instructions, warnings, and precautions for the types of forklift the operator will be authorized to operate
- Differences between the forklift they will be operating and an automobile
- Operation of the engine or motor
- Steering and maneuvering
- Visibility (including restrictions because of loading)
- Fork and attachment adaption, operation, and use limitations
- Vehicle capacity and stability
- Any vehicle inspection and maintenance that the operator will be required to perform
- Refueling and/or charging and recharging of batteries

- Operating limitations
- Any other operating instructions, warnings, or precautions listed in the operator's manual for the types of vehicle that the employee is being trained to operate

Operators are also required to be trained on workplace-specific information such as the type of surface conditions, type of loads to be carried and their stability, load manipulation, stacking and unstacking process, type of pedestrian and/or other traffic to be encountered in the workplace, narrow aisles or other restrictions where the forklift will be operated, any hazardous location or insufficiently ventilated area that the forklift will be operated in, and ramps and/or sloped surfaces where the forklift may be operated.

PERFORMANCE TESTING

After successful completion of a classroom training course, drivers must be given an opportunity to complete practical, hands-on training with the type of forklift they will be operating. The hands-on training may include how to perform a pre-operational inspection; how to operate and move the forklift; how to pick up, travel, set, and stack a load; and refueling and/or charging of batteries. The driver should then be evaluated on proper implementation of training and competency of operation of the forklift.

Operators of forklifts from other employers can provide proof to the new employee. This proof can suffice for the in-class portion, but cannot replace the new operation evaluation.

OBSTACLE COURSE SETUP

To evaluate the driver's skill and competency, an obstacle course should be constructed. Although the obstacle course's length and difficulty is up to the trainer, the course should provide some realistic scenarios the driver will encounter. Some suggested scenarios and obstacles may include the following:

- Boarding and starting the forklift
- Driving forward and backward
- Parking
- Picking up a load and moving forward
- Placing and/or stacking a load
- Driving through narrow or variously sized "aisles"
- Picking up a load and driving in reverse
- Parking and turning off the forklift

The driver's performance on the course should be recorded by the trainer and kept with the driver's training records.

SAFE OPERATING PROCEDURES

Although driving a forklift is unlike driving a typical automobile, many of the driving rules are still the same. For instance, forklift drivers should always drive on the right side of the road or aisle and should not speed or tailgate other vehicles. Additionally, forklift drivers should remember that pedestrians always have the right of way. Finally, forklift drivers should always wear their seatbelt.

CHANGING AND CHARGING STORAGE BATTERIES

The Occupational Safety and Health Administration's Standard 1910.178(g) describes requirements for changing and charging storage batteries. Because handling batteries can pose a health and safety hazard, there must be a designated area with a station for flushing and neutralizing spilled electrolyte. Additionally, because batteries can be heavy, an overhead hoist or other similar method for moving batteries is required. The area must also have fire protection and a protective system of posts or curbs around the charging apparatus to prevent damage. Fumes can also be released during charging; as such, adequate ventilation is necessary.

Because handling and maintaining batteries involves corrosive chemicals and battery acid, additional standards apply to these activities. Staff involved in battery charging and/or changing activities should wear steel-toe boots, chemical-resistant gloves, an acid apron, and eyewear and face protection (29 CFR 1910.13[a][1]). This PPE will reduce injury if acid is splashed or spilled on the operator. Face protection such as safety goggles and face shields should be worn in accordance with the ANSI Z87.1-1989 specifications. Finally, in accordance with OSHA 29 CFR 1910.151, "...suitable facilities for quick drenching or flushing of the eyes and body (i.e., eye/face wash and shower) shall be provided within the work area for immediate emergency use" because the eyes or body of any person may be exposed to injurious corrosive materials during battery handling. According to the American National Standards Act (2004), Emergency Eye Wash and Shower Equipment, the eye/face wash and shower must be within 10 seconds in distance of the hazard and on the same level as the hazard. Operators must ensure that this unit is clearly identified with proper signs and adequate lighting. Portable eye washes can supplement built-in units where needed.

The following are additional practices that should be followed when working with batteries:

- A carbon filter or siphon should be used when handling electrolyte.
- Acid should always be poured into water. Never pour water into acid.
- Check to be sure ventilation caps are functioning. The battery (or compartment) cover(s) must be open to dissipate heat.
- Be sure to properly position and chock tires of the PIV before attempting to change or charge batteries.
- Never smoke in battery charging areas.
- Open flames, sparks, and/or electric arcs should be prevented in battery charging areas.
- Keep metal tools and other metallic objects away from the top of uncovered batteries.

OPERATING AND DRIVING A FORKLIFT

In addition to being proven a competent forklift driver (through successful completion of training and evaluation), the driver should inspect the forklift before it is used each day. According to OSHA standard 1910.178(p), anytime the forklift is observed to be in "need of repair, defective, or in any way unsafe, the PIV [forklift] shall be taken out of service until it has been restored to safe operating condition." In addition, OSHA also states that fuel tanks are not to be filled while the engine is running. Spills should be prevented; however, if a spill occurs, it should be properly cleaned up following appropriate procedures or allowed to completely evaporate before restarting the forklift. Any fuel leaks should be repaired before the forklift is operated. Finally, an open flame (e.g., a lighter) should never be used to check electrolyte levels in storage batteries or gasoline levels in fuel tanks because this could cause an explosion.

When driving a forklift off-site on the road, it is important to remember the rules of the road. In addition to the safe operating procedures discussed in OSHA 1910.178(n), Section 2.7.5 states:

- All traffic regulations shall be observed, including authorizing plant speed limits.
- A safe distance shall be maintained approximately three truck lengths from the truck ahead and the truck shall be kept under control at all times.
- The right of way shall be yielded to ambulances, fire trucks, or other vehicles in emergency situations.
- Other trucks traveling in the same direction at intersections, blind spots, or other dangerous locations shall not be passed.
- The driver shall be required to slow down and sound the horn at cross-aisles and other locations where vision is obstructed. If the load being carried obstructs forward view, the driver shall be required to travel with the load trailing.
- Railroad tracks shall be crossed diagonally wherever possible. Parking closer than 2.4 m [8 ft] from the center of railroad tracks is prohibited.
- The driver shall be required to look in the direction of, and keep a clear view of, the path of travel.
- Grades shall be ascended or descended slowly.
- When ascending or descending grades in excess of 10%, loaded trucks shall be driven with the load upgrade.

HANDLING OF LIQUID PROPANE GAS

Proper procedures should be used when handling or changing liquid propane gas cylinders in the forklift. Refer to the manufacturer's instructions for changing the tank on that specific forklift. However, best practices include the following:

- Park, turn off the forklift, and set the brake.
- Turn off the gas tank (typically by turning the cylinder's valve) and disconnect the cylinder.
- Remove the cylinder using proper lifting techniques.
- Lift and place a new cylinder in place on the forklift.

- Reconnect the clamp and turn the cylinder valve.
- Check for leaks before restarting the forklift.

Storage of forklift cylinders should be in accordance with NFPA's liquefied petroleum gas code (NFPA, 2011). Cylinders must be secure and protected from tampering and damage. They should not be stored in areas with high temperatures or near ignition sources. Finally, when not in use, the service valves on the cylinders should be closed.

CHECKLIST/PRE-INSPECTION FORM SAMPLES

The U.S. Occupational Safety and Health Administration 1910.178(q) requires that all forklifts be examined at least daily or after each shift. Examples of pre-inspection forms that can be used for typical gasoline, diesel, electric, or liquid propane forklifts are provided by OSHA at http://www.osha.gov/dte/library/pit/daily_pit_checklist.html.

MECHANICAL EQUIPMENT AND GUARDING

There are many different types of machinery that can be found in WRRFs, varying from pumps to belt filter presses, all with the potential for moving parts and electrical and stored energy.

Most pieces of machinery have guards that are designed to be in place over the moving parts (OSHA 29 CFR 1910.211 [Subpart O] [OSHA, 2011] addresses standards for guarding machinery). "Guarding" is an engineering design feature that is meant to keep personnel from hazards. The guard encloses the danger. Belts and pulleys, chains and sprockets, rotating coupling, and racks and gears are among some of the common items that should have guards over them. Not all equipment automatically comes with guards installed on them. These may be installed by the owner. Where guards need to be removed for maintenance, LOTO programs are to be used.

Examples of guarding methods include barrier guards, two-hand tripping devices, and electronic safety devices.

Barrier guards, according to OSHA's Machine Guarding e-Tool, are designed to keep the operator's hands and arms from entering the "danger zone," as prescribed by the particular machine. Barrier guards are typically the first point-of-operation safeguard considered for machines. Types of equipment or machinery in WRRFs that typically require barrier guards include couplings, belts, chains, and pumps. Figure 2.16 shows a type of barrier guard called a *barrel style pump guard* (in orange). These are available as an aftermarket guard and fit a variety of different pumps.

Figure 2.16 Barrel-Style Pump Guard (in Orange)

Another common type of guard is referred to as an *interlocked guard.* This type of guard functions (when opened or removed) when the tripping mechanism or power automatically shuts off or disengages. The moving parts of the machine are stopped and the machine does not work until the guard is put back in place. This type of guard may use electrical, mechanical, hydraulic, or pneumatic power or a combination of these to work.

Two-hand tripping devices work by requiring that both of the operator's hands touch the machine operating controls at the same time. Further, the controls are located at such a distance from the machine that the operation stops before the operator touches or reaches into the machine with their hands.

Although the OSHA standard requires guarding, many pieces of equipment are not manufactured with guards, or must be purchased separately. If you purchase equipment or a machine that does not come with a guard, one can be retrofitted.

Safety devices provide an immediate means for stopping a piece of equipment in an emergency (see Figure 2.17). Common devices include electrical lockout-stop switches located adjacent to the equipment, proximity switches, light beam curtains, and mechanical interlocks. Some machines are equipped with an emergency pull cable or emergency stop. When the cable is pulled, or the button is pressed, machine operation is halted and an alarm is generated. All safety devices should be periodically tested as part of the maintenance program.

MOBILE ELEVATED WORK PLATFORMS/LIFTS

Mobile elevated work platforms (MEWPs) refer to a whole group of equipment formerly known as aerial lifts, aerial work platforms, manlifts, powered platforms, and vehicle-mounted work platforms, but still commonly referred to as "lifts." This is referred to in OSHA 29 CFR 1910.67, which requires aerial devices (aerial lifts) acquired on or after July 1, 1975, to be designed and constructed in conformance with the applicable requirements of the American National Standard for "Vehicle Mounted Elevating and Rotating Work Platforms," ANSI A92.2—1969, including appendix, which is incorporated by reference (Grainger, 2020). American National Standards Institute A92.2 has been revised a number of times, most recently as ANSI A92.22-2020, Safe Use of Mobile Elevated Work Platforms (MEWPs) (ANSI, 2020).

TYPES OF MOBILE ELEVATED WORK PLATFORMS

Mobile elevated work platforms are used to move and/or elevate personnel, tools, and material to working positions. Mobile elevated work platforms allow workers to work from a work platform and within guardrails. When operated in accordance with manufacturer guidance and the ANSI standard, MEWPs can be safer than performing work from ladders or scaffolds.

Scaffolds are defined by OSHA as elevated temporary work platforms. They allow people to work in areas that are high off the ground or otherwise inaccessible.

Mobile elevated work platforms pose a number of hazards and risks if not operated properly. Some of the most common include:

- Falling from the MEWP
- Being struck by an object dropped from the MEWP (always barricade the working area beneath the MEWP)

Figure 2.17 Emergency Stop on Machine

- The MEWP tipping over
- Electric shock
- Entanglement hazards
- Contact with the ceiling or other overhead objects
- Crushing of appendages in scissor-lift mechanisms

According to the new ANSI standards, MEWPs are now categorized into two groups: Group A, which are MEWPs with platforms that move vertically but stay inside the tipping lines (like a scissor lift), and Group B, which includes all other MEWPs, but typically boom-type MEWPs where the platform extends past the machine's chassis (like an aerial lift, as shown in Figure 2.18).

OPERATION OF MOBILE ELEVATED WORK PLATFORMS

Only trained and authorized persons are allowed to operate or work on a MEWP. The training certificate must specify the type and group of MEWP they are trained to operate.

A MEWP is to be inspected before operation. The inspection should include fluid levels; leaks; wheels and tires; battery; lower-level controls; horn; gauges; lights; backup alarms; steering; and all lift components including controls, personal protective devices, hydraulic, air, pneumatic, fuel and/or electrical systems, missing or unreadable placards or warnings, outriggers, loose or missing parts, and the guardrail system.

A fall prevention harness is to be worn and the occupants are to be attached to the manufactured anchor point.

A "manufactured anchor point" is an anchor point designed to meet the strength requirements of OSHA. Non-manufactured points such as polyvinyl chloride pipes, vents, and other structures typically aren't strong enough for a harness to be attached to.

Workers must not climb up on the guardrails or climb out of the basket. If a worker cannot reach their work area without climbing up on the rails, then a different or more appropriately sized MEWP should be used.

Figure 2.18 Aerial Lift

Wheel chocks are devices meant to prevent unintended movement. They are a supplementary control used in addition to the equipment's normal braking system.

Surface conditions should be assessed for stability and incline. Outriggers stabilize the MEWP and provide a solid, stable base by distributing the MEWP over a larger area. The outriggers should be positioned on pads or on a solid surface. If on an incline, wheel chocks should be installed. A lift should never be operated on ground that cannot be stabilized or that has been recently backfilled because it can tip over.

A minimum clearance of at least 10 ft [3 m] is to be kept from any overhead power line or other conductive object (i.e., wires, transformers, ducts, and pipes). Regardless of whether or not the lines are believed to be "energized," power lines should always be treated as if they are. This 10-ft [3-m] clearance reduces the chance that the lift will unintentionally come into contact with overhead lines even when the worker's arms are extended upward.

Mobile elevated work platforms have lower controls and upper controls. Lower controls are only used in emergency situations. They cannot control driving or steering the lift. It's always better for the worker to have control of the lift because they can see hazards at their level.

When refueling the MEWP (if applicable), it is important to always have the equipment turned off and the boom or platform lowered and stowed. The operator should make sure there are no open flames. They should take care to avoid putting fuel into the hydraulic tank.

Inspections are required by OSHA by a qualified person. Checklists should be created for both frequent, annual, pre-start, and pre-delivery inspections on MEWPs and should be kept on file in case they are requested by OSHA.

Always refer to the manufacturer's instructions for the specific MEWP. They may have more requirements and guidance for use.

TEST YOUR KNOWLEDGE

1. Guards come standard on all machines and equipment.
   - ☐ True
   - ☐ False
2. What type of crane is shown in this photograph?

   a. Tower crane
   b. Mobile crane
   c. Bridge crane
   d. Gantry crane
3. A MEWP or lift that exceeds the load rating and is operated on a slope, grade, or ramp could tip over.
   - ☐ True
   - ☐ False
4. Who requires guards on equipment?
   a. ANSI
   b. Individual WRRFs
   c. OSHA
   d. The state
5. Working in close proximity to energized power lines or other electrically energized conductors can increase risk of electrocution.
   - ☐ True
   - ☐ False
6. The rated load of the crane should be plainly marked on each side of each crane.
   - ☐ True
   - ☐ False

7. It is only required to inspect cranes and hoists one time per year.
   - ☐ True
   - ☐ False

8. Label the parts of the forklift with the following terms:

   Fork Mast Overhead Guard Counterweight
   Tilt Cylinder Load Backrest

9. Cranes can be modified and load capacity rerated by someone who knows how to operate the crane.
   - ☐ True
   - ☐ False

10. Forklift training only includes classroom training; it's not required to conduct a practical test and be evaluated on the ability to operate.
    - ☐ True
    - ☐ False

11. It is acceptable for operators/occupants of MEWPs to achieve additional height in the platform by climbing on the guardrail or using ladders.
    - ☐ True
    - ☐ False

CHAPTER SUMMARY

SAFETY PROGRAMS AND HAZARD IDENTIFICATION	According to OSHA guidance, the core elements of any safety and health program should contain sections on ■ Management leadership ■ Worker participation ■ Hazard identification and risk assessment ■ Education and training ■ Program evaluation and improvement
HIERARCHY OF SAFETY CONTROLS	■ The Hierarchy of Safety Controls can be used to eliminate or reduce safety and health hazards. ■ Most effective to least effective are • Elimination • Substitution • Engineering controls • Administrative controls • PPE
PERSONAL PROTECTIVE EQUIPMENT (PPE)	■ Most facilities require basic or standard PPE, which includes • Protective clothing, • Head protection, • Eye/face protection, and • Hand and foot protection. ■ Additional or different PPE may be required when handling or cleaning up spilled chemicals.
FALL PREVENTION	■ Fall protection methods and devices are typically divided into two categories: • Those that prevent falls (fall restraint) • Those that arrest falls ■ Examples of fall protection methods and devices include • Rails • Guards • Guardrails • Barriers • Fall-arrest systems • Positioning device or fall-restraint systems • Safety nets • Hole covers • Various work practices and procedures, including those related to the safe use of ladders

LOCKOUT/TAGOUT (CONTROL OF HAZARDOUS ENERGY)

- Lockout/tagout typically involves placing an energy isolation device on a machine or piece of equipment
  - so that it becomes inoperable and
  - most commonly when machines or equipment are being repaired or maintained.
- Isolation may also include placing valve locks or plugging pipes to prevent water from being turned on or entering a certain location.
- In some cases, one lock is adequate.
- Where multiple crews are working on equipment, the main shutoff would have a lockout hasp—a device that would fit through the hole in the main power shutoff and allow for multiple locks to be used
- An LOTO program that meets OSHA specifications requires employee training.
- The following are mandates that an employer must follow to protect employees:
  - Develop, implement, and enforce an energy control program.
  - Use lockout devices for equipment that can be locked out; tagout devices can be substituted instead of lockout devices only if tagout devices can provide the same degree of protection.
  - Ensure that new or overhauled devices are capable of being locked out.
  - If existing equipment cannot be locked out, develop a tagout program.
  - Develop, document, implement, and enforce energy control procedures.
  - Use only LOTO devices authorized for the particular application and ensure that they are durable, standardized, and substantial.
  - Ensure that LOTO devices identify the individual users.

CONFINED SPACES HAZARDS

- Examples of confined spaces found in WRRFs and collection systems include, but are not limited to, manholes, meter vaults, wet wells, valve chambers, and pumping stations.
- The U.S. Occupational Safety and Health Administration defines a *confined space* using three conditional criteria and all three criteria must be met before it is considered a confined space:
  - The space must be large enough and have a configuration where a worker can enter the space.
  - The space has a limited means of access or egress (the way to, through, and away from an exit).
  - The space was not designed for continuous human occupancy.
- A confined space may be classified as either a non-permit-required confined space or a permit-required confined space.
  - A non-permit-required confined space is a confined space that does not contain or have the potential to contain any recognized hazard capable of causing death or serious physical harm.
  - Permit-required confined space entries are expected to be relatively infrequent and involve little to no risk because hazards have been eliminated or controlled.
  - A permit-required confined space entry will include, at a minimum, an
    - attendant,
    - entrant, and
    - entry supervisor.
- Safe confined space entry will address
  - hazardous atmospheres,
  - entry procedures, and
  - atmospheric monitoring.

HAZARD COMMUNICATION

- Hazard communication is typically recognized as the standard for employees' right to know.
- There are three broad components to a written HAZCOM program, each with its own detailed requirements listed in the OSHA standard that outlines specific rules for employers and employees to follow:
  - Labels and other forms of warning for chemicals in the workplace
  - Safety data sheets
  - Employee information and training

CHEMICAL HANDLING, STORAGE, AND RESPONSE

- When handling chemicals, even in small quantities, it is important to know the chemical's hazardous properties and the safest means of handling the chemical.
- Proper chemical storage is important not only for safety, but for quality control.
- U.S. Occupational Safety and Health Administration defines three types of spills:
  - Incidental release
  - Operational spill
  - Emergency response spill
- The OSHA Hazardous Waste Operations and Emergency Response (HAZWOPER) regulation at 29 CFR (Code of Federal Regulations) 1910.120 paragraph (q) outlines specific spill response training requirements.

HEAVY EQUIPMENT & MACHINERY

- Forklifts (PIVs)
- Cranes (bridge and gantry)
  - Always follow manufacturer's directions to minimize risk.
  - Never lift more than the rated weight.
  - Always lift straight up (lifting at an angle may cause a hoist tor crane to tip or drop its load).
  - Clearance must be maintained above and to the side of cranes.
  - Boxes, equipment, or chemicals should not be stored near cranes.
  - Walkways should not be placed in a crane operating zone that would compromise employee safety when the crane is in operation.
  - Only designated personnel are permitted to operate a crane.
  - Crane inspections must be conducted per OSHA regulations.
- Hoists
- Machinery
- Lifts
  - For forklifts powered by electric motors or internal combustion engines, OSHA standards cover
    - fire protection,
    - inspection,
    - maintenance,
    - use,
    - operation, and
    - training.
  - There are multiple classifications of forklifts and they can be fueled by diesel, electrical, gasoline, or liquified petroleum/propane.

HEAVY EQUIPMENT & MACHINERY *(continued)*

- Drivers should
  - be aware and familiar with workplace conditions and pedestrian traffic and be able to properly load and carry materials,
  - be able to properly load and carry materials,
  - remember that pedestrians always have the right of way, and
  - be aware of atmospheric hazards and potentially hazardous or poorly ventilated areas.
- Prequalification of a candidate, training, and performance testing are required to operate a forklift.
- Safe operating procedures relate to
  - operating and driving the vehicle,
  - changing and storing charging batteries,
  - gas handling, and
  - pre-inspection of equipment.

■ Guards
- Installed guards are meant to keep personnel from hazards.
- Belts and pulleys, chains and sprockets, rotating coupling, and racks and gears are among some of the common items that should have guards over them.
- Examples of guarding methods include barrier guards, interlocked guards, two-hand tripping devices, and electronic safety devices.

■ Mobile elevated work platforms
- Refer to a whole group of equipment formerly known as
  - aerial lifts,
  - aerial work platforms,
  - manlifts,
  - powered platforms, and
  - vehicle-mounted work platforms.
- Are used to move and/or elevate personnel, tools, and material to working positions
- Allow workers to work from a work platform and within guardrails
- Pose a number of hazards and risks if not operated properly
- Can only be operated by trained and authorized persons
- Are to be inspected before operation
- Require that operators follow safety procedures related to
  - wearing a proper fall prevention harness
  - operation
  - inspection
  - fueling

References

American Chemical Society. (2018). Guide for chemical spill response planning in laboratories. https://www.acs.org/content/acs/en/about/governance/committees/chemicalsafety/publications/guide-for-chemical-spill-response.html

American Crane. (2021). American Crane website. https://www.americancrane.com/

American National Standards Institute. (2020). *Mobile elevating work platforms* (ANSI/SAIA A92.22-2020).

Crane Repair Company. (n.d.). Which type of overhead crane do you need? https://cranerepairco.com/blog/which-type-of-overhead-crane-do-you-need/

Grainger. (2018). Quick tips #107: Overhead crane safety, 29 CFR 1910.179. https://www.grainger.com/know-how/safety/material-handling-safety/mechanical-lifting-hazards/kh-overhead-crane-safety-107-qt

Grainger. (2020). Quick tips #377: Mobile elevating work platforms. https://www.grainger.com/know-how/safety/working-at-heights/fall-protection-arrest/kh-aerial-work-platform-safety-tips

Institute of Electrical and Electronics Engineers Standards Association. (2002). *IEEE 1584 and 1584A Set - 1584 & 1584A Set arc flash* [Print plus CD-ROM]. IEEE SA - The IEEE Standards Association - Home. https://standards.ieee.org/standard/1584and1584ASet-0.html

National Fire Protection Association. (2011). NFPA 58: Liquefied petroleum gas code (2011 ed.).

National Fire Protection Association. (2021). *Standard for electrical safety in the workplace* (NFPA 70E). https://www.nfpa.org/codes-and-standards/all-codes-and-standards/list-of-codes-and-standards/detail?code=70E

National Institute for Occupational Safety and Health. (2015). Hierarchy of controls. https://www.cdc.gov/niosh/topics/hierarchy/default.html

Page-Bottorff, T. (2018). The habit of safety. https://safestart.com/articles/the-habit-of-safety/

U.S. Occupational Safety and Health Administration. (n.d.-c). *Recommended practices for safety and health programs.* https://www.osha.gov/shpguidelines/management-leadership.html

U.S. Occupational Safety and Health Administration. (n.d.-d). *OSHA quick card, work zone traffic safety.* https://www.osha.gov/Publications/OSHA-work-zone-safety-english.html

U.S. Occupational Safety and Health Administration. (1993). *Letter of interpretation—permit required confined space standard as it applies to certain operations.* https://www.osha.gov/laws-regs/standardinterpretations/1993-10-22

U.S. Occupational Safety and Health Administration. (1998). *Powered industrial trucks; Code of Federal Regulations,* Part 1910, Title 178, Appendix A.

U.S. Occupational Safety and Health Administration. (2004). Personal protective equipment. OSHA 3151-12R. https://www.osha.gov/sites/default/files/publications/osha3151.pdf

U.S. Occupational Safety and Health Administration. (2008). *eTools home: Powered industrial trucks.* https://www.osha.gov/SLTC/etools/pit/index.html

U.S. Occupational Safety and Health Administration. (2011). *Machinery and machine guarding;* OSHA Title 29 CFR Part 1910 Subpart O.

U.S. Occupational Safety and Health Administration. (2019). *Ladders*; OSHA Title 29 CFR Part 1910 Subpart D. https://www.osha.gov/laws-regs/regulations/standardnumber/1910/1910.23

Suggested Readings

Campbell Institute. (n.d.). *Practical guide to leading indicators: Metrics, case studies and strategies.* https://www.nsc.org/Portals/0/Documents/CambpellInstituteandAwardDocuments/WP-PracticalGuidetoLI.pdf

Emory University Environmental Health and Safety Office. (2016). *Compressed air safety.* http://www.ehso.emory.edu/content-guidelines/Toolbox/ToolboxTraining_CompressedAirSafety.pdf

Federal Motor Carrier Safety Administration, A&I Online; Safety Measurement System. http://ai.fmcsa.dot.gov/SMS (accessed October 2020).

Manuele, F. (2021). *Advanced safety management: Focusing on z10 and serious injury prevention*. Wiley-Blackwell.

National Institute for Occupational Safety and Health. (2007). *Pocket guide to chemical hazards*.

National Institute for Occupational Safety and Health. (2020). Total worker health. https://www.cdc.gov/niosh/twh/default.html

The Center to Protect Workers' Rights. (2004). *Biological hazards in sewage and WRRFs hazard alert*. https://www.cpwr.com/sites/default/files/publications/hazsludge.pdf

The Pennsylvania State Association of Township Supervisors. (2014). *Wastewater treatment plant operator certification training, Module 15: The activated sludge process Part 1* (revised July 2014).

Theobald, D. (2016). *Water technology mitigating health hazards for workers at biological WRRFs*. https://www.watertechonline.com/wastewater/article/15550161/mitigating-health-hazards-for-workers-at-biological-wastewater-treatment-plants

U.S. Department of Transportation Federal Highway Administration. (2009). 2009 Edition with Revision Numbers 1 and 2 incorporated, dated May 2012 (PDF). https://mutcd.fhwa.dot.gov/pdfs/2009r1r2/pdf_index.htm

U.S. Environmental Protection Association. (1999). *Wastewater technology fact sheet ozone disinfection*. https://www3.epa.gov/npdes/pubs/ozon.pdf

U.S. Environmental Protection Agency. (2015). *Wastewater technology fact sheet disinfection for small systems*. https://www.epa.gov/sites/production/files/2015-06/documents/disinfection_small.pdf

U.S. Occupational Safety and Health Administration (n.d.-a). Regulations (Standards 29 CFR), 1910. https://www.osha.gov/laws-regs/regulations/standardnumber/1910

U.S. Occupational Safety and Health Administration. (n.d.-b). Regulations (Standards 29 CFR), 1926. https://www.osha.gov/laws-regs/regulations/standardnumber/1926

U.S. Occupational Safety and Health Administration. (2011a). *Hand and portable powered tools*; OSHA Title 29 CFR Part 1910 Subpart P.

U.S. Occupational Safety and Health Administration. (2011b). *Tools – Hand and power;* OSHA Title 29 CFR Part 1926 Subpart I.

U.S. Occupational Safety and Health Administration. (2021). United States Department of Labor general industry standard. 1910. U.S. Occupational Safety and Health Administration. https://www.osha.gov/laws-regs/regulations/standardnumber/1910

Water Environment Federation. (2021). *Safety, health, and security standards for water resource recovery* (Manual of Practice No. 1, 7th ed.); T. Page-Bottorff, Ed. Water Environment Federation.

CHAPTER 3

Physical and Chemical Treatment

Introduction

When primary and secondary treatment are not enough to meet discharge or regulatory limits, tertiary and advanced wastewater treatment processes are often required to further reduce specific contaminants. Tertiary and advanced treatment processes are becoming more common as water reuse becomes mainstream. This chapter covers physical–chemical treatment for facilities with advanced processes. Many of these processes mirror drinking water processes as they are used for nutrient removal, removal of toxic constituents, and for overall effluent polishing. The topics covered in this chapter include chemical delivery systems, coagulation and flocculation, clarification, filtration, granular activated carbon (GAC) filtration, and ion exchange.

COAGULATION AND FLOCCULATION

In the wastewater treatment industry, coagulation involves adding inorganic chemicals, organic polymers, or a combination of the two to wastewater to remove solids or phosphorus. Suspended colloidal particles encountered during advanced treatment are generally negative because of pH, and their mutual repulsion is enough to prevent them from settling.

Colloidal—A mixture in which very small particles of one substance are distributed evenly throughout another substance. The particles are generally larger than those in a solution, and smaller than those in a suspension. Paints, milk, and fog are colloids.

Coagulants destabilize these charges, reducing the repulsion between particles and promoting their settling. Flocculation is similar, except that the chemicals "chain" the particles together into larger particles called *flocs* (aggregates), which makes them easier to remove from the system. In the flocculation step, the chemically treated water is gently stirred to form large flocs of suspended solids or phosphorus precipitates. These particles are then removed via clarification and filtration. If the proper form and quantity of coagulant or flocculant are applied to the wastewater, approximately 90% of the phosphorus and suspended solids typically present in the secondary effluent may be removed.

CLARIFICATION

Depending on the treatment levels required, coagulated and flocculated effluent may enter a clarifier or settling basin, where gravity causes the solids to settle on the bottom. Patented high-rate clarification or ballasted flocculation systems can also remove coagulated–flocculated solids from effluents. A major selling point of these technologies is that they require less space than conventional clarifiers. The main goal of clarification is to remove solids and produce a cleaner effluent.

FILTRATION

Filtration is a process in which particles in a liquid or gaseous fluid are removed by the use of a medium that permits the fluid to pass through but retains the solid particles. In filtration, effluent passes through a media—typically sand, anthracite, or garnet—to remove suspended or colloidal matter. Cloth or fabric media filters such as disk filters may also be used. Filtration is typically a tertiary treatment step located downstream of secondary processes. Its primary goal is removing suspended solids from a secondary effluent, coagulation–flocculation process, or tertiary clarification process effluent. For example, filtering coagulation, flocculation, and clarification effluent can typically reduce suspended solids concentrations from between 3 and 5 mg/L to virtually zero and phosphorus concentrations from between 0.5 and 1 mg/L to 0.1 mg/L or less. An efficient filtration system should also be able to be cleaned (backwashed) using less than about 5% of its influent flow.

GRANULAR ACTIVATED CARBON

Despite secondary treatment, coagulation–flocculation, sedimentation clarification, and filtration, some soluble organic materials may persist in the effluent. Granular activated carbon is a treatment unit used in drinking water and wastewater treatment systems to remove organics. Granular activated carbon systems use the adsorption process to remove organics such as taste and odor compounds, chlorine, and volatile organic compounds (VOCs).

ION EXCHANGE

The ion exchange process is used in both drinking water and wastewater treatment applications. The purpose of ion exchange systems is to remove particles with charged ions that are too small to be removed by conventional clarification and filtration methods. Ion exchange involves a reaction in which a charged ion is exchanged for a similarly charged ion attached to a solid particle in water. Ion exchange is commonly used in water treatment to soften by replacing and removing calcium and magnesium ions and reducing hardness. In wastewater applications, ion exchange is used for the removal of per- and polyfluoroalkyl substances, nitrogen, heavy metals, and total dissolved solids (TDS). In all these applications, water is passed through a bed containing either a naturally occurring zeolite or a synthetically manufactured polymer resin. Both materials contain charged ions selected to react with the ions associated with the removable solid in water.

LEARNING OBJECTIVES

Upon completing this chapter, you will be able to:

- Identify various types of chemical delivery systems (dry feed vs. liquid feed) used in advanced wastewater treatment and their corresponding equipment and components. Perform basic chemical storage calculations.
- Understand the chemical reactions that take place when coagulants and other common chemicals used in water are used. Learn how to determine what water quality changes occur and perform basic alkalinity and acid neutralization calculations.
- Understand the basic principles of design for coagulation, flocculation, sedimentation, and filtration and how these processes are configured in an advanced wastewater treatment train.
- Identify and troubleshoot the various types of mixers and basins associated with these processes. Perform basic power calculations.

Chemical Dosing

FUNDAMENTALS

Chemicals are used in both mainstream (main liquid treatment train) and side stream (any process flow separate from the main train) treatment for a variety of purposes ranging from disinfection (covered in Wastewater Treatment Fundamentals I), coagulation, water softening, and pH adjustment. When operating chemical delivery systems, there are multiple components that come into consideration including dose determination, storage space, and type of feed equipment. Equipment type is determined by the chemical physical state (solid, liquid, or gas). This section covers chemical dose determination for common coagulants, chemical delivery, handling, and storage considerations, and chemical feed equipment and their process control systems. Chemicals used for disinfection are discussed in *Wastewater Treatment Fundamentals I—Liquid Treatment* (WEF & ABC, 2018). More in-depth coverage of chemical handling and polymer and polymer systems for solids handling can be found in *Wastewater Treatment Fundamentals II—Solids Handling and Support Systems* (WEF & ABC, 2021), especially as it relates to chemical receiving, storage, and safety.

PURPOSE AND FUNCTION

Coagulants are commonly used in both wastewater and water treatment facilities. Coagulants either destabilize similarly to charged particles so they will settle more readily or react chemically with compounds to precipitate them. Precipitation and destabilization can both occur in a solution. They are not mutually exclusive. For example, a coagulant added to enhance settling may also precipitate phosphorus. This section mainly focuses on coagulation and not on phosphorus precipitation by metal salts (covered extensively in *Wastewater Treatment Fundamentals I*, Chapter 9).

THEORY OF OPERATION

The amount of coagulant needed depends on time and wastewater characteristics. For example, the aluminum and iron salt concentrations needed for good phosphorus removal are typically proportional to the phosphorus concentration and, to a lesser extent, the water's alkalinity. Lime requirements depend largely on the wastewater's alkalinity.

Although all coagulants typically increase sludge production, aluminum and iron salts add significant quantities of dissolved solids (sulfates, chlorides, etc., as measured by the TDS parameter) to the final effluent. This must be considered if being used as part of a water reuse train. Also, alum and ferric chloride consume alkalinity (about 0.55 mg/L of calcium carbonate [$CaCO_3$] per mg/L of alum and 0.93 mg/L $CaCO_3$ per mg/L of ferric chloride) and may lower pH. Sometimes, pH or alkalinity adjustment are required, which adds chemical costs. In waters with low natural (background) alkalinity, utility staff may want to consider alternatives (e.g., sodium aluminate or aluminum chloride). Alum also enhances heavy metal removals, and the heavy metal concentrations in the precipitated sludge could affect biosolids management options.

If lime is used, it can be recovered via "recalcination," but this process consumes large amounts of energy. High-pH lime coagulation, however, can disinfect the water and precipitate certain heavy metals. As a result, the chemical sludge's heavy metals concentrations could limit the facility's biosolids management options.

Metal salts (e.g., alum and ferric chloride) are commonly used coagulants in wastewater treatment; however, the coagulant should be chosen based on anticipated performance, cost, and material handling constraints. The capital costs for coagulant-addition facilities are similar, but the operations and maintenance costs depend on the wastewater characteristics and desired effluent pollutant levels. Unless there are specific concerns about solids handling and disposal, alum will often be selected over ferric chloride because it is less expensive, less corrosive, safer, and easier to handle. In addition, iron can interfere with UV disinfection.

In addition to metal salts, polymers may be added to promote coagulation and flocculation. A large variety of polymers is commercially available; they may be natural or synthetic, and liquid or dry. Synthetic polymers may be further classified as cationic, anionic, or nonionic based on the type of charge (or lack thereof) on the polymer chain. Polymers may be used alone or with a metal salt as a coagulant aid. In either case, the polymer feed rate is low: typically, less than 5 mg/L and often less than 1 mg/L. The use of polymers reduces sludge volume greatly compared to the volume of sludge produced by conventional coagulants. The equipment used to deliver polymer dose is like metal salts, but smaller. Totes are commonly used for liquid polymer. Dry polymers may also be batched with appropriate equipment.

In addition to inorganic coagulants, organic polymers may also be used. Polymers are high-molecular-weight, water-soluble organic chemicals that can be used as coagulants, settling aids, or filtration aids. They may be cationic, anionic, or nonionic. When used as a filtration aid, a polymer dose of less than 1.0 mg/L typically can increase the strength of the chemical floc and control the depth of floc penetration into the filter. For best results, the polymer should be added directly to the filter influent. However, if polymers are used as settling aids, filtration aids may be unnecessary. Similar polymers may be used for clarification and filtration aids; however, the doses vary greatly. Conventional clarification aid doses are 1 to 10 mg/L for DADMAC and epi-DMA and filter aid doses range from 0.005 to 0.05 mg/L for high molecular weight nonionic polymers (Crittenden, 2012). Higher doses can contribute to filter clogging and excessive backwashing. Also, polymers cannot compensate cost-effectively for improper chemical coagulation or poorly performing secondary processes. When determining the proper polymer dose, a common test procedure is to add an initial dose (typically approximately 1.0 mg/L), observe the effects, and then lower or increase the dose in increments of 0.25 to 0.5 mg/L, as appropriate, until a desired operating range is obtained. Conventional coagulants are covered more thoroughly in Table 10.1 of *Wastewater Treatment Fundamentals Book 2*.

In addition to turndown capability, chemical delivery systems are also selected based on redundancy and capacity. Redundancy for a facility is dependent on the chemical that is used being either critical or noncritical. An example of a critical (non-interruptible) chemical delivery system would be chlorine for disinfection and coagulants. A noncritical (interruptible) system would be chemicals used for water softening, corrosion control, or taste–odor control. Critical systems that cannot stop service require redundancy. This is typically built into the system by having standby pumps or multiple feeders should one fail.

DESIGN PARAMETERS—DETERMINING CHEMICAL DOSE

Phosphorus may be precipitated from wastewater by chemical reaction with most mineral salt coagulations, including various salts of iron and aluminum. The principal source of aluminum in phosphorus precipitation is aluminum sulfate, typically known as alum.

When alum or ferric chloride is added to wastewater, the resulting metallic hydroxide floc can entrap suspended solids and settle slowly, physically sweeping out suspended solids. When alum is added to water, it reacts with bicarbonate alkalinity to form insoluble aluminum hydroxide, which is a gelatinous floc that sweeps out suspended particles as it settles in a clarification basin or is removed via filtration. Similarly, ferric chloride forms insoluble ferric hydroxide, which traps suspended solids and removes them as it settles in clarification basins or is removed by filters.

When metal salts such as aluminum sulfate and ferric chloride are added to water, they hydrolyze (i.e., add OH ions onto their chain). The solubility of aluminum hydroxide ions can be plotted at different pH values using equilibrium constants for hydrolysis reactions. In Figure 3.1, the blue shaded area shows where aluminum hydroxide is insoluble and will precipitate out of solution. The green shaded region is where adsorption also occurs in addition to flocculation.

Adsorption is the adhesion of atoms, ions, or molecules from a gas, liquid, or dissolved solid to a surface. This process creates a film of the adsorbate on the surface of the adsorbent. This process differs from absorption, in which a fluid is dissolved by or permeates a liquid or solid, respectively.

The optimal "sweep flocculation" ranges shown in the equilibrium diagram for aluminum hydroxide in Figure 3.1 are used as "starting points" for jar tests to determine optimum coagulant doses in a wastewater treatment train. This is where you will get the most solids formation and coagulation from your aluminum dose.

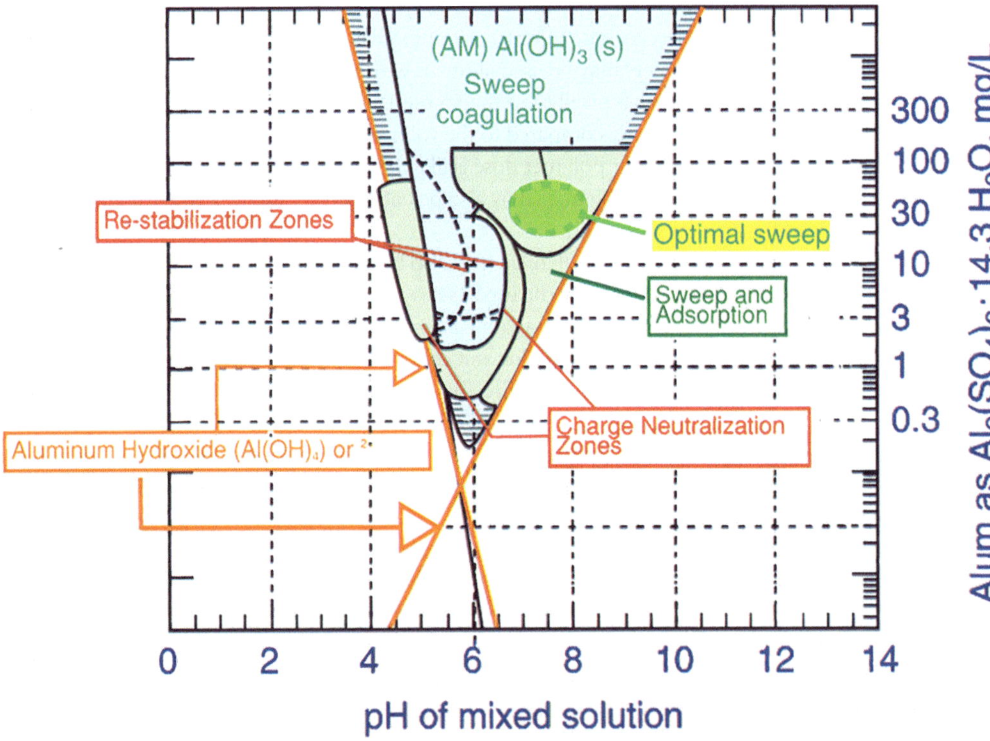

Figure 3.1 Solubility Diagram for Aluminum Hydroxides at 25 °C and Optimal Zones for Coagulation and Flocculation (Adapted from Amirtharajah & Mills [1982] and Crittenden [2012]).

Optimal doses of aluminum sulfate will typically range around 20 to 40 mg/L of $Al_2(SO_4)_3 14H_2O$. Because water quality varies, this dose range for aluminum is simply a starting point. Optimal dose is not always consistent because water quality varies, so jar tests are used to determine optimal doses for a specific water source/process stream. Solubility diagrams also exist for ferric chloride.

Jar tests and alkalinity are discussed in *Wastewater Treatment Fundamentals I—Liquid Treatment* (WEF & ABC, 2018).

The "correct" coagulant dose for enhanced suspended solids removal varies as the wastewater characteristics vary and should be determined based on extensive jar testing and bench-scale or pilot-scale tests.

The most widely used technique is the jar test. In its simplest form, this test involves collecting an effluent sample, transporting it to the laboratory, and then dividing it among several beakers. Each beaker receives a different coagulant dose, and the contents are subjected to various mixing regimes (primarily through varied mixer speeds) in a ganged multiple paddles mixing device. After an appropriate period, stirring is stopped, and the floc is allowed to settle. (The tests may be run multiple times to gauge the effects of different mixing periods.) Operators then determine the optimum coagulant dose based on the clarity of the supernatant after settling. Operators often visually inspect the supernatant to select the best dose, but measuring filtered supernatant turbidity in a laboratory turbidimeter provides a more objective means of selecting the optimum dose. Experience has shown that the jar test is effective, but there are issues with scale-up. Therefore, optimum dosages are often determined in pilot- or full-scale facilities. Figure 3.2 shows a conventional testing apparatus with sludge produced from various concentrations of aluminum sulfate. The left beaker had no coagulant added, and the far-right beaker had too much coagulant added.

DESIGN PARAMETERS—STORAGE AND MATERIALS

Probably the most important criteria for chemical systems are chemical and material compatibility. A rigorous search for chemical/material combinations should be performed and manufacturers of equipment should always be consulted. Polyvinyl chloride and Type 316 stainless steel are commonly used for alum but are incompatible for ferric. Table 3.1 presents a short list of incompatible chemicals used in wastewater treatment.

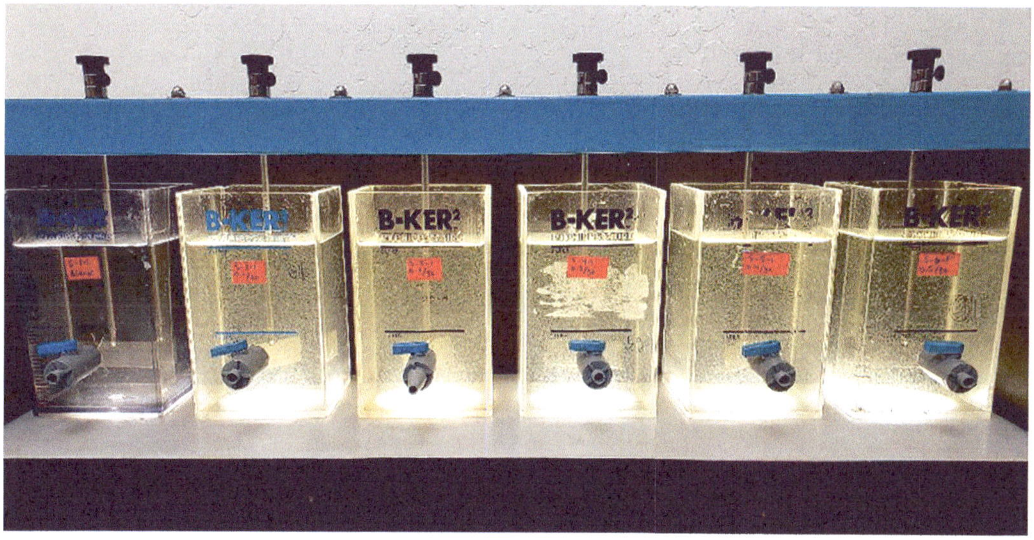

Figure 3.2 Typical Jar Test Apparatus and Dose Determination

Table 3.1 Short List of Incompatible Chemicals Used in Water and Wastewater Treatment (Adapted from Davis, 2010)

Chemicals	Keep out of Contact With:	Reaction	Pipe Material Compatibility
Activated Carbon	Oxidizing agents such as chlorine	Fire	Type 316 Stainless Steel; fiberglass-reinforced plastic, rubber
Aluminum Sulfate	Lime	Exothermic reaction	Type 316 SS; PVC
Ammonia	Chlorine	Exothermic reaction	Steel; Type 316 SS
Ferric Chloride	Potassium permanganate	Chlorine gas formation	PVC
Sulfuric Acid	Strong bases, metals	Exothermic reaction	PVC

TEST YOUR KNOWLEDGE

1. The adhesion of atoms, ions, or molecules to a surface is called:
   a. Adsorption
   b. Absorption
   c. Molecule merging
   d. Fusion
2. What should you never mix with chlorine?
   a. Ammonia
   b. Ferric chloride
   c. Sulfuric acid
   d. Activated carbon
3. What is a typical dose range for aluminum sulfate in a jar test?
   a. 20 to 40 mg/L
   b. 0 to 5 mg/L
   c. 5 to 20 mg/L
   d. 40 to 50 mg/L
4. What is an optimal pH range for aluminum sulfate coagulation?
   a. 0–1
   b. 1–3
   c. 5–6
   d. 6–7
5. What is compatible with sulfuric acid?
   a. Steel
   b. Water
   c. Sulfur
   d. PVC

PROCESS VARIABLES

pH, particle charge (measured by streaming current monitors), TDS, power, alkalinity, turbidity, and feed pump speed are all process variables.

CONTROLLING CHEMICAL DOSING

In a physical–chemical water resource recovery facility (WRRF), the best-designed equipment cannot substitute for proper control of chemical coagulation. Insufficient coagulant doses will produce an excessively turbid effluent. Excessive doses may have the same result, as will inadequate or excessive mixing. Excessive turbidity increases the probability of microorganisms escaping subsequent disinfection because of turbidity's shielding effect. Successful treatment depends on proper coagulation (i.e., proper chemical doses and adequate mixing). Figure 3.3 shows an exploded cutaway of the mixing chamber, typically used for proper mixing. Achieving proper coagulation has challenged operators for many years, and a wide variety of control techniques has been developed. Most involve laboratory or bench-scale tests, the results of which operators then scale up to full-scale facility operations.

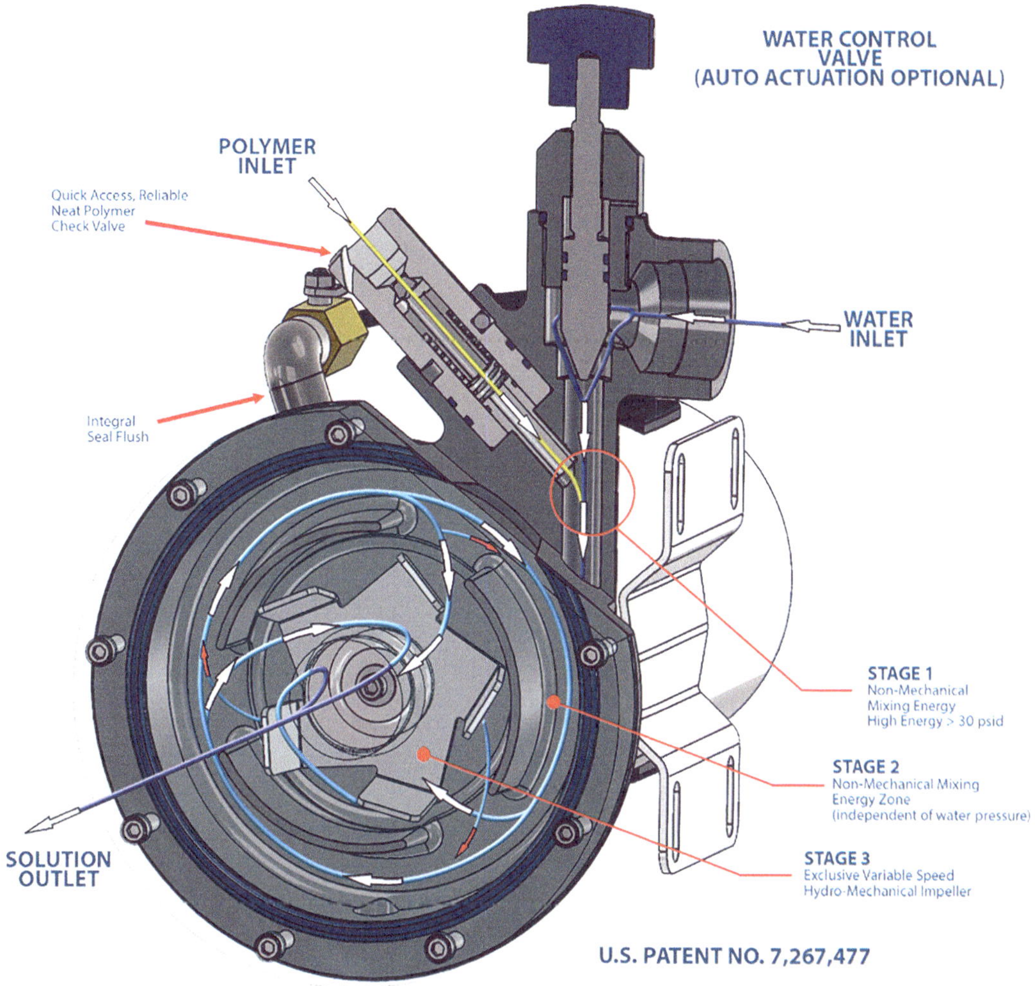

Figure 3.3 Mixing Chamber Cutaway (Provided courtesy of VeloDyne)

Modern WRRF designs enable operators to automatically control the amount of coagulant feed in proportion to facility flow. However, coagulant demand may depend on additional variables. Data from jar tests preceding facility design may indicate a useful correlation between a parameter that can be monitored automatically and the required coagulant dose. For example, the required lime dose may correlate well with pH. If so, it can be controlled automatically via a pH monitor on the settling tank effluent. Alum or ferric chloride doses may correlate well with the raw wastewater's phosphorus content, so the WRRF design may incorporate some means of automatically controlling coagulant doses based on some measure of phosphorus concentration. If the WRRF lacks such controls or if an existing automatic system malfunctions, then the coagulant dose must be controlled via manual tests (e.g., jar tests).

The jar test, however, has limitations. For example, the optimum coagulant dose indicated in a jar test beaker may not necessarily provide optimum results on a WRRF scale because the two mixing efficiencies probably differ. Also, there is an inherent time lag involved; by the time, the optimum chemical dose has been determined, the water involved has already passed by. This time lag can be shortened if the operator is alert to changing conditions that require dose adjustments. Because temperature plays an important role in coagulation, the wastewater samples for jar tests must be collected and analyzed only after all other preparations are made to reduce the possible effect of room temperature on the samples. Valid results demand that WRRF conditions be closely simulated. Dosing solutions or suspensions are prepared from the stock materials used in facility treatment.

The required frequency for jar tests will vary widely from one locale to another. Some wastewater streams may vary little in coagulant demand from day to day or from hour to hour. Others may vary markedly during the day because of the wastewater's major diurnal variations. High turbidity in the chemical settling-tank effluent reliably indicates improper chemical dosing. After checking to be sure that some mechanical failure of the chemical feed system is not at fault, operators must quickly use the jar test to determine the change in coagulant dose required. Even with a dose correction, some time will elapse before a noticeable improvement occurs because the settling tank is already filled with improperly coagulated wastewater. If the filter system is equipped with provisions for adding filter aids, the improperly coagulated settling tank effluent may receive compensating coagulant doses before filtration. This may avoid impairing the WRRF's effluent quality despite possible shortened filter runs.

In terms of system instrumentation and controls, an ideal, automated control system for chemical phosphorus removal would adjust the metal salt, polymer, and supplemental alkalinity doses to maximize phosphorus removal and minimize costs with changing wastewater. However, such systems do not seem to have a consistent history in the industry. Therefore, a common process control system is to manually set and adjust the dose rates based on observed performance. This typically results in the dose being set higher than needed to account for fluctuations in phosphorus loading without violating permit limits at peak conditions. Other methods incorporate a related parameter (e.g., flow, pH, or turbidity) or adjust the dose based on historical and anticipated diurnal loadings to provide a degree of automated or quasi-automated control.

NORMAL PROCESS CONTROL

The most important process control systems during phosphorus and enhanced suspended solids removal are those involving the chemical dose and filter operations. Typical control schemes used in wastewater treatment applications include manual, on-off, feedforward, ratio, feedback, and feedforward/feedback combinations. The most sophisticated methods of control require the use of programmable logic controllers (PLCs) or computers.

Residual control is the simplest type of automated process control. Residual control is typically used in disinfection, where direct feedback is provided on chlorine residual and delivery systems can respond to changes in flow or demand. Unfortunately, residual control cannot be used with coagulants because they leave no easily monitored residual. Flow meters are most commonly used for feedforward control for flow pacing coagulants; then, turbidity is also measured pre- and post-clarification and filtration to ensure the treatment train is operating optimally.

TROUBLESHOOTING

Simple chemical delivery systems such as static mixer very rarely require any troubleshooting. The more complicated the system and the more moving parts generally will require more operator maintenance and changeout of consumables. Table 3.2 provides a summary of troubleshooting for chemical systems and insufficient chemical dosing.

TEST YOUR KNOWLEDGE

1. **Which of the following *are* compatible?**
   a. Alum, copper
   b. Ammonia, chlorine
   c. Sulfuric acid, NaOH
   d. Alum, Type 316 SS
2. **High turbidity is observed post-settling or post-filtration. What is the most likely cause?**
   a. High alkalinity
   b. Proper chemical dose
   c. Clog or tear in chemical feed tubing
   d. A pH of 6.5
3. **What would cause a chemical feed tubing line to rupture at a location away from the peristaltic pumps?**
   a. Pump operating at speeds that are too high
   b. Coagulant or acid has dissolved the tubing
   c. Pump has been started against an open valve
   d. Pump has been started against a closed valve
4. **What is a limitation of the jar test?**
   a. Lack of mixing power
   b. Insufficient coagulant dose
   c. Lack of ability to mimic full-scale chemical dosing lag
   d. Jars are too small

Table 3.2 Probable Causes and Solutions for Troubleshooting Various Chemical Delivery Systems (Adapted from U.S. EPA, 1978)

Indicators	Probable Cause	Check/Monitor	Solutions
■ High turbidity in settling tanks	■ Improper chemical dosage ■ Mechanical failure in feed system	■ Jar test ■ Visual inspection	■ Correct dosage according to jar test results ■ Repair failure in feed system
■ Air slaking occurring during storage of quicklime	■ Adsorption of moisture from atmosphere when humidity is high.	■ Humidity, storage facility not air-tight	■ Make storage facilities air-tight, and do not convey pneumatically
■ Feed pump discharge line clogged	■ Chemical deposits	■ Visual inspection of line	■ Increase dilution water
■ Grit conveyor or slaker inoperable	■ Foreign material in the conveyor	■ Broken shear pin	■ Replace shear pin and remove foreign material from grit conveyor
■ Paddle drive on slaker is overloaded	■ Lime paste is too thick ■ Grit or foreign interfering with paddle action	■ Visual inspection	■ Adjust compression on the spring between gear reducer and water control valve
■ Lime deposits in slurry feeder	■ Velocity too low		■ Maintain continuously high velocity by use of a return line to the slurry holding tank
■ Chemical feed lines ruptures	■ Positive displacement pump has been started against a closed valve/vessel	■ Valve positions	■ Open valves in feed lines before pump is started
■ Pump head of diaphragm pumps clogged	■ Unwanted moisture in dry polymer facility can create "fish-eyes"	■ Dry polymer storage tank and head of pump	■ Clean the pump if clogged; ensure no moisture contamination is entering the storage
■ Increasingly increased polymer dosing rates	■ Lowered active concentration due to aged or expired totes/chemical	■ Perform polymer jar tests with old and new polymer batches	■ If the polymer is aged and has a lowered active concentration, then new polymer of the same type should be replaced

Coagulation and Flocculation

FUNDAMENTALS

Coagulation and flocculation are often the first steps in drinking water treatment and are commonly used as part of an advanced wastewater treatment facility to further treat process streams and remove smaller particles following primary and secondary treatment. Coagulation is essentially the addition of a chemical for the purpose of conditioning suspended, colloidal, or dissolved matter before flocculation. Flocculation is the aggregation of destabilized colloidal particles and subsequent precipitation into "flocs" (Crittenden, 2012).

Turbidity is light scatter caused by colloidal particles that cause water to appear cloudy.

PURPOSE AND FUNCTION

The objective of coagulation and flocculation is turbidity reduction, that is, to turn small particles into large particles that can be removed by clarification or filtration.

For wastewater treatment and resource recovery facilities, this unit process is to treat the water to make disinfection more effective or to help meet nutrient discharge limits. To reduce the eutrophication rate in receiving waters, many dischargers are required to reduce effluent phosphorus or nitrogen concentrations.

Eutrophication is the process by which an entire body of water, or parts of it, becomes progressively enriched with minerals and nutrients. It has also been defined as "nutrient-induced increase in phytoplankton productivity." This leads to dense growth of plant life and death of animal life from lack of oxygen.

Phosphorus is often the growth-limiting nutrient for algae in fresh surface water. Although biological nutrient-removal processes can significantly reduce phosphorus concentrations, chemical phosphorus-

Please refer to *Wastewater Treatment Fundamentals I,* Chapter 9 for biological nutrient removal processes.

removal systems remain common. Some facilities may use a combination of biological and chemical processes (e.g., the Phostrip process) to achieve an effluent total phosphorus concentration of less than 1 mg/L.

Coagulant typically is added to a tertiary process. However, some WRRFs also add coagulant to the activated sludge tank to enhance solids removal in secondary clarification. Other facilities add coagulant to raw wastewater as it enters the facility to enhance removal in primary clarifiers. If coagulants are added to primary treatment, care must be taken not to remove all the phosphorus, which is an essential nutrient for biological growth in secondary treatment. No matter where coagulants are added, chemical coagulation will increase the volume of sludge that must be handled in clarification/clarification processes. Enhanced suspended solids removal processes are essentially identical to chemical phosphorus removal systems. These processes typically include coagulation, flocculation, clarification, and filtration.

THEORY OF OPERATION

Inorganic coagulation such as aluminum sulfate and ferric chloride break down rapidly when mixed with the water to be treated. When metal salts hydrolyze, they also form acid that may affect pH and alkalinity. These chemical reactions, pH, and alkalinity considerations are discussed more in the previous chemical section under determining chemical dosing requirements. The chemistry of coagulation is extremely complex. This section will cover basic conceptual theory of charge neutralization and subsequent flocculation.

Recommended Reading: For an in-depth discussion and additional details on all theoretical mechanisms of coagulation and flocculation, refer to Chapter 9 of *MHW's Water Treatment Principles and Design* (Crittenden, 2012).

Most colloidal particulates are negatively charged. The negative charges repel one another. It's like pushing the negative ends of two magnets together. The two ends can only get so close together. Colloidal particles in solution push apart from one another in the same way. This prevents the particles from settling and from forming larger particles. They are "stable." When inorganic metal salts are added to water, they form positively charged ions. Alum forms Al^{3+}. Ferric chloride forms Fe^{3+}. The hydrolyzed aluminum is surrounded by six water molecules. The negatively attracted oxygen ion binds closer to the positively charged aluminum ion; this leaves the positively charged hydrogen ions facing outward. These molecules can then form an attraction bond with negatively charged colloidal and suspended particles.

Aluminum ions will also form solid hydroxide compounds when introduced to water. The full reaction for aluminum sulfate with water is given below:

$$Al_2(SO_4)_3 14H_2O \leftrightarrow 2Al(OH)_{3\downarrow} + 6H^+ + 8H_2O + 3SO_4^{2-}$$

Adsorbed means stuck to the surface.

When alum reacts with water, it forms a solid-aluminum hydroxide [$Al(OH)_3$]. This solid appears as tiny white snowflakes in the water. These snowflakes and the chemical reactions that produce them don't remove bacteria or phosphorus or organic matter directly. However, they can remove small particles indirectly as they settle. This action is called *sweep* or *sweeping*. Smaller particles are caught by the falling floc and adsorbed. The result is clear water with few colloids. Figure 3.4 shows negatively charged colloids suspended in solution before coagulation (left jar) and settled sludge after coagulation/flocculation (right jar).

Agglomerate means to come together.

Once the particles begin to coagulate, flocculation increases the collisions of coagulated solids and promotes chemical bridging between particles, so they agglomerate to form settleable or filterable flocs.

Brownian motion is the erratic and random movement of microscopic particles in a fluid resulting from their collision with the fast-moving molecules in the fluid. This can also be referred to as *diffusion*.

Whether done in separate basins, before clarification, or within the clarification tank (e.g., flocculating or high-rate clarifiers), flocculation involves prolonged and gentle agitation of coagulated particles to form larger, denser particles. Mixing energy is low. The flocs should never be subjected to mixing more intense than that used for initial flocculation or else the floc will break up.

The two types of flocculation are microscale (small) and macroscale (large). Microscale flocculation is the primary mechanism of aggregation of small particles due to Brownian motion, and macroscale

Figure 3.4 Suspended Colloids and Settled Solids After a Jar Test (Provided courtesy of Advanced Chemical Systems™)

flocculation is the mechanism of aggregation for particles larger than 1 μm. Macroscale flocculation is controlled by fluid velocity gradients created by mixers and differential (gravity) settling.

During flocculation, the coagulated effluent is stirred slowly via air or mechanically driven paddles, causing the solid particles to agglomerate, become heavy, and settle on the tank floor.

Flocculation is similar to train cars lightly bumping into one another end to end. This action connects the couplings between cars and extends the overall length of the train. This occurs at low speed (low velocity). Imagine what would happen if a train car came running down the track at high speed and ran into the back of the train. All that energy would derail several cars and break apart the larger train. Flocculation is similar. We need enough energy to make the connection, but less than the strength of the particles (flocs) overall.

Flocculation basins are often baffled to reduce short-circuiting. Two or three basins in series may also be used. The detention time for each basin typically ranges from 10 to 45 minutes. As the water moves through the process, the stirring becomes gentler, allowing the larger floc particles to collide and bond into larger marcroflocs. The objective of flocculation is to bring particles into contact so that they will collide, stick together, and grow to a size that will readily settle or filter out.

Obtaining the proper speeds for flocculation can be tricky. There must be enough mixing energy to facilitate particle collisions, but too much energy will force the "flocs" to break up back into smaller, non-settleable particles.

DESIGN PARAMETERS

Coagulation and flocculation require two separate components: a method of rapid mixing to disperse chemical and a method of gentle mixing to increase floc size. Rapid mixing typically uses a small basin and turbine or propeller mixer. There are three types of mixers used in flocculation: horizontal shaft, vertical shaft, and hydraulic basins. These are covered in the "Equipment" section below.

The action of coagulation and flocculation mixing depends on the characteristics of the particles and the fluid conditions. The basic design criteria for mechanical mixers are the Camp-Stein velocity gradient (G) and hydraulic retention time (t). Table 3.3 summarizes some typical detention time and velocity gradient parameters for coagulation and flocculation. Notice how flocculation detention times are longer than flash mixing and with lower velocity gradients.

G: Camp-Stein velocity gradient is how fast the water is moving compared to how fast nearby layers of water are moving. G is expressed as "per second" or s^{-1}.

Table 3.3 Typical Detention Time and Velocity Gradient G Values for Mixing and Flocculation in Wastewater (Adapted from Metcalf & Eddy, 2003; Davis, 2010)

Process	Range of Values	
Coagulant Mixing	**Detention Time**	**G Value (s^{-1})**
Typical rapid mixing operations in wastewater	5–30 seconds	500–1500
Rapid mixing for initial dispersion	<1 second	1500–6000
Rapid mixing of chemicals in contact filtration processes	<1 second	2500–7500
Flocculation	**Detention Time**	**G Value (s^{-1})**
Typical flocculation processes used in wastewater treatment	30–60 minutes	50–100
Flocculation before direct filtration	2–10 minutes	25–150
Flocculation in contact filtration process	2–5 minutes	25–200
Low turbidity color removal		20–70
High turbidity solids removal		30–80
Softening, 10% solids		130–200
Softening, 39% solids		150–300

The dispersion and immediate mixing of coagulant into water is called *flash mixing* or *rapid mixing*. Metal salts dissolve in water very quickly (within 10^{-5}s) (Crittenden, 2012). This means that nearly instantaneous and intense mixing is desired for coagulation with a high-velocity gradient. Table 3.4 shows some typical design parameters for turbine and propeller mixers.

For flocculation, much longer mixing times are needed at lower intensities to allow for the aggregation of destabilized particles into larger flocs. Time requirements for flocculation vary depending on other downstream processes. For conventional treatment, which includes clarification basins, typical flocculation times are 20 to 30 minutes; however, for direct filtration, shorter flocculation times of 10 to 20 minutes are typically selected (Crittenden, 2012; Davis, 2010).

The maximum G for paddle flocculators is limited to about 60 s^{-1}. Any higher and severe wear of the shaft bearings will occur. The basic design criteria for mechanical flocculators use velocity gradient and detention time as major design parameters. Flocculation basins are typically divided into various

Table 3.4 Typical Design Parameters for Turbine and Propeller Mixers (Adapted from Metcalf & Eddy, 2003)

Parameter			
Horizontal flow mixing	**Symbol**	**Unit**	**Value**
Velocity gradient	G	1/s	500–2500
Rotational speed	N	r/min	40–125
Ratio impeller diameter to equivalent tank diameter	D/T	Unitless	0.25–0.4
Vertical flow mixing	**Symbol**	**Unit**	**Value**
Velocity gradient	G	1/s	500–2500
Rotational speed	N	r/min	25–45
Ratio impeller diameter to equivalent tank diameter	D/T	Unitless	0.4–0.6

compartments. The velocity gradient is tapered so that the intensity is lower in the subsequent basins. This keeps the floc from breaking back apart. Baffle walls are used to separate basin compartments with the top slightly submerged and space on the bottom for sludge removal. Table 3.5 outlines additional flocculation basin design criteria.

EXPECTED PERFORMANCE

The time required for flocculation depends on the wastewater or effluent characteristics. For example, the South Tahoe (California) Public Utility District lime-coagulated, activated sludge effluent flocculates in 5 minutes in an air-agitated basin. Good flocculation is obtained even without agitation. However, the lime-coagulated, trickling filter effluent at the Orange County (California) Water District's Water Factory 21 is difficult to flocculate. Here, 20 to 30 minutes of careful mixing is needed to form a good settling floc.

EQUIPMENT

Equipment consists of basins, mixers, and flocculators. Flocculation basins are typically concrete and rectangular. Several flocculation basins in series will perform better than one large basin with the same total volume because they provide better mixing and can use different mixing energies. Different types of equipment are used for flocculation. Impeller-style flocculators are small mixers with high speeds. Think of a boat or airplane propeller. Impellers take up less space in a basin and turn at a much higher frequency compared to other flocculator systems.

Impeller-style mixers may require mounting bridges or may be fastened directly to the basin wall. Impeller-style mixers are designed to produce adequate mixing with the lowest turbulence possible. Impeller mixers

Table 3.5 Typical Design Recommendations for Flocculation Facilities (Adapted from Davis, 2010)

Parameter	Recommendation
G	<50 s^{-1}
Basin	
Depth (D)	1 m (3.28 ft) > wheel diameter
Clearance between wheel and walls	.3048–.70104 m (1–2.3 ft)
Wheel	
Diameter	2.98–4 m (9.8–13.1 ft)
Spacing between wheels on same shaft	1 m (3.3 ft)
Spacing between wheel "rims" on adjacent shafts	1 m (3.3 ft)
Paddle Board	
Width (W)	10.16–15.24 m (4–6.0 in)
Length (L)	2.01–3.51 m (6.6–11.5 ft)
Area of paddles/cross section	0.10–0.25
Number per arm	3
Tip speed	0.15–0.97 m (0.5–3.2 ft/s)
CD, L/W = 5	1.2
CD, L/W = 20	1.5
CD, L/W >> 20	1.9
Motor power	1.5–2 × water power
Turndown ratio	1:4

produce a very high tip speed that tends to break down the floc. Once broken down, floc masses will not re-form. Figure 3.5 shows floc breakage due to excess impeller speeds.

At lower mixing speeds (200 rpm), the floc appears to stick together better than at higher impeller speeds (400 rpm).

Paddle wheel flocculators may be compared to the slow-moving paddle wheel of a paddle wheel boat. Paddle wheel flocculators have traditionally been used in the municipal wastewater treatment process, primarily because their slow speed, low-shear mixing method is well suited to the flocculation process in a water treatment basin. Multiple paddles of considerable size are used to gently mix the basin influent allowing for a robust, high-density floc with a high settling capacity, resulting in low turbidity effluent. The large, engineered paddles reach the entire basin, eliminating dead spots and short-circuiting. Figure 3.6 shows an example of two types of mechanical flocculators.

Variable-frequency drive is a type of motor controller that drives an electric motor by varying the frequency and voltage supplied to the electric motor.

Various components of the flocculators include the motor with a variable-frequency drive, a gear reducer, and mounting plate.

Paddle wheel flocculators have a larger blade surface area than conventional mixers. This means that adequate mixing can be achieved at lower speeds. Lower speeds mean fewer rotation and less wear and tear on the gear box.

When a low energy solution is needed, hydraulic baffle wall flocculators (also known as "Alabama" flocculators) have numerous walls with specific spacing to create a flow velocity gradient. They force the water through a serpentine path (see Figure 3.7).

Walls can be made of steel wood, fiberglass, or other materials. Serpentine flows are used or over/under flows. As the flow progresses through the basin (blue arrows), it is mixed by its own velocity gradients. Figure 3.8 shows a photo of a hydraulic flocculator.

Hydraulic flocculators require very little maintenance—they have no moving parts, no expendable wear items, and no electricity cost. Some disadvantages of hydraulic flocculation are that they are difficult to adjust mixing intensity, they create significant headloss across the basin, and the floc particles formed are not as large as other paddle wheel technologies.

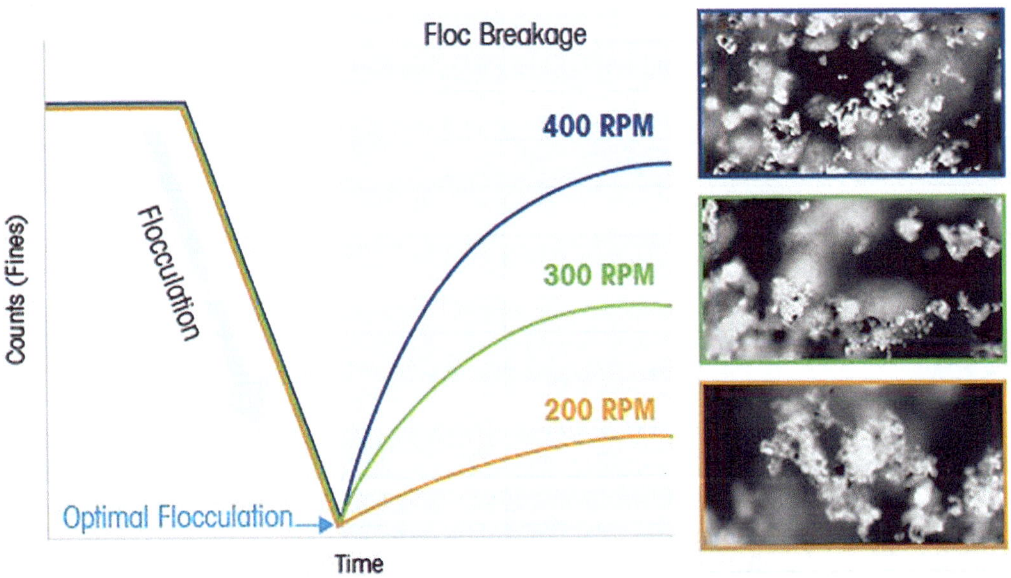

Figure 3.5 Floc Breakage Due to Excess Impeller Speeds (Provided courtesy of METTLER TOLEDO) (https://www.mt.com/global/en/home/applications/L1_AutoChem_Applications/L2_Particle Processing/Formulation_Flocculation.html)

A) Horizontal paddle wheel flocculator

B) Vertical paddle wheel flocculator

Figure 3.6 (A) Horizontal Paddle Wheel Flocculator and (B) Vertical Paddle Wheel Flocculator (Photo provided courtesy of Jim Myers & Sons Inc. [JMS])

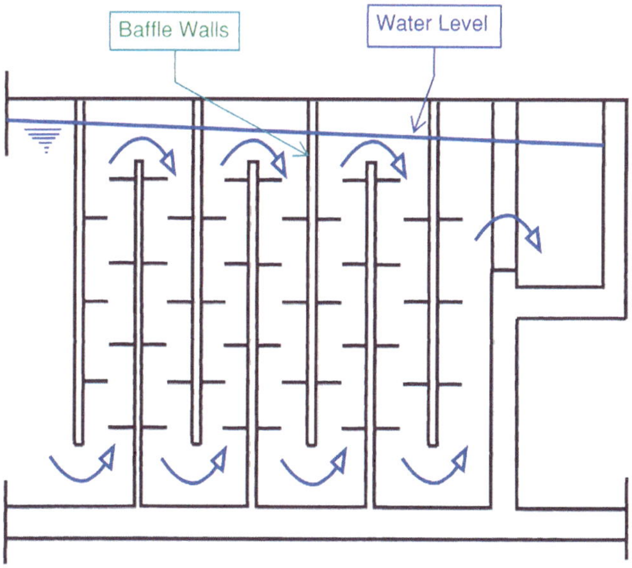

Figure 3.7 Hydraulic Flocculation Schematic

Figure 3.8 A Hydraulic Alabama Flocculation Basin (Upstream and Downstream View) (Photo provided courtesy of Jim Myers & Sons Inc. [JMS])

PROCESS VARIABLES

Process variables include detention time, mixer type, mixer speed, chemical type, chemical dose, and zeta potential. Dosages for chemicals should be based on experience with similar types of waters, or preferably with jar tests and pilot tests.

PROCESS CONTROL

Process control systems typically focus on the ability to maintain adequate coagulant doses. Ideally, an automated control system would dose the chemicals in response to variations in flows and phosphorus loading. In practice, chemical dosing systems are flow paced (i.e., based on influent) if influent phosphorus levels are relatively constant, or they anticipate peak phosphorus loads to avoid permit violations. If jar testing is used to determine chemical doses, the emphasis should be on determining a range of doses that will accommodate the expected variations in flows and phosphorus loading.

Wastewater pH can be measured using online pH measurement systems. A simple feedback loop using the online pH values can then be used to dose caustic or other forms of supplemental alkalinity to replace that consumed by metal salt addition. The amount of operator control available in flocculation is dependent on the type and design of the equipment.

NORMAL PROCESS CONTROL

A common process control practice is to set the doses of coagulant, polymer, and supplemental alkalinity based on historical operating data. In practice, the feed rates are set higher than anticipated to account for variations in the phosphorus load and avoid permit violations at peak conditions. A typical, though not necessarily recommended, mode of operation is to overdose coagulants at average influent concentrations so adequate coagulant concentrations are available at peak influent phosphorus loads.

DATA COLLECTION

Data collection for polished effluent should include process control parameters such as total suspended solids (TSS), phosphorus, alkalinity, and pH monitoring. Typically, online monitoring with periodic grab samples to validate the online measurement may be required to monitor the process and provide feedback control for any supplemental alkali.

MAINTENANCE

Planned maintenance for coagulation and flocculation mostly involves typical maintenance procedures for such equipment as chemical storage tanks, mixing equipment, centrifugal pumps, and chemical feed equipment, which are discussed in other chapters of this book. Flow, pH, and phosphorus monitoring devices should be checked routinely and calibrated periodically based on the manufacturer's recommendations.

Table 3.6 Probable Causes and Solutions for Troubleshooting a Coagulation–Flocculation System (Adapted from U.S. EPA, 1978)

Indicator/ Observation	Probable Cause	Check or Monitor	Solutions
■ Poor floc formation and settling characteristics	■ Chemicals not sufficiently dispersed during rapid mixing ■ Prolonged rapid mixing ■ Improper coagulant dose ■ Flocculator agitators operating at excessive speeds	■ Chemicals not uniformly distributed in rapid mix basin ■ Floc formation breakup in rapid mix basin ■ Jar test ■ Agitator speed	■ Increase speed of rapid mixing device ■ Reduce rapid mixing time ■ Correct coagulant dosage per jar test results ■ Reduce flocculator agitator speed
■ Good floc formation in flocculation basin, but poor settleability in clarifier tank	■ Excessive velocity between flocculation unit and clarifier unit	■ Velocity between flocculation and clarifier units	■ Reduce velocity
■ Clarifier sludge turning anaerobic	■ Sludge blanket present in clarifier ■ Excessive carryover of organic solids from secondary processes	■ Visual inspection ■ Secondary effluent quality	■ Increase sludge removal rate to remove sludge blanket and prevent its formation ■ Correct problems in secondary process

TROUBLESHOOTING

Regardless of the chemical dosing point or whether alum or ferric chloride is the primary coagulant, polymer may also be needed (generally an anionic polymer) to help flocculate solids. If flocculation is inadequate, pin floc, or small flocs not large enough to settle, can develop and result in phosphorus carryover from the primary or secondary clarifier. Flocculation systems are typically easy to maintain, but they can still experience their fair share of troubleshooting. Table 3.6 provides a summary of common problems and solutions.

TEST YOUR KNOWLEDGE

1. **What is the primary goal of coagulation and flocculation?**
   a. To adjust the alkalinity of the wastewater
   b. To create biosolids
   c. To stabilize the colloidal particles
   d. To remove turbidity, bacteria, and nutrients

2. **Colloidal particles and natural organic matter are typically positively charged.**
   ☐ True
   ☐ False

3. **Which is a type of flocculator?**
   a. A high-speed vertical turbine pump
   b. A rapid mixer
   c. A baffled concrete basin
   d. Peristaltic pump

4. **Hydraulic basins can create**
   a. Large flocs
   b. Turbidity
   c. Whirlpools
   d. Solids

5. **Coagulation (rapid mix) requires a high**
   a. Velocity gradient (G)
   b. Detention time
   c. Turbidity
   d. Chlorine demand

6. **Flocculation requires a**
   a. High-velocity gradient
   b. Low detention time (t)
   c. High detention time (t)
   d. High power input

7. **Flocculation requires a**
   a. Low coagulant dose
   b. High velocity gradient (G)
   c. Low velocity gradient (G)
   d. Low detention time (t)

8. **During eutrophication, the water becomes**
   a. Clear
   b. Oxygen sufficient
   c. Oxygen deficient
   d. Muddy

9. **Floc breakage occurs at**
   a. Low chemical dose
   b. High chemical dose
   c. High velocity gradient
   d. Low velocity gradient

10. **Paddle wheels have a large**
    a. Rotational speed range
    b. Mixing area
    c. Power consumption
    d. Footprint

Clarification

FUNDAMENTALS

Clarification is the process of separating suspended particles from water via gravity in large basins or tanks, typically under quiescent conditions. The terms *clarification*, *settling*, and *clarification* are often used interchangeably. Water resource recovery professionals tend to use *clarification*.

PURPOSE AND FUNCTION

Clarification separates solids from liquids via gravity. Clarifiers are used extensively in primary treatment and secondary treatment in conventional wastewater treatment facilities for solids and biochemical oxygen demand removal. Primary and secondary clarification are covered extensively in chapters 4 and 8, respectively, of *Wastewater Treatment Fundamentals I*. This chapter will overlap with some content but will primarily focus on clarification as part of a coagulation–flocculation treatment train in tertiary treatment.

Conventional wastewater clarification basins typically contain mechanically driven scrapers at the bottom of the tank that continuously move the settled sludge to a hopper, from which it is pumped or removed via gravity. The tanks may be either rectangular or circular tanks. The degree of treatment obtained depends largely on the upstream processes (i.e., the type and quantity of chemicals used, mixing times, and the degree of care in process monitoring and control).

THEORY OF OPERATION

CONVENTIONAL SECONDARY CLARIFIER

Conventional clarifier tanks are covered in *Wastewater Treatment Fundamentals I–Liquid Treatment* (WEF & ABC, 2018). The clarifier works by permitting the heavier and larger particles to settle to the bottom of the clarifier. The particles then form a bottom layer of sludge requiring regular removal and disposal. Clarified water then proceeds through several more steps before being sent for storage and use. Figure 3.9 shows a conventional clarifier and ancillary equipment.

UPFLOW SOLIDS CONTACT CLARIFIERS

Upflow solids contact clarifiers provide coagulation, flocculation, and sedimentation in a single (typically circular) steel or concrete tank. These clarifiers are termed "upflow" because the water flows up toward the effluent launders as the suspended solids settle. They are characterized by increased solids contact through internal sludge recirculation. This is a key feature in maintaining a high-clarity effluent and a major difference from horizontal clarifiers. Figure 3.10 shows a solids contact clarifier and its components.

INCLINED PLATE CLARIFIERS

Inclined plate clarifiers work when a solid/liquid stream that has been flocculated enters a tank and flows upward between a pack of inclined plates. The solids fall to the plate surface, where they slide by gravity down to a sludge collection hopper. Figure 3.11 shows a schematic of an inclined plate clarifier.

Ballasted Media—the combination of flocs or suspended solids onto microsand with the aid of a polymer.

Densifying—to increase the density of the solution.

BALLASTED MEDIA HIGH-RATE CLARIFIER

Ballasted media high-rate clarification systems use both gravity and "densifying" solids-contact or ballasted-media flocculation technologies to clarify coagulated and flocculated effluents.

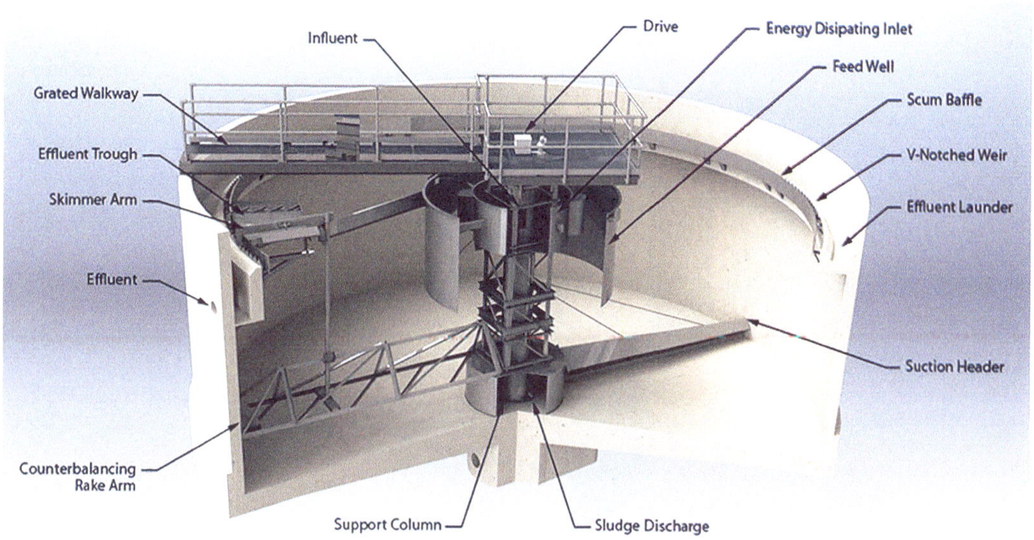

Figure 3.9 **Conventional Clarifier and Equipment** (Provided courtesy of Monroe Environmental)

A major selling point of these technologies is that their high surface overflow rates enable them to have smaller footprints than conventional clarifiers. Figure 3.12 shows a ballasted high-rate clarifier process flow diagram.

In the solids-contact unit's reactor zone, influent combines with reactants and preformed solids from the presettling–thickening zone, flowing up through a draft tube where a specially designed turbine initiates flocculation. As the mixture resettles, its density increases. The internal recirculation rate is up to 10 times that of the influent flow, producing optimum floc density.

The flocculated effluent enters the presettling–thickening zone over a submerged weir. The solids then sink to the bottom of the vessel, where a slow-moving rake helps make them thicker and denser. This maintains solids consistency, while releasing more entrained water. Thickened sludge is periodically cleared from the

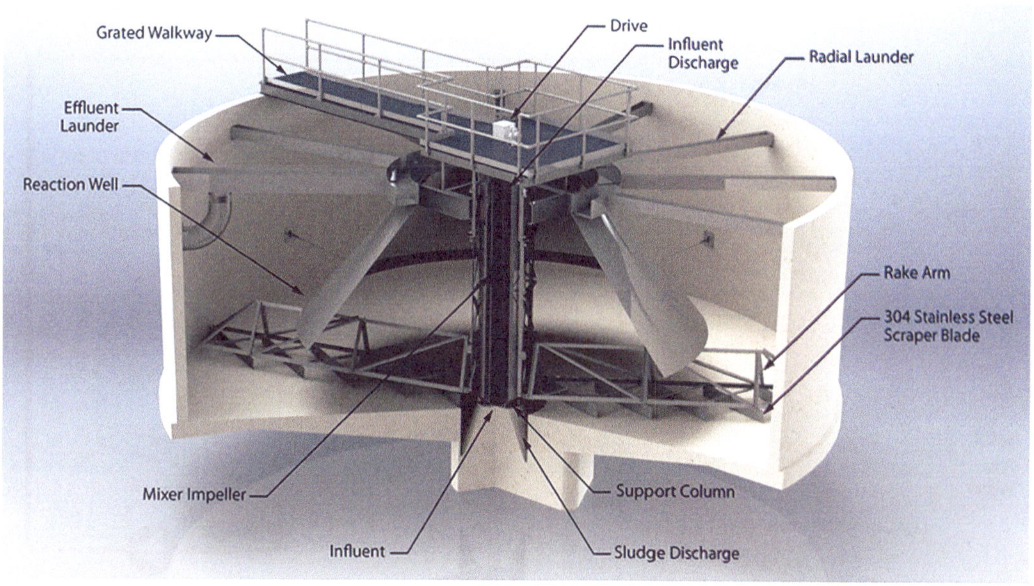

Figure 3.10 **Solids Contact Clarifier and Ancillary Equipment Schematic** (Provided courtesy of Monroe Environmental)

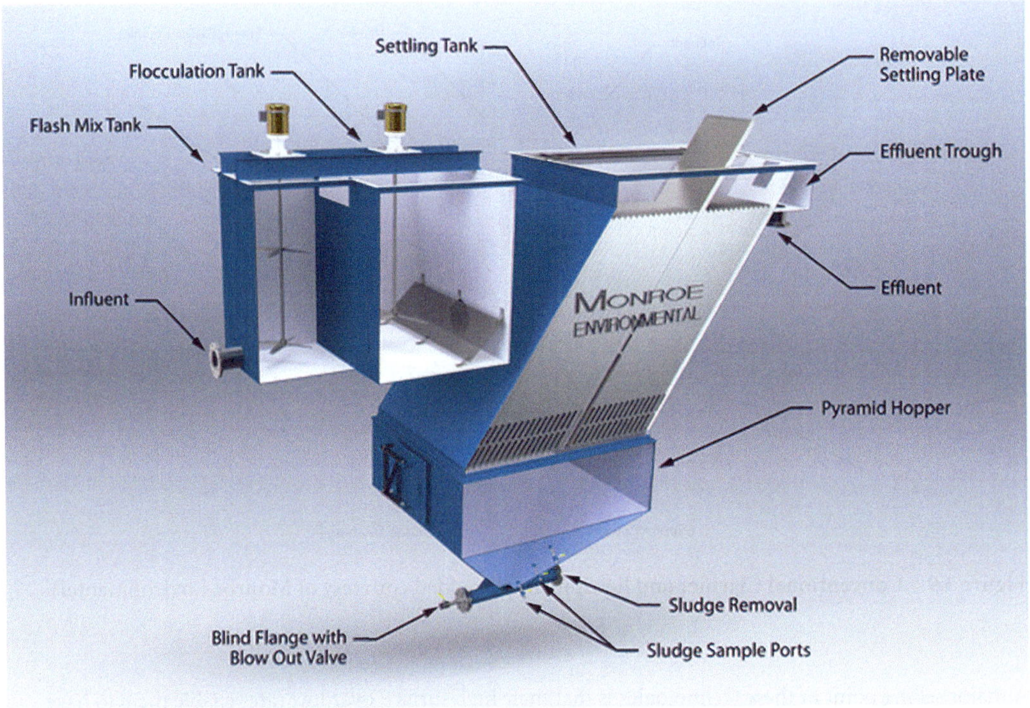

Figure 3.11 Plate Settler and Ancillary Equipment (Provided courtesy of Monroe Environmental)

bottom of the thickener and is typically pumped to a final dewatering mechanism. In the clarification zone, supernatant flows upward through settling plates. The settling plates allow rise rates of up to 0.068 $m^3/m^2{\cdot}s$ (10 gpm/sq ft) for most metal salt coagulation processes—often higher for lime addition processes. The clarified water is uniformly collected in effluent launders above the tubes.

Influent first enters a flash mixing zone, which destabilizes the colloids. Here, wastewater is rapidly mixed with coagulant for about 1 minute. The wastewater then enters the injection tank, where microsand and polymer are added and rapidly mixed for about 1 minute. Then, the water enters a maturation zone, where it is gently mixed for about 4 minutes. This allows the microsand ballasted flocs to grow bigger, so they

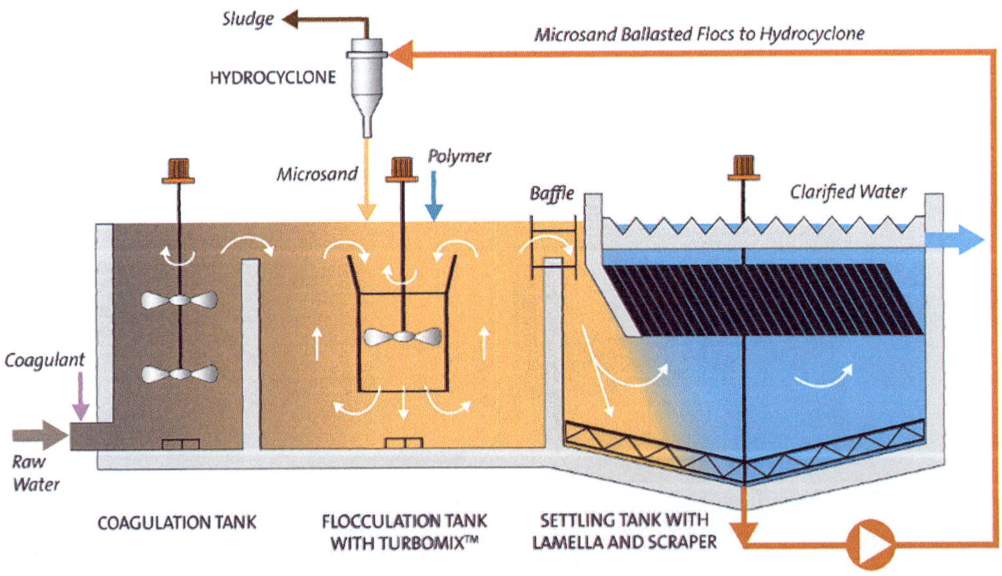

Figure 3.12 Ballasted Media High-Rate Clarification With Sludge Recycle (ACTIFLO®) (Diagram Provided Courtesy of Veolia Water Technologies)

can trap random flocs and settle faster in the clarification tank. The polymer works like a glue to affix the flocs to the microsand. The water then enters the settling tank, where the microsand ballasted flocs settle immediately. The water flows up through settling tubes to collection troughs at the surface and exits the process.

The microsand ballasted sludge is pumped from the bottom of the settling tank to a hydrocyclone, where the microsand is cleaned from the sludge via centrifugal force. The sludge is continuously removed for treatment, and the microsand is re-injected into the system (it is continuously recycled).

Utilities that tried to use upflow, sludge blanket-type clarifiers reported difficulty in maintaining a sludge blanket. The units operated successfully, however, when the surface overflow rate was lowered to that typical of conventional settling tanks and the sludge blanket was significantly reduced or eliminated (essentially converting them to conventional radial-flow basins). The unstable sludge blankets are thought to be primarily caused by variations in the influent's properties. Upflow tanks work better in water treatment, where the influent is relatively uniform and can be pumped into the clarifiers at a more or less constant flowrate.

DESIGN PARAMETERS

Surface overflow rate is the primary design parameter for sizing clarification basins. Overflow rate is defined as flow divided by area, or:

$$\text{Surface Overflow Rate} = \frac{\text{Flow, m}^3\text{/d}}{\text{Area, m}^2}$$

EXAMPLE CALCULATION

What is the overflow rate for a 6-m (20-ft) diameter clarifier with a flow of 19 ML/d (5 mgd)?

$$\text{Surface Overflow Rate} = \frac{\text{Flow}}{\text{Area}} = \frac{1\text{ MGD} \times \left(\dfrac{133\ 680\text{ cu ft}}{\text{mgd}}\right)}{\left(\dfrac{20}{2}\text{ft}\right)^2 \times \pi} = \frac{425.73\text{ cu ft}}{\text{d sq ft}}$$

Other design parameters include length-to-width ratio, length-to-depth ratio, side water depth, and velocity. Table 3.7 summarizes typical design criteria and parameters for horizontal-flow rectangular tanks.

Table 3.7 Typical Design Criteria for Horizontal-Flow Rectangular Tanks (Adapted from Davis, 2010)

Parameter	Units	Value
Type	—	Horizontal-flow rectangular tank
Minimum number of tanks	Unitless	2
Depth (D)	m (ft)	3–5 (9.8–16.4)
Overflow rate	m³/d m² (cu ft/d*sq ft)	40–70 m³/d m² (131–230 cu ft/d sq ft)
Length (L)	Meters (ft)	30 m (wind constraint); (98.4 ft) 60 m (196.85 ft) (chain and flight)
Width (W)	Meters (ft)	6 m max per train (19.685 ft)
L:W		4:1 to 6:1, greater than 6:1 is preferred
L:D		15:1
Velocity	m/s (ft/sec)	0.005–0.018 m/s (.016–.06 ft/sec)

EQUIPMENT

Conventional clarification units consist of large, rectangular, concrete basins divided into two or three sections. Each stage of the clarification process occurs in a single section of the basin. Water movement is horizontal with plug flow through these systems.

In plug flow, the velocity of the fluid is assumed to be constant across any cross section of the pipe perpendicular to the axis of the pipe.

The design is suited to large-capacity basins. Horizontal flow units are still used in some large industrial plants and for clarifying municipal drinking water. The retention time is typically long (6 hours) and is chiefly devoted to settling. Rapid mix is typically designed for 3 to 5 minutes and slow mix for 15 to 30 minutes. This design affords flexibility in establishing proper chemical addition points. Also, such units are relatively insensitive to sudden changes in water throughput.

The long detention time also allows sufficient reaction time to make necessary adjustments in chemical and polymer feed if feed or influent conditions suddenly change. However, for all but very large, treated water demands, horizontal units have high construction costs and require more land space per unit of water capacity compared to circular clarifiers and upflow clarifiers.

UPFLOW SOLIDS CONTACT CLARIFIERS

Upflow clarifiers provide coagulation, flocculation, and clarification in a single (typically circular) steel or concrete tank. These clarifiers are termed *upflow* (see Figure 3.13) because the water flows up toward the effluent launders as the suspended solids settle. Because retention time in an upflow unit is approximately 1 to 2 hours, upflow basins can be much smaller in size than horizontal basins of equal throughput capacity. An overflow rate of 1.71 to 3.06 m/h (0.70 to 1.25 gpm/sq ft) of surface area is normal for clarification. Combination softening–clarification units may operate at up to 3.67 m/h (1.5 gpm/sq ft) of surface area due to particle size and densities of precipitated hardness.

To achieve high throughput efficiency, upflow units are designed to maximize the linear overflow weir length while minimizing the opportunity for short-circuiting through the settling zone. In addition, the two mixing stages for coagulation and flocculation take place within the same clarification tank.

INCLINED PLATE CLARIFIER

An incline plate clarifier is designed to remove particulates from liquids. They are often used in primary water treatment in place of conventional settling tanks. They are also used in industrial water treatment. Unlike conventional clarifiers, they use a series of inclined plates. These inclined plates provide a large effective settling area for a small footprint. Figure 3.14 shows a high-rate solid contact clarifier with plates.

Solid particles begin to settle on the plates and begin to accumulate in collection hoppers at the bottom of the clarifier unit. The sludge is drawn off at the bottom of these hoppers and the clarified liquid exits the unit at the top via a weir.

The most common wastewater application for incline plate clarifiers is as part of tertiary treatment. Inclined plate clarifiers can be integrated to the treatment process or stand-alone units can be used to

Figure 3.13 Solids Contact Clarifier and Ancillary Components (Photo provided courtesy of Jim Myers & Sons Inc. [JMS])

Figure 3.14 High-Rate Solids Contact Clarifier and Ancillary Equipment (Photo provided courtesy of Jim Myers & Sons Inc. [JMS])

increase the flow through existing WRRFs. One option for integrating incline plate clarifiers to existing facilities is for conventional or sludge blanket clarifiers to be upgraded by attaching a bundle of inclined plates or tubes before the overflow in the so-called "clear water zone." Figure 3.15 shows a closeup photo of plate settlers.

This bundling of the plates can increase the settling area by twofold, resulting in a decrease in the solids loading in the overflow.

Table 3.8 summarizes some advantages and disadvantages of different clarification process equipment.

Figure 3.15 Closeup Photograph of Plate Settlers (Photo provided courtesy of Jim Myers & Sons Inc. [JMS])

Table 3.8 Typical Applications for Clarification Process in Advanced Wastewater Treatment (Adapted from Crittenden, 2012)

Typical Applications	Advantages	Disadvantages
Rectangular Basin (Horizontal Flow)		
■ Many municipal and industrial water works	■ More tolerable to shock loads	■ Subject to density flow creation
■ Particularly suited for large-capacity facilities	■ Predictable performance	■ Requires careful design and construction
	■ Easy operation	■ Requires separate flocculation basin
Upflow (Radial Flow)		
■ Small to midsize municipal and industrial water plants	■ Economical compact	■ Short-circuiting
■ Best suited where rate of flow and raw-water quality are constant	■ Easy sludge removal	■ Careful operation
		■ Limitation on practical unit size
Solids Contact Clarifiers		
■ Water softening	■ Good softening and turbidity removal	■ Sensitive to shock loads
■ Flocculation–clarification treatment of raw water that has constant quality and flow	■ Flocculation and clarification in one unit	■ Sensitive to temperature change
■ Plants treating a raw water with low solids concentration	■ Compact	■ Plant-operation dependent

PROCESS CONTROL

Process control systems for chemical clarification units focus on the surface loading or overflow rate, detention time in the basins, and the solids loading rate (the rate of solids feed divided by the clarifier's cross-sectional area, expressed as $kg/m^2 \cdot d$ [lb/d/sq ft]). Because these parameters are typically set during design and are physically limited by the unit's size provided, operators will only have indirect control over these parameters. Some indirect control measures might include using in-facility tankage or equalization basins, system storage in collection systems, careful pump operations to minimize influent surges, and flow distribution among multiple clarifiers to reduce or increase surface overflow rate and detention time.

SLUDGE RECYCLING/WASTING

More solids are generated at WRRFs when chemical addition is practiced. Adding chemicals to settling tanks will also increase sludge generation. For example, when chemicals are added to primary clarifiers for enhanced primary treatment, the additional solids generated will necessitate higher sludge withdrawal rates to prevent other operating problems in the clarifiers. Unless some form of solids contact or floc blanket clarification is desired, chemical sludge is typically removed or wasted out of the system. However, some WRRFs with separate mixing, flocculation, and settling basins recycle settled sludge to either the mixing or flocculation basins to enhance floc formation. The previously precipitated particles in the return sludge may form a group for more rapid agglomeration of colloidal particles in the wastewater. However, if the organic solids in the sludge are unusually high, recycle may impair floc formation.

While metal salt addition will increase the nonvolatile fraction of the mixed liquor suspended solids (MLSS) in an activated sludge facility (requiring higher MLSS values to maintain treatment levels), the major effects of sludge wasting from chemical treatment are typically felt more in the downstream solids-handling processes. In addition to the higher amounts of solids to be handled, chemical sludges are also difficult to thicken and dewater (the reasons why are not yet fully understood). Also, the properties of these sludges and the degree of difficulty in handling them will not be the same from one facility to the next. While alum sludges are typically more difficult to thicken and dewater than ferric chloride sludges, the site-specific nature of sludge merits pilot- or full-scale testing of its effects on solids-handling facilities.

NORMAL PROCESS CONTROL

The clarification process is key to overall WRRF performance. If the settling tank effluent contains high suspended solids concentrations, it will reduce the efficiency of downstream unit processes and perhaps even cause a severe upset, resulting in a complete WRRF shutdown. The important variables when operating a clarification process are upstream coagulation and flocculation, the surface overflow rate, and the sludge withdrawal rate. The surface overflow rate depends on the wastewater flow and the volume of recycle flows (e.g., filter backwash water and sludge processing streams). Operators typically cannot control the inflow, but they can control the timing of the recycle flows. The amount of sludge withdrawn from the basins must be enough to avoid overloading the collection arms, prevent floc carryover, and avoid septic conditions. However, excessive withdrawals will result in a thin sludge, which can hurt downstream solids processing. Excessively thick sludge (greater than 6%) can also have adverse effects, including organics solubilization and sludge blanket subject to washout during hydraulic transients. The downstream thickening process may dictate the desired concentrations for sludge withdrawals.

Hydraulic transient is a sharp pressure surge or wave produced when water flow is forced to stop suddenly or change direction abruptly.

Some other issues to be aware of include:

- Short-circuiting and hydraulic instabilities because of variable influent characteristics or poor influent distribution,
- Density currents or thermal stratification resulting in dead spaces at the surface of the clarifier (when the influent is colder than the tank contents) or at the bottom (when the influent is warmer than the tank contents), and
- Wind effects, in which wind blowing across the water surface may set up circulation cells, reducing net volume and detention time in the tank.

Thermal stratification occurs when the temperature differences between two streams of water cause the colder water to settle and allow the warmer water to float over the colder water.

These issues are site-specific, often difficult to spot, and, once spotted, may be hard to analyze. Dyes or tracers are a simple, practical method for assessing the hydraulic performance of clarification basins. Tracer studies can also be used to determine the success of any corrective measures. Theoretical analysis techniques (e.g., computational fluid dynamics) may also be useful in simulating, analyzing, and resolving such conditions.

Langmuir circulation cells are a series of shallow, slow, counter-rotating vortices that trap water in a separate layer from the remainder of the clarifier.

Furthermore, operators should be aware that they may not be able to resolve these issues operationally. Depending on how often they occur, retrofits to the tanks (ranging from strategically placed baffles to tank covers) may be required to resolve them. Failure to address poor solids removal in the clarification process can increase downstream solids processing costs (primarily due to withdrawing excessively thin sludges).

Computational fluid dynamics is a branch of fluid mechanics that uses numerical analyses and computers to solve problems that involve fluid flows.

DATA COLLECTION

The pH and turbidity of the chemical settling tank influent and effluent should be monitored and recorded continuously. Settleability tests on the flocculator effluent (clarifier influent) should be run daily. For phosphorus removal, composite samples of the settling tank influent and effluent should be tested for phosphorus concentration. The solids concentrations in the sludge pumped from the chemical clarifier should be measured daily, based on grab samples. The sludge blanket depth can also provide valuable operating data.

MAINTENANCE

Planned maintenance programs for clarifiers are not significantly different from those for conventional primary or secondary clarifiers, so programs for conventional units can be used as a guide (see Chapter 4 and Chapter 8 of *Wastewater Treatment Fundamentals I*). However, chemical sludges may have different handling characteristics than conventional primary and secondary sludges. Also, sludges produced by different chemicals will have differing properties. For example, the physical characteristics of alum sludge will be different than those of lime sludge. If lime is used as a coagulant, then calcium carbonate deposits may need to be routinely removed from the rapid mixer shaft and paddles and all lime sludge transporting devices may need routine cleaning. The major mechanical equipment to maintain includes polymer mixing equipment; chemical feeders; rapid mixer motors, gear drives, and shaft bearings; flocculator drive motors, speed reducers, and shaft bearings; settling tank scraper drive motors and speed reducers; sludge pumps; and vibrators on lime storage bins.

TROUBLESHOOTING

Other than issues related to the sludge handling characteristics, troubleshooting for chemical settling tanks is not significantly different than that for conventional primary settling tanks, so guides for conventional units can be used as a starting point (see Chapter 4 and Chapter 8 of *Wastewater Treatment Fundamentals I*). The single biggest troubleshooting issue for chemical settling tanks is typically high effluent turbidity. Probable causes and potential solutions are listed below.

- Inadequate sludge removal—adding chemicals to settling tanks increases sludge generation, so sludge withdrawal rates must be higher than those for an equivalent conventional settling tank.
- Poor floc formation—this could be caused by incorrect chemical dose (run a jar test to determine the appropriate dose); interrupted chemical feed (check that the chemical feeders are operating and the chemical supply is not exhausted); or chemical dispersion being inadequate (adjust rapid mixer speed).
- Floc forming but breaking up before settling—check for excessive turbulence in flocculation zone (check and adjust flocculation mixer speed if necessary) or turbulence at the clarifier inlet or within the clarifier (solutions range from reducing flowrate into unit to installing stilling baffles at strategic locations). Another solution might be using a polymer with the inorganic coagulant to strengthen the floc if this has not already been done.

TEST YOUR KNOWLEDGE

1. Which of the following is a type of clarifier basin?
   a. Alabama flocculator
   b. Serpentine flow
   c. Upflow
   d. Anoxic basin

2. In plug flow, the velocity of the fluid is assumed to be __________ across any cross-section of pipe perpendicular to the axis of the pipe.
   a. Constant
   b. Varied
   c. Supercritical
   d. Subcritical

3. When used as part of an advanced water treatment train (coagulation–flocculation-clarification) what is a primary difference from clarifiers used for primary or secondary treatment in conventional wastewater treatment?
   a. Overflow rate
   b. Detention time
   c. Sludge return and recycle rate and removal

4. The following is a major design parameter for settling tanks:
   a. Overflow rate
   b. Alkalinity
   c. Sludge wasting rate
   d. Coagulant concentration

5. Short-circuiting is an issue with clarification tanks where wind blows across the water surface, reducing net volume and detention time in the tank.
   ☐ True
   ☐ False

6. When a constant water temperature is not ensured in clarification tanks, thermal stratification may occur with cold "dead" zones on the bottom of the tank.
   ☐ True
   ☐ False

7. When poor floc formation is observed, what should be done to determine if it is due to inadequate chemical dosing?
   a. Run a jar test
   b. A timed cone settling test
   c. Alkalinity and pH measurement
   d. Sludge density test

8. Calculate the overflow rate of a rectangular clarifier with a side water depth of 4.572 m (15 ft), length of 18.3 m (60 ft), width of 3.66 m (12 ft), and flow of 7570.8 m^3/d (2 mgd):
   a. 73.15 m/d (240 ft^3/sq ft/day)
   b. 7.56 m/d (24.8 ft^3/sq ft/day)
   c. 372 $m^3/m^2 \cdot d$ (1220 ft/day)
   d. 34.56 $m^3/m^2 \cdot d$ (113.39 ft/day)

9. **Which is a common problem in a settling tank?**
   a. Poor floc formation
   b. Floc breakup and excessive turbulence
   c. Low sludge formation
   d. Inadequate mixing

10. **In an advanced wastewater clarifier as part of a coagulation–flocculation treatment train, why would you not want to waste all of the sludge?**
   a. You do want to turn up the pumps and waste all of the sludge.
   b. Biomass is required in the treatment train to seed the basins in secondary treatment.
   c. Previously precipitated particles in the return sludge may form a nucleus for more rapid flocculation.
   d. It costs too much power to waste all the sludge.

Filtration

FUNDAMENTALS

In advanced wastewater treatment, filtration is used to remove the residual TSS in secondary effluent. Filtration involves passing flow through a bed or layer of sand or other media or a combination of two or three different types of media to remove suspended or colloidal matter. Although the term *filtering* implies a straining mechanism, several other mechanisms occur simultaneously in the filter media and affect the process's efficiency. These mechanisms include clarification (settling) on the filter media, flocculation within the media's pore spaces, and physical and chemical adsorption onto the filter media.

Adsorption and *absorption* mean quite different things. *Absorption* is where a liquid is soaked up into something like a sponge, cloth, or filter paper. The liquid is completely absorbed into the absorbent material. *Adsorption* refers to individual molecules, atoms, or ions gathering on surfaces.

Figure 3.16 shows the three main mechanisms of filtration: straining, adsorption, and absorption.

PURPOSE AND FUNCTION

Filtration removes particulate matter that is suspended in wastewater. The wastewater may be pretreated with coagulants and/or polymer to agglomerate fine particles and colloids upstream of the filter. Wastewater may also be filtered directly without chemical addition. Activated sludge processes, for example, are already well flocculated and may not need chemicals unless phosphorus precipitation is desired. Effluents from trickling filters and rotating biological contactors are not well-flocculated and may need more chemical addition for filtration to be effective.

Filtration removes particulates and also removes nitrogen and phosphorus in the process. Recall that bacteria contain about 10% nitrogen and 1.5% phosphorus by weight. That means, if 30 mg/L of TSS is discharged in the final effluent, 3 mg/L of particulate nitrogen and 0.45 mg/L of particulate phosphorus are also discharged. Filtration, therefore, is a necessary step toward meeting low nitrogen and phosphorus limits.

Occasionally, filtration alone without upstream coagulation and flocculation is used to further reduce suspended solids concentrations in activated sludge effluents. This is less effective for trickling filter and rotating biological contactor effluents because their solids are less naturally flocculent.

THEORY OF OPERATION

A typical filter consists of a holding vessel or tank with an underdrain system and several layers of media. Filter structures (see Figure 3.17) often used for water treatment are readily adaptable to wastewater filtration.

Figure 3.16 Mechanisms of Removal During Filtration

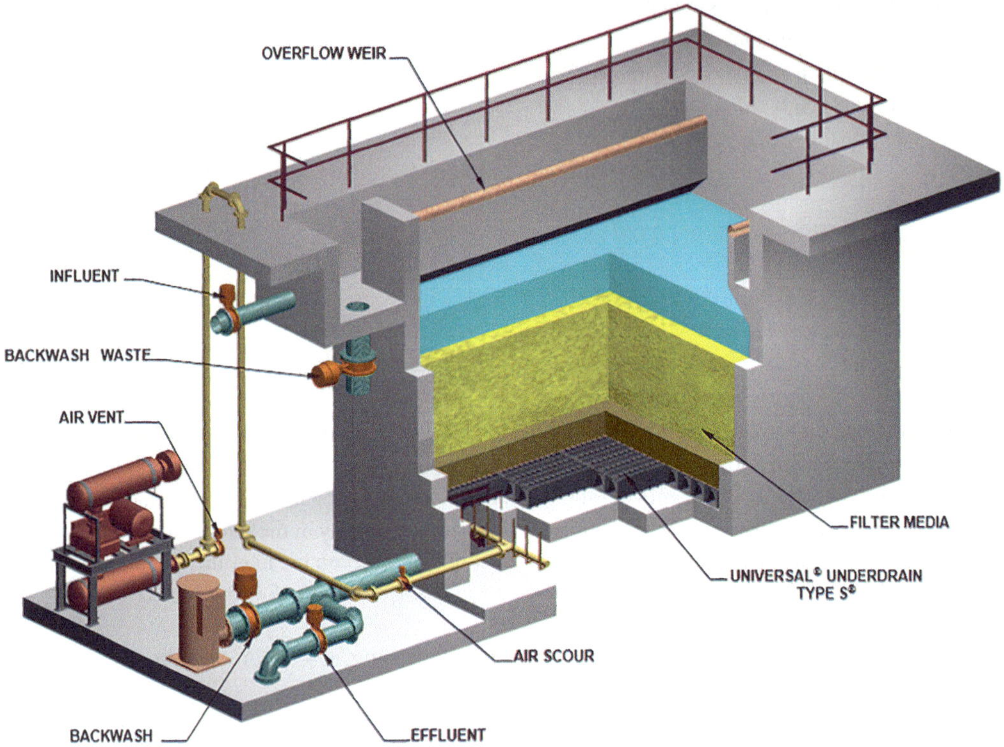

Figure 3.17 Cross Section of a Typical Gravity Media Filter and Ancillary Components (Xylem Inc)

Typical filter appurtenances include the underdrain system and wash water troughs. In conventional downflow filters, the underdrain system supports the filter media and distributes the backwash water. They range from graded gravel beds to plastic and steel nozzles inserted to precast channels, poured in concrete, or affixed on steel plates.

Wash water troughs are typically located at the top of the filter to collect backwash water and convey it to the filter drain. The rectangular troughs are constructed of fiberglass, sheet metal, or concrete and topped with weir plates. Media separator baffles may be placed on the underside of the troughs to minimize loss of filter media.

Wastewater is more difficult to filter than well or surface water because the solids are sticky from biological activity and more highly concentrated.

Dual-media and multimedia conventional filters allow more of the filter depth to be used to capture solids, making wastewater filtration practical. Just as there are different types of filter systems, there are also different configurations. Inline filtration is when coagulant is added to a process and then the stream is sent directly to filtration. "Direct" filtration includes a flocculation step. Conventional filtration is the most commonly used and includes coagulation, flocculation, clarification, and, finally, filtration as part of a treatment train. These processes typically occur downstream of primary and secondary treatment processes at a typical WRRF. Figure 3.18 shows these different configurations.

A multimedia filter is a coarse-to-fine filter; its pore space gradually shrinks as the water moves toward the output point. The number of media grains also typically increases, although the grain size is not uniform because the filtration media have three specific gravities (coal, 1.4; sand, 2.65; and garnet, 4.5). This structure enables removed solids to be stored throughout the depth of the bed rather than only at the surface, greatly extending the length of filter runs between backwashes.

One alternative to media filters is cloth disc filters (see Figure 3.19).

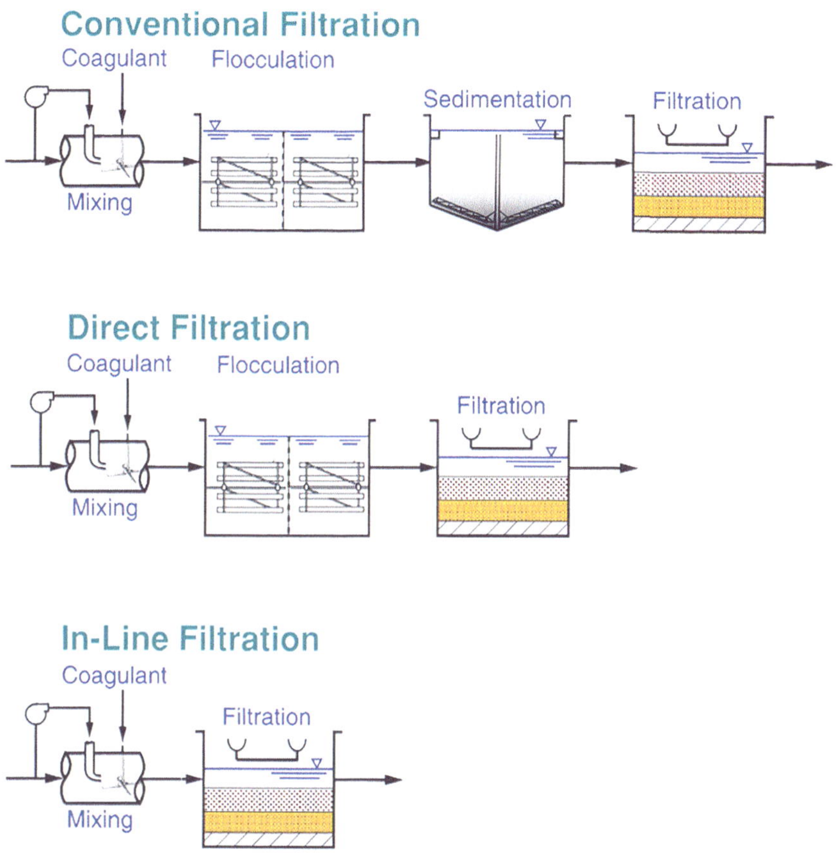

Figure 3.18 Classification of Rapid Filtration by Pretreatment Level (Adapted from Crittenden, 2012)

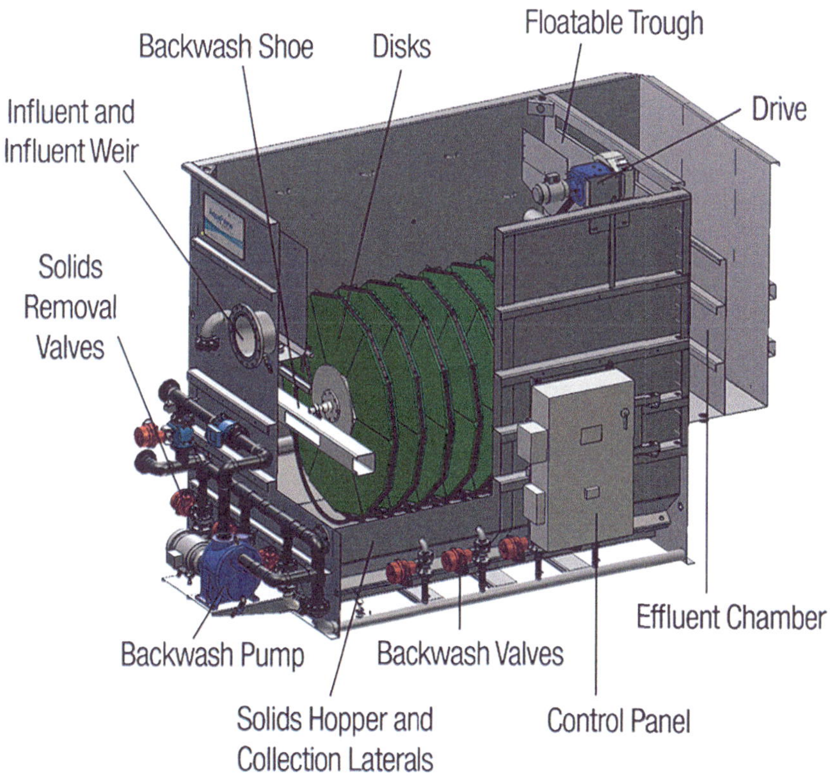

Figure 3.19 Cloth Media Filter Schematic (Provided courtesy of Aqua-Aerobic Systems, Inc.)

In a cloth filter inlet, wastewater enters the tank or basin. Liquid passes through the cloth membrane via gravity. As solids accumulate on the media, a mat is formed, and the liquid level in the tank or basin increases. The filtered liquid enters the internal portion of the disk where it is directed to final discharge. At a predetermined level or time, the backwash cycle will be initiated. Solids are backwashed from the surface by liquid suction from both sides of each disk. During backwash, disks are cleaned in multiples of two, unless a disk rotates slowly, allowing each segment to be cleaned. Backwash water is directed to the headworks. Filtration is not interrupted during this cycle.

In typical shallow-bed sand or diatomite water treatment filters, wastewater tends to quickly "blind" the filter surface. Nevertheless, some shallow-bed single-media filters—designed specifically for wastewater applications—have been successful.

DESIGN PARAMETERS

Filtration is used in addition to regular coagulation and clarification for removal of solids from surface water or wastewater. This prepares the water for use as potable, boiler, or cooling makeup. Wastewater filtration helps users meet more stringent effluent discharge permit requirements.

Quartz sand, silica sand, anthracite coal, garnet, magnetite, and other materials may be used as filtration media. Silica sand and anthracite are the most commonly used types. Table 3.9 presents typical properties of these materials.

The uniformity coefficient is a numerical expression of the variety in particle sizes in mixed natural soils, defined as the ratio of the sieve size through which 60% (by weight) of the material passes to the sieve size that allows 10% of the material to pass. The porosity is a measure of the void spaces in a material and is a fraction of the volume of voids over the total volume, between 0 and 1, or as a percentage between 0% and 100%. Strictly speaking, some tests measure the "accessible void," that is, the total amount of void space accessible from the surface. The hardness is a qualitative ordinal scale, from 1 to 10, characterizing scratch resistance of various minerals through the ability of harder material to scratch softer material. The size and shape of the filter media affect the efficiency of the solid's removal. Sharp, angular media form large voids and remove less fine material than rounded media of equivalent size.

Void—empty space formed around filter media with lots of angles (not spherical).

The media must be coarse enough to allow solids to penetrate the bed for 5.08 to 10.16 cm (2 to 4 in.). Although most suspended solids are trapped at the surface or in the first 2.54 to 5.08 cm (1 to 2 in.) of bed depth, some penetration is essential to prevent a rapid increase in pressure drop.

Pressure drop describes a change in pressure from point A (high pressure) to point B (low pressure). With filters, the higher the drop, the greater the effect on performance. For example, if pressure continues to build up when a substrate is trying to get through a filter, it indicates too much resistance to flow. Pressure drop is undesirable because it can result in loss of energy in the system (higher power costs) and poorer effluent quality.

Sand and anthracite for filters are rated by effective particle size and uniformity. The effective size is such that approximately 10% of the total grains by weight are smaller and 90% are larger.

Grains by Weight—the amount of filter media of a particular size.

Table 3.9 Typical Properties of Filter Media Used in Rapid Filters (Adapted from Crittenden, 2012)

Property	Unit	Garnet	Sand	Anthracite
Effective size	mm	0.2–0.4	0.4–0.8	0.8–2.0
Uniformity coefficient	Uniformity coefficient	1.3–1.7	1.3–1.7	1.3–1.7
Density, p	g/mL	3.6–4.2	2.65	1.4–1.8
Porosity, ε	%	45–58	40–43	47–52
Hardness	Moh	6.5–7.5	7	2–3

Therefore, the effective size is the minimum size of most of the particles. Uniformity is measured by comparison of effective size to the size at which 60% of the grains by weight are smaller and 40% are larger. This latter size, divided by the effective size, is called the *uniformity coefficient*; the smaller the uniformity coefficient, the more uniform the media particle sizes.

Finer sands result in shallower zones for the retention of suspended matter. The most desirable media size depends on the suspended solids characteristics as well as the effluent quality requirements and the specific filter design. In general, rapid sand filters use sand with an effective size of 0.35 to 0.60 mm (0.014 to 0.024 in.) and a maximum uniformity coefficient of 1.7. Coarse media, often 0.6 to 1.0 mm (0.024 to 0.04 in.), are used for closely controlled coagulation and clarification. Table 3.10 shows typical design criteria and parameters for these values.

EXPECTED PERFORMANCE

Filters are capable of achieving a low TSS (<30 mg/L) and turbidity (<2 nephelometric turbidity units [NTUs]) and are a necessary polishing step for wastewater effluent with stringent nitrogen and phosphorus discharge limits. Their performance can be affected by improper upstream pretreatment (coagulation, flocculation, and sedimentation).

FILTER AIDS

Chemical addition is especially important when secondary treatment is followed directly by filtration. If filtration is preceded by distinct coagulation, flocculation, and settling steps, then the filters may not need any chemicals. However, most filter designs typically provide for chemical addition (typically alum and some type of polymer).

Table 3.10 Typical Design Criteria for Dual Media Filters (Adapted from Davis, 2010)

Parameter	Reported Range	GLUMRB* recommendation
Anthracite Coal on top		
Effective size	0.9–1.4 mm (.035–.055 in.)	0.8 m (.0315 in.)
Uniformity Coefficient	1.3–1.8	<1.9
Shape factor	0.46–0.73	
Porosity	0.53–0.6	
Specific Gravity	1.45–1.75	
Depth of medium	0.4–0.6 mm (.015–.024 in.)	
Sand on bottom		
Effective size	0.37–0.7 mm (.015–.027 in.)	0.45–0.55 mm (.018–.22 in.)
Porosity	0.4–0.47	
Specific Gravity	265	
Depth of medium	0.3–0.75 m (11.8–29.5 in.)	>0.6 m (23.6 in.)
Filtration rate	7–20 m/h (23–65 ft/hr)	
Backwash rate	30–60 m/h (98–196 ft/hr)	>24 m/h (78 ft/hr)
Backwash duration	10–20 min	>15 m/h (49.2 ft/hr)
Surface wash rate	4.9–10 m/h (16.1–32.8 ft/hr)	>4.9 m/h (16 ft/hr)
Air scour	0.6–1.5 m^3/min (21–53 cu ft/min)	

*GLUMRB = Great Lakes Upper Mississippi River Board

FILTER BACKWASH

When the filter's pores become clogged, they need to be cleaned. One of the best ways to clean a filter is to backwash it, meaning reversing the flow and increasing the velocity at which water passes back through the filter. This, in effect, blasts the clogged particles off of the filter.

The typical backwash methods for conventional granular media filters include:

- Water backwash with surface and subsurface washers (surface washers are typically used in single-medium filters; subsurface washers, installed at the expected depth of the expanded interface, may be necessary in dual-media or multimedia filters).
- Water backwash with an auxiliary air scour where the air is applied for 3 to 4 minutes before the backwash cycle begins. This is typically used in dual-media and multimedia filters or as an alternative to mechanical surface washers.
- Simultaneous air–water backwash. This is typically used in single-medium unstratified filter beds but is less common in dual-media or multimedia filter beds because the air tends to float anthracite out of the filter.

Backwashing requires conventional filters to be taken offline. The cumulative production downtime can be significant. To circumvent this problem, continuous-backwash granular media filters were developed. These proprietary filters backwash and filter simultaneously, and filter secondary effluents effectively.

Granular media filtration can be done via gravity flow in open concrete or steel structures or under applied pressure in steel pressure vessels. Its operation and control may be readily automated.

EQUIPMENT

Many types of media can be used for filtration. In addition to traditional granular media (typically sand and anthracite), natural or synthetic fiber or fabric, synthetic fuzzy balls and beads, and other filtration media are available. Nevertheless, most WRRFs using filtration use granular media.

GRANULAR MEDIA FILTERS

Backwashing of granular media filters involves several steps. First, the filter is taken offline, and the water is drained to a level that is above the surface of the filter bed. Next, compressed air is pushed up through the filter material. This causes the filter bed to expand, which breaks up the compacted filter bed and forces the accumulated particles into suspension. After the air scour cycle, clean backwash water is forced upward through the filter bed. This continues the filter bed expansion and carries the particles in suspension into backwash troughs suspended above the filter surface. In some applications, air and water streams are simultaneously pushed upward through the granular media, followed by a rinse water wash. Figure 3.20 shows a filter and trough during a backwash operation.

Backwashing continues for a fixed time or until the turbidity of the backwash water is below an established value. At the end of the backwash cycle, the upward flow of water is terminated, and the filter bed settles by gravity into its initial configuration. Water to be filtered is then applied to the filter surface until the headloss setpoint is reached and the backwash cycle needs to be repeated. Figure 3.20 shows a filter in backwash mode.

CONTINUOUS BACKWASH UPFLOW FILTER

The deep bed upflow approach allows the filter to operate continuously because backwashing of the media is an integral part of the filtering process. In addition, the continuous backwash sand filter is easy to operate and allows for mechanical, chemical, and biological filtration processes.

Figure 3.21 and Figure 3.22 present examples of an upflow, deep bed, granular media filter that filters liquid suspensions to the designed effluent quality, while at the same time cleaning the sand bed so the filter does not shut down during the backwash period. Feed water is passed upward through the sand bed and exits the rapid sand filter as clean water. At the same time, sand can be removed from the bottom, cleaned by a simple internal washing system that does not require backwash pumps or storage tanks, and returned to the top of the sand bed. A small portion of the filtered water is used to wash the sand and leaves the filter as a reject stream.

Figure 3.20 Wastewater Filter in Backwash Mode (Xylem Inc)

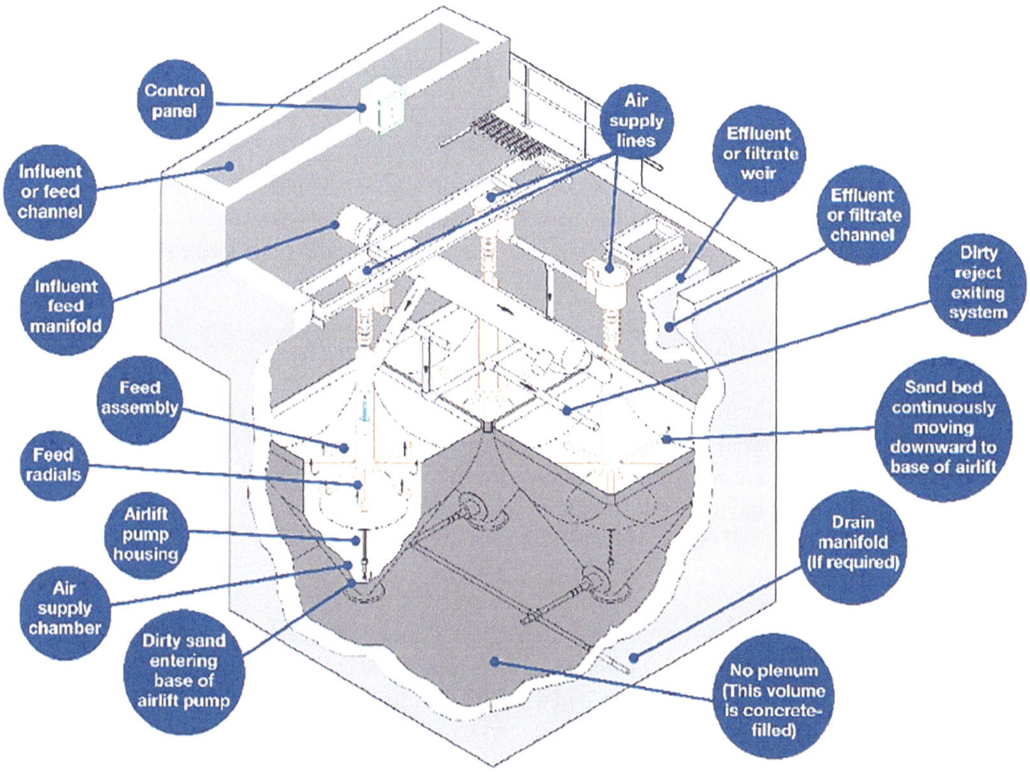

Figure 3.21 Schematic of a Continuous Backwash Upflow Filter (System of Four Filters) (Copyright 2020 Parkson Corporation)

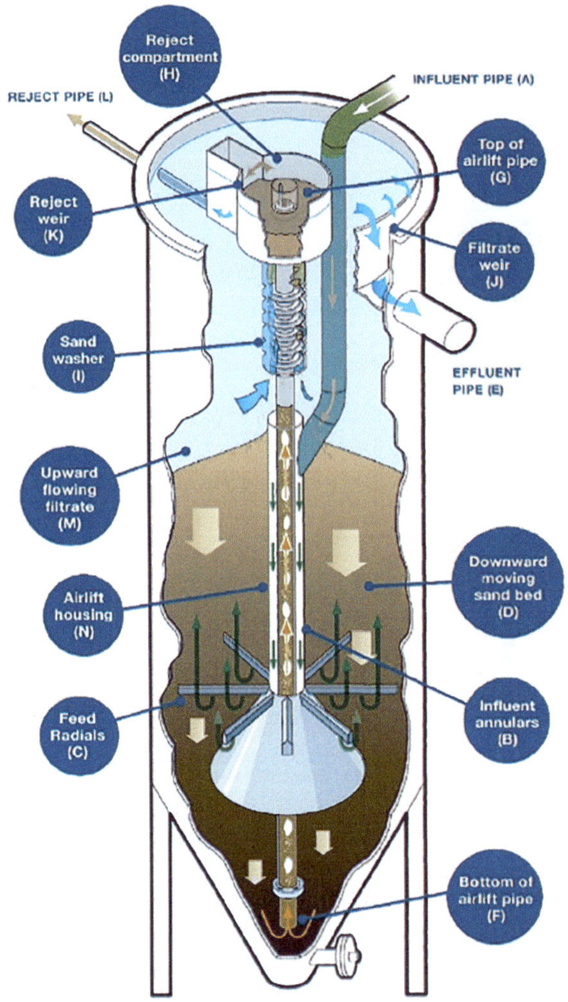

Figure 3.22 Schematic of a Continuous Backwash Upflow Filter (Closeup of One Filter) (Copyright 2020 Parkson Corporation)

During operation, the pressure drop remains low and even throughout all filters. Equal pressure drop ensures even distribution of feed to each filter without the need for splitter boxes or flow controls. Therefore, a typical multiple unit installation can use a common header pipe with feed connections and isolation valves for each filter.

Figure 3.22 shows the process flow for a continuous backwash upflow filter.

Influent feed is introduced at the top of the filter and flows downward through an annular section between the influent feed pipe and airlift housing. The feed is introduced to the bottom of the sand bed through a series of feed radials that are open at the bottom. As the influent flows upward through the downward moving sand bed, organic and inorganic impurities are captured by the sand. The clean, polished filtrate continues to move upward and exits at the top of the filter over the filtrate weir and out through the effluent pipe.

The sand bed containing captured impurities is drawn downward into the center of the filter where the airlift pipe is located. A small volume of compressed air is introduced at the bottom of the airlift, drawing the sand into the airlift pipe. The sand is scoured within the airlift pipe at an intensity of 169 to 254 m/h (100 to 150 standard cu ft/min). The effectiveness of this scouring process is vastly greater than what can be expected in conventional sand filtration backwash. The scouring dislodges any solid particles attached to the sand grains. The dirty slurry is pushed to the top of the airlift and into the reject compartment. From the reject compartment, the sand falls into the sand washer and the lighter reject solids are carried over the reject weir and out the reject pipe. As the sand cascades down through the concentric stages of the washer, it encounters a small amount of polished filtrate moving upward, driven by the difference in water level between the filtrate pool and the reject weir. The heavier, coarser sand grains fall through this small countercurrent flow while the remaining contaminants are carried back up to the reject compartment. The clean, recycled sand is deposited on the top of the sand bed where it once again begins the influent cleaning process and its eventual migration to the bottom of the filter.

SURFACE WASHING

For conventional filters, an effective surface wash ensures that clumps of media and floc do not survive the backwash as mud balls.

The presence of mudballs is an indication of problems in your filter bed. A mudball is the result of solids accumulating in a filter, then combining with filter media to form into a ball of solids. These mudballs block the passage of water, creating higher velocity filtration and uneven distribution.

A typical surface-wash apparatus consists of a revolving pipe with a number of nozzles suspended by a bearing at the pipe's center. These units, positioned 2.5 to 5.1 cm (1 to 2 in.) above the normal surface of the filter, inject wash water through the nozzles (at less than 340 to 690 kPa [50 to 100 psi] of pressure), causing the pipe to rotate. The water supply requirements for these devices vary, but typically range from 0.5 to 0.7 L/m^2·s (0.75 to 1.0 gpm/sq ft).

Typically, surface wash starts approximately 60 seconds before the main backwash pumps are turned on and stops 60 seconds before the end of backwash to allow the bed to reclassify. The nozzles, directed downward approximately 30° from the horizontal, cause agitation throughout the bed depth. The surface wash water must be high quality to avoid plugging the nozzles. A strainer with 2.4-mm (0.09-in.) perforations is often used in the surface wash line as a precautionary measure.

Air–water backwash techniques are an alternative to mechanical surface wash. They involve injecting air through the filter underdrain system. The rising air bubbles break up mud balls in the filter. In dual-media or multimedia beds using anthracite, the backwash water and the air cannot be applied at the same time because the air tends to float the coal out of the filter.

PROCESS VARIABLES

Filtration systems should include instrumentation to monitor turbidity (both single filter effluent and combined filter effluent), headloss, backwash rate, and filter run length.

Backwashing is the reversal of flow through the filters at a higher rate to remove clogged particles from the filters. Backwash run times can be anywhere from 5 to 20 minutes, with rates ranging from 5.4 to 17 L/m^2·s (8 to 25 gpm/sq ft) of filter bed area, depending on the quality of the prefiltered water.

If auxiliary air scour backwash is used, airflow monitoring should also be included.

FLOWRATE

There currently are three principal methods to control flowrates through conventional downflow filters:

1. Constant rate with fixed head,
2. Constant rate with variable head, and
3. Variable rate with fixed or variable head.

The filter run length is the sum of the time the filter is producing water between backwashes. It should include units (hours, minutes, etc.).

In the first method, the flowrate is kept constant by an effluent rate control valve. The valve is nearly closed at the beginning of a filter run and opens slowly as the filter becomes clogged. The run terminates when the valve is fully open.

In the second method, a constant rate is achieved via some type of influent control (e.g., an influent weir box with a starting [driving] water surface elevation above a design terminal headloss). The filter run is terminated when the filter headloss buildup reaches the design terminal headloss.

As filtration proceeds, an increasing amount of pressure, called *headloss* across the filter, is required to force the water through the filter. Headloss should be continuously measured to help determine when the filter should be backwashed.

In the third method, the flowrate is allowed to decline as the headloss builds up. Declining rate systems may be influent- or effluent-controlled.

TYPICAL PROCESS CONTROL

All filters have common elements that enable them to perform the functions discussed below. Filter instrumentation can range from passive systems for monitoring headloss and influent and effluent turbidity to fully automated operations. At the passive monitoring level (typically for smaller facilities), effluent turbidity and headloss values may be monitored manually. Operators take filter units offline and initiate and complete backwashes. At the other end of the range, fully automated filters will include instrumentation and control logic that can take filter units offline, backwash them, and put them back into service with little, if any, operator input. These systems typically include motorized valves, flow meters, differential pressure meters, and turbidity meters, all feeding information or receiving operating commands from a central control unit with a microprocessor-based device (e.g., a PLC).

Filter operation depends on the flowrate through the filter, which, in turn, depends on the headloss through the filter. Effective filtration control should involve continuous monitoring of headloss and influent and effluent turbidity. Lower filter influent TSS (i.e., better upstream secondary process operations) mostly will have slower headloss buildup compared to higher influent TSS operation. Increase run times and decrease backwash needs. Figure 3.23 illustrates the efficiency of particle capture. During the first segment (after backwashing), the effluent turbidity rises to a peak and then falls rapidly. This segment is called *filter ripening* and generally lasts for an hour after backwash. Ripening is the process of media conditioning and occurs as clean media captures particles and becomes more efficient at collecting additional particles. Breakthrough starts to occur as filter effluent turbidity rises.

Breakthrough in filters is similar to a parking garage. When the garage first opens, all potential parking spaces are available to the entering vehicles. The parking garage gradually fills with vehicles. Because people need to get off at different floors or prefer to park in the shade or near an elevator or stairway, the parking garage does not fill up evenly. As time goes on, it becomes more difficult for new vehicles entering the garage to find their "preferred" parking spots. Some drivers will eventually make their way through the entire garage and come out the other side, even though some parking spots still remain. Breakthrough occurs when some particles make their way all the way through the filter. When there are lots of particles coming in, breakthrough happens sooner.

FILLING THE FILTER

When starting up a filter for the first time or filling one that has been drained, special precautions are required to avoid serious damage to the filter. First, the filter must be filled with water. For a conventional filter, operators should slowly pump water up through the filter underdrain system from the backwash supply (essentially, a slow backwash). Rate should be determined by the filter manufacturer. For a

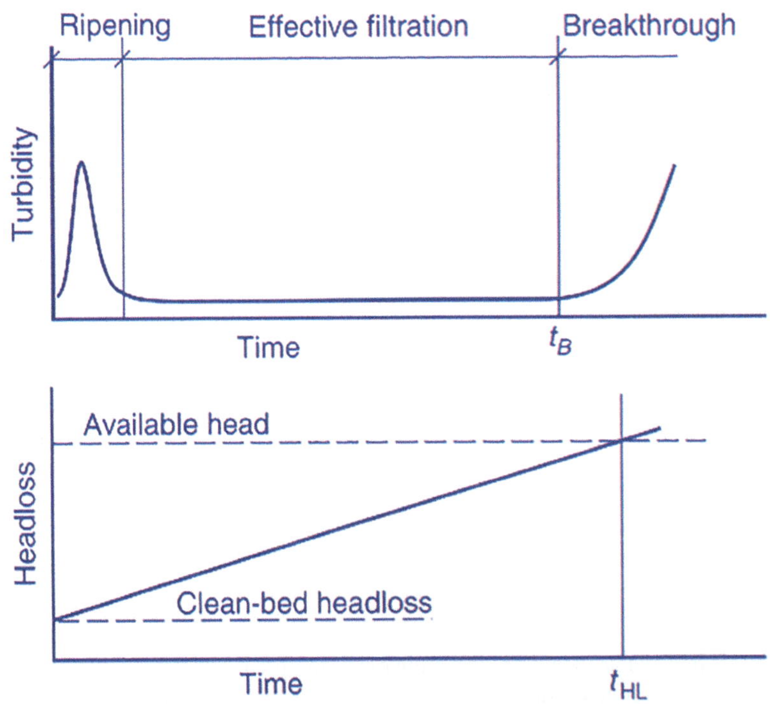

Figure 3.23 Typical Headloss During a Media Filter Run (Adapted from Crittenden, 2012)

continuous backwash upflow filter, filling it from the bottom expels air that would otherwise be trapped in the media. If the filter media are dry, the filter must be filled and allowed to soak for 8 hours before backwashing is attempted. If the filter has been filled from the top, it should not be backwashed—even if a filter cycle has since occurred. Water moving down through a filter will not remove trapped air, which can completely overturn the gravel during backwashing, creating gravel mounds or allowing fine media to be lost from the bed during subsequent filter cycles. In severe cases, the filter media and the gravel may have to be replaced. If the filter has been filled from the top, it should be drained completely, and then operators should initiate a slow backwash.

Whenever a filter drains below the level of the top of the media or a dry filter is being filled, the bottom-filling procedure must be strictly followed to avoid severe damage to the filter media and support gravel.

FILTER AIDS

A key element in making the most of filter performance is using an appropriate amount of filter aid (e.g., polymer or alum). The filter aid improves filtration in several ways:

- Strengthening the floc
- Controlling the depth of floc penetration into the beds
- Improving the clarity of the filtered water
- Increasing the permissible flowrate through the bed

The amount of filter aid needed increases as water temperature decreases, flowrates increase through the filters, and applied turbidities increase. The optimum dosage causes the maximum desired filter headloss to be reached just as turbidity breakthrough is impending. If lime is the primary coagulant, using alum (5 to 20 mg/L) as a filter aid (supplementing polymer) may reduce filter effluent turbidity.

The filter aid coats the grains of the filter media. When a new filter enters service, three or four filter runs may be needed to properly coat the grains. Conversely, three or four filter backwashes may be required to reduce the coating if a filter aid dose is reduced. Operators should proceed with caution when reducing filter aid doses because the residual effects of previous coats on the bed may delay the full response to the reduced dose. Controlling the filter aid dose becomes a relatively simple task once operators learn to interpret the turbidity and headloss recordings.

Modern filtration systems include feed equipment for filter aids (typically polymers).

EFFECTS OF POLYMERS AND POLYMER HANDLING

If the polymer dose is too low (see Figure 3.24, note A), the floc will be fragile and shear and will result in penetration of the filter. This terminates the filter run prematurely because of a breakthrough of effluent turbidity. If the polymer dose is too high (see Figure 3.24, note B), the floc becomes too strong for penetration of the filter, causing a rapid buildup of headloss in the upper portion the filter. The excessive headloss prematurely terminates the filter run. The optimum polymer dose (see Figure 3.24, note C) allows the filter to reach terminal headloss just as its effluent turbidity begins increasing. It prevents premature termination of the filter run because of excessive headloss or turbidity breakthrough, thereby reducing the amount of filter wash water to be recycled through the WRRF.

TURBIDITY MONITORING

Continuously monitoring the effluent turbidity of each filter helps operators maintain the proper application rate of filter aid to each filter unit. If the effluent turbidity exceeds the maximum predetermined allowable value, the filter should be automatically taken out of service and backwashed.

FILTER BACKWASHING

Successful filter operations depend on clean filter media. If the media are not cleaned adequately, biological organisms could accumulate and eventually plug the filter. Adequate backwashing, along with surface wash or air–water backwash, can eliminate plugging. Table 3.11 provides a summary of typical backwash delivery systems.

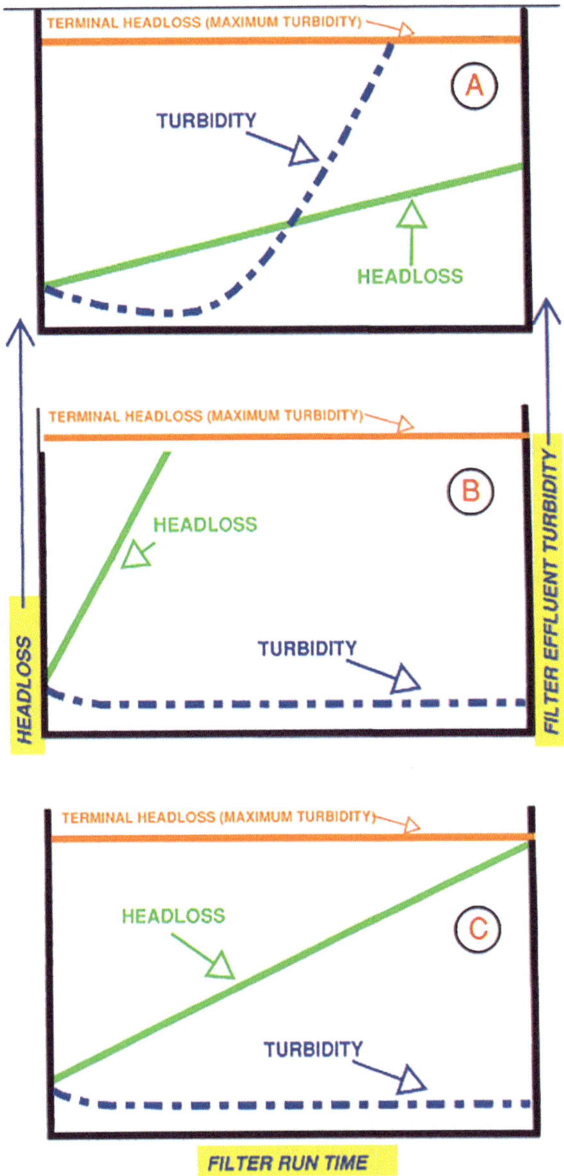

Figure 3.24 Effects of Polymers as Filtration Aids: (A) Inadequate Polymer Dose, (B) Excessive Polymer Dose, and (C) Optimal Polymer Dose (Adapted from WEF, 2017)

Table 3.11 Comparison of Conventional Backwash Delivery Systems (Adapted from Crittenden, 2012)

Water Supply	Backwash Pumps	Elevated Tank	Effluent Channel
Description	Pumps, sized to provide the entire backwash flow, water pulled from finished water clearwell.	Small pumps withdraw water from the filter effluent channel or finished water clearwell and send it to an elevated tank used during backwash.	Filter effluent flows to a common effluent channel, the water elevation of which is controlled by a weir.
Advantages	Provides maximum control over backwash rates.	Only requires small pumps.	No pumps required.
Disadvantages	Large pumps are required.	Backwash flowrate will decline with water level in the tank.	Less control over backwash flowrate; deep filters are required.

High headloss through a filter bed typically indicates the need for backwashing, so headloss through each filter bed must be monitored. After each backwash, operators should note in the daily log the initial headloss at a predetermined flowrate. At a given flowrate, the headloss through a clean filter should be the same (between 0.3 and 0.6 m [1 and 2 ft]). If the initial headloss gradually increases, it may indicate that the backwash is too short to clean the filter media thoroughly and should be lengthened.

Sudden changes in filter flow should be avoided because hydraulic surges tend to cause particles to penetrate the bed. Changes in flow should be gradual, not abrupt. As the filter rate decreases, the length of the filter runs will increase significantly.

Water resource recovery professionals used to believe that the filter media had to expand by 50% for cleaning to be effective. However, optimum scouring of the media particles occurs when media suspension begins because the total shearing force against the media grains equals the weight of the media in water. The shearing force remains constant, no matter how much more the media expansion and wash rates increase. So, a wash rate corresponding to 25% expansion of the anthracite is adequate for a dual-media filter.

How much filter media expand depends on the backwash rate, which, in turn, depends on the water temperature and the filter media's size and specific gravity. (There is an indirect relationship between temperature and water viscosity.) The backwash water needed may be as much as 8% to 10% of the filter influent flowrate, although a well-operated filter should use less than 5% of its influent flow for backwashing. A rise in water temperature from 10 to 20 °C (50 to 68 °F) requires the wash rate to increase about 30% to maintain a given expansion of silica sand. Operators must observe changes in wastewater temperature and adjust the backwash rates accordingly.

For most conventional filters, the optimum backwash duration depends on the filter's use, but typically ranges from 5 to 8 minutes. Highly treated backwash water (e.g., filter effluent or chemically coagulated and settled wastewater) will always have low suspended solids concentrations. Operators should carefully monitor the backwash flow indicator to maintain the desired backwash rate.

RECYCLE OF BACKWASH FLOWS

Because all filter backwash wastewater is recycled and reprocessed, utilities sometimes calculate the percentage of WRRF throughput that is recycled as backwash to indicate the system's efficiency. Typically, if the backwash water percentage is more than 5%, the system's performance is questionable. Questionable performance may stem from any of the following causes:

- Excessive solids carryover from the settling tank
- Excessive filter aid doses with reduced filter runs
- An inoperative filter surface wash system or a surface wash that is too short for the backwash cycle
- Excessive backwash duration.

Smaller WRRFs, those with high diurnal flow variations, and those without equalization storage capability may require backwash water percentages as large as 8% to 10% of influent flow.

Most WRRFs collect filter backwash wastewater in a storage tank so it can be recycled through the treatment processes at a controlled rate. The backwash holding tank also may receive other waste flows to equalize the rates at which these flows are returned to the treatment processes. Some of the tank contents may be returned to the rapid mix unit to reduce chemical usage and improve coagulation because the backwash water typically contains some unreacted or partially reacted coagulant.

DATA COLLECTION

The primary control parameters for filtration are effluent turbidity and filter headloss; filters used as part of phosphorus or nitrogen removal should also have those parameters monitored. Turbidity and headloss are typically continuously monitored via automated instruments. Total suspended solids measure particles. Turbidity (NTU) measures light scatter. You can have high TSS (big chunks) and low turbidity. You can also have low TSS and high turbidity (lots of colloids). Turbidity is the best filter control parameter because it can be measured rapidly and automatically, but TSS measurements must be taken in a laboratory (using a filter and an over). The two parameters can be correlated. Turbidity can be correlated to suspended solids as follows:

$$TSS = (TSSf)(T)$$

where

TSS = total suspended solids (mg/L),
TSSf = factor to convert turbidity readings to total suspended solids, and
T = turbidity (NTU).

The TSS-to-turbidity relationship depends on particle size and other factors. It changes regularly. This means TSSf must be determined routinely—typically weekly. The TSSf is determined by measuring the TSS of a sample and dividing by the turbidity of the sample.

MAINTENANCE

Every day, operators should ensure that rotary arms rotate freely by watching them operate. They should check the calibration of effluent turbidimeters at least once per week. Flow and headloss instrumentation should be checked and calibrated according to the manufacturer's recommendations. Significant mechanical equipment that require maintenance include backwash pumps, surface wash pumps, filter pumps (influent or effluent pumps may be part of the system), filter aid feed pumps, valves, and automatic valve operators.

TROUBLESHOOTING

Filtering wastewater is difficult. The wastewater's solids content varies and could be high if upstream processes are improperly operated. Variable wastewater flow also causes operating difficulties. For example, rapid sand-filter media are widely used in water treatment but not in wastewater treatment because of poor performance when filtering flows with high suspended solids concentrations. Dual-media and triple-media (multimedia) filters are more effective for wastewater treatment. Filters with alternative media or designs (e.g., traveling bridge and continuous backwash filters) have also been used and are becoming more popular in wastewater treatment.

Table 3.12 describes symptoms, probable causes, and potential solutions for troubleshooting a filter system.

TEST YOUR KNOWLEDGE

1. A mechanism of removal during filtration is
   a. Straining
   b. Settling
   c. Fusion
   d. Destabilization

2. Direct filtration is best used when turbidity is less than
   a. 50
   b. 100
   c. 10
   d. 15

3. Conventional filtration (with coagulation, flocculation, clarification) is best used in water with __________ turbidity levels.
   a. Low
   b. High

4. What would not cause the filter headloss to be low?
   a. Inadequate dosage of polymers as filter aid
   b. Coagulant feed system malfunction
   c. Change in coagulant demand
   d. A sudden increase in alkalinity

5. If the percentage of backwash water recycled is consistently high (over 5%) for a filter system, the operator should aim to
   a. Improve solids settling
   b. Increase filter aid dosage
   c. Add more aluminum sulfate
   d. Adjust the pH

Table 3.12 Probable Causes and Solutions for Troubleshooting a Filter System (U.S. EPA, 1978)

Indicators/ Observations	Probable Cause	Check or Monitor	Solutions
Effluent turbidity too high	Filter needs backwashing Inadequate upstream chemical coagulation	Turbidity in excess of 1 NTU Run jar tests; determine coagulant dosage.	Remove filter from service and backwash. Feed proper coagulant dosage.
High headloss through filter bed	Filter needs backwashing	High headloss through filter	Remove filter from service and backwash.
High headloss through filter just backwashed	Insufficient backwash time to clean filter Inoperative surface wash arm or air scour	Initial headloss greater than normal Visually inspect wash arm.	Increase the setting on the backwash timer to provide longer backwash period. Repair surface wash arm or air scour system.
Percentage of backwash water recycled exceeds 5%	Solids carryover to filter basin is too high Filter aid dosages too high Surface wash system not working	Suspended solids concentration Filter aid dosage Visual inspection of surface wash system	Provide better treatment in settling tanks by improving solids settling. Reduce filter aid dosage. Repair surface wash system.
Clogging of filter surface indicated by very rapid increase in headloss after backwash	Rapid accumulation of solids on the top surface of media due to inadequate previous clarification, excessive filter aid dosages, or inadequate surface wash	Rate of headloss buildup	Improve pretreatment or change to dual or mixed-media filters to provide greater porosity at the top of the filter. Reduce or eliminate filter aid dosage to allow particles to penetrate deeper into the bed. Provide adequate surface washing and backwashing.
Short filter runs	High headloss caused by surface clogging	Visual inspection headloss through filter	Replace sand with dual or mixed media. Reduce filter aid dosage. Use polymer as a filter aid to control rate of headloss buildup.
Filter effluent turbidity increases suddenly, but filter headloss is low	Inadequate dosage of polymers as filter aid Coagulant feed system malfunction Change in coagulant demand	Excessive turbidity Chemical feeders Run jar tests	Increase polymer dosage. Repair feeders. Adjust coagulant dosage.
Mud ball formation	Inadequate backwash flowrate and surface wash	Visual inspection	Provide backwash flowrate up to 20 gpm/sq ft and maintain proper auxiliary scour.
Gravel displacement	Introduction of air to filter underdrain in the backwash water	Visual inspection of gravel bed	If displacement is severe, filter media may have to be replaced. Limit the total flow and head of water available for backwash.
Loss of media during backwashing	Excessive flows used for backwashing Auxiliary scour excessive	Backwash rate Backwash program	Reduce rate of backwash flow. Cut off the auxiliary scour 1 to 2 minutes before the end of the main backwash.
Difficult to clean filter adequately in warm weather at normal backwash rates	Decreased viscosity of backwash water due to higher temperatures		Increase backwash rate until desired bed expansion during backwash is achieved.
Air binding causing headloss to increase prematurely	Air bubbles accumulate within bed due to filter influent containing dissolved oxygen at or near saturation levels, or lowered water level and stoppage of flow through filter during preparation for backwashing reduces pressure and releases gases	Rapidly increasing headloss through filter	Provide more frequent backwashing to prevent bubbles from accumulating as quickly. Maintain maximum water depths above the beds.

6. Label the three zones (A, B, and C) on the following chart graph of a filter run:

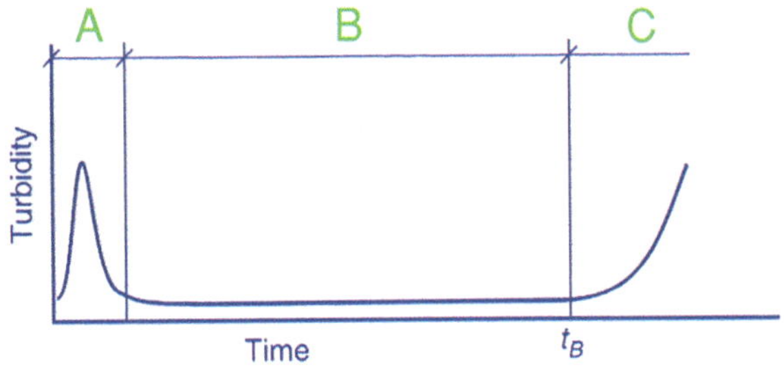

7. An operator observes high headloss through the filter bed, they should
   a. Adjust the coagulant dose
   b. Increase flocculation speed
   c. Decrease the sludge wasting rate
   d. Remove the filter from service and backwash

8. Air bubble accumulation within the filter (and subsequent increasing headloss) is most likely caused by
   a. Malfunctioning diffusers
   b. Excessive coagulant dose
   c. Not enough coagulant dose
   d. Dissolved oxygen saturated filter influent

9. At your facility you have noticed that your filter becomes more difficult to clean during hot summer months and requires more backwash water. This is most likely due to
   a. Excessive algae growth during warm weather
   b. Excessive nutrient discharge from organic runoff during warm weather
   c. A decrease in flowrates due to higher water consumption during summer months
   d. Decreased water viscosity due to higher temperatures

Granular Activated Carbon

FUNDAMENTALS

Granular activated carbon is a tertiary treatment used to removed poorly degraded, soluble organics, commonly called *refractory organics*. The technology uses porous carbon particles with a large surface area that have been heated to activate ions on the particle surface (see Figure 3.25). The thermal process aids the adsorption between the carbon surface area and organics in the water. Table 3.13 presents common types of organics removed through the GAC process.

PURPOSE AND FUNCTION

Granular activated carbon is a treatment unit used in drinking water and wastewater treatment systems that uses adsorption to remove organics such as taste and odor compounds, chlorine, and VOCs. Granular activated carbon has been manufactured to produce a large number of pores within the carbon particles and increase the surface area of the particle. The larger the surface area or attachment area on a single carbon particle, the more effective a carbon particle is at removing soluble organics.

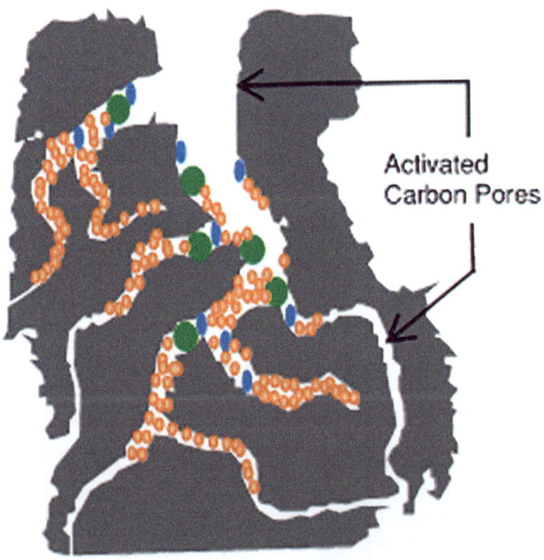

Figure 3.25 Activated Carbon Particle With Attachment Locations

Table 3.13 Organic Compounds Amenable to Removal by GAC (U.S. EPA, 2000)

Class	Example(s)
Aromatic solvents	Benzene, toluene, xylene
Polynuclear aromatics	Naphthalene, biphenyl
Chlorinated aromatics	Chlorobenzene, PCBs, endrin, toxaphene, DDT
Phenolics	Phenol, cresol, resorcinol
	chlorophenols, alkyl phenols
Aromatic amines and high- molecular-weight aliphatic amines	Aniline, toluene, diamine
Surfactants	Alkyl benzene sulfonates
Soluble organic dyes	Methylene blue, textiles, dyes
Fuels	Gasoline, kerosene, oil
Chlorinated solvents	Carbon tetrachloride, perchloroethylene
Aliphatic and aromatic acids	Tar acids, benzoic acids
Pesticides/herbicides	2,4-D, atrazine, simazine, aldicarb, alachlor, carbofuran

Granular activated carbon has a limited attachment area for interaction with soluble particles. Once spent, the carbon must be regenerated, which is defined in the following section titled, "Theory of Operation."

THEORY OF OPERATION

Activated carbon has a large surface area per unit weight, making it an efficient adsorption material. Nonpolar molecules occur when atoms share electrons symmetrically such that the charge of the molecule is neutralized. Large, organic, nonpolar molecules do not dissolve in water (hydrophobic) and will adsorb to solids such as carbon.

Activated carbon systems can use two different types of activated carbon: powder activated carbon (PAC) and GAC. Powder activated carbon is a free powder form that can be added directly to treatment basins, whereas GAC systems house carbon in columns for water to pass through. Powder activated carbon is considered explosive and careful safety precautions need to be made when handling and storing PAC. For this reason, GAC systems are more commonly used.

Granular activated carbon columns can either be fixed bed or moving bed reactors, although fixed bed reactors are more common. The columns are typically downflow, meaning system influent enters the top and moves downward through the media before exiting the column at the bottom. Columns can be used in series to add column height and treatment efficiency to the system. As with most media treatment systems, GAC columns use a backwashing system when headloss or pressure increase across the media becomes too large. Backwashing is completed by pumping flow backwards through a column to remove and collect particles gathered on the media surface.

Over time, GAC may need to be regenerated to reactivate the surface for better attachment. This process is completed by heating or steaming the carbon and returning it to the column for service. Regeneration can be completed on-site with the investment in a multiple hearth furnace or kiln. Alternatively, carbon can be collected and removed for regeneration off-site.

DESIGN PARAMETERS

Typical parameters to consider when designing GAC systems are carbon size, carbon bed depth, hydraulic loading rate, and empty bed reactor time. Table 3.14 lists typical values used for design.

Table 3.14 Typical Design Parameters and Values

Parameter	Values	Units
Common mesh sizes	12 x 40, 8 x 30, 20 x 40	
Empty bed contact time	10–50	minutes
Carbon use rate	24–72 (0.002–0.004)	g/m^3 lb/cu ft
Hydraulic loading for upflow columns	10–20 (33–66)	m^3/h·m^2 ft^3/h·sq ft
Hydraulic loading for downflow columns	5–12 16–39	m^3/h·m^2 ft^3/h·sq ft
Backwash rates	20–50 (66–164)	m^3/h·m^2 ft^3/h·sq ft
Backwash expansion	10–50	percent
Backwash duration	10–15	min
Bed depths	3–12 (10–39)	m ft
Column diameters	0.75–3.6 (2.4–12)	m ft
Depth-to-column ratios	1.5:1–4:1	

Granular activated carbon systems are typically placed after filtration units because the buildup of larger solid materials on the GAC media may restrict pore openings, reducing the effective surface area of the carbon and degrading its adsorptive capacity. Maintaining consistent water quality and flow into the GAC system will help to make the most of backwashing and regeneration times. The rate of adsorption in municipal wastewater typically increases as the pH decreases and is poor at a pH greater than 9.0.

EXPECTED PERFORMANCE

Granular activated carbon system performance is completely dependent on the water quality and quantity from upstream treatment processes. The successful application of GAC to the feed water in the system should consistently have a suspended solids concentration of 20 mg/L or less without any surges in water quality for consistent performance of the unit. Other wastewater constituents in the feed stream such as 5-day biochemical oxygen demand can also adversely affect the carbon adsorption process. Generally, the performance requirements of a GAC system are defined by permitted plant requirements for specific constituents. Granular activated carbon can remove a variety of constituents to varying degrees. The specific performance requirements need to be discussed with the GAC manufacturer, and the manufacturer should provide feedback about the system's design.

EQUIPMENT

Carbon contact columns consist of:

- A contact tank (see Figure 3.26), which can either be an open concrete basin or an enclosed steel vessel.
- A plenum area, which is similar to the underdrain area of a traditional deep bed filter. This is an area where carbon filtered effluent of the contactor is collected before exiting.
- A surface wash system, which is a series of rotating pipe arms installed in the first 15 to 25 cm (6 to 10 in.) of the carbon media. The piping is used to jet high velocity water or air into the carbon media to loosen particles in the heavily inundated upper levels of the media.

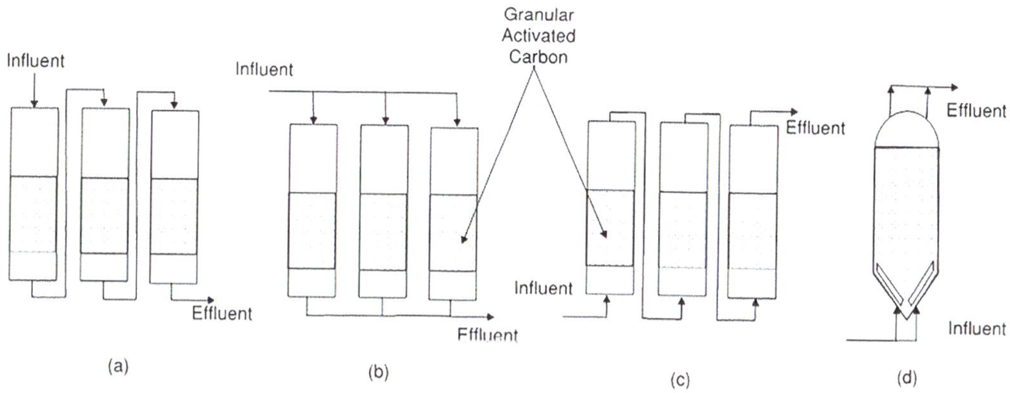

Figure 3.26 Types of Activated Carbon Contactors: (a) Downflow in Series, (b) Downflow in Parallel, (c) Upflow Expanded in Series, and (d) Moving Bed

- Backwash pumps, which are used pump water backward through the media to lift and expand the carbon media and remove particles that have accumulated on the media.
- Carbon slurry lines on the contactors, which are used to remove and replace the used or spent carbon with either regenerated, makeup, or replacement carbon.

Additionally, some systems may be equipped with an on-site carbon regeneration system. The carbon regeneration system includes the furnace used to heat and regenerate the attachment areas on the carbon surface. The furnace will also have to be equipped with the appropriate air scrubbing of after burner systems to treat the air to air quality permit standards. Each furnace will also generally be equipped with a screen conveyor to feed the spent carbon into the furnace. Lastly the regeneration system will include a slurry pump to pump the regenerated carbon back to the carbon contactor column.

Most carbon contact columns are either upflow (i.e., flowing from the bottom to the top of the column or countercurrent) or downflow (i.e., flowing from the top to the bottom of the column) operation. The upflow (countercurrent) approach uses carbon more efficiently and minimizes plugging by larger particulates. Activated carbon will remove oxygen from the air in an enclosed contacting vessel, so operators must carefully check a carbon contactor's oxygen content before entering it.

PROCESS VARIABLES

Typical process variables to monitor for GAC systems are influent and effluent pressure or pressure drop across the system, hydraulic loading rate, and backwash flowrate. Pressure drop across the system is used to determine when backwashing procedures should be initiated across the media. When the pressure drop plateaus, backwashing should be initiated.

PROCESS CONTROL

Process control for GAC systems aims to maintain the removal efficiency of carbon through backwashing and carbon regeneration. Consistently treating high-quality water reduces the required backwashing and regeneration frequencies. So, careful operator attention to primary treatment and secondary treatment systems will have downstream effects on the adsorption process in GAC systems. Backwashing consistently and monitoring the backwashing frequency or changes in backwashing frequency will indicate when it may be time for carbon regeneration.

NORMAL PROCESS CONTROL

As carbon adsorbs organic particles from wastewater, it eventually becomes saturated and breakthrough of particles can occur. The effluent concentrations of the organic compounds to be captured indicate the need to remove exhausted carbon for regeneration. Long-term efficiency of the carbon adsorption and regeneration processes can be aided by routinely calculating the chemical oxygen demand (COD) removed per pound of carbon regenerated. A large decrease in this value through several regeneration cycles indicates that the carbon is becoming fouled, is losing efficiency, and needs to be replaced.

Longer contact times may improve treatment efficiency. This can be accomplished by adjusting the GAC column flow to be in series or by reducing the hydraulic loading rate to the system. If contact times become too long, they could encourage anaerobic conditions in the carbon columns, which may create a corrosive environment due to the favorable atmosphere for the release of hydrogen sulfide.

One approach for minimizing costs and regeneration cycles would be to operate carbon adsorption vessels in series with active units and at least one on standby. By operating columns in series, as shown in Figure 3.27, the first vessel continues to remove organics even though its discharge exceeds the limit, and the second polishes. Once the second vessel experiences a breakthrough event, it is isolated, and the third vessel is opened for operation.

DATA COLLECTION

Data collection and control test requirements are driven by local regulatory requirements. Assuming carbon regeneration is conducted off-site, data collection is straightforward and mirrors that of traditional filtration operations. Granular activated carbon column operation time, headloss via pressure drop, and influent and effluent concentrations are all collected to monitor treatment efficiency of the system.

MAINTENANCE

The beds typically are in column form. Typical contact times are 20 to 40 minutes for domestic wastewaters.

The carbon used for wastewater treatment must be able to withstand the repeated handling required during regeneration and backwashing. Carbon regeneration is completed by a thermal process. Carbon is removed from the unit and heated to oxidize and remove the organic matter collected on the carbon surface. Regeneration can be completed in a rotary kiln or a multiple hearth furnace. Regeneration of carbon can be completed on-site or off-site by removing the carbon slurry from the contactor and hauling it to a facility for regeneration. Five percent to 10% of the carbon may be destroyed as part of the regeneration process. Eventually, the carbon will need to be replaced in lieu of regeneration with virgin carbon.

TROUBLESHOOTING

Table 3.15 describes symptoms, probable causes, and potential solutions for carbon adsorption systems.

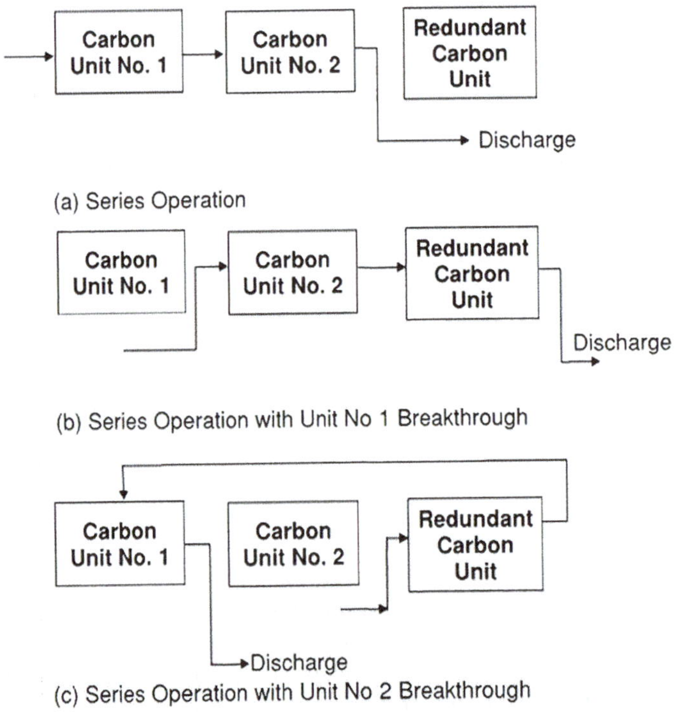

Figure 3.27 Granular Activated Carbon System in Series Operation Schematics

Table 3.15 **Troubleshooting for Soluble COD Removal Processes**

Indication	Probable Cause	Potential Solution
Excessive headloss in carbon beds	Biological growths in columns	Backwash carbon beds more frequently.
	High influent turbidity	Improve upstream treatment.
	Carbon breaking up during regeneration	Remove carbon and wash fine carbon particles from system; if necessary, replace carbon with a different type.
	Inlet or outlet screens plugged	Backflush screens.
Decrease in carbon efficiency	Improper regeneration	Adjust carbon regeneration process or replace carbon.
Hydrogen sulfide being generated in carbon beds (may corrode metal and concrete)	Anaerobic biological activity in carbon beds	Reduce detention time by reducing the number of beds in service or the amount of carbon in a bed.
		Apply 5 to 10 mg/L of chlorine to carbon bed influent.

TEST YOUR KNOWLEDGE

1. Granular activated carbon is regenerated through what mechanism?
   a. Heat
   b. Lime slurry
   c. Digestion
   d. Replacement
2. The pores in GAC particles __________ the surface area of the particle.
   a. Decrease
   b. Agitate
   c. Increase
   d. Destroy
3. Granular activated carbon columns in series __________ total column height.
   a. Increase
   b. Decrease
   c. Double
   d. Halve
4. Granular activated carbon adsorption is typically used as a __________ treatment process.
   a. Primary
   b. Secondary
   c. Tertiary
   d. Preliminary
5. What is the most common GAC contactor configuration?
   a. Uplow
   b. Downflow
   c. Vortex
   d. Braided
6. When particles do not adsorb and make their way through the GAC system, this is called __________.
   a. Pushing
   b. Breakthrough
   c. Non-adsorption
   d. Bleeding
7. The expected performance of a GAC system is completely dependent on what?
   a. The upstream treatment processes
   b. The downstream treatment processes
   c. The disinfection treatment process
   d. The screening process
8. Upflow contactor may also be called what?
   a. Expanded bed
   b. Lifted bed
   c. Downward bed
   d. Raised bed

9. Kilns and multiple hearth furnace equipment are associated with what process in a GAC system?
   a. Regeneration
   b. Backwashing
   c. Testing
   d. Liberation

10. Granular activated carbon systems are effective at removing __________ organics.
   a. Floating
   b. Dissolved
   c. Wet
   d. Particulate

Ion Exchange

FUNDAMENTALS

Ion exchange systems trade one ion for another. Home water softeners are one example of this technology. A resin bed is saturated with one type of ion. In the case of home water softeners, sodium (Na^+) is typically used as an ion (see Figure 3.28). The sodium ions are loosely attached to the resin through chemical bonds. Water is passed through the resin bed. Ions in the passing water that are more strongly attracted to the resin than the sodium ions trade places, or are exchanged, with the sodium ion on the resin. The sodium ions continue with the treated water. In the case of home water softeners, the goal is to attract calcium (Ca^{2+}) and magnesium (Mg^{2+}) ions, because these ions contribute to hardness in water. There are many different resin types available that remove different types of ions and replace them with other, less harmful ions. Examples include resins that remove ammonia, nitrate, phosphates, or heavy metals. Resins may be made of a few different materials. One type is a resin made of naturally occurring materials with ion exchange capabilities such as analcime, chabazite, clinoptilolite, erionite, ferrierite, heulandite, laumontite, mordenite, and phillipsite. These types of resins are called *zeolites*. Many ion exchange manufacturers produce synthetic resins using polymer bases. Both materials contain charged ions selected to react with the ions associated with the desired removable solid in water.

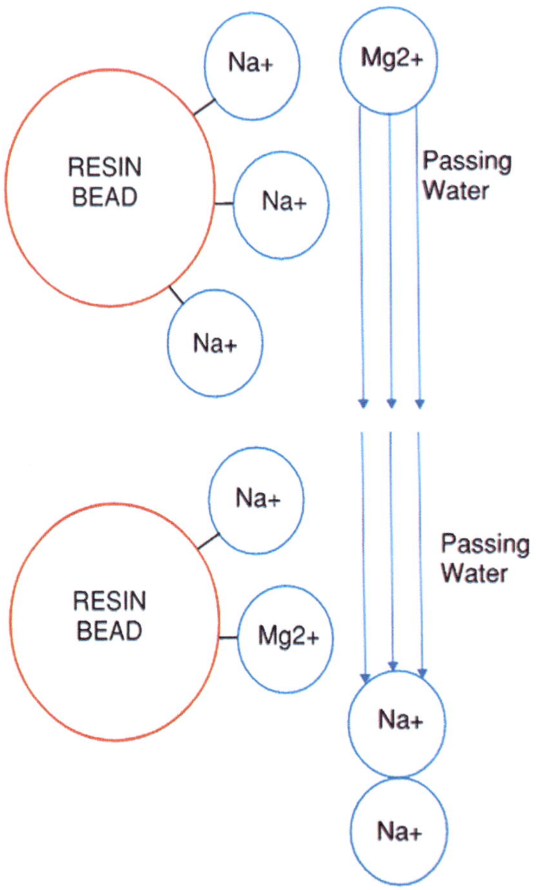

Figure 3.28 Ion Attachment to a Resin Bead. A Positive Ion Passing by in the Water Can Only Replace Another Positive in the Resin. Positive and Negative Charges Don't Mix When It Comes to Trading Ions

Eventually, the resin fills up with the exchange ions and breakthrough occurs. Think back to the parking garage analogy. When this occurs, the resin must be regenerated. Depending on the resin type, regeneration may involve backwashing the resin column with a strong salt solution or strong acid. The collected ions are washed out of the column and are replaced by the ions in the regenerating solution. Ion exchange can only remove dissolved ions and cannot remove solids.

PURPOSE AND FUNCTION

The purpose of ion exchange systems is to remove particles with charged ions that are too small to be removed by conventional clarification and filtration methods. The systems can also be unselective and remove all positively or negatively charged ions or be tailored to a specific preference for a certain ion depending on its size, charge, and structure. Systems are either cation systems, where they exchange positively charged ions, or anion systems, where they exchange negatively charged ions.

THEORY OF OPERATION

Ion exchange is a reversable process that uses sorption methods to remove charged ions from water streams. Sorption includes the physical process of adsorption (i.e., accumulation on media surface) and absorption (i.e., assimilation within media).

Resins are typically made of polymers for wastewater applications. The polymer chains that comprise the resins have several different types of ions on them (see Figure 3.29). The first type of ion is called the *mobile counter ion*, which is the ion that is exchanged for the desired ion to be removed from the wastewater stream. The fixed co-ion is the non-exchangeable ion that has a charge opposite of the counterion that attracts a dissolved ion in water.

As the water flows through an ion exchanger, dissolved ions to be removed are attracted to the fixed co-ions located in the resin beads. The resins are designed such that the desired ions to be removed have a higher affinity for the fixed co-ion than the mobile counterions. This promotes the release of mobile counterions, such as hydroxide, into the water and the attachment of dissolved ions in the water to the resin for removal. Once the resin's mobile counterions have all been exchanged, the resin will no longer work, and the ion exchanger will need to be regenerated and reloaded with mobile counterions.

Exchange resins are typically classified by functional group: strongly acidic cation exchangers, weakly acidic cation exchangers, strongly basic anion exchangers, or weakly basic anion exchangers.

The "strong" in strong acid exchangers does not refer to the physical strength of the resin; rather, it refers to the strength of the ion's charge or a resin's ability to completely break apart into its components, or dissociate, at any pH. For instance, sodium dissociates into one sodium ion and one chloride ion. In other words, the "strong" in the strong acid exchanger refers to the resin's ability to readily give up a proton or positive charge. Some examples of strong acids are hydrochloric acid (HCl) and sulfuric acid (H_2SO_4).

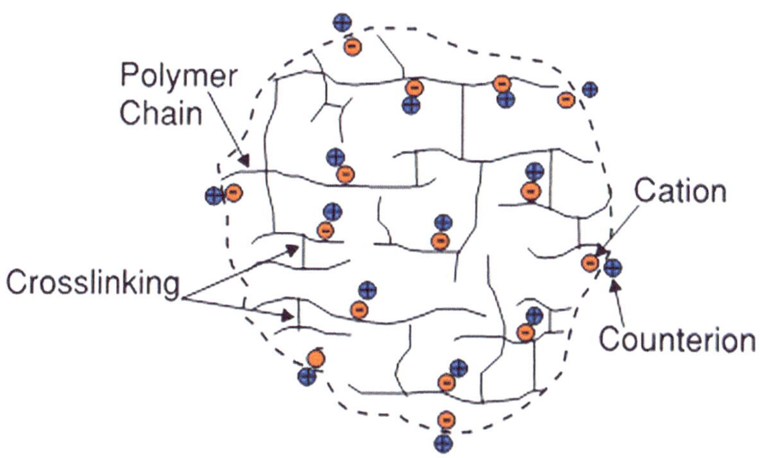

Figure 3.29 Schematic of Organic Cation Exchange Bead

Weak acid exchangers will not readily give up a proton unless the pH is greater than 6. As the pH increases, the capacity of the resin increases, with a maximum capacity typically around pH 10 to 11. An example of a weak acid is acetic acid (CH_3COOH).

Strong base exchangers will readily give up the negatively charged hydroxide ion if the pH is less than 13 or is practically independent of pH. Strong base resins are typically less stable than strong acid resins and can sometimes have a foul smell. An example of a strong base is sodium hydroxide (NaOH). Weak base exchangers will only give up a hydroxide ion (OH^-) if the pH is between 8.3 and 6.7. An example of a weak base is ammonia (NH_3).

DESIGN PARAMETERS

The selectivity of the ion resin indicates the preference a resin has for a specific ion in solution. The greater a resin's selectivity coefficient (K), the stronger the affinity the resin has for a specific ion. Table 3.16 summarizes the typical design criteria and parameters for ion-exchange resins.

The exchange capacity is an important characteristic to evaluate when selecting an ion exchange resin. This is the number of mobile exchange ions the resin has available. It is commonly reported in milliequivalents per gram of dry resin (meq/g).

The pH range is critical for the operation of weak acid or weak base ion exchangers. Keeping the pH in the exchange bed within the correct range directly affects the effectiveness of the resin in removing ions.

Table 3.16 Typical Ranges for Design Data and Criteria for Ion Exchange Resins (Adapted from Davis, 2010)

Parameter	Strong Acid Cation Resin	Strong Base Anion Resin	Unit
Exchange capacity	3.6–5.5	1.8–2.0	Meq/g as $CaCO_3$ or mmol/g
Operating capacity	50–70	40–60	% of exchange capacity
Surface loading rate	400–800 (1312–2624)	400–800 (1312–2624)	m/d ft/d
Backwash rate	12–20 (39–66)	5–7.5 (16–25)	m/hr ft/hr
Regenerate	Lime (NaCl)	Lime (NaCl)	
Regenerate concentration	5–10	2–15	%
Regenerate dose	80–320 5–20	80–320 5–20	kg NaCl/m^3 lb NaCl/cu ft
Parameter	**Range of Values**		
Pressure Drop	35–70 kPa (design) or 5–10 psi, 135 kPa (20 psi) max during operation		
Diameter	<2.5 m (<8.2 ft) for fiberglass tanks <3.5 m (<11.5 psi) for steel tanks		
Bed height of resin	>0.9 m (>3 ft)		
Bed height: diameter	1.5:1 to 3:1		
Expansion of resin bed	<100% Max, <50% Typical		
Column height	<4 m (<13 ft)		

The surface loading rate is the flowrate through the bed per square foot of resin bed surface area. The surface loading rate is used to control the pressure drop across the bed. Similar to filters, the pressure drop across the bed increases as the resin is spent. So, the design value for pressure drop is an average pressure drop allowable.

The operating capacity is a measure of the actual performance of the resin. Operating capacity is always less than the advertised exchange capacity.

It is important to know what the recommended regenerant, dose, and concentration are when selecting an ion exchange resin. The regenerate is the compound added to the ion exchanger to reload the resin with counterions.

The empty bed contact time is the time that water to be treated is in contact with the resin within the ion exchanger. It is calculated by dividing the volume of the empty bed by the flowrate.

The bed height of the resin needs to meet minimum criteria to prevent ion breakthrough. The column height of the exchanger must also meet minimum criteria to allow for expansion of the bed during backwashing. Ion exchange columns can be used in series to increase the total system column height.

EXPECTED PERFORMANCE

The resin used in an ion exchanger is selected to maximize performance of the unit to remove desired constituents. Ion exchange can be used for softening, decarbonization, demineralization, and other special applications. In softening applications, calcium and magnesium ions are exchanged for sodium ions. Regeneration of these systems uses common salts. In decarbonization applications, carbonate (CO_3) components of hardness are removed resulting in partially demineralized water. For decarbonization applications, hydrochloric acid is required for regeneration. Demineralization applications remove all salts that are dissolved in water by using a combined strongly acidic cation exchanger and a strongly basic anion exchanger. Demineralization applications often include a CO_2 tower and use hydrochloric acid for regeneration. Nitrate and heavy metals removal are common applications of ion exchange in wastewater treatment applications. Resins for these applications are designed to target specific constituents for removal.

Heavy metals include arsenic, lead, mercury, cadmium, nickel, chromium, and so on.

The performance of an ion exchange unit is commonly dependent upon the regeneration cycle. Maintaining effective operation by effectively reducing the concentrations of the desired constituents through the ion exchange while minimizing backwash and regeneration cycles is key to efficient performance of ion exchange units.

EQUIPMENT

An ion exchange column consists of:

- An enclosed vessel contact tank: Most ion exchange columns are either upflow (i.e., flowing from the bottom to the top of the column or countercurrent) or downflow (i.e., flowing from the top to the bottom of the column) operation. Countercurrent tanks can use two different processes: the floating bed process and downflow countercurrent process. A floating bed process unit loads in the up-current direction but regenerates in the down-current directions, whereas a down-flow countercurrent process is the opposite. The contactors are made from stainless steel or fiberglass. Tanks are coated internally to avoid deterioration of the vessel from the inside.
- A plenum area that is similar to the underdrain area. This is an area where effluent of the contactor is collected before exiting.
- Backwash pumps are used to pump water backward through the resin to remove particles that have accumulated on the media.
- Other equipment may include pumping, piping, valves, nozzles, pH analyzers, pressure gauges, and transmitters.

PROCESS VARIABLES

Process variables to monitor for ion exchange systems are pH, pressure, empty bed contact time, backwashing flowrates, regeneration dose, and concentration, as described in the "Design Parameter" section of this chapter.

PROCESS CONTROL

Process control of ion exchange systems is like that of conventional filters. Pressure drop across the unit is monitored and the unit is taken offline for backwashing when pressure drop hits maximum setpoints. The backwashing process is typically automated. Monitoring the pH and influent and effluent concentrations will help operators gauge the effectiveness of the treatment system. Regeneration of the resin media and resin cleaning are routine maintenance steps.

TYPICAL PROCESS CONTROL

The operation of ion exchange systems includes four processes: backwashing, regeneration, rinsing, and service. Backwashing is completed after each run through the system. The purpose of backwashing is to remove any suspended particles that have been captured in the resin, such as dirt or silt. Insufficient backwashing is the cause of many issues with ion exchangers. Regeneration is the next step of the ion exchange operation. Regeneration uses chemical addition such as lime or hydrochloric acid to disengage ions that have attached to the resin. Even with the regeneration process, resin will still lose some capacity each regeneration and the resin will need to be completely removed eventually. Based on the resin manufacturer's recommendations, once or twice a year, the regeneration chemical may be changed to do a deeper regeneration.

Rinsing follows regeneration to remove excess regenerate before units are returned back to service. Rinsing always occurs in the same direction as regeneration to promote complete rinsing of regenerate.

Service is when the units are returned to operation after rinsing. Water for treatment is passed through the exchanger to remove ions. As resin adsorbs particles from wastewater, it eventually becomes saturated, and breakthrough can occur. The effluent concentrations of the organic compounds to be captured indicate the need for the cycle to start over again with backwashing.

Longer contact times may improve treatment efficiency. This can be accomplished by adjusting the ion exchange column flow to be in series or by reducing the hydraulic loading rate to the system. One approach for minimizing costs and regeneration cycles would be to operate ion exchange columns in series with active units and at least one on standby. By operating columns in series, the first vessel continues to remove particles even though its discharge exceeds the limit, and the second polishes. Once the second vessel experiences a breakthrough event, it is isolated, and the third vessel is opened for operation.

The waste streams from an ion exchange system include backwashing and regeneration streams. Both of these streams generally contain a brine due to the ion exchanges. Disposal of brine can be difficult, and the final disposal method will be dependent on the concentrations in the waste streams. In some instances, the waste streams can be blended with water and returned to the head of the treatment plant slowly. In cases where the concentrations are too high, the brine may need to be further treated and disposed of.

DATA COLLECTION

Data collection requirements for ion exchange depend on local regulatory requirements for specific constituents to be removed. Monitoring influent and effluent concentrations will indicate to operators when the operation cycle needs to reset and engage backwashing. Online monitoring of regenerant pH will also be required, along with periodic grab samples, to validate the online pH measurement.

MAINTENANCE

Planned maintenance for ion exchange beds involves maintaining component pieces per a manufacturer's recommendation such as centrifugal pumps and chemical feed equipment. Flow, pH, and ammonia monitoring devices should be checked routinely and calibrated periodically based on the manufacturer's recommendations. Longer term maintenance involves resin replacement. Backwashing and regeneration cycles prolong the life of resin, but eventually the resin will need to be replaced at the end of its life cycle or when treatment integrity is compromised due to long-term use.

TROUBLESHOOTING

The following are some ion exchange issues to be aware of and possible solutions:

- Backwashing time and flowrates can be adjusted to make the most of the operations cycle. Removing dirt and silt buildup effectively through the backwashing cycle is a critical step in effective operations, and determining the best backwash rates and times for the waste stream will troubleshoot many treatment effectiveness issues.
- Regeneration with lime slurry alone can be a slow process. The regeneration rate can be increased by adding salt (sodium chloride) to the regenerant solution.
- The high pH regenerant can cause scaling in the zeolite exchange bed and the piping and appurtenances. High backwashing flowrates may be required to remove these deposits, although this should be balanced with the need to minimize media loss during backwashing.

TEST YOUR KNOWLEDGE

1. What type of resin is considered naturally occurring?
   a. Polymorphic
   b. Zeolite
   c. Cationic
   d. Anionic
2. What is the most common material used for ionic exchange regeneration?
   a. Lime slurry
   b. Hydroxide
   c. Polymer
   d. Carbon slurry
3. An anionic exchanger transfers what type of ion?
   a. Positive
   b. Negative
   c. Neutral
   d. Double
4. A strong acid reactor operates best at what pH?
   a. 7
   b. 9
   c. 2
   d. 10
5. The operation capacity is always __________ the exchange capacity.
   a. More than
   b. Less than
   c. Equal to
   d. Subtracted from
6. What step of operations follows regeneration?
   a. Backwashing
   b. Rinsing
   c. Service
   d. Cleaning
7. What is the indication that a backwash needs to occur?
   a. Breakthrough
   b. Temperature change
   c. pH change
   d. Media color change
8. Which of the following applications is ion exchange commonly used for to reduce hardness?
   a. Softening
   b. Demineralization
   c. Decarbonization
   d. Dissociation
9. Empty bed contact time is typically measured in what units?
   a. Minutes
   b. Hours
   c. Million gallons per day
   d. Gallons per minute
10. A resin's selectivity is typically indicated by what symbol?
    a. J
    b. K
    c. Q
    d. M

CHAPTER SUMMARY

Chemical Delivered Systems

- Chemical feeders can be classified into dry feeders or liquid feeders.
- Turndown is a ratio of an instrument's maximum allowable operating span to its minimum operating span. For example, if a pump is capable of a maximum output of 100 gpm and it has a 100-to-1 turndown, then it can be adjusted to a minimum flow of 1 gpm.
- Typical doses of the coagulant aluminum sulfate range from 20 to 40 mg/L and the optimum pH range (lowest solubility) is between 6 and 8.
- As a general rule, approximately 0.5 mg/L of alkalinity (as $CaCO_3$) is required to react with each 1 mg/L of alum, and approximately 1.0 mg/L of alkalinity (as $CaCO_3$) is required to react with each 1 mg/L of ferric chloride.
- If the process water is alkalinity deficient, alkalinity may be added with lime, calcium hydroxide, or other bases.
- Typical maintenance for chemical systems includes changing pumps, gears, and proper tubing maintenance (especially with peristaltic tubing).
- For the most part, if coagulant delivery mechanical systems are operating well and high turbidity is still observed, jar tests should be performed to determine optimum doses with changing water quality.

Coagulation–Flocculation

- Turbidity is the primary process variable for coagulation/flocculation. The more total colloidal material in water the less clear it appears, which correlates to a higher turbidity.
- When inorganic metal salts are added to water, they dissociate to yield positively charged ions such as Al^{+3}. As the particles form bonds with water and come together, they form larger particles or "flocs," which may then be settled out via gravity.
- During flocculation, the coagulated water is stirred slowly via air or mechanically driven paddles, causing the solid particles to agglomerate and settle to the tank floor.
- Paddle wheel flocculators are the most common flocculation systems, but impeller-type mixers are also used.
- When a low-energy solution is needed, hydraulic baffle wall flocculators (also known as Alabama flocculators) have numerous walls with specific spacing to create a flow velocity gradient.
- Normal process control systems typically focus on the ability to maintain adequate coagulant doses.
- Flocculation systems are typically easy to maintain, but they can still experience their fair share of troubleshooting.

Clarification

- Clarification is the process of separating suspended particles from water via gravity in large basins or tanks and is very similar to primary and secondary conventional clarification (refer to *Chapter 4 and Chapter 8 of Wastewater Treatment Fundamentals I*).
- Rectangular horizontal flow tanks are the most common clarification basins. Proprietary ballasted and plate settlers are becoming more common in drinking water and advanced wastewater treatment and are typically designed and produced by manufacturers.
- Compact and relatively economical upflow clarifiers provide coagulation, flocculation, and clarification in a single (typically circular) steel or concrete tank.
- For redundancy, at least two tanks are typically provided in a facility. Process control focuses on the overflow rate of tanks.

$$\text{Surface overflow rate} = \frac{\text{Flow, m}^3\text{/d}}{\text{Area, m}^2}$$

- Common issues related to clarifiers include inadequate sludge removal, poor floc formation, and floc breakup (excessive turbulence).

Filtration

- Filtration involves passing flow through a filtering medium or a combination of two or three media to remove suspended or colloidal matter.
- Dual-media and multimedia conventional filters allow more of the filter depth to be used to capture solids, making wastewater filtration practical. Just as there are different types of filter systems, there are also different configurations.
- Inline filtration is when coagulant is added to a process and then the stream is sent directly to filtration.
- "Direct" filtration includes a flocculation step.
- Conventional filtration is the most commonly used and includes coagulation, flocculation, clarification, and finally filtration as part of a treatment train.
- Normal filter operation depends on the flowrate through the filter, which, in turn depends on the headloss through the filter. So, effective filtration control should involve continuous monitoring of headloss and influent and effluent turbidity.

	■ Settled secondary effluent: TSS (mg/L) = (2.0 to 2.4) × turbidity (NTU) ■ Filter effluent: TSS (mg/L) = (1.3 to 1.5) × turbidity (NTU) ■ Many filter issues can be resolved with cleaning the filter via backwash; however, sometimes it is necessary to adjust upstream coagulant doses.
GAC	■ A tertiary treatment system, similar to filtration, is used to remove organics from wastewater and drinking water. ■ Uses porous carbon that is thermally activated to increase the likelihood of particle attachment on the carbon surface. ■ Backwashing and carbon regeneration are processes used to extend the usable life of the carbon. ■ Using GAC units in series while limiting the hydraulic loading rate improves the treatment efficiency.
Ion Exchange	■ Tertiary treatment system that uses organic or synthetically manufactured polymer resins to remove charged ions in water and wastewater treatment systems. ■ Common applications include softening, nitrogen removal, TDS, and metals removal, among other applications, to remove soluble ions. ■ Typical operation procedures associated with ion exchange treatment include backwashing regeneration and rinsing to improve attachment capacity of the ion exchange resin. ■ Cationic exchangers exchange positive ions, whereas anionic exchangers exchange negative ions.

References

Amirtharajah, A., & Mills, K. M. (1982, April). Rapid-mix design for mechanisms of alum coagulation. *Journal AWWA, 74*(4), 210–216.

Crittenden, J. (2012). *MWH's water treatment principles and design* (3rd ed.). John Wiley & Sons, Inc.

Davis, M. (2010). *Water and wastewater engineering design principles and practice.* Water Environment Federation.

Metcalf & Eddy. (2003). *Wastewater engineering: Treatment and reuse.* McGraw-Hill.

U.S. Environmental Protection Agency. (1978, January). *Field manual for performance evaluation and troubleshooting at municipal wastewater treatment facilities.*

Water Environment Federation & Association for Boards of Certification. (2018). *Wastewater treatment fundamentals I—Liquid treatment.* Water Environment Federation.

Water Environment Federation & Association of Boards of Certification. (2021). *Wastewater treatment fundamentals II—Solids handling and support systems.* Water Environment Federation.

Water Environment Federation. (2017). *Operation of water resource recovery facilities* (7th ed., WEF Manual of Practice No. 11).

Suggested Readings

American Water Works Association & American Society of Civil Engineers. (2005). *Water treatment plant design.* McGraw-Hill.

Association of Boards of Certification. (2017). *Need-to-know criteria: Wastewater collection operator class IV.* Association of Boards of Certification.

Davis, M. (2008). *Introduction to environmental engineering.* McGraw-Hill Companies.

National Fire Protection Association. (2019). *Standard for fire protection in wastewater treatment and collection facilities.*

Spellman, F. (2014). *Water and wastewater treatment plant operations.* CRC Press.

Swain, H. (2020). *TOMCO2 systems water treatment solutions.* https://tomcosystems.com/wp-content/uploads/2019/01/WaterTreatment.PSF_.WhitePaper.pdf

Wahlberg, E. K. (1994). Influence of activated sludge flocculation time on secondary clarification. *Water Environment Research,* 779–786.

Water Environment Federation. (2013). *Automation of water resource recovery facilities.*

Water Environment Federation; American Society of Civil Engineers; & Environmental and Water Resources Institute. (2018). *Design of water resource recovery facilities* (6th ed.; WEF Manual of Practice No. 8; ASCE Manuals and Reports on Engineering Practice No. 76). Water Environment Federation.

Water Environment Federation; American Water Works Association; & International Water Association. (2011). *International standard units for water and wastewater processes* (WEF Manual of Practice No. 6). Water Environment Federation.

CHAPTER 4

Advanced Activated Sludge

Introduction

This chapter assumes that the reader is already familiar with activated sludge. If you have not already completed Chapter 8 ("Activated Sludge") and Chapter 9 ("Nutrient Removal") in *Wastewater Treatment Fundamentals I—Liquid Treatment* (WEF & ABC, 2018), review those chapters first. This chapter addresses more advanced topics and technologies.

LEARNING OBJECTIVES

Upon completing this chapter, you will be able to

- Understand the basic function and differences among various types of microscopes.
- Learn about the protozoa and metazoa commonly found in wastewater treatment and how they can be correlated to treatment conditions.
- Understand the types and operational principles of sequencing batch reactor systems.
- Understand the types and operational principles of moving bed biofilm reactor and integrated fixed-film activated sludge systems.

Microbiology

MICROSCOPIC EXAMINATION

By examining the mixed liquor with a microscope, much can be determined about the activated sludge process. The presence of various microorganisms within the mixed liquor floc can quickly indicate good or poor treatment. It is not necessary to be a skilled microbiologist to be able to identify or count individual species. Instead, it is important to be able to recognize significant groups of microorganisms such as filamentous bacteria; protozoa (amoebas, flagellates, and free-swimming or stalked ciliates); and rotifers. Additionally, the microscope can be used to observe other important factors such as the presence or absence of filaments, floc characteristics, and supernatant characteristics (WEF, 2017b).

Filaments grow in response to different conditions. See *Wastewater Treatment Fundamentals I*, Chapter 8, for discussion of filaments and sludge bulking.

The identification of filamentous microorganisms can help narrow the investigative process during troubleshooting. Knowing whether a filamentous microorganism is, for example, Nocardia, Microthrix, Thiothrix, or Type 0041 is important to pinpoint an activated sludge settling problem. The specific type of filamentous microorganism will be an indication of whether the problem is caused by a low dissolved oxygen (DO) concentration, nutrient-deficient waste, a low food-to-microorganism ratio (F:M), or other conditions. Most water resource recovery facilities (WRRFs) have microscopes that can only determine the presence of excessive filaments, but not the type of microorganism. In many cases, WRRFs seek the assistance of colleges, universities, or commercial laboratories, where proper staining techniques and stronger microscopes are used, for definitive identification of the type of microorganisms present at the facility.

Two types of microscopes are typically used at WRRFs: a high-magnification microscope composed of a one compound lens system and a stereoscopic binocular microscope with two complete-lens systems for producing an image with significantly lower magnification. The term *compound microscope* refers to the high-magnification microscopes using multiple lenses that provide total magnification from 40 to 1000×. Traditionally, these microscopes had a singular ocular lens (eyepiece) and were called *monocular*. Currently, most compound microscopes are binocular and have two ocular lenses that help reduce eyestrain, but do not provide stereoscopic vision. The stereoscopic binocular microscope, also called a *dissection microscope*, has two complete-lens systems (one for each eye) and provides lower magnification, typically from 4 to 60×.

Figure 4.1 shows the monocular compound and stereoscopic binocular microscopes. All compound microscopes contain a basic lens system consisting of an objective lens that produces an image, typically magnified, of the object observed, and an ocular lens that further magnifies the image. The objective and ocular lenses are situated at opposite ends of the microscope's body tube, with the objective lens located closer to the specimen. Magnification is a measure of how big an object looks to the eye, where life-size images are specified as "1." Magnification is written by a number followed by "×" (10× means 10 times life-size). The total magnification of a given compound-lens system is the product of the magnification of the ocular lens (typically 10×) and the magnification of the objective lens (which can be varied). The number of objective lenses on a microscope typically varies from one to four and the lenses are held in place on a revolving nosepiece. A typical combination for magnification on a microscope with four objective lenses is 4, 10, 40, and 100×. The magnification of a particular lens is noted on the side of its barrel. Most often, the 100× objective lens is an oil-immersion lens. If so marked, immersion oil should be used with this lens, with its front lens in contact with the cover glass.

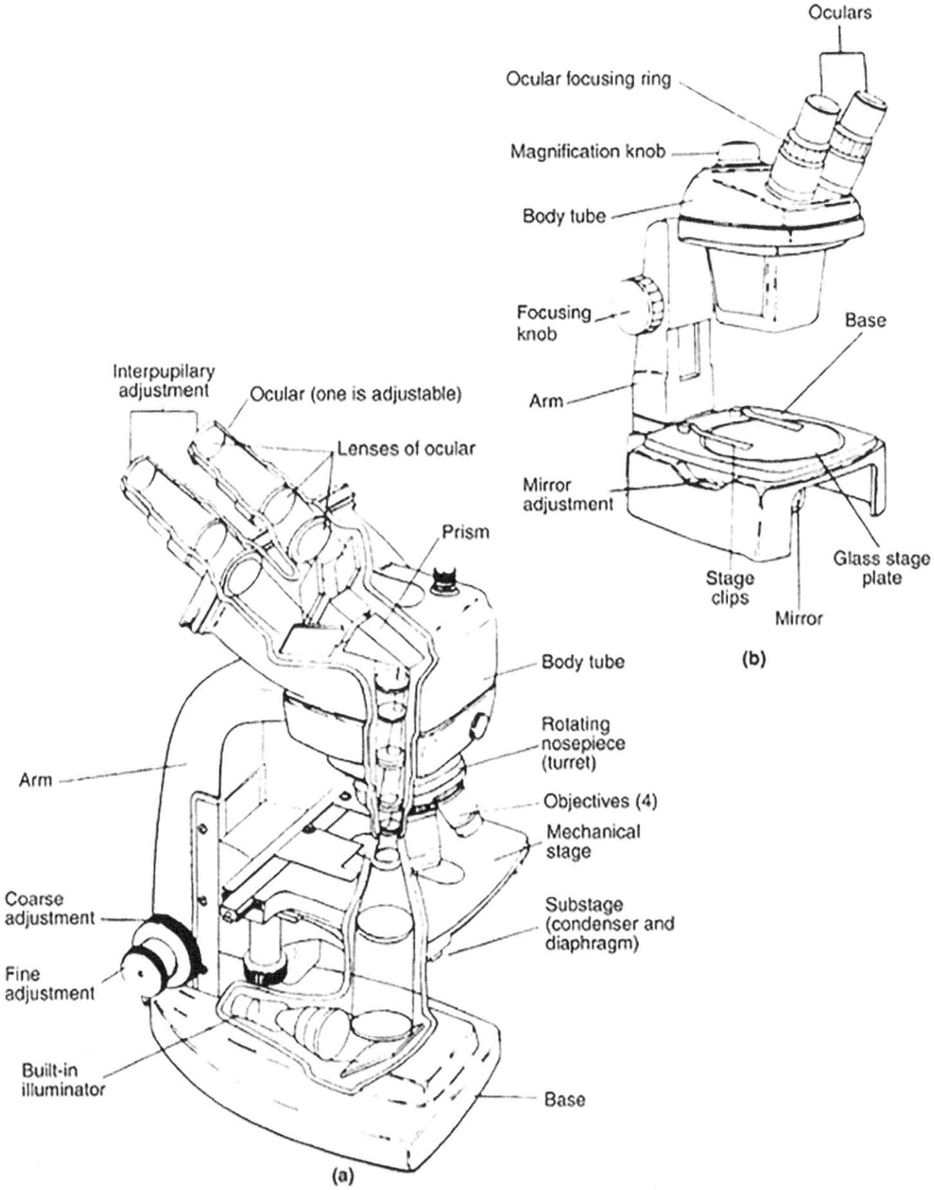

Figure 4.1 Comparison of (a) Compound and (b) Stereoscopic Binocular Microscopes. The Compound Microscope Is Sectioned to Show the Pathway of Light From Illuminator to Eyepiece. (WEF, 2017b)

Focusing a microscope is accomplished by changing the position of the body tube with respect to the position of the stage, the platform on which the specimen to be observed is placed. In light microscopy, the object observed is seen by the light it reflects or transmits. One of the primary differences between the low-magnification stereoscopic microscope and the high-magnification compound microscope is that the stereoscopic microscope typically uses reflected light and the compound microscope typically uses transmitted light.

Although the microscope is a magnifying instrument, magnification alone does not necessarily improve the ability to observe details. The ability of a microscope to allow small objects to be distinguished is more important than its overall magnifying capabilities. The resolving power or resolution of an optical system is its ability to make clear and distinct the separate parts of an object. In other words, it is a measure of how clearly details can be seen with the microscope.

The resolving power of a lens system depends on the angle of the light cone gathered by the objective and the wavelength of light used. The unaided human eye has a resolving power of approximately 0.1 mm, which is determined by the limited number of light-sensitive cells comprising the retina of the eye. A microscope coupled with the human eye, by magnifying an object and increasing the angle through which light is gathered, produces a lens system with a greatly improved resolving power. The theoretical limit of resolution of the light microscope is approximately one-half of the wavelength of the light used. The resolving power of a compound light microscope (with a 10× ocular and 100× objective) with the human eye as a receptor is approximately 0.2 μm (when using immersion oil, with a 100× immersion objective fitted with an oiled condenser top).

One micron meter [μm] equals 1/1 000 000 of a meter or 1/1000 of a millimeter.

The vertical distance that remains in focus at any particular setting is called the *depth of field*. This is a constant value for a given objective lens. If a specimen is thicker than this value, focusing up and down is required to clearly see the area being observed. This technique, called *optical sectioning*, can give valuable information on the three-dimensional shape of a specimen. When used with the fine-line adjustment knobs, the thickness of an object can be estimated. On many microscopes, the fine adjustment knobs are calibrated in measuring units of micron meter (microns).

The low-magnification (4 to 60×) stereoscopic binocular (or dissecting) microscope allows the observation of objects that are too large or too thick to be seen with higher magnifications, but too small to be seen with the unaided eye. It also provides an opportunity to observe these objects in three dimensions. This type of microscope typically consists of two compound-lens systems that are focused simultaneously on the same region of the object; each eye views the field from a different angle. This arrangement provides the observer with binocular vision and increased depth perception with long working distance. Most stereoscopic microscopes feature variable or zoom magnification. This is accomplished by varying the distance between the objective elements and ocular lenses. As with the compound microscope, total magnification is calculated by multiplying the separate magnification of the objective and ocular lenses. In some models, magnification (zooming) is varied by rotating a knob of the microscope or by rotating a ring around the center of the body tube. The lamp is typically positioned so that light reflects off the specimen and enters the lens system. Reflected light is best for color and depth perception.

The compound and stereoscopic binocular microscopes described previously are considered standard equipment by many laboratories. In addition to these two types of microscopes, there are many accessories, modifications, or special microscopes available to yield better contrast or resolution. Several considerations for selecting a microscope (or accessories) include the following: the specimen (size range, live or dead, and stained or unstained) that will be examined; the type of illumination (transmitted or reflected, filtered or unfiltered, and special contrast techniques) that will be necessary; the magnification and resolution requirements of the objective to show the smallest detail of interest; photographic requirements that are necessary; and budget constraints. Table 4.1 compares commonly used microscopes.

The dark-field microscope shows border effects in transparent specimens and greatly improves their contrast, but loss of internal cell details occurs. Phase contrast is of value in the study of transparent microorganisms. It is used routinely (typically by specialized laboratories) for examining such organisms as the filamentous bacteria that cause bulking. Structures in living cells may be seen because greater

Table 4.1 Comparison of Different Types of Microscopes Commonly Used (WEF, 2017b)

Characteristics	Compound Bright-Field Microscope	Dark-Field Microscope	Phase-Contrast Microscope	Fluorescence Microscope
Basic principle	Transmitted light	Transmitted light	Transmitted light	Reflected UV, blue, or green light for excitation; light emitted in form of visible fluorescence
	Optical lenses	Optical lenses	Optical lenses and diffraction plates	
	Darker object on a bright background	Bright object on a dark background	Light waves in phase	Fluorescent object against a dark background
Highest magnification	Visual 1000×	Visual 1000×	Visual 1000×	Visual 1000×
Advantages	Cells can be viewed either alive or dead; simple operation	Cells can be viewed either alive or dead; no staining is necessary	Cells can be viewed either alive or dead; no staining is necessary	Improved specimen visibility; specimens can be rendered selectively visible by stains
	Low cost	High contrast	Moderately high contrast	
Disadvantages	Staining is typically required	Loss of internal details; need strong light	May introduce halos on the image, boundaries, and edges	Typically, not used to observe living cells; high cost

contrast is produced between them and their surroundings. This makes it possible to distinguish structural details that vary only slightly in thickness or light refraction.

PROTOZOA

Protozoa ("little animals") are a large and diverse group of single-cell organisms that are typically much larger than bacteria. Many of them are quite mobile, swimming rapidly through their microscopic world. Others attach themselves to tiny particles and extend into the water around them (see Figure 4.2). They play primary roles in the clarification of effluent because of their predation on dispersed bacterial growth. In many instances, specific protozoa can be easily identified and quantified in wastewater samples. Protozoa may or may not have a test (shell). Their primary function in wastewater is the removal of non-flocculated bacteria and small floc particles. There are some that may use some of the organics in the water as well as other higher life forms. Most of these organisms can be easily viewed with a phase-contrast microscope at 100× total magnification. Viewing using a bright-field microscope may show the organisms, but the level of detail and organism identifying keys may be difficult to see.

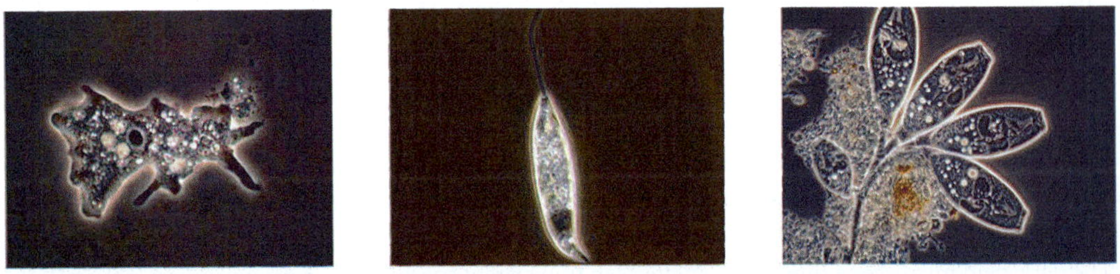

(a) Amoeba (b) Flagellate (c) Ciliate

Figure 4.2 Protozoa in Wastewater (WEF & ABC, 2018) (Reprinted with permission from Steven C. Leach)

The floc size and density along with filament presence have a direct effect on settling, and the mixed liquor's bulk water directly influences the clarity of the discharge water. All of these characteristics are important when performing a microscopic analysis on the wastewater biomass.

Identification of higher life forms is typically combined with floc size and density interpretation and the bulk water examination to understand the health of the biomass. Higher life forms provide insight to the biomass age and stability.

Flowrate and the rate at which sludge is wasted affect the protozoa population in the activated sludge process. Any organism that settles and is returned back to the biological reactor after sedimentation has a much greater chance of survival than an organism that always remains in suspension because it is immediately subject to washout in the effluent. Thus, the activated sludge process selects for organisms that settle and protozoa that attach to the sludge floc. The retention time of the sedimentation tank is also important to the survival of protozoan populations. Microbial solids in the sedimentation tank may reduce the DO concentration.

Spatial and temporal distributions of protozoa and effluent quality are all interrelated. This has led to the proposal that protozoa can be used to indicate facility performance and effluent quality. In some cases, effluent quality may be related to the presence or absence of certain protozoan species. This relationship occurs because while there is a fairly wide range of loading over which an individual protozoan species may live, there is a narrower range of loadings over which it may be present in large numbers. The dominant species present, therefore, can provide an immediate microbiological indication of the loading condition of the facility and of the quality of effluent expected.

In the activated sludge process, grouping the protozoa by their feeding or behavior makes observation easier for the non-specialist without losing too much information. Similarly, it has been shown that certain protozoa that are either difficult to differentiate or are known to have similar ecological roles may be grouped together for monitoring purposes.

As the succession of various microbial populations in an activated sludge unit develops, the quality of the effluent also improves. It is this link between effluent quality and the species of protozoa present that has led to the theory that protozoa are indicators of effluent quality. Table 4.2 summarizes the results of a large-scale survey of resource recovery facilities in the United Kingdom that indicated there was a correlation between the protozoan species structure of the activated sludge and the quality of the effluent (WEF, 2017b). Results indicated that the protozoa found within the mixed liquor were associated with the effluent. For example, it is evident from the table that *Carchesium polypinum* was found principally in facilities that produced good quality effluent (highest frequency of occurrence when effluent 5-day biochemical oxygen demand (BOD_5) ranged from 11 to 20 mg/L), but flagellated protozoa were restricted to facilities producing inferior effluents (effluent BOD_5 above 30 mg/L).

There are several ways in which protozoa affect effluent quality, the most important being the removal of bacteria by predation and enhancement of floc formation. Additional roles that protozoa play in the activated sludge process include growth stimulation of bacteria and healthy flocs and degradation of organic waste.

Table 4.2 Percentage Frequency of Occurrence of Some Protozoa in Facilities Producing Effluents Within the Four Ranges of 5-Day Biochemical Oxygen Demand (WEF, 2017b)

BOD_5 range, mg/L	0–10	11–20	21–30	>30
Vorticella convallaria	63%	73%	37%	22%
Vorticella fromenteli	38%	33%	12%	0%
Carchesium polypinum	19%	47%	12%	0%
Aspidisca cicada	75%	80%	50%	56%
Euplotes patella	38%	25%	24%	0%
Flagellated protozoa	0%	0%	37%	45%

Figure 4.3 summarizes the results of an experiment in which pilot-scale activated sludge processes were operated in the presence and absence of ciliated protozoa. Under protozoa-free conditions, highly turbid effluents of inferior quality were produced; most of the turbidity was a result of large numbers of bacteria suspended in the effluent. In the presence of protozoa, however, the clarity of the effluent was greatly improved and this was associated with a significant decrease in the concentration of viable bacteria in the effluent.

In activated sludge facilities, it has been reported that, in the absence of protozoa, approximately 50% of *Escherichia coli* entering the wastewater are removed by unidentified processes, including natural die-off. When ciliates are present, approximately 95% of the *E. coli* are removed. The continual cropping of bacteria by protozoa not only removes excess suspended bacteria from the wastewater, but also seems to stimulate the growth of healthy bacteria, producing a more active floc. The purification of wastewater, as measured by the rate of oxygen consumption accompanying its degradation, occurs more rapidly and proceeds further in the presence of ciliates and bacteria than when bacteria alone are present. It is thought that predation by protozoa indirectly increases bacterial activity by reducing their number and keeping bacteria in a state of prolonged physiological youth so that their rate of assimilation of organic materials is greatly increased. Effluent quality, as measured by almost all typically used parameters, is thus improved by the activities of protozoa in the treatment process.

A large increase in flagellates compared to recent microbial analysis results likely indicates a recent shock that affected the biomass. The smallest flagellates are the most indicative of recent upsets to a system.

Flagellates are some of the most active protozoa. They can vary in size from 1.5 to 25 μ. They all possess at least one flagellum (tail), and many have more than one. The flagellum is used for both movement and feeding (Figure 4.4a). Some species may reside within a small test (shell) and only expose the flagellum to capture bacteria (Figure 4.4b). They use a lot of energy to feed and require a considerable amount of easily

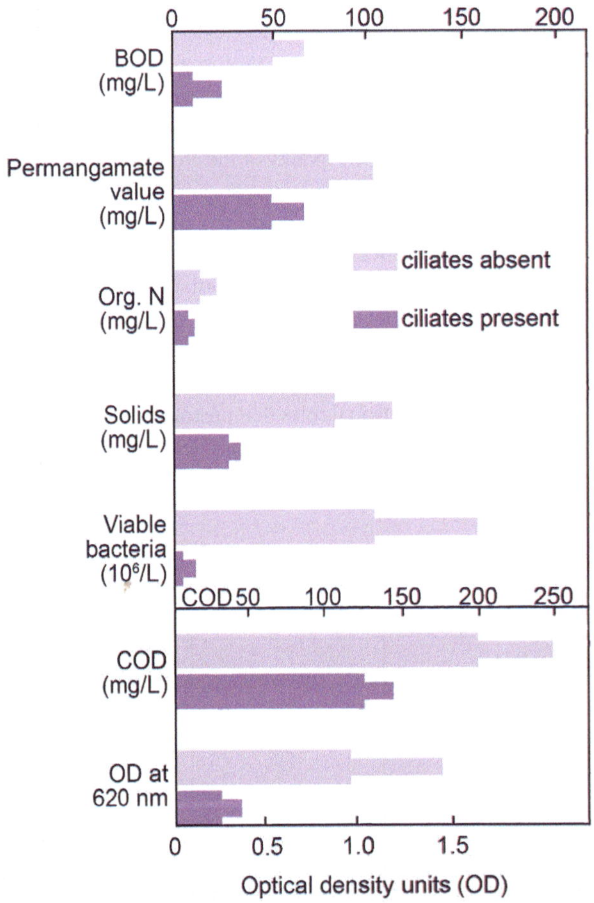

Figure 4.3 Bar Chart of Effluent Quality Issuing From Laboratory-Scale Pilot Facilities Operating in the Presence and Absence of Ciliated Protozoa. The Shoulders on the Bars Indicate Ranges of Variation Between Replicate Facilities (Adapted from Curds, 1992)

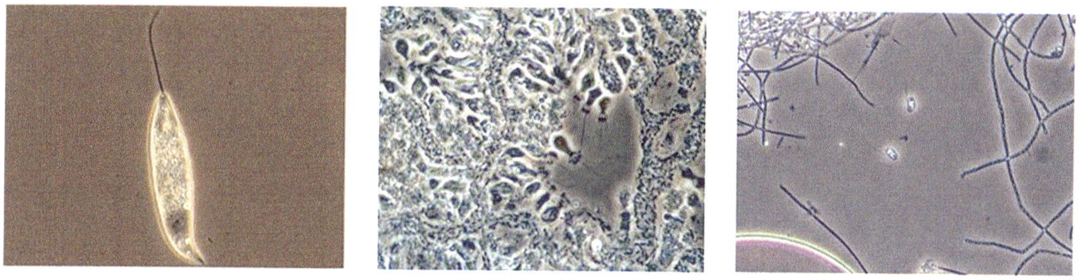

(a) Single-tail flagellate (b) Flagellates within tests (cells) (c) Two small flagellates

Figure 4.4 Flagellated Protozoa (WEF, 2018) (Reprinted with permission from Steven C. Leach)

accessible bacteria, which are primarily found in the mixed liquor bulk water. This is why we see more in a young sludge or after a recent upset. If viewed in high numbers compared to previous mixed liquor examinations, chances are a recent shock has affected the biomass. The smallest flagellates (Figure 4.4c) are the most active and most indicative of recent upsets to a system.

Amoebas are protozoa that use pseudopodia (arm-like, cytoplasm-filled projections of the cell membrane) that are able to change their form for movement and to engulf and ingest pray for digestion. The majority of amoeba will feed on organics, free bacteria, and, in some cases, other protozoa and metazoa (Figure 4.5b shows a rotifer engulfed by a large amoeba). Typically, more amoebas are seen in younger sludge or as a reaction to a recent shock, when the bulk water is loaded with free motile bacteria (Figure 4.5c). Their size ranges from 2 to 3 μ to 5000 μ.

Testate (shelled) amoebae (Figures 4.5a and 4.6a) have been observed in various systems operating in a steady-state and non-steady-state condition (Figure 4.6b). The presence of a crushed shell (Figure 4.6c) may be the result of the microscope slide cover slip and not a toxic shock. Typically, toxins will not break the shell, but will harm the amoeba itself. They are not necessarily the best indicator organisms; however, the color of the shell will suggest how long they have been in the system. Their shell is clear when they are new to the system.

A copper-colored amoeba shell (Figure 4.6d) indicates the amoeba has been in the system for a longer period and chances are that biomass growth conditions have been stable for that period.

Free-swimming ciliates are classified based on how their rows of cilia (tiny hairs) (Figure 4.7a) on the body are arranged. They use these cilia for movement and for gathering bacteria for ingestion. The cilia cover the entire body and, in some organisms, surround the oral region. All ciliates will contain a water expulsion vesicle (WEV) that releases the water either absorbed through the cell wall or ingested with food (Figure 4.7b). This is easily recognizable by observing the opening and closing of the expulsion vacuole. Certain *Paramecium* sp. will have a radiated canal feeding the WEV (Figure 4.7c). The majority of them will divide from 1 to 8 times per day, depending on a number of environmental conditions. Free-swimming ciliates are very active and constantly feeding. The majority will feed on free bacteria found in the bulk water, indicating the presence of non-flocculated bacteria. This would generally indicate a nonstable system; however, free-swimming ciliates have been found in very stable systems as well. Observing these organisms regularly in your biomass may also be a result of infiltration.

Documenting the presence of free-swimming ciliates during optimum performance and poor operations helps the operator detect changes to the biomass.

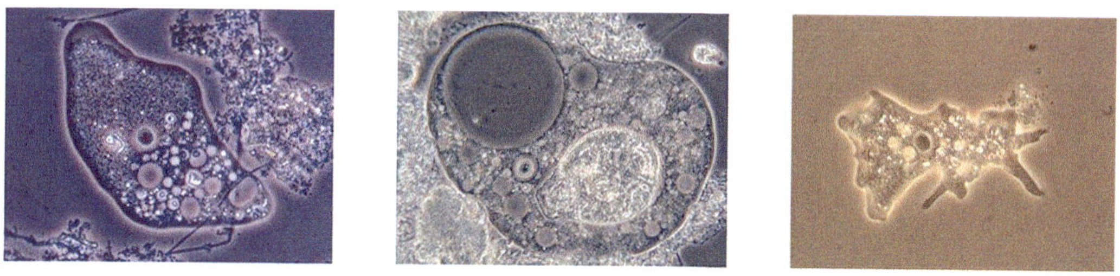

(a) Testate (shelled) amoeba (b) Large amoeba with rotifer inside (c) Naked amoeba

Figure 4.5 Naked Amoeba (WEF, 2018) (Reprinted with permission from Steven C. Leach)

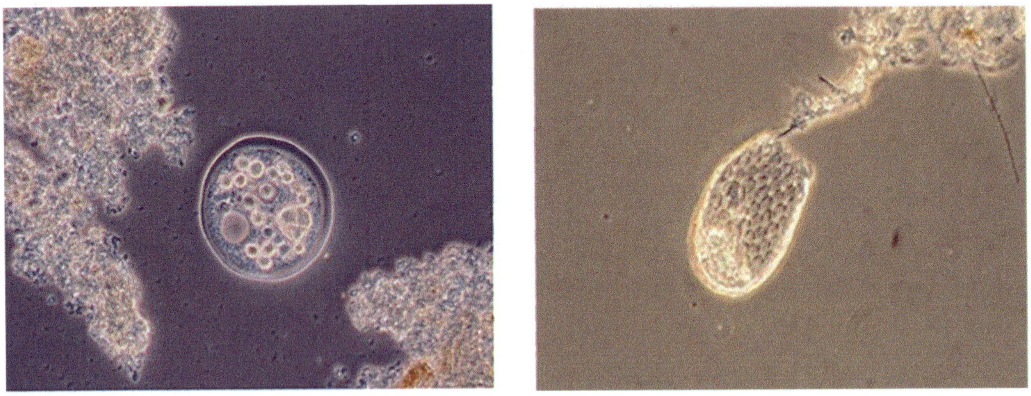

(a) Testate (shelled) amoeba

(b) Testate (shelled) amoeba-*Euglypha*

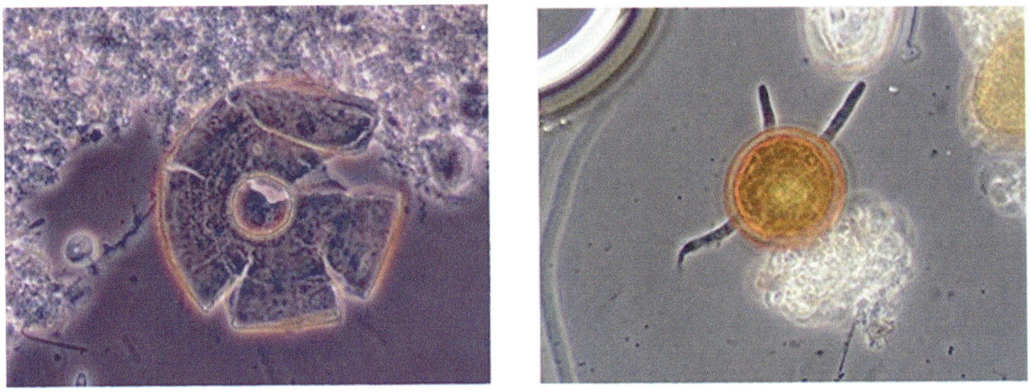

(c) Crushed test (shell)

(d) Testate (shelled) amoeba with indeterminate pseudopodia

Figure 4.6 Testate (Shelled) Amoebae (WEF, 2018) (Reprinted with permission from Steven C. Leach)

Stalked (attached) ciliates are the largest order of ciliates. They include solitary forms (*Vorticella* sp., Figure 4.8a) and colonial forms (*Opercularia* sp., Figure 4.8b). Some attached ciliates may contract along their stalk and others may not. There are a few that even reside within lorica (shell) (*Cothurnia* sp., Figure 4.8c). These ciliates are primarily bacteria feeders. The cilia that circles the oral region creates a current that helps bring bacteria into the mouth cavity. These organisms will attach to floc, inert material, or anything that may provide a base. There are no body cilia unless the organism is experiencing division (Figure 4.8d) or a change in location is needed as a reaction to a change in the environment. They will then develop a trochal band, which is a ring of cilia toward the base of the ciliate that is used to move to an area with a more satisfactory growth environment. This then becomes a "telotroch" (detached ciliate from stalk), seen in Figure 4.8e.

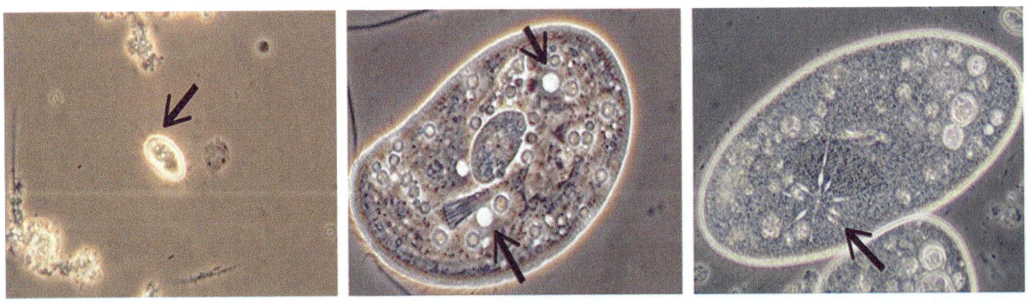

Arrow pointing to cilia state amoeba

Arrows pointing to WEVs

Paramecium with WEV

Figure 4.7 Free-swimming Ciliates (WEF, 2018) (Reprinted with permission from Steven C. Leach)

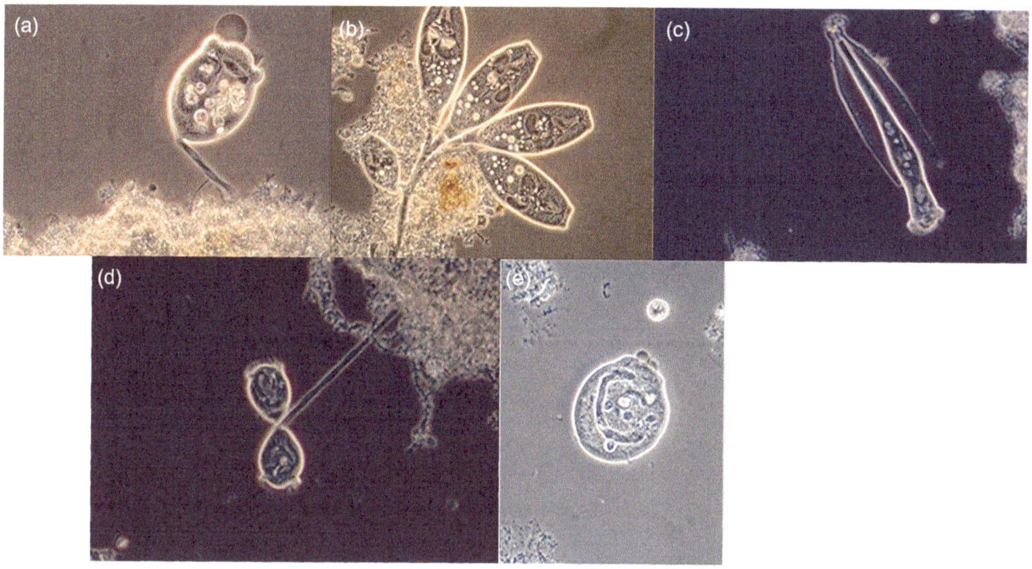

Figure 4.8 Stalk Ciliates: (a) Vorticella (b) Opercularia (c) Cothurnia (d) Daughter Cell on Top (e) Telotroch (Detached Ciliate From Stalk) (WEF, 2018) (Reprinted with permission from Steven C. Leach)

Crawler ciliates are much easier to identify because of the movement and cirri (i.e., fused-together cilia) (Figure 4.9a). The cirri resemble legs as they move along the floc and are much easier to see than cilia. These protists will crawl around the floc (Figure 4.9b) and feed similar to cattle. They do not possess cilia on their body, except near their mouth, and do not swim well (Figure 4.9c), as compared to the free-swimming ciliates. Because these organisms primarily feed crawling on floc, it makes sense that an ideal floc with a relatively smooth surface is present. Crawler ciliates are typically found in a stable biomass.

Suctoria are unique protozoa in that they may have cilia in their younger stage or as free-swimming larvae, but, once they have matured, they lose the cilia and develop tentacles. The tentacles (Figure 4.10a) may move in out from the main body as the organism waits for its prey. These tentacles can both pierce the cell wall of their prey to keep it held while another tentacle will penetrate the captured organism and extract the cytoplasm. Some species will attach through a stalk (Figure 4.10b), whereas others may free-float (Figure 4.10c) through the water.

Suctoria do not like higher ammonia concentrations and are commonly found in a stable nitrifying biomass.

Carnivorous ciliates use toxicysts (small venom-containing capsules) to anesthetize their prey to make it easier to swallow. Some species will even produce pexicysts (which are like tiny harpoons) to attach themselves to their prey. Many of these organisms may be more difficult to identify without further experience, but a few are easily recognizable (Figure 4.11). The presence of carnivorous ciliates is indicative of a stable system.

Table 4.3 compares the sizes of various types of protozoa.

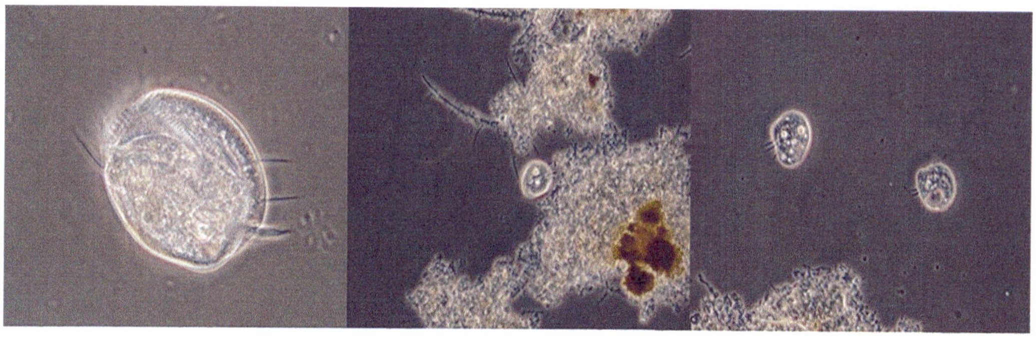

(a) *Euplotes sp.* (b) Crawler ciliate on floc (c) Crawler ciliates

Figure 4.9 Crawler Ciliates (WEF, 2018) (Reprinted with permission from Steven C. Leach)

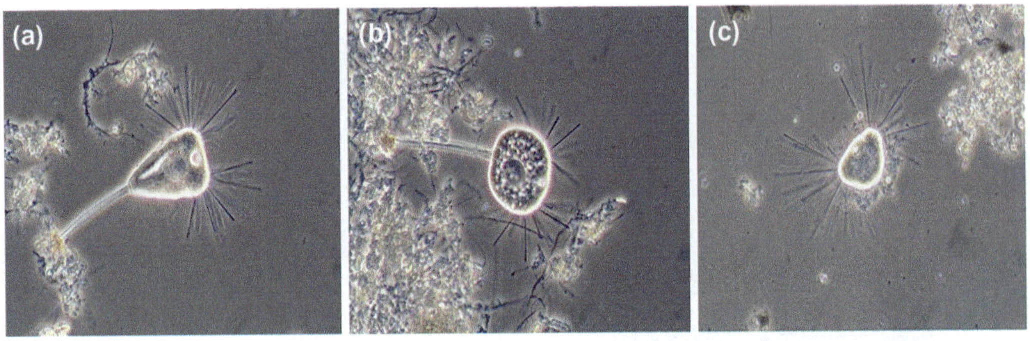

Figure 4.10 Suctoria (WEF, 2018) (Reprinted with permission from Steven C. Leach)

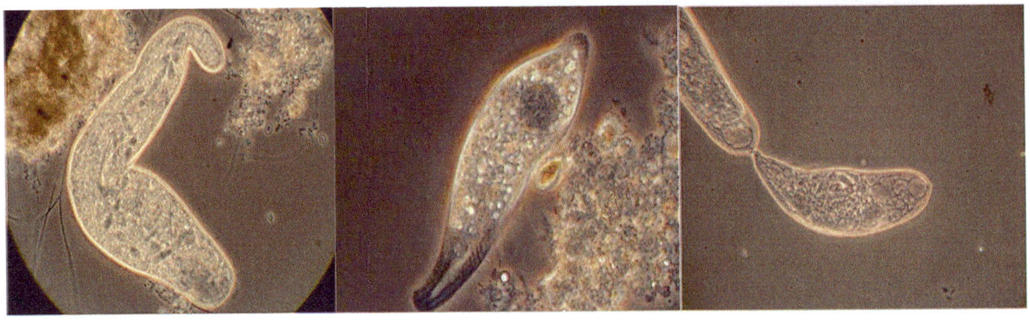

(a) *Spirostomum* (b) *Lionotus* (c) *Blepharisma*

Figure 4.11 Carnivorous Ciliates (WEF, 2018) (Reprinted with permission from Steven C. Leach)

Table 4.3 Succession of Protozoa in an Activated Sludge Facility (Jahn et al., 1979)

Protozoa	Length (μm*)
Small amoebae	30–50
Larger amoebae	1000–5000
Flagellates	5–20
Ciliates	100–350
Long ciliates	1000–3000

*One micron meter [μm] equals 1/1 000 000 of a meter or 1/1000 of a millimeter.

METAZOA

Metazoa is a term that technically applies to all animals composed of more than one cell. They are larger, more complex animals compared to protozoa. Included in the metazoan group are organisms known as *rotifers*, so named because of their rotating crowns of hairlike cilia. Rotifers are fairly common in some wastewater treatment processes and play an active role in the breakdown of organic wastes. Nematodes, water bears, bristle worms, water fleas, and seed shrimp are all metazoa that may be found in a wastewater treatment process. Metazoa may or may not have a digestive tract and central nervous system. The cells in the metazoa seen in wastewater treatment processes are differentiated into tissues, which have various functions within the organism. Metazoa are heterotrophs and consume organic material that may consist of decaying organic matter, bacteria, protozoa, or other metazoa. Most grow slowly and, as a result, are found in older activated sludge or attached-growth conditions. The term "very old sludge" describes their slow growth rate relative to the faster-growing bacteria and protozoa. Most metazoa favor environments low in ammonia, so the systems containing metazoa are typically well nitrified.

The presence of multicellular organisms in a WRRF is most likely the result of their presence in soil or freshwater environments, and their entrance to the treatment process is probably caused by stormwater runoff infiltration in sewer lines or stormwater collected in combined sewer systems. Once within the WRRF, conditions are often favorable for their proliferation as they have been found in significant numbers in various facilities. Frequently, they will proliferate for a time and then their numbers will decrease. It is often impossible to determine exactly where they came from or why they may dominate the facility population for a period of time by simply taking their place as part of the natural food chain occurring among the organisms within the facility.

Rotifers perform many beneficial roles in stabilizing the organic wastes of lagoons and fixed-growth and activated sludge processes. These include stimulating microfloral activity and decomposition, enhancing oxygen penetration, and recycling mineral nutrients. In the activated sludge process, by consuming large quantities of bacteria, rotifers help keep the bacterial population healthy and in an active growing state. Additionally, the turbidity of the secondary effluent is reduced as the rotifers consume non-floc-forming bacteria. Rotifers enhance floc formation by snipping off particles of activated sludge floc to expose new absorptive surfaces for floc formation. This snipping action also permits oxygen to penetrate the floc. Mucus secreted at either the mouth opening or the foot also aids in floc formation, and mucus-coated waste material excreted by the rotifers may act as a bonding site for bacteria.

In aerobic processes, rotifers' consumption of bacteria and solids contributes to biochemical oxygen demand (BOD) reduction. Rotifers living within the slime on the walls of biological reactors may also help to control slime growth and prevent anaerobic conditions within the slime layer (Calaway, 1968).

The types and abundance of rotifers present in any wastewater treatment process are influenced by many factors, including DO, mean cell residence time (MCRT), mixed liquor suspended solids (MLSS), pH, temperature, and trophic level (BOD concentration). The numbers, types, and activity (movement, feeding habits, etc.) of the rotifers present in the treatment process can be observed and related to process performance. Significant changes in rotifer populations may signal changes in wastewater strength and composition.

Typically, not only will there be one to four types of rotifers in any well-operated activated sludge process, but each process will have its own characteristic group of rotifers. By determining the types and relative numbers of the organisms that are characteristic of the process when it is operating correctly, an operator will have a quick means of checking process conditions. For example, if a toxic substance enters an activated sludge process, the rotifers will immediately exhibit decreased activity or die. Typically, this condition may go undetected until there is total process failure. However, if an operator notice a change in the rotifers present, it may be possible to limit the effect of the substance on the entire process.

Within a well-developed activated sludge process, a succession of higher life forms occurs with increased stabilization of organic wastes (Gerardi, 1986). Rotifers are the next higher form of microlife beyond stalked ciliates. They are strict aerobes and are typically found in environments that contain at least 3 mg oxygen/L but may also be found in wastewater with lower oxygen levels.

Rotifers (Figure 4.12) are a group of metazoa with approximately 2500 identified species. The name comes from the Latin words "rota," meaning "wheel," and "ferre," meaning "bearing"—hence, "wheel-bearing animal." The wheels that rotifers have are actually disc-like ciliated structures called *corona* and are their basic characteristic. The corona is constantly moving, bringing food particles into the rotifer or, in some cases, acting as a means of mobility. Unlike protozoa, they are rarely found in large numbers in wastewater treatment processes. Typically, there will only be one to four species of rotifer present in a well-operated WRRF. Rotifers are commonly found in wastewater treatment systems with longer sludge ages but have been viewed in industrial facilities with solids retention times (SRTs) of less than 5 days. Observing and quantifying rotifers contributes to the evaluation of treatment conditions. These animals move either by crawling or swimming and feed primarily on bacteria, detritus, and some protozoa. Rotifers help promote oxygen transfer within the floc by "snipping off the top layer." Rotifers can reproduce sexually or asexually and will have live young or eggs. Daily routine microscopic examination of the rotifers present in the

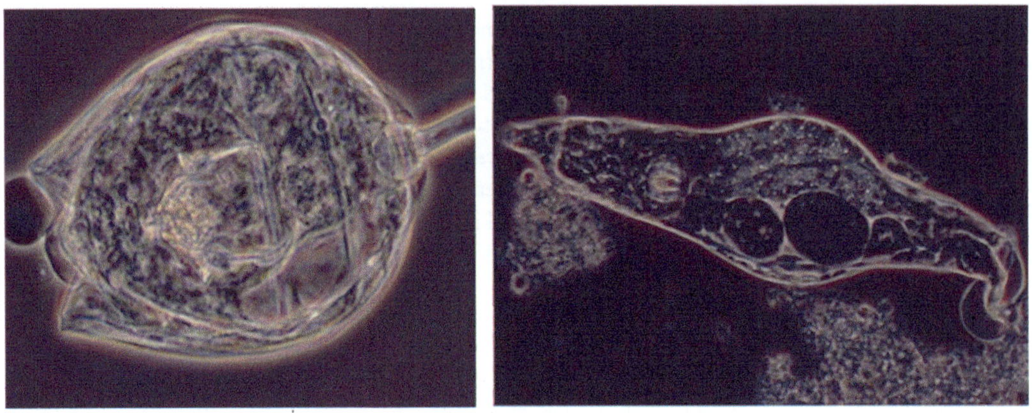

Figure 4.12 Rotifers (WEF, 2018)

wastewater treatment process can provide useful information. By correlating rotifer types, abundance, and activity to laboratory data, an operator can gain valuable insight to operation of the WRRF.

Other metazoa in WRRFs include nematodes (phylum Nematoda), water bears (phylum Tardigrada), bristle worms (phylum Annelida), gastrotrichs (phylum Gastrotricha), water fleas (phylum Cladocera), seed shrimp (phylum Ostracoda), and water mites (phylum Hydracarina). Others have also been observed, although less frequently, such as leeches (phylum Annelida). Nematodes (meaning "thread-like") are the most frequently observed of these metazoa.

Free-living nematodes (Figure 4.13) are macroinvertebrates (i.e., animals lacking a backbone or spinal column) that are found virtually every place there is rotting organic matter, from the soil to lake bottoms. Some nematodes are parasitic (living in the tissues of plants or animals) and those observed in WRRFs are free living (Figure 4.13). Free-living nematodes consume bacteria, yeasts, fungi, and algae. They may be feeding on dissolved or decaying organic matter or feeding on fecal material. Predatory species may consume rotifers, tardigrades, small annelids, or other nematodes.

Most species of nematodes look much alike. They have a cylindrical shape and flexible cuticle, or outer layer. Their bodies have no obvious external features and are cylindrical with tapering ends, including a blunter anterior (head) end and a tapered tail. Free-living nematodes found in WRRFs are microscopic in size, measuring 0.5 to 3.0 mm in length and 0.02 to 0.05 mm in width. They move through the action of longitudinal muscles against the internal pressure of the body fluids and the elastic properties of the cuticle. This allows them to move through soil, mud, and tissues of plants and animals, and also permits the nematode to swim. Nematodes swim inefficiently in a whip-like or rapid thrashing motion, which is often observed in activated sludge.

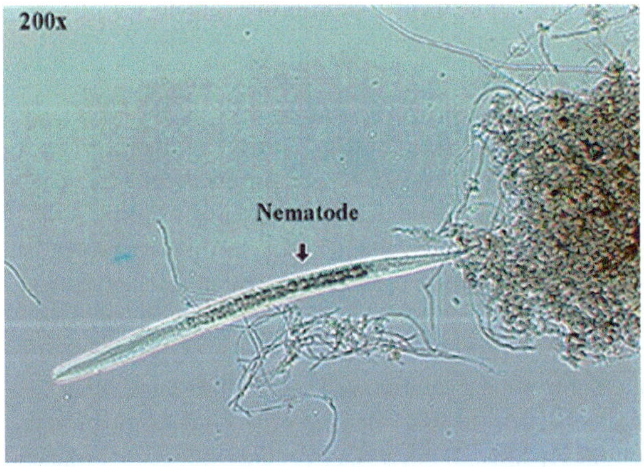

Figure 4.13 Free-Living Nematode (WEF, 2017b) (Reprinted with permission from Ronald G. Schuyler)

Nematodes can be viewed with a bright-field microscope at 100× magnification. Identification and classification can be confusing; however, this typically is not a problem because wastewater professionals are often more interested in the total numbers of nematodes present. Microscopic examination will reveal significant variations in shape and structure. Identification of nematodes is based largely on anatomical features and should be made with an appropriate illustrated key because several different species may occur at the same time within a wastewater treatment system.

Nematodes thrive in aerobic habitats including wastewater treatment processes where DO concentrations are high and food is abundant. They may be found in large numbers in the effluent from WRRFs, particularly activated sludge processes, because of their motility and resistance to chlorination. At the time of this writing, no conclusive correlations have been found between the occurrence of the different genera of nematodes and specific aerobic treatment processes.

Activated sludge processes are continuously seeded with nematodes that are not removed in the primary clarifier. Nematode populations are affected by several environmental factors, including parasitic fungi, DO concentrations, temperature, and wastewater strength and composition. Of the factors influencing nematode populations, changes in wastewater strength and composition have the greatest effect. Food seems to be the primary variable. Nematode numbers increase with greater wastewater strength and accompanying stabilization. Changes in wastewater strength that generate a significant change in the microflora or nematode diet may also generate a change in the types of nematodes present.

Oxygen is supplied to the tissues in the nematode by diffusion. More oxygen is required by a nematode as it grows and matures; oxygen need is at a maximum for adult nematodes. Therefore, in oxygen-poor environments, where sufficient oxygen cannot diffuse to the tissues, activity, growth, and reproduction are reduced. Under anaerobic conditions, nematodes become quiescent. For this reason, they have been used as bioindicators of low DO or anaerobic environments. Warm wastewater temperatures cause decreased nematode numbers, whereas cold wastewater temperatures cause increased numbers. This is at least partially caused by the relationship between oxygen content of water and temperature; cold water is able to hold more DO than warm water.

Most metazoa are found in close association with a substrate. They have traditionally been associated with activated sludge with longer MCRTs. There is some speculation that some actually dominate in the slime coatings found in association with the walls of the biological reactors and in other areas where the organisms would find suitable substrate. As these slime layers slough or are washed off, the organisms are washed into the sludge and can then be seen in the suspended growth environment of the mixed liquor.

All metazoa found in WRRFs have found the environment to be suitable for their survival. They are believed to be washed into the system through infiltration and are able to survive and possibly reproduce in the facility. They simply take their place in the existing food chain and are observed by the operator.

Figure 4.14 shows typical metazoa including bristle worms (Aeolosoma) (Figure 4.14a), gastrotrichs (Figure 4.14b), and nematodes (Figure 4.14c).

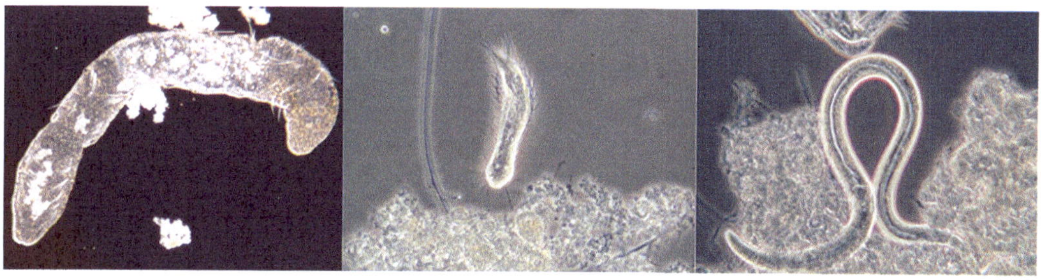

(a) Bristle worm (b) Gastrotrich (c) Nematode

Figure 4.14 Typical Metazoa (WEF, 2018) (Reprinted with permission from Steven C. Leach)

Annelids are an example of nematodes observed in WRRFs. These are segmented worms with a soft, muscular body wall covered by a thin cuticle. The bristle worm is the most frequently observed annelid in WRRFs. Bristle worms are typically less than 5 mm long. The segments are easily observed under a low-power microscope, as are the bristles, or setae, associated with each segment. Pigment globules give them a spotted appearance (Figure 4.15). The color of the pigment varies with the species but is typically red or orange. In large concentrations, they can cause red sludge or pink foam conditions.

Under optimal growth conditions in activated sludge, Aeolosoma have been known to reproduce so rapidly that they turn settled sludge the color of the worm itself and cause excessively fast settling. Systems west of Denver, Colorado, such as West Jefferson Sanitation District, continually fight Aeolosoma takeover. This has also been documented in Canada (WEF, 1995).

CORRELATING PROTOZOA AND METAZOA TO OPERATING CONDITIONS

Correlating a WRRF's operation with higher life form (protozoa and metazoa) presence and abundance will provide a more detailed representation of what you should expect from your own biomass. There are certain organisms that we know become more apparent during higher loading rates, such as flagellates and amoeba. This may also occur with a younger sludge age because these organisms typically reproduce faster than many of the other protozoa and metazoa and require a larger amount of free motile bacteria to survive. As the organics are reduced and more floc becomes present, ciliates such as free-swimmers and crawler ciliates become abundant. This will also progress to stalked ciliates, suctoria, and, as the sludge age increases, the presence of metazoans such as rotifers. The primary key is how this relates to your system, and the only way to use that is to collect your own data with microscopic analysis.

In most properly functioning biological wastewater treatment systems, a hierarchy of microscopic organisms exists. These organisms include bacteria, fungi, algae, protozoa, and metazoa. Together, they are responsible for effective wastewater treatment. Bacteria do the majority of the work by using soluble and particulate organic substrates as sources of carbon and energy.

The growth and performance of the bacteria depend on the growth pressures that constitute their presence, such as their food, temperature, pH, and DO. These changes may be observed in floc formation, presence of filamentous bacteria, or even in the sudden presence of dispersed or free motile bacteria. It is easy for someone with an understanding of basic microscopic observation to recognize the difference between a dense or weak floc. Bacteria can be found as individual cells in the bulk water (free bacteria), as filaments (growing in long chains), or, ideally, as flocculated particles (floc). The dominant form often has a direct correlation to how well the wastewater treatment system is performing. Because identification of most bacteria requires more advanced techniques, using higher life forms in the biomass such as protozoa (the majority are single-celled organisms) and metazoa (multi-cellular) makes a simple microscopic observation much easier. Both these groups of organisms may be used as indicators of steady-state or non-steady-state operation. An operator who can determine what the organism present means to the overall

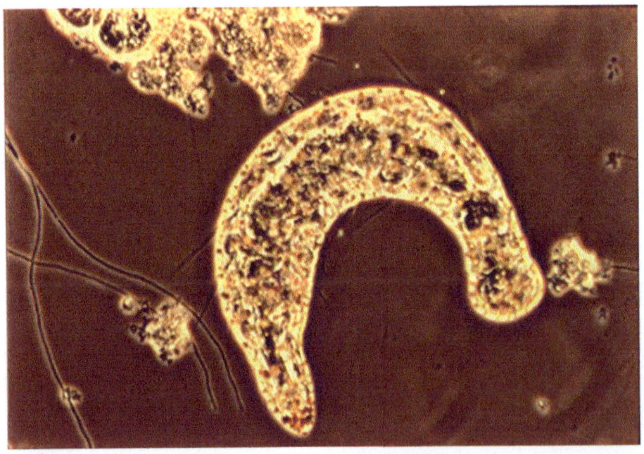

Figure 4.15 Aeolosoma Worm (WEF, 2017b) (Reprinted with permission from Ronald G. Schuyler)

biomass will have a better chance of understanding or preventing significant upsets. The difficulty becomes the understanding and transfer of this knowledge from facility to facility. Not all facilities will have a wide variety of higher life forms, but they may still operate ideally. Just because you do not see crawler ciliates or stalked ciliates and your performance is optimal, there is no reason to panic. The best organisms are often the organisms you see when your facility is running the best. Therefore, observing your biomass when the system is running well will help provide a good baseline of what is common. In addition, the distribution of protozoa can vary from week to week as can be seen in Figure 4.16, which shows an example of typical protozoa successions in an activated sludge process.

Another key in using higher life forms as indicators is diversity. The more diverse their population, the healthier and more stable the overall biomass will be. Protozoa and metazoa are orders of magnitude larger than bacteria and can be easily observed microscopically. The majority of these organisms are affected by BOD and ammonia concentrations, sludge age, and toxic shocks. Many higher life forms flourish and decline in certain environmental conditions. They can also be found in different forms depending on their surrounding conditions. As such, the appearance or lack of these organisms provides a basis for understanding how well the microlife is performing. For example, the majority of protozoa reproduce through binary fission, resulting in two daughter cells. Therefore, the presence and abundance of daughter cells may be a function of ideal environmental conditions. Another example is the presence of protozoan or metazoan cysts. Many higher life forms have the ability to produce a cyst (a protective coating) that enables the organism to withstand unfavorable conditions. An operator who recognizes the presence of cysts may realize that the biomass is experiencing a change, possibly because of an organic shock load. If the cysts are identified quickly enough, operational changes might be made to reduce the effect of the shock. In WRRFs with no influence of industrial waste, it would not be common to see cysts unless it was a function of reproduction. However, the industrial system with variations in substrate or operations could see cysts on a regular basis. An important point to make is that cysts take time to form, so a quick toxin dump or shock to the system will not produce cysts. The change has to be a slower response for an organism to encyst.

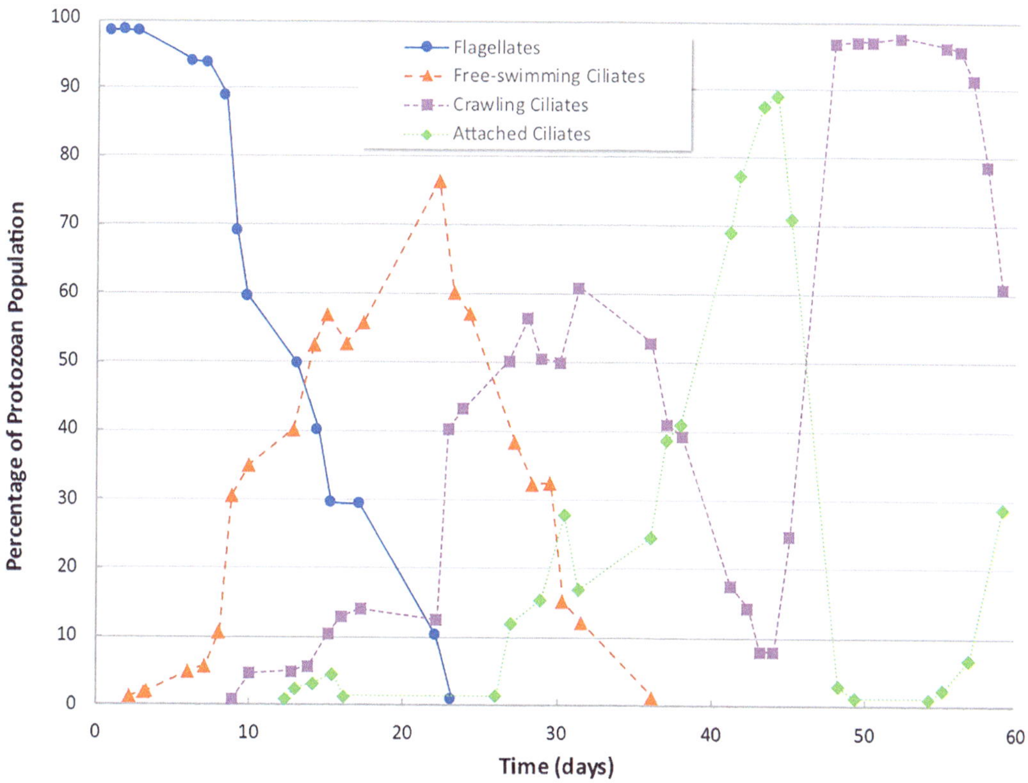

Figure 4.16 Protozoa Successions in an Activated Sludge Facility (Data from WEF, 2017b)

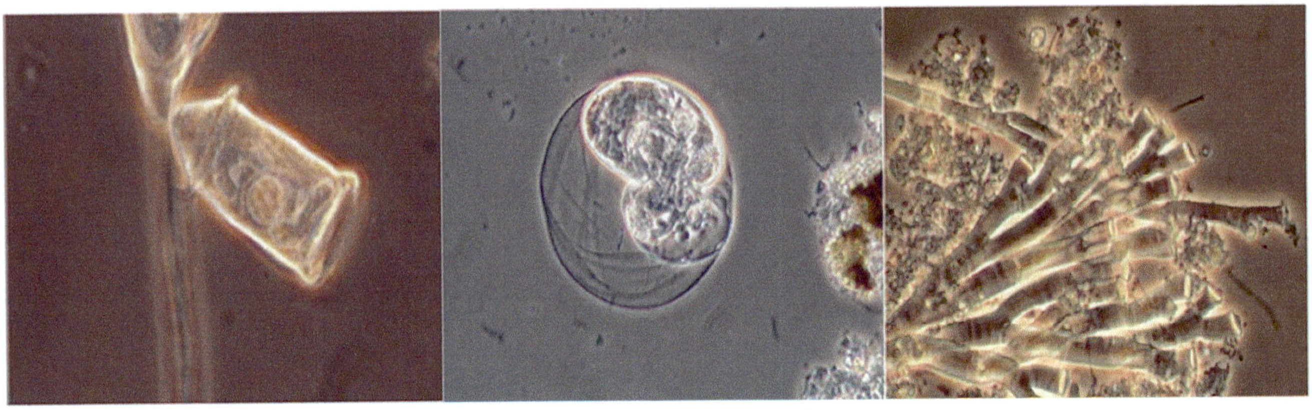

Daughter cell with trochal band | Cyst | Empty stalks

Figure 4.17 Daughter Cell With Trochal Band, Cyst, and Empty Stalks (WEF, 2018) (Reprinted with permission from Steven C. Leach)

A final example is the presence of detached stalked ciliates. If it is a reaction to reproduction, you may be able to view elongated nuclei. If not, chances are the growth pressures have changed slowly and the protist is detaching to find a more favorable environment. If an operator has taken a fresh sample of the biomass and placed it under the microscope with no delay, it is a real-time example of what is happening in the basin at that time. If you observe an entire slide showing detached stalked ciliates (Figure 4.8e) or empty stalks (Figure 4.17), it is very likely the biomass is experiencing a shock to the system and further testing or analysis is needed.

An important point to recognize is that some facilities may never have a large diversity of protozoa or metazoa, yet their treatment system runs very well. Every facility is its own environmental system, with various growth pressures and operating conditions. The presence or absence of higher life forms will not tell you exactly what is wrong with the system but does suggest to you something is happening; therefore, a further in-depth look at the operation may be needed.

TEST YOUR KNOWLEDGE

1. Where should a microscopic exam be performed?
   a. Aeration basin
   b. Clarifiers
   c. Return activated sludge (RAS)
   d. Waste activated sludge (WAS)
2. Focusing a microscope is accomplished by changing the position of the body tube with respect to the position of the stage.
   ☐ True
   ☐ False
3. The resolving power, or resolution of an optical system, is a measure of how many detectable colors can be seen with the microscope.
   ☐ True
   ☐ False
4. Phase contrast microscopy is of value in the study of opaque microorganisms and is routinely used for examining organisms like rotifers and nematodes that cause bulking.
   ☐ True
   ☐ False
5. The objective and ocular lenses of a microscope are situated side-by-side, and the objective lens is always located under the specimen.
   ☐ True
   ☐ False
6. What are the four focus points in performing a microscopic exam?
7. One micron meter (μm) equals 1/1000 of a meter or 1/10 of a millimeter.
   ☐ True
   ☐ False

8. One of the primary differences between the low-magnification stereoscopic microscope and the high-magnification compound microscope is that the stereoscopic microscope typically uses UV light and the compound microscope typically uses refracted light.
   - ☐ True
   - ☐ False
9. What is the difference between protozoa and metazoa?
10. Suctoria are good indicators of nitrification.
   - ☐ True
   - ☐ False
11. Metazoa found in the activated sludge process are believed to enter the system through infiltration of the collection system.
   - ☐ True
   - ☐ False
12. What is the difference between a free-swimming ciliate and a crawler ciliate?
13. Are rotifers protozoa or metazoa?
14. Protozoa can eat all the bacteria in an aeration basin.
   - ☐ True
   - ☐ False
15. A phase contrast microscope will provide more detail than a typical bright-field microscope.
   - ☐ True
   - ☐ False
16. Under proper environmental conditions, what physical characteristic allows bacteria to grow so quickly?
17. Metazoa found in the activated sludge process are believed to enter the system through infiltration of the collection system.
   - ☐ True
   - ☐ False
18. Lower life forms in activated sludge, such as individual bacteria cells, typically provide insight to the biomass age and stability.
   - ☐ True
   - ☐ False
19. The activated sludge process selects for organisms that swim well and consume the bacteria that create the sludge floc.
   - ☐ True
   - ☐ False
20. Nematodes can be found in the effluent of a WRRF due to their resistance to chlorination.
   - ☐ True
   - ☐ False
21. Annelids are segmented worms with soft, muscular body walls.
   - ☐ True
   - ☐ False
22. In high enough concentrations, bristle worms can cause red or pink colored foam conditions.
   - ☐ True
   - ☐ False
23. In activated sludge facilities, it has been reported that when ciliates are present, only 20% to 50% of the *E. coli* are removed.
   - ☐ True
   - ☐ False
24. A WRRF may never have a large diversity of protozoa or metazoa, yet it operates very well.
   - ☐ True
   - ☐ False
25. Aeolosoma organisms have been known to reproduce so rapidly they can change the color of the MLSS and cause rapid settling conditions.
   - ☐ True
   - ☐ False

Sequencing Batch Reactors

INTRODUCTION

Sequencing batch reactors (SBRs) are a variation of the activated sludge process. Sequencing batch reactor systems differ from activated sludge facilities because they combine biological treatment and solids settling and clarification into a single basin, or tank, whereas conventional facilities rely on multiple basins. An SBR is a variable volume fill-and-draw batch activated sludge system. Settling occurs when the air and mixers are turned off and a decanter provides for the withdrawal of treated effluent. According to a 1999 U.S. Environmental Protection Agency (U.S. EPA) report, an SBR is no more than an activated sludge facility that operates in time rather than space. Sequencing batch reactor systems involve introducing a quantity of waste to a reactor, providing an adequate time period for the treatment of the waste, and then discharging a volume of effluent plus waste sludge that is essentially equal to the original volume of waste introduced to the reactor. Sequencing batch reactor basins are equipped with both mixing and aeration systems that are operated independently. Their capability to either "mix only" without introducing oxygen to the reactor, or to mix and supply oxygen simultaneously, provides the operator with a powerful tool to manipulate the SBR; it also provides flexible operational configuration variations to meet the requirements of specific waste treatment applications.

Table 4.4 Typical Parameter Values for Low-Rate and High-Rate Sequencing Batch Reactors (Reprinted with permission from Aqua-Aerobics Systems, Inc.)

Parameter	Low-Rate SBR	High-Rate SBR
Food : mass (1/d)	0.05–0.1	0.15–0.4
Treatment cycle duration (h)	4.8–6.0	4.0
Typical LWL MLSS (mg/L)	4000–4500	2000–2500
Hydraulic retention time (h)	14–24	6–14

Sequencing batch reactors proceed through a series of phases for every cycle and will typically complete 4 to 6 cycles per day per SBR tank for domestic wastewater treatment. The SBR treatment cycle can be adjusted to undergo aerobic, anaerobic, and anoxic conditions to achieve biological nutrient removal (BNR), including nitrification, denitrification, and phosphorus removal. Typically, 50% to 75% of the liquid volume of the SBR containing settled biomass is retained at the end of every cycle, although a higher percentage may be necessary depending on effluent requirements. The minimum liquid level determines the volume of sludge inventory that can be retained in the SBR reactor and the maximum volume of influent that can be accommodated for any individual cycle. Biochemical oxygen demand and total suspended solids (TSS) levels of less than 10 mg/L can be achieved consistently. Total nitrogen limits of less than 3 mg/L can also be achieved by aerobic conversion of ammonia to nitrates (nitrification) and anoxic conversion of nitrates to nitrogen gas (denitrification) within the same tank. Low phosphorus limits of less than 1 mg/L can be attained by using a combination of biological treatment (anaerobic phosphorus absorbing organisms) and chemical agents (metals salts) within the vessel and treatment cycle. Discharge limits that require a greater degree of treatment may necessitate the addition of a tertiary filtration unit following the SBR treatment phase.

Discrete cycles are used during prescribed, programmable time intervals, and MLSS remains in the reactor during all cycles. For conventional systems, there are five steps that are carried out: fill, react, settle, decant, and idle (Figure 4.18). As a result of the batch nature of the process, flow equalization and/or multiple

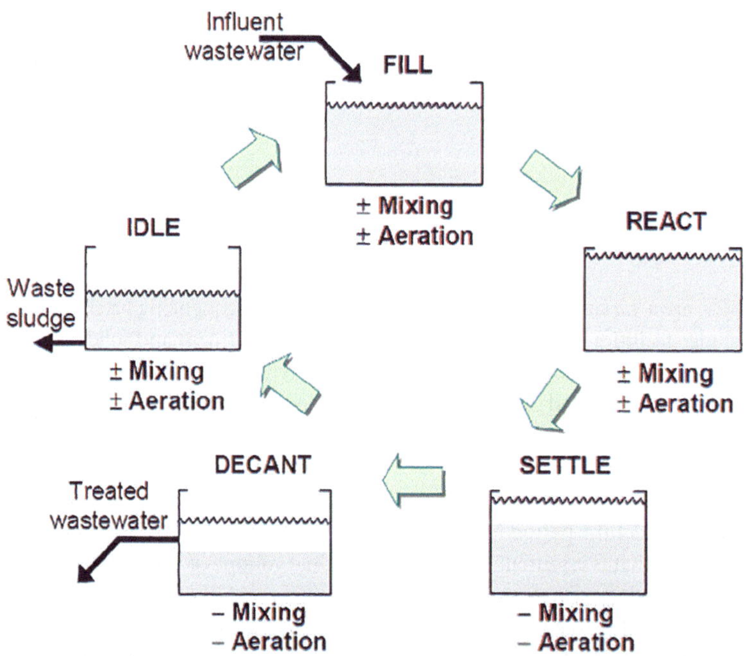

Figure 4.18 Typical Sequencing Batch Reactor Operating Phases (University of Florida, TREEO Center, 2000) (Reprinted with permission from UF TREEO Center)

reactors must be provided to accommodate continuous-flow operation. Mixing and aeration are provided by equipment similar to that found in conventional continuous-flow facilities. Secondary clarifiers are not required. Instead, specially designed decanters are provided in each reactor for effluent withdrawal.

Sequencing batch reactor systems are typically designed for small-to-medium flow applications. According to the Water Environment Federation et al. (2018), more than 1000 SBR-type facilities are operating in the United States. Approximately 80% of these have flows of 4000 m^3/d (1 mgd), or lower, and few, if any, are as large as 40 000 m^3/d (10 mgd). A small number of larger facilities have been constructed outside of the United States, ranging from 190 000 m^3/d or 50 mgd to 490 000 m^3/d or 130 mgd.

FLOW-PACED (FLOW CONTROL) AND TIME-BASED (TIME CONTROL) OPERATION

Under a flow-paced batch system, a facility receives the same volumetric loading and approximately the same organic loading during every cycle. The SBR basin already has stabilized supernatant and settled MLSS in it, which dilutes the batch of incoming influent. Typically, flow-paced systems are designed to handle the average design flow, and the cycles are automatically adjusted to handle higher flows. As the flow changes, the aeration time is adjusted proportional to the flow. Time-based operation provides the flexibility of a fixed cycle regardless of the flow entering the system. The operator has full control of modifying the cycle, as needed, to meet the required effluent. In a time-based operation, the batch integrity is maintained at average flows and maximum day flows. In essence, an adequate decant volume is provided to process the maximum design flow per cycle while maintaining the entirety of each and every phase of operation up to the maximum design flow condition. This results in a system that processes a variable volume per cycle at flow conditions less than the maximum design value. In the event that a period of sustained flow exceeding the maximum design value enters the system, the level sensor system will eliminate some fraction of each affected phase of operation with the exception of the settle phase. Table 4.5 presents a comparison of the two aforementioned operational strategies. Because of the inherent dynamic nature of a batch operating system, for an SBR system to be effective, the facility must have proper monitoring, allow operators to adjust the cycle time, and employ knowledgeable operators who are properly trained to make the necessary adjustments to the cycle (New England Interstate Water Pollution Control Commission, 2005).

SEQUENCING BATCH REACTOR SYSTEM OPERATIONAL PHASES

The description of the operating cycle of a fill-and-draw SBR is useful to understand the SBR treatment process functions. The fill-and-draw SBR operation involves the following five basic steps: fill, react, settle, decant, and idle/sludge waste (Figure 4.18).

Table 4.5 Comparison of Flow Control and Timed Control Strategies (University of Florida, TREEO Center, 2000) (Adapted with permission from UF TREEO Center)

Flow Control	Timed Control
Total cycle length is variable for a variable influent flow to accommodate a fixed volumetric loading.	Total cycle length is fixed.
Fill continues until TWL is reached during each cycle.	Because of the fixed fill duration, the water level typically stops short of TWL.
A cycle may start at any time of the day, depending on the lengths of previous cycles.	The cycle starting times are fixed relative to the 24-hour clock.
MLSS may be routinely measured at TWL.	The bottom water level or, perhaps, the mid-water level, may be used as a reference point for MLSS measurement
Full depth is available for aeration during the react phase.	Partial depth is typically available for aeration during react phase.
React phase is shortened or eliminated to accommodate storm flows.	A special (shorter) storm cycle is used for storm flows.

FILL PHASE

During the fill phase, the basin receives influent wastewater. The influent brings food to the microbes in the activated sludge, creating an environment for biochemical reactions to take place.

Time Duration

The overall length of the fill phase may be determined by a timer (timed control) or by water depth (flow control), as described below. Under the timed control strategy, wastewater flow into the reactor is stopped after a certain period of time has elapsed. Under timed control, the fill phase is ended whether or not the reactor liquid level has reached the top water level (TWL). The alternate control strategy is flow control. In this case, the fill phase continues until the liquid level reaches the TWL. Once the fill phase is completed, the influent wastewater must be directed to either another SBR reactor or to a holding/surge tank.

Mixing and Aeration

Under a static-fill scenario, there is no mixing or aeration while the influent wastewater is entering the tank. Static fill is used during the initial startup phase of a facility, at facilities that do not need to nitrify or denitrify, and during low flow periods to save power. Because the mixers and aerators remain off, this scenario has an energy-savings component. Static fill can be used to enhance biological phosphorus removal during this anaerobic time period. Depending on the manufacturer, the influent wastewater can be fed through a manifold at the bottom of the basin, so that the wastewater is fed directly into the settled sludge blanket.

Under a mixed-fill scenario, mechanical mixers are active, but the aerators remain off. The mixing action produces a uniform blend of influent wastewater and biomass. Because there is no aeration, an anoxic condition is created, which promotes denitrification using nitrate produced during the previous batch. Anaerobic conditions can also be achieved during the mixed-fill phase when the nitrate- and nitrite-nitrogen concentrations approach zero. Under anaerobic conditions, the biomass undergoes a release of phosphorus. This released phosphate is re-absorbed by the biomass once aerobic conditions are reestablished.

Under an aerated-fill scenario, both the aerators and the mechanical mixing unit are activated. The contents of the basin are aerated to convert the anoxic or anaerobic zone to an aerobic zone. No adjustments to the aerated-fill cycle are needed to reduce organics and achieve nitrification. However, to achieve denitrification, it is necessary to switch the aeration off to promote anoxic conditions for denitrification. By switching the aeration on and off during this phase with the blowers, aerobic and anoxic conditions are created, allowing for nitrification and denitrification. Dissolved oxygen should be monitored during this phase, and should not be higher than 0.2 mg/L. This ensures that an anoxic condition will occur during the idle phase.

REACT PHASE

This phase allows for further reduction or "polishing" of wastewater parameters. During this phase, the influent valve is closed and no additional wastewater enters the basin, and the mechanical mixing and aeration units are running. Because there is no addition or organic loading, the rate of organic removal increases dramatically. Most of the carbonaceous BOD removal occurs in the react phase. Further nitrification occurs by allowing the mixing and aeration to continue; the majority of denitrification takes place in the mixed-fill phase. The phosphorus released during static or mixed-fill phases, plus some additional influent phosphorus, is taken up during the react phase. This phase typically ends under aerobic conditions to limit problems with denitrification and phosphorus release during the subsequent settle and decant phases.

Repetitive "on/off" cycling of the oxygen supply during the react phase will typically produce a DO pattern that reveals an increased time requirement for the DO to decline to the zero level; this is due to the progression of treatment and the resulting reduced microbial oxygen uptake rate (OUR). A timer typically fixes the length of the react phase, regardless of whether timed control or flow control is used. Flow-control SBR systems may shorten or skip the react phase when they are receiving storm flows and timed-control systems switch to a special cycle when receiving high flows.

SETTLE PREPARATION AND SETTLE PHASES

At the end of the react phase, some SBR systems incorporate a brief mixing-only phase called *settle preparation* or *settle prep*. During this short 5- to 10-minute period, the aeration blowers are off, but basin mixing is allowed. This gentle stirring action allows bound gases such as carbon dioxide, nitrogen, and excess DO to be released from the activated sludge flocs. Settle prep enhances the ability of the MLSS to settle and compact during the settle phase. Settle prep is also helpful when mixed liquor temperature decreases during colder seasons, making the mixed liquor denser. Cooler water conditions allow higher dissolved gas saturation, further interfering with MLSS settling.

During the settle phase, activated sludge is allowed to settle and compact under quiescent conditions—no flow enters the basin and no aeration and mixing takes place. The activated sludge tends to settle as a flocculent mass, forming a distinctive interface with the clear supernatant. This phase is a critical part of the cycle because, if the solids do not settle efficiently, some sludge can be drawn off during the subsequent decant phase, thereby degrading effluent quality.

DECANT PHASE

During this phase, a decanter is used to remove the clear supernatant effluent. Once the settle phase is complete, a signal is sent to the decanter to initiate the opening of an effluent-discharge valve. There are floating and fixed-arm decanters. Floating decanters maintain the inlet orifice slightly below the water surface to minimize the removal of floatable solids entering the effluent removed during the decant phase. Floating decanters offer the operator flexibility to vary fill-and-draw volumes. Fixed-arm decanters are less expensive and can be designed to allow the operator to lower or raise the level of the decanter. It is optimal that the decanted volume is the same as the volume that enters the basin during the fill phase. It is also important that no surface foam or scum is decanted. The vertical distance from the decanter to the bottom of the tank should be maximized to avoid disturbing the settled biomass. Operators should also monitor the vertical distance from the top of the settled sludge blanket to the decant ports and ensure that the decanter's subsurface ports do not get close enough to draw the lighter settled solids into the effluent flow.

IDLE AND SLUDGE WASTING PHASE

This step occurs between the decant and the fill phases. The time varies, based on the influent flowrate and the operating strategy. During this phase, a small amount of activated sludge at the bottom of the SBR basin is pumped out.

OPERATION OF MULTIPLE SEQUENCING BATCH REACTORS

In dual reactor systems, the total time allotted for all fill phases of operation is equal to the total time designated for the react, settle, decant, and idle, plus sludge waste phases. This allows for a continuous flow to the "system" and the introduction of 100% of the flow to each reactor for 50% of the time in a true batch SBR system (see Figure 4.19). Additional variations of the flow schematic and mode of operation, possibly featuring the direction of 100% of the flow to three reactors for 33.3% or four reactors for 25% of the total treatment cycle time, are also feasible.

Flow is typically entering a treatment system for 24 hours of each day. Each reactor in a dual reactor system would receive flow for a 12-hour period in either a single 12-hour fill phase or in multiple intermittent fill phases that, in summation, total 12 hours. Similarly, with treatment cycles of 6 hours or less of total duration, the react fill and react phases of operation will also be established in combinations that are equal to or less than 50% of the total cycle time. This combination of phase durations permits the use of a common air supply system that may be sequenced from basin to basin to provide aeration during the entire time dedicated to the react fill and react phases or some percentage of that dedicated time. Treatment cycle durations of greater than 6 hours typically use separate air supply systems dedicated for service to each individual reactor.

When possible, fill cycles of individual SBRs should be arranged to spread the peak loads to multiple SBRs and to avoid overloading individual SBRs. For example, an SBR receiving influent from 12:00 A.M. to 6:00 A.M. will see much less BOD and ammonia than an SBR receiving influent from 6:00 A.M. to 12:00 P.M.

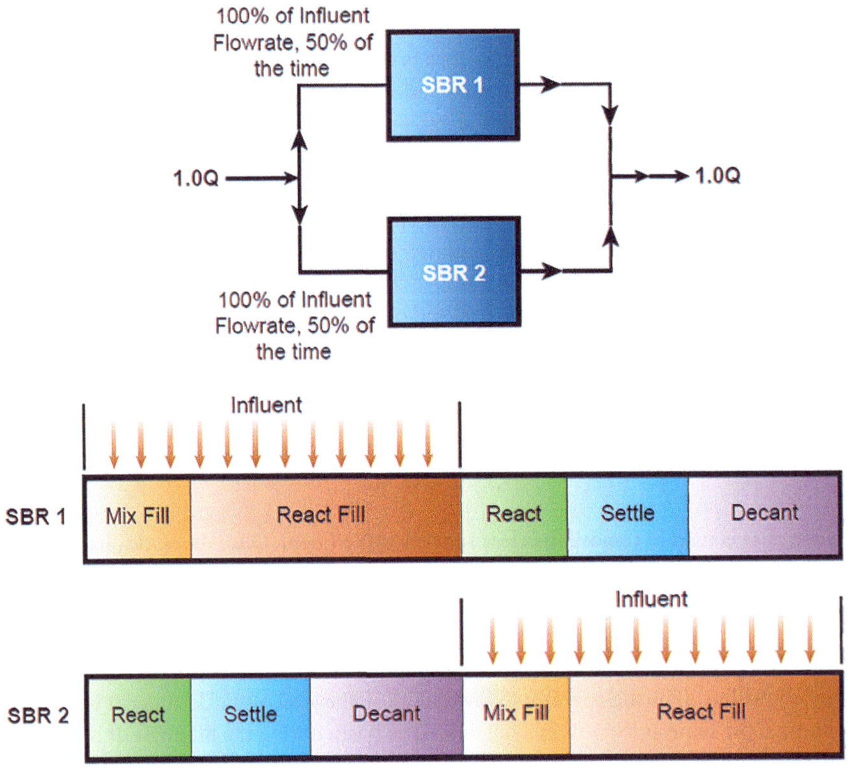

Figure 4.19 Two Sequencing Batch Reactor Basin Systems (Adapted with permission from Aqua-Aerobics Systems, Inc.)

NITROGEN REMOVAL

For nitrogen removal, the fill and react phases are broken down into static fill, mixed fill, and mixed react. Carbon oxidation and nitrification occur in the aerobic react phase and denitrification will happen in the anoxic fill and react phase. Denitrification results from selecting static fill, mixed fill, and mixed react periods that are thus creating anoxic conditions. Intermittent aeration during the react phases, with long enough non-aerated intervals to allow use of all DO, generates the anoxic conditions for denitrification to take place and reduce effluent nitrate- and nitrite-nitrogen values to the range of 0.50 to 2.0 mg/L. Effective denitrification is important when the treatment process is required to achieve effluent total nitrogen concentration values of 3.0 to 5.0 mg/L. In most cases, a highly efficient nitrification design will produce effluent NH_3-N values of approximately 0.50 mg/L. Additionally, in most cases, the requirement for a low organic nitrogen content in the effluent dictates a low organic loading to the treatment system, which results in a greater sludge age and promotes efficient nitrification. A sequence of aerobic conditions followed by anoxic conditions serves to complete the nitrification and subsequent denitrification requirements during the react phase. Figure 4.20 presents an example of this operational strategy.

PHOSPHORUS REMOVAL

As a general principle, the use of a static or mixed fill phase of a specific duration will provide an anoxic to anaerobic environment in the reactor when operating with a design F:M of 0.05 to 0.10 kg/kg·d while treating typical municipal wastewater. Obviously, the ability to manipulate the reactor environment and produce a greater concentration of phosphorus in the waste solids plus the rate of sludge wasting will affect the degree of biological phosphorus removal. Presently, evidence presented by other investigations suggests that an increased organic loading level and the existence of soluble BOD_5 to phosphorus content ratios of 15 to 20:1 in the raw waste may result in a greater degree of phosphorus removal based on attaining a greater concentration of phosphorus within the system cell mass.

The structure of any given treatment cycle allows for the incorporation of an initial mixed fill phase time value that may be set at an appropriate duration to promote microbial release of phosphorus before the subsequent uptake of phosphorus during aerobic conditions.

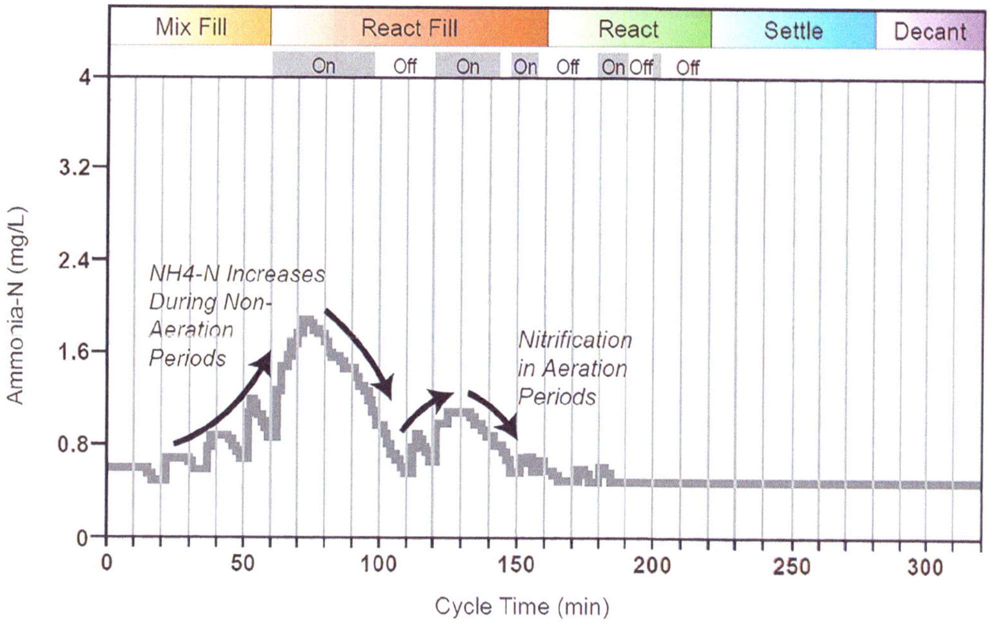

Figure 4.20 Example On-Off Aeration Pattern for Nitrification-Denitrification (Data with permission from Aqua-Aerobics Systems, Inc.)

In an SBR, the anaerobic and aerobic environments required for enhanced biological phosphorus removal (EBPR) are created in the same tank but not at the same time, which is the case in a plug flow reactor. An anaerobic reaction period can be developed during the static or mixed fill phases and during the initial react periods by removing sufficient nitrate following the aeration period. It is also possible to minimize nitrate concentration by incorporating several alternating aerobic and anoxic periods during the react period. Figure 4.21 presents an example of this operational strategy.

Sequencing batch reactors can achieve biological phosphorus and nitrogen removal by programming extended air-off periods to achieve anaerobic conditions during each cycle. Alternatively, a dedicated anaerobic zone and internal recycle can be provided (Figure 4.22).

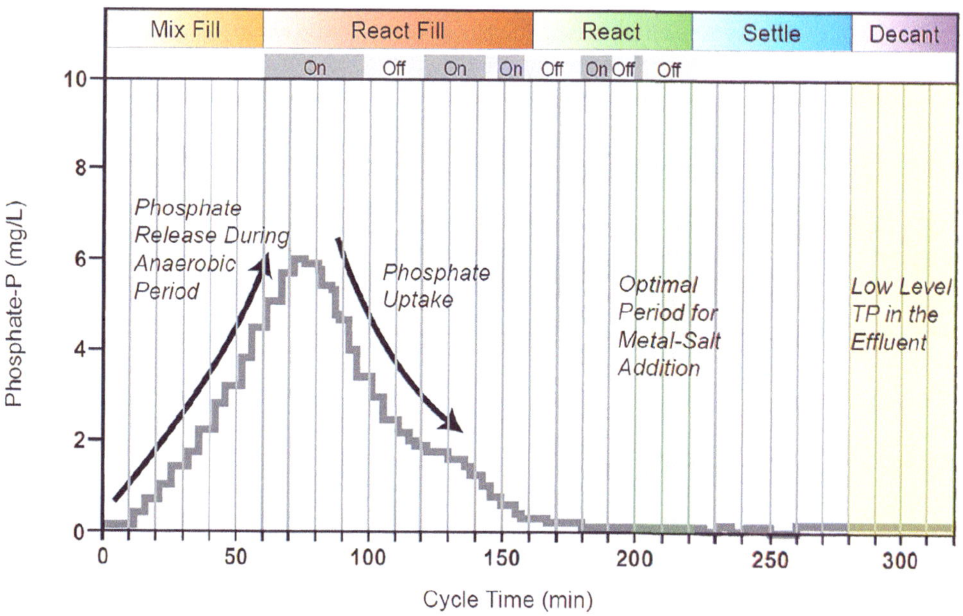

Figure 4.21 Example Operational Strategy for Phosphorus Removal (Data with permission from Aqua-Aerobics Systems, Inc.)

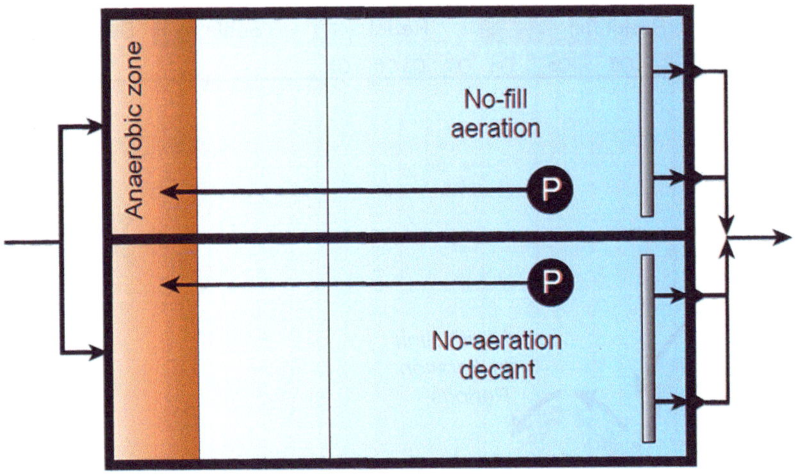

Figure 4.22 Biological Nutrient Removal Sequencing Batch Reactor Configuration With Dedicated Anaerobic Zone and Internal Flow Recirculation for Phosphorus Removal

APPROXIMATION OF STEP-FEED AND CONTACT-STABILIZATION PROCESSES

The SBR fill-and-react cycles can be modified to mimic many activated sludge process configurations. The nitrogen removal and phosphorus removal processes have already been discussed, but step-feed and contact-stabilization processes can also be approximated by delaying feeding the reactor until an RAS aeration period during the react cycle has been completed. Similar to taking samples along the plug flow biological reactor, samples can be taken at different time intervals during the react period and tested for BOD, ammonia, or DO to evaluate the efficiency of the treatment process. Based on these results, aerated and unaerated react times can be adjusted to increase the efficiency of the process.

SEQUENCING BATCH REACTOR EFFLUENT FLOW EQUALIZATION

Both batch inflow and continuous inflow SBR processes discharge effluent during only part of their operational cycle. The effluent flowrate during the decant phase will thus be higher than the influent flowrate. Post-basin effluent equalization smooths out flow variations before downstream processes such as filtration and disinfection (Figure 4.23). By providing storage and a constant smooth flow, effluent filtration and disinfection processes will operate more efficiently. Post-basin effluent equalization also allows downstream processes to be sized smaller because the flow from the basin is metered out and does not hydraulically surge the downstream processes. Effluent equalization also ensures that there are not large variations in operating ranges for the metering pumps and the chlorine analyzers. Ideally, the basin should be of sufficient size to hold a minimum of two decantable volumes. There should be a means of

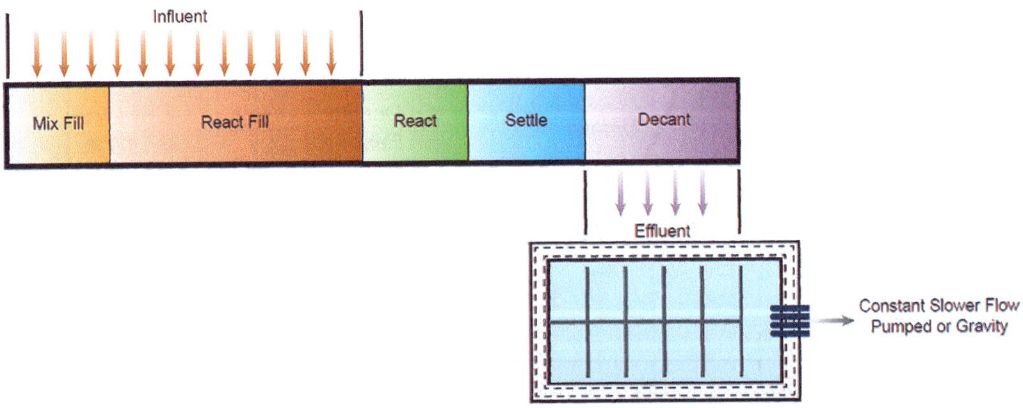

Figure 4.23 Sequencing Batch Reactor Effluent Flow Equalization Tank for Providing Constant Flow to Downstream Effluent Filtration and Disinfection (Reprinted with permission from Aqua-Aerobics Systems, Inc.)

returning the liquid from the post-flow equalization basin to the headworks if a poor decant occurs. These basins should also have a means of removing solids from the bottom of the unit, such as a sloped bottom with a drain or sump.

Note that if the decant time equals the fill time, post-equalization may not be required. This is more common on four-basin SBR systems or larger.

CONTINUOUS-INFLOW SEQUENCING BATCH REACTOR

The continuous-inflow SBR system includes the same equipment, plus a baffle near the influent that minimizes short-circuiting of untreated wastewater (Figure 4.24). This allows a single SBR to receive continuous inflow and bypass the need for multiple SBRs in parallel and/or influent flow equalization.

RELATED SYSTEMS

PRELIMINARY TREATMENT

Sequencing batch reactors generally do not have primary settling tanks; therefore, effective removal or exclusion of grit, debris, plastics, excessive oil or grease, and scum, as well as screening of solids, should be accomplished before the activated-sludge process. Bar screens or mechanical screens should be used to benefit the whole treatment process, and especially the settling phase. Excessive influent debris should not be present to interfere with the solids' ability to settle, thus ensuring a high-quality sludge blanket. Screens also provide protection for the pumps, submerged mixers, diffusers, and instrumentation probes.

Influent flow equalization could be beneficial where significant variations in flowrates and organic mass loadings are expected. Flow equalization is also important if a facility is expected to receive a significant amount of septage or is taking in a significant amount of industrial wastes, so that the septage is not slug loaded and potentially creates a risk of treatment process upset. Influent flow equalization is also very useful when a facility needs to achieve nitrification and denitrification. It is important to note, however, that the size of the influent equalization basin must be carefully considered because an oversized basin can cause negative effects on downstream treatment processes. In the case of a single SBR, the equalization tank is typically large enough to contain all influent flow while the SBR tank is in its react, settle, or decant phases (typically 8 or 12 hours). Coarse-bubble aerators are typically used in the equalization tank to keep wastewater solids suspended and maintain aerobic conditions to minimize odors. In dual basin systems, it also allows for one basin to be taken offline for maintenance or during low load months. Equalization may also allow scum and grease removal at a point before it enters the SBR tank. A way for removing scum, grease, and floatables should be provided in the equalization tank. An equalization tank also allows facilities that must denitrify to maintain a more constant wastewater carbon-to-nitrogen ratio to help ensure that an adequate amount of carbon is available in the denitrification fill phase. Finally, it allows for an equal flow volume into the basin, keeping the F : M fairly stable.

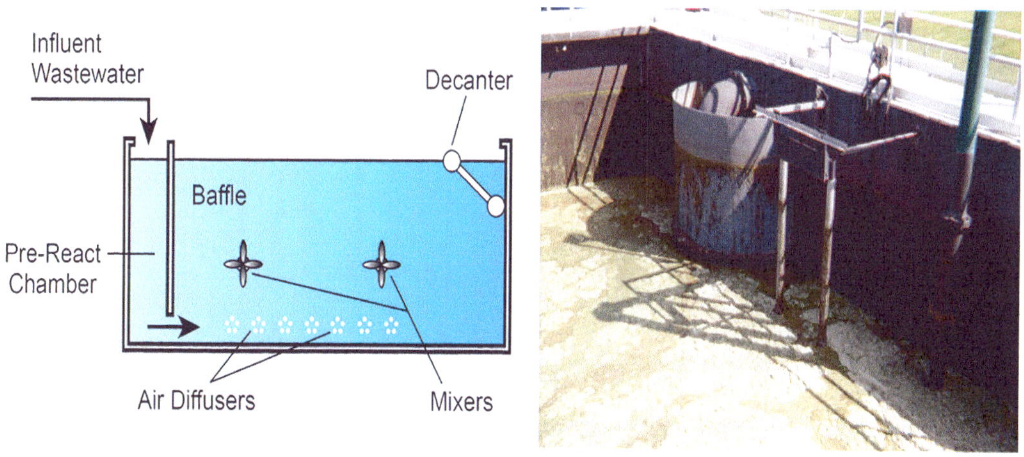

Figure 4.24 Continuous Inflow Sequencing Batch Reactor (University of Florida, TREEO Center, 2000) (Reprinted with permission from UF TREEO Center)

ALKALINITY ADDITION

Nitrification consumes alkalinity, and with a significant drop in alkalinity, the mixed liquor pH will also drop. If the facility influent and the mixed liquor have adequate alkalinity, pH does not change, so it does not need to be raised. Warm weather temperatures and increased mixed liquor temperature encourage nitrification in SBR processes, particularly if they are lightly loaded and there is sufficient aeration. In these cases, it is desirable to operate SBRs so that they also denitrify, even when there is no requirement for nitrogen removal. This allows for recovery of alkalinity through denitrification. When nitrification occurs at SBR facilities, it often occurs during periods of low diurnal flow when a facility may not be staffed. If no testing or chemical addition is available to compensate for an alkalinity decrease, the pH in the SBR unit could drop and cause process upsets.

Chemicals that raise alkalinity, such as sodium bicarbonate and sodium carbonate (soda ash), are recommended over sodium hydroxide. Sodium hydroxide does not raise alkalinity; however, it does raise pH. Ideally, facilities should provide piping for adding alkalinity at both the influent equalization basin and into each SBR basin. It is also desirable to be able to measure alkalinity at each location. Alkalinity addition should be based on the amount measured during the decant phase, not on incoming flow. Alkalinity should be kept in a range of 40 to 70 mg/L as $CaCO_3$ before the decant phase to ensure the nitrification cycle is complete.

The management of both pH and alkalinity are critical to the effective operation of an SBR. Sufficient alkalinity must be present to allow complete nitrification and result in a residual of at least 50 mg/L in the decanted effluent. The pH must be maintained in a manner to prevent it from falling below 7.0 in the reactor basin.

AERATION AND MIXING

Aeration devices are needed to transfer oxygen into the wastewater to allow aerobic degradation of waste and nitrification. The aeration devices must be capable of being cycled on and off. They must be left off during the settling phase and are typically cycled on and off during the idle, fill, and react phases. Aeration devices may be mechanical, diffused air, or jet (University of Florida, TREEO Center, 2000).

A blower is a significant component of the diffused air aeration process. When a single blower per basin is used, it should be sized to provide maximum aeration under worst-case conditions. These conditions typically occur in the summer months, when higher temperatures decrease the amount of oxygen that can be dissolved in wastewater. For facilities that use a single blower per basin, a variable frequency drive should be considered. In a facility configured with only one blower per basin, it is difficult to scale back on the aeration provided. With multiple smaller blowers, units can be taken offline when maximum aeration is not required. This results in electrical cost savings.

The diffusers can be fixed or retrievable for easier maintenance (Figure 4.25). Fine-bubble membrane diffusers are preferable to coarse-air bubble aeration. Fine-bubble diffusers transfer more oxygen to the water due to increased surface area in contact with water. The same amount of air introduced to a big bubble has less surface area in contact with water than an equal amount of air divided into smaller bubbles. The amount of surface area in contact with water is proportional to the amount of oxygen transferred into water. The depth of aerator also plays a part in oxygen transfer due to contact time. The deeper the

Figure 4.25 Fixed-Disc Diffusers With Floating Mixer and Decanter (Left) and Retrievable Tube Diffusers (Right) (Reprinted with permission from Aqua-Aerobics Systems, Inc.)

aerator, the longer it takes for the bubble to come to the surface and the more efficient the oxygen transfer to the liquid.

A jet aerator combines liquid pumping with air diffusion through combined liquid/air jet nozzles (Figure 4.26, left). If the air is turned off, mixing of the tank contents can still be maintained by continuing to run the pump to maintain the recirculated mixed liquor ejected at high velocity from the mixing jet nozzles.

A floating direct drive mixer (Figure 4.26, right) consists of a float-mounted motor that directly drives an impeller mounted on a shaft. This type of mixer maintains constant impellor submergence throughout the range of liquid depth. The unit is typically tethered in position using three cables and can be retrieved for servicing by pulling the unit to the side of the tank. A submerged mixer is typically mounted on a rail at a fixed position in the tank. The entire unit (motor, impeller) remains submerged throughout an operational cycle. Because its position is fixed relative to the tank wall, the mixer's submergence varies depending on the liquid depth. The submerged mixer is typically retrieved using a winch. A platform-mounted mixer typically consists of a motor, gearbox, shaft, and impeller. The shaft must be long enough to keep the impeller submerged throughout the range of liquid depth. The motor and gearbox are readily accessible for servicing.

DECANTERS

Decanters are used to remove treated supernatant after the settling phase is completed. They can either be located at a fixed depth in the tank or a flotation device can support them so that the liquid draw-off always remains close to the surface (University of Florida, TREEO Center, 2000).

The fixed decanter withdraws supernatant down to the bottom water level, which is determined by the design of the SBR tank and cannot be easily changed. During the non-decant phases, a positive air pressure is maintained in the decanter piping within the SBR tank to prevent mixed liquor from entering the decant piping, which would contaminate effluent at the start of the decant phase. A small air pump may be used to ensure maintenance of the pressure. At the beginning of the decant phase, the effluent valve is opened, releasing the air pressure and allowing supernatant to flow through the outlet pipe. In some cases, it is necessary to provide a solenoid-operated relief valve to prevent air locking of the outlet pipe. Fixed decanters allow more lineal feet of decanter, leading to a shorter decant time without disturbing the settled sludge blanket. Fixed decanters have no floating or submersed mechanical parts that can freeze in cold weather. Because the valves and other components of the fixed decanter are installed on the outside of the SBR tank, maintenance is facilitated.

Floating decanters (Figure 4.27) can be set to travel through different ranges of liquid depth, thus allowing the decant volume to be adjusted. A floating decanter uses a submerged circular weir that opens only during the decant phase and draws supernatant from below the water surface. To help ensure that mixed liquor does not enter the decanter during non-decant phases, positive pressure can be maintained in the outlet piping with a small air pump. Floating decanters prevent floating scum and debris from entering

Figure 4.26 Jet Aerator/Mixer Piping and Nozzle (Left) and Floating Direct Drive Mixer (Right) (University of Florida, TREEO Center, 2000) (Reprinted with permission from UF TREEO Center)

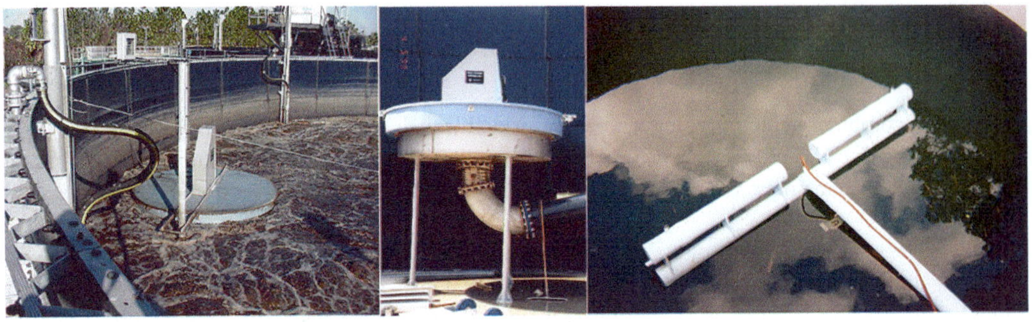

Figure 4.27 Floating Decanters (University of Florida, TREEO Center, 2000) (Reprinted with permission from UF TREEO Center)

the decant line. They also maintain the outlet weir at a constant submergence at all times during the decant phase. Because the depth of withdrawal does not change, the rate of withdrawal remains constant throughout the decant cycle. Floating decanters remove supernatant from 0.3 to 0.6 m (1 to 2 ft) below the bottom water level. This can lead to entrainment of sludge in the effluent in cases where the sludge blanket settles poorly. Floating decanters are susceptible to freezing of the mechanical parts in colder climates and may need to be equipped with heaters.

Care must be taken during design and construction to ensure that the waste sludge suction pipe extends to the bottom of the tank, away from the wall, to maximize the solids concentration that is picked up.

WASTE SLUDGE PUMPS

Waste sludge pumps may be either centrifugal or positive displacement (University of Florida, TREEO Center, 2000). Positive displacement pumps include screw, diaphragm, plunger, airlift, and pneumatic ejector. They may be submerged in the SBR tank or mounted outside the tank wall. The submersible pumps avoid the need to run pipes through the SBR walls. On the other hand, they require the discharge pipe to run over the wall, which can lead to air locking problems. Furthermore, the pump must be removed from the tank by a crane or winch for servicing. The externally mounted waste sludge pumps (Figure 4.28) are readily accessed for service but involve through-the-wall piping.

The smaller the design of a facility, in terms of flow, the more problematic temperature becomes. The smaller the design flow, the greater the likelihood that between 11:00 P.M. and 6:00 A.M. there will be little or no flow. This contributes to a loss of heat from the SBR basins to the surrounding atmosphere. Facilities that have flow throughout this timeframe lose less heat because the incoming flow sustains higher temperatures.

WET/COLD WEATHER ADJUSTMENTS AND OPERATIONAL PLANS

In general, wastewater temperatures are above freezing, but the batch mode of operation exposes the SBR basin to cold winter temperatures. The long, cold ambient air temperature in winter cools the content of a basin below the optimal temperature of 20 to 25 °C. Where practical, basins for small or very small systems or facilities should be housed in a garage-type structure to avoid freezing. Larger basins can also be covered to minimize heat loss. Exposed piping is insulated to protect from freezing and provisions are typically made to prevent the freezing of discharge pipes, decanter valves, and chemical lines and to prevent ice buildup on decanters. Process controllers are a critical component and are placed in dry areas that are not exposed to extreme temperatures.

Figure 4.28 External Wasting Pumps (University of Florida, TREEO Center, 2000) (Reprinted with permission from UF TREEO Center)

Communities that have combined collection systems or that are subject to sanitary sewer overflows during wet weather should consider developing wet weather operating plans or standard operating procedures. Snowmelt, rain, and infiltration and inflow (I&I) can drastically affect the way an SBR functions from a microbial standpoint. Influent oxygen levels as high as 5 mg/L, diluted BOD, and cold sewerage temperatures are all long-term spring occurrences in cold climate regions. The function of the SBR for nutrient removal requires control over the oxygen level during the various SBR phases. Loss of this control due to long periods of excessive I&I can limit the effectiveness of the nutrient-removal process.

CONTROL SYSTEM

Sequencing batch reactor systems rely on automated process logic controllers for their operation. Figure 4.29 presents a typical SBR process logic controller used to control the equipment such as influent flow, blowers, mixers, decanting, and sludge wasting. Figure 4.30 shows close views of an automated SBR valve.

ONLINE PARAMETERS MONITORED BY THE SUPERVISORY CONTROL AND DATA ACQUISITION SYSTEM

Oxidation–reduction potential (ORP), DO, pH, and alkalinity are parameters that should be monitored by the supervisory control and data acquisition (SCADA) system. Manufacturers determine what parameters can be monitored and controlled by the SCADA system. Monitoring of certain parameters is important, and the ability to adjust these parameters from a remote location is ideal. The operator needs to be able to add chemicals to raise the alkalinity and, subsequently, the pH. The setpoint should be an alkalinity value rather than pH-based.

Oxidation–reduction potential measures the electrical potential required to transfer electrons from one compound or element to another compound or element. Oxidation–reduction potential is measured in millivolts, with negative values indicating a tendency to reduce compounds or elements and positive values indicating a tendency to oxidize compounds or elements.

The operator should have the ability to fully control (i.e., modify) the facility-operating parameters such as, but not limited to, cycle times, volumes, and setpoints. Alkalinity monitoring and addition ensures that a pH of less than 7.0 does not occur. For facilities that nitrify and denitrify, ORP monitoring is desirable. Oxidation–reduction potential is the measure of the oxidizing or reducing capacity of a liquid. Dissolved oxygen varies with depth and location within the basin. Oxidation–reduction potential can be used to determine if a chemical reaction is complete and to monitor or control a process.

Operators need the ability to make changes that will modify these readings to achieve appropriate nutrient removal. Because nutrients are also reductants or oxidants, an ORP measurement can find its way to monitor nitrification, denitrification, and phosphorus removal. Ammonia (a reductant) is transformed to nitrate (an oxidant) with the corresponding increase in the ORP value. During denitrification, the elimination of nitrate reduces the ORP value.

General ranges are as follows: carbonaceous BOD (+50 to +250), nitrification (+100 to +300), and denitrification (–50 to +50). An ORP below –50 mV typically indicates anaerobic conditions. Figure 4.31

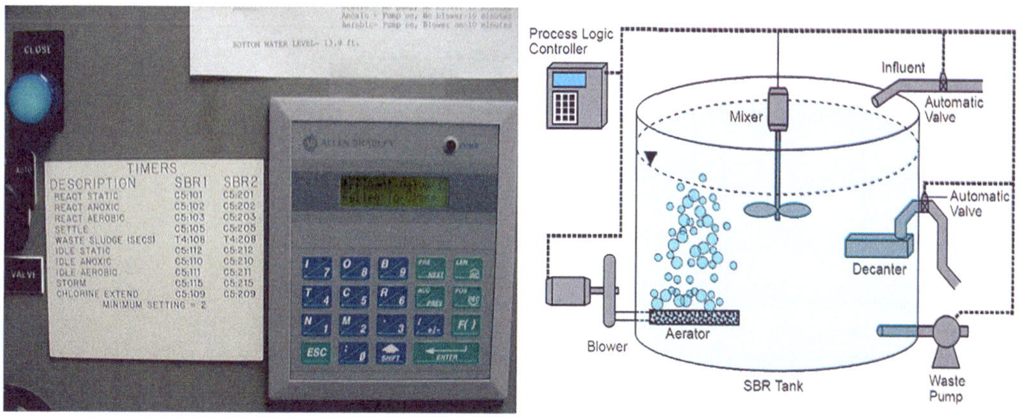

Figure 4.29 Typical Sequencing Batch Reactor Process Logic Controller and Components Automated Components (University of Florida, TREEO Center, 2000) (Reprinted with permission from UF TREEO Center)

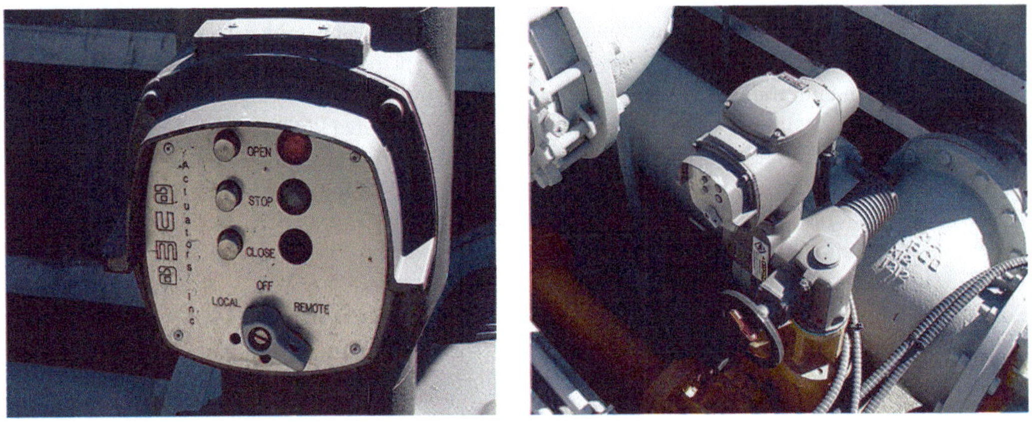

Figure 4.30 Automated Valve (Two Views) (University of Florida, TREEO Center, 2000) (Reprinted with permission from UF TREEO Center)

shows the ORP ranges for various biological reactions. As demonstrated in Figure 4.31, reaction environments do not exist in isolation, rather, they function over ranges of ORP. For EBPR, the anaerobic environment generally begins around –50 mV because fermentation is possible at that ORP. In practice, however, most EBPR facilities consider anaerobic conditions beginning below –150 mV to be optimal for effective phosphorus release in the anaerobic zone, and at least +150 mV in the aerobic zone to ensure effective uptake. Following the rate of change provides better information than tracking the actual value because, for example, the starting point can change from batch to batch. Typically, readings from various instruments (sometimes with exactly the same sensor type and transmitter) differ by a significant margin (50 to 100 mV), even though the sensors are in the same wastewater and even if all of the sensors show identical readings in an ORP standard (WEF, 2013).

Online DO meters are very useful in SBR operation. They allow operators to adjust blower times to address the variable organic loads that enter the facility. Lack of organic strength reduces the react time during which aeration is needed to stabilize the wastewater. Dissolved oxygen probes can be used to control the aeration-blower runtime during the cycle, which, in turn, reduces the energy cost of aeration.

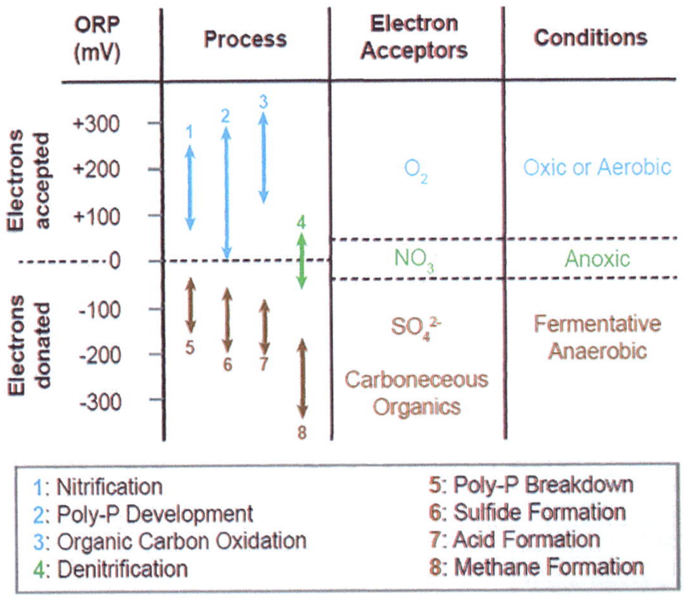

Figure 4.31 Oxidation–Reduction Potential and Metabolic Processes. Adapted from: "Oxidation Reduction Potential on Anaerobic System," River Bend Labs: A Division of Chemtron Corporation, October 25, 2013 (WEF, 2018)

It is desirable to locate DO, pH, and/or ORP probes in a place that can be reached easily by operators for cleaning and calibration. The facility operator should have the knowledge and the ability to program the SCADA system to increase or decrease blower speed. Allowing the operator to adjust the blower speed, through the SCADA system, gives the operator much more control over the DO in the SBR.

COMPOSITE SAMPLING

As with all WRRFs, SBR samples are collected and analyzed for both process control and compliance reporting. Sampling locations must be carefully considered. Sequencing batch reactors that use influent-equalization basins have more representative flow-paced composite samples because the discharge is consistent in volume. In other words, flow equalization and true batch reactions allow for easier composite sampling because the same volume is entering and exiting the basin during each cycle.

SOLIDS RETENTION TIME AND SLUDGE WASTING

Ensuring an adequate SRT is very critical to the SBR BNR process. The design SRT for nitrifying systems is based on the aeration time during the cycle, not the entire cycle time. Sludge wasting should occur during the idle cycle to provide the highest concentration of MLSS. The facility should be operated on pounds of MLSS and not concentration. Sludge from the SBR basins can be wasted to a digester and/or holding tank for future processing and disposal. The digester tank and sludge holding tank capacity should be sized appropriately, based on the sludge treatment and disposal method. Supernatant from the sludge digester and/or holding tank should be returned to the headworks or influent equalization basin so that it will receive full treatment. The facility should be designed so that the supernatant volume and load do not adversely affect the treatment process. A high-level alarm and interlock should be provided to prevent sludge-waste pumps from operating during high-level conditions in the digester and/or holding tanks. Controls should be provided to prevent overflow of sludge from digester tanks and/or holding tanks.

TROUBLESHOOTING

In general, the first step to troubleshooting is to determine if the problem is related to hydraulics, mechanics, processes, or a combination of these factors. When a problem or situation is about to occur, or if it has already surfaced, it is important to take logical and methodical steps to improving the process. The following seven areas are key to troubleshooting; they are critical for narrowing the search for solutions to most problems:

- Thoroughly know the process being evaluated.
- Know all facility flow patterns, including forward flow and sidestreams.
- Know the facility's design parameters and how actual loadings compare to design values.
- Know all maintenance procedures, including equipment maintenance considerations and staff responsibilities.
- Know how to recognize an abnormal condition.
- Know what alternatives are available when trouble develops.
- Know the amounts and characteristics of any industrial waste that may be discharged to the facility.

Specifically for SBRs, when addressing a process upset the operator should first eliminate mechanical, circuitry, and electrical problems as the source of the problem. Areas an operator should investigate include

- Power failures
- The logic controller (hardware or software), timer settings for idle cycle, and fill react cycle
- Communication pathway between the logic controller and the individual equipment
- The failure of the actual equipment (pumps, motors, floats, level sensors, solenoids, control panel, limit switches, etc.)
- Automatic valve failure (partially or completely closed valves)
- Air leaks in air distribution system
- Clogged or broken diffusers, if applicable
- Clogged jet mixer, if applicable
- The fouling of air filters

Process control and troubleshooting tests are summarized in Table 4.6. Once mechanical problems are eliminated as a source of the identified problem, the operator must troubleshoot the process biological and chemical components. Table 4.7 and the following discussion identify the most common biological and chemical problems associated with SBR facility operation. In each case, a problem or observation is indicated. The conditions that might cause the particular problem or observation are then detailed and the various causes that may contribute to the condition are discussed. The various process control analyses required to troubleshoot the problem or observation are then identified. Finally, the control or corrective actions needed to correct the problem are examined.

Table 4.6 Process Control and Troubleshooting Tests (University of Florida, TREEO Center, 2000) (Reprinted with permission from UF TREEO Center)

Parameters	Data Required/Analysis	Unit
	Organic Loading	
COD	Colorimetric analysis	mg/L
BOD, carbonaceous BOD	Bioassay	mg/L
Total organic carbon (TOC)	Colorimetric analysis	mg/L
	Solids Inventory	
MLSS	Gravimetric analysis	mg/L
Mixed liquor volatile suspended solids (MLVSS)	Gravimetric analysis	mg/L
Centrifuge spin	Volumetric analysis	%
	Solids Inventory Calculations	
MCRT	MLSS; waste activated sludge (WAS) TSS; WAS flow	days
Sludge age	AT%; CL%; WAS%	days
F:M	MLVSS; #BOD/#COD·d	#BOD/#MLVSS·d
	Sludge Quality	
SSV_x	Physical analysis	mL/L or %
SSV_5	Physical analysis	mL/L or %
Supernatant TSS or turbidity	Gravimetric or scattered light	mg/L or nephelometric turbidity unit
Depth of blanket	Physical measurement	m (ft)
Microscopic analysis	Visual analysis	N/A
	Sludge Quality Calculations and Nutrients	
NO_3-N, NO_2-N	Colorimetric or electrometric analysis	mg/L
NH_3-N	Colorimetric or electrometric analysis	mg/L
PO_4-P	Colorimetric analysis	mg/L
	Troubleshooting Analyses	
OUR	Analysis	mg O_2/L·h
Specific oxygen uptake rate (SOUR)	OUR; MLVSS	mg O_2/L·h
pH	Electrometric analysis	Standard units
ORP	Electrometric analysis	mV
Alkalinity	Titrimetric analysis	mg/L

Table 4.7 Sequencing Batch Reactor Troubleshooting Chart (University of Florida, TREEO Center, 2000) (Reprinted with permission from UF TREEO Center)

Sequencing Batch Reactor Troubleshooting Chart				
Problem or Observation	**Condition**	**Process Control Analysis**	**Possible Causes**	**Control Action**
I. Loss of solids from reactor due to a high blanket	Poor sludge settling velocity and compaction	SSV_x, SSV_5, sludge volume index (SVI), diluted SSV_x, microscopic examination, NH_3-N, COD, DO, SOUR	■ Glutting (old sludge)	■ Decrease MCRT.
			■ Classic bulking (young sludge)	■ Increase MCRT.
			■ Filamentous bulking	■ Identify conditions contributing to filamentous growth and correct. See comments in narrative below.
			■ Slime bulking	■ Add nutrients.
			■ Foam trapping	■ Optimize pretreatment removal of oil and grease.
			■ Highly nitrified or oxidized sludge	■ Increase anoxic cycle, reduce aerobic cycle.
			■ Toxicity	■ Isolate or split flow, identify source of toxic influent and eliminate, increase aeration cycle, increase MCRT.
			■ High organic loading	■ Short-term, increase aerobic cycle; long-term, increase MCRT.
II. Rapidly settling blanket leaving particulate. Difficulty in maintaining waste concentration	Rapid sludge settling velocity and compaction	SSV_x, SSV_5, SVI, F:M, SOUR	■ Low F:M	■ Increase F:M by decreasing MLVSS.
III. Turbid or cloudy effluent, disinfection problems	A. High effluent BOD or total solids	MLSS, MLVSS, DO, pH, temperature, influent COD or TOC, influent NH_3-N, DO, SOUR	■ Low MLSS or MLVSS	■ Increase MLSS/MLVSS.
			■ Low DO, temperature or pH	■ Increase aeration cycle in fill react, increase MLSS, add alkalinity.
			■ High organic loading	■ If long-term, increase MLSS/ MLVSS and aeration cycle.
			■ High nitrogenous loading	■ If long-term, increase MLSS/ MLVSS and aeration cycle.
			■ Toxicity	■ Isolate or split flow, identify source of toxic influent and eliminate, increase aeration cycle, increase MCRT.
	B. High effluent NH_3-N (Incomplete nitrification)	Influent and process NH_3-N, influent and process alkalinity, pH, temperature, SOUR, DO	■ Influent NH_3-N overload	■ Increase aerobic cycle.
			■ Low DO	■ Increase aerobic cycle.
			■ Low temperature	■ Increase aerobic cycle.
			■ Inadequate aerobic retention time	■ Increase aerobic cycle.
			■ Low pH or alkalinity	■ Add alkalinity.
			■ Low MLVSS (nitrifiers)	■ Increase MLVSS.
			■ Toxicity	■ Isolate or split flow, identify source of toxic influent and eliminate, increase aeration cycle, increase MCRT.

Table 4.7 Sequencing Batch Reactor Troubleshooting Chart (University of Florida, TREEO Center, 2000) (Reprinted with permission from UF TREEO Center) (*Continued*)

Sequencing Batch Reactor Troubleshooting Chart				
Problem or Observation	**Condition**	**Process Control Analysis**	**Possible Causes**	**Control Action**
IV. High-effluent TSS	Individual particle washout	Effluent and recycle TSS or turbidity, F/M, microscopic exam, SOUR	■ Pin floc—low F:M	■ Increase waste cycle, decrease MLSS.
			■ Pin floc—denitrification	■ Increase waste cycle, decease MLSS, increase anoxic cycle.
			■ Pin floc—solids recycle	■ Optimize solids handling.
			■ Straggler floc—high F:M	■ Decrease waste cycle, increase MLSS, increase aeration cycle.
			■ Straggler floc—filamentous	■ Identify filamentous organism (see filamentous control above).
			■ Individual bacterial cells in effluent	■ Decrease waste cycle, raise MLSS, increase aeration cycle, if toxicity, remove source of toxic influent.
V. High-effluent NO_3-N	High effluent NO_3-N	NO_3-N, pH, TOC or COD	■ Lack of or inadequate anoxic conditions	■ Increase anoxic cycle (may require decreasing oxic cycle).
			■ Lack of or inadequate carbon source	■ Add carbon (methanol or acetic acid).
			■ Low pH, temperature or MCRT	■ Add alkalinity, increase MCRT.
VI. Difficulty in maintaining chlorine residual	Chlorine (Cl_2) residual fluctuation, no chlorine residual	Cl_2 residual, supernatant NH_3-N, NO_2-N, turbidity or TSS	■ Incomplete nitrification/denitrification resulting in high NO_2-N in supernatant	■ High NO_2-N in supernatant will result in increased demand. Optimize nitrification and denitrification processes.
			■ High TSS in supernatant	■ High TSS in supernatant will result in increased demand. See Problems I, III, IV.
			■ Reducing agents in supernatant	■ Reducing agents such as H_2S, Fe, Mn in supernatant. Investigate source and eliminate. Increase chlorine feed rate to overcome demand.
VII. High fecal coliform values	Sufficient chlorine (Cl_2) residual, but high fecal coliform values	Supernatant TSS, free and total Cl_2 residual, supernatant NH_3-N, theoretical and actual chlorine contact chamber detention time	■ Excessive TSS in supernatant	■ High TSS in supernatant can result in "blinding" of disinfection process. See Problems I, III, IV.
			■ Short-circuiting of chlorine contact chamber	■ Calculate the theoretical detention time. Conduct dye testing to determine actual detention time.
			■ Chloro-organic compounds	■ If there is no NH_3-N in effluent but organic nitrogen is present, then false residual may be present due to formation of chloro-organic compounds. Use free chlorine to establish residual not total chlorine. Reduce aeration cycle to de-optimize nitrification rate.

Table 4.7 Sequencing Batch Reactor Troubleshooting Chart (University of Florida, TREEO Center, 2000) (Reprinted with permission from UF TREEO Center) (*Continued*)

Sequencing Batch Reactor Troubleshooting Chart				
Problem or Observation	**Condition**	**Process Control Analysis**	**Possible Causes**	**Control Action**
VIII. Foam	Excessive foam or scum on surface of SBR, flow equalization tank or chlorine contact chamber	Microbiological examination, NO_3-N, C-N-P ratio, SRT, oils and grease, DO	■ Excessive filamentous bacteria	■ The presence of hydrophobic filamentous bacteria may lead to excessive scum and foam.
			■ Denitrification	■ Denitrification can result in sludge and foam on surface of SBR.
			■ Nutrient deficiency	■ Foam may also indicate a possible nutrient deficiency. This type of foam may be due to bacteria producing a natural polymer when subjected to nutrient deficient conditions for an excessive period of time.
			■ SRT	■ Both too low and too high an SRT can cause foam problems.
			■ Fats, oils, or grease	■ Fats, oils, and grease and other non-degraded surface-active organics can cause foam problems.
			■ Overaeration	■ Excessive (DO > 4.0 mg/L) may cause foaming.

Corrective actions should be kept to a minimum. This means not making too many changes at once because that may create additional problems. In addition, because the activated sludge process is a biological process, adequate time must be allotted for corrective actions to take effect. In other words, do not make rapid changes after a corrective action is implemented. Each SBR operator needs to establish this optimal operation zone not only for the facility, but for each reactor in the facility. As Table 4.7 demonstrates, different operators may choose different strategies. Ideal operating parameters are those that produce the best treatment at the least cost. Ideal operating parameter ranges are facility and even reactor specific and, in many cases, must be determined by trial and error. When process upsets and problems occur, having established operating parameters makes it much easier for the operator to return to normal operating conditions.

TEST YOUR KNOWLEDGE

1. Sequencing batch reactor basins will typically complete one cycle per day.
   - ☐ True
   - ☐ False
2. Settling of the MLSS occurs in a separate basin outside of the SBR.
   - ☐ True
   - ☐ False
3. Nitrification is encouraged inside an SBR system during periods of increased mixed liquor temperatures.
   - ☐ True
   - ☐ False
4. Sequencing batch reactor basins are equipped with both mixing and aeration equipment that operate simultaneously.
   - ☐ True
   - ☐ False

5. In the conventional SBR design concept, there are three steps carried out during a batch.
   - ☐ True
   - ☐ False

6. Process upsets due to low pH in the SBR can occur if monitoring of process alkalinity is not performed.
   - ☐ True
   - ☐ False

7. During an aerated-fill cycle, both the influent and decant valve are in the open position.
   - ☐ True
   - ☐ False

8. During the anoxic phase of treatment, the DO should be monitored and should be always greater than 0.2 mg/L.
   - ☐ True
   - ☐ False

9. The idle and sludge wasting steps occur between the mixed-fill and aerated-fill phases.
   - ☐ True
   - ☐ False

10. It is recommended that a residual value of 50 mg/L of alkalinity be maintained in the decanted effluent.
    - ☐ True
    - ☐ False

11. The effluent flowrate from an SBR will be higher than the influent flowrate.
    - ☐ True
    - ☐ False

12. A sequence of anaerobic and anoxic phases followed by settling will ensure complete nitrification.
    - ☐ True
    - ☐ False

13. For SBR effluent flow equalization, the equalization basin is sized to hold a maximum of one-half the decantable volume.
    - ☐ True
    - ☐ False

14. For efficient alkalinity addition, sodium hydroxide is preferred over sodium bicarbonate or sodium carbonate.
    - ☐ True
    - ☐ False

15. For nitrogen removal, the fill and react phases include static fill, mixed fill, and mixed react.
    - ☐ True
    - ☐ False

16. To avoid overloading an individual SBR, the fill cycles should be arranged to spread the daily peak loading into multiple SBR basins.
    - ☐ True
    - ☐ False

Integrated Fixed-Film Activated Sludge and Moving Bed Biofilm Reactor Systems

INTRODUCTION

Integrated fixed-film activated sludge (IFAS) and moving bed biofilm bioreactor (MBBR) processes have become increasingly popular technologies for WRRFs trying to increase nitrification and total nitrogen removal capacity. Although both systems can be designed and built as new processes, they are often implemented as upgrades to existing activated sludge facilities.

The IFAS (also known as *hybrid*) process uses either fixed or free-floating media that can be made of plastic, foam, or fabric. The fixed media are anchored within the reactor and typically resemble a structured plastic filter media, net or mesh, or bundles of ropes. Media are installed in an activated sludge basin to encourage the establishment of biofilm in the tank, thus increasing the solids inventory in the basin. The IFAS systems have been shown to achieve 100% nitrification with 50% or less of aeration volume compared to conventional activated sludge systems.

The key to the IFAS process is the increased total biomass inventory that can be achieved by combining the attached biomass with the suspended MLSS. The amount of biomass that can be retained on the different types of media depends on multiple factors, including loading rate, DO concentration, media type, temperature, mixing, suspended activated sludge biomass concentration, and overall SRT. Because the attached biomass is retained in the activated sludge basin and not sent to the clarifiers, use of IFAS technology can increase the capacity of the activated sludge system in the same tank volume without significantly affecting the clarifiers.

Table 4.8 presents a brief description of different types of fixed and free-floating media that are currently in use in IFAS systems. Integrated fixed-film activated sludge systems are often implemented to improve nitrification capacity at existing facilities without increasing the solids load rate on secondary clarifiers.

Table 4.8 Types of Media Used in Integrated Fixed-Film Activated Sludge Systems (WEF, 2017a)

Media Type	Advantages	Disadvantages
	Fixed Media	
Fabric web type	Simple and quick to install, low initial cost, low maintenance, no risk of media loss	Prone to fouling (rags and red worm infestations), accumulation and trapping of debris, may require primary treatment upgrade
Rope type	Simple and quick to install, low initial cost, low maintenance, no risk of media loss	Prone to fouling (rags and red worm infestations), media breakage and entanglement, field assembly required
Polyvinylchloride sheet (tricking filter media)	Simple and quick to install, low initial cost, low maintenance, no risk of media loss	Structural media may impede mixing and air distribution, excess biomass accumulation may cause clogging, rags can cause clogging
	Free-Floating Media	
Polypropylene finned cylinders	Excellent mixing, staged implementation, may convert to MBBR	Potential for media loss (abrasion and washout), fouling of aeration system and screens, difficult to maintain the aeration system, need to account for media distribution

Moving bed biofilm reactor processes are similar to free-floating media IFAS systems, except there is no RAS recycle and the process relies almost entirely on biofilm activity. The key advantage of MBBR processes is their ability to retain slow-growing organisms and to maintain a high equivalent SRT. For this reason, MBBRs are being implemented as separate centrate treatment technologies that use anaerobic ammonia oxidation (anammox) biomass and deammonification processes. These types of systems require specialized biomass to be retained in the reactor for long SRTs (>30 days).

MOVING BED BIOFILM REACTORS

INTRODUCTION

Moving bed biofilm reactors are treatment systems with a buoyant free-moving plastic biofilm carrier media that require mechanical mixing or aeration for uniform distribution throughout the tank. The process includes a submerged biofilm reactor and liquid–solids separation unit. The installations include several process configurations and effluent water quality standards for carbon oxidation, nitrification, and denitrification. Moving bed biofilm reactors can treat wastewater to meet effluent quality standards ranging from secondary treatment (30 mg/L TSS and 30 mg/L BOD_5 monthly average) to more stringent nitrogen limits (advanced wastewater treatment standard total nitrogen less than 3 mg/L). The first MBBR installed in Norway has been routinely inspected and, reportedly, plastic biofilm carrier wear had not been observed after 15 years of continuous operation. Benefits of MBBRs include the following (WEF et al., 2018):

- Requires a smaller tank volume than a clarifier-coupled activated sludge system.
- Biomass retention is clarifier independent. Therefore, solids loading to the liquid–solids separation unit is reduced significantly compared to activated sludge systems.
- It is a continuous flow process and does not require a special operational cycle for biofilm thickness control.
- Multiple basin compartments can be configured in series for nutrient removal without the need for intermediate pumping or return activated sludge pumping to accumulate mixed liquor.
- It can be coupled with a variety of different liquid–solids separation processes including clarifiers, dissolved air flotation, ballasted flocculation, and membranes.
- It is suitable for retrofitting existing activated sludge basins.

MOVING BED BIOFILM REACTOR COMPONENTS

General Description

An MBBR may be a single reactor or several reactors in a series. Typically, each MBBR has a length-to-width ratio (L:W) in the range of 0.5:1 to 1.5:1. The MBBRs contain a plastic biofilm carrier volume (carrier fill) ranging from 25% to 67% of the liquid volume. Sieves typically are installed with one MBBR wall and allow treated effluent to flow through to the next treatment step while retaining the free-moving plastic biofilm carriers. Carbon-oxidation, nitrification, or combined carbon-oxidation and nitrification MBBRs use a diffused aeration system to uniformly distribute plastic biofilm carriers and meet process oxygen requirements. Plastic biofilm carriers in denitrification MBBRs are distributed uniformly by mechanical mixers. Plastic biofilm carriers must be removed before draining and servicing or repairing air diffusers.

Preliminary and Primary Treatment

As biofilm reactors, MBBRs require proper preliminary treatment. Robust screening and grit removal are used to prevent sieve blinding and long-term accumulation of rags, plastics, and sand in the tank. Once accumulated, these materials are difficult to remove. Manufacturers typically recommend no larger than 6- to 12-mm screen spacing if primary treatment is also provided and fine screens with 3-mm spacing for MBBR installations without primary treatment. Scum must also be removed from the system because of its potential to blind the media-retention sieves. Tertiary or add-on MBBR processes receiving wastewater that received significant upstream treatment do not require additional screening.

Plastic Biofilm Carriers

The biofilm carriers are typically extruded or molded from either virgin or recycled high-density polyethylene. Table 4.9 summarizes characteristics and manufacturers of several commercially available plastic biofilm carriers. The carriers are slightly buoyant with a specific gravity between 0.94 and 0.96 g/cm^3. As a result, both clean and biofilm-covered plastic biofilm carriers float in still water. In operating MBBRs, they are uniformly distributed throughout the bulk of the liquid due to the bulk mixing provided by the aeration system, liquid recirculation, or mechanical mixing. Biofilms primarily develop on the protected surface inside of the plastic biofilm carrier. For this reason, the specific surface areas of plastic biofilm carriers listed in Table 4.9 exclude areas that are not inside the plastic carrier. Biofilm carriers have the following parameters characteristic of the specific plastic biofilm carrier:

- **Bulk specific surface area:** biofilm area per unit volume of plastic biofilm carriers (square meter of biofilm per cubic meter of plastic biofilm carrier)
- **Bulk liquid volume displacement:** liquid volume displaced per unit volume of plastic biofilm carriers (cubic meter of liquid displaced per cubic meter of plastic biofilm carrier)

These parameters can be referenced to the basin volume by multiplying them by the carrier fill percentage, resulting in the following calculated parameters for each selected carrier fill value:

Table 4.9 Typical Ranges of Integrated Fixed-Film Activated Sludge Parameters (WEF et al., 2018)

Type of System	Media	Media Fill Volume Percentage*	Biofilm-Specific Surface Area (m^2/m^3)	Recommended MLSS (mg/L)	Minimum Aerobic Hydraulic Retention Time (h) at 12°C
Activated sludge	None	0	0	3000	7
IFAS, fixed bed	Bioweb, Accuweb	70–80	50–100	3000	5
IFAS, sponge MBBR	Linpor, Captor	20–40	100–150	2500	4
IFAS, plastic carrier MBBR	K1 (Kaldnes), Entex, ActiveCell (Hydroxy)	20–65	150–3002	3500	4

*External volume of frame for cord-type fixed-bed media; external volume of cuboids or cylinders with biofilm for moving bed media. Note that fill volume fraction is not the fraction of liquid volume in the activated sludge tank displaced by the media.

- **Net specific surface area:** biofilm area per unit basin volume (square meter of biofilm per cubic meter of basin volume)
- **Net liquid volume displacement:** liquid volume displaced per unit basin volume (cubic meter of liquid displaced per cubic meter of basin volume)

For example, if a plastic biofilm carrier has a 500-m^2/m^3 bulk specific surface area, then the net specific surface area at 50% carrier fill is 500 $m^2/m^3 \times 50\% = 250$ m^2/m^3. Similarly, for a plastic biofilm carrier having a characteristic 0.15-m^3/m^3 bulk liquid volume displacement, the net liquid volume displacement at 50% carrier fill is 0.15 $m^3/m^3 \times 50\% = 0.075$.

Figure 4.32 illustrates how biofilm thickness on plastic carriers vary depending on reactor conditions. Biofilm thickness does not become excessive because of the turbulent motion of the carriers in the reactor. Therefore, effective surface area reduction resulting from increasing biofilm thickness is not a significant factor.

The plastic biofilm carriers maximize protected bulk specific surface area and preserve an adequate open space through which significant advective flow can pass. Conditions in an MBBR promote development of relatively thin and dense biofilms. Sieves are used for retaining the plastic carriers within the MBBR basin. Larger plastic biofilm carriers allow for the sieves to be constructed with larger openings which reduces hydraulic head loss at the sieves. Figure 4.33 presents examples of plastic Bio-Carrier media.

Media Retention Sieves

Plastic biofilm carriers are retained in an MBBR by vertically configured flat sieves (see Figure 4.34) or horizontally configured cylindrical sieves (see Figures 4.35 and 4.36). Aerobic zones typically contain cylindrical sieves and anoxic zones contain flat-wall sieves. The cylindrical sieves extend horizontally into the upward-flowing air bubbles imparted by the diffuser grid. As a result, the air scours accumulated debris from the sieve surface. Water movement from the mechanical mixers is insufficient to dislodge debris accumulated on the wall sieve. Therefore, flat-sieve scour occurs in a denitrification MBBR with a sparging air-header. Removing the debris retained on a sieve aids in maintaining hydraulic throughput. For systems with high approach velocity that causes media migration, air sparge systems on the media retention sieves are also provided.

Sieves used to retain media, and their supporting structural assemblies, if required, typically are constructed with stainless steel. The sieve may be wedge-wire or mesh with approximately 6-mm spacing but can be procured as perforated plates with 5- to 6-mm diameter orifices. Two sieve configurations typically are used in MBBRs: horizontally configured cylindrical sieves (carbon oxidation/nitrification MBBRs) and vertically mounted wall sieves (denitrification MBBRs). Horizontal sieves are attached to the reactor wall by cast-in-place wall thimbles or by inserting wall sleeves through poured or core-drilled holes in the reactor wall. The cylindrical sieves are typically submerged 35% to 65% of the sidewater depth.

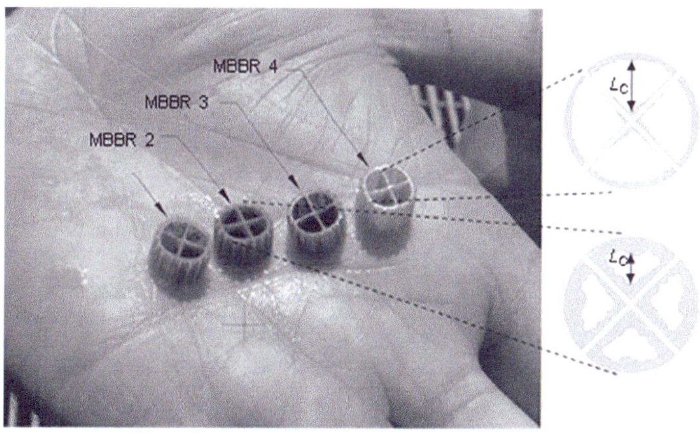

Figure 4.32 Photograph of Biofilm Carriers Taken From Moving Bed Biofilm Reactors in Series and Illustrative Renderings of How Biofilm Thickness on the Carriers Can Vary From Reactor to Reactor (WEF et al., 2018, from Boltz et al., 2009)

Figure 4.33 Examples of Bio-Carrier Media (a) K1; (b) K3; (c) ABC-5; (d) W-1; (e) W-2 (Reprinted with permission from Chandler Johnson)

Figure 4.34 Flat Media Retention Sieve in Anoxic Reactor—Midwest City, Oklahoma (Reprinted with permission from Chandler Johnson)

Figure 4.35 Tubular Sieves in Aerobic Reactors at Ladysmith, British Columbia (Reprinted with permission from Chandler Johnson)

Figure 4.36 Tubular Sieves in Integrated Fixed-Film Activated Sludge Reactor With Air Sparge on Second Row of Media Retention Sieves Also Showing Foam-Migrating Flat Sieves at Water Surface Level (Reprinted with permission from Chandler Johnson)

Vertically mounted flat panel sieves are attached to wall-fixed brackets. The brackets extend outward from the wall and create an approximately 0.15- to 0.3-m space between the sieve panels and the MBBR wall. Forward flow passes through the sieves, then through the void space between the sieve panels and the wall, and finally into the next treatment step through the concrete wall openings located at the liquid surface. A full concrete wall is used to divide MBBRs to provide structural support for the sieve brackets and panels and to eliminate back mixing from the downstream MBBR.

The sieve area is defined based on the allowable hydraulic headloss across the MBBR wall. However, typical sieve design allows for a maximum 50- to 100-mm headloss (at the peak hydraulic flow, which is typically measured as the peak hour flow) across each sieve-containing wall. Proper sieve design is primarily related to system hydraulics. Key sieve related parameters are

- **Sieve loading rate:** wastewater flowrate (including recirculation streams) applied per unit screen area ($m^3/m^2{\cdot}h$)
- **Approach velocity:** wastewater flowrate (including recirculation streams) in an MBBR divided by the reactor cross-sectional area ($m^3/m^2{\cdot}h$)
- **Length-to-width ratio:** 0.5 : 1 to 1.5 : 1 (MBBR L : W > 1.5 may be possible, but also may result in plastic biofilm carrier migration). The approach velocity must be below the recommended limit.

Based on the design hydraulic headloss across each MBBR sieve-containing wall, the corresponding sieve loading rate is selected by the MBBR manufacturer using empirical criteria that accounts for sieve material and the influence of plastic biofilm carrier fill. Sieve loadings are typically 50 to 60 m/h, but values up to 85 m/h have been applied with sufficiently low approach velocity. Typically, cylindrical sieves have a 406- to 610-mm (16- to 24-in.) diameter and a length of 3.6 m (12 ft).

The approach velocity should be less than 30 to 35 m/h under peak flow conditions. Higher approach velocities will cause the media to migrate with the flow and accumulate on the sieves. The result is reduced sieve hydraulic throughput, increased hydraulic headloss, poor oxygen transfer in the zone where media accumulates, and likely an oxygen deficiency because of incomplete use of the MBBR volume. When an approach velocity greater than 30 m/h cannot be avoided (as in the data from operating facilities shown in Figure 4.37), it may be prudent to reduce the sieve loading rate (Leaf et al., 2011).

Provisions for filling and draining an MBBR train are necessary. Small wall openings with screens typically are installed near the floor of the reactor to allow for equalization of the water level between reactors

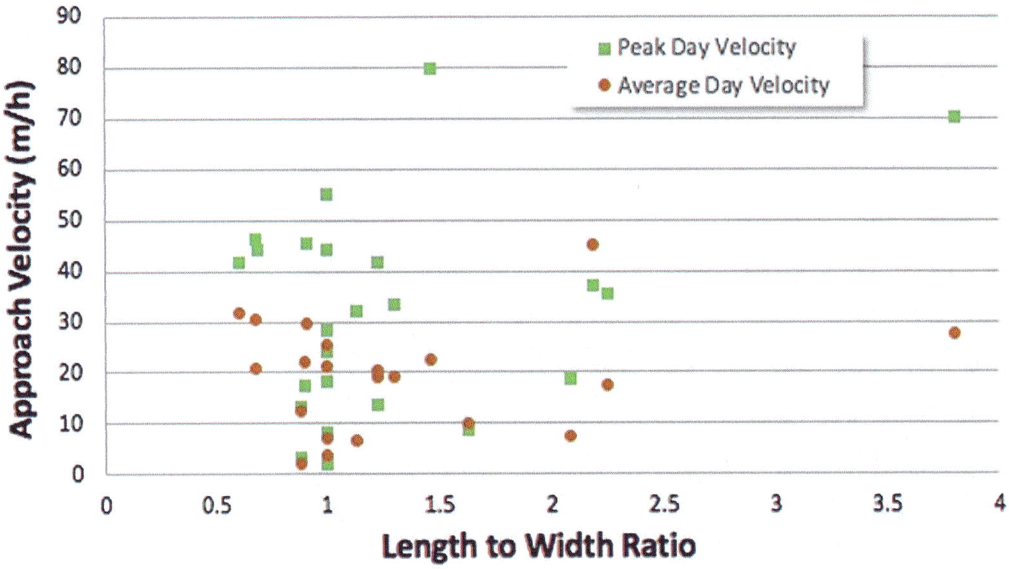

Figure 4.37 Approach Velocity Versus Length-to-Width Ratio of MBRB and Integrated Fixed-Film Activated Sludge Zones (Reprinted with permission from Chandler Johnson)

during fill and drain periods. When possible, the tank should be drained without removing the plastic biofilm carriers; for example, water can be withdrawn from a location immediately downstream of the last MBBR wall or by using an underdrain system.

Aeration System

Diffused air is applied to aerobic MBBRs. The airflow applied through a network of air piping and diffusers that are attached to the tank bottom has the dual purpose of meeting the microbial oxygen requirements and uniformly distributing the plastic biofilm carriers throughout the aerobic MBBR. To promote uniform distribution of the plastic biofilm carriers, the diffuser grid layout and drop pipe arrangement provide a rolling water circulation pattern, as illustrated in Figure 4.38. Fine- and coarse-bubble diffusers have been used in free-floating plastic biofilm carrier reactors.

Coarse-bubble diffusers typically used in MBBRs are plastic or stainless-steel pipes with circular orifices along the underside (Figure 4.39). These coarse-bubble diffusers are less affected by scaling and fouling because of the large dimension and turbulent airflow through the discharge orifice. As a result, coarse-bubble diffusers require less maintenance than fine-bubble diffusers. The coarse-bubble diffusers are designed with a structural end support that enables them to withstand the weight of plastic biofilm carriers when the MBBR is taken out of service and drained.

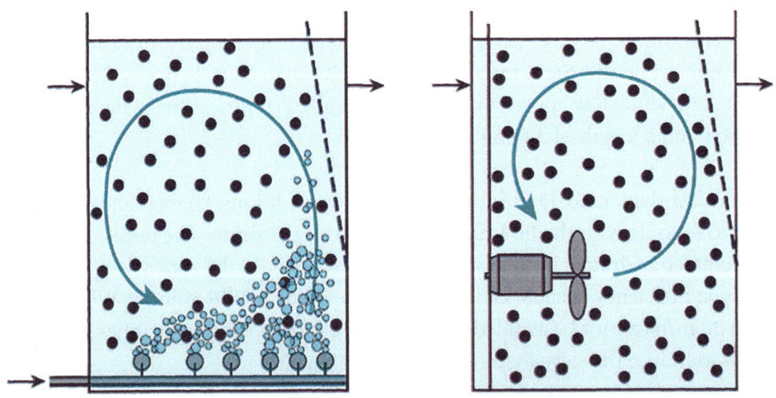

Figure 4.38 Rolling-Water Circulation Pattern Induced by a Diffused Aeration System (Left) and Mechanical Mixers (Right) (Adapted from WEF et al., 2018)

Figure 4.39 Full Floor Coverage Coarse-Bubble Diffusers—Broomfield, Colorado (**Reprinted with permission from Chandler Johnson**)

Coarse-bubble diffusers produce bubbles with a diameter of 6 to 12 mm (compared to a 2-mm diameter typical of fine-bubble diffusers). These coarse bubbles rise rapidly and produce a rolling-water circulation pattern. Multiple drop pipes with individual valves for modulation provide added flexibility to induce a rolling pattern. Coarse-bubble diffusers typically used in MBBRs are 25-mm diameter stainless-steel pipes with 4- to 5-mm diameter orifices that are spaced approximately 50 mm apart along the underside of the diffuser pipe. The coarse-bubble diffuser orifice is smaller than the plastic biofilm carrier to avoid air-pipe and orifice plugging. Clean water oxygen transfer efficiency design values for MBBRs with coarse-bubble diffusers are 3.0% to 3.5% per meter of water submergence. The MBBR-specific coarse-bubble air diffusers have been designed with a 0.8-alpha factor and 0.95-beta factor (Johnson & Boltz, 2013).

When using fine-bubble diffusers, biofilm carriers must periodically be removed to service or replace diffusers. The bubble-rise velocity characteristic of fine-bubble diffusers does not create the rolling-water circulation pattern required to uniformly distribute the plastic biofilm carriers. Therefore, the diffuser-grid layout may require modification to promote a rolling-water circulation pattern. The fine-bubble diffuser configuration required to promote a rolling-water circulation pattern has a negative effect on oxygen transfer efficiency.

Mechanical Mixing

Denitrification MBBRs use mechanical mixers to agitate the activated sludge and to uniformly distribute plastic biofilm carriers. The mechanical mixers may be platform mounted (dry motor) or rail-mounted submersible (wet motor) units. Typically, these mixers agitate liquids with biological suspended solids concentrations up to 10 000 g/m^3. Submersible mechanical mixers typically have a maximum 120-rpm impeller speed and a minimum of three blades per impeller. These features minimize the damage of the impeller and the plastic carriers from abrasion as the plastic carriers are propelled through the impeller. A dry motor mechanical mixer specially designed to withstand the abrasive MBBR environment uses a stainless-steel backward-curve propeller with a round bar welded along its leading edge to avoid damage to the plastic biofilm carriers and impeller wear. The mixer has a large-diameter impeller with a fairly low rotational speed (90 rpm at 50 Hz and 105 rpm at 60 Hz). Because of the plastic biofilm carriers' buoyancy, the mixers are located near the water surface but should not create an air-entraining vortex because DO inhibits denitrification. A slight negative inclination of mixer orientation instead of a fully horizontal inclination helps maintain the rolling-water circulation pattern and uniformly distribute plastic biofilm carriers. It is especially important to use rail-mounted submerged mixer units to facilitate access to the mixer when maintenance is required. These specially designed mechanical mixers are typically sized to input 25 W/m^3.

Solids Separation

The MBBR process performance is dependent of a liquid–solids separation unit. Biomass accumulation in an MBBR is, however, independent of the settler. Therefore, the MBBR process offers considerable flexibility in terms of the type of process that can be used for liquid–solids separation. Typically, the suspended solids concentration in the MBBR effluent stream is at least an order of magnitude lower than typical of activated-sludge bioreactors. As a result, a variety of different solids separation processes have been paired with MBBRs. The MBBR can be combined with compact, high-rate solids separation technologies such as dissolved air flotation and ballasted flocculation. Tertiary units may discharge directly to a filtration unit or coagulation/flocculation/sedimentation (with lamella plate settlers) basins. Circular or rectangular secondary clarifiers also may be used (especially in the case of retrofit application in which the clarifiers already exist).

PROCESS CONFIGURATIONS

Relevant considerations when selecting a MBBR configuration include site-specific treatment objectives, wastewater characteristics, site layout, existing basin configuration (if a retrofit), system hydraulics, existing treatment scheme (if applicable), and the potential to retrofit existing tanks. Figure 4.40 illustrates carbon oxidation, nitrification, and denitrification MBBR flow sheets. Although the process mechanical features of an MBBR are typically consistent, biofilms grown in carbon oxidation, nitrification, denitrification, and combined carbon oxidation and nitrification MBBRs are variable and depend on local environmental conditions.

Biodegradable soluble organic carbon is quickly consumed in an MBBR. Particulate organic matter is entrapped by a biofilm; subsequently, a portion of the entrapped biodegradable organic particulates is hydrolyzed and the resulting soluble organic carbon used. The remaining entrapped particulate exits the biofilm before being hydrolyzed with detached biofilm fragments. Suspended organic particles exit the MBBR in the effluent stream with the detached biofilm fragments and are conveyed with the sieved liquid to the downstream liquid–solids separation unit.

MOVING BED BIOFILM REACTOR STARTUP AND MAINTENANCE

Plastic biofilms carriers typically are delivered to the site in sacks of known volume. The carriers are introduced to the MBBR simply by elevating the sacks and allowing the plastic biofilm carriers to fall into the tank. A recessed impeller pump can be used to transfer the plastic carriers out of the water-filled MBBR to perform aeration system maintenance. The MBBR aeration or mixing system should remain engaged while the pump transfers carriers to temporary storage. Ideally, a dedicated basin or container must be available for interim plastic biofilm carrier storage to ensure that the original carrier fill volume is returned to the emptied tank. If the carrier fill is transferred to another MBBR for temporary storage, then the best way to restore original carrier fill is to pump the plastic carriers back into the emptied tank, drain both MBBRs, and measure the average bed height at several locations. The plastic carrier bed depths should be equal.

It is common for light foam to float on the water surface during MBBR startup and while mature biofilm is developing. If necessary, a defoaming agent may be used. It is important that the defoaming agent is compatible with the plastic carriers by consulting with the manufacturer. Even when agitated, plastic biofilm carriers have a propensity to float when introduced to the water-filled tank but disappear from the water surface within a few days. Carbon oxidation likely occurs after 2 to 15 days; ammonia-nitrogen oxidation occurs after approximately 4 weeks but may take 60 to 120 days to reach an approximate steady state. Approximately 4 to 6 weeks may be required before nitrate/nitrite-nitrogen removal occurs because the denitrifying biofilms will not develop until sufficient nitrate-nitrogen is present.

INTEGRATED FIXED-FILM ACTIVATED SLUDGE

INTRODUCTION

Integrated fixed-film activated sludge facilities are those that incorporate a fixed-film process component (a biofilm growth) within activated sludge basins. The IFAS facilities operate with a level of MLSS concentration similar to conventional activated sludge (CAS) systems with a sludge age that improves sludge settleability and provides carbonaceous removal, nitrification, and/or BNR. The IFAS facilities may incorporate three types of biofilm process media or carriers within the activated sludge basin(s): fixed-bed

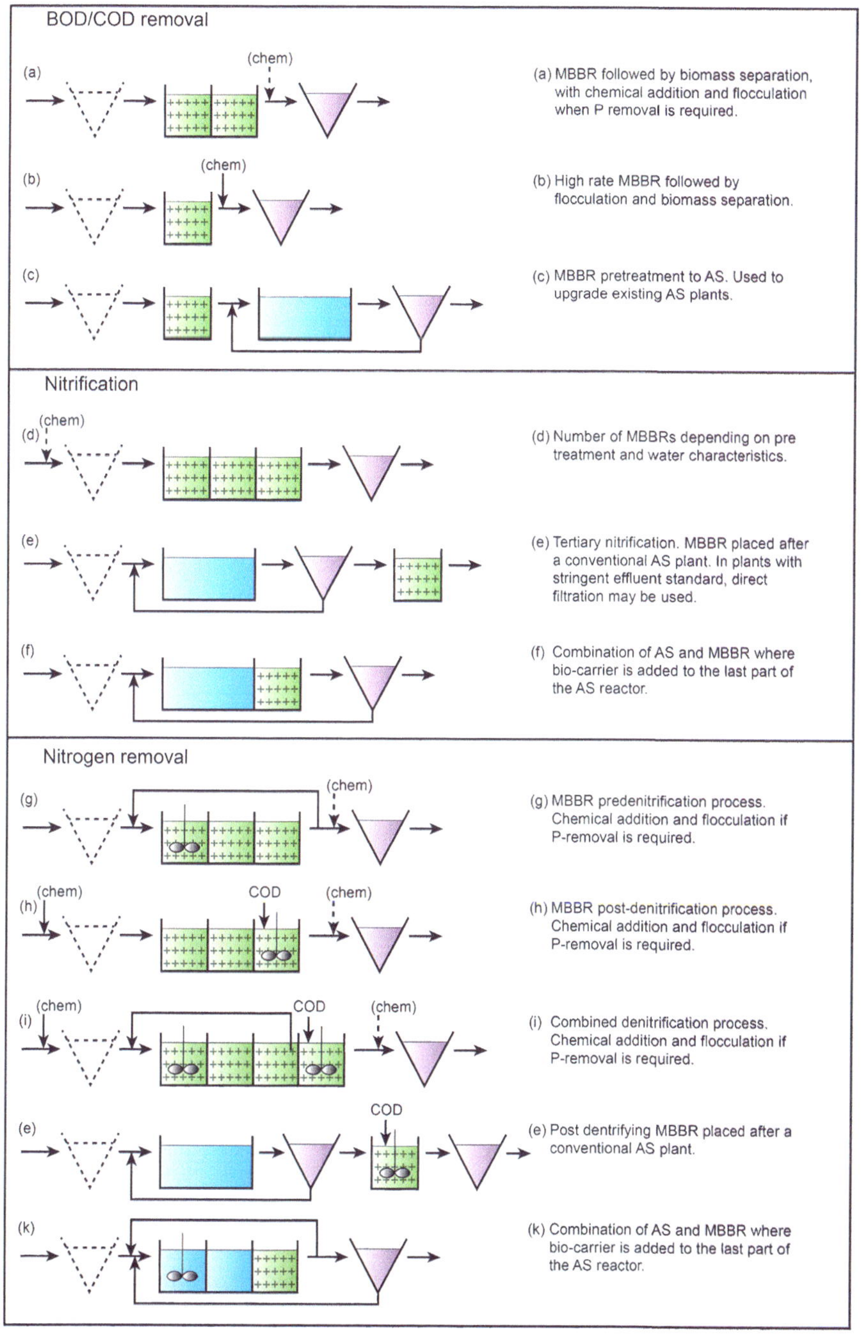

Figure 4.40 Typical Process Flowsheets for Carbon Oxidation, Nitrification, Combined Carbon Oxidation and Nitrification, and Denitrification in a Moving Bed Biofilm Reactor. Rectangular Tanks With Crosses Are Moving Bed Biofilm Reactors and Rectangular Tanks With Mixers Are Anoxic (Ødegaard, 2006) (Reprinted from *Water Science and Technology, 53(9): 17–33*, with permission from the copyright holders, IWA Publishing)

media, free-floating plastic media that move within the bed, and sponge-type carriers that move within the bed. Fixed-bed media are held in frames (Figure 4.41a). Free-floating media are held within the basins or zones (anoxic or aerobic) using screen or sieve assemblies (Figure 4.41b). Sponge-type biofilm carriers are typically allowed to move longitudinally within the aerobic zone and are recycled or pumped back to the front of the aerobic basins using the process's media recycle system (Figure 4.41c). The typical range of MLSS incorporated to IFAS processes is from 1500 to 3500 mg/L, and the upper limit is determined by the secondary clarifier capacity. Therefore, when designing IFAS facilities, designers must decide whether

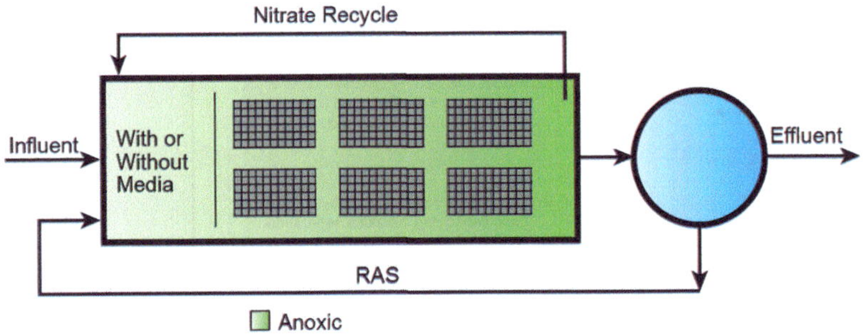

(b) Typical Layout of Cord or Web Fixed IFAS System

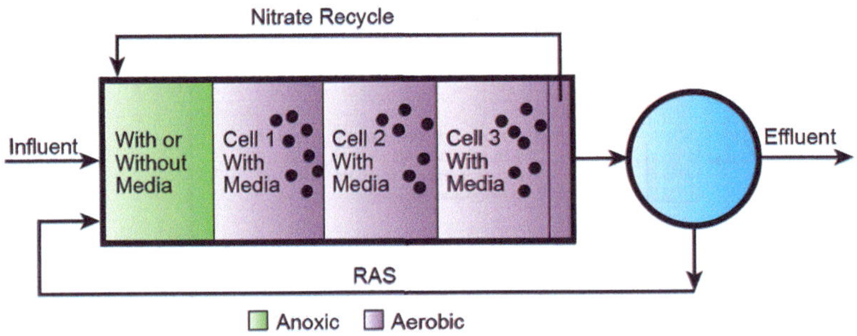

(b) Typical Layout of Plastic Carrier IFAS System

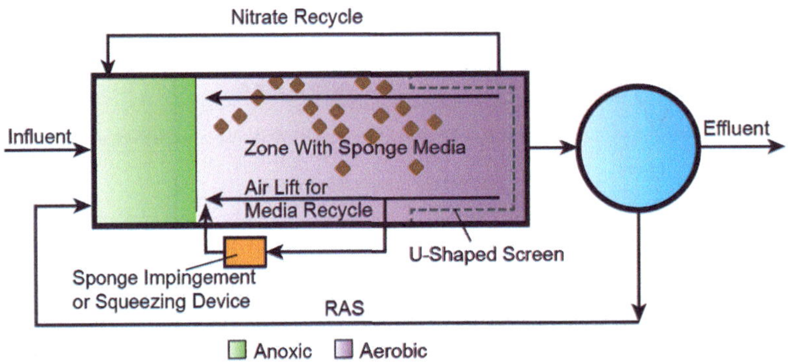

(c) Typical Layout of Sponge Media IFAS System

Figure 4.41 **(a) Typical Layout of Cord or Web Fixed-Bed Medial Integrated Fixed-Film Activated Sludge System. (b) Typical Layout of Plastic Carrier Media Integrated Fixed-Film Activated Sludge System. (c) Typical Layout of Sponge Media Integrated Fixed-Film Activated Sludge System.** (WEF, 2011).

to construct new secondary clarifier infrastructure or to limit the amount of work the MLSS is able to provide, balancing reliance between suspended and fixed-film (biofilm) growth. Additionally, the designer must consider SRTs that will allow for adequate sludge-settling characteristics.

The principle of the IFAS system is to enhance BOD removal and nitrification above the removal that could have been achieved using MLSS in a suspended-growth-only reactor of the same volume.

The biofilm flux rate is the rate of transport of a particular substrate or electron acceptor across the liquid-biofilm interface. Typical units of flux are in grams of substrate removed per square meter of biofilm surface per day or kilograms per 1000 square meters per day. The biofilm surface is the surface area of biofilm that develops on the biofilm media or carriers. It is not the surface area of bare media or carriers.

Flux rates can be in terms of parameters such as chemical oxygen demand (COD), DO, ammonium-nitrogen, oxidized-nitrogen, volatile suspended solids (VSS), and inert solids. It is important to understand the factors that control biofilm flux rates. The flux rate for a substrate such as COD increases with the concentration of substrate COD and the concentration of electron acceptors (DO, oxidized-nitrogen forms) in the mixed liquor (bulk liquid outside the biofilm). Biofilm thickness, biomass density in the biofilm, and the thickness of the stagnant liquid layer (boundary layer) also affect flux rate.

A higher bulk liquid concentration can increase concentrations inside the biofilm. As the thickness of the stagnant liquid layer decreases, as may be observed at higher intensities of mixing or a more open media structure, the concentrations of substrate in the biofilm can increase as well as the substrate flux rate. Additionally, a greater biofilm thickness increases the depth into the biofilm to which the substrate can be used. There is also an increase in the flux rate if electron acceptors are available in these deeper layers. A higher biomass density (MLVSS of the biofilm, in milligrams per liter) also generates an increase in the COD flux rate.

In addition to the biofilm flux rate, the removal that takes place per unit of tank media volume depends on the biofilm-specific surface area (square meter of biofilm surface/cubic meters of tank volume) and the media fill fraction. The removal per unit of tank volume (kg/m³·d) is

$$\frac{\text{Pollutant, kg}}{\text{Tank vol., m}^3\text{/d}} = \frac{\text{Pollutant, kg}}{\text{Biofilm surf. area, m}^2\text{/d}} \times \frac{\text{Biofilm surf. area, m}^2}{\text{Biofilm carrier, m}^3} \times \frac{\text{Biofilm carrier, m}^3}{\text{Tank op. vol., m}^3}$$

The biofilm media or carrier fill fraction is the fraction of the activated sludge basin volume occupied by the media or carriers. At 100% fill, the value of media fill fraction equals 1.0. For fixed-bed media, the media fill fraction is the fraction of the tank floor and height covered by media frames. For plastic and sponge-type free-floating media, the fill fraction is the fraction of the empty basin volume occupied by the carriers. The biofilm specific surface area at 100% media fill fraction depends on several factors. These include

- Type of media used. The specific surface area at a certain fill volume fraction increases from fixed-bed media to sponge moving-bed media to certain types of plastic cylinders (Table 4.9).
- The thickness of the biofilm on the media increases with an increase in organic substrate (soluble biodegradable COD) levels in the bulk liquid. In those situations, the thickness can be decreased by more vigorous mixing to induce a higher rate of biofilm shear.
- The extent of bare media surface that is covered by the biofilm. This can change with the external COD concentration.

Table 4.9 also shows the biofilm-specific surface areas and typical fill fractions applied in IFAS systems for secondary treatment. For free-floating media, the upper limit on fill fractions is determined by the in-basin aeration system or anoxic mixing equipment's ability to create completely mixed (moving bed) conditions. Factors that affect the ability to create these conditions include carrier fill fraction, carrier buoyancy, airflow per unit basin volume or floor area, mixing power in anoxic zones, and the basin's hydraulic end-wall approach velocity, that is, the velocity of the wastewater and all recycle streams toward the end wall of the IFAS basin(s). The end-wall approach velocity is determined by dividing the total hydraulic throughput under peak conditions by the wetted end-wall area. For plastic free-floating media, a typical acceptable peak flow end-wall approach velocity is 30 to 40 m/h.

If the biodegradable soluble COD within the IFAS basins containing media or carriers is higher than 15 to 25 mg/L, then the biofilm will increase in thickness unless the shear (mixing or air scour) is increased for the type of media. If the shear is not changed, then specific surface area may remain constant over a certain range and then decrease. The range over which the surface area remains constant will depend on the type of media (WEF & ASCE, 2018).

For plastic free-floating media, growth on the outer surface of media, which is otherwise barren, can add to the specific surface area, and the increase in thickness reduces the biofilm specific surface area inside the carrier. For cord media, the increase in the concentration of biodegradable soluble COD in the cell may result in development of a tubular biofilm surface enveloping the loops rather than a serrated biofilm surface. This results in a reduction in effective surface area. Biofilm media and carrier design influence the changes in specific surface area with biodegradable soluble COD for a given level of mixing.

As with activated sludge systems, removal of COD per unit of tank volume in the IFAS system increases with increases in MLSS, concentrations of substrate (biodegradable soluble COD), and DO (or oxidized-nitrogen) in the mixed liquor. For the same tank volume, an increase in the suspended-growth SRT increases the concentration of MLVSS. Increasing SRT and subsequent increases in MLVSS can result in a decrease in the biofilm surface area required for removal of organics and for nitrification. The SRT needed for nitrification in an activated sludge process is a function of the maximum specific growth rate (which is related to temperature, the reactor DO concentration, and pH [U.S. EPA, 2010]).

Interaction Between Suspended Biomass and Biofilm Mass

An increase in removal by the suspended growth, as observed at a higher suspended growth SRT, decreases the soluble COD concentrations at any location along the aerobic zone. This decreases the COD uptake rates for the biofilm at that location. For the same SRT, an increase in temperature has a similar effect. Conversely, reducing the SRT or lowering the temperature has the opposite effect, and the rate and the fraction of COD removal in the biofilm will increase.

One can take advantage of the available clarifier capacity to operate the system with a higher MLVSS and SRT and reduce the amount of biofilm required. This may allow for a change in the type of media required to a lower biofilm specific surface area. It also may allow for use of a smaller aeration basin volume compared to an activated sludge system or an MBBR.

The data in Figure 4.42 and Figure 4.43 are from a 3-year case study from the Broomfield, Colorado, Wastewater Treatment Plant. Figure 4.43 shows the change in biofilm during the year (seasonal variation) and seeing higher biofilm mass in the winter. Integrated fixed-film activated sludge systems are more resilient to variability in wasting rates because the biofilm is these systems acts as an insurance policy and will do work when the MLSS cannot. Should the system see a higher-than-normal wasting, the biofilm will kick in and do more work until wasting is back under control and allows the MLSS to do its workload. showing very stable RAS, mixed liquor recycle (MLR), and WAS rates that were only changed once per month (if at all). Figure 4.43 shows that the effluent ammonia-nitrogen concentration was predominantly stable within this 3-year period.

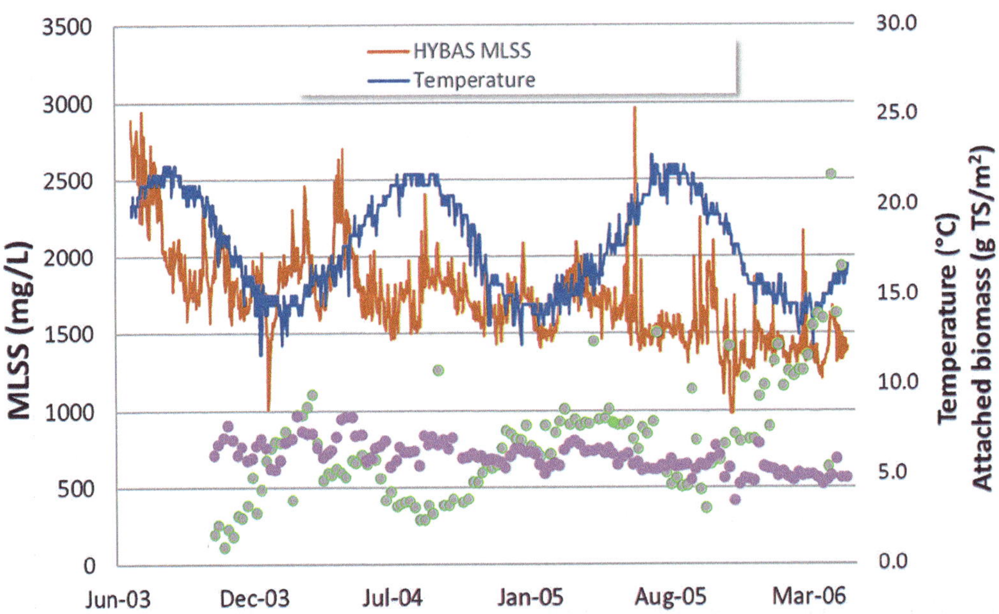

Figure 4.42 Full-Scale Pilot Test Mixed Liquor Suspended Solids and Attached Biomass Data (Data with permission from Chandler Johnson)

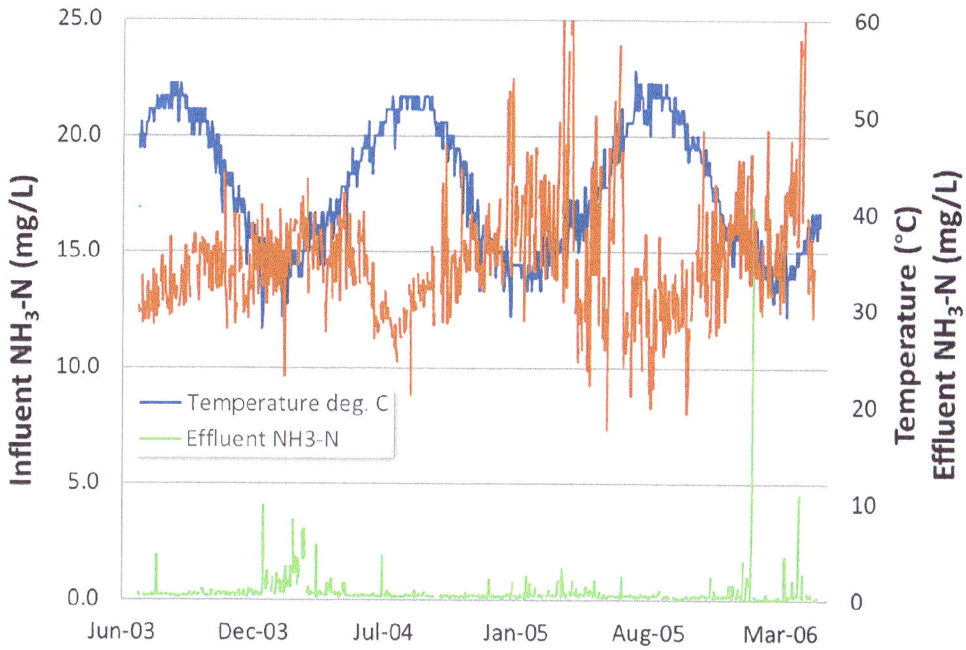

Figure 4.43 Full-Scale Pilot Test Influent/Effluent Ammonia-Nitrogen and Temperature Data (Data with permission from Chandler Johnson)

As with standard activated sludge systems, operators will still need to look at SVIs to understand how well the MLSS is settling. Integrated fixed-film activated sludge systems, in general, have shown good settling rates and typically have SVI values less than 120 mL/L.

Interaction Between Heterotrophs and Nitrifiers and Solids Retention Time Control

Understanding the interaction between heterotrophs and nitrifiers (autotrophs) is a challenge in IFAS systems. As with activated sludge systems, factors that increase soluble biodegradable COD levels in the bulk liquid (from 5 to 20 mg/L) result in a substantial increase in growth rates of heterotrophs preferentially overgrowth of nitrifiers. This decreases the fraction of nitrifiers in both the suspended growth and the biofilm and decreases the nitrification rate in each.

In an activated sludge system, a reduction in SRT close to the washout SRT of nitrifiers decreases the nitrifier biomass until nitrification is lost; addition of a biofilm carrier typically allows nitrifiers to survive. The nitrifier biomass in the biofilm will depend on the soluble biodegradable organic concentration and DO levels in the bulk liquid outside the biofilm.

When nitrifiers are present in the biofilm of the IFAS system, biofilm sloughing seeds the mixed liquor with nitrifiers. At SRTs approaching washout SRT, this sloughing and seeding can significantly increase the fraction of nitrifiers in the MLSS compared to an activated sludge system operating at the same SRT. Therefore, the presence of biofilm helps increase nitrification rates per unit of tank volume. If nitrification in the biofilm is not compromised during a facility upset or suspended growth process washout, then the time required for an IFAS system to recover full nitrification is reduced compared to a system with activated sludge only, due to seeding.

Operation of IFAS systems is similar to that of conventional activated sludge, meaning the use of SRT and wasting strategies are implemented. For IFAS systems, an equivalent SRT is calculated by dividing the total biomass (in the biofilm and mixed liquor) by the daily wasting rate.

The SRT of the suspended growth is typically within 3 to 5 days such that the mixed liquor will have good sludge-settling characteristics, but such that nitrification preferentially occurs in the biofilm under winter conditions; seeding from the biofilm into the mixed liquor allows partial nitrification in the suspended growth during summer conditions.

Assuming that the attached microorganisms' mass is in the order of 15% of the bulk mixed liquor microorganisms, IFAS systems should be operated within ±15% of the design aerobic MLSS SRT. If desired, testing to determine the actual amount of attached biomass on the media can be done. This could give a better understanding of the actual microbial mass in the system and the actual total (attached plus suspended microbial mass) available SRT.

Based on computer modeling results, as SRT decreases, more of the nitrification takes place in the biofilm (Figure 4.44). According to these results, at 15 °C, equal nitrification in the biofilm and mixed liquor takes place at an SRT of 4 days. Figure 4.45 includes a partial listing of design aerobic MLSS SRT (not including the fixed biomass) for full-scale facilities. As is shown, many facilities have design aerobic SRTs in the 2- to 5-day range. In those facilities with low aerobic design SRTs, the bulk of nitrification takes place on the bio-carriers instead of the MLSS.

The weight of the attached biomass can be calculated as the difference between the weight of the media with biomass and the cleaned media without biomass, according to the following procedure:

EXAMPLE: Measuring the Attached Biomass for Use in Integrated Fixed-Film Activated Sludge Solids Retention Time Calculation

Preparation and sample collection

1. Label the aluminum pans.
2. Place the elements in a jar of water.
3. Move the media gently to remove any suspended solids from the elements. Drain the water from the jar while keeping elements in the jar.

Determination of media weight with biomass attached

4. Count out 10 elements and place in an aluminum pan in one layer. Record the number of elements counted out.
5. Dry the pan with media to constant weight in an oven at 105 °C (1 hour).
6. After drying to constant weight, place the pan with media in a desiccator to cool to room temperature (0.25 hour).

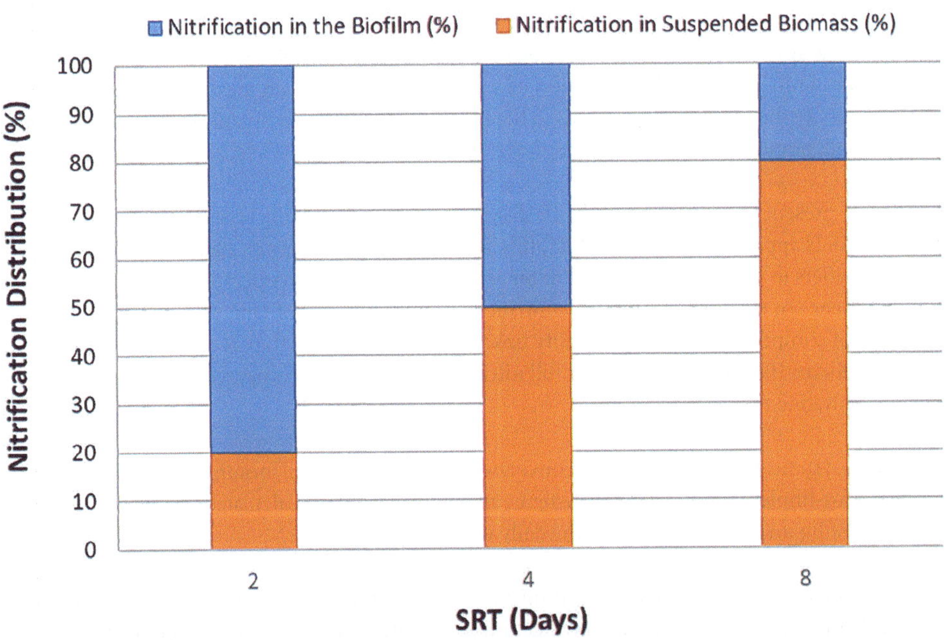

Figure 4.44 As Solids Retention Time Decreases, More of the Nitrification Takes Place in the Biofilm Based on Modeling Results at 15 °C (WEF et al., 2018)

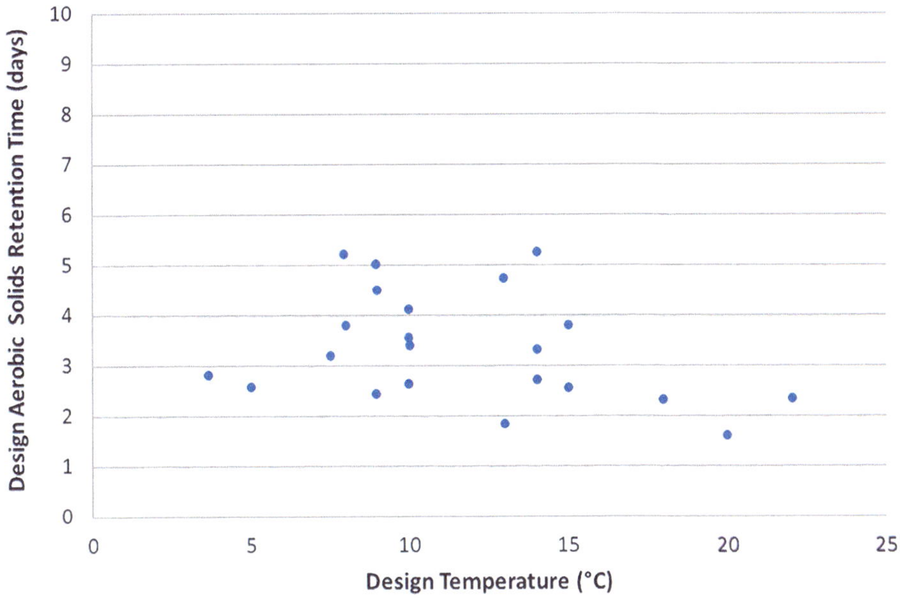

Figure 4.45 Partial List of Integrated Fixed-Film Activated Sludge Facilities' Aerobic Mixed Liquor Suspended Solids Retention Time Versus Temperature (Data with permission from Chandler Johnson)

7. Accurately weigh and record the weight of the media.

$$\text{Weight}_{\text{media with biomass}}\ (\text{g}) = \text{Weight}_{\text{pan with media}}\ (\text{g}) - \text{Weight}_{\text{pan tare}}\ (\text{g}) \tag{4.1}$$

Removal of biomass from media

8. Place the media elements in a 500-mL wide-mouthed Erlenmeyer flask containing concentrated hydrogen chloride to obtain pH of 2.
9. Soak the carriers for 15 minutes.
10. Pour off the choric acid solution while retaining the media elements. Rinse the elements with deionized water by partially filling the flask.
11. Place the media in Erlenmeyer flask containing NaOH (pH of 10) for 15 minutes.
12. Wash the carriers with deionized water to remove the biomass from the media.
13. The media is clean now. If not, repeat steps 7 through 9.

Determination of cleaned media weight

14. Dry the pan with media to constant weight in an oven at 105 °C (1 hour).
15. After drying to constant weight, place the pan with media in a desiccator to cool to room temperature (0.25 hour).
16. Accurately weigh and record the weight of the clean media.

$$\text{Weight}_{\text{cleaned media}}\ (\text{g}) = \text{Weight}_{\text{pan with cleaned media}}\ (\text{g}) - \text{Weight}_{\text{pan tare}}\ (\text{g}) \tag{4.2}$$

Total attached biomass calculations

17. Calculate the weight of the attached biomass as the difference between the weight of the media with biomass (Step 6) and the cleaned media without biomass (Step 15).

$$\text{Weight}_{\text{attached biomass}}\ (\text{g}) = \text{Weight}_{\text{media with biomass}}\ (\text{g}) - \text{Weight}_{\text{cleaned media}}\ (\text{g}) \tag{4.3}$$

18. Calculate the unit biomass (biomass per square meter surface area available) by dividing the biomass weight (step 17) by the number of elements sampled (step 5) and using standard conversions for number of elements per cubic meter and square meters of surface area per cubic meter:

$$\text{Unit biomass}\left(\frac{8}{m^2}\text{carriers}\right) = \frac{\text{Weight}_{\text{attached biomass}}(g) \bullet 1000\frac{mg}{g} \bullet 149\ 000\frac{\text{carriers}}{m^3}}{\text{No. of carriers sampled} \bullet \frac{650\ m^2}{m^3}} \quad (4.4)$$

CONSIDERATIONS FOR INTEGRATED FIXED-FILM ACTIVATED SLUDGE SYSTEMS

Power Demand

Coarse-bubble aeration typically is used to ensure an equal distribution of the buoyant media throughout the mixed liquor column and to reduce the need for and frequency of access to aeration basins. Coarse-bubble aeration is not as efficient as fine-bubble aeration. In addition, some suppliers prescribe high DO concentrations, further increasing aeration requirements and power demand.

Red Worms

As mentioned above, the fixed-bed media process may require higher aeration rates to maintain sloughing rates causing high DO concentrations. This higher DO concentration and fixed static media may result in the growth of red worms in the media. The worms are predators of the biofilm, and their presence means a reduction in the biofilm available to perform treatment. The worms may affect system capacity and performance. In addition, if aeration is lost or access to the tank is required, the red worms may die and become a source of odors, which may be difficult to clean up and remove.

Storage of Media

The moving and fixed-media products occupy large volumes within the aeration basin. During construction, contractors will need a significant site area for staging. During operation, if access is needed to an aeration tank, then some method of storage of the media or transfer to and from an adjacent tank is required. Also, free-floating media and sponges are typically delivered in plastic sacks, and sunlight will degrade the integrity of the sacks with long-term outdoor storage. Contractors should cover the sacks with tarps or store them in shipping containers or within warehouse(s) until installation.

Foaming

There is a risk of foam accumulation in basins and zones designed to retain free-floating media, which may not allow foam to leave the basin or zone. Foam accumulation may get worse with increasing SRT. If the facility tends to accumulate foam, either seasonally or because of changes in the influent characteristics, then it may be beneficial to limit the suspended growth SRT and use more biofilm media instead. This is not a concern in fixed-bed or sponge-type systems because they do not use in-basin screens to retain media. The IFAS systems appear to be prone to *Nocardia*-form foaming, particularly with nitrogen-removal configurations (WEF et al., 2018). Every method recommended for foam control should be available for use by operators of these systems including surface chlorine sprays, surface wasting, RAS chlorination, redirection of all scum and surface waste to prevent reseeding, and defoaming polymer.

Hydraulic Limitation and Loss of Media

The mechanical components are designed with similar design standards for both MBBR and IFAS systems. Each of these process mechanical components influences the hydraulic throughput of MBBR and IFAS systems. The instances of hydraulic failures and media loss are greater with IFAS systems compared to MBBR. There are six IFAS systems that have been subjected to a hydraulic failure, which resulted in plastic biofilm carrier loss (WEF et al., 2018). These failures were public and under the scrutiny of news media. The anecdotal evidence suggests that the hydraulic failures have mostly occurred during construction. Therefore, under these circumstances, limited information exists to describe the conditions that ultimately resulted in the hydraulic failure and biofilm carrier loss.

Leaf et al. (2011) gives accounts of five case studies that illustrate hydraulic failures that resulted in plastic biofilm carrier loss from IFAS processes, with the goal of identifying general mechanisms of hydraulic failure and biofilm carrier loss and recommending applicable design guidelines to avoid hydraulic failure and biofilm carrier loss. These case studies include IFAS installations at

- Broomfield, Colorado
- Raisio, Finland
- Groton, Connecticut
- Hooksett, New Hampshire
- Mamaroneck, New York

The following modes of failure and design guidelines were identified (WEF et al., 2018).

Flat Screen Failure

A loss of flat screen hydraulic throughput is a mode of system hydraulic failure. The absence or inefficiency of air scour device can contribute to the retention of screen accumulation. Plastic biofilm carriers or trash accumulation on the screen surface due to insufficient screen area or excessive approach velocity, which causes hydraulic throughout and substantially reduced and peak hydraulic flows, may result in basin overflow and plastic biofilm carrier loss.

Cylindrical Screen Failure

Cylindrical screens, if not reinforced, could lead to structural failure, which causes plastic biofilm carriers to migrate to the downstream unit processes. Trash accumulation due to insufficient pretreatment causes hydraulic throughput to substantially reduce in addition to peak hydraulic flows that may result in basin overflow and plastic biofilm carrier loss. Biofilm carrier accumulation on the screen surface resulting from insufficient screen area causes too high media fill fraction, improper depth of submergence, and insufficient underneath aeration intensity. Combined with excessive approach velocity, this causes reduction of hydraulic throughput and peak hydraulic flows and may result in basin overflow and plastic biofilm carrier loss.

Avoiding Hydraulic Limitation and Media Loss

Good preliminary treatment for screening and grit removal to prevent inert material accumulation within the bioreactors, as needed. Design features to overcome hydraulic limitations inherent to the IFAS process are as follows:

- Redirecting flow perpendicular to normal basin flow scheme
- Bypassing wet weather flows around the IFAS zone
- Split internal mixed liquor flow using a series of pumps
- Process control with instrumentation to overcome hydraulic limitations during wet weather conditions

Design of Retention Screen

- Typical headloss of 50 to 150 mm across each screen-containing wall at the peak hydraulic flow (McQuarrie & Boltz, 2011)
- Hydraulic loading rate typically less than 60 $m^3/m^2 \cdot h$ under all flow conditions
- Cylindrical screen submergence of 35% to 65% of the reactor sidewater depth
- Cylindrical screen with structural support to resist forces exerted by the plastic biofilm carrier

MAINTENANCE

As described, the major in-basin components are the Bio-Carrier media, aeration system, and media retention screens, and they are designed to have minimal maintenance. The Broomfield, Colorado, IFAS system was started up in November 2002 and has used the same aeration system, media retention sieves, and media (18 years of operation) with little to no maintenance on the in-basin components. Items like blowers, anoxic mixers, instrumentation, and any other rotating equipment used for RAS/WAS/MLR all need to have specific service, which would be similar to that of any standard activated sludge facility.

TEST YOUR KNOWLEDGE

1. Which of IFAS or MBBR is a stand-alone system with no RAS?

2. What control parameter is used for BOD and nitrification?

3. An MBBR can only be used for post-denitrification.
   - ☐ True
   - ☐ False

4. What is typical aerobic SRT range for IFAS systems?

5. The key to the IFAS process is the increased total biomass inventory that can be achieved by combining the attached biomass with the suspended MLSS.
   - ☐ True
   - ☐ False

6. Integrated fixed-film activated sludge systems are often implemented to improve nitrification capacity at existing facilities.
   - ☐ True
   - ☐ False

7. Fabric, web-type media used in IFAS systems is difficult to install but is not prone to fouling by rags.
   - ☐ True
   - ☐ False

8. Polypropylene finned cylindrical shaped media provides excellent mixing and can be converted to an MBBR system.
   - ☐ True
   - ☐ False

9. It can be difficult to maintain the aeration system when using free-floating media.
   - ☐ True
   - ☐ False

10. A difference between IFAS and MBBR systems is the lack of RAS recycle with MBBR systems.
    - ☐ True
    - ☐ False

11. Integrated fixed-film activated sludge media are commonly made of plastic, foam, or fabric material.
    - ☐ True
    - ☐ False

12. Polyvinyl sheet IFAS media have a high initial cost and are maintenance intensive.
    - ☐ True
    - ☐ False

13. In MBBR systems, biofilms primarily develop on the protected surface inside of the plastic biofilm carrier.
    - ☐ True
    - ☐ False

14. Integrated fixed-film activated sludge and MBBR treatment facilities cannot be incorporated to existing facilities and must be constructed as stand-alone unit processes.
    - ☐ True
    - ☐ False

15. In a denitrifying MBBR reactor, the biofilm media is uniformly distributed by mechanical mixers.
    - ☐ True
    - ☐ False

16. Screens with mesh sizes of 20 to 35 mm are used for retaining the plastic carriers within the MBBR basin.
    - ☐ True
    - ☐ False

17. Tertiary or add-on MBBR processes receiving wastewater that received significant upstream treatment do not require additional screening.
    - ☐ True
    - ☐ False

18. Because the specific gravity of MBBR media is between 0.94 and 0.96 g/cm^3, both clean and biofilm-covered plastic biofilm carriers will sink in calm water.
    - ☐ True
    - ☐ False

19. Because of the turbulent motion of the carriers in the reactor, the biofilm thickness will become excessive.
    - ☐ True
    - ☐ False

20. Automated wiping systems are used to remove accumulated debris from cylindrical sieves.
    - ☐ True
    - ☐ False

21. Integrated fixed film activated sludge systems are limited in the amount of nitrification that can be achieved.
    - ☐ True
    - ☐ False

22. An MBBR system installed in Norway that had been routinely inspected showed excessive wear on the plastic biofilm carrier in less than 5 years of continuous operation.
    - ☐ True
    - ☐ False

23. Coarse-bubble diffusers used in MBBR systems are prone to scaling and fouling because of the large orifice dimension, and therefore require more maintenance than fine-bubble diffusers.
    - ☐ True
    - ☐ False

CHAPTER SUMMARY

MICROBIOLOGY

- Higher life forms typically provide insight to the biomass age and stability.
- Protozoa include amoebas, flagellates, and ciliates.
- A large increase in flagellates compared to recent microbial analysis results likely indicates a recent shock that affected the biomass. The smallest flagellates are the most indicative of recent upsets to a system.
- A large increase in flagellates compared to recent microbial analysis results likely indicates a recent shock that affected the biomass. The smallest flagellates are the most indicative of recent upsets to a system.
- Documenting the presence of free-swimming ciliates during optimum performance and poor operations helps the operator detect changes to the biomass.
- Suctoria do not like higher ammonia concentrations and are commonly found in a stable nitrifying biomass.
- Metazoa are animals composed of more than one cell. They are larger and more complex compared to protozoa. Included in the metazoa group are rotifers. Rotifers play an active role in the breakdown of organic wastes. Nematodes, water bears, bristle worms, water fleas, and seed shrimp are all metazoa that may be found in a wastewater treatment process.
- Most metazoa have been associated with activated sludge with longer MCRTs.
- Every facility is its own environmental system, with various growth pressures and operating conditions. The presence or absence and type of higher life forms will not tell you exactly what is wrong with the system but can be combined with a further in-depth look at the operation to optimize treatment conditions.

SEQUENCING BATCH REACTORS

- Sequencing batch reactor systems differ from activated sludge facilities because they combine biological treatment and solids settling and clarification into a single basin, or tank. An SBR is an activated sludge system that operates in time rather than space.
- There are two SBR main control approaches: flow paced (flow control) operation and time based (time control) operation.
- The fill-and-draw SBR operation involves five basic steps of fill, react, settle, decant, and idle/sludge waste.
- Nitrogen and phosphorus removal can be accomplished using SBRs.
- Because of their on-off feed pattern, SBRs present some unique operational challenges. For example, the smaller the design of a facility, in terms of flow, the more problematic temperature may become due to the loss of heat from the SBR during cold nights without wastewater addition to sustain the temperature.
- The operator should have the knowledge and ability to fully control (i.e., modify) the facility-operating parameters, such as (but not limited to) cycle times, volumes, and setpoints.

IFAS AND MBBR SYSTEMS

- Integrated fixed-film activated sludge and MBBR processes have become increasingly popular technologies for WRRFs typically trying to cost-effectively increase cold weather nitrification and total nitrogen removal capacity.
- An MBBR is a treatment system with buoyant free-moving plastic biofilm carrier media that require mechanical mixing or aeration for uniform distribution throughout the tank. The process includes a submerged biofilm reactor and liquid–solids separation unit.
- Integrated fixed-film activated sludge facilities are those that incorporate a fixed-film process component (a biofilm growth) within activated sludge basins. The IFAS facilities operate with a level of MLSS concentration similar to CAS systems with a sludge age that improves sludge settleability and provides carbonaceous removal, nitrification, and/or BNR.
- When nitrifiers are present in the biofilm of the IFAS system, biofilm sloughing seeds the mixed liquor with nitrifiers and significantly increases the fraction of nitrifiers in the MLSS compared to an activated sludge system operating at the same SRT. The time required for an IFAS system to recover full nitrification is reduced compared to a system with activated sludge only, due to seeding.
- Operation of IFAS systems is similar to that of conventional activated sludge, meaning the use of SRT and wasting strategies are implemented. For IFAS systems, an equivalent SRT is calculated by dividing the total biomass (in the biofilm and mixed liquor) by the daily wasting rate.

References

Boltz, J. P., Johnson, B. R., Daigger, G. T., Sandino, J., Elenter, D. (2009). Modeling integrated fixed film activated sludge (IFAS) and moving bed biofilm reactor (MBBR) systems II: Evaluation. *Water Environment Research, 81,* 555–575.

Calaway, W. T. (1968). The metazoa of waste treatment processes—Rotifers. *Journal of the Water Pollution Control Federation, 40,* 412.

Curds, C. R. (1992). *Protozoa in the water industry.* Cambridge University Press.

Gerardi, M. (1986). An operator's guide to protozoa. *Public Works, 7,* 44.

Jahn, T. L., Bovee, E. C., & Jahn, F. F. (1979). *How to know the protozoa* (The Pictured Key Nature Series, 2nd ed.). William C. Brown Publishers.

Johnson, C., & Boltz, J. P. (2013). Aeration system design in integrated fixed-film activated sludge (IFAS) and moving bed biofilm reactors (MBBR) using stainless steel pipe diffusers, manifold, and down pipes. *Proceedings of the 86th Annual Water Environment Federation Technical Exposition and Conference* [CD-ROM]; New Orleans, Louisiana, Oct 5–9; Water Environment Federation.

Leaf, W., Boltz, J. P., McQuarrie, J. P., Menniti, A., & Daigger, G. T. (2011). Overcoming hydraulic limitations of the integrated fixed-film activated sludge (IFAS) process. *Proceedings of the 84th Annual Water Environment Federation Technical Exhibition and Conference* [CD-ROM]; Los Angeles, California, Oct. 15–19; Water Environment Federation.

McQuarrie, J. P., & Boltz, J. P. (2011). Moving bed biofilm reactor technology: Process applications, design, and performance. *Water Environment Research, 83,* 560–575.

New England Interstate Water Pollution Control Commission. (2005). Sequencing batch reactor design and operational considerations. Available for download at https://neiwpcc.org/

Ødegaard, H. (2006). *Innovations in wastewater treatment: The moving bed.* IWA Publishing.

U.S. Environmental Protection Agency. (2010). *Manual: nutrient control design* (EPA-600/R-10-100). U.S. Environmental Protection Agency, Office of Research and Development, National Risk Management Research Laboratory.

University of Florida, TREEO Center. (2000). *Sequencing batch reactor operations and troubleshooting manual.*

Water Environment Federation. (1995). Earthworms in a lively plant. *Water Environment Technology, 7*(6), 14.

Water Environment Federation. (2011). *Biofilm reactors* (Manual of Practice No. 35).

Water Environment Federation. (2013). *Automation of water resource recovery facilities* (4th ed.; Manual of Practice No. 21).

Water Environment Federation. (2017a). *Operation of water resource recovery facilities* (7th ed.; Manual of Practice No. 11).

Water Environment Federation. (2017b). *Wastewater biology: The microlife* (3rd ed.).

Water Environment Federation. (2018). *Activated sludge and nutrient removal* (Manual of Practice No. OM-9).

Water Environment Federation, American Society of Civil Engineers, & Environmental and Water Resources Institute. (2018). *Design of water resource Recovery Facilities* (6th ed.; WEF Manual of Practice No. 8/ASCE Manuals and Reports on Engineering Practice No. 76). Water Environment Federation.

Water Environment Federation & Association of Boards of Certification. (2018). *Wastewater treatment fundamentals I–Liquid treatment.* Water Environment Federation.

Suggested Readings

AECOM/Metcalf & Eddy, Inc. (2014). *Wastewater engineering: Treatment and resource recovery* (5th ed.). Tchobanoglous, G.; Burton, F.; Stensel, D.H.; Abu-Orf, M.; Bowden, G.; & Pfrang, W. (Eds.). McGraw-Hill.

Eikelboom, D. H. (2000). *Process control of activated sludge plants by microscopic investigation.* IWA Publishing; p 102.

Grady, C. P. L. Jr., Daigger, G. T., Love, N. G., & Filipe, C. D. M. (2011). *Biological wastewater treatment* (3rd ed.). CRC Press.

Jenkins, D., Richard, M., & Daigger, G. (2004). *Manual on the causes and control of activated sludge bulking, foaming, and other solids separation problems* (3rd ed.). Lewis Publishers.

Minchin, E. A. (2006). *Protozoa microbiology and guide to microscopic identification.* Merchant Books.

Rittmann, B. E., & McCarty, P. L. (2000). *Environmental biotechnology: Principles and applications.* McGraw Hill.

Society of Protozoologists. (2000). *The illustrated guide to the protozoa* (2nd ed.). Allen Press.

Stevens, G. (2015). BNR process using RAS anaerobic zone. WEF Nutrient Symposium; San Jose California, July 26–28; Water Environment Federation.

Water Environment Federation. (2011). *Nutrient removal* (Manual of Practice No. 34).

Water Environment Federation. (2013). *Operation of nutrient removal facilities* (Manual of Practice No. 37).

Water Environment Federation. (2018). *Glossary of water resource recovery facility design terms.* accesswater.org

Water Environment Federation. (2021). *Energy in water resource recovery facilities* (2nd ed.; Manual of Practice No. 32).

CHAPTER 5

Nontraditional Disinfection

Introduction

This chapter discusses nontraditional disinfection methods such as ozone and peracetic acid being used at water resource recovery facilities (WRRF). Ozone was incorporated into WRRFs that used the pure oxygen activated sludge process for biological treatment. At these facilities, excess pure oxygen was used as the feed source to generate the ozone. More recently, ozone disinfection is being used more at WRRFs because it can result in low bacteria counts and remove contaminants of emerging concern (CECs).

Refer to Chapter 10 of *Wastewater Treatment Fundamentals I—Liquid Treatment* (WEF & ABC, 2018) for detailed information about chlorine and UV.

Peracetic acid (PAA) is an effective disinfectant for removing fecal coliforms and *Escherichia coli*. It can be an affordable alternative to chlorine when WRRFs have disinfection byproducts as part of National Pollutant Discharge Elimination System (NPDES) permit requirements. Peracetic acid has become a disinfection system that more WRRFs are implementing.

Other disinfection methods such as on-site generation of chlorine, hydro-optic disinfection UV (HOD UV), and UV C light-emitting diode (UVC LED) are being implemented at some WRRFs because of potential localized, minimized supply chain issues. These technologies can be used economically at small and point-of-use WRRFs.

Chlorine and ultraviolet (UV) light have been the primary choice for the majority of wastewater disinfection operations in the United States.

Finally, the concept of combining two disinfection processes like chlorine and UV to achieve increased levels of microorganism removal is discussed as an overview of the operation of advanced oxidation processes (AOPs) to achieve microbial disinfection and removal of CECs.

LEARNING OBJECTIVES

Upon completing this chapter, you will be able to:

- Understand the operational practices needed to successfully operate an ozone disinfection process.
- Demonstrate calculations for determining ozone dose for various water quality conditions.
- Troubleshoot common mechanical and process control problems for an ozone disinfection system.
- Understand principles necessary for operation of a PAA disinfection system.
- Overview operational practices for on-site generation of chlorine gas, HOD UV, and UVC LED.
- Overview operational practices using multiple disinfection technologies and AOPs.

Purpose and Function

This chapter will focus on operation and maintenance activities associated with nontraditional disinfection practices. The disinfection process destroys and inactivates bacteria and pathogens in the treated effluent. The sections in this chapter have been divided accordingly to discuss nontraditional disinfection approaches that are available in the wastewater industry. Although the actual number of utilities using ozone in the United States is quite limited (fewer than 20), it is growing due to its ability to remove many CECs. Peracetic acid has been included in this chapter because the number of operating facilities is approaching 50 in the United States; as such, PAA is becoming an accepted approach for achieving basic-level disinfection NPDES discharge requirements. Additionally, this chemical has successfully been used in Europe to disinfect wastewaters to both basic and reuse requirements for many years.

Disinfection allows for the destruction and inactivation of bacteria and pathogens before the treated effluent is discharged to a receiving stream.

TEST YOUR KNOWLEDGE

1. **Disinfection will destroy and inactivate which type of pollutants?**
   a. Biochemical oxygen demand (BOD)
   b. Total suspended solids (TSS)
   c. Metals
   d. Bacteria and pathogens

2. **Name a nontraditional disinfection technology.**
   a. Chlorine gas
   b. Sodium hypochlorite
   c. UV
   d. Ozone

3. **Which disinfection has become an accepted form of wastewater disinfection over the past several years?**
   a. Chlorine gas
   b. Sodium hypochlorite
   c. UV
   d. PAA

Ozone Disinfection

THEORY OF OPERATIONS

Ozone disinfects wastewater by oxidizing the cell wall, cellular bonds, and nucleic material of bacteria, viruses, protozoa, and other microorganisms. Ozone produces radical byproducts that also impart damage and destruction of microorganisms as well as CECs.

The effectiveness of ozone for disinfection depends on several factors, including target pathogen, the pathogen's exposure to ozone, particulate shielding, and temperature. Ozone is an unstable gas and must be generated on-site and then transferred into the treated water. The relationship between dosage, demand, and residual for an oxidizing disinfectant can be described by the following:

$$\text{Dose} = \text{Demand} + \text{Residual}$$

Once transferred, a large portion of the initial ozone concentration rapidly decays within the first few seconds ("instantaneous demand"); the remaining concentration ("residual") decays more slowly, with a typical half-life of less than 30 seconds in municipal wastewater effluent. However, because ozone decays so quickly, and much of the actual cT (a measure of the effectiveness of a disinfection process) is accounted for before measurement is practical, treatment trains should be designed using site-specific bench and pilot studies. Operations staff need to ensure that instrumentation is availble and operational to measure the ozone residual.

For a given pathogen, inactivation performance is partially a function of cT, which is, in turn, determined by ozone kinetics. Ozone is unstable in aqueous solution, and will proceed along one of two reaction pathways: the direct pathway or the indirect pathway. Relatively few substrates react directly with ozone, so the direct pathway is the slower and more selective of the two. However, in some cases, ozone will form hydroxide radicals (.OH) or other radicals in its reaction products. These radicals react quickly and non-selectively with neighboring molecules, sometimes forming new radicals and perpetuating the reaction cycle. Ozone demand can be calculated based on the ozone concentration determined during pilot testing. An example calculation is as follows:

$$\text{Ozone} = \text{Ozone dose from testing} \times \text{Flowrate}$$

This indirect pathway greatly increases the ozone reaction rate and range of potential substrates. The removal of bacteria and, therefore, the final effluent concentraton of bacteria is based on the residual of ozone (mgL) times the detention time (cT).

$$cT = c\ (\text{residual concentration}) \times \text{contact time (minutes)}$$

so

Residual concentration = 0.1 mg/L
Contact time = 10 minutes
cT = residual concentration (0.1 mg/L) × Contact time (10 minutes) = 1 mg-min/L

Unlike other chemical feed systems, the feed rate (kg/h or lb/d) for ozone cannot be measured directly; the ozone feed rate is calculated from other measured parameters. Specifically, ozone production is calculated from the mass flowrate and the ozone concentration according to the following formula:

$$\text{Production rate (kg/h or lb/d)} = \text{gas flow (m}^3\text{/min or scfm)} \times \text{ozone concentration (\%)} \times 1.2$$

Ozone is fed as a gas; therefore, the volume depends on both pressure and temperature. In addition, ozone is produced as a mixture of ozone and oxygen. Both of these conditions make measurement of the production rate difficult and may lead to inaccurate calculations of the ozone feed rate. The calculation is done in and is then displayed on the control system.

The main components of the ozone disinfection system include the following (see Figure 5.1):

- Liquid oxygen (LOX) supply (LOX or sidestream)/storage and feed system
- Ozone generation facilities
- Sidestream injection system
- Ozone contactor system
- Ozone offgas destruct system
- Ozone quenching system
- Sampling and monitoring system

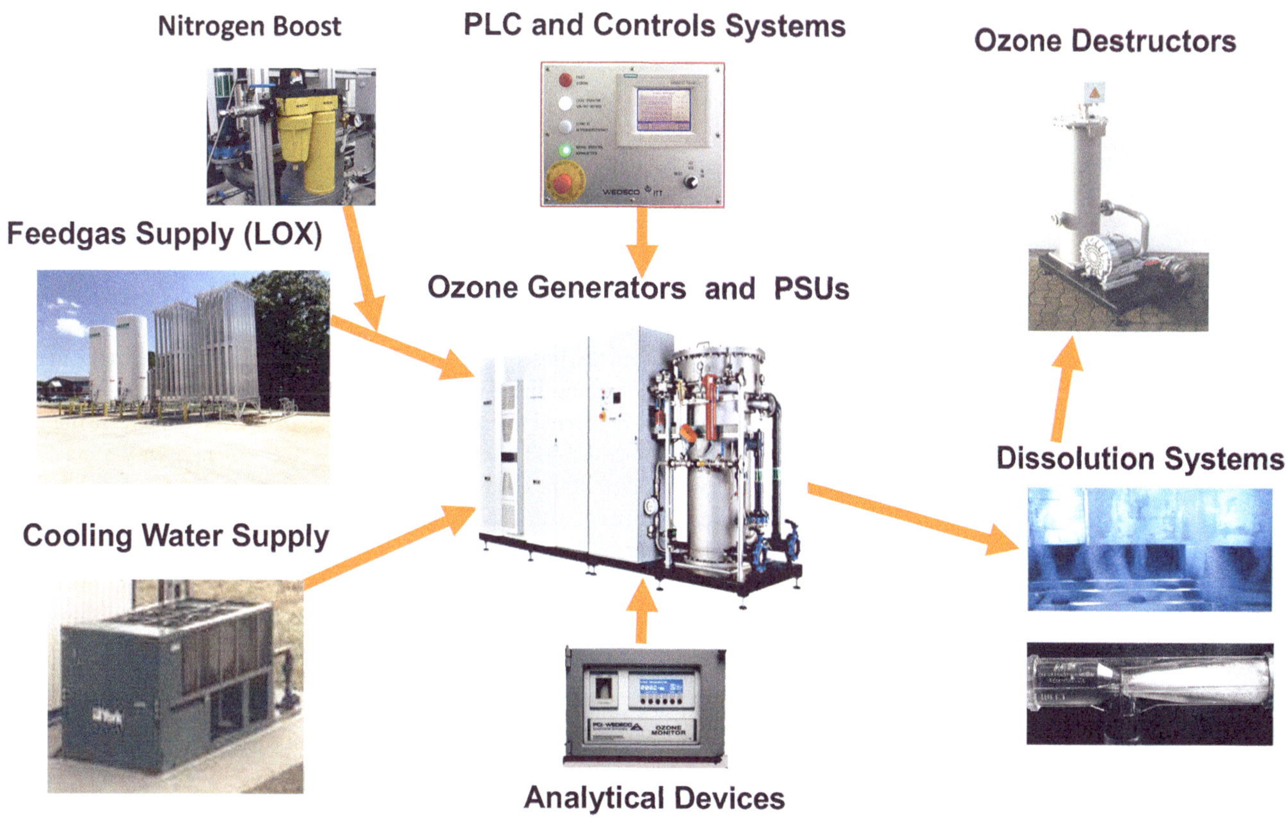

Figure 5.1 Ozone Disinfection System Key Components (Permission of Xylen) Oxygen Supply (LOX or Pure Oxygen) Storage and Feed System (Graphic courtesy of Xylem, Inc.)

The LOX and ozone quenching agent (i.e., calcium thiosulfate) bulk storage tanks are generally located outside and next to the contactors. The LOX system and the ozone generators, power supply units, and associated instrumentation and controls are housed inside a building. The ozone sidestream injection, contactors, offgas destruct units, and sampling and monitoring systems are located at the contactors.

Ozone is produced in an ozone generator by using oxygen or air as a feed gas. At WRRFs, that can come from the oxygen supply system (if the WRRF is a pure oxygen activated sludge plant) or a separate supply like LOX. The feed gas is passed through electrodes carrying a high voltage (6000 to 20 000 V), which produces a corona. The feed gas is converted to ozone, and the efficiency depends on the type of feed gas—air (1% to 4%) or oxygen (1% to 10%). The use of pure oxygen yields ozone between 4% and 12%, by weight. Approximately 80% to 95% of the energy will be converted to heat and must be removed at the ground electrode, through cooling water. The operational variables are the applied power, efficiency of the generator, flow of feed gas, and temperature.

SIDESTREAM INJECTION SYSTEM

The feed gas treatment must be designed to remove moisture from the feed gas. The recommended moisture concentration in the generator is <1 mg/L (<1 ppm) by weight. For the ozone to do its work of disinfection and oxidation, it must be brought into the water and dispersed as finely as possible. This is generally accomplished through fine-bubble diffusers located in baffle chambers or in a turbine-type contactor via sidestream injection. The selection of either using diffusers or sidestream injection is dependent on the dose. Diffusers may be more efficient in the transfer of ozone when higher doses may be required.

Sidestream systems with venturi injectors are also becoming increasingly popular. Increasingly, sidestream injection is used for ozonation, which involves splitting off a portion of the main flow into a side stream. Ozone is injected to this side stream and then the side stream is mixed back into the main flow. The sidestream injection system is composed, at a minimum, of a booster pump, a venturi, and nozzles for introducing the side stream back into the main flow. A sidestream injection system may also include degassing equipment and inline mixers.

COOLING SYSTEM

Finished facility water is generally used in ozone cooling systems at WRRFs in cases where packaged chillers are not used. When finished facility water from nonpotable systems is used, this can result in changes to material selection of the cooling systems. Finished water can be more aggressive, such as beneficial reuse reverse osmosis or operational/maintenance temperatures (such as more frequent heat exchanger cleaning) on account of increased fouling due to more rapid biofilm formation and accumulation in nutrient-laden tertiary filtrate.

CONTACTOR/CONTACT CHAMBER

An ozone contactor has several compartments, in series, with bubble diffusers at the bottom. In the first compartment, the water flows downward against the rising bubbles, and, in the second compartment, the water flows upward. The chambers are covered to prevent the escape of ozone and to increase the partial pressure of the ozone in the contactor. Additional chambers follow to ensure necessary contact time between the ozone and the water. Each of the chambers has sampling ports, so that the ozone concentration in each chamber can be determined. This capability is needed to calculate the product of the concentration and detention time, to obtain the required cT value.

OFF-GAS SYSTEM

The offgas from ozone contactors generally exceeds the Occupational Safety and Health Administration limit of 0.1 mg/L (0.1 ppm) by volume; therefore, the remaining ozone has to be recycled or destroyed. The offgas is first passed through a demister, which traps small water droplets on stainless steel mesh. The gas is then heated and passed through a destruction unit, which contains a catalyst to speed up the process. The power requirement is between 1 kW and 3 kW per 100 scfm (3 m^3/min) of gas flow.

TROUBLESHOOTING

Tables 5.1 through 5.5 are troubleshooting tables related to LOX storage, vaporizer, ozone generator, ozone contactor, processes. Process problems can occur due to the calculated feed rate being different than the actual feed rate as a result of instrument or sampling errors. The operator can be led to believe the current ozone feed rate is correct and, therefore, that the ozone dose is incorrect. However, the feed rate calculation may be incorrect. If the dissolved ozone residual is consistently high or low, the first remedial action is to confirm the feed rate calculation. Both the mass flow instrument and the high concentration analyzer are difficult to calibrate and subject to fouling.

EXPECTED PERFORMANCE

If water quality issues are accounted for within the design, ozone can be very effective in the removal of bacteria and pathogens. Ozone is experiencing a reinterest because of its ability to remove some types of CECs. The challenge is that if the ozone dose determined through pilot testing exceeds 5 mg/L, then there is potential to form bromated disinfection byproducts.

TEST YOUR KNOWLEDGE

1. **Which of the following statements about ozone is correct?**
   a. Ozone can remove contaminants of emerging concern.
   b. Ozone is practical if the facility's effluent contains high levels of suspended solids, chemical oxygen demand, or organic matter.
   c. Ozone is a less complicated disinfection method that chlorine and requires a more complex contact system.
   d. Costs for ozone are low because of the equipment and power requirements.

2. **The type and quantity of the target pathogen are the only factors that influence the ability of ozone to disinfect.**
   ☐ True
   ☐ False

3. **What are the two common methods to introduce ozone in the wastewater for disinfection?**
   a. Sidestream injection and mixers
   b. Sidestream injection and diffusers
   c. Mixers and diffusers
   d. Direct injection and mixers

Table 5.1 Liquid Oxygen Storage Process Troubleshooting

Condition	Possible Cause	Remedial Action
High ambient oxygen level alarm	Incorrect alarm setpoint	Verify alarm setpoints are correct.
	Oxygen leakage	Check all oxygen piping and joints for leaks.
	LOX tank filled improperly	Review the fill procedure with the LOX supplier to ensure the tricock valve is wide open and the tank is not over-pressurized during the fill.
	Multifunction oxygen analyzer instrument	Check calibration of oxygen monitor. Calibrate as required.
Liquid oxygen tank low-level alarm	Low tank level	Check records to determine when tank was filled last. Refill if required.
	Multifunction level analyzer	Check level transmitter for proper operation. Calibrate as required.
	System leakage	Check all piping for leaks.
Liquid oxygen tank pressure low	Low tank level	Check records to determine when tank was filled last. Refill if required.
	Malfunctioning pressure building coil	Check pressure building coil valves are open and pressure regulator setpoint is 517 kPa (75 psig).
	Multifunction pressure analyzer	Check pressure transmitter for proper operation.
	System leakage	Check all piping for leaks.
Liquid oxygen tank pressure high	Multifunction pressure analyzer	Check pressure transmitter for proper operation.
	Pressure buildup in the system	Check that pressure builds, setpoint is correct, the economizer line valves are open, and the tank is not isolated and LOX can flow freely to the vaporizers.

Table 5.2 Process Troubleshooting for Vaporizers

Condition	Possible Cause	Remedial Action
Gaseous oxygen (GOX) vaporizer discharge temperature low	Online time setpoint too high	Reduce cycle time to value within the specified range.
	Defrost problems	Troubleshoot vaporizer for defrost mode problems such as outlet valve remaining open.
Gaseous oxygen vaporizer discharge temperature high	Incorrect operation setpoint	Check that GOX heater is set up for proper operation.

Table 5.3 Process Troubleshooting for Ozone Generators

Condition	Possible Cause	Remedial Action
Gaseous oxygen header pressure low	Low GOX flow mass	Confirm proper GOX flow.
	Inlet valve incorrect position	Confirm the pressure-reducing valve (PRV) setting and verify valve is operating in accordance with the design.
	Header pipe leakage	Correct piping leaks.
	Pressure instrument malfunction	Recalibrate the pressure transmitter.
	GOX header pressure as sensed by pressure switch has triggered GOX header pressure low alarm.	Confirm the pressure switch setpoint and verify sensor is operating in accordance with the design.
Gaseous oxygen header pressure high	High GOX flow mass	Confirm proper GOX flow.
	Inlet valve incorrect position	Confirm the PRV setting and verify valve is operating in accordance with the design.
	Pressure instrument malfunction	Recalibrate the pressure transmitter.
Oxygen inlet high pressure	Flow path blockage	Verify flow control valves are in normal operating positions.
	Malfunctioning pressure regulator	Confirm upstream pressure regulating valve is operating properly.
	Alarm at ozone generator	Confirm the alarm setpoint and verify sensor is operating in accordance with the design.
Oxygen inlet low pressure	Flow path blockage	Verify flow control valves are in normal operating positions.
	Malfunctioning pressure regulator	Confirm upstream pressure regulating valve is operating properly.
	Depleted oxygen level in system	Verify LOX supply is not empty.
	Alarm at ozone generator	Confirm the alarm setpoint and verify sensor is operating in accordance with the design.
GOX inlet low flow	Flow path blockage	Verify flow control valves are in normal operating positions.
	Malfunctioning pressure regulator	Confirm upstream pressure regulating valve is operating properly.
	Depleted oxygen level in system	Verify LOX supply is not empty.
	Alarm at ozone generator controller interface	Confirm the alarm setpoint and verify sensor is operating in accordance with the design.
Product ozone analyzer low concentration/failure	Malfunctioning ozone analyzer	Verify analyzer has power and gas flow.
	Incorrect ozone concentration setpoint	Verify ozone concentration controller is set up for auto mode with proper setpoint.
	Low oxygen level	Verify oxygen flowrate is at proper value
	Malfunctioning ozone generator	Troubleshoot ozone generator for low ozone production per manufacturer's O&M manual
	Alarm at ozone generator controller interface	Confirm the alarm setpoint and verify sensor is operating in accordance with the design.

Table 5.3 Process Troubleshooting for Ozone Generators (*Continued*)

Condition	Possible Cause	Remedial Action
Ozone outlet temperature high	Temperature instrument malfunction	Verify a flow path for cooling water through the operating generator and that a shell cooling water pump is operating.
	High generated ozone concentration	Verify ozone concentration controller is set up for auto mode with proper setpoint.
	High oxygen flowrate	Verify oxygen flowrate is at proper operating value.
	High oxygen inlet temperature	Check that oxygen inlet temperature is below the maximum rated value.
	Cooling water temperature above maximum rating	Check that cooling water temperature is below maximum rating and flow is correct.
	Alarm at ozone generator	Confirm the alarm setpoint and verify sensor is operating in accordance with the design.
Inverter temperature high	Inverter cabinet temperature high	Verify cabinet air conditioner is operating properly
	Inverter cabinet temperature high	Confirm the cooling water temperature is below maximum rating and flow is correct
	Inverter cabinet temperature high	Verify cooling water flow path through the power supply unit equipment and that a cooling water pump is operating.
	Alarm at ozone generator	Confirm the alarm setpoint and verify sensor is operating in accordance with the design.
Inverter fault	Alarm at ozone generator	Confirm the alarm setpoint and verify sensor is operating in accordance with the design.
Cooling water flow low	Air trapped in cooling system	Remove excessive from the air cooling water system.
	Water leakage	Correct piping leaks.
	Water path blockages	Verify flow control valves are in the normal operating positions.
	Alarm at ozone generator	Confirm the alarm setpoint and verify sensor is operating in accordance with the design.
Cooling water outlet temperature high	High temperature water supply	Check cooling water supply for high temperature.
	Water path blockages	Verify flow path for cooling water through the shell and check that flowrate is adequate.
	Incorrect valve position	Confirm cooling system valve position.
	High ozone concentration	Verify ozone generator is operating below 10% ozone concentration.
	Temperature instrument malfunction	Check temperature switch setpoint and calibration.
	Alarm at ozone generator	Confirm the alarm setpoint and verify sensor is operating in accordance with the design.
High ambient ozone level alarm	Ozone leakage	Verify low-rate vent fans are on and dampers are open.
		Shut down and purge ozone system; allow room to clear of ozone.
		Check all ozone piping and tubing joints for leaks.
		Check ozone generator covers and high voltage bushing connections on shell for leaks.
	Ambient ozone monitor for ozone area triggers oxygen leak alarm	Confirm the alarm setpoint and verify sensor is operating in accordance with the design.
Yellow stains in ozone generator	Particulates in generator	Check oxygen filter downstream of vaporizers for element/cartridge failure.
	High dew point nitric acid in generator	Verify maintenance of supplemental air system and confirm providing air with low dew point
		Check for proper dew point of process oxygen –51 °C (–60 °F).
		Confirm the maintenance log to verify ozone generator is properly cleaned.

Table 5.4 Ozone Contactor Process Troubleshooting

Condition	Possible Cause	Remedial Action
High dissolved ozone concentration	Low water flowrate in the system	Verify proper process water flowing through the system.
	Incorrect dosage setpoint	Verify contactor dosage setpoint is correct.
	Malfunction ozone analyzer	Check analyzer operation and calibrate as required.
Low dissolved ozone concentration	High water flowrate in system	Verify process water flowrate through the system.
	Incorrect dosage setpoint	Verify that the contactor dosage setpoint is correct.
	Malfunctioning ozone analyzer	Check analyzer operation and calibrate as required.
Poor treated water quality	Incorrect ozone dosage	Check for proper ozone production setting.
	Poor supplied water quality	Check quality of process water entering the system.
	Low ozone transfer efficiency	Check transfer efficiency of ozone dissolution equipment.
	Excessive scum buildup in system	Check contactor for excessive scum buildup on the interior walls.

Table 5.5 Ozone Destructor Process Troubleshooting

Condition	Possible Cause	Remedial Action
Destructor heater over-temperature	Clogged gas path	Verify path through the destructor(s) and that unit is above minimum flow.
	Offgas blower	Confirm offgas blower is operating.
	Incorrect temperature setting	Check that temperature setpoint is correct on temperature controller and that controller is operating properly.
Destructor differential pressure high	High operating flowrate	Verify destructor(s) is operating at or below flowrate.
	Malfunctioning pressure gauge	Check that differential pressure gauge is operating properly and that switch setpoint is correct.
Destructor vent gas ozone high	High operating flowrate	Verify ozone destructor is operating within normal flowrate.
	Incorrect destructor heat setting	Check that operating destructor has heat temperature setpoint –1 °C (30 °F) above its inlet gas temperature. Adjust as required. Check operating destructor(s) has its temperature controller set up for auto mode.
	High ozone dosage applied to contactors	Check ozone offgas concentration is at or below 0.5%.
	Malfunctioning ozone analyzer	Switch to another ozone destructor; troubleshoot destructor for excessive ozone outlet.
Condensate in destructor	Moisture in destructor	Check demister pad for proper installation and cleaning.
	Improper destructor heater operation	Check destructor inlet heater for proper operation.
	Clogged destructor drain line	Verify condensate drain system is operating correctly.

Emerging Disinfection Methods

PERACETIC ACID DISINFECTION

THEORY OF OPERATION

Peracetic acid (CH_3COOH, PAA) is a strong oxidant that exists as an equilibrium mixture of water, hydrogen peroxide, and acetic acid. With an oxidation potential nearly as high as that of ozone, PAA is a highly reactive chemical solution that can both disinfect and oxidize contaminants in wastewater and drinking water. Furthermore, due to its reactivity, PAA does not persist in environmental spaces, decomposing into hydrogen peroxide and acetic acid. Although highly reactive, PAA solutions are generally stable and can maintain strength if kept in the proper conditions (i.e., low temperatures).

Disinfection (i.e., inactivation of microorganisms) and oxidation (i.e., degradation of pollutants) properties of PAA are mainly attributed to the formation of reactive oxygen species that may occur through numerous reactions. Specifically, disinfection by PAA occurs according to reactions that happen with microorganism cell walls, cell membranes, or metabolic functions.

The effectiveness of a disinfectant depends on the dose (i.e., how much is used) and the contact time. Peracetic acid is no exception. The contact time for a WRRF is determined by the design engineer when they size the contact chamber. Operators can influence the contact time by placing more chambers into service or removing them from service.

In general, the higher the dose of disinfectant used, the less time required to disinfect. For PAA, a dose of 2 to 3 mg/L is typically used at a contact time of 10 to 15 minutes. Minimum contact time may vary from one state to the next with design regulations.

OPERATIONAL PARAMETERS

Because PAA disinfection efficacy is site specific, system components for PAA systems are site specific as well.

Overall, proper sizing, application points, drainage, and maintenance are all key elements to ensure the desired disinfection performance.

> Peracetic acid should be fed into the contact basin at commercially available strength (i.e., 15%).

> For flexibility and maintenance, it is recommended that a contact tank should have two or more sections available.

MIXING DESIGN

One critical component that lies within the contact basin and helps to maximize use of the contact basin is the mixing component. There are three types of mixing that can be considered for use in PAA contact basins: hydraulic, static, and mechanical.

Static mixing occurs by adding PAA upstream of a fixed component such as a static mixer or baffle wall. The advantages are low maintenance requirements and low pump energy requirements. The disadvantages are that there is reduced efficiency as low flows, and pressure drops may reduce hydraulic capacity in the facility.

> Static mixing uses pressure drops across fixed components to generate mixing energy.

Mechanical mixing uses mechanical energy imparted by moving components to achieve a homogenous effluent stream. There are two types of mechanical mixing devices available for use: induction mixing and mechanical mixers. In induction mixing, a motor is submersed in the feed stream and PAA enters the unit as PAA is mixed with influent waters. In mechanical mixing, PAA is added upstream of the mixer and then mixes with the influent stream as the flow passes the mixer. The advantages are lower headloss compared to static and hydraulic mixing, good mixing over various flowrates, and complete mixing in properly sized tanks. The disadvantages are maintenance requirements, capital costs, and energy costs.

Hydraulic mixing makes use of the facilities' hydraulic profile to generate a mixed effluent. In ideal circumstances, this mixing can be achieved with already installed system elements such as a weir or flume. Peracetic acid is added immediately upstream of the hydraulic element. The advantages are low maintenance, completely mixed effluents, and that no additional energy is required. The disadvantages include dependence on headloss in existing hydraulic profile and being less efficient at lower flows.

STORAGE

Peracetic acid can be stored either inside or outside based on its properties. As such, there are applicable codes that must be followed. In general, PAA storage and feed buildings are classified as a Group H-3 and H-4 occupancy if they exceed storage limits allowed by the 2012 International Building Codes, 2012 International Fire Codes, and the 2012 International Mechanical Codes.

Definition: Group H-3, H-4 International Building Code classification for chemical storage. The "H" represents high hazard, and the number indicates severity of the hazard (1 being the highest, 5 being the lowest).

There are several codes that specify the maximum allowable quantities in each area (determined by sprinkler access), which can increase to 100% if sprinklers are present. Furthermore, specifications for ventilation systems, emergency power and alarms, spill control and prevention, and spacing of storage components are also required and given in the 2012 International Building Codes. For outdoor storage, weather protection, distance from public access, and secondary spill containment and management specifications are also provided, with not all being required, but recommended.

After determining where the PAA will be stored, the net key consideration is how much is to be stored. To determine appropriate storage volumes, minimum inventory, regulatory, and bulk storage considerations should be discussed. Minimum inventory volume should be based on the amount necessary to meet desired or regulated disinfection efficacy at average flows and average dose for at least 2 weeks plus delivery time. Inventory calculations should also include the total amount needed to cover peak flow events, dosage required for daily peak flow, and the number of sequential days the peak event happens. Another factor to consider is total space requirements on-site and within the storage tank themselves. Bulk storage tanks should be sized such that the working volume is 125% of the delivery volume, plus 10% heel, plus 10% head space at the minimum.

Minimum inventory is the basis for total storage volume, not the entire storage volume!

Peracetic acid is generally stored in high-density polyethylene and passivated stainless steel tanks. Bulk storage tanks are equipped with venting and pressure relief components that allow breathability and prevention of contamination, operator exposure to vapors, and damage from over-pressurization. Tanks also come equipped with a discharge nozzle. Piping connections to the nozzle must not create backflow conditions.

There are not many PAA suppliers compared to other disinfectants; thus, considerations should be given to delivery delays.

At a minimum, PAA storage tanks are fitted with some form of temperature monitoring system, PAA level indicators, and leak detection (for double-wall tanks only).

Typically, discharge nozzles are located at the bottom of the tank to create a flooded suction for feed pumps.

PROCESS MONITORING AND CONTROL

Similar to other chemical disinfectant systems, typical parameters that are monitored include flow, temperature, and PAA level, with the addition of vapors as well. Additionally, pressure sensors, storage temperature, chemical storage level, and secondary containment alert systems are all recommended to help ease maintenance and provide flexibility in processes. Chemical feed pump controls and calibration are critical to ensuring proper disinfection occurs, determining usable life, and determining when tubing or pump changes need to occur.

Flow-pacing can be used to deliver successful PAA doses; however, because water quality can be variable at WRRFs, other process controls may be necessary. Integral contact time dose pacing is another process control technique for PAA usage as is residual feedback systems. Completely online, the residual feedback system reacts to eliminate differences between the measured residual and desired PAA setpoint in the system. While nice, a drawback is the lag time that occurs in the system due to the placement downstream. The lag time, combined with rapid changes in flowrates, can result in over- and underestimates of the appropriate feed rates.

Feed-forward process control is another control technique that allows the operator to proactively dose the system in the proper way by taking into account PAA demand using statistical analysis of PAA residual as a function of different water quality parameters.

It is of utmost importance for operators to be fully aware of their system and make detailed observations when doing checks and rounds. Specifically, operators must observe and record chemical liquid and gauge levels, unusual noises, and leaks (refer to Tables 5.6 and 5.7).

Table 5.6 **Operator Checklist for Small Peracetic Acid Feed Systems** (WEF, 2020)

Item	Yes	No	Comments
What is the effluent flow?			
What is the PAA residual?			
Are there any tank leaks?			
Is there adequate chemical in storage?			
Do chemical totes need to be changed?			
Are process valves set properly?			
Are there leaks at feed pumps? If yes, which?			
What is the pump output speed?			
Are there any unusual noises coming from the pumps?			
Any housekeeping needs?			
Any maintenance needs?			

ANCILLARY SYSTEMS

Ancillary systems are systems that are key to storage and management of PAA. These systems include chemical transfer/unloading stations, air compressors/blowers, sump pumps, and spill management systems. Should these systems not be maintained properly, problems including corrosion, PAA contamination, or even worse service interruptions could happen, presenting safety and health concerns.

TROUBLESHOOTING

Like other chemical feed systems, PAA is most effective when it is properly dosed, has the correct contact time, and efficiently inactivates microorganisms. Many factors including pH, hydraulic considerations, dosing strategies, contact time, temperature, and organic and inorganic matter concentrations can play

Table 5.7 **Operator Checklist for Large Peracetic Acid Feed Systems** (WEF, 2020)

Item	Yes	No	Comments
What is the effluent flow?			
What is the PAA residual?			
Are there any bulk storage tank leaks?			
Is there adequate chemical in bulk storage?			
Are unloading transfer pumps ready for operation and functioning properly?			
Are process valves set properly?			
Are there leaks at feed pumps? If yes, which?			
What is the pump output speed?			
Are there any unusual noises coming from the pumps?			
Is the bulk storage tank leak and level instrumentation and other PAA instrumentation working properly?			
Any housekeeping needs?			
Any maintenance needs?			

Table 5.8 Peracetic Acid System Troubleshooting

Factor	Possible Issue	Action to Take
Electrical/instrumentation issues	Power failures; poor power quality; faulty wire connections	Backup power generation; regular electrical connection checks
Chemical dosing	Air locking; chemical aging; low residual	Offgas valves; chemical assay
Hydraulics	Short circuiting; blockages	Equipment cleaning; flow regulating
Upstream process effects	Poor BOD removal; inorganic coagulants for phosphorus removal	
Biofilm formation	Biofilm growth inside of storage tanks	Elevate disinfectant dose

a large role in the efficacy. Often, an issue with one of these factors can lead to issues with another factor. As such, issues with the following parameters can likely be attributed to one of the following categories outlined in Table 5.8.

TEST YOUR KNOWLEDGE

1. **The three types of mixing that can be considered for use in PAA contact basins are:**
   a. Hydraulic, static, and mechanical
   b. Hydraulic, diffused air, and mechanical
   c. Jet mixing, diffused air, and static
   d. Diffused air, mechanical, and jet mixing

Other Emerging Disinfectants

ON-SITE GENERATION OF CHLORINE GAS

Because of supply chain issues, on-site generation of chlorine gas has emerged as a cost-effective alternative to chlorine gas. On-site generation systems are increasingly popular for wastewater disinfection and include both unseparated and separated electrolyzer systems. One emerging operational change is to use on-site generation of a hypochlorite system to generate chlorine gas. It should be noted that the generation of chlorine gas on-site is not regulated by the Clean Air Act Risk Management Plan.

Although the product of each system is the chlorine gas, the electrolysis method can also result in variations of the hardness concentration and other properties of the feed material. Because crystallized salt is dissolved and used for electrolysis in brine systems, control of the calcareous components may be achieved using water softening or by selecting the desired quality of the crystallized salt. Figure 5.2 shows an on-site chlorine gas generation system.

On-site generation of chlorine gas can reduce the safety issues associated with the use of chlorine gas.

HYDRO-OPTIC DISINFECTION ULTRAVIOLET LIGHT

Hydro-optic disinfection UV is a system with total internal reflection-based design that, when coupled with the comprehensive monitoring of such critical parameters as flowrate, UV lamp intensity, and ultraviolet transmission percentage, allows the system to achieve and maintain the specified UV dose. Figure 5.3 presents a system schematic of HOD UV. The HOD UV technology measures four critical parameters in real time to maintain a specified UV dose: percent ultraviolet transmittance (% UVT), flowrate, UV lamp intensity (kW), and apparatus (consisting of total internal reflection and dose pacing). The system uses a proprietary total internal reflection (TIR)-based design that, when coupled with the comprehensive monitoring of critical parameters, allows the system to achieve and maintain the specified UV dose. The system's patented TIR technology, which is similar to fiber optic science, recycles UV light energy within the HOD UV chamber. This is especially important given that, in traditional UV

Figure 5.2 On-Site Generation of Chlorine Gas System (**Klorigen OSG/Anchorage Water and Wastewater Utility**)

systems, metal adsorbs or "detracts" the UV dose the closer it gets to metal, whereas the TIR enhances the UV dose. Application of HOD UV has demonstrated success in dechlorinating water and is just now receiving interest from wastewater facilities. Simply put, the UV photons are effectively lengthened and provide a greater opportunity to inactivate microorganisms and decompose the free chlorine. Hydro-optic disinfection UV has been applied in industrial applications for dechlorination and biofilm control.

Based on their use at industrial facilities, HOD UV systems are capable of achieving NPDES basic level disinfection criteria. Where the system has been used, operations staff have found it to be very reliable. This system has not received conditional acceptance by the Division of Drinking Water—California Department of Health Services.

Hydro-optic disinfection UV is typically applied in industrial applications for dechlorination and biofilm control.

ULTRAVIOLET LIGHT-EMITTING DIODE SYSTEMS

Ultraviolet C light-emitting diode systems (see Figure 5.4) based on a solid-state technology generally tend to operate with low electrical consumption and less relative space compared to conventional technology. A comparison of UVC LED to traditional UV systems applied to wastewater indicates that there were significant cost savings for this technology. Ultraviolet C light-emitting diode systems have similar

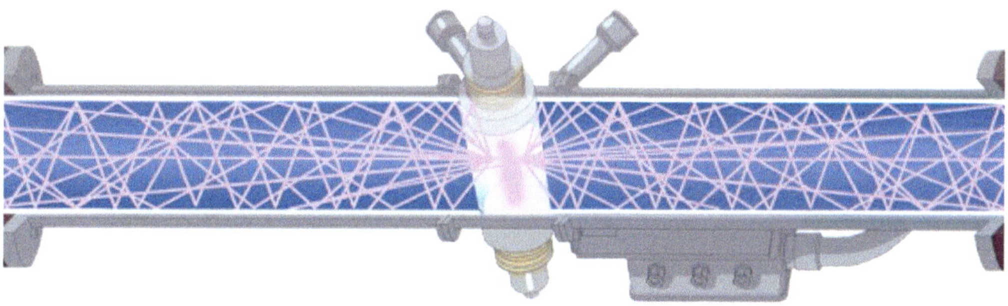

Figure 5.3 Hydro-Optic Disinfection Ultraviolet Light System (**Reprinted with permission from Atlantium Technologies**)

Figure 5.4 Ultraviolet C Light-Emitting Diode Disinfection System (Photo courtesy of Aqua-Aerobic Systems, Inc.)

challenges to those of more traditional UV systems. These systems have been more applicable at the point-of-use/point-of-entry market. The systems used in these markets are more "plug and play"; so, when the unit fails, it is replaced by another unit.

Testing conducted on UVC LED systems have shown these systems to be able to achieve basic level disinfection at very low flowrates 5 to 54 m^3/d (1 to 10 gpm) based on effluent quality.

MEMBRANES—MEMBRANE BIOREACTORS

Membrane bioreactor processes can be effective at removing pathogens.

Disinfection often is differentiated from liquid–solid treatment processes, including clarification, tertiary filtration, and membrane treatment. However, these processes can be effective at removing pathogens, resulting in the ability to decrease the size of downstream disinfection processes. Additional regulatory changes will be needed to allow for membrane bioreactors to be used for disinfection.

PASTEURIZATION

Pasteurization has been conditionally approved as a system that can achieve Tittle 22 high-level disinfection requirements.

Pasteurization uses the exhaust heat from an engine-driven generating unit as a heat source to inactivate pathogens. Pasteurization recently was approved for Title 22 disinfection in California, joining chlorine and UV as approved technologies. The process also results in the co-generation of electricity. Developed and named after Louis Pasteur in 1865, pasteurization is the process of heating liquids to destroy bacteria, protozoa, molds, and yeast. The effectiveness of the process is heavily dependent on liquid temperature and exposure time. Pasteurization has been used, for a long time, by the food industry to inactivate microorganisms in products like milk, eggs, and wine. Other applications include the treatment of wastewater sludge and disinfection of wastewater.

TEST YOUR KNOWLEDGE

1. An advantage of UVC LED on wastewater disinfection is its
   a. Ability to be "plug and play" for some point-of-use applications
   b. Higher energy use
   c. Shorter lamp life
   d. Low cost

2. On-site generation of chlorine gas has gained acceptance due to
   a. Ease of use
   b. Supply chain issues resulting in short supply and higher costs
   c. Similar chemical distribution system
   d. Cost

3. Hydro-optic disinfection UV has been successfully used for
   a. Wastewater disinfection
   b. Combined sewer overflow disinfection
   c. Reuse disinfection
   d. Dechlorination

Combined Disinfection Methods and Advanced Oxidation

The combination of two or more disinfection processes in the treatment train has some advantages. As previously discussed, the combination of UV and various oxidants can result in AOPs to remove trace organic pollutants. Additionally, multiple processes in series can provide an additional margin of safety through additional log removals; if a pathogen is not inactivated as it passes through the first process, the next process may inactivate it. Finally, combined processes may provide complementary pathogen inactivation, such that a disinfection process that provides effective inactivation in certain classes of pathogens, but not others, is complemented by another process that is known to be effective for the other pathogens.

In water reclamation systems, advanced chemical-oxidation processes are increasingly used to remove trace organic pollutants. However, they have not yet realized widespread use in wastewater treatment for disinfection. These processes use high doses of UV light and chemical oxidants, such as ozone or peroxide. Examples include the following:

A combination of two or more disinfection processes in the treatment train can provide advantages if the treatment facility is producing fit-for-purpose water.

- Peroxone (ozone + hydrogen peroxide)
- Ozone + UV
- Peracetic acid + UV
- Performic acid (PFA) + UV
- Chlorine + UV
- Hydrogen peroxide + UV
- Hydrogen peroxide + metal catalyst + UV
- Fenton's reagent
- Sunlight + metal catalyst

Use of these technologies results in a high level of disinfection, in addition to chemical oxidation, because the doses required for chemical oxidation are higher than for disinfection.

The development of the dose is for an advanced oxidation system is a function of the contaminant being removed and, therefore, bench scale and demonstration testing will need to be conducted before the equipment is selected.

TEST YOUR KNOWLEDGE

1. **Combined disinfection for systems include**
   a. PAA + chlorine
   b. Ozone + PAA
   c. Chlorine + UV
   d. PAA + PFA

2. **Advanced oxidation processes are used to remove**
   a. Trace organic pollutants
   b. Inorganics
   c. Heavy metals
   d. Chlorine residual

3. **What would be a reason to install a combined disinfection system?**
   a. Disinfection of fit-for-purpose water
   b. Economics
   c. Ease of operation
   d. Least amount of maintenance

CHAPTER SUMMARY

OZONE	■ All pathogens are target constituents in addition to some CECs. ■ Required dose is most influenced by the concentration and type of dissolved organic matter present, carbonaceous alkalinity, pH, and temperature. ■ Dose varies with target pathogen and must be determined through site-specific bench- and pilot-scale studies. ■ Formation of disinfection byproducts, especially bromate and aldehydes, is of concern. ■ Followed by a tertiary biological treatment step to remove BOD associated with natural organic matter. ■ Human exposure to ozone must be addressed promptly.
PAA	■ All pathogens are target constituents. ■ For secondary treated effluent, contact time is 10 to 15 minutes with a dose of 2 to 3 mg/L, dependent on effluent quality. ■ Required equipment for storage and dosing is similar to hypochlorite systems.
ON-SITE GENERATED CHLORINE	■ All pathogens are target constituents. ■ Contact time is 15 minutes at peak flow or maximum pumping rate. ■ Used when substantial quantities of stored sodium hypochlorite are required on-site and not available via chemical suppliers. ■ Various system configurations are available, and each has specific O&M considerations.
HOD UV LIGHT	■ All pathogens are target constituents in addition to biofilm control and de-chlorination. ■ Key operational parameters are flowrate, UV lamp intensity, and % UVT. ■ Generally applied for industrial wastewater treatment.
UVC LED	■ All pathogens are target constituents. ■ Low electrical consumption and less relative space required compared to conventional UV systems results in significant cost savings. ■ On a rapid development schedule. ■ Demonstrated for low flowrates for wastewater 5 to 54 m^3/d (1 to 10 gpm).
COMBINED DISINFECTION PROCESSES AND ADVANCED OXIDATION	■ Targets specific pathogens and trace organic pollutants. ■ Unit processes may be combined in series to achieve desired pathogen removal. ■ Results in a high level of disinfection.

References

Water Environment Federation. (2020). *Peracetic acid disinfection: Implementation considerations for water resource recovery facilities.*

Water Environment Federation & Association of Boards of Certification. (2018). *Wastewater treatment fundamentals I—Liquid treatment.*

Suggested Readings

Gottschalk, C., Libra, J. A., & Saupe, A. (2010). *Ozonation of water and waste water: A practical guide to understanding ozone and its applications* (2nd ed.). Wiley-VCH.

Leong, L. Y. C., Kuo, J., & Tang, C. (2008). *Disinfection of wastewater effluent—Comparison of alternative technologies.* Water Environment Research Foundation.

Metcalf & Eddy, Inc./AECOM. (2014). *Wastewater engineering: Treatment and resource recovery* (5th ed.). McGraw-Hill.

National Water Research Institute & American Water Works Association Research Foundation. (2003). *Ultraviolet disinfection guidelines for drinking water and reuse.* U.S. Environmental Protection Agency.

Rakness, K. L., Najm, I., Elovitz, M., Rexing, D., & Via, S. (2005) *Ozone in drinking water treatment: Process design, operation and optimization.* American Water Works Association.

U.S. Environmental Protection Agency. (1986). *Design manual: Municipal wastewater disinfection* (EPA-625/1-86-021).

U.S. Environmental Protection Agency. (1986). *Quality criteria for water.* U.S. Environmental Protection Agency, Office of Water Regulations and Standards.

von Sonntag, C., & von Gunten, U. (2012). *Chemistry of ozone in water and wastewater treatment: From basic principles to applications.* IWA Publishing.

Water Environment Federation. (2017). *Operation of water resource recovery facilities* (7th ed., WEF Manual of Practice No. 11).

CHAPTER 6

Membranes

Introduction

Membranes are thin, semi-permeable barriers. They are used to remove contaminants from water and wastewater. Semi-permeable membranes are more permeable to some things (like water) and less permeable or impermeable to other things (like salt or bacteria).

The size of the membrane pores determines what can pass through and what is held back. Membranes do not degrade contaminants like activated sludge; they physically remove them from the water. They can be compared to secondary clarifiers in an activated sludge process. As a result, membrane processes will always generate a waste stream that contains all the contaminants that were in the feedwater. Throughout this chapter, we will be using the following terms to discuss membrane systems:

Pores are openings or holes in the membrane.

Feedwater—The water entering the membrane system.

Concentrate—Water that does not pass through the membrane. This water contains all of the contaminants that were unable to pass through the membrane. The concentrate stream may be from 10% to 50% of the feedwater flow. The concentrate is sometimes referred to as *retentate* because it is retained by the membrane. This waste stream must receive further treatment or be properly disposed of.

Permeate and filtrate—Water that passes through the membrane. This water has fewer contaminants than the feedwater and makes up 50% to 90% of the feedwater flow. Permeate and filtrate are also called product water. "Filtrate" is more common for membrane filtration and "permeate" is more common for reverse osmosis.

Flux—Flux is the amount of water that passes through a specified area of membrane per unit time. It is also called the permeate flux. It is expressed as liters per square meter of membrane surface area per hour ($L/m^2{\cdot}h$) or gallons produced per square foot of membrane surface area per day (gpd/sq ft).

Water recovery—The percentage of feedwater that becomes permeate or filtrate.

Membranes are commonly represented using a diagram like the one shown in Figure 6.1. The letter "Q" is used to symbolize flow. The letter "C" is used to symbolize concentration. Subscripts are used to designate which flow and concentration is being discussed. For example, Q_F is the feedwater flow.

Membrane processes are used for several purposes in wastewater treatment. First, membranes can replace the secondary clarifier in the activated sludge process. In this case, the overall biological/membrane process is known as a *membrane bioreactor* (MBR). Second, membranes can be used for advanced removal of particles to further reduce the turbidity following secondary clarification. This process is known as *membrane filtration*. It is used for tertiary treatment, typically when the facility effluent is used for potable or nonpotable reuse. Third, membranes can be used to remove dissolved species from water, like salts and trace organics such as disinfection byproducts, industrial solvents, pharmaceuticals, and personal care products. This process is known as *reverse osmosis*. It is most commonly used when the facility effluent is used for potable reuse.

Potable reuse is the intentional recycling of highly treated wastewater into a municipal water supply. Potable reuse can be indirect potable reuse, meaning that the treated water passes through an aquifer or reservoir before entering the drinking water system, or direct potable reuse, meaning the treated water will be piped directly to the drinking water system with perhaps only storage tanks in between.

The types of membranes that remove particles are different from the membranes that remove dissolved species. Many aspects of the design and operation for each are different. Membrane filtration and MBR processes both remove particles, so the design and operation are similar and are presented together in this chapter. These processes use either microfiltration or ultrafiltration membranes. A variety of reverse osmosis membrane products are available, including seawater reverse osmosis (SWRO), brackish water

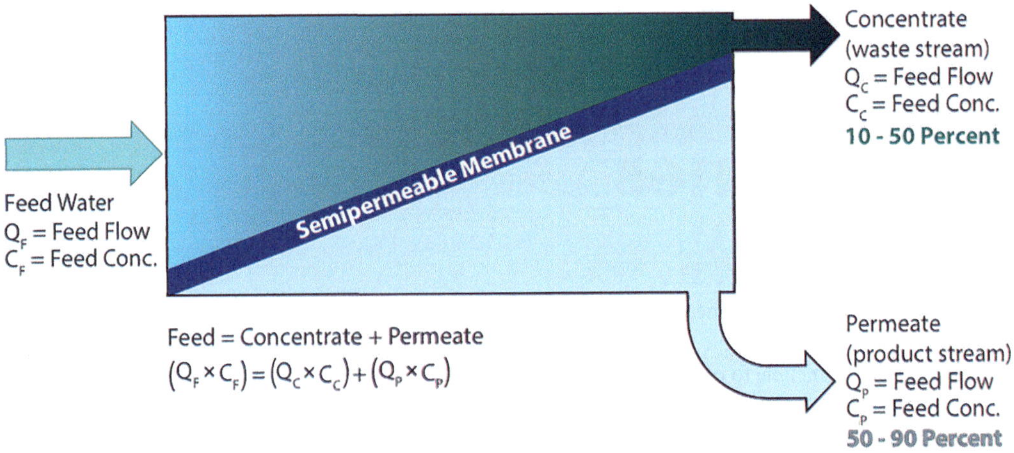

Figure 6.1 Typical Representation of a Membrane Process

reverse osmosis (BWRO), and nanofiltration membranes. Brackish water reverse osmosis membranes are the most common type used in wastewater applications. Nanofiltration membranes are not effective at removing the trace organics that must be removed for potable reuse, and SWRO membranes operate at much higher pressure and are more expensive to operate.

LEARNING OBJECTIVES

Upon completing this chapter, you will be able to:

- Explain the purpose of membrane processes in wastewater treatment.
- Explain the specific application for using each type of membrane product.
- Explain the fundamentals of how membrane systems remove contaminants from water.
- Calculate key process variables for membrane systems.
- Identify potential causes and strategies to prevent or minimize membrane fouling.
- Describe the main components of membrane filtration and reverse osmosis systems.
- Identify and be able to calculate parameters used to track membrane system performance.
- Identify the instrumentation used to monitor membrane processes.
- Develop a daily operations routine to maintain stable membrane system operations.
- Describe typical membrane maintenance requirements.
- Explain how to troubleshoot membrane systems.
- Describe how to handle common chemicals and other hazards related to membrane systems.

Purpose and Function

Membranes are used for several purposes in wastewater treatment. The selection of the membrane treatment technology depends on the type and form of contaminants. Membrane filtration and MBR are primarily used to remove particles. Salts and other dissolved substances can pass through microfiltration and ultrafiltration membranes. Reverse osmosis membranes have pores that are so small, only water molecules can pass through them. Everything else, including salts and dissolved substances, is held back (rejected) by a reverse osmosis membrane. Figure 6.2 presents a summary of the contaminants that can be removed by each class of membranes.

The purpose of the membranes in a membrane bioreactor is to separate the mixed liquor suspended solids (MLSS) from the wastewater effluent—the same purpose as secondary clarifiers. Thus, membranes replace secondary clarifiers in an MBR. Membranes are more effective at removing solids than clarifiers and can achieve an effluent turbidity less than 0.2 nephelometric turbidity units (NTUs) and total suspended solids (TSS) less than 1 mg/L (WEF, 2012). Eliminating secondary clarification can allow the activated sludge

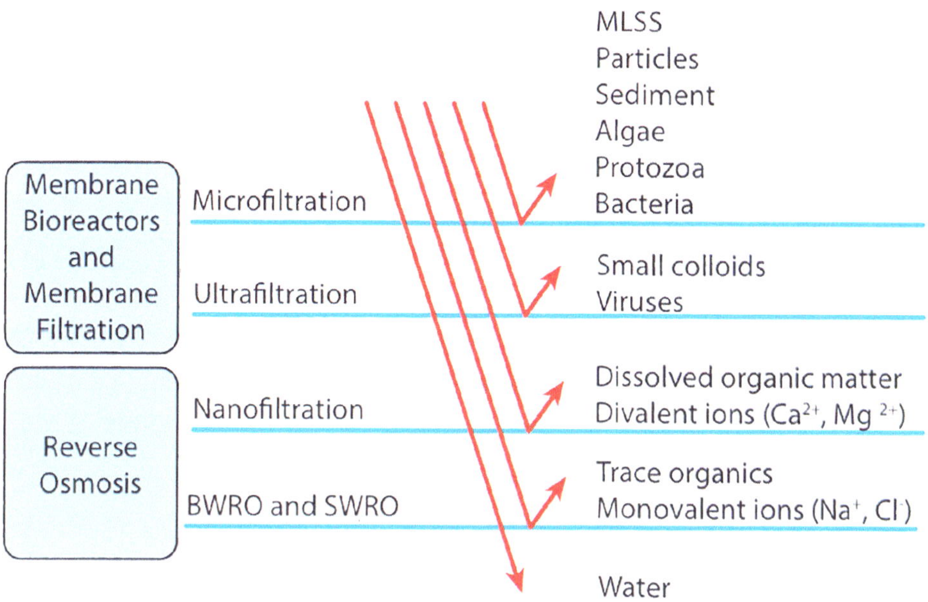

Figure 6.2 Types of Contaminants That Can Be Used by Each Class of Membrane Product Used in Wastewater Treatment

process to operate at higher MLSS concentrations, shorter hydraulic retention times (HRTs), and longer solids retention times (SRTs). These changes result in less sludge production and higher quality effluent with respect to lower turbidity, TSS, biochemical oxygen demand, and bacteria concentrations. Membrane bioreactors can use less space than conventional activated sludge systems because the membranes are more compact than secondary clarifiers. The MBR membrane modules can be placed in the activated sludge tank or in a separate tank. More information about MBRs is available in *Membrane Bioreactors* (WEF, 2012).

The purpose of membrane filtration for tertiary treatment is to provide a higher quality effluent that can be used for other purposes, like wastewater reuse. Membrane filtration will often be the first process in an advanced treatment train for potable reuse. In addition to lower turbidity, membrane filtration is excellent at removing pathogens, which is a key requirement for potable reuse. Microfiltration and ultrafiltration membranes can get credit for a log removal value (LRV) of 4 for *Giardia* and *Cryptosporidium*.

Log removal value is the measure of how well *Giardia*, *Cryptosporidium*, and viruses are removed in a treatment system. Log removal value is often described as the "number of nines." 1 LRV = 90% removal, 2 LRV = 99% removal, 3 LRV = 99.9% removal, 4 LRV = 99.99% removal, and so on.

The purpose of reverse osmosis in an advanced treatment train is to remove trace chemicals that may be in treated wastewater. A primary concern is trace organics, such as pharmaceutical, personal care products, industrial chemicals, and disinfection byproducts. Reverse osmosis does excellent removal of many dissolved chemicals, particularly charged, hydrophobic, and high-molecular-weight molecules. Reverse osmosis is not as effective at removing low-molecular-weight, neutral chemicals. Reverse osmosis will effectively remove total organic carbon (TOC) and is a required part of the treatment train for potable reuse in places that limit the effluent TOC concentration (like California in the United States).

Hydrophobic means a chemical that doesn't like being in water.

TEST YOUR KNOWLEDGE

1. A key advantage of membrane processes is that they do not generate a waste stream.
   - ☐ True
   - ☐ False

2. When a membrane system replaces the secondary clarifier in an activated sludge process, the resulting process is called a:
   a. Microfiltration
   b. Membrane bioreactor
   c. Reverse osmosis
   d. Wastewater recycling

3. Which of these types of membranes is best for removing disinfection byproducts?
   a. Microfiltration
   b. Ultrafiltration
   c. Nanofiltration
   d. Brackish water reverse osmosis

4. A membrane bioreactor can have a higher MLSS concentration, shorter HRT, and longer SRT than a conventional activated sludge system with a secondary clarifier.
   ☐ True
   ☐ False

5. What is membrane flux?
   a. The holes in a membrane that allow water to pass through
   b. A chemical to clean pipes before soldering
   c. The amount of water that passes through a specified area of membrane per unit time
   d. The membranes used in an activated sludge system

6. Which membrane has pores so small that only water molecules pass through?
   a. Membrane bioreactor
   b. Reverse osmosis
   c. Microfiltration
   d. Nanofiltration

7. What is potable reuse?
   a. The intentional recycling of highly treated wastewater into a municipal water supply
   b. The removal of pathogens from wastewater
   c. The reuse of wastewater for irrigation of parks and green spaces
   d. The reuse of wastewater for industrial purposes

Theory of Operation

MECHANISMS OF REMOVAL

Membrane filtration removes particles by straining and reverse osmosis removes dissolved material by dissolution and diffusion.

Membrane filtration and reverse osmosis use different mechanisms to remove contaminants, as shown in Figure 6.3. Microfiltration and ultrafiltration membranes have pores that are smaller than the particles or pathogens to be removed. Removal occurs by straining of the particles at the surface of the membrane. Particles that are larger than the pores are effectively 100% removed by the membrane. However, 100% removal can never be guaranteed because of imperfections in the membrane, equipment limitations, and fiber breakage that occurs as membranes age. As a result, microfiltration and ultrafiltration membranes typically receive credit for LRV = 4, which corresponds to 99.99% removal. Because particles are physically strained from the water, removal is generally independent of operating conditions like feed pressure or temperature.

Reverse osmosis membranes are often described as not having pores through which liquid water can flow. Instead, water and contaminant molecules dissolve into the membrane material, travel across the membrane, and re-liquify on the permeate side. The membrane has a greater attraction for water than

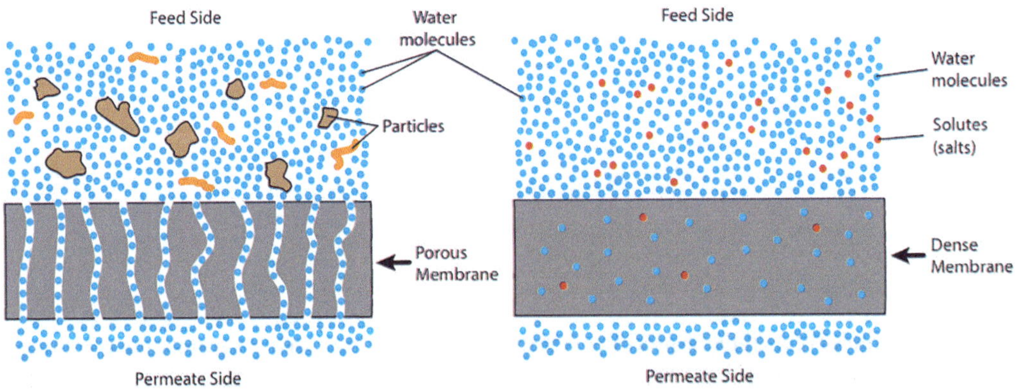

Figure 6.3 Mechanism for Removal of Contaminants (left panel: membrane filtration relies on straining, in which the particles to be removed are strained at the surface because they are too big to fit in the pores; right panel: reverse osmosis relies on a difference in how much water and contaminants enter the membrane material and how fast they move through the material)

contaminants, so the water enters the membrane and travels across much faster. Removal occurs because of a fast movement of water and slow movement of contaminants. Because contaminants are able to enter the membrane material along with water, reverse osmosis cannot achieve the high removal that can membrane filtration. Reverse osmosis can achieve 99% removal or higher of ionic species like sodium and chloride and large molecules like TOC, but may have lower removal of small, uncharged molecules. Furthermore, the removal efficiency can be affected by operating conditions like pressure, temperature, and feed concentration.

Ionic species are chemicals that have a charge in water. The charge can be positive, like sodium (Na^+), or negative, like nitrate (NO_3^-).

OSMOTIC PRESSURE IN REVERSE OSMOSIS

The performance of reverse osmosis membranes is affected by osmotic pressure. Osmotic pressure is caused by the flow of water through a semi-permeable membrane from a dilute solution into a concentrated one. In a solution, equilibrium is achieved when the concentration of the contaminants is the same everywhere. If a drop of dye is placed in a glass of clean water, the dye molecules will slowly move by diffusion until they spread throughout the whole glass and the water is evenly colored, as shown in Figure 6.4(a). The drop of dye was originally a place of high concentration and the clean water a place of low (actually, zero) concentration. As the dye spreads out, the concentration in the clean water increases and the concentration in the droplet of dye decreases, until the concentration is the same everyplace.

Osmosis is a similar process to diffusion, but with the addition of a semi-permeable membrane. Suppose that instead of placing a droplet of dye directly into a glass of water, the dye is placed inside a bag made of a semi-permeable membrane, and then the bag is sealed and immersed in the glass of water, as shown in Figure 6.4(b). The semi-permeable material of the bag allows water but not dye molecules to flow through. The solution still wants to reach equilibrium with the same concentration of dye everyplace in the glass. Because the dye cannot pass out of the bag to increase the concentration of dye in the water, the water will pass into the bag to dilute the concentration of dye within the bag. As water passes into the bag, the bag will swell. If the bag cannot stretch much, the pressure inside the bag will increase. The amount of water that will try to enter the bag will depend on the original concentration of dye in the bag: a higher concentration of dye results in more water trying to enter the bag to dilute it, which results in higher pressure.

As the pressure inside the bag builds up, the pressure acts as a driving force to push water out of the bag. At some point, the force pushing water into the bag due to the difference in concentration will be equal to the pressure pushing water out of the bag. At that point, the system is said to be in equilibrium and the flow of water into the bag will stop. The pressure required to achieve equilibrium is known as *osmotic pressure.* The osmotic pressure of a solution is related to the concentration of the dye. The higher the concentration, the higher the pressure will be.

Osmotic pressure is the pressure resulting from water passing through a semi-permeable membrane in an attempt to dilute a concentrated solution.

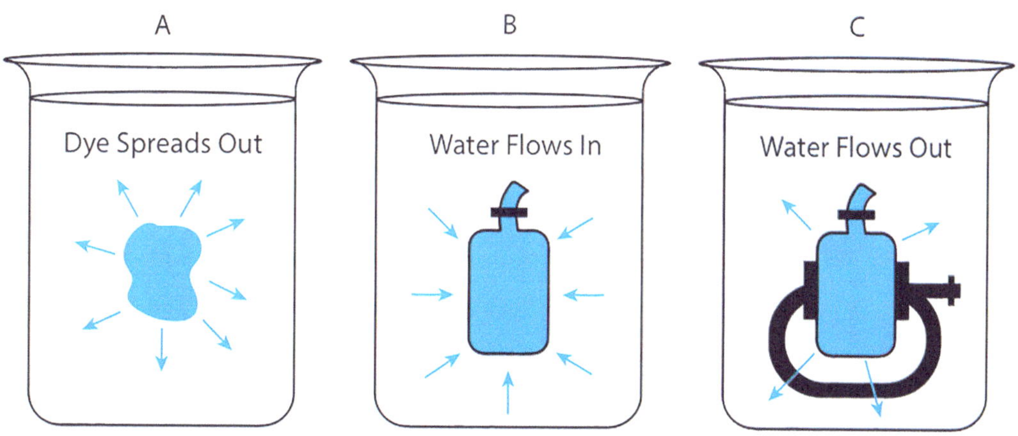

Figure 6.4 **(a) Diffusion: Dye Spreads Throughout the Water Until the Concentration Is the Same Everywhere; (b) Osmosis: Water Flows Into the Semi-Permeable Membrane Bag, Diluting the Dye Solution in the Bag and Causing the Bag to Swell; and (c) Reverse Osmosis: Pressure Causes the Water in the Bag to Flow Out, Producing Additional Fresh Water**

Reverse osmosis, as the name indicates, is simply osmosis in reverse. Osmosis causes water to flow from the clean water outside the bag to the concentrated solution inside the bag, making the concentrated solution more dilute. If the bag is squeezed with pressure equal to the osmotic pressure, no water will flow in either direction. However, if the bag is squeezed with pressure greater than the osmotic pressure, as shown in Figure 6.4(c), the excess pressure will force water inside the bag to flow out. The water flows from the concentrated solution into the fresh solution, making more fresh water. Thus, reverse osmosis is a treatment process that produces fresh water by applying pressure to a concentrated solution to force water to pass through a semi-permeable membrane. The pressure must be greater than the osmotic pressure.

MEMBRANE FOULING AND SCALING

An important issue in membrane systems is membrane fouling. Membrane fouling occurs when the water flow through the membrane decreases because of materials that deposit, grow, or accumulate at the membrane surface or within the membrane itself. Fouling controls how often membrane systems must be backwashed or cleaned. Failure to manage fouling increases operation and maintenance (O&M) costs and may increase how often the membranes need to be replaced. Additional details about membrane fouling are available in *Membrane Bioreactors* (WEF, 2012) and *MWH's Water Treatment: Principles and Design* (Crittenden et al., 2012).

MEMBRANE FILTRATION

Fouling in membrane filtration and MBRs depends on how it restricts water flow or how (and whether) it can be removed. Foulants can reduce water flow by collecting at the membrane surface or by plugging the pores, as shown in Figure 6.5. The collection of particles and suspended material at the membrane surface is known as *cake fouling*. This cake layer consists of material that is larger than the membrane pores so that it is trapped and collects on the membrane surface.

Material that is smaller than the membrane pores can enter the pores but become trapped within the membrane. This material can consist of small colloids, dissolved material, and microbial cells. This type of fouling is called *pore blocking*. The activated sludge process produces many different types of organic compounds. Some of these are very large organic molecules that can stick to membrane surfaces and contribute to pore blocking.

The second way to describe fouling in membrane filtration is by whether it is *physically reversible, chemically reversible*, or *irreversible*. Physically reversible fouling is loosely attached and can be removed by backwashing the membrane. The cake layer and some loosely attached material in the pores can typically be removed by backwashing. Chemically reversible fouling is caused by materials that cannot be removed by backwashing, but can be removed by chemical cleaning. The material that adsorbs so strongly that it cannot be removed even by chemical cleaning is called *irreversible fouling*. Irreversible fouling can build up slowly over time, reducing the permeate flowrate of the membrane system. Over a long period of time, typically years, the irreversible foulants can accumulate to the point that the membranes must be replaced.

Backwashing is pumping water through the membrane in reverse to clean it. Backwashing typically includes an air scouring step to help break up the foulants.

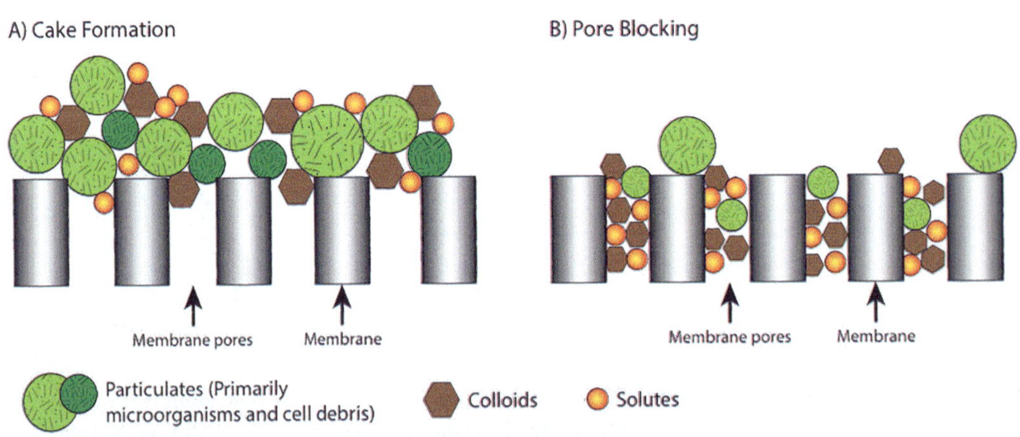

Figure 6.5 Membrane Filtration Fouling Mechanisms: (a) Cake Formation and (b) Pore Blocking (Adapted from WEF, 2012)

REVERSE OSMOSIS

Fouling in reverse osmosis systems can be caused by four things: (1) accumulation of particles; (2) oxidation of soluble metals; (3) precipitation of inorganic salts, which is known as *scaling*; and (4) adsorption of organic and biological matter.

Reverse osmosis membrane elements, which will be described later in this chapter, have very narrow feed channels that are held open by a feed channel spacer. If the feedwater contains a significant amount of turbidity, it will be captured by the feed channel spacers and collect in the feed channel. The reverse osmosis industry uses a parameter called the *silt density index* (SDI) to measure the amount of particulate matter that can contribute to fouling. Reverse osmosis systems do not have a backwash cycle, so it is important to pretreat the water to reduce the SDI to below the manufacturer's specifications before it enters the reverse osmosis system. Manufacturers typically require an SDI of less than 3. Membrane filtration is often placed ahead of reverse osmosis in advanced treatment systems and provides excellent pretreatment that minimizes the SDI and particulate fouling.

The silt density index (SDI) is a test to measure the amount of particulate matter in the water that might contribute to particle fouling. The SDI does not have units.

Some water sources can contain dissolved trace metals like iron and manganese. Exposing these metals to oxygen can create insoluble compounds like iron oxide (rust) and manganese dioxide (MnO_2). These precipitate and form particles that can foul reverse osmosis membranes. This type of fouling is not common in wastewater systems because the effluent from an activated sludge or MBR process is typically well oxygenated.

Fouling from scaling is a key factor that limits the recovery of water in reverse osmosis systems.

The third type of reverse osmosis fouling is scaling. Scaling is caused by the precipitation of sparingly soluble salts. As clean water passes through the membrane to form permeate, dissolved species stay on the feed side of the membrane and become more concentrated. While this happens, they may reach their saturation limit, which is the point at which they are too concentrated to stay in solution. An example of reaching a saturation limit is adding sugar to tea until no more sugar will dissolve. Typical salts and minerals that might cause problems are calcium carbonate, calcium sulfate, barium sulfate, and strontium sulfate. In wastewater systems, calcium phosphate can be a problem. As water is converted to permeate, the remaining water continually becomes more concentrated, such that the limit on how much fresh water can be produced is limited by the point at which scaling will occur. Scaling can be controlled by adding acid and scale inhibitors to the feedwater. Ultimately, the fouling caused by scaling is what limits the maximum recovery (i.e., the ratio of product water to feedwater) in a reverse osmosis system.

Viscosity is the property that describes how easily a liquid flows. A liquid like water has a lower viscosity and flows easily, whereas honey or motor oil have a higher viscosity and do not flow as easily as water. When water gets colder, the viscosity increases and it doesn't flow through the membrane as easily, so the permeate flow is lower.

The fourth and final type of fouling is organic and/or biological fouling. This fouling is caused by microorganisms and organic matter in the feedwater. Microorganisms can attach and grow on the membrane surface in the feed channel. Organic matter is soluble and can pass through membrane filtration pretreatment. Once through, organic matter can clog the feed channels in a way similar to particulate fouling. Disinfection can minimize biological growth in the reverse osmosis system, but typical reverse osmosis membranes are not resistant to chlorine. Free chlorine (bleach) cannot be used. The disinfectant of choice in reverse osmosis systems is chloramines.

Flux is the flow of water through the membrane surface area and is described in more detail in the next section. It has units of gallons per square foot per day (gpd/sq ft) or liters per square meter per hour (LMH).

EFFECT OF TEMPERATURE AND PRESSURE

As fouling occurs, the production of treated water will decline. Otherwise, more pressure will be needed to produce the same flow. Monitoring flow and pressure is the key way to know when to backwash or clean a membrane system. Unfortunately, other operating conditions affect the permeate flowrate. Specifically, an increase in pressure increases flow. An increase in water temperature will also increase flow because the water is less viscous at higher temperatures. In reverse osmosis systems, an increase in the total dissolved solids (TDS) concentration of the feedwater will decrease permeate flow because of the increase in osmotic pressure. It is hard to determine how much fouling is happening if the amount of fouling, pressure, temperature, and feed TDS concentration are all changing at the same time.

The specific flux of a new, clean membrane is called the *membrane permeability*. The specific flux has units of liters per square meter per hour per bar ($L/m^2 \cdot h\text{-}bar$) in SI units and gallons per square foot per psi (gpd/sq ft/psi) in English units. Note that 1 bar = 100 kPa.

Procedures have been developed to normalize the permeate flux to standard operating conditions so that the effect of fouling can be separated from the change in performance resulting from changes in operating conditions. A parameter called the *specific flux* is defined as the permeate flux divided by the transmembrane pressure (TMP) for membrane filtration or net driving pressure for reverse osmosis. Equations for calculating specific flux are presented in the "Process Variables" section later in this chapter.

The transmembrane pressure is the feed pressure minus the filtrate or permeate pressure.

TEST YOUR KNOWLEDGE

1. **Which one of the following statements is true?**
   a. Reverse osmosis membranes have larger pores the microfiltration membranes.
   b. The mechanism for particle removal in membrane filtration is straining at the surface of the membrane.
   c. Reverse osmosis membranes are able to achieve complete removal of all dissolved compounds.
   d. Particle removal in membrane filtration is significantly affected by operating conditions like feed pressure and temperature.
2. **Osmotic pressure helps water flow through a reverse osmosis system so that the water flow will be faster when the osmotic pressure is higher.**
   - ☐ True
   - ☐ False
3. **The cake layer that forms on the surface of a microfiltration or ultrafiltration membrane can be removed by backwashing.**
   - ☐ True
   - ☐ False
4. **Which type of reverse osmosis fouling is most responsible for limiting the water recovery in reverse osmosis systems?**
   a. Accumulation of particulate matter
   b. Oxidation of reduced metals
   c. Scaling from precipitation of insoluble minerals
   d. Adsorption of organic and biological matter
5. **Calculating the specific flux helps identify whether fouling is occurring.**
   - ☐ True
   - ☐ False
6. **Which type of reverse osmosis fouling is measured with the SDI?**
   a. Accumulation of particulate matter
   b. Oxidation of reduced metals
   c. Scaling from precipitation of insoluble minerals
   d. Adsorption of organic and biological matter
7. **Which of these salts or minerals can precipitate and cause scaling on reverse osmosis membranes?**
   a. Sodium chloride
   b. Calcium phosphate
   c. Free chlorine
   d. Calcium tetrachloride

Design Parameters

Important factors in the design and operation of a membrane systems include:

- The size of contaminants it can remove,
- How well it can remove them,
- What fraction of the feedwater is recovered as product water,
- What pressure it operates at,
- How much membrane surface area is needed,
- How often it must be backwashed, and
- How often it must be cleaned.

The membrane surface area depends on the product water flowrate (i.e., how much water must be produced) and the permeate flux (i.e., how fast water can pass through the membrane). Table 6.1 shows the typical values for these membrane system parameters. In membrane filtration, the TMP is the difference between the feed pressure and permeate pressure. In reverse osmosis, the osmotic pressure also needs to be taken into account. The overall pressure available to push water through a reverse osmosis membrane is known as the *net driving pressure* (NDP).

Expected Performance

MEMBRANE FILTRATION

Because membrane filtration removes particles, protozoa, and bacteria by straining, it can achieve a very high level of removal as long as there are no defects or damage to the equipment, such as holes or tears in the membrane. Typically, membrane filtration will be rated to achieve LRV = 4 for protozoa. If any membrane fibers break (discussed later), the filtration performance can decrease. Filter runs typically last

Table 6.1 **Typical Values of Key Operating Parameters for Membrane Systems**

	Membrane Filtration	Reverse Osmosis
Rejection rating (what contaminants are removed from the feedwater by the membrane)	Microfiltration: 0.1–0.45 μm Ultrafiltration: 0.01–0.1 μm	>99% sodium chloride (NaCl) removal
Removal efficiency	LRV = 4 or greater	Charged species: >99% Neutral species: 0–99%
Typical water recovery	90%–95%	50–80%
Permeate flux	34–102 LMH (20–60 gpd/sq ft)	14–24 LMH (8–14 gpd/sq ft)
Feed pressure	21–210 kPa (3–30 psi)	Potable reuse: 500–1000 kPa (73–145 psi) Seawater reverse osmosis: 6000–8500 kPa (870–1230 psi)
Filter run duration (backwash frequency)	15–90 minutes	N/A
Cleaning frequency	Maintenance clean or maintenance wash: weekly to monthly Clean-in-place: 1–3 months	Monthly to yearly

Maintenance cleans and washes are described more in the "Maintenance" section later is this chapter.

15 to 90 minutes, followed by a short backwash that lasts for a minute or so. In some MBR systems, the filtration cycle is followed by a short period of relaxation (i.e., no filtrate production) instead of a true backwash. Filtrate is used for backwashing, which reduces the total water produced slightly. The net filtrate production (called the *water recovery*) is typically about 90% to 95% of the feedwater. Typical cleaning frequency is every 1 to 3 months, and the time between cleaning can be improved with better quality feedwater. The difference between backwashing and cleaning will be explained later in this chapter.

REVERSE OSMOSIS

The reverse osmosis membranes used in wastewater applications are typically BWRO membranes. These membranes are typically rated for >99% sodium chloride (NaCl) removal. Although reverse osmosis membranes remove all kinds of ionic species, the performance tests done by manufacturers only include NaCl. The parameter monitored for performance is conductivity. The conductivity in the permeate will be 95% to 99% less than the conductivity of the feedwater. A second parameter of interest is TOC. A well-functioning reverse osmosis system should be able to remove the TOC to below 0.25 mg/L. Reverse osmosis is effective at removing many pharmaceuticals and personal care products but may have more difficulty removing disinfection byproducts and industrial solvents. Performance will decline over time as the membranes age.

The water recovery for reverse osmosis systems typically ranges from 75% to 85% when used in wastewater applications. Reverse osmosis systems are not backwashed, but cleaning must be done monthly to annually.

TEST YOUR KNOWLEDGE

1. The water recovery from membrane filtration systems is higher than reverse osmosis systems.
   - ☐ True
   - ☐ False

2. Reverse osmosis systems are backwashed more frequently than membrane filtration systems.
   - ☐ True
   - ☐ False

3. What can cause bacteria to get through a membrane?

a. Operating at too low of a pressure
b. Backwashing too frequently
c. Not enough backwashing
d. Tears or holes in the membrane

4. What chemical do manufacturers use to test the performance of reverse osmosis membranes?

a. Sodium chloride
b. Calcium phosphate
c. Chlorine (bleach)
d. Pharmaceuticals

Process Variables

Monitoring a membrane system involves a lot of parameters. Some of the most important ones are described in this section. Typically, these are measured and calculated by the facility's data acquisition system. Many operators will never need to calculate these parameters themselves. However, some lead operators like to calculate these parameters in a spreadsheet so they have a better understanding of what is going on. Example calculations are shown here to help lead operators understand the calculations. Other operators will benefit from the explanations about what the parameters are.

MEMBRANE FILTRATION

The fraction of the contaminants that are removed by the membrane system is called *rejection*, commonly expressed in percent. Rejection can be used to describe the removal of suspended and dissolved materials. In reverse osmosis, rejection of TDS and electrical conductivity is commonly monitored as a measure of the membrane's performance. Rejection is calculated as

$$R = \left(1 - \frac{C_p}{C_f}\right) \times 100 \tag{6.1}$$

Where

R = percent rejection and
C_p, C_f = permeate and feedwater concentrations, mg/L.

For example, if the feed nitrate concentration is 4.5 mg/L and the permeate nitrate concentration is 0.062 mg/L, then the percent rejection is

R = [1 – 0.062 mg/L/4.5 mg/L] × 100 = 98.6%.

When the rejection is very high (>99.9%), the rejection is calculated using a special mathematical function called a *logarithm*. The rejection is then called the *LRV*. The logarithm converts the percent rejection into a number the reflects how many 9's are in the rejection, such that 99% is LRV = 2, 99.9% is LRV = 3, 99.99% is LRV = 4, and so forth. The equation for LRV is

$$\text{LRV} = \log(C_f) - \log(C_p) = \log\left(\frac{C_f}{C_p}\right) \tag{6.2}$$

Where

LRV = log removal value, dimensionless.

For example, if viruses are measured as 56 000 virus/L in the feed and 2 virus/L in the permeate, then the log removal is calculated as

LRV = log(56,000) – log(2) = 4.75 – 0.30 = 4.45.

Log removal value is the way to report the rejection of microorganisms, bacteria, and viruses. Typically, membrane filtration will use LRV and reverse osmosis will use percent rejection.

Water recovery is the net permeate (reverse osmosis) or filtrate (membrane filtration) flow divided by the feedwater flow. Reverse osmosis systems always have a concentrate stream, so feedwater recovery is based on the ratio of permeate flow to feed flow. Because microfiltration and ultrafiltration systems use filtrate for periodic backwashing, the water used for backwashing is taken into account:

$$r = \left(\frac{Q_p}{Q_f}\right) \times 100 = \left(\frac{V_f - V_{bw}}{V_f}\right) \times 100 \tag{6.3}$$

Where

r = recovery, in units of percent;
Q_p, Q_f = permeate and feed flowrates, m^3/d (gpm);
V_f = volume of water fed to membrane over filter run, m^3 (gal); and
V_{bw} = volume of water used during backwash, m^3 (gal).

For example, if a reverse osmosis system has a feed flowrate of 900 gpm and a permeate flowrate of 730 gpm, the recovery is calculated as

r = (730 gpm / 900 gpm) × 100 = 81%.

If a membrane filtration system produces 4800 gal of filtered water during a filter run and uses 180 gal of that filtrate during the backwash, the recovery is calculated as

r = (4800 gal – 180 gal)/4800 gal × 100 = 96.3%.

Flux is the rate at which water is passing through the membrane surface area. Membrane systems operate best when operated within a specific range of flux. A high flux means each module can produce more treated water, but the membranes can foul rapidly when the flux is too high. Thus, the system will have been designed for the flux that minimizes the size of the system while also minimizing the rate of fouling and associated downtime for maintenance, such as chemical cleanings. Flux is calculated from

$$J = \frac{Q}{A} \tag{6.4}$$

Where

J = volumetric water flux through membrane, L/m^2·h or m/s (gallons per square foot per day, gpd/sq ft);
Q = flowrate, L/h (gpm); and
A = membrane surface area, m^2 (sq ft).

For example, if a membrane filtration skid has 50 membrane elements that each have a membrane surface area of 66 sq ft and the skid produces 80 gpm, the flux is calculated as

J = (80 gpm)(60 min/hr)(24 hr/d) / (50 elements)(66 sq ft per element) = 35 gpd/sq ft.

To determine the total system membrane surface area, it is necessary to know the membrane surface area of each membrane element and the number of membrane elements in the system. This information will be available in the membrane manufacturer specifications and the operations manual. The flux through the membrane can change when the operating conditions change. Specifically, the flux will increase when either the feed pressure increases or the temperature increases. The increase in temperature will decrease the viscosity of the water, which will allow it to flow through the membrane pores more easily. To determine if membranes are fouling, the change in flux due to pressure and temperature must be factored out, which can be done by calculating the flux at standard conditions. The standard temperature is 20 °C.

Calculating the flux at standard conditions involves normalizing the flux for temperature, and then dividing by the TMP to determine the specific flux, as follows:

$$J_s = J_m(1.03)^{T_s - T_m} \qquad (6.5)$$

Where

J_m, J_s = flux at measured and standard temperatures, L/m²·h (gpd/sq ft), and
T_m, T_s = measured and standard (20 °C) temperatures, °C.

For example, if a membrane filtration membrane skid has a flux of 35 gpd/sq ft at 17 °C, the standard flux at 20 °C is calculated as

J_s = 35 gpd/sq ft × $(1.03)^{(20\,°C - 17\,°C)}$ = 35 gpd/sq ft × 1.0927 = 38.2 gpd/sq ft

The TMP is the net pressure that forces water through a membrane filter and is calculated by subtracting the permeate pressure from the feed pressure.

$$J_{sp} = \frac{J_s}{TMP} \qquad (6.6)$$

Where

J_{sp} = specific flux at standard temperature, L/m²·h·bar (gpd/sq ft/psi), and
TMP = transmembrane pressure, kPa (psi).

For example, if a membrane filtration membrane skid has a standard flux of 38.2 gpd/sq ft at a TMP of 11.2 psi, the specific flux at 20 °C is calculated as

J_s = 38.2 gpd/sq ft / 5.6 psi = 6.8 gpd/sq ft/psi.

REVERSE OSMOSIS

The fraction of the contaminants that have been removed by the membrane system is called *rejection*. Rejection in reverse osmosis is calculated using the same equation (Equation 6.1) that was presented for membrane filtration. Rejection can be calculated for bulk parameters like conductivity or individual species like nitrate.

Feedwater recovery can be calculated using Equation 6.3 based on ratio of permeate flow to feed flow.

The osmotic pressure is proportional to the concentration of the dissolved species in the solution and depends on the temperature and a constant called the *ideal gas constant*. Waters with higher TDS concentrations have higher osmotic pressure. Typical operators will never need to calculate the osmotic pressure. The equation for osmotic pressure is presented here for lead operators who want to develop their own spreadsheet to track performance:

Moles are a way of keeping track of large numbers of molecules. Shoes come in a pair, eggs come in a dozen, and molecules come in a mole. A dozen eggs has 12 eggs. A mole of molecules has 6.022 × 1023 molecules. The "Chemistry Review" section in Chapter 9 of *Wastewater Treatment Fundamentals I* describes the procedure for converting milligrams to moles.

$$\Pi = CRT \qquad (6.7)$$

Where

Π = osmotic pressure, kPa;
C = concentration of dissolved species, mol/L;
R = ideal gas law constant, 8.3145 L-kPa/mol-K; and
T = temperature (K).

The concentration in the equation for osmotic pressure is in molar units. To convert a concentration from mass units (mg/L) to molar units (mol/L), it is necessary to convert from milligrams to moles. In potable reuse systems, it is common to measure TDS of the solution and calculate the molar concentration assuming that the TDS is composed of sodium chloride. Sodium chloride dissociates into 2 ions. Every ion or molecule must be accounted for in osmotic pressure. The molecular weight of sodium chloride is 58.45 g/mol. This means the molar concentration for a solution with TDS = 500 mg/L is

500 mg/L × (2 ions in NaCl)/(58.45 g/mol × 1000 mg/g) = 0.017 mol/L.

The temperature is reported in Kelvin. To convert from Celsius to Kelvin, add 273.15 to the temperature in Celsius. Thus, the osmotic pressure of feedwater with 500 mg/L TDS at 20 °C is

0.017 mol/L × (0.083145 L-bar/mol-K) × (20 + 273.15) K = 0.42 bar.

Reverse osmosis permeate flux is calculated using the same equation as membrane filtration (J = Q/A) and is corrected for temperature using the same equation as well (Equation 6.4 and Equation 6.5). To calculate specific flux, however, it is necessary to divide the temperature-corrected flux by the NDP instead of the TMP, as follows:

$$J_{sp} = \frac{J_s}{NDP} \tag{6.8}$$

Where

J_{sp} = specific flux at standard temperature, L/m²·h·bar (gpd/sq ft/psi);
J_s = flux at standard temperature, L/m²·h (gpd/sq ft); and
NDP = net driving pressure, kPa (psi).

The NDP is the pressure that actually causes water to pass through the membrane and is proportional to permeate flow. Pressure in the permeate line reduces the available pressure, as does the osmotic pressure:

$$NDP = \Delta P - \Delta\Pi = (P_F - P_p) - (\Pi_F - \Pi_p) \tag{6.9}$$

Where

ΔP = difference in pressure between the feed and permeate streams, kPa (psi);
ΔΠ = difference between the feed and permeate osmotic pressure, kPa (psi);
P_F, P_p = feed and permeate pressures, kPa (psi); and
Π_F, Π_p = feed and permeate osmotic pressures, kPa (psi).

It is typically not necessary for operators to calculate the NDP manually. The NDP is typically calculated by the reverse osmosis control system software directly from online instruments.

TEST YOUR KNOWLEDGE

1. What is the conductivity rejection for a reverse osmosis system if the feed conductivity is 1855 µS/cm and the permeate conductivity is 27.6 µS/cm?
   a. 96.5%
   b. 97.5%
   c. 98.5%
   d. 99.5%

2. In a challenge test for a membrane filtration system, the *Cryptosporidium* concentration is measured as 172 000 oocysts/L in the feed and 19 oocysts/L in the permeate. What LRV is achieved?
   a. 1.28
   b. 3.96
   c. 5.24
   d. 6.51

3. What is the water recovery of a reverse osmosis system if the feed flowrate is 475 gpm and the permeate flowrate is 360 gpm?
   a. 24%
   b. 32%
   c. 76%
   d. 132%

4. What is the water recovery of a reverse osmosis system if the feed flowrate is 1960 m^3/d and the permeate flowrate is 2580 m^3/d?
   a. 24%
   b. 32%
   c. 76%
   d. 132%

5. A reverse osmosis system has 24 pressure vessels; each pressure vessel has 7 membrane elements, and each membrane element has 37.1 m^2 of membrane surface area. What is the permeate flux if the permeate flowrate is 3250 m^3/d?
   a. 0.0217 L/m^2·h
   b. 21.7 L/m^2·h
   c. 152 L/m^2·h
   d. 521 L/m^2·h

6. What is the flux for a membrane filtration skid if the skid produces 275 gpm and has 94 membrane elements that each have a membrane surface area of 66 sq ft?
   a. 46 gpd/sq ft
   b. 51 gpd/sq ft
   c. 61 gpd/sq ft
   d. 64 gpd/sq ft

7. What is the specific flux at 20 °C for a membrane filtration system that has a flux of 72.6 LMH when the water temperature is 13 °C and the TMP is 84 kPa (0.84 bar)?
   a. 70 LMH/bar
   b. 75 LMH/bar
   c. 89 LMH/bar
   d. 106 LMH/bar

8. What is the specific flux at 20 °C for a membrane filtration system that has a flux of 41 gpd/sq ft when the water temperature is 26 °C and the TMP is 11.5 psi?
   a. 3.0 gpd/sq ft/psi
   b. 3.6 gpd/sq ft/psi
   c. 4.3 gpd/sq ft/psi
   d. 34 gpd/sq ft/psi

9. What is the NDP of a reverse osmosis system that is operating with a feed pressure of 107 psi, permeate pressure of 9.1 psi, feed osmotic pressure of 11.3 psi, and permeate osmotic pressure of 0.81 psi?
   a. 85.8 psi
   b. 87.4 psi
   c. 97.1 psi
   d. 108 psi

Equipment for Membrane Filtration and Membrane Bioreactors

This section covers the equipment and components commonly found in membrane filtration systems, which include microfiltration and ultrafiltration membranes. Microfiltration and ultrafiltration membranes are also used in membrane bioreactors. The equipment commonly used in nanofiltration and reverse osmosis will be described in a separate section.

HOLLOW FIBER MEMBRANES

Membrane filters typically use hollow fiber membranes, but also come in flat sheet configurations. Hollow fiber membranes are molded as hollow tubes. The opening that runs through the center of the fiber is called the *lumen*. Figure 6.6 shows a cross-section of a hollow fiber membrane.

In dead-end filtration, all of the feedwater is filtered through the membrane. In an inside-out configuration, water is fed to the lumen and filtrate is collected on the outside of the membrane. In an outside-in configuration, the feedwater surrounds the membrane and filtrate is collected from the lumens.

Filtration occurs through the semi-permeable membrane in either an outside-in or inside-out configuration, as shown in the Figure 6.7. In an outside-in configuration, the feedwater surrounds the membrane fibers and water passes through the fiber wall into the lumen. Particles and pathogens are retained on the outside. The clean water, or filtrate, is collected in the lumen. In an inside-out configuration, the feedwater is in the lumen and the rejected materials and pathogens are trapped on the inside of the lumen as the water passes outward through the fiber wall. The filtrate surrounding the fiber is collected. Membrane filtration uses "dead-end filtration," meaning all of the feedwater is filtered through the membrane and becomes filtrate.

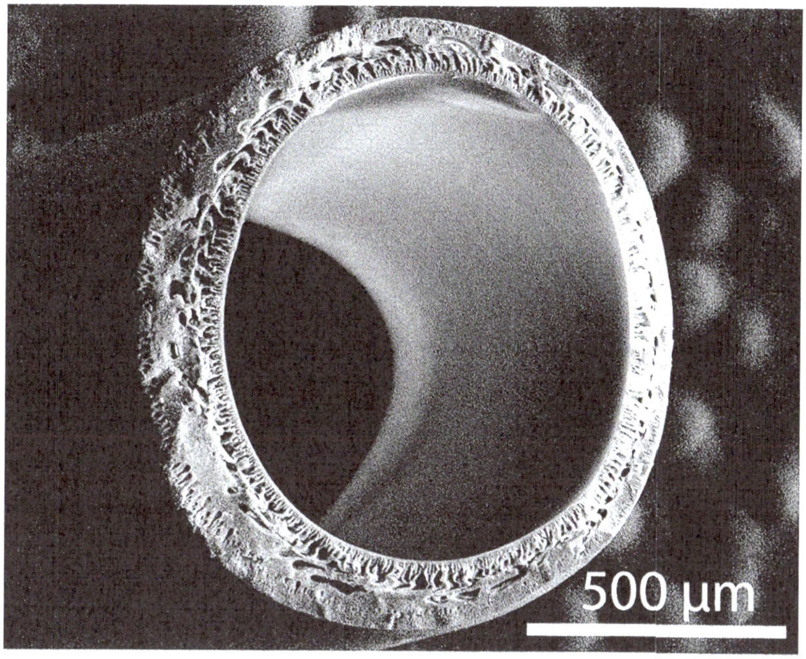

Figure 6.6 Cross-Section of Hollow Fiber Membrane
Wikimedia Commons. From "SEM Cross-Section of a Polysulfone Hollow Fiber Membrane Fabricated by Nonsolvent-Induced Phase Separation," by RobertsBiology, 2019 (https://commons.wikimedia.org/wiki/File:Polysulfone_hollow_fiber_membrane_SEM_cross-section.png). CC BY-SA 4.0. In the public domain.

There are differences between microfiltration and ultrafiltration membranes. Arguably, the main difference is the pore size. The membrane structure contains pores where straining occurs. The size of the pores is not homogeneous throughout the membrane. A range of opening sizes exists throughout the membrane. Microfiltration membranes typically have pore sizes from 0.1 to 0.45 µm, whereas ultrafiltration membranes have pore sizes from 0.01 to 0.1 µm. Fibers are packed into modules as a fiber bundle. Fiber bundles can range between several hundred fibers to over 10,000 fibers. The number of fibers that are bundled is typically higher for ultrafiltration membranes than for microfiltration membranes. With respect to the constituents that are rejected, microfiltration membranes filter out particles, sediment, algae, protozoa, and bacteria, whereas ultrafiltration membranes can also filter out small colloids and viruses in addition to those materials (see Figure 6.2).

Ultrafiltration has a smaller pore size and a larger fiber bundle size and can reject more materials than microfiltration membranes.

A homogeneous material's composition and properties are the same throughout.

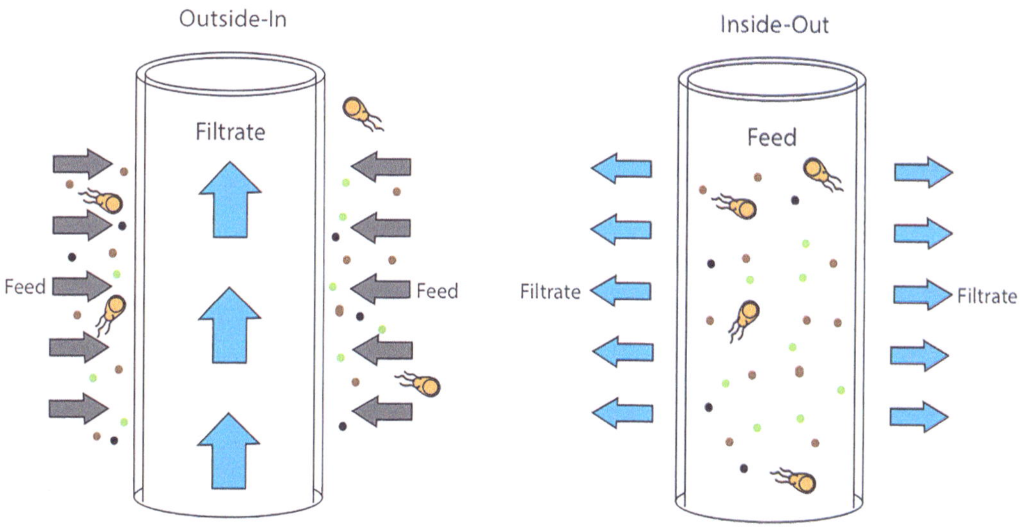

Figure 6.7 Outside-In and Inside-Out Hollow Fiber Configurations

Polymeric materials are made up of polymers. Polymers are large molecules made up of a series of building blocks. An example of a polymer is polypropylene $(C_3H_6)_n$. The subunit, C_3H_6, is repeated for "n" number of times to form the molecule.

Improvements to membrane materials have led to greater chemical resistance and higher permeability, resulting in more efficient operations.

Microfiltration and ultrafiltration membranes are typically made from ceramic or polymeric material. Table 6.2 presents a summary of examples of polymeric and ceramic membrane materials. Polypropylene membranes were commercialized in the 1980s. Many full-scale systems still use polypropylene membranes. Additional membrane materials have been developed since then. These have higher permeability and chemical resistance than polypropylene membranes. Membranes with higher permeability have lower operating costs because more filtrate can be produced with less energy. Of the new polymeric membrane materials, polyvinylidene fluoride (PVDF) and polytetrafluoroethylene have the best chemical resistance, which allows for longer service life and more effective chemical cleanings.

Polymeric membranes can be categorized into isotropic and anisotropic membranes, as shown in Figure 6.8. The type is dependent on how the membrane is manufactured. Isotropic (or symmetric) membranes have the same mechanical properties throughout the membrane and the filtration barrier exists along the depth of the membrane. Anisotropic or asymmetric membranes have a filtration "layer," with the rest of the fiber depth acting as a support structure.

Ceramic membranes are significantly more expensive than polymeric membranes but have longer life expectancy and chemical resistance. Ceramic membranes can be categorized as isotropic or anisotropic membranes and can come as fibers or tubular membranes.

FLAT SHEET MEMBRANES

An alternative to hollow fiber membranes is flat sheet membranes. Flat sheet membranes are used in submerged configurations for MBRs. Submerged configurations are discussed below. In a flat sheet configuration, two adjacent sheets are sealed together to form an envelope, with a spacer inside. The envelopes are arranged vertically with space between them for the MLSS to flow. A suction pump draws the water through the membrane. In other words, water flows from the MLSS outside the envelope to the filtrate inside the membrane (i.e., outside-in configuration). Flat sheet membrane modules look similar to submerged hollow fiber modules and work the same way. Submerged modules are described more in the next section.

SUBMERGED CONFIGURATION

In a submerged configuration, the membranes are suspended in open-to-atmosphere basins or tanks, as shown in Figure 6.9. The feed water surrounds the membranes and water is filtered in an outside-in setup. The filtrate is collected in a filtrate header, which is typically connected to a filtrate pump. The filtrate pump provides the suction for outside-in filtration to occur. Hollow fiber membranes can be thought of as thousands of tiny straws sipping the feed water out of the basin. There are cavitation limitations on the filtrate suction pump. This limits the operating range for TMP. It is lower for submerged configurations than for pressurized configurations. The typical TMP operating range for submerged configurations is on the order of 21 to 41 kPa (3 to 6 psi).

For additional discussion on pump cavitation, see the pump chapter in *Wastewater Treatment Fundamentals II.*

In a submerged configuration, filtrate may be collected from one end of the module or from both ends. Fibers are bundled into modules and are fixed on each end with a resin material. When water is filtered through the hollow fiber, the filtrate will travel in the lumen. On the ends where filtrate is collected, membrane fibers are not plugged with resin to allow filtrate to flow out of the lumen and into a common header.

Table 6.2 Summary of Polymeric and Ceramic Membrane Materials

Polymeric	Ceramic
Polyethersulfone (PES)	Silicone carbide
Polyvinylidene fluoride (PVDF)	Zirconium
Teflon (PTFE)	Titanium dioxide
Polyvinyl chloride (PVC)	Aluminum oxide
Polypropylene (PP)	

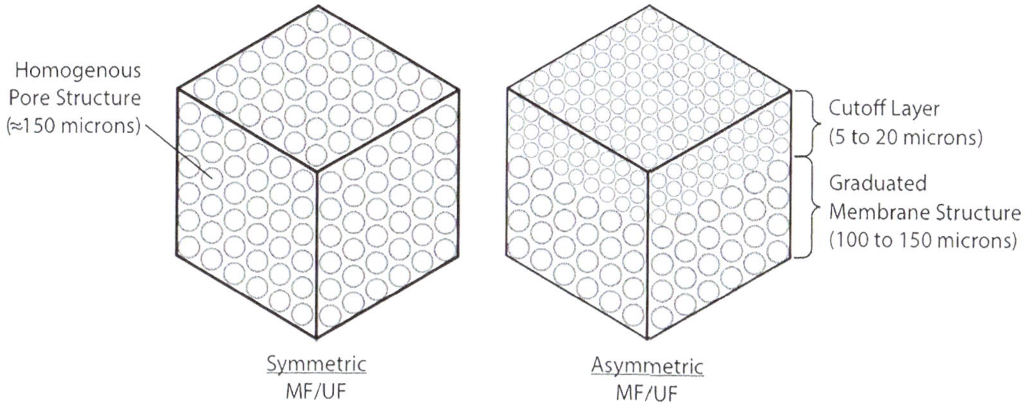

Figure 6.8 Comparison of Symmetric and Asymmetric Membranes (Adapted from U.S. EPA, 2005)

Submerged configuration membranes can be found in both MBR and advanced treatment applications. When used in drinking water applications, the membranes are suspended in a low solids concentration solution. In wastewater applications, the membranes may be submerged in mixed liquor as part of an MBR system or in filtered wastewater (post-secondary clarification) for either municipal wastewater treatment or for recycled water applications.

In MBR applications, aeration and flow recycling are important to minimize fouling and maintain membrane permeability. When submerged in mixed liquor, aeration provides scouring of the membrane surface. Bubbles traveling up to the water surface move the fibers and scrub off easily detached materials form the membrane surface. Aeration prevents excessive accumulation of sludge on the membrane surface and helps maintain permeability. Flow recycling is an important parameter when the submerged membranes are located in a separate tank from the aeration basin, which is a common configuration. A portion of the flow going into the membrane tank is recycled back to the biological reactors. The recycle ratio is set as a constant multiplier of either the influent flow or filtrate flow. A higher recycle ratio results in a lower solids concentration in the membrane tank and a lower recycle ratio results in a higher solids concentration. Operating at a lower recycle ratio can lead to greater solids buildup on the membrane fiber due to higher solids concentration in the tank and lower crossflow velocity.

Aeration and flow recycling are important parameters for controlling fouling in MBR applications.

For all submerged applications, the process water must undergo screening upstream of the membranes to remove larger particles and debris. Larger particles and debris can damage membranes. It is desirable

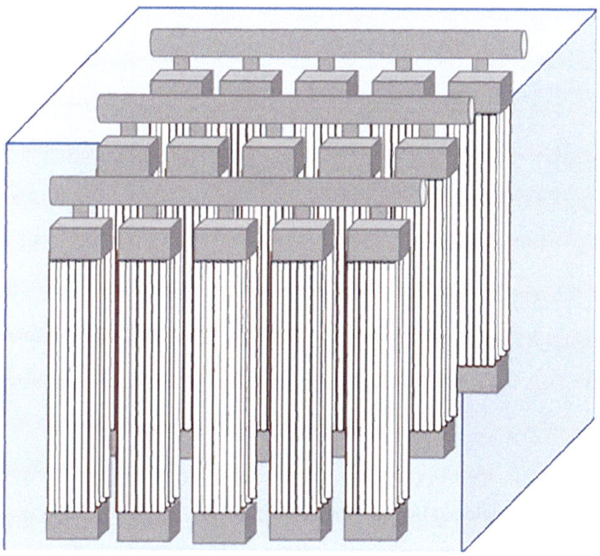

Figure 6.9 Schematic of Submerged Membrane Filtration Configuration

for screening to occur immediately upstream of membranes, although this is not always possible for all applications. In MBR applications, the process water is screened upstream of the biological reactors (in headworks facilities) instead of immediately upstream. Typically, 1- to 3-mm fine screens are used in wastewater applications due to the higher concentration of solids and potential presence of debris. For drinking water applications and lower solids applications (i.e., filtered wastewater and recycled water), smaller size screens can be used (100 to 300 μm). It is also important to keep membrane tanks covered to prevent foreign materials from entering the tank and damaging membranes.

PRESSURIZED CONFIGURATION

In a submerged configuration, the membranes are bundled into modules that must operate in an outside-in configuration. In pressurized configurations, both outside-in and inside-out membrane configurations are possible (Figure 6.10). Typically, a low-pressure feed pump directs the feed into the pressurized module. The feed and the filtrate flow may enter and leave either through one end of the module or through side ports. Figure 6.10 depicts modules with side ports for both inside-out and outside-in configurations.

The feed can either enter the module through a side port, as shown in Figure 6.10, or through one end of the module. For inside-out configurations, the lumens act as the feed channel and the filtrate leaves through a side port or through a channel in the potting resin. For outside-in configurations, the feedwater enters the module either through a side port or through feed channels in the potting resin surrounding the membrane fibers, and the filtrate leaves through the lumens. Some modules do not use side ports, and one end of the module is designated as the feed end and the other as the filtrate end. Filtrate is collected in a common header from all of the modules. The operating range for TMP is higher for pressurized configurations than for submerged configurations. The TMP operating range is manufacturer- and system-dependent; however, some systems can operate with TMPs greater than 210 kPa (30 psi).

Pressurized membrane configurations can be found in both drinking water and wastewater applications with low solids concentrations. For wastewater applications, pressurized systems can be used in municipal wastewater and recycled water applications. In municipal wastewater applications, pressurized systems are not used to treat mixed liquor, but are used to further treat filtered wastewater.

Pressurized systems are typically dead-end filters. This means there is no flow recycling like there may be for submerged systems. The process water must undergo screening upstream of the membranes to remove larger particles and debris. Larger particles and debris can damage membranes. Because there are lower solids and a lack of debris in the membrane feed, a finer filter can be used as pretreatment. Typically, 100- to 300-μm cartridge filters are used for this purpose.

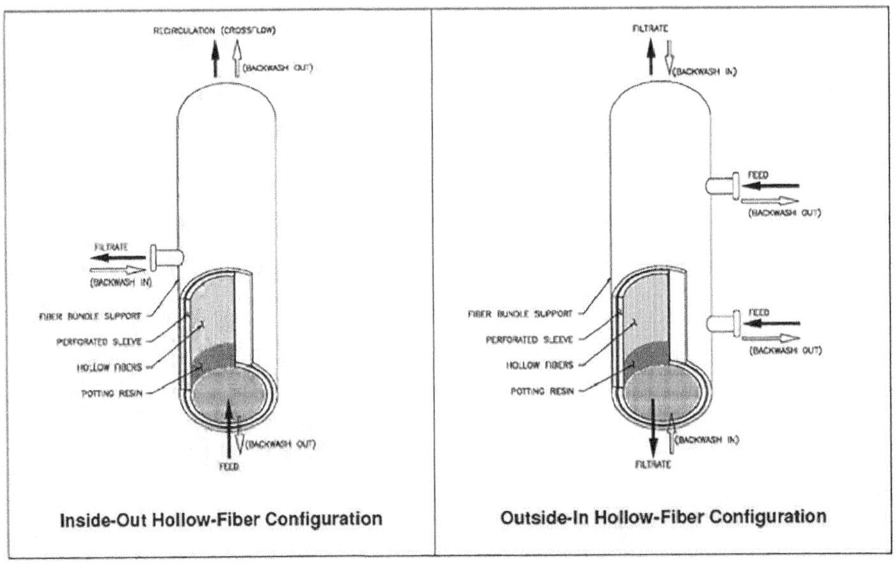

Figure 6.10 Pressurized Inside-Out and Outside-In Hollow Fiber Configurations (U.S. EPA, 2005)

SKID CONFIGURATION

A membrane filtration system is composed of several components. A train/skid/rack is a group of modules that share common valving and can be isolated as a group. The terms *train*, *rack*, and *skid* are synonymous. Table 6.3 provides descriptions of typical components.

Table 6.3 Summary of Membrane Filtration System Components

Component	Description
System train/skid/rack	■ A *rack* is defined as a stage that shares common valving ■ Provides structure to mount system components like modules, piping, etc.
Membrane modules	■ Contains membrane fiber bundles
Drum screen	■ Installed upstream of biological reactors ■ Used for MBR systems ■ Typically sized ■ Prevents damage to membranes from larger particles
Strainer	■ Installed upstream of membranes ■ Not used for MBR systems ■ Filter out particles in the 100- to 300-μm size range ■ Prevents damage to membranes from larger particles
Compressed air system	■ Provides air for air scouring during backwash and pressure decay tests and can be used for air-actuated valves ■ Provides constant air scouring for MBR systems ■ Comprised of air compressor, air receiver tank, and air lines
Filtrate pump	■ Used for MBR systems ■ Provides suction to pull process water into lumens
Chemical cleaning system	■ System may be used for clean-in-place and/or enhanced flux maintenance ■ Comprised of the following: • Cleaning pump • Chemical pumps and solution • Heater • Cleaning tank
Backwash system	■ Consists of a reverse filtration pump that transfers backwash water to membrane modules ■ Can also be used to flush membranes with filtrate before resuming production
Water quality monitors	■ Provides information on system conditions and performance • Feed and filtrate turbidity • pH • Temperature
Instrumentation	■ Monitors process parameters and is used as control variable • Pressure • Flow

TEST YOUR KNOWLEDGE

1. **If removal of viruses is desired, microfiltration membranes should be used.**
   - ☐ True
   - ☐ False
2. **Isotropic membranes have the same mechanical properties throughout the membrane.**
   - ☐ True
   - ☐ False
3. **Which of the following is true for an outside-in membrane configuration?**
   a. Air scour applies air so that it travels from the outside of the hollow fiber to the inside of the hollow fiber.
   b. The lumen is plugged with resin on both potted ends.
   c. Water is filtered from the outside of the hollow fiber to the inside and materials are rejected at the membrane surface.
   d. Filtrate is collected on the outside of the hollow fiber membranes.
4. **Which of the following is true for a submerged system versus a pressurized system?**
   a. Submerged systems have a narrower operational TMP range than pressurized systems.
   b. Submerged systems operate with a feed pump to draw water into the lumens.
   c. Both submerged and pressurized systems are dead-end filters.
   d. Submerged systems can come in both outside-in and inside-out configurations.
5. **Membrane bioreactor systems use a drum screen upstream of biological reactors to remove larger particles.**
   - ☐ True
   - ☐ False
6. **Which component is not commonly found in membrane filtration systems?**
   a. Strainer/drum screen
   b. Turbidimeter
   c. Compressed air system
   d. High-pressure feed pump

Process Control for Membrane Filtration and Membrane Bioreactors

Process control of membrane filtration relies primarily on installed online instrumentation. This section describes the instrumentation commonly used in membrane filtration processes for process control and operation.

INSTRUMENTATION

The instruments in a membrane filtration system may be separated into two categories: process and water quality. Examples of process instruments include pressure transmitters and flow transmitters. Water quality instruments measure parameters like turbidity, pH, and temperature. These instruments are used for a variety of membrane filtration processes for monitoring and process control. The following sections describe the role of instruments in process monitoring for various membrane filtration processes.

PRODUCTION

During production, feedwater is filtered through the membrane and filtrate is produced. Typically, there is a filtrate flow setpoint that is used to control valves and/or pumps to meet this setpoint. The filtrate flow meter provides feedback to the system if modulation of valves/pumps is needed to meet the setpoint. Flow is an important parameter to monitor because it provides information on the flux of the system. Flow meters can also provide totalizer information, which can be used to calculate recovery of the system.

Integrity testing is a diagnostic test used to evaluate if there are membrane breaches that would result in loss of the membrane's ability to remove target constituents like pathogens and solids. Direct and indirect monitoring methods are required for evaluating the integrity of membranes when used as a pathogen barrier (i.e., a unit process that reduces/removes/inactivates pathogens through treatment).

Pressure transmitters are typically installed on the filtrate lines. Pressurized systems also typically incorporate a pressure transmitter on the feed line. Submerged systems typically do not have a feed pressure transmitter; however, the feed pressure can be calculated using the tank level to evaluate the head pressure on the membranes. The feed and filtrate pressure values are used in the TMP calculation, which is an indicator for the fouling status of the membranes. Flow and pressure information can be used to calculate the specific flux of the system, which is an indicator for the production efficiency of the system.

INTEGRITY TESTING

According to U.S. state and federal regulations, both indirect and direct monitoring are required for the membrane filtration system when used to remove pathogens. Performing a pressure decay test (PDT) satisfies the direct monitoring requirement and monitoring the filtrate turbidity continuously satisfies the indirect monitoring requirement.

During a PDT, the membrane rack is pressurized with air to meet a target start pressure setpoint. During this step, airflow is measured and valves control the flow of air to prevent excessive airflow that may damage the membranes. During pressurization, the air control valve may alternate between opening and closing to maintain the target pressure. After pressurization, there will be a stabilization step to allow for the pressure to stabilize. After a set amount of time has elapsed for stabilization, the test start pressure is logged and the subsequent decrease in pressure is measured over the designated amount of time for the test. At the end of the test period, the test end pressure is logged. The membrane rack is depressurized as the final step of the PDT. Next, the rate of decay is calculated. The rate of decay is the amount of pressure lost per unit of time. The rate of decay is used in further calculations to evaluate the pathogen LRV for the tested rack. If the pressure decay rate (PDR) is less than the system's upper control limit (UCL), the rack may be placed back into queue for production. If the PDR exceeds the system's UCL, there is a possibility of a breach in the membrane integrity. In this scenario, the rack would be placed offline for further troubleshooting. Figure 6.11 presents the steps that may be taken to troubleshoot a potential membrane breach.

The filtrate turbidity is monitored continuously throughout the PDT and satisfies the indirect monitoring requirement. The frequency of monitoring must be at least once every 15 minutes to satisfy the "continuous" requirement in U.S. Environmental Protection Agency (U.S. EPA) regulations. If the filtrate turbidity exceeds 0.15 NTU for two consecutive readings for a period exceeding 15 minutes, a direct integrity test (i.e., PDT) must be performed. The PDT performed in response to the elevated turbidity is a separate test from the required daily PDT for each rack. If the PDT passes, it is possible that the turbidity meter requires maintenance or calibration.

FOULING

Transmembrane pressure is the main parameter for tracking fouling for hollow fiber membranes. Transmembrane pressure is the difference between the feed pressure and filtrate pressure. Most systems use pressure transmitters on both the feed and filtrate lines. For submerged systems, there may not be a pressure transmitter on the feed. As noted earlier, feed pressure can be calculated from water depth. When

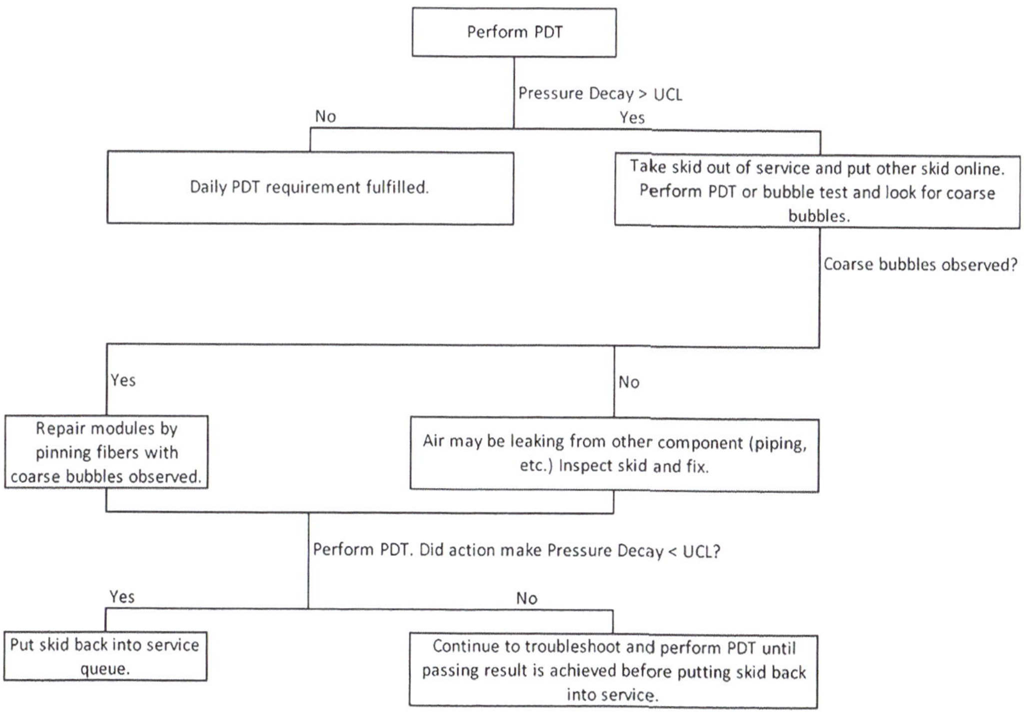

Figure 6.11 Flowchart for Pressure Decay Tests

evaluating TMP, the pressure measured by the filtrate pressure transmitter while the system is in standby (i.e., no filtration) can be used as the feed pressure for the TMP calculation.

Chemicals can be used to remove foulants and restore membrane performance. Maintenance cleans can be used to address short-term fouling and CIPs can be used to address more severe long-term fouling.

Transmembrane pressure can be used to evaluate short- and long-term fouling trends. If the TMP increases significantly during the filtration cycle (e.g., 14 to 21 kPa [2 to 3 psi]), an accelerated rate of fouling is occurring in the short term. If the TMP value increases during the filtration cycle, the following backwash should lower the TMP value. Under normal operating conditions, increasing TMP accumulated during filtration will be mostly corrected by the following backwash cycle. If the TMP value does not decrease, the backwash may not be effective for the type of fouling taking place. If TMP is stable for long periods of time (e.g., weeks) and increases slowly throughout that time, the rate of fouling is considered low. Maintenance cleans can be used to clean the membranes periodically. The TMP before and after the maintenance cleans should be recorded to evaluate if the maintenance clean was effective. Eventually, membrane systems must be cleaned with a stronger regime and a clean-in-place (CIP). It is intended to remove the accumulation of foulants/scale on the membranes. The concentration of the CIP solution and duration of CIP can be adjusted to accommodate different extents of fouling. The TMP should be tracked to evaluate when a CIP is necessary; the TMP before and after each CIP should be recorded to evaluate the efficacy of the clean. The "Maintenance" section contains more information on evaluating fouling trends.

TEST YOUR KNOWLEDGE

1. **What instruments are used to monitor the production cycle?**
   a. Feed pressure transmitter, flow meter, and filtrate pressure transmitter
   b. Airflow rate, turbidimeter, and flow meter
   c. Feed and filtrate pressure transmitter
   d. Feed pump speed, filtrate valve percent opening, and turbidimeter

2. **For a membrane system providing pathogen protection, if the filtrate turbidimeter is greater than 0.15 NTU for a period longer than 30 minutes, the membrane is considered intact.**
   ☐ True
   ☐ False

3. **Which of the following statements about the PDR is not true?**
   a. The PDR is calculated by dividing the loss of pressure divided by the test period.
   b. If the PDR is greater than the UCL, the membranes are intact.
   c. Per U.S. EPA regulations, the PDR should be measured once a day to satisfy the direct monitoring requirement.
   d. The PDR can be used to evaluate if there is a potential breach in a membrane skid.

4. **What parameter is used to evaluate the effectiveness of chemical cleans?**
   a. TMP
   b. Turbidity
   c. Filtrate pressure
   d. Feed flow

Operation of Membrane Filtration and Membrane Bioreactors

Collecting operational data and visually inspecting system and components are known as *rounds*. Daily rounds are one of the most important activities for monitoring the membrane filtration process. Through daily rounds, operators capture important process parameters while the system is stable and not transitioning between processes (e.g., coming out of backwash) to evaluate if the production target is being met, the cleanliness of membranes, and membrane integrity.

The main operational activities are geared toward maintaining and confirming the stability of the membrane system. *Stability* refers to the system's ability to maintain consistent production while meeting all treatment objectives. The following sections describe common routine and preventive practices for operating membrane filtration systems.

NORMAL/DAILY OPERATION

The daily tasks for an operator are centered on monitoring process performance and maintaining stable operations. Confirming production targets, membrane integrity, and the fouling state of membranes are all aspects of maintaining stable membrane filtration operations.

While doing rounds, operators also physically inspect the system for signs of abnormalities (e.g., leaks, suspect instrument values, etc.) that are not observable on a monitoring screen. Any abnormalities are noted and reported if further action is needed. Operators confirm through daily rounds that scheduled activities, such as daily PDTs, have been performed. Table 6.4 summarizes critical membrane filtration monitoring parameters that are recommended to be tracked during daily rounds.

Table 6.4 Daily Rounds for Tracking Membrane Filtration Parameters

Process	■ Feed pressure ■ Filtrate pressure ■ Feed flow ■ Feed chloramines concentration (mg/L) ■ Membrane tank level (submerged systems)
Production	■ Filtrate flow ■ Recovery ■ Total filtrate volume produced
Membrane Integrity	■ PDT PDR ■ *Cryptosporidium* LRV (not for MBRs) ■ *Giardia* LRV (not for MBRs) ■ Filtrate turbidity
Fouling Status	■ TMP ■ Specific flux

The data collected during daily rounds may be used for further data analysis and compared with historical trends to evaluate the status and condition of the membrane filtration system. Corrective maintenance can sometimes be identified by studying historical trends before small problems get worse. Table 6.5 presents a summary of parameters and how they may be used to diagnose system status.

STARTUP

System startup protocols are typically provided by the membrane manufacturer and should be followed to ensure that none of the components are misused or damaged as a result of improper startup. Examples of some common startup practices are as follows:

1. Perform integrity test in factory and repeat on-site to confirm membrane integrity.
2. Operate system at low and high end of flow ranges.
3. Confirm system is able to operate at design flux stably for a long period of time (e.g., 30 days).
4. Test maintenance systems (backwash, maintenance clean, CIP).

Table 6.5 Parameters Used to Diagnose Membrane Filtration System Status

Parameter	Analyses of Historical Trends
TMP	■ Observe rate of increase and evaluate if fouling rate is an issue. ■ If fouling rate is an issue, may consider increasing frequency of enhanced flux maintenance. ■ If system is approaching max TMP, schedule time for CIP.
PDR	■ Monitor PDR trend. If stable, no issue. If increasing, consider performing bubble test to evaluate if there are defects.
Filtrate turbidity	■ If turbidity is trending upwards, turbidimeter body/cell may need to be cleaned. If after cleaning readings are still high and/or increasing, there may be membrane defects. Cross check with PDR trend to see if there is an integrity issue.

SHUTDOWN AND MEMBRANE STORAGE

System shutdown protocols are typically provided by a membrane manufacturer and should be followed to ensure that none of the components are misused or damaged as a result of improper shutdown. The procedures for storage will depend on whether short- or long-term storage is desired. Short-term storage can range from 1 day to a week. The duration range for what is considered short-term storage may differ between manufacturers and the manufacturer's definition should be used for each specific application. If the desired storage period is longer than the specified short-term duration range, long-term storage protocols should be used. The "Maintenance" section provides some common practices for membrane storage.

TEST YOUR KNOWLEDGE

1. **What type of information can be determined with daily rounds?**
   a. Type of foulants on the membrane
   b. Time until membrane failure
   c. All routine maintenance of analyzers have been completed
   d. Membrane integrity

2. **Short-term storage of membranes can range from 1 day to 1 month.**
   ☐ True
   ☐ False

3. **Filtrate turbidity is the only parameter that can be used to evaluate the integrity of membranes.**
   ☐ True
   ☐ False

Data Collection, Sampling, and Analysis for Membrane Filtration and Membrane Bioreactors

Data collection, sampling, and analysis are important for monitoring the performance of membrane filtration systems. Data can be collected on a supervisory control and data acquisition (SCADA) system for short- and long-term trends. Historians may also be used to generate reports for specific time periods at a specified data frequency. Operators may also collect data on their daily rounds during stable and normal operations to ensure that the system is functioning as intended. Table 6.6 summarizes key parameters that should be tracked on SCADA and included in reports and daily rounds. This list is not comprehensive of all parameters and should be modified to best suit each application.

Analysis of collected data can provide insight to process performance and is a valuable tool for troubleshooting. Some parameters are used to confirm that the system is operating as intended; these include the air scour flowrate, filtrate flowrate, membrane tank level (submerged systems only), and backwash flowrates. Other trends can be analyzed further for additional insights on performance and to inform decision-making. Transmembrane pressure can provide both short- and long-term trends on fouling and evaluate if cleanings need to be performed to better control fouling. Figure 6.12 presents an example of how TMP may increase over time. The TMP should decrease after chemical cleans if the clean was effective. If there is little recovery of TMP, it is possible that a more aggressive cleaning regime is needed or there is irreversible fouling.

Pressure decay rates from PDTs can be tracked to evaluate the condition of membranes. As membranes age, the PDR will slowly increase. Monitoring the PDR can prompt membrane repairs before issues become serious. The filtrate turbidity trend may also be monitored to evaluate maintenance needs to maintain stable monitoring. If turbidity spikes after backwashes or changes in flow, the readings may be subject to interference from particles in the meter body; this is not necessarily reflective of the actual water quality.

Starting from least aggressive and more frequent to most aggressive and least frequent, chemical cleaning methods in membrane filtration systems include: CEBs performed daily or during every backwash cycle, maintenance cleans, maintenance washes, or enhanced flux maintenance cleans (EFM) performed daily to weekly, and CIP performed monthly to annually.

Maintenance for Membrane Filtration and Membrane Bioreactors

MEMBRANE FOULING

Membranes systems will gradually accumulate foulant and scaling materials that cannot be removed by the backwash and air scour cycle. Therefore, some form of routine chemical cleaning is typically required. Cleaning regimens are typically site-specific based on actual expected performance at the design flux and

Table 6.6 **Summary of Key Parameters for Data Collection for the Membrane Filtration Process**

Parameter	SCADA	Reports	Daily Rounds	Calculated
Feed pressure	×		×	
Filtrate pressure	×		×	
Feed flow	×		×	
Filtrate flow	×		×	
Flux	×	×	×	×
Feed chloramines concentration	×		×	
Membrane tank level (submerged systems)	×		×	
Recovery	×	×	×	×
Total filtrate volume produced	×	×	×	×
Air scour flowrate	×			
Backwash flowrate	×			
Runtime	×		×	
TMP	×	×	×	×
Specific flux	×			×
PDT results	×	×	×	×
Pathogen LRV	×	×	×	×
Feed turbidity	×		×	
Filtrate turbidity	×	×	×	

the expected range of water quality conditions. Organic and biological foulants are typically addressed with the use of sodium hypochlorite. In cases of persistent and heavy organic fouling, caustic soda may be added with the hypochlorite to raise pH above 10 for more effective cleaning. Backwashes can be performed with the addition of sodium hypochlorite at an approximate target concentration of 30 to 50 mg/L providing a short cleaning regime. This can be performed daily or during every backwash cycle. Some manufacturers refer to this practice as *chemically enhanced backwash* (CEB). A less frequent but stronger chemical cleaning is a form of CEB with a higher target residual chlorine of 100 to 500 mg/L

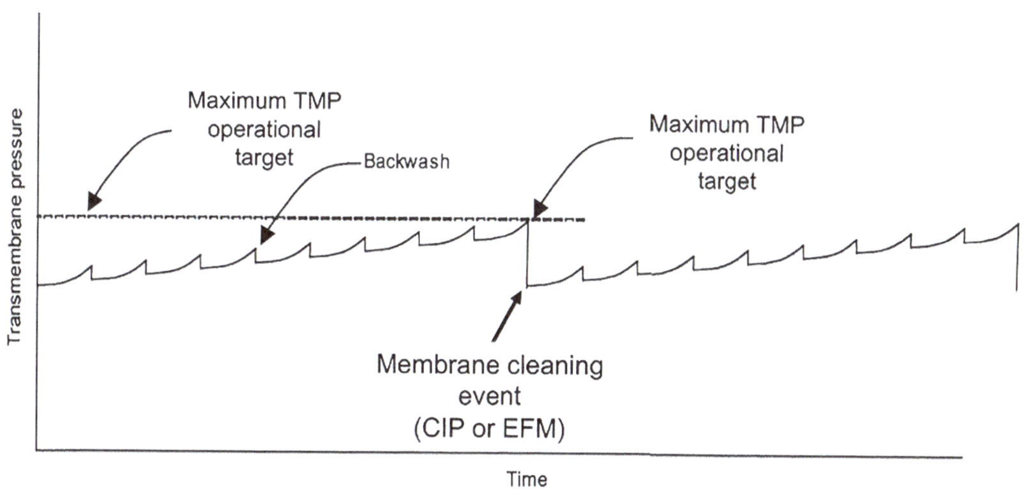

Figure 6.12 **Short- and Long-Term Transmembrane Pressure Trends**

with a short soak (e.g., 20 minutes). This is referred to as enhanced flux maintenance (EFM), which is sometimes also called maintenance clean or maintenance wash. It is typically an automated cleaning sequence programmed to occur from 1 to 3 times a week to 2 times every month. Enhanced flux maintenance may provide sufficient performance recovery to eliminate the need for CEBs and decrease the frequency of longer CIP chemical cleaning cycles. In fact, EFMs are commonly used and may provide a good tradeoff in terms of operational availability of the system and minimizing operating costs relative to operating for longer periods of time between CIPs and allowing TMP to build up to higher levels, which causes higher energy use. Clean-in-place processes rely on hypochlorite concentrations that are specific to the membrane manufacturer's recommendations but are commonly in the range of 1000 to 1500 mg/L for submersible PVDF membranes and 3000 to 5000 mg/L for pressurized PVDF membranes. Clean-in-place is typically performed every 1 to 3 months in reuse applications and less frequently in drinking water applications, especially if supplemented by EFM regimens.

Inorganic scaling, such as that associated with iron and manganese, can be addressed with citric acid cleaning. Typically, a citric acid CIP is carried out every 3 to 6 months and, in some cases, annually. However, in some applications, such as when an iron-based coagulant is added upstream of the membrane filters, more frequent cleaning may be required. Typically, a CIP may use a 2% wt citric acid solution, whereas EFM may use 0.1% to 0.2% wt (1000 to 2000 mg/L) citric acid solution. Generally, the frequency of cleaning is based on actual performance and can be adjusted to provide both preventive and effective cleaning addressing both early onset and heavy fouling events. Manufacturers generally recommend the use of a systematic cleaning regimen to prevent accumulation of irreversible fouling. Systematic CIP sequences can be triggered based on accumulated runtime, calendar interval, or a certain filtrate production volume.

A *2% wt* refers to chemical concentration in units of percent weight (% wt). A 2% wt is equivalent to a concentration of 20 000 parts per million.

STORAGE

System shutdown protocols are typically provided by a membrane manufacturer and should be followed to ensure that none of the components are damaged as a result of improper shutdown. The procedures for storage will depend on whether short- or long-term storage is being performed. Short-term storage can range from 1 day to 1 week. The duration range for short-term storage may differ between manufacturers and the manufacturer's definition should be used for each specific application. If the desired storage period is longer than the specified short-term duration range, long-term storage protocols should be used. Common requirements for membrane storage include the following:

a. Perform CIP before storage
b. Membranes must remain moist
c. Store away from direct sunlight
d. May have temperature range requirement for storage
e. Pickle membranes with chemical solution (chlorine or sodium bisulfite) and periodically check strength of chemical solution

For downtime of 1 week to 1 month, the system may be soaked in a mild solution with chlorine residual (e.g., 20 to 50 mg/L as Cl_2). For longer periods, pickling with 1% sodium bisulfite is generally recommended; however, it is always best to consult manufacturer-specific recommendations.

FIBER PINNING

Fiber pinning is performed to restore membrane integrity. Typically, this involves removing a membrane module from the rack and disassembling the endcap. Broken fibers must first be located using a module tester, which is done by supplying a low-pressure air supply (e.g., <70 kPa [<10 psi]) using a pressure regulator and submerging the filtrate end of the module in water. Typically, this is accomplished by placing the module upright and removing the endcap. Next, a small amount of water is added directly onto the top of module (if it has edges) or a module-specific adaptor is used for the pressure-testing individual modules. Broken fibers are revealed by the presence of bubbles. The air pressure can be modulated to pinpoint the precise location of any broken fibers. Typically, metal pins supplied by the membrane manufacturer are glued into the broken fibers. This process is known as *pinning*. Pinning essentially takes those individual fibers out of service. Because each module typically contains thousands of bundled fibers together, losing a small percentage of fibers due to pinning will not significantly affect membrane module performance.

Typical manufacturing warranty is to have no more than 0.5% of broken fibers over an estimated life of 10 years for either microfiltration or ultrafiltration polymeric membrane modules. The number of hollow fibers in commercial membrane modules may range from 3000 to 7000 modules, 0.5% of which equates to 15 to 35 broken fibers over the lifetime of the membrane.

OTHER MAINTENANCE

Membrane systems have several components besides membrane modules, including but not limited to:

- Feed or filtrate pump
- Backwash pump
- Variable frequency drives (VFDs)
- Strainers
- Compressed air system for air-actuated valves and pressure decay testing
- Air scour system consisting of air blowers (or compressed air may be used depending on system size)
- Air dryer unit
- Air receiving tank(s)
- Flow control valves
- Open/close valves
- Solenoid valves
- Pressure regulators
- Flow meters
- Pressure transmitters
- Temperature transmitters
- Turbidity meters.

Specific maintenance intervals for each subcomponent should be obtained from the manufacturer's O&M manuals.

TEST YOUR KNOWLEDGE

1. **Which of the following are examples of commonly used chemical cleaning methods for membrane filtration systems?**
   a. Backwashing and air scour
   b. Enhanced flux maintenance and CIP
   c. Taking units out of service and letting them rest
   d. Reducing the flux at which membranes operate
2. **What is the typical frequency for maintenance washes?**
   a. Monthly to annually
   b. Daily to weekly
   c. Weekly to monthly
   d. Every backwash cycle
3. **What chemicals are typically used for cleaning hollow fiber membranes?**
   a. Caustic and sulfuric acid
   b. Sodium hypochlorite and citric acid
   c. Detergent solutions and hydrochloric acid
   d. Organic solvent and water
4. **The process of repairing broken hollow fibers is called?**
   a. Pinching
   b. Parking
   c. Pinning
   d. Plucking
5. **What procedures should be taken for membrane storage?**
   a. Membranes should be drained of process water
   b. Membranes should be taken off and wrapped
   c. Membranes do not require special procedures for storage
   d. Membranes should be soaked in a mild chlorine solution or with a preservative such as sodium bisulfite

Troubleshooting for Membrane Filtration and Membrane Bioreactors

MEMBRANE FOULING

The fouling rate of membrane systems may change due to changes in feed water quality. Total organic carbon can indicate organic fouling. Total organic carbon can increase due to minor upsets in upstream biological treatment. In addition, systems that use secondary effluent as feedwater may occasionally receive higher levels of TSS, which cause more fouling. Generally, once TSS concentrations return to normal levels, the fouling rate will decrease and performance should stabilize with backwashes and not require chemical cleaning. Monitoring feedwater turbidity using an online meter can be used to verify that changes in TSS concentration return to normal levels. However, once organic fouling forms on the membrane, it is harder to remove with only backwashes and chemical cleaning is generally needed. It is useful to periodically measure TOC, iron, manganese, TSS, and turbidity levels in the feedwater to establish normal expected ranges for these water quality parameters and to correlate with normal long-term membrane performance. Fouling events are typically exhibited by a rapid decline in temperature-corrected specific flux and increase in TMP. During fouling events, grab samples should be collected to determine which water quality parameters are outside of the normal range. Organic fouling can be addressed by chemical cleaning with sodium hypochlorite, combined with supplemental addition of sodium hydroxide. Iron- and manganese-based scales typically take longer to develop; however, they can be significant if iron and manganese levels are in the range of 0.2 mg/L or higher. This scaling can be mitigated by using chemical cleaning with citric acid more frequently.

Biological fouling of membranes can occur when there is an insufficient amount of monochloramine residual in the water. This is especially applicable in reuse applications. Combined chlorine residual can be easily verified through testing with the commonly used N,N-diethyl-p-phenylenediamine-based methods using commercially available test kits, handheld meters, or laboratory methods. Typically, combined chlorine residual of 2 to 5 mg/L (as Cl_2) is effective to control the onset of biological fouling. If the residual is too low, it should be increased. Any accumulated biological fouling typically is effectively addressed with sodium hypochlorite cleaning.

Combined chlorine typically refers to the monochloramine residual. However, when chlorine and ammonia are added to water, some dichloramine may also be formed, and so *combined chlorine* or *chloramines* refers to the mixture of mono- and dichlroamine. Typically, trichloramine does not form in most applications, unless a Cl_2-to-nitrogen ratio of 5 is exceeded or pH of the water is very acidic.

TROUBLESHOOTING FILTRATE TURBIDITY

Filtrate turbidity is typically used as a performance indicator as well as continuous indirect integrity monitoring between daily membrane integrity tests. Turbidity meters correlate light scattering from the sample stream to the presence of solids and colloids in the water. Results are reported as NTUs. Over time, turbidity readings may increase due to sloughing of biofilms that form in sample lines, accumulation of dust particles, or membrane failures. Typically, the maintenance requirements are minimal on the meters; however, periodic flushing of the sample lines with a higher flowrate can be helpful to minimize buildup of biofilms in the sample line. In addition, air bubbles can result in false turbidity readings. While some commercial turbidity meters may have an algorithm to correct turbidity readings due to the presence of bubbles, other meters may use simple devices such as a bubble trap. A bubble trap generally consists of a larger diameter pipe than the sample line. Flow enters from the top, allowing the sample flow to be slower than the speed of bubble rise (Figure 6.13).

TROUBLESHOOTING FAILED PRESSURE DECAY TESTS

The PDT is used for membrane integrity testing. It is designed to produce a narrow range of acceptable pressure decay, which is the difference between start and end pressure over a defined hold test duration, typically 5 minutes. The PDR (measured as kPa/min [psi/min]) is compared to the system's UCL. This is

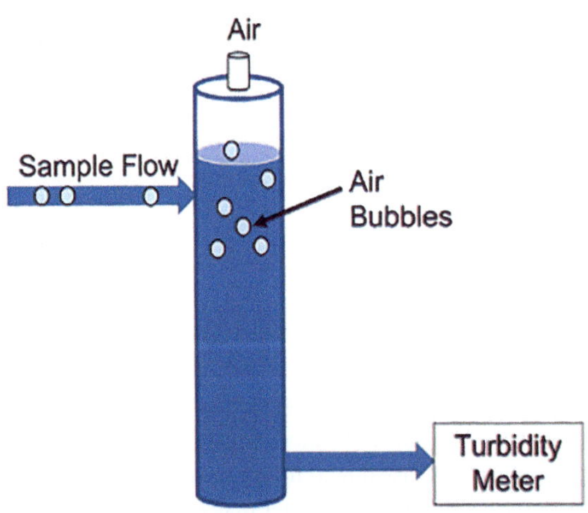

Figure 6.13 Schematic of a Bubble Trap Apparatus

a system-specific parameter corresponding to 4-log removal of pathogens. A failed membrane integrity test can be caused by mechanical leaks in the system and/or failed membrane fibers. If the PDR exceeds the UCL, no credit can be claimed for pathogen removal. Pathogen removal credit is a must in drinking water treatment and potable water reuse applications; therefore, a failed PDT requires immediate troubleshooting. The first step in troubleshooting the PDT is to inspect the failed train or unit for any obvious signs or air leaks through careful inspection of fittings on the air supply to the membrane modules (if applicable) or common header. A solution with soap can be used as a bubble test to aid in this inspection. Over prolonged use, plastic fittings may become looser due to vibrations in the system or they may become cracked due to exposure to the elements. In some cases, membrane module connections to the membrane rack may develop a leaky O-ring that would need to be replaced. Once the system has been inspected for mechanical leaks on the feed side of the membranes, the next step is to repeat it while observing the system. Each module should be visually inspected for the presence of coarse air bubbles at the top of the module. Alternatively, a sonic tester may be used to "listen" for the presence of leaks on individual modules and to pinpoint the approximate location where the leak is occurring (e.g., at the top, middle, or bottom). Any modules with coarse bubbles and/or leaks identified by sonic testing would need to be identified and removed for further testing using an individual module pressure tester.

TEST YOUR KNOWLEDGE

1. **Which of the following water quality parameters should be checked when troubleshooting suspected fouling of the membrane systems?**
   a. Feed ammonia, nitrite, and nitrate
   b. Feed TSS, TOC, and turbidity
   c. Iron, manganese, and pH
   d. Total coliforms, fecal coliforms, and chlorine residual
2. **Which of the following water quality parameters should be checked when troubleshooting suspected scaling of the membrane systems?**
   a. Iron and manganese and other metals
   b. TOC, ammonia, and TSS
   c. Total coliforms, fecal coliforms, and chlorine residual
   d. Temperature, pH, and electrical conductivity
3. **What are the typical causes for drifting or abnormally high turbidity readings on the online turbidity meters?**
   a. Seasonal changes in temperature, pH, and electrical conductivity
   b. Buildup of biofilms in the sample lines and the presence of air bubbles
   c. Loss of calibration
   d. Defective meter parts
4. **What steps should be taken if a membrane filtration system fails a required daily pressure decay test in a potable reuse facility?**
   a. Taken immediately out of service for troubleshooting of a failed PDT and repair of the cause
   b. Allowed to return to production and scheduled for maintenance at the next convenient time
   c. Stop production and order replacement of the membranes in the failed train
   d. Perform a backwash and EFM to restore membrane performance

Equipment for Reverse Osmosis

Reverse osmosis systems consist of several main components, which are further described and illustrated in subsequent sections. These include

- Spiral-wound elements
- Pressure vessels
- Feed pumps
- Cartridge filters
- Energy recovery devices
- Chemical skids for pretreatment chemicals
- Clean-in-place skid
- Flush skid

The following subsections provide details of the primary reverse osmosis system equipment.

SPIRAL-WOUND ELEMENTS

The basic production unit of a reverse osmosis system is the spiral-wound element, which is designed to contain a large amount of membrane material in a compact space. Figure 6.14 presents a schematic of a spiral-wound element. Spiral-wound elements consist of layers of membrane sheets (called "leaves") that are glued along three edges to form an envelope. The unglued edge is sealed around the perforated central tube. The purpose of the central perforated tube is to collect the product water (i.e., permeate) that passes through the membrane sheets. The term *permeate*, rather than *filtrate*, is used to define the product water produced by a reverse osmosis system. A common 20-cm (8-in.) -diameter spiral-wound element consists of approximately 20 leaves (U.S. EPA, 2005). These leaves together provide a total membrane surface area of 37.1 m^2 (400 sq ft) per 20-cm (8-in.) membrane element (most common). Ten-centimeter (4-in.) and 40-cm (16-in.) elements are also used, although they are less common. Permeate only has to pass through one leaf. Once it is inside the envelope, it travels through a permeate carrier that routes the permeate to the central perforated tube. Between each leaf there is a spacer made of a mesh-type material, known as the *feed spacer*. It creates a pathway for feedwater to travel efficiently across the surface of the membrane sheet and exit on the other end as concentrate. To compactly assemble a single reverse osmosis element, the leaves and feed spacers are spirally wound together and encased within a rigid outer wrap (typically made of fiberglass) and capped on either end with a hard plastic cap (i.e., anti-telescoping device) that prevents unwinding of the membrane sheets due to axial forces from cross-flow filtration. The direction of feedwater flow is often printed on the outer shell of every spiral-wound element. It is important to install the elements in the correct flow direction if there is one. More recently, some reverse osmosis membrane manufacturers are producing bi-directional flow membranes, although such membrane configurations are less common. Figure 6.15 shows multiple elements side-by-side upon removal from service, a cross-section view of a 20-cm (8-in.) element, and an uncased 20-cm (8-in.) spiral-wound element.

A mesh-like material that separates the semi-permeable membrane layers and creates flow passages in a spiral-wound module.

A rigid plastic fitting found on the concentrate side of a pressure vessel to prevent the physical deformation of the membrane elements due to high differential pressure.

Cross-flow filtration is the application of water at high velocity perpendicular to the surface of a membrane. Cross-flow filtration results in differential pressure in which the membrane, support, and spacer layers are displaced axially (i.e., in the direction of the feed flow) from the center.

A reverse osmosis semi-permeable membrane consists of several layers cast on top of each other in a structure known as a *thin-film composite*. Figure 6.16 shows an example of the layering for a typical thin-composite semi-permeable membrane. The uppermost layer (facing the feedwater) is known as the *active layer* and is extremely thin (<2 μm, or 10× thinner than a human hair). The active layer must be extremely thin to achieve an effective water flow through the membrane because the resistance to flow through such a membrane is inversely proportional to its thickness. The active layer is responsible for allowing water to pass through while stopping the dissolved contaminants (which are called solutes). In other words, this active layer is the "workhouse" that makes reverse osmosis membranes able to produce near-ultrapure product water. The active layer is commonly made of polyamide and is supported by a thicker polysulfone porous layer (thickness range of 30 to 40 μm), which is cast onto a polyester nonwoven support fabric (approximately 100 μm in thickness). The direction of flow is very important for reverse osmosis membranes. The feed flow must be applied to the polyamide active layer such that permeate is collected on the opposite side. Because of this structure, reverse osmosis membranes cannot be backwashed like low-

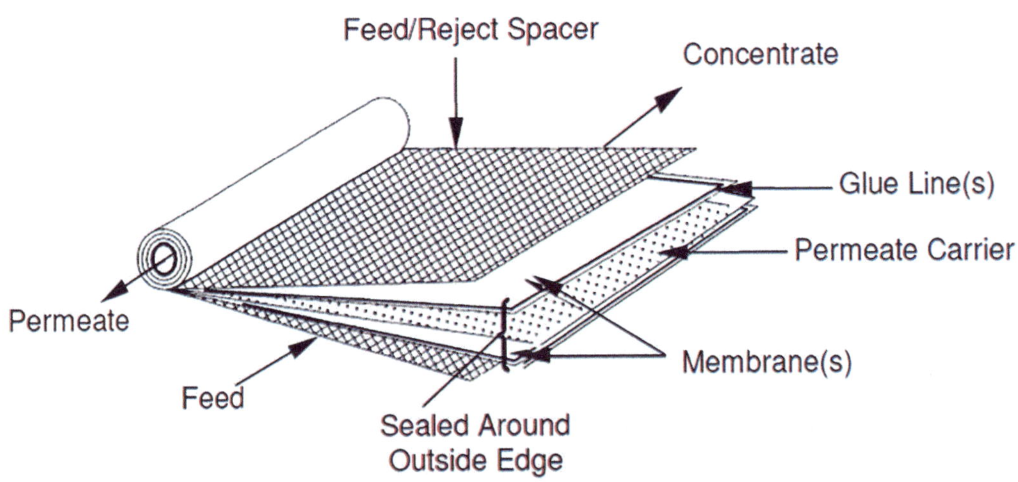

Figure 6.14 Schematic Showing Components of a Spiral-Wound Reverse Osmosis Element (U.S. EPA, 2005)

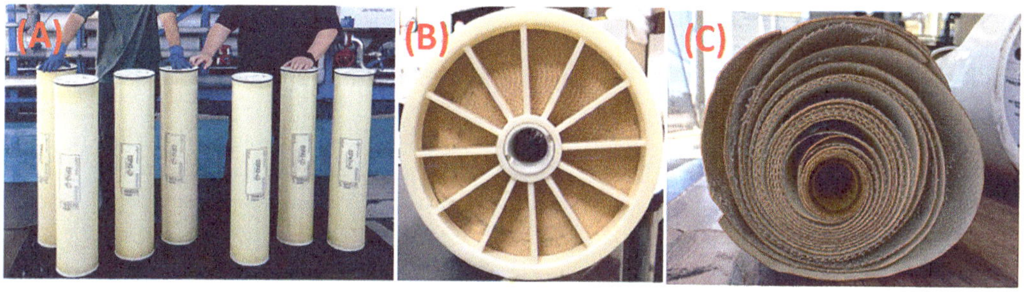

Figure 6.15 **Panel A: Seven 20-cm (8-in.) Reverse Osmosis Elements upon Removal from Service, Panel B: Cross-Section View of a 20-cm (8-in.) Membrane Element, Panel C: Demo Uncased 20-cm (8-in.) Membrane Spiral-Wound Element**

pressure filtration membranes. If flow is applied in the opposite direction, there is a risk of separating the reverse osmosis membranes layers, permanently damaging the membrane.

A shortcoming of polyamide-based membranes is that they are extremely sensitive to strong oxidants like free chlorine and ozone. Even small residuals of free chlorine (≥0.1 mg/L) will permanently damage this layer of the membrane. When damage occurs, the rejection of dissolved solutes (e.g., TDS) will dramatically decrease—permanently. Thus, it is important to have instrumentation to monitor reverse osmosis feedwater for these oxidants and to automate system shutdowns when they are detected. Although reverse osmosis membranes do not tolerate strong oxidants, they are moderately compatible with weaker oxidants such as chloramines. Chloramines do degrade polyamide membranes over time so it is important to minimize exposure as much as possible. Because of this sensitivity to oxidants, chlorinated feedwater is commonly dechlorinated for reverse osmosis applications. Polyamide-based membranes are stable over a pH range of 3 to 11 and have general resistance to bacterial degradation.

Even small residuals of free chlorine will permanently damage reverse osmosis membranes and it is important to shut down the system if any chlorine is detected in the feedwater.

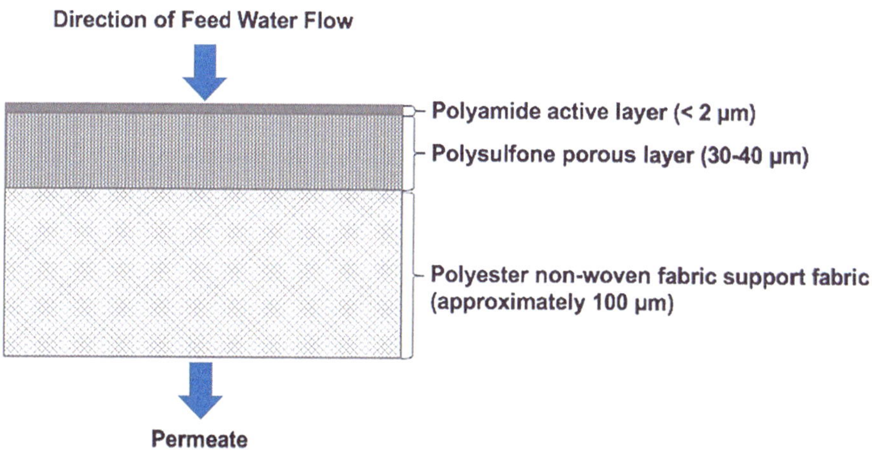

Figure 6.16 **Schematic Showing Layers of a Thin-Film Composite Reverse Osmosis Membrane**

TEST YOUR KNOWLEDGE

1. Similar to low-pressure filtration membranes, reverse osmosis membranes can be backwashed.
   - ☐ True
   - ☐ False

2. How many membrane sheets must feedwater pass through to become permeate in a spiral-wound reverse osmosis element?
   a. Only one
   b. Two to four
   c. At least 10
   d. All of them

3. What are some common properties of polyamide-based spiral-wound reverse osmosis elements?
   a. Sensitive to free chlorine and ozone
   b. Very resistant to bacterial degradation
   c. Operationally stable when pH is above 11
   d. Consists of a single-layer membrane

4. What is the purpose of a feed spacer for a reverse osmosis element?
   a. Helps separate out large debris from entering the membrane
   b. Separates the membrane layers and creates a flow passage for feedwater
   c. Provides structure to the membrane that would otherwise collapse
   d. Spaces to space out feedwater from permeate

PRESSURE VESSELS

Pressure vessels are rigid cylindrical housings that contain multiple spiral-wound elements in series. Figure 6.17 shows a number of these components from a 20-cm (8-in.) membrane element reverse osmosis system. Pressure vessels are pressurized by pumping feedwater to one end of the vessel in combination with a partially closed valve on the other end. This way, the driving pressure forces water through the membrane to produce permeate throughout the membrane surface housed in the pressure vessel. The remaining untreated feedwater exits through the other end of the vessel as concentrate. Pressure vessels are often made of reinforced fiberglass to withstand high operating pressures (e.g., >2100 kPa [>300 psi]). They are generally designed to house between four and eight elements, with six and seven elements being the most common. A number of critical components are needed to neatly and tightly house multiple elements inside pressure vessels to prevent leaks and/or unintentional passage of feedwater into the permeate collection tubes. These include seals (e.g., O-rings), interconnectors, endcaps, thrust cones, shims, and retaining rings. Figure 6.18 provides a schematic of a pressure vessel and identifies the major critical components. Table 6.7 provides a description and function of the pressure vessel components.

TEST YOUR KNOWLEDGE

1. Pressure vessels house multiple spiral-wound elements in series.
   ☐ True
   ☐ False

2. Which seals/O-rings prevent unintentional passage of feedwater into the permeate channel?
   a. Permeate interconnector O-ring
   b. Brine seal
   c. Endcap O-ring
   d. Permeate interconnector and endcap O-rings

3. What is the purpose of retaining rings in the context of reverse osmosis system pressure vessels?
   a. To provide a tight seal between spiral-wound elements
   b. To keep the endcaps from sliding outward
   c. To prevent backflow within the pressure vessel
   d. To connect the permeate tubes of adjacent pressure vessels

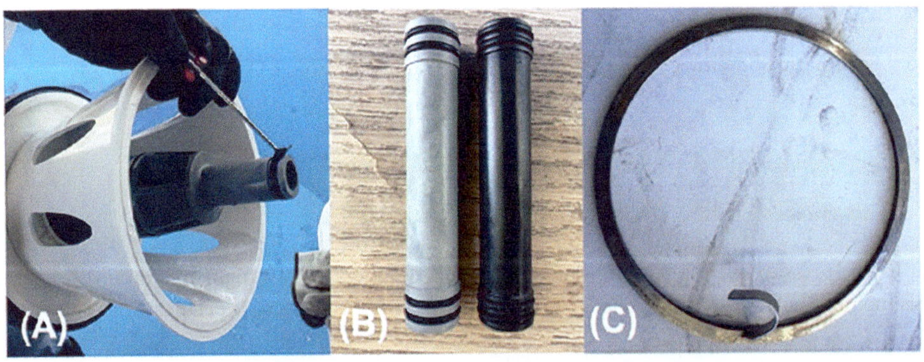

Figure 6.17 Key Components for a 20-cm (8-in.) Pressure Vessel Reverse Osmosis System. Panel (A) Endcap Adapter With Thrust Cone Containing Two Seals, Panel (B) Two Permeate Interconnector With Two seals on Either Side, Panel (C) Retaining Ring

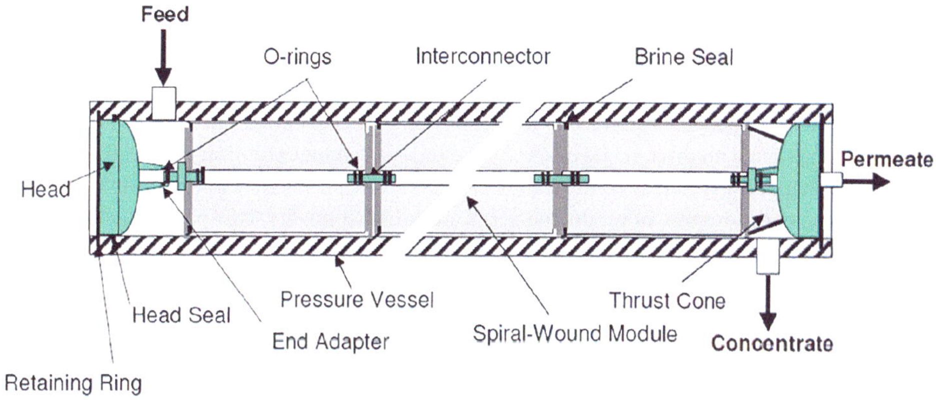

Figure 6.18 Schematic Showing Pressure Vessel and Pressure Vessel Components (U.S. EPA, 2005)

STAGES AND TRAINS

Because each pressure vessel contains only six or seven spiral-wound elements, it is common for reverse osmosis systems to have multiple pressure vessels in parallel. A group of pressure vessels in parallel is known as a *stage*. In a stage, the feedwater is distributed equally so that each pressure vessel produces virtually the same amount of product water (i.e., permeate) and nontreated water (i.e., concentrate) in terms of both flow and water quality—assuming all spiral-wound elements are equal. The permeate and concentrate from all vessels within a stage are each collected into a common manifold for distribution/disposal (or further treatment) of the end products.

Table 6.7 Major Components for Reverse Osmosis Pressure Vessels

Component	Function
Pressure vessel	Rigid cylindrical housings served to pressurize and tightly encase multiple spiral-wound elements in series, typically 6 to 7.
Spiral-wound element	Contains multiple membrane sheets that are spirally wound into a cylindrical rigid casing. Feedwater flows tangential to the membrane sheets (i.e., cross flow). These elements fit snugly into pressure vessels of the same diameter.
Permeate interconnector	A rigid tube that connects the permeate tube of two spiral-wound membranes in series. Each permeate interconnector has either one or two O-rings on either end that provide a seal to prevent passage of feedwater into the permeate tube.
Endcap	A rigid part provided by the pressure vessel manufacturer that is placed on either end of each pressure vessel to seal off the pressure vessel from the exterior. The diameter of the endcap must match the diameter of the pressure vessel. Endcaps have O-rings to prevent passage of feedwater into the permeate tube.
Retaining ring	A metal, ring-shaped part that keeps the endcaps from sliding outward from the pressure vessel. Design varies by pressure vessel manufacturer. The correct installation of the part is extremely important to prevent safety hazards.
Brine seal	Rigid or rubber-like seal that is pre-installed on spiral-wound elements. Located on the feed side of spiral-wound elements, they prevent or minimize flow of feedwater along the outer shell of the elements. This part does not provide a critical seal between the permeate channel and the feed/brine, but is important to maintain proper hydraulics throughout the pressure vessel.
Thrust cone	A cone-shaped rigid spacer placed between the last element and endcap located on the concentrate side of the pressure vessel. The thrust cone helps prevent sliding of the elements and avoid blocking of the concentrate exit port.
Shims	Rigid plastic spacer-like parts, available in a variety of lengths. *Shimming* adds a physical spacer between the feed side and first element to more tightly pack all elements together to minimize shifting of elements and connections during startup and shutdown sequences.

It is also common for reverse osmosis systems to contain multiple stages in series. Subsequent stages are designed to either produce additional permeate by treating the concentrate from the previous stage (thus, increasing recovery) or to further treat the permeate from the previous stage (known as two-pass or multipass systems). Figure 6.19 shows two reverse osmosis system configurations, with Panel A showing a multistage system with three stages in a 4 : 2 : 1 vessel array configuration, and Panel B showing a two-pass system with a 2 : 2 vessel array configuration. The permeate production will be lower for a two-pass configuration at the expense of producing permeate of higher quality. Two-pass configurations are sometimes used for seawater desalination applications because the feedwater is at a sufficiently high concentration such that two passes through membranes in series is necessary to reach the target permeate concentration. In contrast, multistage configurations are able to increase water recovery by treating the concentrate from the first stage in a subsequent stage. Multistage systems containing two to three stages are most common for recycled water and brackish groundwater applications. A common characteristic of multistage configurations is the decreasing number of pressure vessels in each subsequent stage. The reduction in pressure vessels is necessary because the concentrate flowrate declines as permeate is generated and the velocity in the feed channels declines as the flowrate declines, which causes solutes to build up near the membrane surface. If a single stage has approximately 50% water recovery, the concentration flowrate (which is the feed to the next stage) is 50% of the feed flowrate. Thus, it would be necessary to reduce the number of pressure vessels by half to account for the fact that the feed flowrate

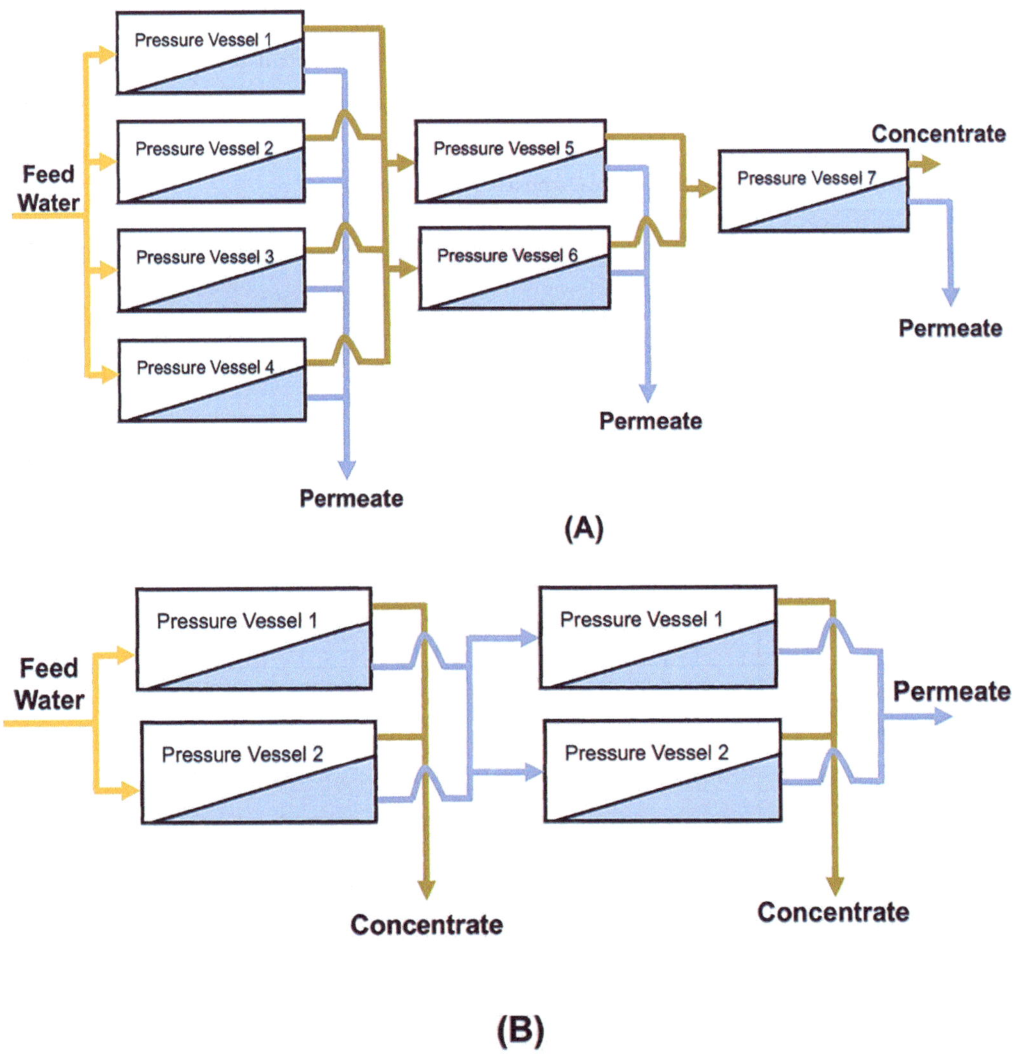

Figure 6.19 Schematic Showing Common Reverse Osmosis System Configurations. Panel (A) Multistage System With Three Stages in 4 : 2 : 1 Vessel Array Configuration, Panel (B) Two-Pass System With a 2 : 2 Vessel Array Configuration

to the second stage is half of the feed flowrate to the first stage. As a result, a general rule for multistage configurations is to have half as many pressure vessels in each subsequent stage (e.g., Stage 2) compared to the stage that precedes it (e.g., Stage 1).

A *reverse osmosis train* is defined as a group of pressure vessels that share common valving that can be isolated for inspection, repair, and cleaning. Each train is typically designed to have a dedicated feed pump (U.S. EPA, 2005). A *skid* is another term used to describe a reverse osmosis train. Figure 6.20 provides examples of two 1890-m^3/d (0.5-mgd) reverse osmosis trains manufactured on a common skid to operate in parallel, one being a three-stage reverse osmosis train and the other being a two-pass reverse osmosis train.

TEST YOUR KNOWLEDGE

1. Which is the best array configuration to reach highest recovery?
   a. Two-pass system
   b. Single-stage system
   c. Multistage system
   d. Single pressure-vessel system
2. What is the vessel array configuration of a reverse osmosis train consisting of 100 vessels in Stage 1, 50 vessels in Stage 2, and 25 vessels in Stage 3?
   a. 4:2:1
   b. 1:2:4
   c. 100:50:25
   d. 25:50:100
3. What is the best configuration of pressure vessels for a three-stage system containing a total of 140 pressure vessels?
   a. 20:40:80
   b. 70:70
   c. 80:40:20
   d. 140:70:35
4. How is a reverse osmosis train defined?
   a. Having at least two or more vessels per stage
   b. A bank of pressure vessels that share common valving
   c. A railway wagon filled with reverse osmosis membrane elements
   d. A reverse osmosis system consisting of multiple stages

Figure 6.20 Two 1890-m^3/d (0.5-mgd) Reverse Osmosis Trains in Multistage Configuration Mounted on a Common Skid

REVERSE OSMOSIS PUMPS

Reverse osmosis feed pumps provide the driving pressure to produce permeate. Reverse osmosis feed pumps are sometimes configured as two pumps in series: a transfer (booster) pump followed by a high-pressure pump. The transfer pump provides sufficient head to the high-pressure pump for it to operate efficiently. Booster pumps are typically horizontal centrifugal pumps, whereas high-pressure pumps are commonly vertical multistage pumps. More recently, horizontal centrifugal pumps have been used as high-pressure pumps. The high-pressure pump is often controlled by a VFD that is used to vary pump speed and flow to target a permeate flow setpoint. Figure 6.21 shows a 56-kW (75-hp) vertical, multistage, high-pressure pump used for a 1890-m^3/d (0.5-mgd) reverse osmosis system used for recycled water production. For such an application, pressures greater than 690 kPa (100 psi) are common. Seawater desalination applications operate at much higher pressures (e.g., up to 8500 kPa [1230 psi]).

A VFD allows a pump to be operated at a set speed. A VFD is often used for pumps that operate at a flow setpoint.

Multistage reverse osmosis systems sometimes have an additional pump located between stages, known as an *interstage booster pump*. These pumps are located on the feed to the second or third stage to increase the pressure above the concentrate pressure from the previous stage. This increases the permeate flow in the latter stage. The higher pressure is necessary to increase flux and permeate flow. The higher osmotic pressure that occurs in latter stages results from the increased concentration of solutes in the concentrate stream. Higher feed pressure is needed to overcome the higher osmotic pressure. Multistage pumps are smaller than the feed pumps because there is less flow.

CARTRIDGE FILTERS

Cartridge filters are used in reverse osmosis systems to prevent suspended solids in the feedwater from reaching the membrane elements. When the reverse osmosis system is preceded by a dedicated pretreatment process such as membrane filtration, cartridge filters would seem redundant. However, they are still important. Without them, debris in the reverse osmosis feed tank could get through and damage the membranes. Cartridge filters are often placed between the booster pump and high-pressure pump. The filter housing (shown in Figure 6.22[a]) is typically sized to fit multiple cartridge filters (shown in Figure 6.22[b]). Cartridge filters often have a nominal pore between 1 µm and 5 µm. The cartridge filters must be replaced periodically, typically because of an increase in differential pressure across the filter housing and/or time interval.

Cartridge filters get their name from the removable filters (cartridges) they contain. Individual cartridges are cumulable items that require regular replacement.

Figure 6.21 A 56-kW (75-horsepower) Multistage Vertical High-Pressure Pump for Two 1890-m^3/d (0.5-mgd) Reverse Osmosis Trains Used for Recycled Water Production

Figure 6.22 Panel (a) Cartridge Filter Housing. Panel (b) Cartridge Filters

ENERGY RECOVERY DEVICES

Energy recovery devices (ERDs) are used to recover a portion of the residual concentrate pressure to increase the feed pressure to another stage. The most basic type of ERD is a mechanical turbocharger that resembles those used in modern cars. An ERD makes it possible to recover residual pressure by using leftover concentrate pressure through the use of a turbine. It is important to note that the concentrate and feed water do not make contact with each other. Rather, it is the turbine that transfers available energy from one to the other. Figure 6.23 provides a turbocharger-like ERD installed on a three-stage reverse osmosis system.

A more sophisticated ERD couples a pump with a mechanical turbine. These devices combine the benefits of a VFD-driven pump and energy recovery through a mechanical turbine. These devices are more expensive and are typically only used when the pressure increase from the turbocharger alone is not sufficient to achieve the feed pressure needed for the stage being considered. Energy recovery devices are particularly relevant for high-pressure reverse osmosis applications (like seawater desalination) because the available pressure in concentrate streams is very high. These devices can help offset a portion of the power demand. Figure 6.24 provides a schematic of a two-stage reverse osmosis system and the location of previously discussed components, including the ERD.

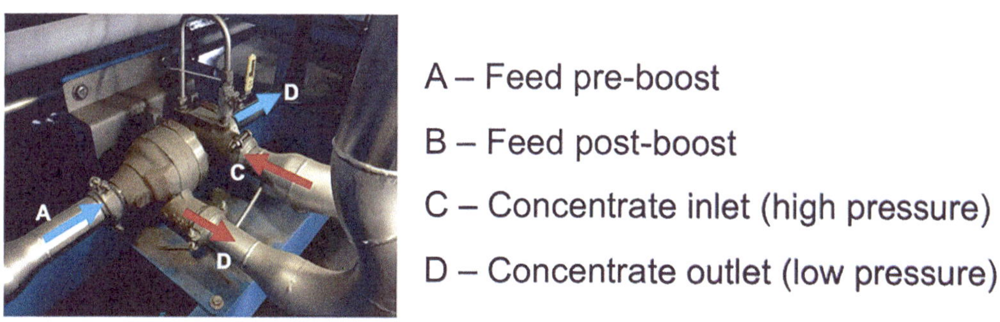

Figure 6.23 Example of an Installed Turbocharger-Type Energy Recovery Device to Recoup a Portion of the Residual Concentrate Pressure

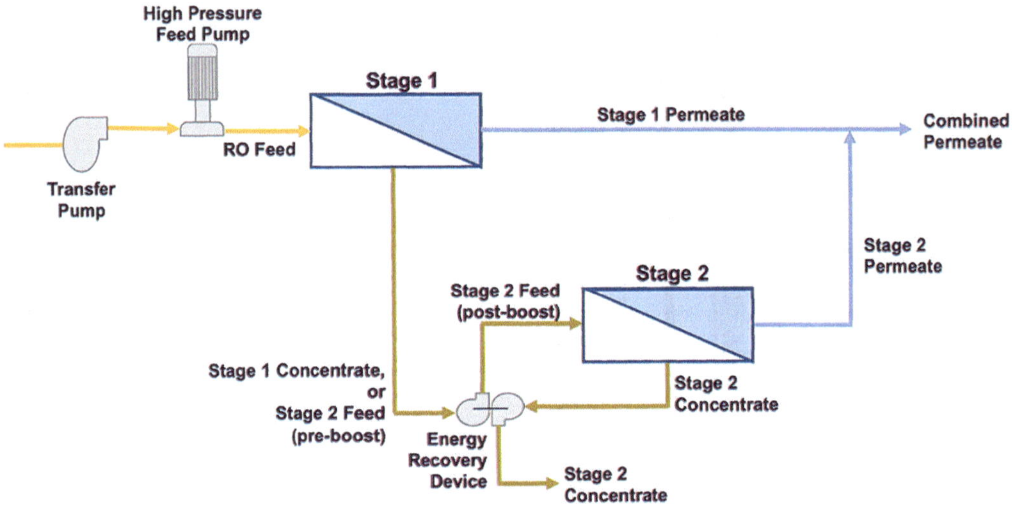

Figure 6.24 Schematic of a Two-Stage Reverse Osmosis System With Feed Pumps and Energy Recovery Device

Scaling is caused by the deposition and/or precipitation of inorganic compounds (phosphate, calcium, silica, iron, aluminum). Scaling will plug up membrane surface, ultimately causing a decrease in production. Scaling can be irreversible if not addressed appropriately. Scaling can be controlled by a combination of acid and anti-scalant in the correct doses.

CHEMICAL SKIDS FOR PRETREATMENT CHEMICALS

Reverse osmosis membranes are prone to fouling and scaling. Figure 6.25 shows a scanning electron microscope image of phosphate-based scaling. Such scaling can decrease membrane performance and make it more difficult (i.e., higher pressure) to maintain production. It is common for reverse osmosis systems to have dedicated chemical skids to dose pretreatment chemicals to the reverse osmosis feed. The main pretreatment chemicals are acid (e.g., sulfuric acid) and anti-scalant. Acid increases the solubility of certain inorganic compounds that can otherwise precipitate onto the membrane surface, eventually causing a decrease in water. Acid addition systems typically operate using a feed pH setpoint so the acid is dosed to reach its setpoint. Reverse osmosis feed pH setpoints are site-specific and depend on water quality. Membrane manufacturers, anti-scalant vendors, and/or operators/engineers with experience in reverse osmosis operation can determine the appropriate pH setpoint based on the water quality. Anti-scalants are proprietary blends of chemicals that inhibit crystallization and reduce the rate of precipitate formation of sparingly soluble salts. Anti-scalant vendors typically specify the dose required to limit scale formation. Some vendors offer water quality-based modeling to determine the right product and dose. The anti-scalant dose can be determined by testing, often together with the acid. The correct balance of acid and anti-scalant becomes especially important when operating at higher feed water recovery (e.g., ≥85%).

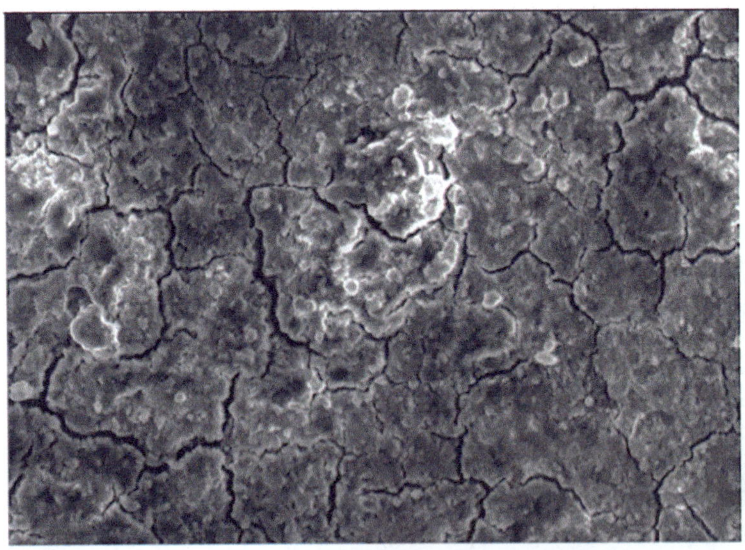

Figure 6.25 Microscope Imagery of Phosphate-Based Scaling

Residual chloramines can prevent or diminish biological fouling but, as discussed earlier, can degrade the polyamide layer of the reverse osmosis membrane. For recycled water/reuse applications, a reverse osmosis feed chloramine dose of 2 to 3 mg/L is common, although less is generally preferred. Chloramines are often added upstream of the reverse osmosis system to also prevent biological fouling of upstream membrane processes (e.g., membrane filtration). Another chemical used during reverse osmosis operation is sodium hydroxide, which is typical for desalination applications to enhance boron removal for multi-pass configurations.

(Biological) Fouling is a general term to denote a decrease in membrane performance caused by the deposition of suspended or dissolved solids on the membrane surface. A type of fouling is biological fouling, which is caused by the accumulation of biological and/or organic material on the membrane surface. Biological fouling also causes the membrane surface to be plugged, eventually decreasing production. Biological fouling can be controlled by proper pretreatment (e.g., membrane filtration) and/or the use of a mild oxidant/disinfectant compatible with reverse osmosis membranes, such as chloramines.

CLEAN-IN-PLACE SKID

Even though pretreatment chemicals maintain steady water flow and prevent scaling/fouling, reverse osmosis membranes still slowly accumulate a small degree of fouling/scaling that needs to be removed with periodic cleans. These periodic cleans are performed when a single reverse osmosis train is offline and help restore performance. Clean-in-place skids consist of pumps, tanks, valving, piping, and a heater, as shown in Figure 6.26. The pump and tank must be sized in accordance with membrane manufacturer cleaning recommendations. Clean-in-place cleans are performed by recirculating and soaking a chemical solution targeted to remove foulants and/or scalant of concern. It is important to note that the membranes are not removed from the pressure vessels to perform a CIP. Hence, the name "clean-in-place" procedure. Typically, a "low pH" clean uses citric acid and a "high pH" clean uses sodium hydroxide (i.e., caustic soda). Cleaning regimens and cleaning tips are further detailed in membrane manufacturer bulletins as well as in the "Maintenance" section.

PERMEATE FLUSH SKID

To protect the reverse osmosis membrane during system shutdowns, it is common to flush the system with reverse osmosis permeate before shutdown. Allowing membranes to sit in feedwater can cause the system to become septic and/or cause passive membrane fouling. Permeate flushes are performed using a dedicated reverse osmosis flush skid, which consists of a flush pump, permeate tank, and necessary valving and piping. The permeate flush duration must be long enough to displace feed/concentrate throughout the system. The flush skid tank is often sized to contain enough permeate to flush all reverse osmosis trains in the event of a systemwide shutdown. In such an event, the trains are commonly flushed one by one instead of simultaneously, so only one flush pump is used at a time.

TEST YOUR KNOWLEDGE

1. **Energy recovery devices allow a reverse osmosis system to recoup part of the residual concentrate pressure.**
   - ☐ True
   - ☐ False
2. **What is the purpose of a CIP skid?**
   a. To periodically clean the membranes by recirculating heated chemicals
   b. To further treat permeate that is not up to spec
   c. To clean the reverse osmosis skid after the membranes are removed
   d. To restore rejection of target compounds
3. **Why is flushing with permeate important upon reverse osmosis system shutdowns?**
   a. To increase recovery
   b. To avoid passive fouling and septicity
   c. To increase performance upon restart
   d. To reduce recovery

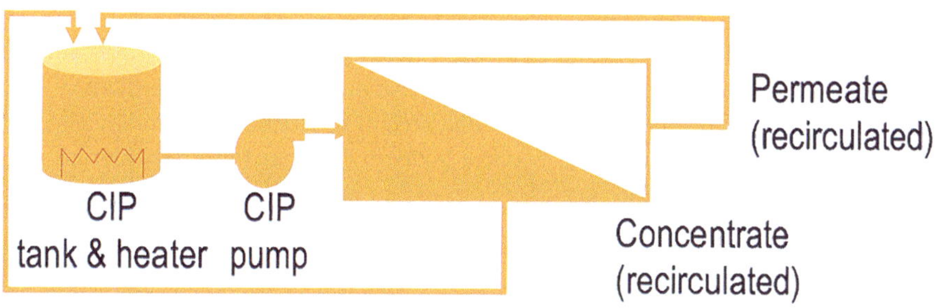

Figure 6.26 Main Components of a Reverse Osmosis Clean-in-Place Skid

Process Control of Reverse Osmosis

Process control of reverse osmosis systems relies primarily on installed online instrumentation. This section describes the instrumentation used during the operation of reverse osmosis systems.

INSTRUMENTATION

Monitoring instrumentation, such as pressure transmitters, temperature sensors, and flow meters, is part of reverse osmosis systems. In addition, certain water quality monitors, such as TOC, electrical conductivity, oxidation–reduction potential (ORP), and pH sensors and analyzers are commonly used for reverse osmosis systems. The following sections describe the role of instruments in process monitoring for various reverse osmosis processes.

PRODUCTION

During production, feedwater flows through the vessels under pressure and produces permeate. Typically, a permeate flow setpoint is used to control valves and/or feed pumps to meet this setpoint. The permeate flow meter provides feedback to the system if modulation of valves or pumps is needed to meet the setpoint. Flow is an important parameter to monitor because it provides information on the flux of the system. Flow meters are also located on the concentrate stream of reverse osmosis systems, which can be used to calculate recovery of the system. Recovery can be adjusted by modulating the concentrate valve so the concentrate flow meter is used as a reference to reach the target recovery.

Pressure transmitters are typically located in the feed, permeate, and concentrate lines. Multistage system will have pressure transmitters on each stage. When monitored routinely, feed pressure provides an indication of the extent of fouling in the system. It is important to note that temperature has a direct effect on feed pressure, which must be considered. Other pressure-related parameters include specific flux and net driving pressure, although these require additional parameters (such as flow and conductivity) and the use of spreadsheet for calculating purposes. These parameters are used as indicators for the fouling status of the membranes.

Conductivity is commonly monitored throughout a reverse osmosis system. It is used to calculate conductivity rejection as well as fouling status parameters, as discussed. Conductivity rejection trending is used to monitor the status of membranes as they can eventually lose rejection over years of operation, which can influence when membranes need to be replaced. Conductivity rejection is also used to monitor membrane integrity because a sudden increase in permeate conductivity would be indicative of compromised integrity (failed O-ring, damaged membrane). Conductivity is used for specific flux calculation as a normalization parameter, where both feed and permeate conductivity account for a degree of osmotic pressure that must be accounted for.

Total organic carbon analyzers are common for reverse osmosis systems that need to monitor product water quality carefully (e.g., potable reuse). Total organic carbon analyzers are typically located on the reverse osmosis feed and combined reverse osmosis permeate, which can be used to calculate TOC rejection. Similar to conductivity, a sudden increase in permeate TOC concentration can indicate compromised integrity.

Because reverse osmosis systems can operate based on feed pH setpoint, reverse osmosis feed pH monitoring is common for reverse osmosis systems. During production for a system that operates on a feed pH setpoint, pH monitoring ensures that sufficient acid is dosed to the reverse osmosis feed. Alarms would be triggered if pH were to drift above/below the desired setpoint.

Oxidation–reduction potential monitoring is used to ensure the oxidation potential of the feed water is not beyond a set value. Oxidation–reduction potential is influenced by the presence of chloramines in the feedwater as well as other stronger oxidants, such as free chlorine. The presence of free chlorine would cause an increase in the ORP in the feedwater, which can cause a system shutdown to protect the membranes from oxidation.

Operation of Reverse Osmosis

The main reverse osmosis system operational activities are geared toward maintaining and confirming stability of the system. *Stability* refers to the system's ability to maintain consistent permeate production while meeting all treatment objectives. The following sections describe common routine and preventive practices when operating reverse osmosis systems.

NORMAL/DAILY OPERATION

The daily tasks for an operator focus on monitoring process performance and maintaining stable operations. Confirming permeate targets, system integrity, and the fouling/scaling state of system are all aspects of maintaining stable reverse osmosis operations. Daily rounds are one of the most important routine activities for monitoring the reverse osmosis process. Through daily rounds, operators capture important process parameters while the system is stable to evaluate if the production target is being met, the status of fouling degree, and system integrity. While doing rounds, operators also physically inspect the system for signs of abnormalities (e.g., leaks, suspect instrument values, etc.) that are not observable on a monitoring screen. Any abnormalities are noted and escalated if further action is needed. Through daily rounds, operators also confirm that scheduled activities have been performed. For example, are both acid and anti-scalant being dosed appropriately? Table 6.8 summarizes a recommendation of critical reverse osmosis monitoring parameters that should be tracked during daily rounds.

Reverse osmosis systems are typically operated at constant flux (i.e., constant permeate production). If changes to key parameters (such as recovery) are made, it is important to monitor the system to ensure operational stability. For multistage systems, it is important to monitor performance on a per-stage basis in addition to overall performance. A drop in overall performance may be affected by a single stage and can be more easily identified by single-stage tracking. In this case, the low-performing stage can be appropriately addressed (e.g., by a CIP procedure) rather than changing overall system parameters. As an example, the overall flux for a multistage system will remain constant if the system is operated on an overall permeate setpoint, although, over time, one (or more) stage will slowly foul due to slight scaling and/or biofouling. Because of fouling, one stage will have a harder time sustaining production at system pressure and part of the permeate flow will be made up by the other stages that have not fouled. This will

Table 6.8 Summary of Key Parameters for Data Collection for the Reverse Osmosis Process

Process	■ Feed pressure ■ Concentrate pressure ■ Permeate pressure ■ Feed conductivity ■ Concentrate conductivity (for each stage) ■ Permeate conductivity (for each stage)
Production	■ Feed flow ■ Concentrate flow ■ Permeate flow ■ Recovery
System integrity	■ Conductivity rejection ■ Permeate TOC (if available)
Fouling status	■ Net driving pressure ■ Specific flux
Pretreatment chemical	■ Anti-scalant dose ■ Chloramines dose ■ Acid dose

increase the flux of that stage. To this extent, it is important to track fluxes on a routine basis through daily rounds to verify that fluxes of individual stages are not exceeding membrane manufacturer limits. Figure 6.27 shows the effect of a how a lead stage makes up for the lack of production from a trail stage to maintain overall flux constant.

Another aspect of long-term reverse osmosis membrane operation is conductivity rejection. Although reverse osmosis membrane performance is relatively consistent over a range of feed water quality, over time, reverse osmosis membranes slowly lose their ability to reject solutes compared to when they are first installed. The decrease in rejection occurs due to the slow degradation of the polyamide layer of reverse osmosis membranes from general operational wear and tear. This may be exacerbated by the presence of oxidants, as discussed earlier. When membrane oxidation occurs, a general reduction in conductivity rejection and increased permeability (plotted as specific flux) is observed. A noticeable reduction in conductivity rejection can be observed as early as a few years since membrane install and is often noticeable after 5 years of membrane operation. A reduction in conductivity can be the primary cause for replacing membranes.

STARTUP

System startup protocols are typically provided by a reverse osmosis system supplier or membrane manufacturer and should be followed to ensure that none of the components are misused or damaged as a result of improper startup. Documents provided by these entities should be reviewed upon system startup.

MEMBRANE CLEANINGS

Periodically, reverse osmosis membranes must be cleaned with a chemical solution to restore performance. These are typically performed on the same skid where membranes are used for normal operation, and thus are known are CIP procedures. The frequency of CIP procedures is site-specific and may occur monthly to yearly. It is common for multistage reverse osmosis systems to undergo a stage-specific cleaning schedule, with later stages often requiring more frequent cleans than leading stages. Per membrane manufacturer guidelines, CIP procedures should be commenced when

- Normalized differential pressure increases more than 20%, or
- Normalized permeate flowrate decreases by more than 10%, or
- Normalized salt passage increases by more than 20%.

> Clean-in-place procedures are performed to restore membrane performance; the system is not in production during such procedures. A chemical solution is recirculated through the membranes using a dedicated CIP skid. These procedures help dissolve and dislodge scalants and foulants that accumulate on the surface of reverse osmosis membranes over time.

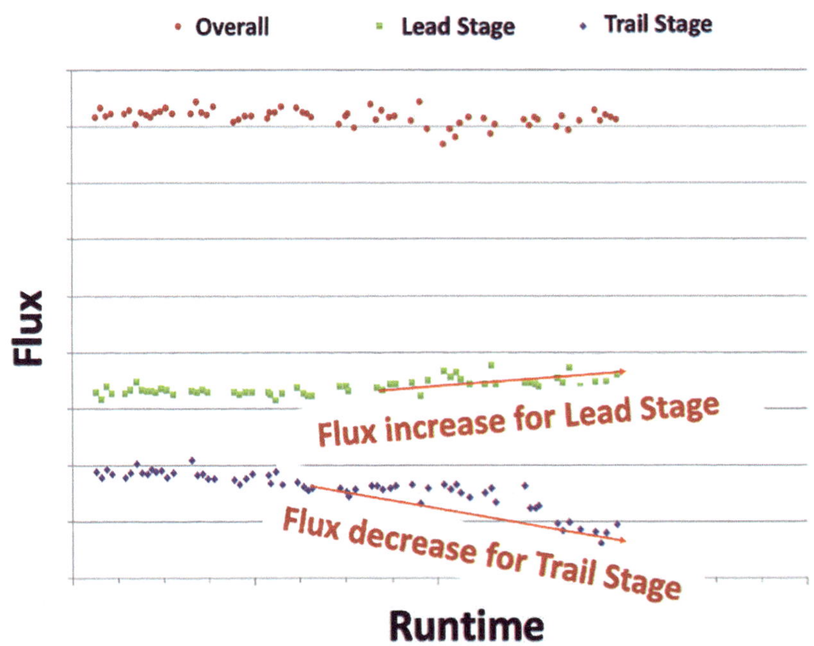

Figure 6.27 Effect of Multistage on Overall Flux (Typical Unit: gpd/sq ft) as a Result of Runtime (Typical Unit: Hours) Resulting in Long-Term Fouling

Another general rule to consider is cleaning when specific flux decreases by more than 20% since the last cleaning cycle. The "Maintenance" section of this chapter contains additional details for reverse osmosis membrane CIP procedures (such as chemicals used and cleaning strategies).

SHUTDOWN AND MEMBRANE STORAGE

System shutdown protocols are typically provided by the membrane manufacturer and should be followed to ensure that none of the components are misused or damaged as a result of improper shutdown. As earlier discussed, the use of permeate flushes helps prevent passive fouling. The procedures for storage will depend on whether short- or long-term storage is desired. Short-term storage can range from 1 day to a week. The duration range for what is considered short-term storage may differ between manufacturers and the manufacturer's definition should be used for each specific application. If the desired storage period is longer than the specified short-term duration range, long-term storage protocols should be used. The "Maintenance" section provides some common practices for membrane storage. A sodium bisulfite (SBS) solution is often used to store membranes for longer periods time. Membrane storage can occur within the skid or by bagging individual elements.

Bagging of reverse osmosis membranes is needed to properly store them for future use. A preservative solution is added during the bagging process. The preservative is generally SBS. The bag is then vacuum sealed.

TEST YOUR KNOWLEDGE

1. Reverse osmosis is typically operated at constant feed pressure and variable flux.
   - ☐ True
   - ☐ False
2. What type of activity is done during daily rounds for reverse osmosis systems?
   a. Initiate permeate flushes to restore performance
   b. Perform maintenance of online instrumentation
   c. Collect samples for compliance purposes
   d. Verify drawdowns of related chemicals, such as anti-scalant
3. What is a proper way to store reverse osmosis membranes?
   a. Follow storage protocols from membrane manufacturer
   b. Store in a chlorine solution of at least 1000 mg/L
   c. Store in an acid solution of pH below 2
   d. Store in a warehouse after membranes have fully dried in the field
4. At what frequency should reverse osmosis membranes be cleaned using CIP procedures?
   a. Frequency is site-specific, and may occur monthly to yearly
   b. Every time the reverse osmosis system is shut down for maintenance
   c. Every 24 hours to meet compliance requirements
   d. Reverse osmosis membranes should not be cleaned as cleaning may damage them

Data Collection, Sampling, and Analysis for Reverse Osmosis

An organized data log is key to interpreting historical reverse osmosis performance. Consistently logging data allows operators to identify important reverse osmosis trends. For example, trends help operators predict when chemical cleans (e.g., CIPs) should be triggered. In general, data collection should happen at least daily for reverse osmosis systems. Apart from keeping track of the performance, the log sheets are also valuable tools for troubleshooting and may be needed in the cases of warranty claims. Table 6.8 identifies parameters to be collected and entered into a logbook. Figure 6.28 demonstrates key trends related to reverse osmosis performance. Panel A shows an increase in feed pressure, which increases permeate flux as well as salt rejection, to an extent. Panel B shows how higher temperature lowers the viscosity of water, which, in turn, increases permeate flux. Conversely, salt rejection will decrease because membrane structures are less "tight" at higher temperatures. Panel C shows that increasing recovery will eventually lower permeate flux and salt rejection. Panel D shows that an increase in feed concentration (e.g., higher TDS) will lower permeate flux due to increased osmotic pressure and will decrease salt rejection due

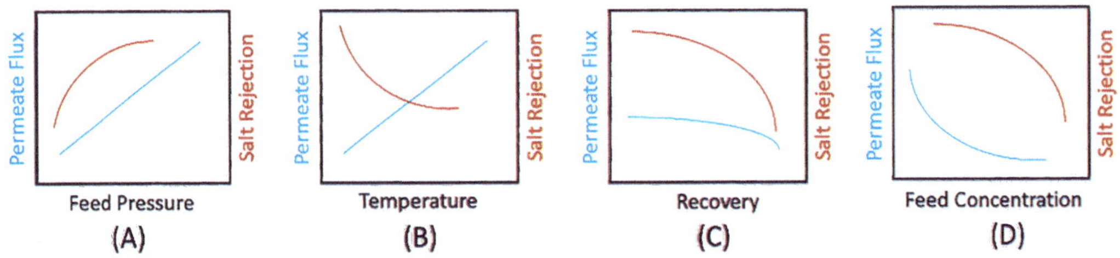

Figure 6.28 Key Trends Related to Reverse Osmosis Performance

to increased salt diffusion across the membrane. Other key trends to track and plot for reverse osmosis performance include

- Feed pressure
- Temperature-corrected specific flux
- Flux
- Recovery
- Feed and permeate conductivity
- Salt passage

Reverse osmosis trends are often better visualized on a longer-term scale of at least a few months (i.e., >3 months), as shown in Figure 6.29. This is because day-to-day changes tend to be very small. Longer term trends make it possible to interpret big picture trends, such as fouling, and visualize frequency of cleans for a given system. There can be precipitous drops in performance (e.g., specific flux, salt rejection) that occur over the course of few days and weeks. Such precipitous changes are often indicative of problematic operations. For example, a membrane integrity failure may show up as a sudden decrease in salt rejection. Rapid scaling (or fouling) can be indicated by a dramatic decrease in specific flux. In addition to visualizing performance on a long-term basis, it is also valuable to plot variables based on the time membranes have been in operation (runtime). Runtime may be in hours, days, or months. Runtime counters are typically available from feed pumps where runtime is only accumulated when the pump is online. Plotting in runtime is beneficial in that it only displays data when the system is operational and avoids gaps in charts due to downtime.

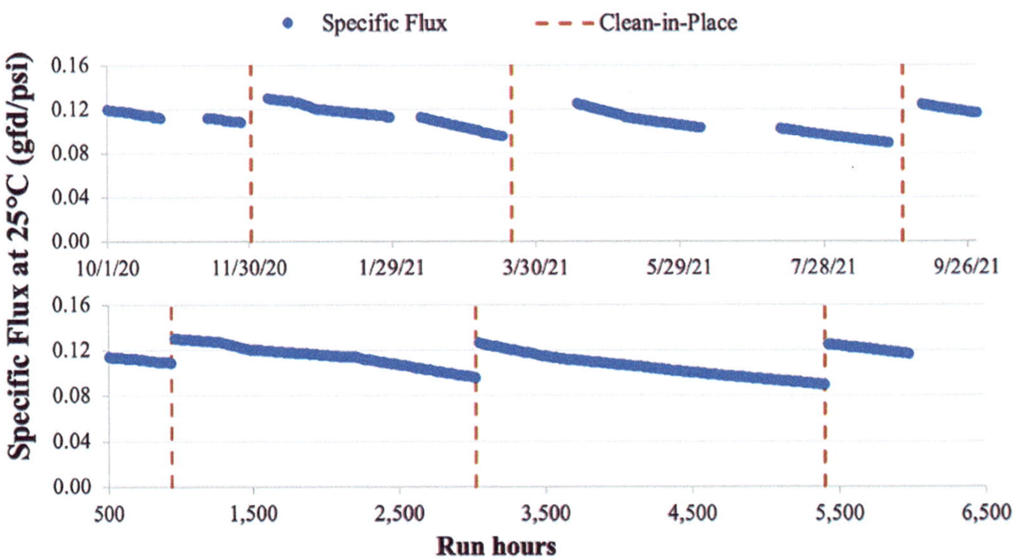

Figure 6.29 Identical Data Plotted as Run Hours (Bottom) and by Date (Top). Plotting in Run Hours Helps to "Smooth Out" Data Visualization due to Periodic System Shutdowns.

Maintenance for Reverse Osmosis

MEMBRANE FOULING

Fouling of reverse osmosis membranes will gradually occur and require chemical cleaning. This is the only membrane maintenance necessary for reverse osmosis (and nanofiltration) membranes. Similar to membrane filtration membranes, organic and biological fouling may occur, especially in applications with relatively high TOC concentrations and inadequate levels of chloramines to suppress formation of biological films on membrane surfaces. Unlike membrane filtration, a sodium hypochlorite solution cannot be used for reverse osmosis and nanofiltration membranes because it will degrade the separation layer material. Reverse osmosis membrane manufacturers specify maximum exposure tolerance to free available chlorine (e.g., 1000 ppm*h). Additionally, feed water chlorine levels must be below detection limits (<0.02 mg/L). Instead of hypochlorite, sodium hydroxide solutions are used for removing organic, biological, and colloidal silica from membrane surfaces. Both reverse osmosis and nanofiltration membranes consist of thin-film composite membranes using polyamide polymer layered on top of supporting polymer film (see Figure 6.16). These membranes are susceptible to damage from high and low pH exposures. As a general rule, a mild CIP solution is typically attempted first to try to restore membrane performance. An example of a mild CIP solution is using sodium hydroxide to target a pH of 10. A high pH cleaning is typically followed by low pH cleaning to remove scalants like calcium phosphate, calcium carbonate, and other sparingly soluble minerals. In some applications, cleaning order may be reversed using low pH cleaning first, followed by high pH cleaning. Proprietary cleaning agents may be used to target specific scalants and foulants. There are numerous commercial products that may provide better recovery of membrane performance than using manufacturer or industry recommended generic chemicals.

Membrane chemical exposure tolerances are commonly expressed in units of concentration multiplied by time (e.g., unlike reverse osmosis membranes, microfiltration/ultrafiltration membranes typically have exposure tolerance to free chlorine on the order of 1 000 000 ppm*hr). Refer to the reverse osmosis membrane manufacturer's technical service bulletins and O&M manuals for recommended cleaning protocols, including choice of cleaning chemical and concentration, membrane pH, and temperature limits to minimize membrane damage, and specifications for minimum CIP solution volumes; recirculation flowrates per pressure vessel; and recirculation and soak durations.

STORAGE

Short-term storage during normal shutdown is not necessary if the membranes are properly flushed with reverse osmosis permeate. However, when a system must be idle due to changes in seasonal water demand, manufacturer recommendations for long-term storage should be followed. Common practices for reverse osmosis membrane storage include the following:

- Membranes must remain moist.
- Membranes should be stored away from direct sunlight to prevent potential drying, which can lead to cracking and delamination.
- Membranes may have a certain temperature range requirement for storage.
- Pickling membranes with chemical solution (such as SBS) and periodically checking the strength of the chemical solution is the most effective method to preserve membranes for weeks or months at a time.

Preservation of reverse osmosis membranes without their removal from the pressure vessels can be done by preparing a pickling solution. Pickling solutions typically consist of 1% wt sodium bisulfite. Pickling procedures can use the CIP tank and CIP pump to perform a recirculation through the entire system for a short duration (e.g., 15 to 30 minutes). After injecting the pickling solution, the system should be isolated by closing inlet and outlet valving to allow membranes to stay soaked in the pickling solution. Manufacturers of reverse osmosis membranes recommend checking the stability of SBS monthly by measuring pH. Sodium bisulfite effectively acts as a biocide agent, controlling formation of biofilms, and as a reducing agent, preventing membranes from being oxidized by oxygen.

There are many chemicals besides SBS that may be used to inactivate biological film growth. However, it is important to consult with membrane manufacturers for compatibility of the biocide agent to prevent any potential membrane damage.

SEAL REPLACEMENTS

Pressure vessels contain several O-ring seals, as mentioned in previous sections. As shown in Figure 6.17 and Figure 6.18, these include endcap seals, endcap permeate adaptor seals, permeate interconnector O-rings, and pressure vessel seals. Commonly, reverse osmosis membranes are supplied with spare plastic interconnectors and O-rings; however, these O-rings can be made of cheaper materials such as polyethylene and can degrade over time. Ethylene propylene diene monomer is a more resistant and relatively inexpensive rubber material that may provide longer service life. Viton, which is a fluoropolymer, provides the best resistance to long-term exposure to chloramines and cleaning chemicals. Viton parts can be supplied for the majority of the type of seals used in commercial pressure vessels.

OTHER MAINTENANCE

Reverse osmosis (and nanofiltration) membrane systems have other components besides pressure vessels and membrane elements, including, but not limited to the following:

- Transfer pump
- High-pressure pump
- Permeate flush pump
- Clean-in-place pump
- Variable-frequency drives
- Motorized (e.g., typically electric) open/close and flow control valves
- Pressure gauges/indicators and transmitters
- Temperature meters
- Conductivity meters
- Flow meters

Each component requires its own separate maintenance program and interval that can be obtained from the manufacturer's O&M manual.

TEST YOUR KNOWLEDGE

1. **Reverse osmosis membranes can be cleaned with citric acid and sodium hypochlorite using similar protocols for microfiltration or ultrafiltration membranes.**
   - ☐ True
   - ☐ False

2. **What is the typical frequency for chemical cleaning of reverse osmosis membrane systems?**
   a. Daily to weekly
   b. Weekly to monthly
   c. Every backwash cycle
   d. Monthly to annually

3. **What method should be performed during prolonged downtime of the reverse osmosis membrane systems?**
   a. Draining the reverse osmosis system
   b. Pickling with a biocide agent, such as SBS
   c. Soaking reverse osmosis system with chlorine solution
   d. Performing a CIP

4. **What is the common cause of failure of O-ring seals in a reverse osmosis system?**
   a. Extreme temperatures during operation
   b. Continuous exposure to combined chlorine
   c. Physical damage due to movement of water through the system
   d. Biological degradation

Troubleshooting for Reverse Osmosis

MEMBRANE FOULING AND SCALING

Fouling of reverse osmosis and nanofiltration membranes typically occurs gradually over time. However, it can become accelerated due to changes in feed water quality, malfunctioning equipment such as failed flow control valves, or failures in pretreatment, such as loss of anti-scalant or sulfuric acid chemical dosing. Changes in feed water quality may occur because of upsets in the upstream biological treatment due to storm or peak flow events or they may occur because of equipment and process failure, resulting in water quality that has significantly higher TOC concentration, higher total phosphate and orthophosphate, or other potential scalants such as iron and manganese. This can lead to an accelerated rate of fouling and shorten the CIP interval. In the absence of obvious causes experienced at the upstream treatment processes, changes in the key foulants can be verified by laboratory analysis of grab samples and should be performed as part of routine monitoring. Generally, sudden changes in water quality parameters are not expected and any short-term events may not necessarily promulgate in a decreased membrane performance.

Loss of anti-scalant feed or sulfuric acid can result in more rapid fouling caused by scaling of sparingly soluble minerals such as calcium carbonate, calcium phosphate, and silica scale. If not addressed quickly, rapid scaling due to failed chemical feed will require a chemical cleaning to restore membrane performance; however, this could also lead to accumulation of irreversible fouling. Chemical feed system calibration should be routinely verified (e.g., weekly) with a manual drawdown test at the pump to ensure proper dosing in accordance with setpoints. Grab samples to measure pH should be performed to verify pH adjustment before and after chemical dosing locations and compared to online meter reading (if applicable) that is used for automatic adjustment of the chemical feed rate.

Particulate and colloidal fouling can occur if there is loss of prefiltration to at least 5 µm. This type of fouling may first appear on the cartridge filters that are used, in some cases, together with membrane filtration pretreatment. The presence of higher particle and colloid concentrations can be verified by measuring SDI. The SDI method passes sample water through a 0.45-µm filter at 210 kPa (30 psi) over an approximately 20-minute test duration. Time to fill 500 mL of filtered water is measured initially (time = 0 min) and after 5 minutes, 10 minutes, and 15 minutes. The value at 15 minutes (SDI_{15}) is calculated based on the relative increase in the time it takes to filter the same amount of water after continuous filtration. Typically, the SDI_{15} value should be less than 3.0, and values above 3.0 are considered poor. In such situations, performance of upstream filtration should be investigated in addition to reviewing typical fouling of the cartridge filters. As cartridge filters foul, more pressure will be needed to push the same amount of flow through them. The condition of cartridge filters can be tracked with differential pressure (i.e., difference of inlet and outlet water pressure) and elevated fouling will typically be accompanied by higher differential pressure, which indicates replacement is needed.

TROUBLESHOOTING INCREASES IN REVERSE OSMOSIS PERMEATE CONDUCTIVITY

As part of normal operations and routine monitoring, each of the reverse osmosis system units (or trains) should undergo routine verification of permeate conductivities from each stage and each vessel. Permeate conductivities will gradually increase across all stages as part of the normal membrane aging process. Vessel conductivities should be tracked historically and compared between other vessels in the same stage and other trains to identify any outliers that are outside normal distribution. As described in earlier sections, pressure vessels may contain six or seven membrane elements that are connected with permeate interconnectors. Each interconnector has O-rings that provide a seal between feed and permeate streams. The O-ring material can be susceptible to chemical damage from the continuous use of chloramines as well as cleaning solutions.

Vessels with significantly higher permeate conductivities, such as the one highlighted in Table 6.9, can be inspected manually by opening the vessel and extracting all of the membranes and subjecting each to a wet test to measure sodium chloride rejection to determine if membranes are intact. Vessels can be inspected further using vessel probing techniques to identify the location of failures within the pressure vessel. Probing is done by inserting a sampling tubing directly into the individual vessel's permeate sample port and taking conductivity readings while pushing the sampling tubing further into the vessel to identify any breaches due to specific element position in the vessel (see Table 6.10 or example results). Once the location of failure is identified (# 4 in Table 6.10), the vessel can be taken offline for manual inspection by extracting the membranes and inspecting the O-ring conditions on the permeate interconnector and/or replacing the suspected membrane element with a spare. A failed membrane element can be tested with an apparatus to check rejection of sodium chloride.

Table 6.9 Example of Vessel Conductivities From Stage 1 Consisting of 25 Vessels

Stage 1	Electrical Conductivity of Permeate (uS/cm)				
Vessels No. 1–5	14	12	16	11	15
Vessels No. 6–10	11	13	12	15	16
Vessels No. 10–15	13	15	60	13	17
Vessels No. 16–20	10	14	17	12	16
Vessels No. 21–25	12	11	16	15	14

Note: Highlighted cell in yellow represents an outlier vessel that should be investigated further for a potential O-ring seal breach using a technique called *vessel probing*.

Table 6.10 Example of Results From Vessel Probing

Element Position	# 1	# 2	# 3	# 4	# 5	# 6	# 7
Electrical Conductivity (uS/cm)	12	13	11	334	18	14	16

Note: Highlighted cell in yellow represents an outlier element position that should be investigated further for potential O-ring seal breach by taking the element out of vessel and inspecting permeate interconnector seals.

TEST YOUR KNOWLEDGE

1. **What are common causes for fouling of reverse osmosis membranes?**
   a. Particulate and colloidal matter
   b. Pore plugging
   c. Growth of biological films and deposits of organics
   d. Accumulation of salt and clay deposits

2. **What are common causes for scaling of reverse osmosis membranes?**
   a. Loss of chemical feeds such as anti-scalant agent or sulfuric acid
   b. Too little of chloramines residual
   c. Too high of TDS
   d. Feed water pH too low

3. **What methods can be used for troubleshooting reverse osmosis permeate conductivity increases?**
   a. Measurement of conductivity in feed and permeate to calculate removal
   b. Measurement of conductivities from individual vessels and from vessel probing
   c. Check the SDI of the feed water
   d. Check chloramines residual and increase if too low

Safety Considerations

TYPICAL CHEMICALS USED WITH MEMBRANE FILTRATION

Typical chemicals used with membrane filtration include sodium hypochlorite, ammonium hydroxide, citric acid, sodium hydroxide, SBS, and strong inorganic acids such as sulfuric or phosphoric acids. Table 6.11 provides a summary of commonly used chemicals, their purpose, and addition points. Always consult chemical safety data sheets (SDSs) for any chemical before use for important information on hazards associated with the chemical. Safety data sheets also contain information on the proper personal protective equipment. In addition, identify the closest emergency eyewash station and/or emergency shower for each work area containing chemicals. Sodium hypochlorite (NaOCl), typically 12% to 15% wt, is used for the chloramination of the feedwater to control biological fouling and during the maintenance cleans (or washes) and CIP events. Sodium hypochlorite is corrosive, a strong oxidizer, and a basic solution (pH 11.5 to 13). Ammonium hydroxide (NH_4OH), also known as aqua ammonia, is typically 19% to 30% wt. Aqua ammonia is used for adding free ammonia to form chloramines in the feedwater for biological fouling control. Ammonium sulfate ($[NH_4]_2SO_4$), 40% wt, is another chemical gaining in popularity that may be used as a source of ammonia for chloramination.

Always refer to the SDSs before starting to work with any chemicals to understand chemical hazards, permissible short- and long-term exposure limits, personal protective equipment requirements, spill response procedures, material incompatibility, firefighting, ecological effects, and disposal measures.

Ammonium hydroxide solutions offgas over time, producing toxic ammonia fumes. Care should be taken when working in direct contact with the bulk liquid. Consult SDSs for proper personal protective equipment. Respirators and/or self-contained breathing apparatus are typically required when working with ammonium hydroxide. Citric acid, typically 50% wt, is used for chemical cleanings such as maintenance cleans and CIP for removal of inorganic scalants (e.g., iron and manganese). For maintenance cleaning, a solution of 0.1% wt (1000 mg/L) may be used, whereas CIP solutions are typically performed with 2% wt solution. Citric acid is a weak organic acid that is nonflammable and mildly reactive with some metals. Sodium hydroxide (NaOH), typically 25% to 50% wt, is a corrosive liquid with pH 14. Sodium hydroxide is used for increasing the pH of NaOCl-based CIP solutions to increase effectiveness of chemical cleaning for removal of organics and for neutralization of spent CIP waste

Table 6.11 Typical Chemicals Used With Membrane Filtration and Reverse Osmosis Systems

Chemical	Common Name	Formula	Typical Strength	Purpose	Addition Points
Sodium hypochlorite	Bleach	$NaOCl$	12%–15% wt	Chloramination to control biofouling	Membrane filtration feed
Ammonium hydroxide	Aqua ammonia	NH_4OH	19%–30% wt	Chloramination to control biofouling	Membrane filtration feed
Ammonium sulfate	N/A	$(NH_4)_2SO_4$	40% wt	Chloramination to control biofouling	Membrane filtration feed
Citric acid	N/A	$C_6H_8O_7$	50% wt	Chemical cleans	CIP tank
Sodium hydroxide	Caustic	$NaOH$	25%–50% wt	Chemical cleans	CIP tank
Sodium bisulfite	SBS	$NaHSO_3$	40% wt	Quenching chlorine residual/membrane preservative	CIP tank
Anti-scalant agent	Scale inhibitor or threshold inhibitor	Proprietary	100% wt	Controlling sparingly soluble minerals	Reverse osmosis feed
Sulfuric acid	N/A	H_2SO_4	93%–98% wt	Controlling calcium carbonate and calcium phosphate scales	Reverse osmosis feed

solutions. Sodium bisulfite ($NaHSO_3$), typically 40% wt solution, is used for quenching residual chlorine in the makeup water before using citric acid or in the spent NaOCl CIP solutions before disposal.

TYPICAL CHEMICALS USED WITH REVERSE OSMOSIS

Typical chemicals used with reverse osmosis systems include citric acid, sodium hydroxide, sulfuric acid, anti-scalant agent (also known as *threshold inhibitor* or *scale inhibitor*), and generic and proprietary cleaners. Table 6.11 provides a summary of commonly used chemicals, purpose, and addition points. Chemical SDSs for all chemicals should be consulted for important information on hazards associated with the chemical and proper personal protective equipment. Of note, sulfuric acid spill should never be diluted with water due to the strong exothermic reaction that occurs. Reverse osmosis membranes can be damaged by improper use of chemical cleaning solutions. Manufacturer-specific information should be consulted to ensure chemical cleaning solutions are within acceptable ranges and compatible with the reverse osmosis membranes. Citric acid, typically 50% wt solution, is used for "low pH" cleaning protocols to remove buildup of scalants such as calcium phosphate and calcium carbonate. Similar to the membrane filtration systems, a 2% wt citric acid solution, pH 2 to 3, is commonly used. Sodium hydroxide, typically 25% wt, is used for "high pH" cleans that target removal of organics, biofilm, and certain scalants such as silica. Typically, a mild CIP solution with pH 10 to 10.5 can be tried first to restore membrane performance. More aggressive chemical cleaning may be needed after prolonged use of membranes that accumulate significant fouling. Typically, CIP solutions that use the minimum and maximum allowable pH in terms of membrane compatibility are considered aggressive. In some cases where citric acid cleans are ineffective, inorganic acids may be added to lower the pH of the cleaning solution. Care should be taken when using concentrated acids when preparing the CIP solutions.

Examples of inorganic acids include hydrochloric acid (HCl), sulfuric acid (H_2SO_4), phosphoric acid (H_3PO_4), and nitric acid (HNO_3).

ELECTRICAL AND MECHANICAL HAZARDS

Components of membrane systems typically include feed or filtrate pumps and/or backwash pumps that use 480-V power supplies. Control panels, such as the operator interface terminals or human-machine interface, typically use single-phase power sources. Care should be taken when troubleshooting and isolating electrical circuits to prevent injury or equipment damage. Clean-in-place systems also typically contain electrical water tank heaters that can overheat and cause fire if used improperly. Equipment that contains electrical components, such as an immersed electrical heater, should be labeled and contain a warning sign of electrical hazards.

A number of mechanical hazards such as rotating shafts, motorized valves, and other components with moving parts are common in membrane systems. All equipment should be properly locked out and tagged out before any work to prevent injuries and equipment damage. In membrane filtration systems, care should be taken when working with automatic strainer filters that use pressurized water and motorized cleaning mechanisms. Before maintenance, strainers need to be isolated from the pressurized water source in addition to deenergizing at the power source. Reverse osmosis systems use transfer pumps, high pressure pumps, booster pumps, and flush pumps. These pumps may require periodic maintenance and should be properly isolated before any work. Always consult the O&M manuals to review specific hazards associated with the equipment, including proper shutdown and isolation procedures.

CLEAN-IN-PLACE SYSTEMS AND WASTE DISPOSAL

Spent CIP solutions may have too high or too low pH and/or contain chlorine residual. Commonly, the waste solutions are neutralized to adjust pH to 7 and quenching any residual chlorine occurs before either disposal of waste solutions or return of waste streams to the influent flow for treatment. Caustic may be added to raise pH of the citric acid solution. Citric acid may be used to neutralize high pH solutions containing caustic. Sodium bisulfite may be used to quench chlorine residual. Specific discharge requirements should be reviewed in advance and neutralization procedures followed to meet compliance with these requirements. Consult local regulations before discharging spent chemical solutions to the sanitary sewer.

TEST YOUR KNOWLEDGE

1. What are examples of typical CIP chemicals for membrane filtration systems?
    a. Sulfuric acid and caustic
    b. Caustic and sodium hypochlorite
    c. Sodium hypochlorite and citric acid
    d. Citric acid and caustic

2. What are examples of typical chemicals used for reverse osmosis systems to control scaling?
    a. Sulfuric acid and caustic
    b. Caustic and sodium hypochlorite
    c. Sulfuric acid and anti-scalant agent
    d. Citric acid and caustic

3. What documentation should always be reviewed before working with any chemicals?
    a. Chemical's certificate of analysis
    b. SDSs
    c. O&M manual
    d. Chemical compatibility guide

4. What is commonly done to prevent electrical hazards and injuries when servicing equipment?
    a. Turn off the equipment and shut off the breaker.
    b. Display a caution tape around working area.
    c. Deenergize at power source and lock out tag out.
    d. Consult SDSs.

5. What is commonly done before disposal of spent CIP solutions?
    a. Record temperature and conductivity before disposal.
    b. Neutralize by adjusting pH and adding dechlorinating agent.
    c. Dilute solution and send to drain.
    d. No neutralization is necessary.

CHAPTER SUMMARY

	Membrane Filtration	Reverse Osmosis
PURPOSE	■ Water purification technology ■ Removes particles, sediment, algae, protozoa, and bacteria. Some membranes can remove small colloids and viruses as well. ■ Does not remove dissolved constituents ■ Used for nonpotable and potable reuse	■ Water purification technology ■ Removes large majority of solutes, including TDS, ions, organics, and pathogens ■ Used for ocean desalination, water reuse, and brackish water desalination
THEORY OF OPERATION	■ Removal of particles occurs via straining at the surface of the membrane. ■ Particles that are larger than the pores are effectively 100% removed by the membrane. ■ Membranes operate as dead-end filters in which all of the feedwater passes through the membrane. ■ Water passing through the membrane is called *permeate*. ■ Membranes slowly lose performance due to fouling and must be cleaned to restore performance.	■ Passage of water through a polyamide membrane via diffusion ■ System pressure must exceed feed osmotic pressure to generate permeate via reverse osmosis. ■ Reverse osmosis system operated as cross-filtration ■ Water passing through the membrane is called *permeate* and untreated water exiting on the other side of the membrane is called *concentrate* ■ Membranes slowly lose performance due to fouling and must be cleaned to restore performance.
DESIGN PARAMETERS	■ Design flux depends on application. Can range from 34–102 LMH (20–60 gpd/sq ft). ■ Recovery typically 90%–95% ■ Clean-in-place typically expected every 3 months	■ Design flux depends on application. Typically, between 14–24 LMH (8–14 gpd/sq ft). ■ Design recovery depends on application. Can exceed 90% for certain reuse applications and be limited to 50% for seawater desalination. ■ Vessel array configuration depends on a number of interrelated design parameters
PROCESS VARIABLES	■ Production is controlled using a filtrate flow setpoint. Systems are typically run at constant flux. ■ Transmembrane pressure is the difference between the feed and filtrate pressure. ■ Chloramines are typically dosed upstream of membrane systems to control biological fouling.	■ Feed pressure depends on a number of variables, including permeate flow setpoint, temperature, recovery, feed concentration. ■ If flux is kept constant, a decrease in temperature-corrected specific flux (LMH/bar or gpd/sq ft/psi) indicates a decrease in performance (i.e., more pressure needed to maintain production). ■ Acid and anti-scalant can be added to control scaling, potentially allowing for higher recovery and/or longer runtimes between cleans.
EQUIPMENT	■ Upstream screening (strainers or drum screen) ■ Membrane modules ■ Air scouring system ■ Chemical cleaning system ■ Backwash system	■ Transfer pumps ■ High feed pumps ■ Pressure vessels ■ Cartridge filters ■ CIP skid ■ Permeate flush skid ■ Acid/anti-scalant chemical skids

EXPECTED PERFORMANCE	■ Reduce turbidity to <0.2 NTU ■ Achieves 4 LRV for *Giardia* and *Cryptosporidium*	■ 90%–99% reduction in TDS ■ High rejection of inorganics, organics, and pathogens ■ Consistent performance over for a broad range of feedwater quality
MAINTENANCE	■ Daily observation for visual and audible indications of malfunction ■ Regular maintenance should be performed in accordance with membrane manufacturer's recommendations. ■ Perform maintenance cleans and CIPs to restore performance per manufacturer's recommendations. ■ Maintain an inventory of spare parts.	■ Daily observation for visual and audible indications of malfunction ■ Regular maintenance should be performance in accordance with membrane manufacturer's recommendations. ■ Clean-in-place procedures with compatible products to restore performance ■ Maintain an inventory of spare parts.

References

Crittenden, J. C., Trussell, R. R., Hand, D. W., Howe, K. J., & Tchobanoglous, G. (2012). *MWH's water treatment: Principles and design*. John Wiley & Sons.

U.S. Environmental Protection Agency. (2005). *Membrane filtration guidance manual.*

Water Environment Federation. (2012). *Membrane bioreactors* (WEF Manual of Practice No. 36).

Suggested Readings

Toray. (2020). *Operation, maintenance, and handling manual for membrane elements.* https://www.toraywater.com/knowledge/pdf/HandlingManual.pdf

U.S. Environmental Protection Agency. (2001). *Low-pressure membrane filtration for pathogen removal: Application, implementation, and regulatory issues* (EPA-815/C-01-001). https://cfpub.epa.gov/si/si_public_record_report.cfm?Lab=OGWDW&TIMSType=&count=10000&dirEntryId=143146&searchAll=&showCriteria=2&simpleSearch=0&startIndex=30001

Water Environment Federation. (2016). *WET operator essentials—Membrane applications in wastewater.* https://www.wef.org/login/?ReturnUrl=/globalassets/assets-wef/direct-download-library/members-only/operator-essentials/2010-2011/wet---operator-essentials---membrane-bioreactors---jan11.pdf

Water Environment Federation. (2016). *WET operator essentials—Membrane bioreactors.* https://www.wef.org/login/?ReturnUrl=/globalassets/assets-wef/direct-download-library/members-only/operator-essentials/2016/wet-operator-essentials---membrane-applications-in-wastewater---feb16.pdf

Water Environment Federation. (2018). *The water reuse roadmap.*

CHAPTER 7

Water Reuse

Introduction

The increased use of advanced treatment at today's water resource recovery facilities (WRRFs) is due to the greater need of water reuse to replace our traditional freshwater supplies. This chapter describes the current state and ongoing trends for water reuse in the United States. We cover the wide spectrum of reuse from nonpotable reuse (such as golf course irrigation) to potable reuse (such as treated water augmentation). Each type of reuse is performed in many different ways. The drivers behind the advanced treatment processes you are working with will be discussed here and will go well with the knowledge gained from the other chapters.

LEARNING OBJECTIVES

Upon completing this chapter, you will be able to:

- Understand the differences between the major kinds of water reuse.
- Understand the types of source waters used for water reuse.
- Identify the important environmental and public health regulations for water reuse.
- Understand the level of treatment required for different forms of water reuse.
- Identify the water quality targets for different forms of water reuse.
- List the methods used to ensure potable reuse is safe for public health.

Water Reuse Overview

There are many reasons to reuse water and there are all kinds of ways to perform water reuse. For example, irrigation is one type of nonpotable reuse, and drinking water is one type of potable use. Either way, water reuse is a valued asset for sound water management.

HISTORY OF WATER REUSE

Fresh water has been, and continues to be, a vital resource for all communities. As the number of people in the United States has increased, the demand for fresh water, in turn, has increased. In the recent past, between the 1930s to 1970s, the primary method to meet these needs was to build more reservoirs and dams for water storage as well as aqueducts to redistribute the water supply. Building this type of water infrastructure, however, began to decrease in the 1980s due to concerns for the environment and a lack of unclaimed freshwater flow. Methods to ensure enough water supply for growing populations shifted to both water conservation and water reuse.

At the same time, wastewater systems were also being built with the primary goal of disposing of wastewater in ways that would not contaminate the freshwater supply. Environmental concerns for wastewater disposal became a main public concern. The Clean Water Act of 1972 was passed as a result. This act was the driver for the construction of most of our WRRFs to achieve primary and sometimes secondary levels of treatment. Water resource recovery facilities today need to meet increasingly stringent water quality regulations to protect the environment. For this reason, some level of advanced treatment exists now at many facilities.

There are some 15 000 wastewater utilities in the United States that treat approximately 121 120 ML/d (32 bil. gpd) of wastewater, most of which is released back into environment.

As a whole, both our water and wastewater infrastructure are now being used and considered for use in meeting the country's water demand through the link of water reuse. Recent severe drought conditions in the southwestern United States have persisted longer than in the past, prompting agencies to include the potential effects from climate change in their long-term planning for meeting current and future drinking water demands. When faced with such challenges, seeing wastewater as a resource rather than a waste stream has been a major shift by the public and industry to adapt to water scarcity concerns. This change in mindset will continue to increase the reliability of our water supply systems in the face of large-scale drought, population growth, and a limited supply of new water sources.

CURRENT TRENDS IN WATER REUSE

Current trends in water reuse may be split into nonpotable and potable reuse, based on whether the water will be consumed by humans (Figure 7.1). Nonpotable reuse may be further broken down into centralized and decentralized nonpotable reuse. Centralized nonpotable reuse is the most common and oldest form of water reuse and includes uses such as irrigation of golf courses and publicly owned parks and landscaping. These so-called "recycled water systems" are often identified by purple-colored pipe to distinguish from traditional wastewater or drinking water piping and uses. Decentralized nonpotable reuse is most commonly referred to as "on-site reuse" and encompasses smaller and localized treatment systems at the building or dense community itself. Because this application is still at a nonpotable reuse stage, the buildings require dual plumbing for use of recycled water for toilet flushing, clothes washing, and irrigation. The key benefit of on-site reuse is the reduction of potable water demand and wastewater discharge.

Reuse of captured stormwater is also on the rise in the United States, but differs from the water reuse types described in this chapter because it is not of wastewater origin. As a result, treatment requirements for stormwater reuse differ greatly from those of wastewater reuse. If stormwater is combined with wastewater for treatment, then the wastewater treatment requirements will govern.

Potable reuse has been gaining in popularity since the early 2000s. One key reason for this is that planned potable reuse can use our existing water infrastructure; it does not require a new "purple pipe" recycled water distribution system to offset potable water demand. Historically, potable reuse has been practiced by groundwater recharge through surface spreading and groundwater injection. Both are considered examples

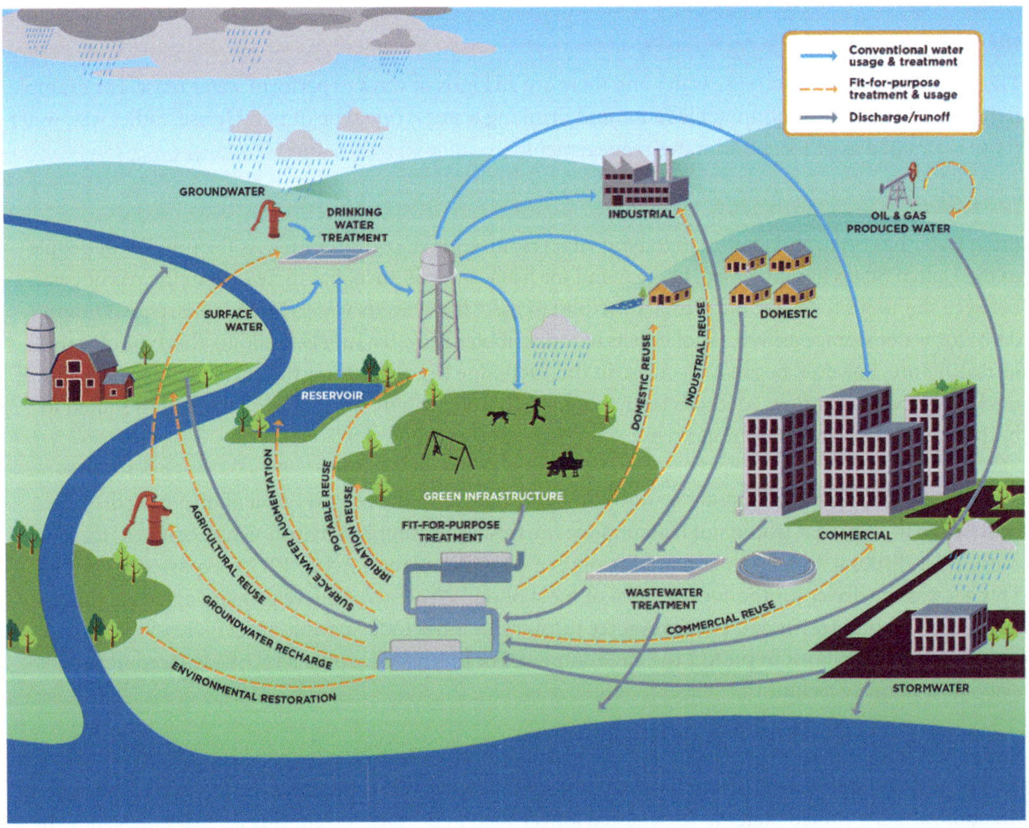

Figure 7.1 Examples of Water Reuse Schemes (U.S. EPA, 2021)

of indirect potable reuse. As will be further discussed in this chapter, more direct forms of potable reuse, such as surface water augmentation, raw water, and treated water augmentation are currently being considered by regulators, the public, and utilities to allow more sustainable uses of potable water.

For all types of water reuse, the "fit-for-purpose" treatment concept is becoming a common theme. Fit-for-purpose allows different treatment technologies to be used to meet the needs of different uses of recycled water. For example, recycled water for golf course irrigation does not need the same level of treatment as recycled water that will be used for potable supply.

"Fit-for-purpose" treatment refers to the design of treatment systems for specific uses of water.

DE FACTO POTABLE REUSE

De facto potable reuse refers to the use of surface water (such as a river or reservoir) as a drinking water supply that is downstream of wastewater discharges. The concept of de facto potable reuse makes one think about how to best reduce adverse effects of wastewater discharges on downstream communities. During drought conditions, these effects can be increased and pose both public health and other environmental concerns.

The term *de facto potable reuse* was coined in a 2013 paper by Rice, Wutich, and Westerhoff (2013) that looked at how often wastewater occurred in water supply sources. The study covered a period from 1980 to 2008 and concluded that de facto reliance on wastewater was on the rise. The study also found that wastewater made up between 7% to 100% of supplies for drinking water treatment plants in the study. The industry has recognized wastewater as an important water resource, with several states developing regulations to enable wider and safer potable water reuse. Later sections in this chapter will be about *planned* potable reuse as opposed to de facto reuse.

POTABLE WATER REUSE

Planned potable reuse is at the forefront of water reuse in the United States. Planned potable reuse occurs when a water resource of wastewater origin is treated with engineered processes and/or the environment to adequate standards that protect public health. Figure 7.2 presents a map of potable reuse projects in the United States. Much of the advanced treatment facility needs are driven by the end use of the water for human consumption. These drivers include the level of treatment and reliability to address the potential for failures. Potable reuse is often split into "indirect potable reuse" and "direct potable

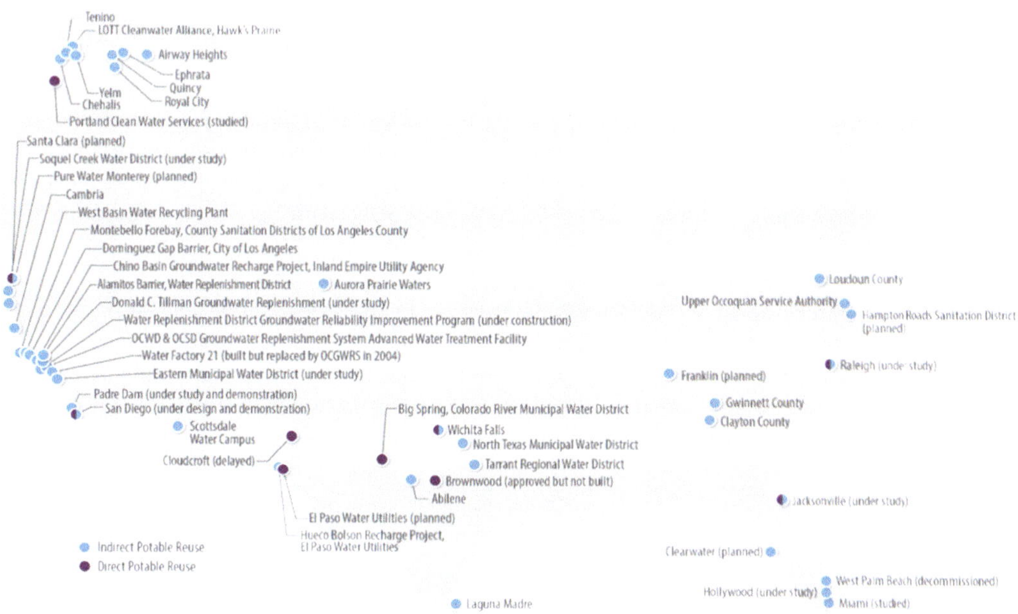

Figure 7.2 Potable Reuse Projects in the United States from U.S. EPA's Potable Reuse Compendium (U.S. EPA, 2017)

reuse." The key distinction between the two is that an indirect potable reuse project has an environmental buffer to provide greater response time for failure. Key features of a potable reuse project, in contrast to a nonpotable reuse project, are the level of treatment needed to meet pathogen reduction requirements, multiple barriers of treatment, and monitoring. The concept of reliability for a potable reuse project also drives how the treatment facility will be operated.

According to the National Water Reuse Action Plan (U.S. EPA, 2020) developed by the United States Environmental Protection Agency (U.S. EPA), several states anticipate major increases in potable water demand in the near future. For example, the Florida Potable Reuse Commission expects adding 4163 ML/d (1.1 bil. gpd) of fresh water to meet projected needs by 2035. In California, population growth of 5 million is expected by 2050, which may increase the need for extra water supplies by 3406 to 20 439 ML/d (900 mgd to 5.4 bil. gpd). Potable reuse is expected to provide major portions of the water supply in these two states to meet future water demands.

TEST YOUR KNOWLEDGE

1. **Decentralized nonpotable reuse is most commonly referred to as "on-site reuse."**
   - ☐ True
   - ☐ False

2. **Which of the following describes "de facto" potable reuse?**
   a. The planned use of recycled water for potable uses
   b. The planned use of recycled water for irrigation
   c. The use of a surface water that is affected by upstream wastewater discharge for potable supply
   d. The use of stormwater for a potable supply

3. **The concept of "fit-for-purpose" treatment refers to:**
   a. Recycled water treatment to meet effluent guidelines defined by U.S. EPA
   b. Offsetting the use of potable water with advanced treated recycled water
   c. Treatment of recycled water to meet specific water quality objectives based on the intended uses of water
   d. Recycled water treatment for industrial reuse purposes

Secondary and Tertiary Treated Municipal Wastewater

TYPICAL WATER QUALITY

Treated secondary effluent water quality generally can be classified as "nitrified" or "non-nitrified" effluent. Non-nitrified effluent contains free ammonia and typically has higher concentrations of total organic carbon (TOC). In nitrified effluents, most of the nitrogen is in the form of nitrate, with concentrations ranging from 40 to 50 mg/L as nitrogen. If the secondary process also provides partial denitrification, nitrate concentrations normally range from 10 to 20 mg/L as nitrogen. Nitrified and partially denitrified wastewaters generally meet most standards established for nonpotable reuse but need disinfection to be fully compliant.

In 2012, U.S. EPA provided national guidelines on water reuse to offset the use of potable water and included water quality treatment objectives that should be met (U.S. EPA, 2012). Table 7.1 shows examples of expected ranges of water quality parameters on influent, primary, secondary, and tertiary effluents. Another water quality parameter to consider when evaluating wastewater for water reuse is total dissolved solids (TDS). In some cases, TDS of reclaimed water can be above 1000 mg/L (typically TDS is 500 to 1500 mg/L). In this case, partial demineralization may be needed to reduce the levels of TDS and minimize effects of salinity on plant health. Potential scaling in downstream uses is also a reason for partial demineralization. Specific water quality and disinfection objectives are discussed in later sections of this chapter based on intended end use.

Partial demineralization is a process used to remove TDS from water by removing hardness and salts (e.g., Ca^{2+}, Mg^{2+}, Na^{+}, K^{+}, Cl^{-}, F^{-}, Br^{-}, SO_4^{2-}, HCO_3^{-}, HPO_4^{2-}, HPO_4^{2-}, NO_3^{-}).

Table 7.1 Source Water Quality and Treated Water Quality (adapted from Asano et al., 2007, & U.S. EPA, 2012)

Water Source	Primary Influent	Primary Effluent	Secondary Effluent	Tertiary Effluent
Chemical Oxygen Demand (mg/L)	250–800	150–350	40–80	<25
Biochemical Oxygen Demand (mg/L)	110–350	100–250	<25	<5
Total Suspended Solids (mg/L)	120–400	10–100	<10	<10
Total Nitrogen (mg/L)	20–70	40–80	10–20	10–20
Total Phosphorus (mg/L)	4–12	3–5	<3	1–3
Turbidity (NTU)	20–100	10–50	<2.5	<2
TOC (mg/L)	80–260	30–50	<20	5–15
pH	6–8	6–8	6–8	6–8
Alkalinity (mg/L $CaCO_3$)	100–400	100–300	<200	80–200
Total Coliforms (MPN/100 mL)	>1 000 000	10 000–1 000 000	10 000–30 000	100–30 000

PATHOGEN CONCENTRATIONS

Pathogenic microorganisms such as enteric virus and protozoa (e.g., *Giardia* and *Cryptosporidium*) are found in wastewater. Indicators such as total and fecal coliforms are often used for monitoring the presence of human pathogens and to establish performance criteria for water reuse. As potable reuse becomes more common, there is a growing effort to understand the concentrations of pathogens in both raw wastewater and secondary and tertiary effluents.

COMPOUNDS OF EMERGING CONCERN

Compounds of emerging concern (CECs) include a broad group of nonregulated compounds for which the fate, transport, and health relevance are not well understood. Examples of CECs include pharmaceuticals and personal care products (PPCPs) such as medications, shampoos, lotions, and makeup; endocrine disrupting compounds (EDCs); contrast agents (such as those used for x-rays); herbicides (to kill weeds); pesticides (to kill insects); and industrial and artificial compounds (e.g., Teflon). For nonpotable reuse, the presence of CECs does not pose significant health concerns. However, for potable water reuse, CECs are viewed as a concern. Many CECs do not have established regulatory limits or health references and are only detected in the low part per trillion to part per billion concentrations. Compounds of emerging concern are moderately to poorly removed by conventional activated sludge treatment and require advanced water treatment using reverse osmosis membranes, advanced oxidation, ozone, and/or granular activated carbon for removal.

Fate: The "fate" of a chemical refers to how the chemical will change and degrade over time. Transport: The "transport" of a chemical refers to how the chemical will move within a system. Will it adsorb to other compounds? Does it travel in water? Does it travel in air? Health relevance: The "health relevance" of a chemical refers to its effect on human health. For many chemicals, we do not yet know how or at what doses they affect human health.

NUTRIENTS AND MINERALS

Generally, reclaimed waters are rich in nitrogen, phosphorus, and carbon. High levels of nutrients can lead to algal blooms, which can adversely affect the environment where discharged. Cyanotoxins can be produced from algal bacteria and contaminate drinking water sources. U.S. EPA has developed effluent guidelines that are implemented through the National Pollutant Discharge Elimination System (NPDES). The Clean Water Act has also established total maximum daily loads (TMDLs) for impaired water bodies. Recycled water can be treated to lower nutrient concentration in receiving bodies to improve water quality and provide remediation of impaired water bodies. Phosphorus is often the limiting nutrient and may be addressed further by advanced treatment to minimize effects to receiving water bodies.

Endocrine disrupting compounds include industrial chemicals such as those used in some hydraulic fluids, lubricants, and plastics. Endocrine disrupting compounds can alter how the endocrine system works. The endocrine system is made up of the body's hormones and regulates all of the biological processes, including the development and function of the brain, nervous system, and reproductive system.

IMPORTANCE OF SOURCE CONTROL PROGRAMS

Source control programs are used to minimize the effects of industrial discharges to WRRFs and prevent release of recalcitrant (i.e., difficult to treat) compounds into the environment. Source control programs are especially important for potable water reuse to minimize organic loading to the reclamation facility and minimize any potential adverse health effects. Water and wastewater agencies that are subject to an NPDES permit are required to establish local limits for industrial discharges that may adversely affect wastewater treatment or cause discharge permit violations.

TEST YOUR KNOWLEDGE

1. **Nitrified effluent contains free ammonia and typically has higher concentrations of TOC.**
   - ☐ True
   - ☐ False

2. **High concentrations of nutrients in secondary and tertiary effluents can lead to algal blooms.**
   - ☐ True
   - ☐ False

Cyanotoxins are toxic byproducts of cyanobacteria. The most common examples of cyanotoxins include microcystines, cylindrospermopsin, anatoxins, and saxitoxins. Contact and ingestion of water with cyanotoxins present can lead to skin irritation, allergic reactions, upset stomach, vomiting, and diarrhea, as well as potential liver and kidney damage and respiratory arrest.

Impaired water bodies are those that may be used for drinking water supplies, recreational purposes, or agriculture that are contaminated by discharged streams containing pollutants. U.S. EPA assists states, territories, and authorized tribes in development of TMDLs for these water bodies that establish a maximum limit of pollutant discharged to preserve and restore water quality.

Nonpotable Water Reuse

TYPICAL USES OF RECYCLED WATER

Nonpotable water reuse includes a number of applications from the early days of water reuse. Historically, the largest uses of potable water supply were for both agriculture and irrigation uses. As the value of the pristine freshwater supplies increased, people looked for ways to reduce demand for potable water by using fit-for-purpose treatment. Industrial water use was also considered a "low hanging fruit" for early water reuse, and source waters for these applications shifted from potable water supplies to WRRF discharges and other innovative methods. On-site reuse is one of the most up-and-coming forms of nonpotable reuse. Public acceptance for dual plumbing uses has increased over time. Current on-site reuse practices are supported by the guidance resources developed by the National Blue Ribbon Commission for On-Site Non-potable Water Systems (NBRC). Over the years, NBRC, led by participating municipalities, worked closely with utilities, the public, heath regulators, and other agencies to develop resources for establishing on-site water reuse practices that effectively address the concerns of closer contact with recycled water with treatment and monitoring that is protective of public health.

AGRICULTURE

Water reuse for agriculture is viewed as a more sustainable source of water during drought conditions. Recycled water is also rich in nitrogen and phosphorus, which reduces the fertilizer use and can increase crop yields for more efficient use of crop lands. Key public health considerations for agricultural use of recycled water include whether the crop is a food crop or a non-food crop (such as livestock feed). If the crop is a food crop, regulators will also consider if the recycled water contacts the edible part of the crop or not. These distinctions have made it possible to use highly treated WRRF effluents for many agricultural uses of recycled water at the secondary or tertiary level of treatment possible for this large industry. In terms of public health, the key water quality concerns for recycled water have been pathogen concentrations, as regulated by bacterial indicators such as total and fecal coliform. Disinfection with free chlorine, for example, is generally how these public health concerns are addressed. Other non-public health requirements include TDS, chloride, sodium, hardness, boron, and other elements that may affect the productivity of the crop, but do not affect public health. U.S. EPA grants states the ability to provide specific practices to use recycled water to irrigate food crops. Typically, these practices include frequent monitoring to ensure pathogenic organisms are effectively inactivated in the recycled water used on food crops. Food crop irrigation with recycled water is practiced in Oregon, Washington, California, Arizona, New Mexico, Texas, Florida, Pennsylvania, and Hawaii. Recycled water is also used for irrigating non-vegetable crops such as cotton, fodder, or for crops that must be processed further before human consumption.

IRRIGATION

Key public health considerations for the use of recycled water for irrigation closely follow those considerations for agricultural uses. Depending on the potential for human contact, the recycled water is generally required to be treated to a tertiary disinfected level. For example, it is used to irrigate highway medians, golf courses, and parks. In this case, the crop is often grass and, sometimes, water quality drivers are based on those characteristics key for healthy grass. Often, a golf course may require lower levels of TDS and sodium due to sensitivity of certain grass types, but these grasses benefit from the increased nitrogen and phosphorus load of the recycled water. Irrigation continues to be a key use of recycled water in many communities.

INDUSTRIAL

Industrial uses of recycled water are broad and varied and can include boiler feed, cooling tower use, dust control during construction, and street sweeping. The specific use will dictate most water quality requirements. Most of these uses do not have the potential for contact with people, and so the water quality requirements for public health safety are generally less than that of agriculture or irrigation. However, some utilities choose to meet a higher level of treatment (e.g., tertiary filtration and disinfection) to meet potential contact uses because there may not be enough economy of scale for specific fit-for-purpose treatment. In this case, the utility could then use the recycled water for both irrigation and an industrial use such as street sweeping.

RECREATIONAL

Reclaimed water can be used for recreational purposes such as creating artificial ponds, lakes, and wetlands; making artificial snow; and augmenting river and stream flows. Artificial ponds and lakes can provide aesthetic improvements to the recreational zones within a community. Adding reused water to rivers or streams can enhance aquatic or riparian habitat, offset the effects of drought conditions, and reduce water quality impairment. Water quality and treatment requirements for river or stream augmentation are often based on the designated use of the water and the enhancement and protection of native aquatic life. An NPDES permit is typically required and includes whole effluent toxicity testing, biological monitoring, and chemical monitoring. Wetlands provide many important functions such as supporting natural habitats, flood protection, and enhancement of water quality. For recreational snowmaking, recycled water must be filtered, treated for removal of nutrients to limit runoff to surface streams during snowmelt, and disinfected to achieve at least 4-log reduction of virus.

A *riparian habitat* is the area between a moving source of water (i.e., river, stream) and land, sometimes called the "riverbank." Riparian habitats are thriving ecosystems that supply food, nutrients, and shelter for a large and diverse group of plants and animals. Riparian areas also provide natural groundwater recharge, protect waterways from excess nutrients and sediment, and help control flooding.

ON-SITE

On-site water reuse means it is decentralized, typically implemented by larger buildings and campuses. On-site nonpotable reuse can provide beneficial uses such as toilet flushing and irrigation, offsetting the use of potable water and decreasing discharge flows into the sewer collection system. On-site reuse is being considered by densely populated municipalities, with newer buildings including "dual plumbing" that allows for separate potable and nonpotable lines throughout the building structure. This practice uses an on-site nonpotable water reuse system (ONWS), which captures, treats, stores, and distributes recycled water to different fit-for-purpose uses.

Decentralized water reuse refers to systems that treat water from a specific source (i.e., manufacturing complex, farm), within a close vicinity to the source. Decentralized water reuse systems are not part of a community's centralized water/ wastewater systems (i.e., municipal sewer system).

On-site nonpotable water reuse systems can come in different sizes and configurations and are commonly small-scale, packaged automated systems operated and managed by the building managers. As shown in Figure 7.3, source wastewaters may include blackwater, stormwater, rainwater, condensate, and foundation

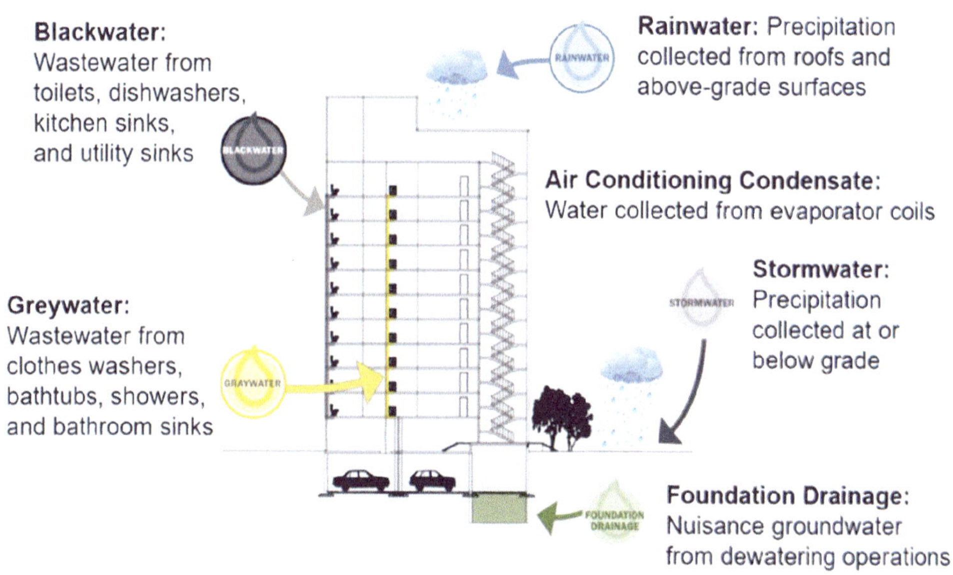

Figure 7.3 Potential Source Waters for On-Site Reuse (U.S. EPA, 2018)

drainage. Blackwater is the most challenging source water from a treatment and health risk perspective. Other sources, such as condensate or even greywater, may require less treatment. The ONWS must provide adequate treatment for a particular use. On-site nonpotable water reuse systems are relatively new, and treatment guidelines are currently being developed for different end uses. An ONWS may be subject to International Plumbing Code, Uniform Plumbing Code, National Sanitation Foundation standards, and/or specific local municipality guidelines as well as permit requirements from state agencies.

OVERVIEW OF REGULATORY GUIDELINES FOR RECYCLED WATER

FEDERAL GUIDELINES FOR RECYCLED WATER

The Clean Water Act of 1972 defined levels of wastewater treatment that avoid eutrophication and gross degradation in the environment. Primary treatment for separation of solids from the liquid stream was key. Secondary treatment was necessary for reduction of biochemical oxygen demand (BOD). Today, water reuse applications often require treatment beyond that provided by primary and secondary treatment. U.S. EPA has played an important role in providing guidelines for water reuse and driving progress through legislation and funding.

STATE AND LOCAL GUIDELINES FOR RECYCLED WATER

States have regulated recycled water use as necessary for their needs and have primarily focused on human contact versus nonhuman contact uses. Regulations may be based on treatment technique, in which a certain type of treatment process must be used. Regulations may also be water quality based, in which the recycled water must meet certain levels of bacterial indicators or nutrients.

OVERVIEW OF TREATMENT AND WATER QUALITY OBJECTIVES

U.S. EPA's *Guidelines for Water Reuse* (2012) provides a thorough overview of the uses of recycled water as well as treatment and water quality objectives. Table 7.2 provides examples of treatment and water quality objectives for recycled water based on intended uses. Generally, recycled water has to be treated to higher standards when used for nonpotable uses that have potential human exposure and even higher standards for potable reuse applications, as will be discussed in later sections of this chapter (Figure 7.4).

Secondary treatment and disinfection are minimum requirements for agricultural uses such as non-food crops, processed food crops, and restricted urban irrigation uses as well as for industrial and recreational reuse applications. For agricultural food crops, tertiary filtration before disinfection is required. Nonpotable reuse with groundwater recharge requires secondary treatment before injection. Potable reuse via groundwater recharge by surface spreading requires not only secondary, tertiary filtration and disinfection, but also soil aquifer treatment to meet the drinking water standards once water reaches the water table at the saturated zone. The compliance point is defined at the end of the vadose or the unsaturated zone (see sidebar).

> Soil aquifer treatment combines physical, chemical, and biological treatment of water as it percolates through soil, sand, silt, and rocks until it reaches the saturated zone. Once water reaches the saturated zone, all the voids and porous media are full of water. The unsaturated zone, also known as the "vadose zone," is where treatment takes place.

TYPICAL TREATMENT TRAIN CONFIGURATIONS

Typical treatment train configurations include headworks or screening, primary treatment, and secondary treatment. Tertiary treatment varies in method and can include additional solids removal via filtration through cloth filters, sand filters, or granular media filters. Tertiary treatment may also include disinfection, typically through free chlorine contact, combined chlorine contact, or ultraviolet disinfection.

ADVANCED WATER TREATMENT

Advanced water treatment for recycled water may involve membrane filtration treatment either for solids removal (such as micro- or ultrafiltration membrane processes) or dissolved solids removal (such as electrodialysis reversal, nanofiltration, or reverse osmosis). Because this is often driven by crop or landscape irrigation, the removal of TDS may be done for a portion of the flow, and then blended back, in a fit-for-purpose method.

RECYCLED WATER DISTRIBUTION SYSTEM

RECYCLED WATER STORAGE

Like any distribution system, recycled water distribution and storage has its own key considerations. Because recycled water has a higher potential for biogrowth than typical potable supplies, managing detention time in recycled water storage reservoirs or tanks is important to ensure the recycled water

Table 7.2 **Examples of Treatment and Water Quality Objectives for Recycled Water Applications** (adapted from U.S. EPA, 2012)

Reuse Application	Treatment	Water Quality Objectives
Agricultural non-food crops or processed food crops and restricted urban irrigation	■ Secondary ■ Disinfection	■ pH 6.0–9.0 ■ ≤30 mg/L BOD ■ ≤30 mg/L TSS ■ ≤200 FC/100 mL ■ 1.0 mg/L Cl_2 residual
Agricultural food crops	■ Secondary ■ Tertiary filtration ■ Disinfection	■ pH 6.0–9.0 ■ ≤10 mg/L BOD ■ ≤2 NTU ■ Non-detectable FC ■ 1.0 mg/L Cl_2 residual
Urban irrigation (Unrestricted)	■ Secondary ■ Tertiary filtration ■ Disinfection	■ pH 6.0–9.0 ■ ≤10 mg/L BOD ■ ≤2 NTU ■ Non-Detectable FC ■ 1.0 mg/L Cl_2 residual
Industrial reuse for cooling towers	■ Secondary ■ Disinfection ■ Coagulation and filtration (optional)	■ pH 6.0–9.0 ■ ≤30 mg/L BOD ■ ≤30 mg/L TSS ■ ≤200 FC/100 mL ■ 1.0 mg/L Cl_2 residual ■ Additional requirements based on recirculation ratio
Recreational reuse, including uses such as for creating wetlands, augmenting natural streams	■ Secondary ■ Disinfection ■ Additional requirements may be applicable	■ pH 6.0–9.0 ■ ≤30 mg/L BOD ■ ≤30 mg/L TSS ■ ≤200 FC/100 mL ■ 1.0 mg/L Cl_2 residual ■ Variable additional requirements
Nonpotable reuse groundwater recharge	■ Primary at minimum for spreading ■ Secondary at minimum for injection	■ Site specific and use specific
Potable reuse groundwater recharge by spreading	■ Secondary ■ Tertiary filtration ■ Disinfection ■ Soil aquifer treatment	■ Meet drinking water standards after percolation through vadose zone ■ ≤2 mg/L TOC ■ ≤2 NTU ■ pH 6.5–8.5 ■ 1.0 mg/L Cl_2 residual ■ Non-detectable TC
Potable reuse groundwater recharge by injection and surface water augmentation	■ Secondary ■ Tertiary filtration ■ Disinfection ■ Advanced treatment	■ Meet drinking water standards ■ ≤0.5 mg/L TOC ■ ≤2 NTU ■ pH 6.5–8.5 ■ 1.0 mg/L Cl_2 residual ■ Non-detectable TC ■ Variable additional requirements

FC = fecal coliforms, TC = total coliforms, NTU = nephelometric turbidity units, TSS = total suspended solids

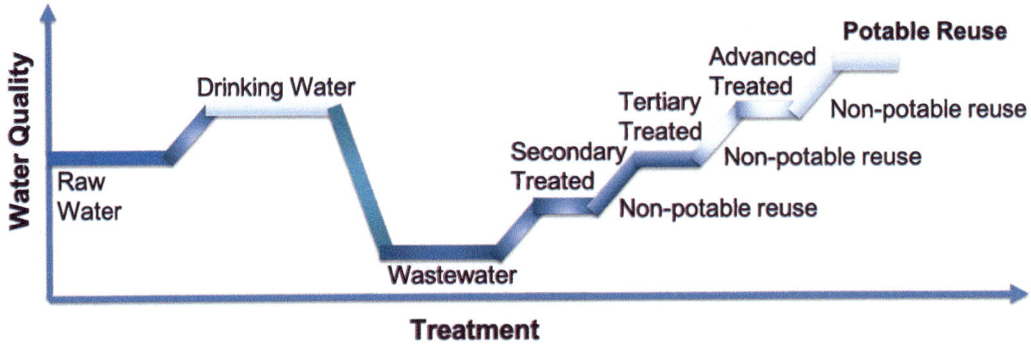

Figure 7.4 Water Quality and Treatment (adapted from U.S. EPA, 2012)

quality does not degrade during storage. Strategies for controlling biogrowth and other water quality issues include aeration, limiting detention time, adding a disinfectant such as chlorine, adding an algaecide such as copper sulfate, and other methods. For engineered storage tanks, the tank material and coating must be chosen carefully to prevent corrosion from the stored water.

DISINFECTANT RESIDUAL MANAGEMENT

Recycled water is generally required to have a chlorine residual that must be maintained through distribution for uses that may result in human contact, such as irrigation of public parks. Chlorine demand is expected to be greater for recycled water. As a result, maintaining the required residual level may require chemical dosing booster stations within particularly large recycled water distribution systems.

EXAMPLES OF NONPOTABLE WATER REUSE PROJECTS

A number of nonpotable reuse projects have been highlighted by the industry in the past several decades. These projects continue to operate once the infrastructure is in place, speaking to the sustainability of nonpotable reuse. It is common to see purple pipe in many cities where potable water costs are increasing due to limited local supplies, urban growth, and agricultural demands.

Key examples of nonpotable reuse projects include Monterey One Water, a pioneer of agricultural reuse in Monterey, California, that now has a large portfolio of reuse applications including planned groundwater recharge. The reuse facility treats 113 ML/d (30 mil. gal) of reclaimed water from four sources including wastewater, industrial processing water, crop drainage water, and stormwater runoff. Approximately 4856 ha (12 000 ac) of farmland in northern Monterey County use recycled water for irrigation of food crops such as lettuce and strawberries. In 2020, Monterey One Water completed commissioning of a 19 ML/d (5 mgd) planned groundwater recharge facility.

Irrigation use in cities continues to be common in many states. West Basin Municipal Water District operates a 151 ML/d (40 mgd) recycled water facility that offsets the use of potable water equivalent to 80 000 households. The district pioneered industrial uses for recycled water with five different customer-specific recycled waters for large refineries and other industrial facilities in their district.

On-site water reuse is being developed to new levels in cities across the nation ranging from New York, Chicago, Atlanta, Portland, Seattle, San Francisco, and Los Angeles. These include examples of individual residential and commercial buildings, technology centers, university and college campuses generating recycled water on-site for irrigation, cooling towers, toilet flushing, wetlands, and steam and chiller plants (William J Worthen Foundation, 2018).

The City and County of San Francisco is a densely populated area where on-site reuse has become both economically and culturally favorable. San Francisco Public Utilities Commission (SFPUC) developed a nonpotable water program that was added to the local Health Code in 2012 allowing for the collection, treatment, and beneficial use of recycled water in individual buildings and at the district scale. In 2015, SFPUC formally began to require that all new construction projects of 23 234 m^2 (250 000 sq ft) or more install and operate an on-site nonpotable water system (Article 12C to the San Francisco Health Code)

(SFPUC, 2020). A few examples of such projects include the 181 Fremont building, which has an estimated 3630 m^2 (420 300 sq ft) of office spaces and 68 residential units. The 181 Fremont building has an on-site reuse system using a membrane bioreactor that supplies 4.9 mil. ML (1.3 mil. gal) annually of recycled water to offset potable use for toilet flushing (SFPUC, 2020). The Living Machine system at 525 Golden Gate Avenue uses a combination of treatment steps, including flow equalization and recirculation tanks for feeding a wetland system. The Living Machine also includes filtration to 5 micron and ultraviolet light disinfection to supply 0.2191 L/s (5000 gpd) of recycled water for toilet and urinal flushing.

Hassalo on 8th Wastewater Treatment & Reuse System is an urban development project in Portland, Oregon's, busy Lloyd neighborhood of office buildings and residential units. The Hassalo on 8th is a four-block on-site reuse system that treats up to 2 L/s (45 000 gpd) using trickling filters, constructed wetlands, and disinfection for a variety of beneficial uses such as toilet flushing, irrigation, cooling tower use, and excess recycled water pumped to dry wells for groundwater recharge.

The WaterHub at Emory University in Atlanta, Georgia, has an on-site reuse capacity to treat up to 17 L/s (400 000 gpd), or nearly 40% of the campus water needs, for steam, chiller plants, and residential hall toilet flushing. Similar to the Living Machine in San Francisco, the WaterHub uses reciprocating wetlands that consist of adjacent biological cells that contain plants and rock media. The treatment process uses anaerobic, anoxic, and aerobic zones that are filled with wastewater on a recurrent basis, similar to tides, that supports nitrification and denitrification. The system involved approximately 12 months of construction and acclimation, and although it's fully automated, a certified wastewater treatment operator provides oversight of the system process and coordinates mechanical and maintenance needs.

The Solaire Building in New York City is a multi-residential housing building with 35 582 m^2 (383 000 sq ft) that includes an on-site reuse system that treats 1 L/s (25 000 gpd) of recycled water for cooling and toilet flushing. Solaire Building's on-site system is the first such reuse system in New York City and was commissioned in 2013. The Solaire Building's system consists of a membrane bioreactor followed by ultraviolet light and ozone disinfection.

TEST YOUR KNOWLEDGE

1. **Historically, the largest use of potable water supply was for industrial uses.**
   - ☐ True
   - ☐ False
2. **Recycled water that comes in contact with the edible part of a crop will likely need to be disinfected before use.**
   - ☐ True
   - ☐ False
3. **Tertiary treatment may include:**
   a. Coarse bar screen
   b. Trickling filter
   c. Chlorine disinfection
   d. Grit removal
4. **Recycled water has a higher potential for biogrowth during storage and distribution than typical potable supplies.**
   - ☐ True
   - ☐ False

Potable Water Reuse

Planned potable reuse is when recycled water is used for human consumption and can be implemented directly or indirectly (Figure 7.5). Recycled water can be placed back into the environment (indirect potable reuse) before being reused for human consumption, or it can be introduced directly to a public water system (direct potable reuse). Generally, each form of potable reuse has different requirements based on how direct the pathway is from wastewater to drinking water. The shorter the time the product water spends between the advanced WRRF and the drinking water system, the more direct it is considered to be. More direct forms of potable reuse often require more treatment and monitoring to ensure public health protection.

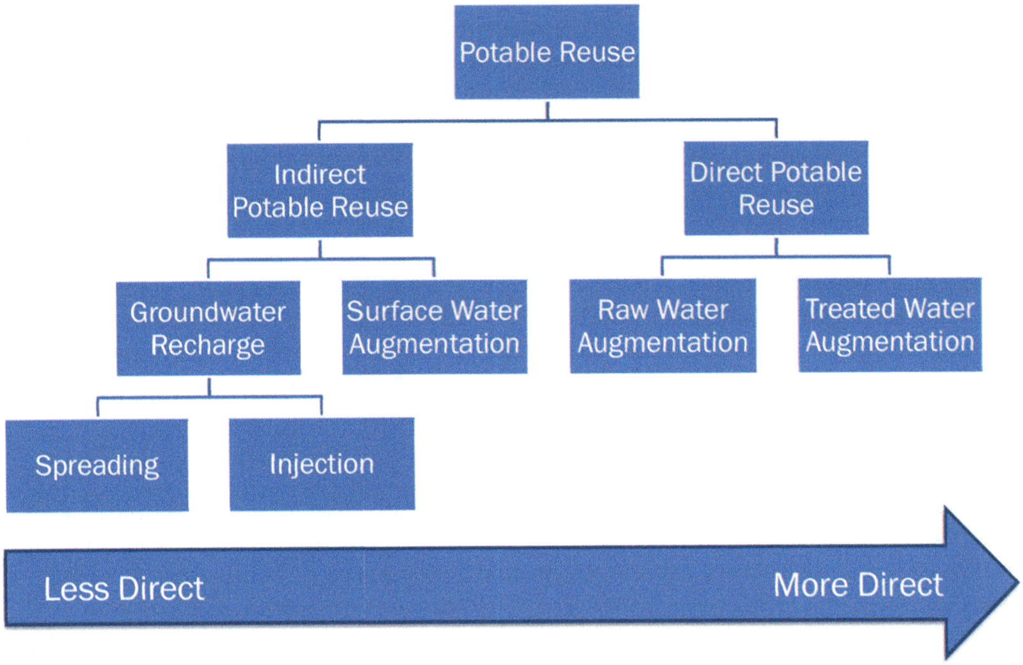

Figure 7.5 Forms of Potable Reuse

Indirect potable reuse is the planned placement of recycled water into an environmental storage system (such as an aquifer or reservoir) that is used as a source of domestic drinking water supply for a public water system.

POTABLE REUSE WITH AN ENVIRONMENTAL BUFFER

When recycled water is treated, placed back into the environment, and then used as a source of drinking water supply, it is considered indirect potable reuse.

In potable reuse, environmental buffers provide multiple benefits to protect public health. For example, water that is placed in an aquifer may benefit from soil aquifer treatment, and water that is placed in a reservoir benefits from the dilution and mixing provided in the natural system. In every case, the placement of recycled water into an environmental buffer provides valuable time referred to as "response time" for the water utility to react to protect public health in the event of a treatment failure.

Indirect potable reuse is often a convenient way to introduce recycled water into a drinking water system by using existing natural barriers that currently provide comfort to regulators and consumers alike. However, there are challenges associated with placing recycled water into the environment. These challenges include additional environmental permitting of discharges and risk of recontamination of the highly advanced treated water, among others.

GROUNDWATER RECHARGE

Through groundwater recharge (or replenishment), existing groundwater basins are augmented with reclaimed water as a way to increase the reliability of water supply. Other drivers for groundwater recharge are to reduce, stop, or even reverse declines of groundwater levels (Asano et al., 2007). For this type of potable reuse application, groundwater can be recharged via surface spreading or direct injection to the groundwater aquifer. In surface spreading, reclaimed water is applied to a spreading basin where it then infiltrates and percolates through the soil layer referred to as the "vadose (unsaturated) zone" before reaching the groundwater (aquifer). Surface spreading has the benefit of added treatment as the reclaimed water percolates through the vadose zone, referred to as "soil aquifer treatment." Direct injection (also known as subsurface application) consists of direct conveyance of reclaimed water into an aquifer through groundwater injection wells. In general, the reclaimed water is highly treated before injection, often to drinking water standards. Additionally, direct injection is particularly effective in creating a barrier of fresh water that prevents seawater intrusion into the aquifer of coastal communities. Figure 7.6 illustrates the flow of reclaimed water for these two types of groundwater recharge.

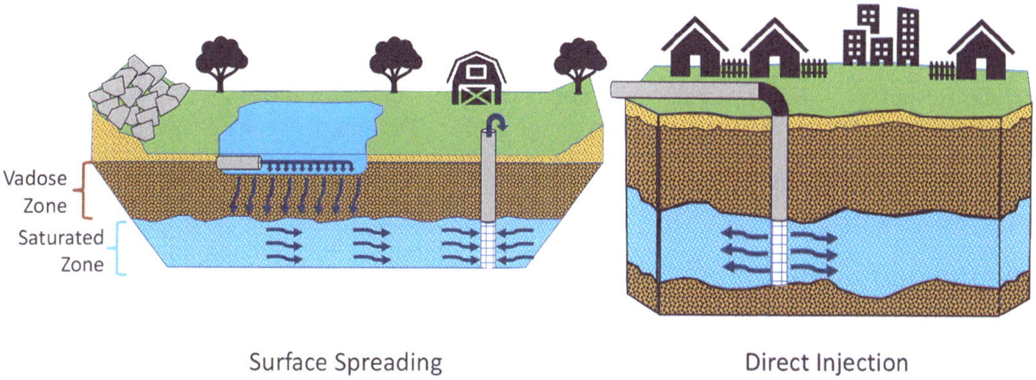

Figure 7.6 Methods to Achieve Planned Groundwater Recharge

SURFACE WATER AUGMENTATION

Reservoirs and rivers that serve as a source water for drinking water treatment plants act as a convenient way to introduce recycled water into a potable water system, and this is considered surface water augmentation.

For regions that rely on water imported from many hundreds of miles away as a water supply, augmenting local surface waters with recycled water increases the reliability of the water supply. Figure 7.7 illustrates the flow of water in a planned surface water augmentation project.

POTABLE REUSE WITHOUT AN ENVIRONMENTAL BUFFER

In the United States, most potable reuse projects to date include an environmental buffer such as a groundwater aquifer or a drinking water reservoir. However, in some cases, potable reuse occurs without an environmental buffer, and this is defined as direct potable reuse.

Environmental buffers provide valuable benefits such as natural attenuation mechanisms (through soil aquifer treatment, dilution, and mixing) and time to respond to treatment failures. Without the environmental buffer, direct potable reuse projects must account for the loss of valuable response time through other means such as a source control program, enhanced treatment, monitoring, and well-planned responses.

RAW WATER AUGMENTATION

Direct potable reuse can take many forms, but California has defined terms for two primary types. The first is raw water augmentation, which is defined as "the planned placement of recycled water into a system of pipelines or aqueducts that deliver raw water to a drinking water treatment plant that provides water to a public water system." Figure 7.8 illustrates the flow of water in a raw water augmentation project.

TREATED WATER AUGMENTATION

The second type of direct potable reuse identified by California is treated water augmentation, which is defined as "the planned placement of recycled water directly into a finished water distribution system of a public water system." Figure 7.9 illustrates the flow of water in a treated water augmentation project. The primary difference between raw water augmentation and treated water augmentation is the inclusion of the drinking water treatment plant in a raw water augmentation project.

Surface water augmentation, often referred to as "reservoir augmentation" or "reservoir water augmentation," is the planned placement of recycled water into a raw surface water reservoir used as a source of domestic drinking water supply for a public water system, or into a constructed system conveying water to such a reservoir.

Direct potable reuse is the planned use of recycled water that is introduced directly to a domestic drinking water supply for a public water system or into a constructed system conveying water to such a system, without the use of an environmental buffer.

Raw water augmentation is the planned placement of recycled water into a system of pipelines or aqueducts that deliver raw water to a drinking water treatment plant that provides water to a public water system.

Treated water augmentation is the planned placement of recycled water directly into a finished water distribution system of a public water system.

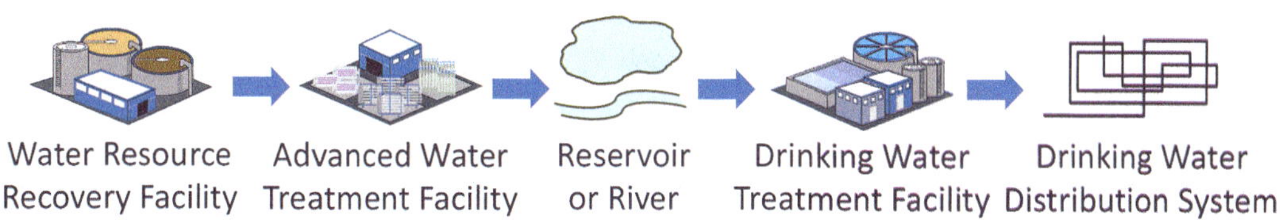

Figure 7.7 Surface Water Augmentation

Figure 7.8 Raw Water Augmentation

PUBLIC HEALTH AND SAFETY CONSIDERATIONS

Potable reuse poses new challenges for protecting public health and safety. In drinking water treatment, control of pathogenic organisms is accomplished by primary disinfection and filtration. Disinfection frameworks are based on the concept of concentration and time of exposure to a disinfectant such as chlorine, chlorine dioxide, ozone, or ultraviolet light. U.S. EPA provides a framework for achieving a certain log reduction of pathogens based on established dose response curves (U.S. EPA, 2006). Filtration processes also provide pathogen removal and use turbidity as a surrogate to claim a certain log inactivation. For nonpotable reuse, typically a chlorine contact time of 450 mg/L·min is required following a filtration process to meet disinfection requirements for 5-log inactivation of poliovirus. Using treated wastewater as a new source water for drinking water requires the public water system to consider the following:

- What additional precautions should be taken to ensure public health protection?
- Is there additional monitoring that needs to occur to ensure proper treatment?
- What happens in the event of a treatment failure or upset?

These questions are important to consider with wastewater as a source because wastewater will have a higher concentration of pathogenic microorganisms and chemical contaminants than a traditional source water. Pathogenic organisms, such as virus, are present at concentrations ranging from 10^5 to 10^9 in number/L, whereas protozoa and bacteria are present at up to 10^5 number/L (U.S. EPA, 2012). This higher loading of pathogenic microorganisms as well as a higher concentration of chemical contaminants requires greater control and consistency of treatment. To protect public health, one of the key aspects that should be considered is the reliability of the potable reuse system.

In potable reuse, "reliability" is defined as the ability of a potable reuse system to provide water that meets or exceeds the public health protection provided by existing drinking water supplies.

Reliability can be achieved by providing adequate treatment and controlling failures through failure prevention and failure response. The "4Rs" framework provides an excellent summary of the concepts that are imperative to achieving reliability in potable reuse (Pecson et al., 2015). Reliability is the first of the 4Rs. The remaining 3Rs—redundancy, robustness, and resilience—are described in the following sections.

REDUNDANCY

Redundancy is the use of measures beyond minimum requirements to ensure that treatment goals are more reliably met or that performance can be more reliably demonstrated.

Redundancy provides failure prevention. By providing treatment beyond the minimum requirements, the probability of producing water that does not meet those requirements decreases. For example, in drinking water applications, a drinking water treatment plant may be required to provide 4-log reduction of virus. In many cases, the drinking water treatment plant will not rely on one treatment process to meet this requirement, but will instead have two different types of treatment processes that provide the necessary degree of protection (a concept often referred to as "multiple barriers"). This way, if one of those processes is not functioning, the drinking water treatment plant can still produce water that is safe for public consumption. The concept of redundancy and multiple barriers allows for more operational flexibility because if one process is not performing well, it allows the plant to continue to operate and produce water.

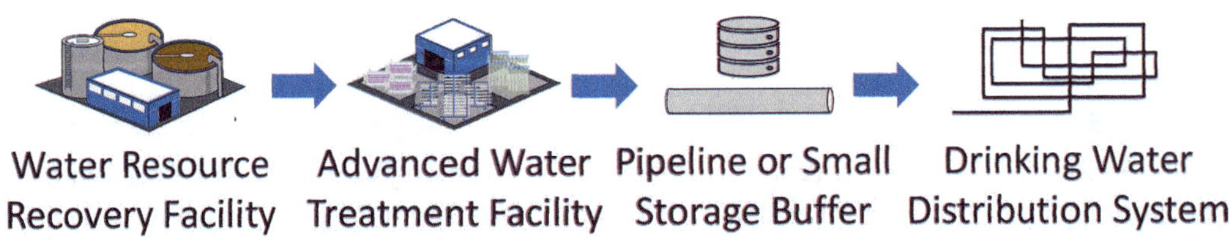

Figure 7.9 Treated Water Augmentation

ROBUSTNESS

Robustness also provides failure prevention. Having a robust system provides protection against many different contaminants. In potable reuse, the source water (wastewater) can contain a wide variety of pathogenic microorganisms and chemicals. A robust system will include a diversity of treatment mechanisms to provide protection against the wide variety of contaminants. For example, in a traditional full advanced treatment train, the treatment processes have different mechanisms that allow for removal of contaminants: membrane filtration removes contaminants physically through size exclusion, reverse osmosis removes contaminants physically through size exclusion and charge exclusion, and the ultraviolet/ advanced oxidation process (AOP) removes contaminants through ultraviolet light inactivation and oxidation. These different mechanisms provide protection against different-sized pathogens and chemicals with different sizes, structures, characteristics, and bonds. They also provide protection against contaminants that are still unknown.

Robustness is the ability of a potable reuse system to address a broad variety of contaminants and resist catastrophic failures.

RESILIENCE

Resilience provides failure response. A resilient potable reuse system is one that can respond to a failure when it occurs and prevent the undertreated water from being distributed. In potable reuse, this may include a variety of strategies including high-frequency monitoring with alarms, highly trained operators who can detect and respond to a failure quickly, the ability to divert water before it reaches the consumer, or some form of retention time to allow for more time to respond in the event of a failure. A resilient system is one that anticipates failures and has an appropriate response to them when they occur.

Resilience is the ability of a potable reuse system to successfully eliminate or minimize the effects of treatment process failure.

MONITORING AND REPORTING

Monitoring the water quality produced by a potable reuse system is crucial to ensuring that treatment processes are performing as expected and are producing water that is safe for public consumption. Monitoring requirements for potable reuse are similar to drinking water standards because, ultimately, water produced in a potable reuse system becomes drinking water.

Potable reuse projects monitor for pathogenic microorganisms and key chemical constituents. While requirements may vary from state to state, continuous monitoring of surrogate parameters used to demonstrate removal of pathogens is often required because of the large potential effect that pathogens can have on public health. Chemical monitoring tends to be less frequent (not continuous), but may be weekly, monthly, or quarterly depending on the type of chemical. Chemical monitoring follows drinking water standards, and potable reuse projects are likely to be required to meet primary and secondary maximum contaminant levels (MCLs), action levels, and any state requirements such as notification levels in California. They may also be required to monitor priority pollutants and other chemicals such as CECs.

RESPONSE TO OFF-SPEC WATER

In the event that water quality does not meet requirements or treatment processes do not perform as required, the water may be classified as "off-spec." Depending on the contaminant that is not meeting the regulatory requirement, there may be a chance to do a confirmatory sample to confirm that the water is, indeed, off-spec before taking additional action. However, once the water is confirmed to be off-spec, action should be taken immediately to remove the water as a source water to a potable system.

Off-spec water is water that has not met one or more regulatory requirements.

To remove off-spec water from the system, there are several options. The first is to divert the water before it reaches the potable system. For indirect potable reuse projects, diversion would occur before reaching the environmental buffer, if possible. For direct potable reuse, diversion would occur before the water reaching a drinking water treatment plant or before it enters the drinking water distribution system. Projects should plan for diversions and know what types of treatment excursions trigger a diversion. For diversions to be effective at protecting public health, monitoring must be frequent enough to be able to catch the pulse of off-spec water before it reaches its destination.

For indirect potable reuse projects, one option for responding to off-spec water is to decouple the drinking water system from the potable reuse system. For example, in a surface water augmentation project, if a drinking water treatment plant has an alternative water source, the drinking water treatment plant could stop pulling water from the reservoir that may have received off-spec water. This is an attractive option for a project that needs time to run additional water quality analysis to confirm whether the water is truly off-spec.

Finally, if off-spec water is detected and diverting or temporarily holding the water is not feasible, the advanced water treatment facility may be forced to shut down. Shutting down can be an expensive endeavor and, depending on regulatory requirements, may take coordination and approval from the regulatory agency to be able to restart the plant.

TEST YOUR KNOWLEDGE

1. **Direct injection is the most effective groundwater recharge method to create a barrier to prevent seawater intrusion to a confined aquifer.**
   - ☐ True
   - ☐ False
2. **The difference between indirect potable reuse and direct potable reuse is the use of an environmental buffer.**
   - ☐ True
   - ☐ False
3. **What are the "4Rs" in the potable reuse framework?**
   a. Reliability, redundancy, recalcitrant, resilience
   b. Reliability, redundancy, robustness, resilience
   c. Responsibility, redundancy, robustness, resilience
   d. Reliability, reinvestigate, robustness, resilience

Groundwater Recharge

BACKGROUND

Nearly one third of the potable water provided by public agencies in the United States is from groundwater sources (National Research Council [NRC], 2012). Because the natural replenishment of groundwater occurs slowly, groundwater supplies are often used at a rate greater than can be replenished naturally. This lowers the groundwater levels over time, and leads to groundwater depletion. To increase groundwater supply and reliability, agencies can artificially recharge groundwater basins. This practice has been increasing over the last several decades. This section answers the following questions:

- How can planned groundwater recharge be achieved?
- What are the benefits of groundwater recharge?
- What is the current regulatory state for groundwater recharge?
- What are examples of existing groundwater recharge projects?

Unplanned groundwater recharge may occur through the land discharge of wastewater where the disposal rates exceed the rates of evaporation and plant uptake (Asano et al., 2007). Unplanned groundwater recharge may also occur from leaking wastewater collection systems and unlined storage basins. Unplanned groundwater recharge may impair groundwater quality and should be avoided.

PLANNED GROUNDWATER RECHARGE

Groundwater recharge using reclaimed water has been practiced in the United States Since the 1960s (Asano et al., 2007). As identified by Bouwer (1978) and Asano and Wasserman (1980), the purposes of this form of planned reuse include the following:

- To reduce, stop, or even reverse the decline of groundwater levels;
- To protect underground fresh water from saltwater intrusion for coastal aquifers;
- To store surface water, runoff, and reclaimed water for future use; and
- To mitigate potential problems of land subsidence.

There are different ways to achieve planned groundwater recharge for water reuse (Figure 7.10), as further described in the following subsections.

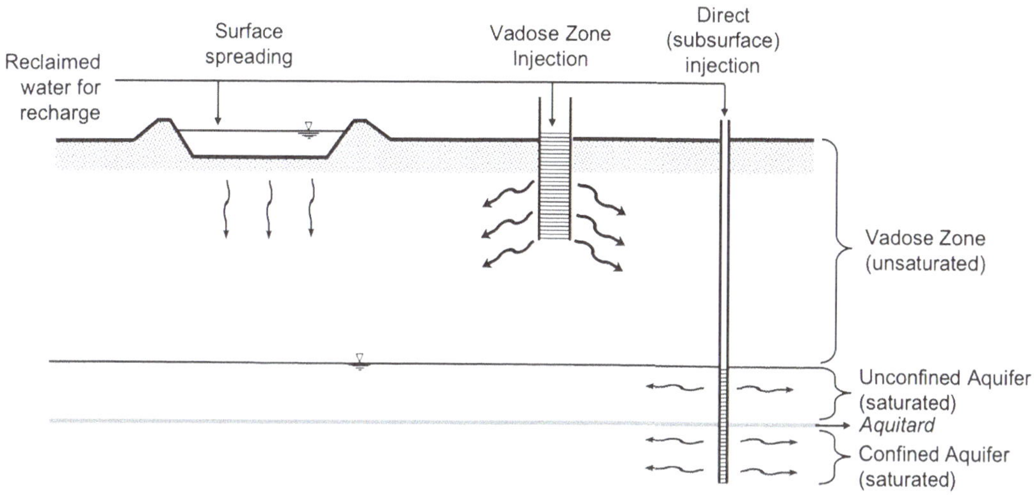

Figure 7.10 Planned Groundwater Recharge Methods

SURFACE SPREADING

Surface spreading is the simplest, oldest, and most widely applied method of groundwater recharge (Todd, 1980). In this form, reclaimed water is applied to a large spreading basin allowing the water to infiltrate and percolate through the vadose (unsaturated zone) before reaching the groundwater aquifer (Figure 7.11). Surface spreading basins require little maintenance and are thus a favored groundwater recharge method. Surface spreading also provides added treatment benefits as the reclaimed water percolates through the vadose zone before reaching the aquifer. However, more land is required to perform surface spreading in comparison to direct injection. The land must also have suitable soil characteristics. In addition, surface spreading is not effective in creating a seawater intrusion barrier for coastal aquifers.

VADOSE ZONE INJECTION

A vadose zone is a dry well installed in the unsaturated zone above the water table. Vadose zones are primarily used where aquifers are deep and construction of direct injection wells is cost-prohibitive. The mechanism of injection is to recharge water through large-diameter boreholes allowing the water to enter the vadose zone. Although infiltration rates are relatively high (similar to direct injection rates), vadose zone wells cannot be backwashed and a severely clogged well may become permanently out of commission. As such, reliable pretreatment is important to maintain the performance of vadose zone wells.

DIRECT INJECTION

Direct injection, also known as "subsurface injection," is achieved when reclaimed water is conveyed directly into an aquifer. This type of groundwater recharge is considered higher maintenance and more energy intensive because it requires injection wells. In addition, direct injection typically requires more

Rio Hondo Spreading Ground

San Gabriel Spreading Ground

Figure 7.11 Aerial View of Surface Spreading Grounds in California (U.S. EPA, 2009)

highly treated reclaimed water because it does not benefit from the additional treatment provided by percolation through the vadose zone that occurs for surface spreading projects. However, direct injection is more effective in creating a freshwater barrier against saltwater intrusion for coastal aquifers.

SELECTION OF GROUNDWATER RECHARGE METHOD

There are many site-specific variables that affect the design and selection of the most appropriate groundwater recharge method. If a confined (saturated) aquifer is being considered, then direct injection is the only feasible method. For unconfined (unsaturated) aquifers, the depth of the aquifer will be key to selecting the most appropriate method. In general, direct injection becomes more expensive as the depth of the groundwater increases. Land price and availability are also important considerations. Other considerations include changes to groundwater levels, such as groundwater mounding. Figure 7.12 provides a sample decision tree for the selection of a groundwater recharge method.

OVERVIEW OF WATER QUALITY OBJECTIVES

When practicing groundwater recharge for potable purposes, it is essential to improve water quality before surface spreading or direct injection. Surface spreading often requires at least disinfected wastewater treated to secondary or tertiary standards before land application. In contrast, direct injection will typically require more treatment, such as reverse osmosis and advanced oxidation before injection to eliminate most water quality concerns (Asano et al., 2007). Water quality concerns for groundwater recharge that must be addressed for potable reuse application include:

- Microbiological (pathogenic) quality
- Total mineral content (TDS)
- Inorganic constituents prone to precipitation (such as phosphates)
- Toxic constituents (such as heavy metals)
- Nutrients
- Trace organic constituents

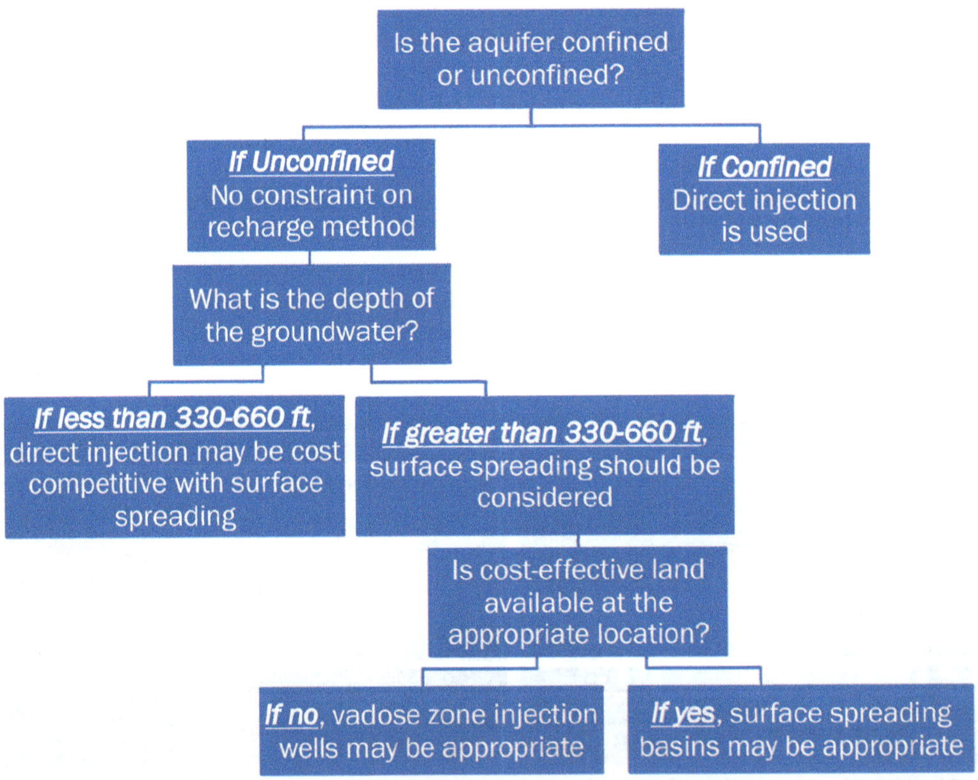

Figure 7.12 Sample Decision Tree for Selection of Groundwater Recharge Method (adapted from U.S. EPA, 2012)

These requirements can change with time such that an exhaustive description of all the requirements will not be covered in detail in this chapter. Microbiological quality concerns refer to the potential presence of pathogens that may be present in the recycled water because requirements for groundwater recharge pretreatment vary considerably and are often state-specific and project-dependent; in some cases, non-disinfected water may be allowed for recharge. When recharge water comes in contact with soil and rocks in the aquifer, metals and radionuclides may dissolve into the groundwater; therefore, it is important to adjust water quality of recycled water before discharge. Trace organic constituents present a broad range of organic compounds of potential concern, including potential disinfection byproducts such as trihalomethanes and haloacetic acids. In many cases, wells used for injection are classified as Class V by the U.S. EPA, and some states such as California and Florida require that the injected water must meet primary and secondary drinking water standards before injection. Typical water quality parameters used for groundwater recharge regulations include total nitrogen, nitrate, nitrite, TOC, pH, iron, total coliform bacteria, and others.

OVERVIEW OF REGULATORY GUIDELINES

FEDERAL GUIDELINES AND REGULATIONS FOR GROUNDWATER RECHARGE

There are currently no federal regulations directly governing water reuse practices in the United States (U.S. EPA, 2012). As such, states maintain the primary regulatory authority for practice of water reuse and have adopted rules and developed programs for water reuse to ensure that water reuse projects are designed, constructed, and operated in a manner protective of the environment, other beneficial uses, and human health (U.S. EPA, 2012).

As part of the Safe Drinking Water Act of 1974, U.S. EPA also established a national program to prevent groundwater recharge activities that endanger drinking water sources. As a result, U.S. EPA has promogulated a series of underground injection control (UIC) regulations to protect underground sources of drinking water. Even though states often have their own regulations for groundwater recharge, they do not supersede federal regulations that prohibit endangerment of underground sources of drinking water. Specifically, the promogulated UIC regulation states:

> No owner or operator shall construct, operate, maintain, convert, plug, abandon, or conduct any other injection activity in a manner that allows the movement of fluid containing any contaminant into underground sources of drinking water, if the presence of that contaminant may cause a violation of any primary drinking water regulation under 40 CFR part 142 or may otherwise adversely affect the health of persons. (Underground Injection Control Program, 1983)

Drinking water sources such as groundwater aquifers are a limited and precious resource that may be contaminated by municipal and/or industrial waste if underground injection is not executed properly. It must be ensured that groundwater recharge projects will not transmit contaminants to our drinking water aquifers that will make them unsafe to use.

STATE AND LOCAL GUIDELINES FOR RECYCLED WATER

Local regulations for recycled water vary considerably by state and region. For the purposes of this chapter, a brief discussion is provided by U.S. EPA regional divisions (Table 7.3). Regulatory information provided in this chapter is not exhaustive.

OVERVIEW OF RECLAIMED WATER PRETREATMENT

Before recharge, municipal wastewater typically receives a minimum of conventional secondary treatment, but may also receive additional filtration (e.g., dual-media filter) followed by disinfection. Depending on the regulatory framework (state and federal) as well as the utility' infrastructure, further treatment may be necessary. Further treatment may include membrane filtration, reverse osmosis, and ultraviolet disinfection (which may be used as an AOP). In California, for example, it is necessary to include what's referred as "full advanced treatment" for subsurface injection applications. Full advanced treatment consists of reverse osmosis and advanced oxidation (Advanced Treatment Criteria, 2014). Reverse osmosis provides removal of a broad range of dissolved organic and inorganic constituents as well as pathogenic organisms, such as virus, protozoa, and bacteria. The AOP must demonstrate sufficient oxidation of select indicator compounds, often used by combining ultraviolet processes with hydrogen peroxide (i.e., ultraviolet/H_2O_2) or hypochlorous acid (i.e., ultraviolet/HOCl). Membrane filtration is often included as pretreatment to the reverse osmosis system as well as to provide additional pathogen removal for protozoa and bacteria. Chapter 6 further discusses the use of membranes for water reuse. For recharge occurring via surface spreading, the soil itself provides additional treatment to purify water before reaching the groundwater (Trussell et al., 2018).

Table 7.3 Summary of Recycled Water State Regulations by U.S. EPA Region (adapted from U.S. EPA, 2012)

Region	Summary
Northeast (Regions 1 and 2)	■ Regulatory requirements and guidelines have not been implemented in most states of this region. ■ *Massachusetts*: Promogulated reuse regulations in 2009. Requires TOC limit for discharge. ■ *New Hampshire*: No regulations, but has developed identified design criteria for reuse of reclaimed water. Projects approved on case-by-case basis. ■ *Rhode Island*: Developed guidelines in 2007 for four categories: (1) restricted irrigation, (2) unrestricted irrigation, (3) non-contact cooling water, and (4) agricultural reuse for non-food crops. Established water quality criteria, signage, and setback distances for these categories. ■ *Vermont*: Adopted rules for indirect discharge. Requires that land-based discharge be considered before approval of surface water discharge. ■ *New York*: Isolated uses of reclaimed water for cooling. Roadmap developed to promote water reuse for other applications. ■ *New Jersey*: *Technical Manual for Reclaimed Water for Beneficial Reuse* codified in 2009. Establishes requirements for feasibility studies, developing and implementing reuse for two areas—public access and restricted access. ■ *Connecticut and Maine*: Installations approved on a case-by-case basis.
Mid-Atlantic (Region 3)	■ *Delaware*: Permits administered by the Delaware Division of Water, primarily agricultural irrigation. State regulations require advanced treatment for unrestricted access reuse. Reuse intended for irrigation of crops for human consumption is not allowed without further processing. ■ *District of Columbia*: No regulations or guidelines, but considers projects on a case-by-case basis. ■ *Pennsylvania and Maryland*: Both have guidelines for water reuse. Maryland 2010 guidelines require zero nitrogen addition to groundwater for new permits. ■ *Virginia*: Adopted new regulations for water reuse in 2008. Allows on-site treatment and reuse of reclaimed water. Allows indirect potable reuse. ■ *West Virginia*: No regulatory requirements and guidelines have been implemented at the time of publication.
Southeast (Region 4)	■ *Alabama and Georgia*: Each have guidelines governing various aspects of reuse. ■ *Kentucky*: Does not have regulations or guidelines for reuse. ■ *Mississippi*: Has regulations that cover potential for restricted urban reuse, agricultural reuse for non-food crops, and industrial reuse. ■ *South Carolina*: Has regulations governing reuse. Wastewater facilities that apply to discharge to surface waters must conduct an alternatives analysis to demonstrate that water reuse is not economically or technologically feasible. ■ *Tennessee*: Allows reclaimed water to be distributed for land application reuse by industrial customers, commercial developments, golf courses, recreational areas, residential developments, and other nonpotable uses. ■ *Florida*: Has one of the more mature water reuse programs in the nation. Forty-nine percent of treated wastewater is reclaimed and reused (Florida Department of Environmental Protection, 2012). Laws enacted in 2008 prohibit ocean discharge of treated wastewater by 2025, except as a backup to a reuse system. In most instances, reuse is the addition of highly treated reclaimed water to a surface water body or aquifer for indirect potable reuse. ■ *North Carolina*: Has established reuse regulations that classify systems as conjunctive or non-conjunctive. A "conjunctive reclaimed water system" refers to a system in which beneficial use of reclaimed water is an option and reuse is not necessary to meet the wastewater disposal needs of the facility. A "non-conjunctive reclaimed water system" typically has evolved from land disposal system permits and refers to a system in which the reclaimed water utilization option is required (or dedicated) to meet the wastewater disposal needs of the facility and no other disposal or utilization options are available.
Midwest and Great Lakes (Region 5 and 7)	■ The Midwest and Great Lakes states are beginning to develop regulations and guidelines for water reuse. ■ *Illinois, Indiana, Iowa, Michigan, Missouri, Nebraska*: Have water reuse regulations. ■ *Kansas, Minnesota, Ohio*: Have water reuse guidelines. ■ *Wisconsin*: No regulatory requirements and guidelines have been implemented at the time of publication.

Table 7.3 Summary of Recycled Water State Regulations by U.S. EPA Region (adapted from U.S. EPA, 2012) (*Continued*)

South Central (Region 6)	■ *Louisiana*: No regulatory requirements and guidelines have been implemented at the time of publication. ■ *Arkansas*: Has adopted regulations for land disposal with a provision for irrigation of forage and non-contact crops using reclaimed water. ■ *New Mexico*: Provides guidance that identifies four different qualities of reclaimed water, Class 1A being highest quality for unrestricted urban use. Class 1A uses treatment that removes colloidal material and color that can interfere with disinfection. Class 1B, 2, and 3 are based on secondary treatment processes. Spray irrigation of food crops is not allowed, although surface irrigation with Class 1B or 1A is allowed without contact with edible portions of crops. ■ *Oklahoma*: Has proposed and adopted new water reuse regulations. Establishes four categories of reclaimed water (Categories 2 through 5). Each category has a different level of treatment and permitted uses. Category 2—unrestricted access irrigation exclude application on food crops that could be eaten unprocessed and on processed food crops within 30 days of harvest. Category 3—reclaimed water, the regulations also exclude use on athletic fields with potential for skin-to-ground contact. ■ *Texas*: First regulations created in 1997. Reclaimed water use is regulated by the Texas Commission on Environmental Quality. Must meet limits for *Enterococci* based on type of reclaimed water (Type 1 limit is 4 CFU/100 mL for unrestricted use; Type 2 limit is 35 CFU/100 mL for restricted use).
Mountains and Plains (Region 8)	■ *Colorado*: Regulations for reuse identify three qualities of reclaimed water: Classes 1, 2, and 3, with Class 3 being the highest water quality and requiring secondary treatment filtration and disinfection for use in unrestricted urban applications. ■ *Montana*: Established gray water rules in 2007 and updated in 2009. ■ *North Dakota*: Has guidance on water reuse for a number of categories, including urban, agricultural, industrial, and groundwater recharge. ■ *South Dakota*: Has guidelines on reuse of reclaimed water for irrigation of food and non-food crops. ■ *Utah*: Allows on-site use of reclaimed water inside a treatment plant boundary for landscape irrigation, washdown, and chlorination system feed water. Also allows alternate disposal methods of land application and reuse of either Type I (potential human contact) or Type II (human contact unlikely). ■ *Wyoming*: No regulatory requirements and guidelines have been implemented at the time of publication. Agriculture irrigation, for both food and non-food crops, is addressed on a case-by-case basis as part of construction permitting regulations.
Pacific Southwest (Region 9)	■ *Arizona*: Reclaimed water regulations have evolved since initial adoption in 1972. Current regulations address reclaimed water permitting, requirements for reclaimed water conveyances, reclaimed water quality standards, and allowable end uses. Arizona established five qualities of reclaimed water from A+ to C, with A+ being the highest quality. Class A+ reclaimed water in Arizona receives secondary treatment followed by filtration, disinfection, and nitrogen reduction to less than 10 mg/L total nitrogen. ■ *California*: Current regulations in California related to water reuse are complex and have been in a state of continual flux as water districts and utilities look to expand their use of reclaimed water. Criteria for water recycling was first identified in 1969 by the Porter-Cologne Water Quality Control Act. Recently, the California Division of Drinking Water has developed potable reuse regulations for Groundwater Replenishment in 2014 (surface and subsurface application) and Surface Water Augmentation in 2018. These regulations outline a number of requirements such as treatment processes, TOC limits, minimum retention times, and demonstration of log removal for regulated pathogens (e.g., *Cryptosporidium*, *Giardia*, and enteric virus). By 2023, the division of drinking water is required to adopt uniform water recycling criteria for direct potable reuse through raw water augmentation by the end of 2023. ■ *Hawaii*: All water reuse projects are subject to review and approval by the Hawaii State Department of Health Wastewater Branch. Guidelines for reclaimed water were released in 1993. The guidelines define three classes of reclaimed water, R-1, R-2, and R-3. R-1 is the highest quality (treated effluent that has undergone filtration and disinfection) and can be used for spray irrigation without restriction on use. R-2 is disinfected secondary effluent and uses are subjected to specific restriction and controls. R-3 is the lowest quality of reclaimed water (undisinfected secondary) whose uses are severely limited. ■ *Nevada*: Has both guidelines and regulations. The guidelines describe criteria to be required in engineering plans for irrigation reuse projects.

Table 7.3 Summary of Recycled Water State Regulations by U.S. EPA Region (adapted from U.S. EPA, 2012) (*Continued*)

Pacific Northwest (Region 10)	■ *Alaska*: No regulatory requirements and guidelines have been implemented at the time of publication. Alaska is exploring decentralized water and wastewater treatment for rural homes. ■ *Idaho*: Has both regulations and guidelines for reuse of municipal and industrial wastewater. U.S. EPA is responsible for issuing and enforcing all NPDES permits in Idaho. ■ *Oregon*: Has requirements for use of reclaimed water for beneficial purposes. Facilities are required to manage and operate reclaimed water projects under a water reuse management plan. ■ *Washington*: In 1992, state legislature passed the Reclaimed Water Act, which encourages development and use of reclaimed water. The state recognizes reclaimed water as an important mechanism to reduce discharge of treated wastewater into Puget Sound.

MONITORING AND OPERATION OF WATER RESOURCE RECOVERY FACILITIES

OPERATION AND MAINTENANCE OF SURFACE SPREADING GROUNDS

For surface spreading, to maximize infiltration, it is pivotal to keep the surface ground wetted and to keep soil unclogged. The quality of the reclaimed water should not inhibit infiltration. Typically, spreading grounds are subdivided into smaller basins that can be alternately filled or dried to allow for maintenance of basins not in use. The cycle length of spreading basins is dependent on soil conditions, development of a clogging layer, and distance to the groundwater table. Algae accumulation can clog the bottom of the basins and reduce infiltration rates. Algae can be prevented by upstream nutrient removal or by reducing detention time of the reclaimed water in the basins. Summer months can be particularly challenging for controlling algae because growth rates increase under higher temperature and higher solar intensity. Maintenance of basins involves cleaning the bottom of the basins by scraping the top layer of the soil. The basin is often dried for maintenance, which allows for aeration and drying of the clogged layer. This procedure helps improve infiltration rates for the next wetting cycle.

Intrinsic means to belong to naturally, and a tracer is a distinct parameter that may be used to track something. In this case, an intrinsic tracer is some parameter that is naturally occurring in groundwater that may be distinctly and sufficiently detected downstream.

GROUNDWATER TRAVEL TIME

One consideration for groundwater recharge projects is managing the travel time of reclaimed water before it is removed from the aquifer again for use at the downstream well. Travel times can range from days to years, depending on a number of factors (e.g., groundwater flow, recharge flow, soil characteristics, distance from recharge location to downstream well). Figure 7.13 provides a schematic representing travel time from groundwater recharge via surface spreading to recovery from a downstream well. Certain jurisdictions require minimum groundwater travel time as a method to provide pathogenic crediting. Because travel times are project and site-specific, the verification of travel times is often required. These can be achieved through "tracer tests" using non-toxic tracers as well as groundwater numerical modeling. For tracer tests, tracers can be "added" to the receiving water or be "intrinsic" to recharge water. An added tracer is intentionally added to the recharge water for the purpose of measuring it once combined with the receiving groundwater. An intrinsic tracer is present in the recharge water at levels different than the receiving groundwater. In either case, the tracers must be non-reactive, have measurable characteristics that are different than the receiving groundwater, and be detectable in the downstream gradient relative to recharge location to determine travel time. Tracer tests are typically performed once a project first begins recharging reclaimed water to verify that earlier groundwater numerical modeling results were accurate. Tracer tests must be approved by the local regulatory body to obtain pathogenic credits (if applicable). In California, for example, projects are given 1 log of virus reduction credit for 1 month of demonstrated groundwater travel time verified by a tracer test using an added tracer (Pathogenic Microorganism Control, 2014). Certain jurisdictions allow projects to rely on numerical modeling alone to demonstrate groundwater travel time. In this case, only partial pathogen credit is typically granted relative to modeled travel time. Recycled water groundwater recharge projects in California must have a minimum groundwater travel time of 2 months.

Regulatory bodies will often use "pathogenic credits" as a measure of treatment efficacy. This is where a certain treatment process (i.e., soil aquifer treatment, ultraviolet) is found to reliably achieve at least a certain log removal of a specific pathogen (i.e., *Cryptosporidium*, *Giardia*, virus). Regulatory bodies will then specify the log reduction (or pathogenic) credits required for specific recycled water uses (i.e., groundwater recharge, potable reuse) that treatment facilities must meet using their desired treatment train.

One log of virus reduction credit means that a certain process has been found to reliably achieve at least 1-log (90%) reduction of virus and, therefore, was awarded a 1-log reduction credit by a regulatory body. Log reduction is calculated using the following formula: log reduction = $-\log_{10}\left(\frac{\text{Contaminant after treatment}}{\text{Contaminant before treatment}}\right)$. One log is 90% reduction, 2 log is 99%, 3 log is 99.9% reduction, and so on.

WATER QUALITY MONITORING

Recharge projects are strictly regulated per federal and state regulations. Recharge projects are subject to complex water quality monitoring to ensure protection of public health and the environment. In addition, water reclamation plants practicing groundwater recharge must ensure performance through various

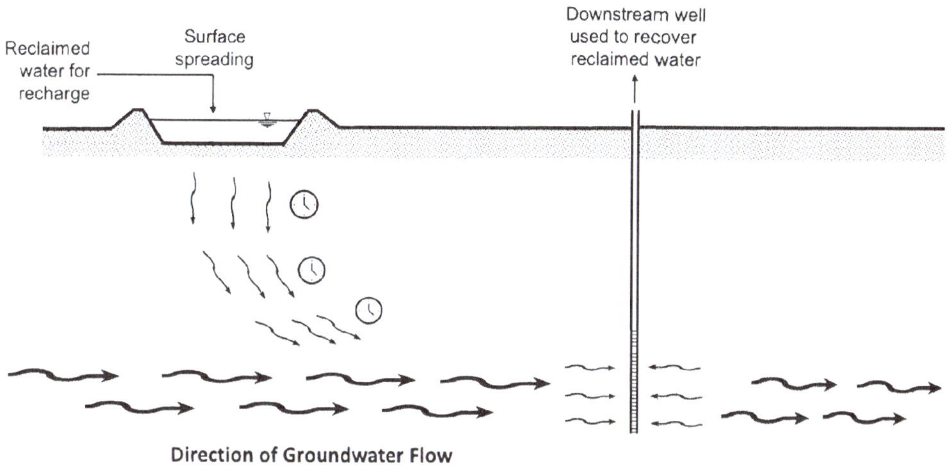

Figure 7.13 Groundwater Travel Time From Point of Recharge to Downstream Well

in-plant analyzers and laboratory tests (e.g., turbidity, TOC, chlorine residual, pH). In general, direct injection requires more stringent water quality monitoring than surface spreading applications because it is understood that percolation of surface spreading provides additional treatment.

On a less frequent basis (monthly to quarterly), groundwater recharge projects are subject to more extensive monitoring, including primary MCLs and action levels, and in California, notification levels as well. Secondary MCLs are often monitored annually; priority toxic pollutants and other CECs are monitored quarterly to annually.

PATHOGENIC CONTROL MONITORING

Like other potable reuse applications, demonstrating pathogen removal for groundwater recharge projects is also essential. Although requirements for pathogen reduction may vary from state to state, most projects must provide reduction of enteric viruses, *Giardia*, and *Cryptosporidium* (V/G/C), and sometimes bacteria. In California, for example, groundwater recharge projects (surface or subsurface applications) must provide sufficient treatment to achieve a cumulative log reduction of at least 12-log enteric virus reduction, 10-log *Giardia* reduction, and 10-log *Cryptosporidium* reduction. Specific to California, treatment must be provided by at least three separate treatment trains, where each process may be credited with no more 6-log reduction and no less than 1-log reduction (Pathogenic Microorganism Control, 2014). This is done to provide treatment redundancy and to reduce effects of treatment failures. To be granted log-removal credits, the project must demonstrate ongoing monitoring using the pathogenic microorganism of concern or a microbial, chemical, or physical surrogate parameter(s) that verifies the performance of each treatment process's ability to achieve its credited log reduction. In addition, retention time in the groundwater can also provide pathogen log-removal credit. Up to 6 log of virus log can be granted for a groundwater retention time of 6 months. If retention times exceeds 6 months, the project will be credited with 10-log *Giardia* cyst reduction and 10-log *Cryptosporidium* oocyst reduction.

Surrogate parameter means a microbial, chemical, or physical property that has a proven direct correlation to the pathogenic microorganism or compound of concern. Surrogate parameters are typically easier and cheaper to evaluate and can be used to monitor removal of pathogens/CECs or indicate treatment failure.

COMPLIANCE REPORTING

Routine monitoring and sampling that takes place during operation of a groundwater recharge project eventually makes its way into a routine compliance report that is reported in some form to a regulatory agency. Such reporting takes place to evaluate compliance with effluent limitations or performance goals as established in the project's permit.

In California, groundwater recharge projects must provide annual reports to notify public water systems and drinking water well owners with downgradient sources potentially affected by the groundwater recharge plant. The report shall include:

In this instance, "downgradient sources" refers to the drinking water wells (sources) that the recharged groundwater is flowing toward.

1. A summary of the plant's compliance status with the monitoring requirements;
2. A list of any violations of the project's permit and groundwater recharge regulations along with a summary of corrective action;

Diluent water in this scenario refers to an additional water source (i.e., lake, reservoir) used to dilute or reduce the contribution of recycled municipal wastewater in a groundwater replenishment project.

3. Any detections of monitored chemicals or contaminants and any observed trends in the monitoring wells and diluent water supplies;
4. Information pertaining to the vertical and horizontal migration of the recharge water plume;
5. A description of any changes in the operation of any unit processes or facilities;
6. A description of any anticipated changes, along with an evaluation of the expected effect of the changes on subsequent unit processes;
7. The estimated quantity and quality of the recycled municipal wastewater and diluent water to be applied for the next calendar year;
8. A summary of the measures taken to comply with wastewater source control and the effectiveness of implementing these measures; and
9. Increases in diluent water relative to recharge wastewater effluent during the previous calendar year and increases anticipated for the next calendar year.

Compliance reporting is specific to governing agencies and facilities. Thus, it is important to be familiar with all aspects of the permit to ensure all of the proper reporting is completed at the required frequency.

EXAMPLES OF GROUNDWATER RECHARGE PROJECTS

There are several groundwater recharge projects throughout the world that are planned or are currently in operation. Table 7.4 provides an overview of some notable projects, the treatment technologies used, and the ultimate use of the recharged water.

TEST YOUR KNOWLEDGE

1. **Algae can cause reduced infiltration rates in spreading basins.**
   - ☐ True
   - ☐ False
2. **Full advanced treatment consists of reverse osmosis and advanced oxidation.**
   - ☐ True
   - ☐ False
3. **When should direct injection be a considered method for groundwater recharge?**
   a. The aquifer is unconfined
   b. The depth of the aquifer is less than 300 ft
   c. There is available land to perform surface spreading
   d. The depth of the aquifer is greater than 800 ft
4. **Which is the most accurate method to estimate the travel time of reclaimed water recharged to the groundwater?**
   a. Added tracer
   b. Intrinsic tracer
   c. Numerical modeling
   d. Darcy's law

Surface Water Augmentation

BACKGROUND

Surface water augmentation, often referred to as "reservoir augmentation" or "reservoir water augmentation," is the planned placement of recycled water into a raw surface water reservoir used as a source of domestic drinking water supply for a public water system, or into a constructed system conveying water to such a reservoir.

De facto is a Latin phrase that translates to "in fact." This term is used to describe things that occur in reality, but are not be recognized officially.

Although there are still only a few planned surface water augmentation projects, de facto surface water augmentation has been occurring for years. When a WRRF discharges to a river, that water is carried downstream and is often used as a source of supply for a drinking water treatment facility. The National Research Council estimated in 2012 that of the 1.4 ML/s (32 bil. gpd) of wastewater effluent discharged, 876 424 L/s (20 bil. gpd) were discharged to surface waters (NRC, 2012). Although most de facto reuse scenarios produce water that meets drinking water regulations, the benefits of planned surface water augmentation include better treatment and monitoring to ensure public health protection.

Table 7.4 Examples of Groundwater Recharge Projects

Project Location	Capacity	Recharge Method	Description of Treatment	Ultimate use of Recharged Water
Orange County Water District's Groundwater Replenishment System[1] (California, USA)	100 mgd (130 mgd after final expansion)	Surface spreading and direct injection	MF, RO, UV/H_2O_2	Prevent seawater intrusion and augment potable water; operation began in 2008.
West Basin Municipal Water District[1] (California, USA)	40 mgd	Direct injection	O_3, MF, RO, UV/H_2O_2	Potable water, other uses of purified water, including irrigation, industrial cooling, boiler feed water; operation began in 1995.
Inland Empire Utilities Agency[1] (California, USA)	7 mgd dedicated to recharge	Surface spreading	Primary, secondary, and tertiary treatment for nutrient removal, soil-aquifer treatment	Potable water. Recharge began in 1999, increasing to current recharge capacity.
City of Tucson[2] (Arizona, USA)	5.8 mgd	Surface spreading	Tertiary effluent (dual-sand filtration), soil-aquifer treatment	Nonpotable reclaimed water (primarily irrigation); operation began in 1989.
St-André Aquifer recharge[2] (Flanders, Belgium)	1.5 mgd	Surface spreading	Ultrafiltration, RO, UV disinfection, soil aquifer treatment	Potable water; operation began in 2002.

MF = membrane filtration, mgd = million gallons per day, O_3 = ozone, RO = reverse osmosis, UV/H_2O_2 = UV light AOP using hydrogen peroxide. Project information obtained from utilities' websites.
[1]U.S. EPA, 2017, [2]U.S. EPA, 2012

OVERVIEW OF REGULATORY GUIDELINES

FEDERAL GUIDELINES FOR RECYCLED WATER

National Pollutant Discharge Elimination System

Discharges of treated wastewater to a reservoir require an NPDES permit because reservoirs are considered "waters of the United States." The NPDES permit program falls under the Clean Water Act and addresses pollution through the regulation of point sources. An NPDES permit dictates the requirements for water quality, monitoring, and reporting as well as other provisions that ensure protection of both the ecosystem's water quality as well as human health.

Priority Pollutants List

The Priority Pollutants List was developed in 1977 and contains a list of compounds that are regulated by U.S. EPA (U.S. EPA, 1977). The goal of the Priority Pollutants List (and its predecessor the Toxic Pollutants List) was to identify and regulate toxics in waterways. Water produced by a surface water augmentation project will need to consider the concentrations of the compounds on the Priority Pollutants List and ensure that when the water is discharged to the environment, that the concentrations of these compounds do not exceed effluent limitations set in the NPDES permit.

STATE AND LOCAL GUIDELINES FOR RECYCLED WATER

In 2018, California promulgated regulations for surface water augmentation that specify requirements for treatment, monitoring, reservoir dilution, and reservoir retention time as well as requirements for planning between the agencies involved. Most states do not have specific regulations for surface water augmentation, but may permit such a project on a case-by-case basis.

BASIN PLAN OBJECTIVES

For California, water that is discharged to a surface water may also be subject to the water quality objectives of the area's Basin Plan. Basin Plans establish (a) beneficial uses within the region, (b) water quality objectives necessary to protect those uses, (c) how protections are implemented, and (d) monitoring that is required to ensure water quality standards are met.

OVERVIEW OF REQUIRED TREATMENT PROCESSES

TYPICAL TREATMENT TRAIN CONFIGURATIONS

Defining a "typical" treatment train for surface water augmentation is challenging because only a few projects exist in the United States and only California has developed specific regulations and requirements for surface water augmentation. The primary goals of a system treating municipal wastewater to drinking water standards are to reduce the concentrations of pathogens and chemicals to an acceptable level. There are multiple possible treatment trains that could achieve these goals, but one of the standard treatment trains considered is membrane filtration followed by reverse osmosis and an ultraviolet light/AOP. This treatment train can meet the requirements of full advanced treatment as defined by the California recycled water regulations. In California, full advanced treatment of an oxidized wastewater is required for surface water augmentation projects. Table 7.5 provides other examples of treatment trains for surface water augmentation.

POSTTREATMENT

After the final treatment process, the water will undergo posttreatment. In a surface water augmentation project, the water from the advanced treatment facility will flow into a river or reservoir that is a source of supply for a drinking water treatment plant. Goals of posttreatment include: (a) dechlorination, (b) drinking water treatment plant treatability (effective TOC and turbidity removal), (c) distribution system compatibility, and (d) improving aesthetic quality. Each of these goals is explained in more detail in the following paragraphs.

It is important that the water is dechlorinated before discharge to the environment to protect aquatic life. Often, the dechlorination facility will be near the point of discharge into the receiving stream or water body. This allows any chlorine residual to continue to provide disinfection protection throughout the conveyance to the reservoir or river.

Depending on upstream treatment processes, the water may need to be conditioned to ensure that it will still be treatable at the drinking water treatment plant through processes like coagulation, flocculation, and filtration that are meant to remove TOC and turbidity. Drinking water treatment plants are designed to treat natural surface waters; however, if reverse osmosis is used in the advanced water treatment train, the quality of the source water for the drinking water treatment train will change and will contain much lower concentrations of TDS. To ensure that the drinking water treatment plant processes such as coagulation, flocculation, and filtration can achieve their treatment goals, the advanced treated water may need to be conditioned to achieve an optimal pH or alkalinity.

Additionally, the corrosion potential of the water should be considered. For example, reverse osmosis removes many minerals (including calcium and magnesium) and drops the pH of the water, which makes it corrosive and taste unnatural. To ensure the water does not cause corrosion of conveyance pipelines or the drinking water distribution system and to ensure it is aesthetically pleasing to the consumer, the advanced treated water should be conditioned.

Treatment to condition the water may include the addition of carbon dioxide (CO_2), lime [CaO or $Ca(OH)_2$], sodium hydroxide (NaOH), and/or calcium chloride ($CaCl_2$). Another treatment technique used for conditioning is air stripping to remove CO_2 gas, which increases the pH of the product water.

COMPLIANCE MONITORING AND REPORTING

To ensure that water produced by a surface water augmentation project is protective of public health, monitoring of chemical contaminants and pathogenic microorganisms is required.

WATER QUALITY MONITORING

Requirements for water quality monitoring may vary from state to state; however, because the product water will become drinking water, federal drinking water quality standards should be met before the water being discharged to a surface water. In California, for example, water quality must meet the following before discharge to a reservoir:

- Primary MCLs,
- Secondary MCLs,
- Notification levels, and
- Action levels for lead and copper.

In addition, the water quality should be monitored for priority pollutants and other CECs.

The frequency of monitoring will vary state by state, but generally, primary MCLs and action levels should be monitored monthly to quarterly, secondary MCLs should be monitored annually, and priority toxic pollutants and other CECs should be monitored quarterly to annually. In California, reduced monitoring frequency is possible if the water quality is consistently meeting water quality requirements. Methods for analyzing water quality should be drinking water methods. Method reporting limits should be able to detect the parameter below the regulatory limit.

In addition to these federal drinking water standards, the water quality may also need to meet some environmental discharge requirements. In a reservoir, this may include limits on nutrients like nitrogen and phosphorus to prevent algae growth.

PATHOGENIC MICROORGANISM CONTROL MONITORING

In potable reuse, and all drinking water applications, pathogenic microorganisms pose an acute health risk to the public. Just one exposure event could lead to an illness, so monitoring for pathogen reduction is imperative. Whereas requirements for pathogen reduction may vary from state to state, most projects must provide reduction of V/G/C and sometimes bacteria. In California, for example, surface water augmentation projects must provide 8-log reduction of enteric virus, 7-log reduction of *Giardia*, and 8-log reduction of *Cryptosporidium* before discharging to a reservoir (8/7/8 of V/G/C) and contain no more than 1%, by volume, of recycled municipal wastewater that was delivered to the surface water reservoir during any 24-hour period. Coupled with the additional treatment required by the Long Term 2 Enhanced Surface Water Treatment Rule at a drinking water treatment plant—4/3/2-log reduction of V/G/C—the total pathogen reduction provided by a surface water augmentation project is 12/10/10-log reduction of V/G/C. Figure 7.14 presents a summary illustration of these requirements.

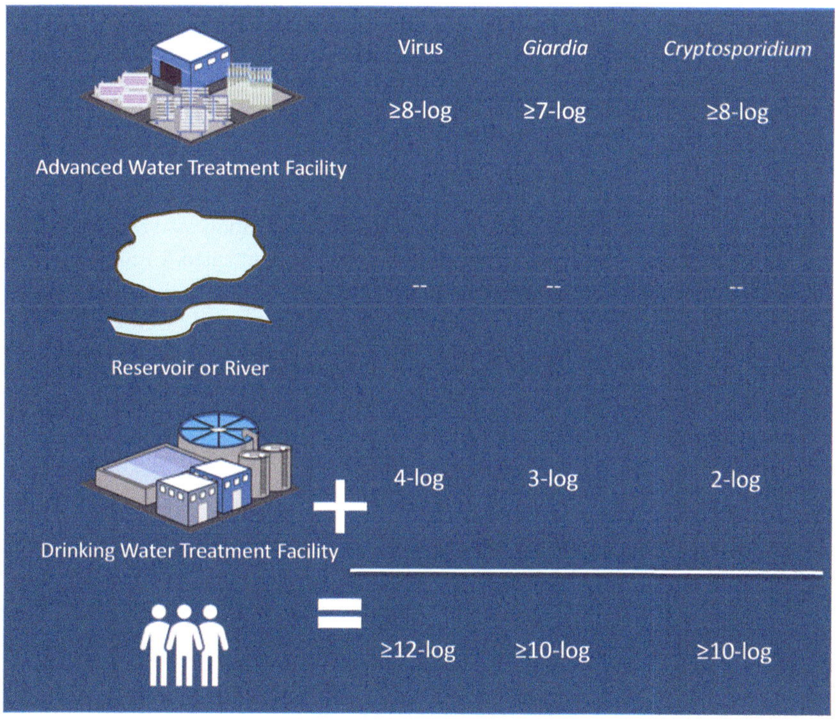

Figure 7.14 Summary of Pathogen Reduction Requirements for Surface Water Augmentation in California

Some states may require varying degrees of pathogen reduction based on the size of the reservoir or river being used. For example, in California, if a reservoir provides ≥100 : 1 dilution of a 24-hour pulse of recycled water, no additional pathogen reduction is required beyond 12/10/10 of V/G/C. However, if a reservoir provides <100 : 1 dilution, but ≥10 : 1 dilution of a 24-hour pulse of recycled water, an additional log reduction of each pathogen (V/G/C) is required. In addition to pathogen reduction requirements based on dilution, California has also specified that additional log reduction of pathogens may be required depending on the retention time of the reservoir. If a reservoir provides <4 months of retention time, an additional 1-log reduction of V/G/C may be required. For example, a project whose reservoir provides <100 : 1 dilution and has a retention time <4 months may be required to provide 10/9/10-log reduction of V/G/C before discharging to the reservoir, resulting in a total log reduction of 14/12/12 of V/G/C after consideration of the pathogen reduction provided at the drinking water treatment facility.

Monitoring pathogen concentrations in water is often impractical because methods for quantifying pathogens take too long or are prohibitively expensive. In lieu of measuring pathogen concentrations, surrogate indicators or established frameworks can be used to determine how well a treatment process is removing certain pathogens. For example, rather than directly measuring the reduction of *Cryptosporidium*, ozone concentration and time could be used to determine expected log reduction through an ozone system based on established frameworks. For a reverse osmosis system, reduction of TOC could be used to determine the log reduction of pathogens through the unit process. Using surrogate indicators allows for more continuous monitoring of each unit process to ensure it is achieving the desired amount of pathogen reduction.

COMPLIANCE REPORTING

Water quality data must be collected and reported in some form to a regulatory agency. Reporting requirements differ from state to state. Although monitoring may occur continuously, the regulatory agency may only ask for a report monthly. Each regulatory agency will specify how the data should be reported. Should you report average values? Do you only need to report exceedances? Each state will have their own requirements for what type of data needs to be included.

Data are often collected via electronic databases. For example, in California, some data and reports can be transmitted electronically through the California Integrated Water Quality System Project; however, some reports must be emailed. It is important to read the permit carefully to ensure all of the proper reporting is completed.

SURFACE WATER RESERVOIR MANAGEMENT

MONITORING FLOWS

Reservoirs are often carefully managed to ensure proper water levels. With augmentation from an advanced water treatment facility, the water agency must also monitor the flowrate of this new source into the reservoir. In California, where regulations require compliance with both a dilution requirement and a retention time requirement, the reservoir flows and levels must be carefully documented. To comply with the dilution requirement, hydrodynamic modeling of the reservoir needs to occur to develop an operations strategy for the reservoir. The operations strategy will account for different flowrates of advanced treated water, different reservoir levels, and periods of reservoir maintenance or drawdown that may affect dilution and retention time. In California, the retention time of a reservoir is calculated by taking the volume of the reservoir at the end of a calendar month and dividing it by the total flow that came out of the reservoir during that month. To calculate the retention time, monitoring of the flows out of the reservoir must also occur.

MAINTAINING RESERVOIR WATER QUALITY

In addition to monitoring flows in and out of the reservoir, the water quality of the reservoir should also be monitored and maintained. Some reservoirs support fish species and may have other beneficial uses, such as recreation, that should be maintained. Water quality in the reservoir and compliance with environmental regulations need to be considered, in addition to public health protection, when designing for the finished water quality of an advanced treatment facility. Regulatory agencies may require monitoring of the reservoir before and after augmentation to ensure that the reservoir water quality is maintained.

Environmental discharges may have limits for nitrogen and phosphorus because these are nutrients that influence algae growth. These nutrient limits may be in terms of a TMDL or an instantaneous concentration. Although these nutrient limits may vary from project to project, it is important that they are monitored and controlled within the reservoir to maintain the water quality.

COORDINATION AND RESPONSE PLANS

Surface water augmentation projects require coordination between wastewater and drinking water agencies. The wastewater agencies are providing the source water that the drinking water agencies will distribute. If an upset occurs on the wastewater side, there must be communication with the drinking water side to alert them of the size and type of upset. Should they stop distributing reuse water? How long was the upset? Both agencies (wastewater and drinking water) should be prepared for situations like this. In California's surface water augmentation regulations, there is a requirement for a "joint plan." The joint plan describes the actions that each agency will take to ensure the project is successful and lays out the contingency plans for when something fails, breaks, or needs to be taken out of service for maintenance. It is the responsibility of both agencies to ensure that the water produced is always protective of public health.

EXAMPLES OF SURFACE WATER AUGMENTATION PROJECTS

There are several surface water augmentation projects throughout the world that are planned or are currently in operation. Table 7.5 provides an overview of these projects and the type of treatment they provide.

Table 7.5 Examples of Surface Water Augmentation Projects

Project Location	Capacity	Description of Treatment	Description of Project
San Diego, California, USA[1]	30 mgd	Water Reclamation: Preliminary and primary treatment followed by 4-stage Bardenpho with biological nitrogen removal and granular media filtration. Advanced Water Treatment: Ozone disinfection followed by biologically activated carbon, membrane filtration, reverse osmosis, ultraviolet light/AOP with sodium hypochlorite, chlorination, and dechlorination.	Reclaimed water augments Miramar Reservoir, which is a drinking water supply for the Miramar Water Treatment Plant; expected to be online in 2025.
Santee, California, USA[1]	12.5 mgd	Water Reclamation: Preliminary and primary treatment followed by 4-stage Bardenpho with biological nitrogen removal and tertiary filtration. Advanced Water Treatment: Micro- or ultrafiltration, reverse osmosis, ultraviolet light advanced oxidation with sodium hypochlorite, chlorination, and dechlorination.	Reclaimed water augments Lake Jennings Reservoir, which is the drinking water supply for the R.M. Levy Water Treatment Plant; expected to be online in 2025.
Upper Occoquan, Virginia, USA[2]	54 mgd	Water Reclamation: Preliminary and primary treatment followed by complete mixed activated sludge with biological nitrogen removal. Advanced Water Treatment: Lime precipitation and two-stage recarbonation with intermediate settling followed by multimedia filtration, granular activated carbon adsorption, chlorination, and dechlorination.	Reclaimed water is blended in the Occoquan Reservoir, which is a drinking water supply for the Fairfax Water Treatment Plant; operating since 1978.
Langford, United Kingdom[2]	10.5 mgd	Water Reclamation: Primary and secondary treatment is provided at the Chelmsford Sewage Treatment Works. Advanced Water Treatment: Biological nitrification-denitrification followed by chemical phosphorus removal and UV disinfection.	Treated reclaimed water is discharged in the River Chelmer 8 km upstream of the drinking water supply; operating since 2003.

[1]U.S. EPA, 2017, [2]U.S. EPA, 2012

TEST YOUR KNOWLEDGE

1. There are federal regulations established for planned surface water augmentation projects.
   - ☐ True
   - ☐ False

2. Posttreatment is an important consideration for surface water augmentation projects to ensure the water does not cause corrosion of conveyance pipelines or the drinking water distribution system.
   - ☐ True
   - ☐ False

3. Water should be dechlorinated before being released to a reservoir or river.
   - ☐ True
   - ☐ False

Direct Potable Reuse

As described previously, direct potable reuse is the planned introduction of recycled water directly into a public water system or into a raw water supply immediately upstream of a water treatment plant. Without an environmental buffer, direct potable reuse projects must provide project elements that compensate for the benefits provided by the environmental buffer (namely response time). As the proximity of the recycled water and the consumer becomes closer, additional protections may be required to protect public health.

REGULATORY GUIDELINES

In the United States, there are no federal regulations for direct potable reuse, and no states have developed regulations. However, regulators in California are required to develop regulations for direct potable reuse by the end of 2023, and other states are also considering developing direct potable reuse regulations. In the absence of regulatory guidelines, some states allow projects on a case-by-case basis. Although there are no regulations in place, there have been many documents developed that provide guidelines and frameworks for direct potable reuse.

The following public documents provide guidance on some of the key considerations for a direct potable reuse project:

- 2017 *Potable Reuse Compendium* (U.S. EPA, 2017)
- 2012 *Guidelines for Water Reuse* (U.S. EPA, 2012)
- *A Proposed Framework for Regulating Direct Potable Reuse in California* (State Water Resources Control Board [SWRCB], 2019)
- *Expert Panel Report: Evaluation of the Feasibility of Developing Uniform Water Recycling Criteria for Direct Potable Reuse* (Olivieri et al., 2016)
- *Guidelines for Direct Potable Reuse in Colorado* (National Water Research Institute [NWRI], 2019)

All of the guidance for direct potable reuse includes similar themes. Each document discusses the importance of multiple diverse barriers to provide protection from pathogenic microorganisms and chemical contaminants as well as clear monitoring and response plans to ensure that the distributed water will protect public health. These themes are in line with the 4Rs discussed previously: reliability, redundancy, robustness, and resilience.

Other key considerations for direct potable reuse beyond treatment of pathogens and chemicals include wastewater source control; operator training; aesthetic considerations; public acceptance; and the technical, managerial, and financial capacity of the project owner.

The 10^{-4} infections per person per year means that pathogens in potable water should not cause more than one infection within a population of 10,000 people per year. Disability adjusted life years (DALYs) represent the loss of an equivalent full year of health due to both premature death and years of healthy life lost due to disability. In this case, 10^{-6} DALYs mean that pathogens in potable water should not cause the loss of more than one healthy year (365 days) within a population of 1 million people per year.

PATHOGENIC MICROORGANISM CONTROL

Direct potable reuse projects must ensure that the concentration of pathogens is reduced to an acceptable risk level before distribution. There are several risk targets that have been used previously for pathogen exposure including: 10^{-4} infections per person per year and 10^{-6} disability adjusted life years. In California, the pathogen reduction requirements for indirect potable reuse were based on a tolerable risk of 10^{-4} infections per person per year and resulted in a requirement for 12/10/10-log reduction of V/G/C before public consumption. For direct potable reuse, California is considering setting treatment requirements based on a daily risk target of 2.7×10^{-7} infections per person per day (SWRCB, 2019). A shift from an annual risk framework to a daily risk framework would require greater consistency in treatment because days in which treatment may not be meeting requirements could no longer be buffered out by the days in which treatment may be exceeding requirements. Treatment would need to consistently meet performance requirements to ensure pathogen concentrations are always being reduced enough to protect public health. This may result in a requirement or preference for redundancy in treatment of pathogens. If 12/10/10-log reduction of V/G/C is required to meet the risk goal, additional treatment may be considered or required to ensure that even if a treatment process is underperforming, 12/10/10-log reduction of V/G/C can still be met.

CHEMICAL CONTROL

Direct potable reuse projects will be required to meet federal drinking water standards before distribution. Without an environmental buffer, an additional consideration for direct potable reuse projects is the control of pulses of chemicals that may come through the system. Environmental buffers provide some diluting or buffering of pulses of chemicals that may pass through advanced treatment. Without that buffer, direct potable reuse projects will need to compensate by providing additional treatment, blending or dilution, enhanced monitoring, and/or enhanced source control. In addition, because chemical contaminants have such a wide diversity of structure, charge, and size, additional robust treatment systems may be required to provide additional protection for unknown chemicals.

IMPORTANCE OF SOURCE CONTROL

One key aspect of a direct potable reuse project is wastewater source control. The existing national pretreatment program is designed to prevent contaminants from being discharged into collection systems that may pass through the treatment provided at traditional WRRFs. The national pretreatment program targets industrial users that may discharge unwanted chemicals into a sewershed. Although the national pretreatment program is a solid foundation for source control, additional measures may be needed to enhance the source control program. Some additional components of an enhanced source control program may include enhanced local limits, enhanced evaluation of dischargers including an inventory and database of all industrial users in a sewershed, and potential monitoring of bulk parameters (e.g., electrical conductivity, pH, temperature, etc.) or specific chemicals (e.g., volatile organic compounds) within a sewer system to provide early warning for any potential issues (NWRI, 2020). Source control should be considered the first barrier in a direct potable reuse project and can provide protection by preventing unwanted contaminants from entering the system.

MONITORING AND RESPONSE

Direct potable reuse projects lose valuable response time with the loss of an environmental buffer. Environmental buffers provide time—often several months—to respond to potential treatment excursions. Direct potable reuse projects have much less time—on the order of minutes or days—to respond to failure events before water is distributed.

To ensure that water quality is consistently meeting requirements, high frequency monitoring should be used when possible. High frequency monitoring of each treatment process will ensure that processes are performing as designed and will allow a system to detect potential treatment excursions rapidly. If a treatment excursion is detected, the response will also need to be rapid to prevent off-spec water from being distributed.

To provide additional response time, a direct potable reuse project may consider using an engineered storage buffer to hold produced water before distribution. Depending on the size of the project, this may be a viable option and would provide a utility time to react to a treatment excursion, prevent distribution of off-spec water, and provide an alternate water source. Regardless, utilities will need to have detailed response plans to ensure off-spec water is not distributed. When an excursion is detected, the water will need to be diverted, and the treatment plant may need to shut down until the issue can be resolved.

OTHER CONSIDERATIONS

In addition to treatment, monitoring, and response, there are additional considerations for direct potable reuse projects. Direct potable reuse projects will require skilled operators that are able to respond quickly to issues that arise. Operators will need to be trained in advanced water treatment and will need to understand both wastewater and drinking water requirements. Although potable reuse projects have not yet required a special certification for operators of a potable reuse system, states may require operators to have experience with advanced water treatment processes.

The California–Nevada section of the American Water Works Association and the California Water Environment Association partnered to develop the Advanced Water Treatment Operator Certification program. It launched in 2019 and provides an avenue for both wastewater operators and drinking water operators to gain a certification for advanced water treatment processes. For more information of the certification program, visit awtoperator.org.

Operators will be key to ensuring potable reuse projects protect public health, and utilities will need to have sufficient technical, managerial, and financial capacity to ensure a project is sustainable. This includes having the ability and resources to operate and maintain critical infrastructure.

A final consideration for direct potable reuse is the aesthetic quality of the water. Posttreatment of the water should aim to match the mineral content of existing drinking water. Wastewater is typically warmer than traditional source waters, so direct potable reuse systems may need to consider options for cooling the water before distribution. Aesthetic quality affects public perception of the water quality and is important to address.

BUILDING PUBLIC ACCEPTANCE

Potable reuse has an "ick" factor for the public. The thought of drinking water that has been sourced from our wastes can make public acceptance of potable reuse challenging. All potable reuse projects need to have strong community outreach programs to inform the public about the goals and benefits of a potable reuse system. In addition to community outreach, agencies pursuing potable reuse may consider operating a demonstration facility to build confidence in the ability of the proposed treatment to produce safe water before building a full-scale facility. The demonstration facility provides multiple benefits including opportunities for public education through tours, operator experience with advanced treatment processes, and data collection to ensure the processes will protect public health.

Some projects may consider forming an independent advisory panel of experts to provide further confidence in the project. Review by an independent panel of experts can bolster public confidence.

EXAMPLES OF DIRECT POTABLE REUSE PROJECTS

There are several direct potable reuse projects throughout the world that are planned or are currently in operation. Table 7.6 provides an overview of these projects and the type of treatment they provide.

TEST YOUR KNOWLEDGE

1. **There are established federal regulations for direct potable reuse.**
   - ☐ True
   - ☐ False

2. **Source control could be considered the first barrier in a direct potable reuse project and can provide protection by preventing unwanted contaminants from entering the collection system.**
   - ☐ True
   - ☐ False

Table 7.6 Examples of Direct Potable Reuse Projects

Project Location	Capacity	Description of Treatment	Description of Project
Big Spring, Texas, USA[1]	2.5 mgd	Water Reclamation: Primary and secondary treatment followed by filtration. Advanced Water Treatment: Microfiltration followed by reverse osmosis and advanced oxidation with ultraviolet light and hydrogen peroxide.	Reclaimed water augments a raw water pipeline that brings water directly to a drinking water treatment plant; operation began in 2013.
Windhoek, Namibia[2]	5.5 mgd	Water Reclamation: Primary and secondary treatment blended with reservoir water. Advanced Water Treatment: Pre-ozonation, coagulation and flocculation, dissolved air flotation, dual media filtration, ozonation, biologically activated carbon, 2-stage granular activated carbon, ultrafiltration, and chlorine disinfection.	Secondary effluent is blended with reservoir water, and then advanced treated and distributed; installed in 2002.

[1]U.S. EPA, 2012, [2]U.S. EPA, 2017

CHAPTER SUMMARY

SECONDARY AND TERTIARY TREATED MUNICIPAL WASTEWATER	■ Non-nitrified effluent contains free ammonia and typically has higher concentrations of TOC. In nitrified effluents, the majority of the nitrogen is in the form of nitrate. ■ Concentrations of pathogens, CECs, nutrients, and minerals should be considered when evaluating secondary and tertiary treated municipal wastewater for water reuse applications. ■ Wastewater source control is important for ensuring public and environmental health protection.
NONPOTABLE WATER REUSE	■ Nonpotable reuse includes applications for agriculture, irrigation, and industry and reduces the demand for potable water. ■ On-site reuse is a form of decentralized nonpotable reuse that can be used in large buildings or on campuses. ■ Treatment requirements for nonpotable applications may vary depending on the amount of human contact involved in the end use.
POTABLE WATER REUSE	■ Planned potable reuse is the planned use of recycled water for human consumption. ■ Indirect potable reuse includes the use of an environmental buffer such as a groundwater aquifer, surface water reservoir, or river. ■ Direct potable reuse is a form of potable reuse that does not include an environmental buffer. ■ Strategies to consider when implementing potable reuse include reliability, redundancy, robustness, and resilience.
GROUNDWATER RECHARGE	■ Groundwater recharge can be implemented through surface spreading, vadose zone injection, or direct injection. It is a form of indirect potable reuse. ■ Treatment requirements for groundwater recharge may vary depending on state or local regulations as well as the type of groundwater recharge (spreading vs. injection). ■ Groundwater travel time can be used to obtain pathogen removal credits for certain recharge applications. ■ Direct injection is effective at forming a barrier to prevent seawater intrusion of the groundwater of coastal communities.
SURFACE WATER AUGMENTATION	■ Surface water augmentation, often referred to as "reservoir augmentation" or "reservoir water augmentation," is the planned placement of recycled water into a raw surface water reservoir used as a source of domestic drinking water supply for a public water system, or into a constructed system conveying water to such a reservoir. It is a form of indirect potable reuse. ■ Treatment requirements for surface water augmentation may vary depending on state or local regulations as well as the amount of retention time and dilution/mixing provided by the surface water. ■ Surface water augmentation projects require communication between the wastewater agency and the drinking water agency to ensure protection of public health.
DIRECT POTABLE REUSE	■ Direct potable reuse is the planned introduction of recycled water directly into a public water system or into a raw water supply immediately upstream of a water treatment plant. ■ Direct potable reuse projects have much less time than indirect potable reuse projects to respond to failure events before water is distributed. ■ Direct potable reuse projects may require higher levels of treatment than indirect potable reuse projects to account for the loss of response time. ■ All potable reuse projects should include a public outreach program to build public acceptance.

References

Advanced Treatment Criteria, 22 C.C.R. § 60320.201. (2014).

Asano, T., & Wassermann, K. L. (1980). Groundwater recharge operations in California. *Journal of the American Water Works Association, 72*(7), 380–385.

Asano, T. (2006). Water reuse via groundwater recharge. *International Review for Environmental Strategies, 6*(3), 205–216.

Asano, T., Burton, F. L., Leverenz, H. L., Tsuchihashi, R., & Tchobanoglous, G. (2007). *Water reuse: Issues, technologies, and applications.* McGraw-Hill.

Bouwer, H. (1978). *Groundwater hydrology*. McGraw-Hill.

Florida Department of Environmental Protection. (2012). *2011 reuse inventory*. http://www.dep.state.fl.us/water/reuse/docs/inventory/2011_reuse-report.pdf

National Research Council. (2012). *Water reuse: Potential for expanding the nation's water supply through reuse of municipal wastewater.* The National Academies Press.

National Water Research Institute. (2019). *Guidelines for direct potable reuse in Colorado* (NWRI-2019-03).

National Water Research Institute. (2020). *Enhanced source control recommendations for direct potable reuse in California* (NWRI-2020-07).

Olivieri, A., Crook, J., Anderson, M., Bull, R. J., Drewes, J. E., Haas, C. N., Jakubowski, W., McCarty, P. L., Nelson, K. L., Rose, J. B., Sedlak, D. L., & Wade, T. J. (2016). *Expert Panel report: Evaluation of the feasibility of developing uniform water recycling criteria for direct potable reuse.* National Water Research Institute.

Pathogenic Microorganism Control, 22 C.C.R. § 60320.108. (2014).

Pecson, B. M., Trussell, R. S., Pisarenko, A. N., & Trussell, R. R. (2015). Achieving reliability in potable reuse: The four Rs. *Journal of the American Water Works Association, 107*(3), 48–58.

Rice, J., Wutich, A., & Westerhoff, P. (2013). Assessment of de facto wastewater reuse across the U.S.: Trends between 1980 and 2008. *Environmental Science & Technology, 47*(19), 11099–11105.

San Francisco Public Utilities Commission. (2020). *Onsite water reuse program guidebook.* https://www.sfpuc.org/sites/default/files/documents/OnsiteWaterReuseGuidebook2020.pdf

State Water Resources Control Board. (2019). *A proposed framework for regulating direct potable reuse in California* (2nd ed.).

Todd, D. K. (1980). *Groundwater hydrology*. John Wiley & Sons.

Trussell, B., Trussell, S., Qu, Y., Gerringer, F., Stanczak, S., Venezia, T., Monroy, I., Bacaro, F., & Trussell, R. (2018). A four-year simulation of soil aquifer treatment using columns filled with San Gabriel Valley sand. *Water Research, 144,* 26–35.

Underground Injection Control Program, 40 C.F.R. § 144 *et seq*. (1983).

U.S. Environmental Protection Agency. (1977). *Reuse of municipal wastewater for groundwater recharge.*

U.S. Environmental Protection Agency. (2006). *Ultraviolet disinfection guidance manual for the final long term 2 enhanced surface water treatment rule* (EPA 815/R-06-007).

U.S. Environmental Protection Agency. (2009). *Recycled water for groundwater recharge: Innovative recharge projects and source water implications.* Presentation by Ted Johnson, PG, CHG. Water Replenishment District of Southern California. U.S. EPA Achieve Document.

U.S. Environmental Protection Agency. (2012). *Guidelines for water reuse* (EPA 600/R-12-618).

U.S. Environmental Protection Agency. (2017). *Potable reuse compendium* (EPA 810/R-17-002).

U.S. Environmental Protection Agency. (2018). *On-site non-potable water reuse research. Technical brief (EPA 600/F-18-342).*

U.S. Environmental Protection Agency. (2020). *National water reuse action plan—Collaborative implementation (Version 1)* (EPA 820/R-20-001).

U.S. Environmental Protection Agency. (2021). *Basic information about water reuse.* https://www.epa.gov/waterreuse/basic-information-about-water-reuse#basics

William J Worthen Foundation. (2018). *Onsite non-potable reuse: Practice guide.* https://www.collaborativedesign.org/water-reuse-practice-guide

Suggested Readings

American Society of Civil Engineers Value of Water Campaign 2020. *The economic benefits of investing in water infrastructure.*

Florida Potable Reuse Commission, WateReuse Florida, Florida Section of the American Water Works Association Water Utility Council, Water Research Foundation, & WateReuse Association. (2020). *Framework for the implementation of potable reuse in Florida.* WateReuse Association.

General Requirements, 22 C.C.R. § 60320.100. (2014).

U.S. Water Alliance. *National blue-ribbon commission for onsite non-potable water systems.* http://uswateralliance.org/initiatives/commission

CHAPTER 8

Characterization and Sampling of Sludge

Introduction

It is important for operators to be familiar with different types of sludge, their characterization, and proper sampling methods to operate and control water resource recovery facilities (WRRFs). The waste generated in different treatment processes is termed *sludge* that contains both liquid and solids. The sludge generated from primary and secondary clarifiers has a significant amount of organic matter and requires further treatment and processing before disposal. The term *biosolids* is used to connotate the primary organic solid product of treatment that meets U.S. Environmental Protection Agency or other applicable criteria for beneficial use. Anything preceding that point in the wastewater process is generally called *solids.*

This chapter provides operators with information on characterization and sampling of sludge to operate WRRFs effectively.

LEARNING OBJECTIVES

Upon completing this chapter, you will be able to:

- List the types of wastewater sludge for domestic WRRFs.
- Understand the characteristics of sludge from different treatment processes.
- Identify which sludge parameters need to be analyzed.
- Explain how sludge samples can be collected, prepared, and preserved.

Characterization of Sludge

TYPES OF SLUDGE

PRIMARY SLUDGE

Primary sludge is suspended solids and organics collected by gravitational settling in a primary clarifier before biological treatment. It occurs after influent screening and grit removal and consists of settled wastewater contaminations. The composition of this sludge depends on the characteristics of the influent. Primary sludge consists of organic matter, such as feces, vegetables, fruits, textiles, paper, and so on. Primary sludge includes 50% to 70% of the suspended solids that enter the WRRF in the influent. Typical solids concentrations in raw primary sludge are 3% and 6%. The portion of volatile solids varies from 60% to 85% (WEF & ABC, 2018). In some facilities, primary sludge is generated from chemical precipitation and is then called *chemically treated primary sludge.*

SECONDARY SLUDGE

The secondary sludge comes from the secondary biological treatment process, which uses microorganisms to break down organic matter in the wastewater. The microorganisms feed on the biodegradable material in the wastewater in the aeration tank and then flow into a secondary clarifier, where flocculated solids settle out as secondary sludge.

Flocculants help to remove suspended solids from wastewater through settlement by aggregating contaminants into "flocs" that float to the surface of the water. The settled solids are then called *flocculated solids.*

Sludge From Activated Sludge Processes

The activated sludge process is the most commonly used suspended growth process in WRRFs. The activated sludge process produces a large quantity of microorganisms capable of aerobic stabilization of organic material in wastewater. The activated sludge exists normally in the form of flocs, which include living and dead biomass, adsorbed organic matter, and minerals. A conventional activated sludge process consists of an aeration tank, a secondary clarifier, and a recycle system (Figure 8.1). A fraction of the

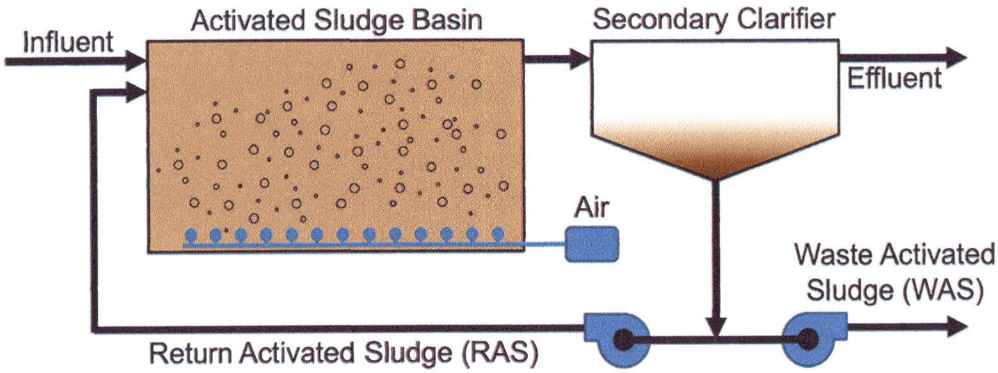

Figure 8.1 Activated Sludge Process (WEF & ABC, 2018) (Reprinted with permission by Indigo Water Group)

sludge settled at the bottom of the secondary clarifier is returned to the aeration tank and it is called return activated sludge (RAS). The excess biomass, which settles at the bottom of the clarifier, is known as *waste activated sludge* (WAS) and is removed for further treatment and subsequent disposal.

Raw wastewater is retained in the aeration tank (Figure 8.2) in a well-mixed environment, where microorganisms responsible for organic matter degradation are kept in suspension. During the aeration, the microorganisms in the aeration tank stabilize the organic matter. In this process, part of the organic matter is synthesized into new cells and part is oxidized to produce energy. After the aeration tank, mixed liquor suspended solids (MLSS) flows into the secondary clarifier.

An aeration tank is the biological reactor where air (or oxygen) is injected to the MLSS.

Sludge From Fixed-Film Processes

Fixed-film systems (also called *attached-growth systems*) are biological wastewater treatment processes (trickling filter, biotower, rotating biological contactor, etc.) that also generate biological sludge. In this system, biological films are attached to inert media such as rocks, plastics, woods, or other natural or synthetic solid materials. Sludge from this process is similar to activated sludge, but the fixed-film

Figure 8.2 Aeration tank of an activated sludge system

sludge contains biomass coming from aerobic, anaerobic, and anoxic zones and increases the sludge retention time.

Sludge From Dissolved Air Flotation

Sludge from dissolved air flotation (DAF) systems contains proteins, fats, and fibrous materials that could not be removed by simple mechanical means. Sludge from DAF systems is also called *float* and is skimmed off of the top. Sludge concentration in a DAF unit typically ranges from 3.5% to 4% solids (Hasan, 2017; WEF & ABC, 2021). If operated properly, the DAF system can achieve at least 85% solids recovery without chemical addition. Chemicals such as polymer and coagulants can improve solids recovery to greater than 90%; however, solids concentration may not improve significantly (Hasan et al., 2017; WEF & ABC, 2021). This study hypothesized that polymer demand is directly correlated with soluble protein and polysaccharide concentration of the sludge, whereas the cake total solids are related to the extracellular polymeric substances (EPS). Figure 8.3 shows a schematic of a typical DAF thickener.

THERMALLY HYDROLYZED SLUDGE

Thermal hydrolysis occurs when sludge is heated above 120 °C for a certain period of time (typically between 20 and 30 minutes) with the purpose of increasing the performance of anaerobic digestion. The reported benefits of thermal hydrolysis relate to increased digestion loading rate due to altered rheological properties, improved biodegradation of sludge (especially activated sludge) due to the solubilization of otherwise particulate substrate and enhanced dewaterability (De Clippeleir et al., 2016; Hasan, 2017). Rheology changes are observed in the form of decreased viscosity when compared to non-hydrolyzed sludge at the same solids concentration. Given that the heat is provided by live steam, systems are operated at high solids concentrations in the range of 16% to 22% total solids in the feed to the thermal hydrolysis system (Figure 8.4), with a hydrolyzed sludge around 11% total solids due to dilution with the steam.

ANAEROBICALLY DIGESTED SLUDGE

Anaerobic digestion is a biological process that stabilizes organic matter in the absence of oxygen. During anaerobic digestion, sludge is biodegraded to water and biogas, reducing the total amount of solids leaving the system. Anaerobically digested sludge is often less odorous than raw material, although it may contain high ammonia and hydrogen sulfide concentrations (WEF et al., 2012). The remaining organic matter, nitrogen, and phosphorus in the digested sludge can be used as a soil amendment and fertilizer.

Characteristics of anaerobically digested sludge depend on the feed sludge and the types of pretreatment used. Feed sludge can be primary sludge only, primary blended with secondary sludge, high-strength substrate (such as fats, oils, and grease), or food waste.

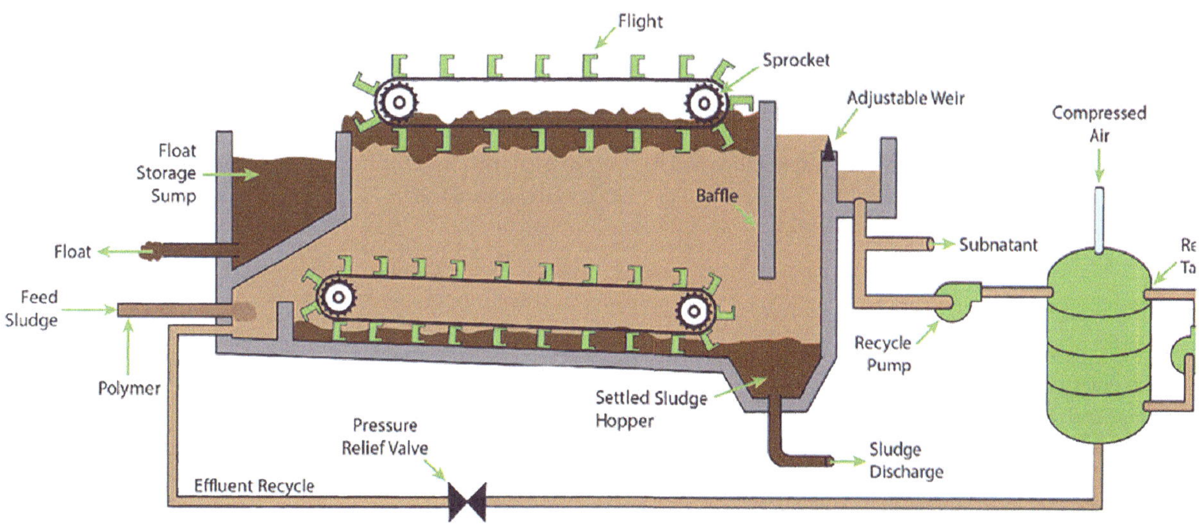

Figure 8.3 Diagram of a Dissolved Air Flotation Thickener (WEF & ABC, 2021) (Reprinted with permission by Indigo Water Group)

Figure 8.4 The Thermal Hydrolysis System (Hasan, 2017)

AEROBICALLY DIGESTED SLUDGE

During aerobic digestion, biomass and other organic solids are aerated and then converted into carbon dioxide, water, nitrite/nitrate, nitrogen gas, and stabilized organic compounds. This process extends the decomposition of solids to a point where the volatile portion is sufficiently low to permit the land application of the biosolids. This process is typically used to stabilize secondary sludge (WEF et al., 2012). For a more detailed description of aerobic digestion and related regulatory requirements, refer to Chapter 3 of *Wastewater Treatment Fundamentals II—Solids Handling and Support Systems* (WEF & ABC, 2021).

Characterization of aerobically digested sludge focuses on process performance and regulatory requirements. Some of the key process parameters are the specific oxygen uptake rate (SOUR), pathogen reduction, total and volatile solids reduction, alkalinity and pH, and quality of the soluble fraction in terms of nitrogen and phosphorus as this may affect liquid process performance if the liquid portion of the digested sludge is returned to the head of the facility.

TEST YOUR KNOWLEDGE

1. **What is the typical removal efficiency for settleable solids in a primary clarifier?**
   a. 10% to 15%
   b. 20% to 50%
   c. 40% to 60%
   d. 95% to 99%

2. Anaerobically digested sludge can be less ________ than aerobically digested sludge even it contains high ammonia and hydrogen sulfide.
   a. Odorous
   b. Viscous
   c. Decomposed
   d. Heated

3. Thermal hydrolysis of sludge is operated at a temperature above
   a. 50 °C
   b. 80 °C
   c. 120 °C
   d. 35 °C

4. Sludge concentration in a DAF unit typically ranges from
   a. 3% to 5%
   b. 5% to 8%
   c. 8% to 10%
   d. Above 12%

SLUDGE CHARACTERISTICS

Sludge consists of organic and inorganic solids present in the raw wastewater and removed in the primary settling tanks, plus organic solids produced during secondary treatment of wastewater and removed in the secondary settling tanks. The characteristics and the quantity of sludge produced depend on the character of the raw wastewater and the treatment process used.

TOTAL SUSPENDED SOLIDS AND TOTAL SOLIDS

Total suspended solids (TSS) are the quantification of particulate solids, excluding both organic and inorganic soluble solids. Conversely, total solids are the quantification of solids both in soluble and in particulate form (both organic and inorganic). Total solids content of sludges is in the design and process control of WRRFs. Total suspended solids are commonly monitored to evaluate the degree of pollution and serve as a key process control parameter for wastewater treatment operation. Both TSS and total solids content are measured by means of the gravimetric procedure. To calculate TSS and total solids, the solids concentration of a sample can be found directly by weighing the sample before and after drying at 103 °C.

VOLATILE SUSPENDED SOLIDS AND VOLATILE SOLIDS

Volatile suspended solids (VSS) are the quantification of particulate organic solids, excluding soluble solids and inorganic solids. Conversely, volatile solids are the quantification of organic solids, both in soluble and particulate form. Both VSS and volatile solids are estimated (by weight loss) by burning the sample at 500 ± 50 °C to volatilize the organics. Figure 8.5 shows the physical fractionations of solids in sludge. Chapter 9 ("Laboratory Procedure") of *Wastewater Treatment Fundamentals II—Solids Handling and Support Systems* (WEF & ABC, 2021) contains detailed explanations of methods to conduct total solids, volatile solids, TSS, and VSS tests.

SLUDGE VOLUME INDEX

The sludge volume index (SVI) is the volume, in milliliters, occupied by 1 g of sludge after 30 minutes of settling. It is used to monitor settling characteristics of activated sludge and other biological suspensions. Although SVI is not supported theoretically, experience has shown it to be useful in routine process control. Sludge volume index is calculated by the following equation:

$$\text{Sludge Volume Index (SVI), mL/g} = \frac{\text{SSV30}\frac{\text{mL}}{\text{L}} \times 1000\ \text{mg/g}}{\text{MLSS, mg/L}} \quad (8.1)$$

Where

SSV30 = settled sludge volume after 30 minutes and
MLSS = mixed liquor suspended solids concentration.

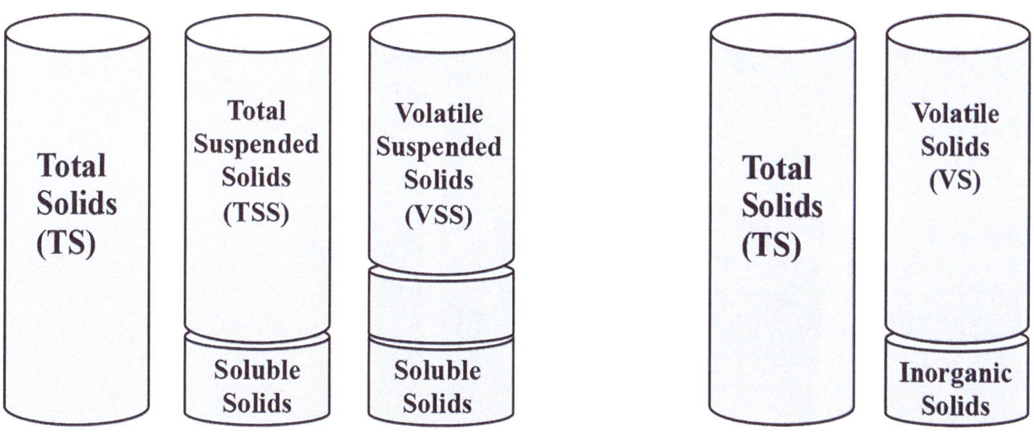

Figure 8.5 Physical Fractionation of Total Solids in Sludge

If the SVI is above 200 mL/g, it is an indication of sludge bulking. Sludge bulking typically occurs when the sludge fails to separate out in the sedimentation tanks/clarifiers. The main cause of sludge bulking is the growth of filamentous bacteria. To avoid sludge bulking, remedies are required that include bypassing the flow, increasing the recycle ratio, adding lime and chlorine, and increasing aeration to the tank. In determining the SVI, precision is limited by the precision achieved in the suspended solids measurement, the settling characteristics of the suspension, and variables associated with the measurement of the settled sludge volume. Figure 8.6 shows the SVI test using an Imhoff cone.

CHEMICAL OXYGEN DEMAND

The measurement of chemical oxygen demand (COD) in sludge is generally related to the determination of the COD-to-VSS ratio (COD : VSS) to cross-check the process model calibration to compare between observed and model-predicted values. The COD : VSS is expressed as mg COD/mg VSS and the typical range of this monitoring parameter is 1.45 to 1.55 mg COD/mg VSS over a range of solids retention time (SRT) activated sludge systems (Melcer, 2004). The measurement of COD : VSS is simple because it only requires measurement of VSS and COD on the sludge sample. The COD in sludge includes particulate COD (pCOD), colloidal COD (cCOD), and soluble COD (sCOD), where pCOD contains biomass concentration as well as particulate substrates. The COD : VSS ratio is calculated using the following equation:

$$\text{COD : VSS} = (\text{pCOD in sludge sample})/\text{VSS} = (\text{COD measured on unfiltered sludge sample} - \text{COD measured on sludge sample filtered through 1.5-}\mu\text{m filter})/\text{VSS} \tag{8.2}$$

ODOR

Odors from any type of sludge can lead to public complaints and, in some cases, to permit violations. An odor is a mixture of volatile molecules present at very low concentrations in the inhaled air that, upon coming in contact with the human sensory system, is able to stimulate an anatomical response. Odors are recognized via the memory effect of previously experienced smells, thus accounting for the high subjectivity of the odor perception. Therefore, measuring odors in any form represents a challenge due to the low concentrations of compounds at which an odor can be perceived.

In wastewater and sludge, odors are monitored using several approaches. Some of these use human sensory panels (olfactometry); others seek to simulate the human olfactory system (e.g., electronic noses), whereas still others are based on concentration measurements (e.g., hydrogen sulfide) in comparison to measured detection thresholds and monitoring process performance of an odor control system.

Figure 8.6 Sludge Volume Index Test Using an Imhoff Cone

NUTRIENTS

Similar to the COD : VSS, nitrogen (N)/VSS and phosphorus (P)/VSS content in sludge can be monitored to cross-check the process model calibration to compare between observed and model-predicted values. The ratio of P : VSS has significance on modeling and analyzing enhanced biological phosphorus removal systems to assess phosphorus mass balance and removal performance. N : VSS and P : VSS can be calculated using the following equations:

$$\text{N:VSS} = (\text{total Kjeldahl nitrogen [TKN] on unfiltered sludge sample} - \text{TKN on sludge sample filtered through a 1.5-}\mu\text{m filter})/\text{VSS} \quad (8.3)$$

$$\text{P:VSS} = (\text{total phosphorus on unfiltered sludge sample} - \text{total phosphorus on sludge sample filtered through a 1.5-}\mu\text{m filter})/\text{VSS} \quad (8.4)$$

The TKN and total phosphorus can be measured using commercially available test kits.

EXTRACELLULAR POLYMERIC SUBSTANCES

Bacteria in activated sludge systems produces glue-type substances called *extracellular polymeric substances* (EPS), which helps them to bind together to produce larger flocs (aggregate of bacterial cells). Bioflocculation is essential for effective separation of microorganisms from the treated effluent in the downstream clarifiers, and this is directly linked with EPS content and composition (Laspidou & Rittmann, 2002; Rahman et al., 2016). The key responsibilities of EPS in microorganisms are as follows:

- To aggregate bacterial cells to form larger flocs.
- To protect microorganisms against harmful environments such as toxic substances contained in the wastewater and sudden changes of pH.
- To remove organic matter (especially particulate and colloidal COD) through adsorption process. The adsorption mechanism is assumed to entrap both biodegradable and unbiodegradable (inert) organic compounds from wastewater using EPS excretion outside of the cell surface. Extracellular polymeric substances are also capable of removing other pollutants, nutrients, and minerals from wastewater.

Typically, EPS are made of protein and polysaccharides (carbohydrate) organic substrates, where protein is the dominant component. Extracellular polymeric substances can be measured in MLSS samples of activated sludge systems to understand process and operational performance.

There are several operational parameters that can affect EPS production such as SRT, dissolved oxygen (DO), the food-to-microorganism ratio, and so on.

SPECIFIC OXYGEN UPTAKE RATE

The SOUR is a facility control parameter that is defined as consumption of oxygen per unit of VSS concentration. It is typically expressed as milligrams of oxygen consumed per gram of VSS (mg O_2/g VSS) present in the activated sludge system. The SOUR can be measured quickly in a laboratory or on-site using typical laboratory equipment such as a biochemical oxygen demand (BOD) bottle, DO sensor, DO meter, magnetic stir bar, and magnetic stir plate. The measurement of SOUR has several benefits to facility operation:

- Influent COD biodegradability evaluation
- Identification of toxic or inhibitory substance present in the wastewater
- Determination of oxygen demand for an aeration tank
- Evaluation of operational stability and condition of a sample

The SOUR is typically calculated using the oxygen uptake rate (mg O_2/L/h) divided by the MLSS concentration (mg VSS/L). This provides a quick measurement of total oxygen demand required for an aeration tank. Table 8.1 illustrates typical ranges of SOUR in an activated sludge system and their potential indication in process conditions.

Table 8.1 Typical Ranges of SOUR Measured in Activated Sludge Treating Municipal Wastewater (Adapted from APHA et al., 2017)

Type of SOUR	Description	SOUR (mg O_2/g VSS)
Low SOUR	Indicates that the activated sludge system has too much MLSS concentration, or the presence of toxic substances in wastewater may induce low SOUR	<12
Normal SOUR	Indicates good BOD removal performance and well-settled sludge in the final clarifiers	12 to 20
High SOUR	Indicates that there is not enough MLSS concentration present in the activated sludge system for the incoming BOD loading to treat	>20

OXIDATION–REDUCTION POTENTIAL

Oxydation–reduction potential (ORP) analyzers measure the water's ability to oxidize (or reduce) organic matter (e.g., sludge) or compounds (WEF et al., 2021). The electric potential is measured in voltage. Oxidation–reduction potential values in wastewater systems and sludges are between –300 to +300 mV. As such, aerated systems will have a positive ORP (oxidizing), whereas negative ORP values describe a system that lacks oxygen (reducing) in the form of molecular (O_2) or other species, such as nitrate (NO_3).

HEAVY METALS AND TOXIC ORGANICS

Heavy metals such as zinc, copper, nickel, lead, cadmium, chromium, and mercury can be found in municipal sludge and have negative effects when used for agricultural land application. Many toxic substances are found in WRRFs that can inhibit the biological treatment process activity. These contaminants are harmful to humans and aquatic ecosystems and can be adsorbed onto the sludge. Other contaminants, such as per- and polyfluoroalkyl substances, endocrine disruptors (nonylphenols), pesticides, dimethyl sulfoxide, and quaternary ammonium compounds, are potential contaminants that are observed in WRRF sludge and have a negative effect on biological treatment processes. Thus, sludge characterization in activated sludge systems may require identifying inhibition activity and other problems.

TEMPERATURE, pH, ALKALINITY

Temperature and pH are common parameters measured in sludge samples and the methods are similar to those used in liquid samples, often relying on probes and multiparameter field instruments. Along with alkalinity (commonly measured as mg of $CaCO_3$/L), pH and temperature are important parameters for monitoring process stability in activated sludge as well as in aerobic and anaerobic digestion.

TEST YOUR KNOWLEDGE

1. At what SVI value of sludge sample represents sludge bulking of an aeration tank?
   a. 100 mL/g
   b. 125 mL/g
   c. >150 mL/g
   d. 130 mL/g

2. At what temperature is a sludge sample burnt in a furnace to measure VSS?
   a. 550 °C
   b. 103 °C
   c. 250 °C
   d. 400 °C

3. Which parameter indicates a quick measurement of total oxygen demand required for an aeration tank?
   a. BOD
   b. COD
   c. SSOUR
   d. TTSS

4. Which parameter helps bacteria to bind together to create flocs?
   a. COD
   b. SVI
   c. pH
   d. EPS

5. Total suspended solids and total solids concentration of a sample can be found directly by weighing the sample before and after drying at
   a. 103 °C
   b. 153 °C
   c. 503 °C
   d. 450 °C

Sampling of Sludge

TYPES OF SAMPLING

Sludge sampling from different liquid treatment processes at intermediate sampling points within the facility is critical to better understand process performance. There are two types sampling: grab sampling and composite sampling. Each of these is described in the subsections below.

GRAB SAMPLING

Grab sampling is a single sample collected from a particular location and time. It represents the characteristics of the wastewater stream at the time of grab sample collection. Grab sampling is typically suitable for a wastewater stream when the composition does not vary over time. However, grab samples always provide an instant representation of the wastewater source because wastewater characteristics can change in any short period of time. Grab sampling requires manual labor and minimal equipment and apparatuses during sample collection. However, it can be time-consuming if there is regular sampling required, especially for a large-scale sampling campaign. It is always recommended that grab samples be collected from a well-mixed location to represent the whole wastewater stream composition. Grab sampling is typically applicable for collecting samples from activated sludge systems, where the process condition does not change significantly.

COMPOSITE SAMPLING

Composite sampling is a combination of multiple grab samples collected over a period of time. This sampling technique is more reliable than grab sampling because it captures the variation of wastewater composition over time. In WRRFs, composite sampling is typically conducted for a 24-hour period. Composite sampling can be categorized into two types: automated and manual composite sampling. Automated composite sampling eliminates human error and labor cost from manual sampling.

For sludge characterization from the liquid treatment process, grab sampling is more preferrable than composite sampling because the process conditions are mostly under steady-state, well mixed, and slower process change conditions.

SAMPLING SIZE AND FREQUENCY

Site-specific conditions apply when determining the analysis and frequency to be performed on samples of the different types of sludges. The sample size needed to accommodate processing such samples is dependent of the type of parameters to be analyzed. In general, parameters and frequency respond to the need for process control, permitting, and planning. Daily samples of aerobic digestion feed sludge for analysis of TKN, for example, may not be warranted given the complexity and significance for the process; conversely, daily sampling for total solids analysis in the same process could be justified as a minimum for process monitoring.

SAMPLING POINT SELECTION

The sludge sampling and measurement plan from the liquid treatment process typically consists of routine monitoring (weekly sampling) and seasonal monitoring (short-duration campaigns conducted on a seasonal basis). Both the routine monitoring plan and seasonal monitoring plan include collection of data from instrumentation around the facility.

Routine monitoring is typically conducted on a weekly basis as part of the facility's regular operating procedures. The intent of routine monitoring is to understand trends in the facility's performance. Seasonal monitoring is typically developed to understand seasonal trends affecting the facility's

Table 8.2 Typical Sludge Sampling Point in a WRRF

Parameter	Units	Primary Clarifier Underflow	Aeration Basin Mixed Liquor	RAS/WAS	Solids Holding Tank	Aerobic Digester	DAF – WAS
Collection Method & Timing →		Grab (Special)	Grab, Online[1] (Regularly)	Grab (Regularly)	Grab (Special)	Grab (Special)	Grab (Special)
Water Quality Samples							
Total COD	mg/L						
Soluble COD	mg/L		√				
TSS	mg/L		√	√	√		
VSS	mg/L		√	√	√		
Ammonium-N	mg/L as N						
Nitrate-N	mg/L as N		√				
Nitrite-N	mg/L as N		√				
Total P	mg/L as P	√				√	√
Ortho P	mg/L as P					√	
SVI	mL/g		√				
Total Solids	%	√				√	√
Volatile Solids		√					√
Operational Parameters							
Flow	MGD	√		√			
DO	mg/L		√				
Temperature	C		√				
pH	—		√				

Legend: COD – Chemical oxygen demand, TSS – Total suspended solids, VSS – Volatile suspended solids, P – Phosphorus, SVI – Sludge volume index, DO – Dissolved oxygen, RAS – Return activated sludge, WAS – Waste activated sludge, DAF – Dissolved air flotation
[1]Online parameters are typically DO, temperature, pH, and total solids.

performance. The sludge characterization is critical for the underflow of the primary clarifier, aeration basin, RAS/WAS from the secondary or final clarifier, aerated sludge holding tank, aerobic digesters, and DAF systems. Table 8.2 summarizes the sludge characterization parameters that can be collected from a WRRFs and may require a special sampling campaign.

SAMPLE COLLECTION PROCEDURE

Different sampling type and volume may be suggested depending on the unit process being monitored. A sampling protocol should be developed in which collection procedure is detailed for each sampling point and each parameter to be analyzed (WEF et al., 2012). Minimum considerations during the collection of representative samples are the use of appropriate sampling containers, sampling location, sampling devices (e.g., sludge judge), and personal protective equipment as well as clear labeling and a chain of custody.

SAMPLE PREPARATION AND PRESERVATION

Sample preparation and preservation depend heavily on the type of analysis to be performed in a sample and the type of sample (grab or composite). It is paramount that the integrity of the sample is maintained during periods between sample collection and analysis. Short holding periods are strongly recommended;

when this is not feasible (e.g., 24-hour composite samples), provisions must be made for preserving samples (WEF et al., 2012). Common techniques for sampling preservation include cooling and storing at 4 °C and acidification; these methods, again, depend on the parameters to be analyzed.

TEST YOUR KNOWLEDGE

1. **Which sampling method is typically suitable for sludge when the compositions do not vary over time?**
   a. Composite
   b. Grab
   c. Diluted
   d. Monitoring

2. **In the clarifier, the composite sampling is typically made for a period of**
   a. 6 hours
   b. 12 hours
   c. 18 hours
   d. 24 hours

3. **Sludge characterization is critical for following what treatment process?**
   a. Overflow of primary clarifier
   b. Overflow of secondary clarifier
   c. Underflow of final clarifier
   d. Grit removal effluent

4. **Sludge samples should be stored at a temperature of**
   a. –4 °C
   b. 0 °C
   c. 4 °C
   d. 25 °C

CHAPTER SUMMARY

TYPES OF SLUDGE	■ Primary sludge consists of a high portion of organic matters accounting for 50% to 60% of the suspended solids. ■ Chemically hydrolyzed sludge includes the sludge from alkaline hydrolysis and acid hydrolysis. ■ Thermally hydrolyzed sludge increases the performance of anaerobic digestion. ■ Anaerobic digested sludge can be from primary sludge only, or blended with secondary sludge, high-strength substrate (such as fats, oils, and grease), or food waste.
SLUDGE CHARACTERISTICS	■ If SVI is above 150 mL/g, it is an indication of sludge bulking. ■ The COD in sludge includes pCOD, cCOD, and sCOD, where pCOD contains biomass concentration as well as particulate substrates. ■ Typically, EPS are made of protein and polysaccharides (carbohydrate) organic substrates, where protein is the dominant component. ■ Specific oxygen uptake rate provides a quick measurement of the total oxygen demand required for an aeration tank.
SAMPLING OF SLUDGE	■ Grab samples provide an instant representation of wastewater source. ■ Composite sampling is typically conducted for a 24-hour period. ■ A sampling protocol is developed where collection procedure is detailed for each sampling point. ■ Common techniques for sampling preservation include cooling and storing at 4 °C and acidification.

References

American Public Health Association, American Water Works Association, & Water Environment Federation. (2017). *Standard methods for the examination of water and wastewater.* American Public Health Association.

De Clippeleir, H., Miranda, M., Zhang, Q., Hasan, M., Al-Omari, A., Passarelli, N., … Debarbadillo, C. (2016). Optimization of belt filter press dewatering of thermally hydrolyzed biosolids: A Blue Plains case study on balancing desired cake solids with filtrate quality. *Proceedings of the 89th Annual Water Environment Federation Technical Exhibition and Conference.* https://doi.org/10.2175/193864716819715608

Hasan, M. (2017). *Evaluation of dewatering characteristics of wastewater sludge from various treatment processes.* The George Washington University. https://www.proquest.com/openview/ae016e9705768dfdfe29961e03c1d2eb/1?pq-origsite=gscholar&cbl=18750&diss=y

Hasan, M., Zhang, Q., Riffat, R., Al-Omari, A., Murthy, S., Higgins, M. J., & De Clippeleir, H. (2017). Mechanistically understanding the dewatering fundamentals: Impact of biological systems and thermal hydrolysis on cake total solids and polymer demand. *Proceedings of the Water Environment Federation.* https://doi.org/10.2175/193864717821495537

Laspidou, C. S., & Rittmann, B. E. (2002). A unified theory for extracellular polymeric substances, soluble microbial products, and active and inert biomass. *Water Research.* https://doi.org/10.1016/S0043-1354(01)00413-4

Melcer, H. (2004). *Methods for wastewater characterization in activated sludge modelling.* IWA Publishing.

Rahman, A., Meerburg, F. A., Ravadagundhi, S., Wett, B., Jimenez, J., Bott, C., ... De Clippeleir, H. (2016). Bioflocculation management through high-rate contact-stabilization: A promising technology to recover organic carbon from low-strength wastewater. *Water Research.* https://doi.org/10.1016/j.watres.2016.08.047

Water Environment Federation & Association of Boards of Certification. (2018). *Wastewater treatment fundamentals I: Liquid treatment.* Water Environment Federation.

Water Environment Federation & Association of Boards of Certification. (2021). *Wastewater treatment fundamentals II: Solids handling and support systems.* Water Environment Federation.

Water Environment Federation, Water Environment Research Foundation, & U.S. Environmental Protection Agency. (2012). *Solids process design and management.* Water Environment Federation.

CHAPTER 9

Management of Solids

Introduction

Solids management describes how we use or dispose solids that have been removed from wastewater, processed, and are ready to leave the water resource recovery facility (WRRF). The solids are typically biosolids, but could also be scum, grease, screenings, grit, or ash. The Water Environment Federation (WEF), along with the United States Environmental Protection Agency (U.S. EPA), define biosolids as "primarily organic solids produced by wastewater treatment processes that can be beneficially recycled" (U.S. EPA, 2018). Before treatment, this material is called "wastewater solids" or "sludge." To be classified as biosolids, the material must meet state and U.S. EPA criteria for beneficial use.

LEARNING OBJECTIVES

Upon completing this chapter, you will be able to:

- Describe the basic regulatory structure governing biosolids treatment to ensure compliance.
- Identify the physical, chemical, and biological characteristics important in assessing biosolids quality.
- Recognize the handling measures that can affect the final biosolids quality.
- Discuss management options for developing a strong biosolids disposal and reuse plan.

Purpose and Function

Operators should be familiar with not only the processes for biosolids treatment, but also reuse options. This chapter will outline the types of solids along with their characteristics and how to manage them effectively.

Types of Solids

Water resource recovery facilities receive and generate many types of solids, including ash, scum, grease, screenings, grit, and sludge (e.g., raw, treated, primary, secondary, tertiary, combined, and chemical). Some also receive septage from on-site treatment systems and liquid sludge from other WRRFs. (For more information on solids, see Book II.) U.S. EPA has mandated requirements to help ensure that biosolids are processed, handled, and treated in a manner that minimizes potential risk to human health. Based on the treatment methods, U.S. EPA has divided biosolids into "Class A" and "Class B" designations (U.S. EPA, 2020b). The treatment processes proposed for Class A biosolids eliminate pathogens, including viruses, whereas treatment requirements for Class B biosolids do not completely eliminate pathogens in the biosolids. As a result, the use of Class B biosolids may require a permit from U.S. EPA with restrictions on land application, crop harvesting, and public access to allow time for pathogen degradation.

Water resource recovery facilities also generate solid wastes, including paper, garbage, automotive residuals, equipment oils and greases, chemical-impregnated rags, lab wastes, office wastes (e.g., printer cartridges and copy- and fax-machine chemicals), light bulbs, batteries, and barrels. These wastes must be disposed of or recycled according to local regulations. (Given that WRRFs are supposed to protect the health of both the public and the environment, they should recycle as much of their wastes as possible [Fischelli, 1992].)

REGULATORY CONSIDERATIONS

Each WRRF has its own unique Biosolids Management Plan based on the applicable laws, regulations, and local ordinances. A summary of federal and state regulatory frameworks for biosolids is provided below.

FEDERAL REQUIREMENTS

Under Section 405(d) of the Clean Water Act (CWA), U.S. EPA established numeric limits and management practices to prevent adverse effects of chemical and microbial pollutants to human health during disposal of sludge and issued regulations and standards currently found in 40 Code of Federal Regulations (CFR) Part 503 (U.S. EPA, 2020b). U.S. EPA also revisits these regulations every 2 years to identify and include any additional pollutants that may occur in biosolids that could be harmful to human health and the environment (U.S. EPA, 2020b).

Any WRRF that generates sludge and performs land application and/or incineration with the owners and operators of surface disposal sites must comply with the requirements under 40 CFR Part 503. 40 CFR Part 503 is a self-implementing rule, meaning the requirements of Part 503 must be met even in the absence of a permit. The Part 503 rule describes the general requirements, requirements for land application, placement on a surface disposal site, and pathogen and vector attraction reduction and incineration, with information pertaining to who the rule applies to, exclusions from the rule, site-specific permits, and other relevant information (U.S. EPA, 2020b). Both Class A and Class B biosolids have specific treatment requirements for pollutants, pathogens, and vector attraction reduction governed by the Part 503 rule.

STATE REGULATIONS

In addition to federal rules, the WRRF must also confirm compliance with state and local regulations for disposing of biosolids. These state regulations may be more stringent than the federal Part 503 rule and may regulate certain types of biosolids more strictly than the federal regulations. Associated state and local permitting authorities will need to be consulted to determine compliance specifications regionally. Sometimes, the National Pollutant Discharge Elimination System (NPDES) permit program that is authorized to state governments by U.S. EPA to oversee compliance with CWA includes a biosolids component to it. To date, nine states have been authorized to implement biosolids programs as part of their NPDES program (U.S. EPA, 2020b). If a state has its own governing rules for handling biosolids, the owner/operator is required to follow the most stringent portions of both the federal and state rules (U.S. EPA, 2018). U.S. EPA maintains a comprehensive list of major legislative mandates, fact sheets, biennial review reports, and updates to existing regulations that have been developed on biosolids management since 1976 (U.S. EPA, 2020b). This list will serve as a key resource for operators, especially when the regulations are constantly evolving.

SITE PERMITS

In conformance with the Part 503 rule, any person owning a WRRF must apply for a permit covering biosolids use or disposal standards. Owning or operating a WRRF here refers to any facility that generates, changes the quality of, or provides final disposition of solids as outlined in 40 CFR Part 503 (U.S. EPA, 2018). Biosolids incinerator owner/operators and certain surface disposal sites with unique site conditions are required to have site-specific pollutant limits in their permits. In most cases, Part 503 requirements will be incorporated into the NPDES permits issued to publicly owned treatment works (POTWs) and WRRFs. For non-NPDES facilities, permitting authorities of U.S. EPA regions and approved states may issue "biosolids-only" permits covering applicable Part 503 requirements. Part 503 has a specific list of exclusions that covers areas of biosolids management for which there are no requirements under this rule; these are summarized below. Part 503 does not include requirements for

- Treatment of biosolids before final use or disposal except for operational parameters that must be met under Part 503 pathogen and vector attraction reduction requirements;
- The selection of the biosolids use or disposal practice because it is a local decision;
- Incineration of biosolids with other wastes;
- Placement of biosolids on land for 2 years or less;
- Industrial sludge generated during industrial wastewater treatment with or without combined domestic wastewater;
- Use or disposal of sludge identified as hazardous in accordance with 40 CFR part 261;
- Use or disposal of sludge with a concentration of polychlorinated biphenyls (PCBs) ≥50 mg/kg of total solids (dry weight basis);
- Use or disposal of ash generated during the firing of sludge in a biosolids incinerator;

- Use or disposal of grit or screenings generated during preliminary treatment of domestic wastewater;
- Use or disposal of sludge generated during the treatment of either surface water or groundwater used for drinking water; and
- Use or disposal of commercial septage, industrial septage, a mixture of domestic septage and commercial septage, or a mixture of domestic septage and industrial septage.

POLLUTANT LIMITS

40 CFR Part 503.13 establishes pollutant limits for biosolids based on the type of management practice. There are three categories of biosolids application for which pollutant limits in the form of cumulative loading rate of pollutants and monthly average or ceiling concentrations of individual pollutants are set, including land application, surface disposal, and incineration (U.S. EPA, 2021a). The pollutants regulated in the land application of biosolids are arsenic, cadmium, copper, lead, mercury, molybdenum, nickel, selenium, and zinc; Table 9.1 presents the limits set for each metal (U.S. EPA, 2021a).

Characterization

The physical, chemical, and biological characteristics of biosolids affect both the use of biosolids and the public's perception of biosolids. For example, while federal and state regulators base land-application requirements on biosolids' pollutant and pathogen concentrations, a land-application site's neighbors may base their acceptance on the material's consistency and odor.

Biosolids should be sampled for metals and organics after they have been processed. The analytical results will describe the actual material to be used or disposed of. Once the chemical and biological standards have been met, the biosolids characteristics that will affect biosolids management options are nutrient concentrations that can possibly affect surface water quality; micronutrient concentrations, which can affect soils and plant growth; and concentrations of various chemical and metal constituents (e.g., mercury and sulfur), which can affect an incinerator's air pollution controls.

It is recommended that WRRFs keep a large database containing multiple years' worth of chemical analysis results, including not only those required by permits, but also any additional testing that has been conducted on-site or at land-application sites, particularly those for metals, organics, and pathogens. These data help with establishing a track record of compliance and can aid in public education related to

Table 9.1 Regulatory Determination and Pollutant Limits for Land-Applied Biosolids (U.S. EPA, 2021a).

Pollutant	Ceiling concentration, milligrams per kilogram (dry weight basis)	Cumulative pollutant loading rate, kilograms per hectare	Monthly average concentrations, milligrams per kilogram (dry weight basis)	Annual pollutant loading rate (kilograms per hectare per 365-day period)
Arsenic	75	41	41	2.0
Cadmium	85	39	39	1.9
Copper	4300	1500	1500	75
Lead	840	300	300	15
Mercury	57	17	17	0.85
Molybdenum	75			
Nickel	420	420	420	21
Selenium	100	100	100	5
Zinc	7500	2800	2800	140

biosolids. The data collected should be within quality control and quality assurance parameters, including external checks. To demonstrate consistency, the data should present stable or decreasing trends in metals and organics concentrations. If this is not achievable, then a response plan should be developed by the WRRF.

Also, an active pretreatment program is recommended to minimize unwanted constituents in the influent of the WRRF. Such constituents tend to end up in the solids, and the presumption may be that the resulting biosolids should not be land applied if these constituents might cause the soil to need remediation for certain uses. To demonstrate that a WRRF incorporates best management practices (BMPs) into their biosolids management program, staff should

- Do necessary sampling and analysis and demonstrate that regulatory standards are being met;
- Do additional sampling and analysis, including priority pollutants and possibly others on a regular basis, because telling the public that the facility's biosolids meet all federal and state standards does not always inspire confidence; specifying how the biosolids surpass those standards does;
- Maintain an effective pretreatment program (even if there are no industries in the facility's service area) and an active pollution prevention program to help all contributors reduce their use of pollutants;
- Develop and use a written sampling plan and chain-of-custody form;
- Review every analysis, including those done by a contract lab, to verify that the appropriate test was done, the results are accurate, and the appropriate quality control measures were taken;
- Review tests performed on-site based on established quality control procedures; and
- Ensure that the biosolids look innocuous. For instance, pieces of fast food wrappers, condiment pouches, or personal sanitary products should not be seen in the material.

A successful land-application program is based on the careful consideration of multiple factors (e.g., the biosolids to be applied, the suitability of proposed land-application sites, and the estimated application rates [which determine the amount of land required]). When biosolids are used in agriculture, regulations and crop needs typically determine which biosolids characteristics are required (e.g., pathogen content, vector attraction, and nutrient content [typically nitrogen, phosphorus, and/or calcium]). Site conditions and economics (e.g., distance to land-application sites, application equipment typically in use) also may determine whether the biosolids should be applied in liquid or cake form.

PHYSICAL CHARACTERISTICS

The textures and particle sizes of biosolids are as varied as the processes used to create them. Heat-dried biosolids are typically granular, and can range in size from a fine dust, to irregular or spherical particles the size of a small ball bearing, to larger, "noodle-like" pieces up to a few inches long. Composts have a texture similar to that of peat, but depending on how they were made, may contain some wood fibers or wood chips. Lime-stabilized solids may be clay-like or soil-like, depending on moisture content and other factors. Aerobically and anaerobically digested biosolids may be liquid or dewatered into a clay-like material.

PATHOGENS AND VECTORS

Pathogen content of wastewater treatment solids depends on the level of treatment they have received. Raw primary solids contain the most pathogens due to lack of treatment. In contrast, biosolids that meet Part 503 Class A standards have low pathogen concentrations, which can, in some cases, even be below detection limits. Concentrations of specific pathogens (e.g., helminths) depend on the sources in a sewershed as well as the climate.

Before they can be land applied, biosolids must meet Class A or B pathogen-reduction standards (as well as relevant vector attraction reduction and pollutant concentrations). Land-application sites that receive Class B biosolids have additional management restrictions (e.g., waiting periods before specific crops can be grown) (U.S. EPA, 2018).

Part 503 requires that Class B biosolids contain less than 2 million colony-forming units (most probable number [MPN]) of fecal coliforms per gram of dry biosolids. Class A biosolids must contain less than 1000 MPN/g of fecal coliforms or less than 3 MPN/4 g of Salmonella bacteria and meet one of six alternatives. One alternative includes testing for enteric viruses and helminth ova (U.S. EPA, 2003). Larger

Six Alternatives for Meeting Class A Pathogen Requirements (U.S. EPA, 2003)

1. Thermally treated biosolids
2. Biosolids treated in a high-pH—high-temperature process
3. Biosolids treated in other processes (requires enteric virus and helminth ova testing)
4. Biosolids treated in a process to further reduce pathogens (PFRP)
5. Biosolids treated in a process equivalent to a PFRP as determined by the permitting authority.

WRRFs may have the personnel, equipment, and accreditation to handle bacterial testing on-site, whereas smaller WRRFs may send samples for bacterial testing to commercial laboratories with appropriate accreditation. Virus and helminth ova testing is nearly always performed by commercial laboratories with appropriate accreditation.

Land-applied biosolids also must meet vector attraction reduction (VAR) requirements via thermal drying, volatile solids reduction (VSR), oxygen uptake rate, time and temperature, or pH increases (U.S. EPA, 2003). Nine VAR options are based on process parameters or testing; three are based on land management (e.g., soil incorporation or injection). To qualify for Exceptional Quality (EQ; see sidebar), the generator must meet one of the process-based VAR options.

For details on pathogen and VAR requirements, methodologies, monitoring, and testing, see *Environmental Regulations and Technology: Control of Pathogens and Vector Attraction in Sewage Sludge* (also called the "White House document") (U.S. EPA, 2003).

METALS

Biosolids contain a wide range of chemical elements, many of which are trace elements, also called heavy metals. This is misleading because the density of some trace elements (e.g., boron and arsenic) is too low to be considered a heavy metal. Likewise, not all trace elements behave as metals in the environment. The ones with metallic chemical behavior are those that are cations in the environment (e.g., copper [Cu^{21}], cadmium [Cd^{21}], and lead [Pb^{21}]). Other trace elements are anions in the environment (e.g., arsenic [AsO_4^{32}], molybdenum [MoO_4^{22}], and selenium [SeO_4^{22}]). This distinction is important because cationic trace elements are more bioavailable when soil pH is low, whereas anionic trace elements are more bioavailable when soil pH is high.

When developing 40 CFR 503, U.S. EPA conducted a risk assessment on a number of trace elements in biosolids and developed a Hazard Index, which considered various pathways of exposure (e.g., direct ingestion of biosolids, plant uptake, and uptake by soil organisms) from trace elements in land-applied biosolids (U.S. EPA, 1995). The agency eliminated two trace elements (fluorine and iron) based on the Hazard Index. Then, regulators developed acceptable concentrations for the remaining 10 elements based on a pathway analysis. The ceiling concentration is the maximum safe concentration for biosolids land applied at agronomic rates.

A lower concentration also was established for what was later called EQ biosolids. The intent of the concentrations for EQ biosolids is that these levels would be safe no matter how much biosolids were applied to land. It is important to note that while biosolids may meet the metals levels associated with EQ biosolids, they cannot be considered EQ biosolids without meeting the Class A pathogen requirements as well.

Part 503 originally regulated 10 trace elements: arsenic, cadmium, chromium, copper, lead, mercury, molybdenum, nickel, selenium, and zinc. Lawsuits filed against U.S. EPA resulted in the removal of chromium as a regulated metal and the removal of the EQ limit for molybdenum.

When Part 503 was made know publicly, there were a number of municipal WRRFs whose biosolids failed to meet the EQ limits for one or more elements. Most, however, met the ceiling concentrations so their biosolids could be land applied, but there was a limit on the cumulative loading to a given site (U.S. EPA, 1995). By the late 1990s, industrial pretreatment was so successful that most WRRFs could meet EQ limits. Because of this, trace elements in biosolids are no longer considered an important limitation in land-application programs.

NUTRIENTS

Biosolids contain various concentrations of macro- and micronutrients. The nitrogen content in biosolids ranges from 1.0% (dry weight) for Class A alkaline-stabilized biosolids to up to 6.0% (dry weight) for heat-dried biosolids. Primary and waste activated solids mostly contain organic nitrogen (in the form of protein) while up to one third of total nitrogen in aerobically and anaerobically digested biosolids is ammonia. Biosolids contain little nitrate.

Vectors and Vector Attraction

Vectors are organisms such as rodents and insects that can spread disease by carrying and transferring pathogens (U.S. EPA, 2003). Vectors, including flies, mosquitoes, fleas, rodents, and birds, can transmit pathogens to humans and other hosts physically through contact or biologically by playing a specific role in the life cycle of the pathogen. Reducing the attractiveness of biosolids to vectors reduces the potential for transmitting diseases from pathogens in biosolids (U.S. EPA, 2003).

Vector Attraction Reduction Options (U.S. EPA, 2003)

Option 1: Meet 38% reduction in volatile solids content.
Option 2: Demonstrate VAR with additional anaerobic digestion in a bench-scale unit.
Option 3: Demonstrate VAR with additional aerobic digestion in a bench-scale unit.
Option 4: Meet a specific oxygen uptake rate for aerobically digested biosolids.
Option 5: Use aerobic processes at greater than 40 °C for 14 days or longer.
Option 6: Alkali addition under specified conditions.
Option 7: Dry biosolids with no unstabilized solids to at least 75% solids.
Option 8: Dry biosolids with unstabilized solids to at least 90% solids.
Option 9: Inject biosolids beneath the soil surface.
Option 10: Incorporate biosolids into the soil within 6 hours of application to or placement on the land.
Option 11: Cover biosolids placed on a surface disposal site with soil

(continued)

or other material at the end of each operating day (note: only for surface disposal).
Option 12: Alkaline treatment of domestic septage: to pH 12 or above for 30 minutes without adding more alkaline material.

Note: Options 1–8 and 12 are the process-based options; options 9–11 are based on land management.

Agronomic rates are determined by the nutrients in the biosolids as compared to the nutrients needed by the crops to be grown. This is typically done based on nitrogen, though phosphorus is also important. The purpose is to reduce the excess nutrients that could run off from a site and affect ground and/or surface waters.

"Exceptional Quality biosolids" refers to biosolids that meet Class A pathogen-reduction requirements, meet VAR requirements [503.33(a)(1) through (8)] at the same time or after pathogen reduction is met, and have low concentrations of regulated pollutants (503.13, Table 3).

The total phosphorus content in biosolids typically ranges from 0.9% to 3.1% (dry weight). This includes both organic phosphorus and various forms of inorganic phosphorus. Water resource recovery facilities that use iron and aluminum salts for phosphorus removal will have solids that contain inorganic phosphorus in the form of iron and aluminum phosphates. Water resource recovery facilities that use lime for dewatering or disinfection will have solids that contain inorganic phosphorus in the form of calcium phosphates. Iron, aluminum, and calcium phosphates are relatively insoluble, which will limit the availability of phosphorus in soils receiving repeated applications of biosolids. Biosolids also contain small amounts of potassium—typically 0.1% to 0.3% (dry weight)—and are not a significant source of potassium for crops. Most of the potassium in biosolids is in a water-soluble form.

In addition, biosolids contain various concentrations of other macro- and micronutrients (e.g., calcium, magnesium, sulfur, iron, manganese, as well as the regulated elements [copper, molybdenum, selenium, zinc]). Some plant species also require nickel for growth. The plant availability of these nutrients depends on the major chemical phases in the biosolids. Elements like copper and zinc are known to be complexed with the biosolids organic matter, and most sulfur in biosolids is in the form of protein. Iron and manganese exist as relatively insoluble oxides.

OTHER CONSTITUENTS

Biosolids contain many organic and inorganic chemical compounds that come from different sources. Some are discharged into the collection system as household chemicals (e.g., surfactants and conditioning agents used in detergents) and pharmaceuticals and personal care products (e.g., hormone prescriptions) and become concentrated in biosolids. Biosolids also contain natural materials (e.g., silica and aluminosilicate clays) that enter the collection system as sediments. For quantitative estimates, U.S. EPA has conducted three national sludge surveys, with the most recent in 2009. These survey results present national estimates of the concentrations of more than 500 pollutants in sludge, including metals; dioxins and dioxin-like compounds; inorganic ions; certain organics (e.g., polycyclic aromatic hydrocarbons, semivolatiles); polybrominated diphenyl ethers (flame retardants); and pharmaceuticals, steroids, and hormones (U.S. EPA, 2009).

In the case of alkaline-stabilized biosolids, lime and lime-containing chemicals (e.g., alkaline coal ash, cement kiln dust, and lime kiln dust) are added to biosolids for disinfection. The lime reacts with water in biosolids to form calcium hydroxide, which has a pH of 12 to 12.5 and adds value to the biosolids because alkaline-stabilized biosolids often are used as a substitute for agricultural limestone.

TEST YOUR KNOWLEDGE

1. **Which of the following is the ultimate goal of the U.S. EPA regulations related to biosolids?**
   a. Minimize cost
   b. Speed of treatment
   c. Maximize the quantity treated
   d. Protect human health

2. **In biosolids treatment, different classes reflect different levels of treatment. Which class represents the most stringent level?**
   a. Class A
   b. Class B
   c. Class EQ
   d. Class Z

3. **The primary federal regulation pertaining to biosolids is**
   a. 40 CFR 503
   b. The Clean Water Act
   c. Vector Attraction Reduction Law
   d. National Pollutant Discharge Elimination System

4. **How do state regulations relate to federal regulations?**
   a. State regulations can be less stringent.
   b. State regulations can only be more stringent.
   c. The state cannot make its own regulations.
   d. States are always responsible for making separate regulations.

5. **Which of the following is one of the pollutants regulated in biosolids under the Part 503 rule?**
   a. Phosphorus
   b. Potassium
   c. Selenium
   d. Nitrogen

6. **An active pretreatment program is implemented to**
   a. Minimize unwanted constituents in the influent
   b. Reduce odors in the facility
   c. Remove pathogens
   d. Reduce the flowrate through the facility

7. **Which of the following are essential to a successful land-application program?**
   a. All testing on the biosolids must be done by outside laboratories.
   b. The biosolids must be in the form of a fine dust.
   c. The type of crops and nutrient needs of those crops should be considered.
   d. The site must have fencing of at least 10 ft in height.

8. **Class A and B biosolids must meet which additional requirements besides the pathogen requirements?**
   a. Surface water quality criteria
   b. Vector attraction reduction
   c. Crop application rate standards
   d. Carbon reduction criteria

Handling

While all wastewater solids and septage must be treated to meet the relevant federal, state, and local regulations, differences in processing methods can result in biosolids with widely varying consistencies, which will affect how they can be transported and beneficially used. Handling methods can greatly affect the consistency of biosolids; screw conveyors to move the biosolids can turn them into a paste that is difficult to apply via manure spreaders, whereas belt conveyors have less effect on the consistency.

For example, two modifications to the solids treatment process at the Allegheny County Sanitary Authority in Pittsburgh, Pennsylvania, had unexpected results. Originally, the biosolids were dewatered via belt presses and then mixed with lime. This produced a crumbly mixture that was easily applied via standard manure spreaders. When facility staff replaced the lime-addition equipment with mixers and screw conveyors, the biosolids became a "pudding" that was difficult to spread. However, when staff replaced the belt presses with centrifuges, the material reverted back to the crumbly, spreadable mixture. Each WRRF sludge is unique, so there needs to be some trial and error to determine what solids handling equipment will provide a specific solids consistency.

CHEMICAL ADDITIVES

Any compound added during wastewater treatment typically ends up in the biosolids and may affect its physical, chemical, or handling characteristics. It also may affect the biosolids' permitted use and disposal options.

Lime and sodium bicarbonate, which are added during anaerobic digestion to control pH, should not be problematic; they only increase the concentration of compounds that occur naturally in biosolids and are typically harmless. Likewise, the acids and caustics used to control pH in aerobic digesters should not affect the biosolids' ability to meet regulatory standards. Calcium, on the other hand, can change the soil chemistry and nutrient uptake at land-application sites, so agronomists may recommend occasional soil tests (e.g., once every 3 years) to check the macro- and micronutrient levels at land-application sites.

Agronomists study crops and soil science, including how the two interact. They are experts in nutrients and how crops grow along with climate effects on crops (American Society of Agronomy, https://www.agronomy.org/about-agronomy).

Some WRRFs use organic polymers or inorganic chemicals to thicken or dewater sludge. Some regulators are concerned that nitrogen-based polymers may decompose to produce amines and other intensely odorous compounds. However, research has shown that the polymer may adsorb to particulate, making it less biodegradable and significantly less reactive—effectively, an inert material (Dentel et al., 2000). The portions of the polymer that do break off are typically innocuous and biodegradable. So, polymer additions should not prohibit the biosolids from being land applied safely.

Solids-conditioning chemicals (e.g., lime, ferric chloride, ferrous sulfate, ferrous chloride, and aluminum sulfate [alum]), which may be added before or after thickening or dewatering processes, should not affect the biosolids' ability to meet regulatory standards. Some WRRFs use potassium permanganate to control odors; this also should not affect the biosolids' ability to meet regulatory standards.

TREATMENT PROCESSES

Most sludge stabilization methods (e.g., anaerobic digestion, aerobic digestion, composting, lime stabilization, thermal treatment, and heat drying) produce land-applicable biosolids, but their products' characteristics and required handling methods will vary. Anaerobic digestion, for example, produces biosolids with low levels of organics and bacteria. However, the material may be somewhat odorous and difficult to dewater.

Aerobic digestion, which is mostly used at small WRRFs, produces a land-applicable biosolids that may be harder to thicken than other types of biosolids.

Composting, when done properly, produces a relatively dry, biologically stable, odor-free biosolids that may be stored outside without developing odors or attracting insects. Some WRRFs sell this material, thereby reducing solids-handling costs. Composted biosolids typically look better and emit fewer odors than other types of biosolids.

Some state regulators question the quality of lime-stabilized biosolids due to historical concerns with stability of the pH of the resulting materials and the potential for odors if the pH does not hold as indicated. Research has shown, however, that if enough alkaline (or lime and heat) is added, then the resulting biosolids meet the pathogen-reduction and vector-attraction-reduction standards for EQ biosolids and are less likely to be odorous because the odorous gases are driven off during processing and can be collected and treated then.

Thermal treatment processes destroy pathogens, improve the resulting biosolids' thickening and dewatering characteristics, and greatly reduce their volume. Although odors are produced during treatment, the final biosolids produced have little odor (as long as the material remains dry). However, the biosolids may have higher metals concentrations and could spontaneously combust. Dewatering thermally treated biosolids by mechanical drying methods or drying beds may create odor problems and may be costly if covers are needed to contain odors or, if outside temperatures require them, to maintain the internal temperatures.

Incineration destroys organic matter, reduces the sludge volume, and produces an ash that may be beneficially used if its metals levels do not exceed regulatory standards. At temperatures well above 816 °C (1500 °F), the ash may become a hard frit that can be used as a fill, roadbed material, or ceramics ingredient. This option may be attractive to larger WRRFs that have difficulty finding enough land for composting or land application.

Emerging sludge-stabilization technologies include a microwave technology and methods involving quicklime and heat. For more information on additional sludge-stabilization methods, see Chapter 10, "Additional Stabilization Methods," of this manual.

Management

USE OPTIONS

Biosolids that meet the requirements of U.S. EPA's Standards for the Use or Disposal of Sewage Sludge (40 CFR 503), as well as state and local standards, can be beneficially used. Biosolids management should incorporate holistic waste management/reuse, value-added product development, and enhanced product performance. Recommended options should lead to simple and safe management alternatives that focus on reusing resources.

LAND APPLICATION

The most common method for using biosolids is land applying it on farmland. In 1988 (before ocean disposal stopped), 33% of all wastewater solids was land applied, according to the preamble of 40 CFR 503. A more recent U.S. EPA study gives a rough estimate that 60% of biosolids are now land applied. Before a WRRF can begin spreading (or injecting) biosolids on farmland, however, all federal, state, and local paperwork must be completed and approved. Some states require every site to be permitted, whereas others require the biosolids generator to obtain a general permit and maintain a register of individual sites. To obtain a general permit, the biosolids generator may need the following:

- An open and ongoing relationship with regulators (they should be familiar with the WRRF and its operations because they have toured the site at the facility staff's invitation);
- A database of biosolids test results (at least 3 years' worth, but preferably since facility startup because regulators will be more comfortable about letting a facility land apply biosolids if extensive physical, chemical, and pathogenic data demonstrate its safety);
- A history of the WRRF's operations and a detailed description of its solids processes;
- A quality control and quality assurance plan, standard operating procedures for each analysis, and chain-ofcustody forms or a logbook of all samples;
- A public relations plan and ongoing outreach efforts to inform the public about the biosolids management program;
- A written description of all land-application activities and personnel, along with the contact information for the WRRF staff and regulators who should be notified if a question arises or emergency occurs;
- A spill prevention and control plan, a contingency plan, and a long-range plan for future improvements to the biosolids management program; and
- A written procedure (and form or logbook) for handling complaints from each land-application site's neighbors.

Some states have training programs for biosolids generators and land appliers. Biosolids personnel should take such programs (whether mandatory or optional) because they will learn how to interact with regulators and what their expectations are. Biosolids personnel also should become active members of biosolids-related professional organizations (e.g., WEF and its member associations, and regional associations), which are sources of useful information, help, and support. In addition to committees devoted to biosolids issues, WEF and its Member Associations also publish technical materials.

Once a WRRF has met the general permit requirements, its personnel can establish the land-application program by doing the following:

- Find suitable farmland. The site must have enough acres to justify the expenditure of time and resources that go into obtaining a permit. The vicinity must be checked for wetlands, wind direction, remoteness, location of neighbors, local ordinances on biosolids use, and the general feeling about biosolids.
- Determine which farmers want the material, explain the approval requirements, verify ownership, obtain the necessary data and signatures, and notify the neighbors and any local officials (e.g., township or borough supervisors, county officials, conservation districts, and regional state regulators);
- Contact all of the relevant biosolids regulators; prepare the necessary paperwork; do the required field work (e.g., soil sampling every 10.1 ha [25 ac] or so) and lab tests; procure aerial photographs of the area and mark the fields, boundaries, and all water-related landmarks (e.g., waterways, wetlands, water supplies, and wells); determine the delivery truck's travel route; and complete any other work required by local agencies;
- Inform local agencies and the public about the benefits of beneficially using biosolids (especially if land application or the WRRF is new to the area), ensure that all program-related personnel (e.g., generators, transporters, and land appliers) understand the benefits of biosolids and can explain them properly to the public, and maintain an ongoing local outreach program for the benefit of anyone who inquires about land application; and
- Inspect the land-application sites regularly to verify that the work is done correctly (e.g., the trucks are not causing problems, mud is not tracked onto roads, odors are controlled, the site is well-maintained, and any unspread biosolids are neatly stored) and that a WRRF representative will be able to answer all questions during regulatory inspections.

LAND RECLAMATION

The use of residuals, largely municipal biosolids, has been successfully used to restore disturbed soils (acid strip mines, coal refuse piles, etc.) for more than 25 years (Alvarez-Campos & Evanylo, 2019; Brown et al., 2014; Haering et al., 2000; Sopper, 1992). Land reclamation is similar to land application except that the biosolids-application rates are higher because the application is not just to provide nutrients for the current growing season but to establish long-term growth on disturbed and nutrient-poor soils (Basta, 2000; Brown et al., 2014; Lu et al., 2012; Nicholson et al., 2018). Biosolids alone or in mixtures with other amendments (e.g., lime, woody material, fly ash, saw dust) have also been used to correct specific problems (Brown & Chaney, 2000; Brown et al., 2014). Restoring soil properties also contributes to vigorous vegetative growth, which is vital to land reclamation (U.S. EPA, 2007). Biosolids also have successfully reclaimed rangeland in New Mexico by providing organic content that helps retain moisture in the soil. Lime-stabilized biosolids work well on strip-mined ground where the pH may be low across the site—the lime content of the biosolids provides the pH adjustment needed to allow grasses to grow—but other types of biosolids are also effective as long as appropriate pH adjustments are made. The material's organic matter and slow-release nitrogen will provide a good base for most seed mixtures. However, if the seed mixture includes warm-season grasses, which develop slowly (over 2 to 3 years), they can be choked by faster-growing fescue and other grasses; thus, multiple applications of biosolids at lower rates may be necessary to avoid this problem. Biosolids, biosolids compost, and yard debris compost have been used to restore land to previous wetland characteristics. The nutrient-rich organic material provides an excellent growth medium for native wetland plant species, while stabilizing slopes (Northwest Biosolids, 2018).

Biosolids have also successfully revegetated capped cells at landfills. Once a cell has been filled, it is typically covered with a geomembrane or filter fabric covered with soil to keep water from infiltrating into the cell and to minimize leachate from the cell. The biosolids are then applied in the same manner as that used for reclaiming strip-mined land.

COMMERCIAL PRODUCTS

Some WRRFs (on their own or with a private company) compost biosolids for use in landscaping and gardening. Biosolids that can be bagged and sold may be used by the public for their gardens and potted plants. Products such as engineered soils, ornamental horticultural fertilizers, synthetic coal, turf grass, biochar, and activated carbon, sourced all or in part from biosolids, are available on the commercial market.

INCINERATION

Some WRRFs choose to incinerate biosolids because they are in colder areas and need an effective management method in winter when land application is not feasible. Some facilities in large metropolitan areas incinerate solids to avoid the odor complaints related to hauling Class B biosolids long distances through city neighborhoods. The incineration process is equipment- and energy-intensive and requires air pollution control devices; the leftover ash must also be used or disposed of. (For more information on incineration processes, see Chapter 10, "Additional Stabilization Methods," of this manual.)

Some WRRFs recover the heat generated during incineration and use it to heat other treatment processes, to generate electricity, or to make steam for heating or use in other treatment processes. Incinerator ash is typically landfilled, but it could be beneficially used if its metal concentrations are within regulatory limits (i.e., all pathogens were burned off during incineration). Biosolids incinerator ash is similar in physical and chemical characteristics to coal ash and can be used in applications in which coal ash is permitted. Incinerator ash may be acidic, neutral, or alkaline, depending on the solids characteristics. Also, incinerator ash will contain more phosphorus than coal ash. Although ash has historically been landfilled, interest in beneficial use of this material is increasing (Dominak, 2005). Beneficial use options for ash include:

- Landfill—Ash can be used alone or blended with soil and used as a landfill cover.
- Fill material—Ash can be used as fill material for excavations. One utility, for example, has a contractor using the material to fill old solids lagoons.
- Soil amendment—In some areas (e.g., those with high clay soils), incinerator ash may be used as a soil additive that produces a soil that handles more easily, allows better drainage and airflow, and includes some valuable minerals.
- Concrete fly ash—Ash has been used as a fly ash substitute in concrete mixes.
- Asphalt additive—Ash has been used as a mineral filler and fine aggregate in asphalt mixes. New Jersey's Department of Solid and Hazardous Waste has permitted the use of incinerator ash for this purpose.

To qualify for most of these uses, the ash must meet certain criteria and be available on demand at specified volumes. Not all WRRFs can meet such requirements, and they must compete with other industries that produce large volumes of suitable ash. Use options for incinerator ash tend to be site-specific. Utilities should pursue all avenues available for using ash. State departments of transportation should be contacted to determine requirements for using incinerator ash as a fly ash substitute in concrete or as a mineral filler or fine aggregate substitute in asphalt mixes. If the incinerator ash is approved for use in mix designs by the state department of transportation, then a considerable market can be opened for incinerator ash.

HEAT DRYING AND OTHER THERMAL PROCESSES

Heat drying produces a biosolids product that can meet all U.S. EPA and most state requirements for unlimited distribution as a soil conditioner or fertilizer. Most, though not all, drying systems are designed to meet the time and temperature requirements for Class A pathogen reduction. Details on the heat drying processes can be found in Chapter 10, "Additional Stabilization Methods," of this manual. The dryers typically also meet VAR standards simply because they routinely dry biosolids to 90% solids or greater. Heat drying does not concentrate pollutants in the biosolids; the dry weight concentrations in both feedstock (dewatered cake) and product (heat-dried biosolids) will be nearly identical. So, if the dewatered cake met the pollutant concentrations in 40 CFR 503's Table 3, the heat dryers routinely produce EQ biosolids, which are suitable for marketing as a soil conditioner or fertilizer.

The primary benefit associated with heat drying has historically been volume reduction. The potential for producing fertilizer with a serviceable nutrient content is another aspect that makes heat drying attractive. Heat drying typically preserves the nutrient content of the feedstock solids, except for nitrogen. Typically, heat drying removes all but a small portion of the soluble ammonia-nitrogen from dewatered biosolids. It does not remarkably change the dry weight concentrations of organic nitrogen or most of the other inorganic elements.

Water resource recovery facility managers should look at examples like the Milwaukee Metropolitan Sewerage District, which has produced and sold a heat-dried biosolids since about 1926, and at Houston, Texas, which has produced and distributed heat-dried biosolids since the 1950s. Both organizations developed successful markets for their products and, as a result, sales revenues play an important role in their budgets. However, such successes must be put into context; Milwaukee and Houston invested many years in developing and servicing their markets. New heat-drying facilities are unlikely to fully emulate these marketing programs until they have been operating for a number of years. Likewise, as the number of heat-drying facilities increases, there is likely to be a downward pressure on the prices of these products. Further, dryers require a lot of power, and increasing energy costs may negatively affect the economic feasibility of this option. Different drying technologies generate products of differing quality; revenues will depend on the needs and desired characteristics of the customer.

This biosolids management option is a proven technology in which odors can be contained and controlled. The resulting biosolids pellets have much less volume and weight than the influent solids and are easily handled, conveyed, and stored. They can be delivered to consumers in bulk or in bags or other containers.

The disadvantages of heat drying and pelletizing include the dust's potential to combust and the potential for overheating and fires. Also, the equipment is expensive, complex, and maintenance-intensive, requiring qualified operators. Air emissions control equipment also is required, especially because drying certain types of solids can result in more odorous pellets and quantities of dust. Solids also may be treated via the following thermal processes:

- Phased thermophilic digestion;
- Heat pasteurization, which produces a marketable fertilizer;
- Residuals gasification, which produces liquid, gaseous, and hydrogen-derived fuels; and
- Treating biomass thermochemically to produce a low- to medium-grade heat content gas.

DISPOSAL OPTIONS

Water resource recovery facilities typically decide whether to use or dispose of biosolids based on the cost of each evaluated option. Sludge that will ultimately be landfilled may need as much treatment as if it were to be land applied, so the related labor and trucking costs may be the overriding factors. There are two types of landfill sites that typically accept solids. One is a monofil, which is a landfill that only accepts stabilized or unstabilized municipal wastewater solids. The other is a co-disposal landfill, which

Landfill Specifics

Leachate and methane gas: Landfill gas is a natural byproduct of the decomposition of organic material in landfills and is composed of roughly 50% methane (the primary component of natural gas), 50% carbon dioxide (CO_2), and a small amount of non-methane organic compounds. (U.S. EPA, https://www.epa.gov/lmop/basic-information-about-landfill-gas).

Groundwater monitoring: Systems must be installed to collect samples from the uppermost aquifer (i.e., defined as the geological formation nearest the natural surface that is capable of yielding significant quantities of groundwater to wells or springs) both up-gradient and down-gradient of the landfill. The samples from the up-gradient wells show the background concentrations of constituents in the groundwater, while the down-gradient wells show any groundwater contamination caused by the landfill. The required number of wells and the spacing and depth of wells is determined on a site-specific basis (U.S. EPA, https://www.epa.gov/landfills/requirements-municipal-solid-waste-landfills-mswlfs#groundwater)

Vectors and nuisances: Operating practices, including compacting and covering waste frequently with several inches of soil, help reduce odor; control litter, insects, and rodents; and protect public health (U.S. EPA, https://www.epa.gov/landfills/municipal-solid-waste-landfills).

accepts both residuals and municipal solid waste. Landfills are subject to federal, state, and sometimes local regulations, which require liners; daily and final covers; collection of leachate and methane gas; groundwater monitoring; control of odors, vectors, and nuisances; and caps and other closure methods.

DIVERSIFICATION AND CONTINGENCY PLANS

Water resource recovery facilities do not just process wastewater, and solids management is not a small part of the operation. It requires at least as much time, effort, and financial commitment as wastewater does. So, all WRRFs should have a contingency plan for unexpected events (e.g., severe weather). Planning for regulatory changes, however, is another matter. Facility staff should establish a good working relationship with regulators and keep abreast of the regulatory changes under development. Changes that will result in more solids treatment will involve more money and equipment, which can take months or years to secure, install, and start up.

Diversification and contingency plans are highly dependent on the conditions surrounding a WRRF, including both geographic and economic considerations. An example would be a facility that performs the following split of treatment, energy recovery, and reuse:

- Fifty percent is incinerated, producing steam for use in various facility processes as an energy recovery method;
- Forty percent is land applied on farms and strip mines by contractors; and
- The rest is landfilled.

This would work in a facility located within city limits, surrounded by businesses and residences so they cannot store solids on-site because of potential odors. This diversity enables the WRRF to operate efficiently and without odors, regardless of maintenance needs or weather conditions.

FACTORS AFFECTING THE DECISION

The most important factor in choosing between different biosolids management options is the applicable laws, regulations, and local ordinances. Check carefully: even if land application is technically allowed, obtaining permits and site approvals can be difficult and time consuming. The regulators who permit generators and approve sites may have many duties and, therefore, little time for these tasks or for complaints from organizations that vehemently oppose land application. In some states, local governments have established local ordinances that impose fees and requirements that double the cost and effort of land application. The current trend is that regulators want WRRFs to produce odorless Class A biosolids.

Another important factor is whether to treat the facility's solids to meet Class A or Class B biosolids requirements. Typically, Class A treatment processes cost more than Class B processes. Most land-application programs use Class B biosolids, which work well on farms and strip mines and are acceptable to farmers and miners. Class A biosolids have lower pollutant and pathogen levels, and can be sold or given away to the public. (The sale price typically does not cover the entire production cost, but does help defray it.)

Other factors that affect the choice of biosolids management options include:

- The WRRF's location;
- Local climate;
- Past solids management practices (everyone is more comfortable with something they know);
- Distance from available beneficial use sites;
- Support of local government officials;
- Estimated costs;
- Available funding; and
- Local landfill availability and pricing.

Each option has advantages and disadvantages, and facility staff should avoid making the decision based solely on cost. Switching from a long-used method to a new one can require significant training investments, especially if the old method was simple (e.g., landfilling) and the new one is complex (e.g., composting or heat drying). Water resource recovery facility personnel also need to decide whether to manage solids inhouse or to outsource the work.

See Figure 9.1 below on the controlling factors related to biosolids reuse.

OUTSOURCING

Outsourcing is the process of hiring another entity to perform some of an organization's work. It enables WRRF personnel to decide how much of the solids management work they want to do and how much they prefer to let someone else do. For example, the WRRF could produce the biosolids and then hire a contractor to land apply it (e.g., perform all the work related to permitting, transporting, land applying, reporting, and communicating with the public and regulators). Alternatively, the WRRF could hire a contractor to handle all of the solids-management processes.

One advantage of outsourcing is that the facility does not have to expand its workforce. Also, facility personnel can rely on the expertise of a biosolids management specialist rather than "starting from scratch" themselves. The facility can also avoid having to invest in a fleet of specialized vehicles (e.g., tractors, spreaders, and front-end loaders), or—if the entire solids-management program is outsourced—any solids or biosolids handling, processing, or storage facilities.

One disadvantage of outsourcing is that the WRRF is still responsible for its contractor's actions, because as far as regulators are concerned, the biosolids generator remains their owner. So, facility personnel should be familiar with all aspects of the contractor's work, monitor the operations, and confirm that the program is effective and meets all requirements. Facility personnel also should maintain a suitable public outreach program and keep in contact with contractor staff, beneficial-use site owners, and regulators.

Once WRRF personnel have decided to hire a solids management contractor, they need to draft an agreement that covers all usual and exceptional matters affecting the services that they expect the contractor to perform. The bid specifications or request for proposal should be developed by a team that includes legal counsel, environmental scientists, engineers, and operations and maintenance staff, as well as input from regulators. (Most outsourcing WRRFs are willing to share their bid specifications.) Any contractor hired to manage a WRRF's solids should be well known in their field and have experienced staff, a good compliance record, and a good work history.

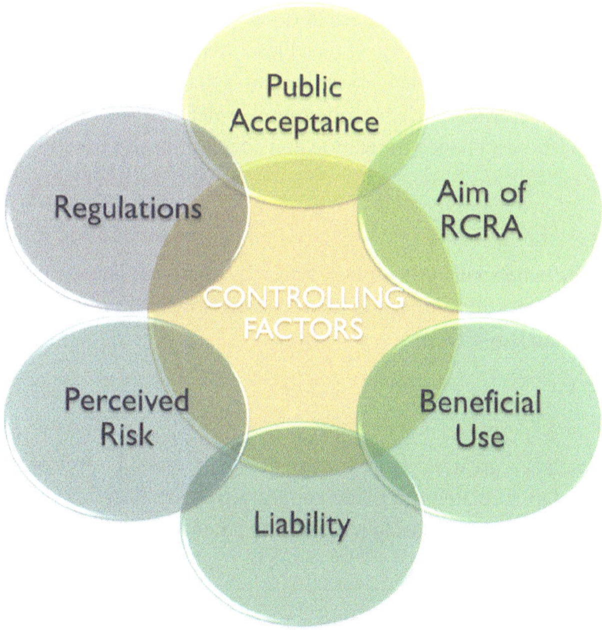

Figure 9.1 Controlling Factors Related to the Beneficial Reuse of Biosolids

ODORS AND PUBLIC ACCEPTANCE

Another important factor is the problem of odor complaints. Some complaints may be justified. Others will have nothing to do with the WRRF's biosolids; rather, they may be the result of odors from neighbors down the road. The complaints may even be more general: "We don't want your sewage sludge here." Justified or not, these complaints will cost the WRRF personnel much time and energy combating their effects on regulators, the news media, and public relations.

Although odor complaints cannot be completely avoided, WRRF personnel can address the problem by

- Producing a product that is as odorless as possible;
- Monitoring conditions at beneficial use sites (e.g., remoteness, wind direction, possibility of inversions, humidity, topography, and occurrence of holiday weekends);
- Encouraging regulators to educate the public about biosolids;
- Supporting environmental programs in schools and teaching students about biosolids;
- Establishing and maintaining public outreach programs on biosolids;
- Inviting local government and media representatives to tour the sites so they become familiar with biosolids and their benefits for farms and strip mines; and
- Recordkeeping.

Water resource recovery facility personnel should keep the required records for as long as the regulations dictate. All permits, lab analyses, and regulatory reports should be maintained in a well-organized, easily accessible manner for regulators or third-party verifiers when they inspect the solids facilities. Facility personnel should also set up an archive system. For example, any paper files should also be available via on-site digital storage and/or servers off-site so the hardcopies are backed up, but the data remain accessible.

REGULATORS AND INSPECTIONS

U.S. EPA and states can issue permits for biosolids management. For example, some biosolids permits are issued through the NPDES. 40 CFR Part 503 is a self-implementing rule. This means that the requirements of Part 503 must be met even if a permit has not been issued. An enforcement action can be taken against a person or WRRF who does not meet the requirements of Part 503 even when that person or WRRF does not have a permit for the use or disposal of sludge. The state and regional coordinators can be found through the U.S. EPA website (U. S. EPA, 2021b).

ENVIRONMENTAL MANAGEMENT SYSTEMS AND INTERNATIONAL ORGANIZATION FOR STANDARDIZATION 14001

Water resource recovery facilities and solids contractors are encouraged to develop environmental management systems (EMSs) to improve their production and use of biosolids. International Organization for Standardization (ISO) 14001 is the primary standard for the EMS and requires that an organization consider all environmental issues relevant to its operations such as air pollution, water and wastewater issues, waste management, soil contamination, climate change mitigation and adaptation, and resource use and efficiency.

OTHER SOLIDS

Other solids separated from wastewater may be beneficially used but are typically landfilled. The use of grit is being explored; it is washed to remove organics and slime, and then separated by size into sand and gravel for use as aggregate in asphalt mixtures and sub-base course materials.

Safety

The risk of worker exposure to pathogens in solids is greatest before treatment. With Class B biosolids, there is some uncertainty during treatment and immediately after land application of the biosolids. Because the concentration of pathogens declines through natural processes, the potential for pathogen exposure decreases over time.

Workers could be exposed to pathogens and irritants while working with Class B biosolids during the period when public access is restricted. Previous work by the National Institute of Occupational Safety and Health (NIOSH) has indicated any incidents of illness in workers in all phases of biosolids operation occurred when compliance was lacking, in particular inconsistent awareness, provision, and use of protective equipment and hygiene practices appropriate for handling Class B biosolids (or biosolids that do not comply with U.S. EPA standards). Water resource recovery facilities should provide a basic level of protection, including appropriate measures from those listed below:

1. Wash hands thoroughly with soap and water after any contact with solids.
2. Avoid touching face, mouth, eyes, nose, or open sores and cuts while working with solids.
3. Wash your hands before you eat, drink, or smoke and before and after using the bathroom.
4. Eat in designated areas away from solids-handling activities.
5. Do not smoke or chew tobacco or gum while working with solids.
6. Use barriers between skin and surfaces exposed to solids.
7. Remove solids from footgear before entering a vehicle or a building.
8. Keep wounds covered with clean, dry bandages.
9. Thoroughly but gently flush eyes with water if solids contact eyes.
10. Change into clean work clothing on a daily basis and reserve footgear for use at the worksite or during solids transport.
11. Do not wear work clothes home or outside of the work environment.
12. Use gloves to prevent skin abrasion.

In addition, NIOSH recommends providing appropriate protective equipment, hygiene stations, and training.

PERSONAL PROTECTIVE EQUIPMENT

Appropriate personal protective equipment (PPE) should be provided for all workers likely to have exposure to solids. The choices of PPE include goggles, splash-proof face shields, respirators, liquid-repellent coveralls, and gloves. Face shields should be made available for all jobs in which there is a potential for exposure to spray or high-pressure leaks, or aerosolized solids during land application. Management and employee representatives should work together to determine which job duties are likely to result in this type of exposure, to conduct appropriate on-site monitoring, and to determine which type of PPE is needed in conjunction with a qualified safety and health professional. If respirators are needed, a comprehensive program would include respirator fit-testing and training (or retraining).

HYGIENE AND SANITATION

Handwashing stations with clean water and mild soap should be readily available whenever contact with solids occurs. In the case of workers in the field, portable sanitation equipment, including clean water and soap, should be provided. Cabs should be wiped down and cleaned of residual mud (or settled dust) frequently to reduce the potential for exposure to solids.

TRAINING

Periodic training on standard hygiene practices for workers handling solids should be conducted by qualified safety and health professionals to cover issues such as the following:

- Frequent and routine handwashing (the most valuable safeguard in preventing infection by agents present in solids), especially before eating or smoking;
- The proper use of appropriate PPE, such as coveralls, boots, gloves, goggles, respirators, and face shields;
- The removal of contaminated PPE and the use of available on-site showers, lockers, and laundry services;
- Proper storage, cleaning, or disposal of contaminated PPE;
- Instructions that work clothes and boots should not be worn home or outside of the immediate work environment;
- Prohibition of eating, drinking, or smoking while working in or around solids; and
- Procedures for controlling exposures to chemical agents that may be in biosolids.

REPORTING

Workers should be trained to report potentially work-related illnesses or symptoms to the appropriate supervisory or health care staff. This may aid in the early detection of work-related health effects (U.S. Department of Health and Human Services et al., 2002).

Costs

The costs of various solids treatment processes delineated in Table 9.2 are based on the results of a study conducted in 2013 and 2014 (Sandberg & Reimers, 2013). From this work, it is noted that the cost of Class A treatment of sludge (disinfection) ranges from $500 to $1,300 per dry ton. This cost includes the processes of thickening, dewatering, disinfection, and stabilization. These costs are based on 2013 dollars and required the assistance of many reports and literature from 2005 and up. The costs include both capital and operation and maintenance costs, which include both thickening and dewatering. It should be noted that these costs vary depending on the region of the United States and Canada. The energy required by these processes is taken into account but may not include the final product costs such as marketing. Table 9.3 describes the distribution across the United States of the different types of disposal along with the average costs (Peccia & Westerhoff, 2015).

TEST YOUR KNOWLEDGE

1. **Which of the following are required by federal regulations for compliance?**
   a. A quality control and quality assurance plan
   b. Standard operating procedures for each analysis
   c. Chain-of-custody forms or a logbook
   d. An approved permit

2. **Once a WRRF has met the general permit requirements, its personnel can initiate a land-application program by doing which of the following:**
   a. Determine which farmers want the material. The site must have enough acres to justify the expenditure of time and resources that go into obtaining a permit.
   b. Have private meetings with only the farmer; the public does not need to know about the land application.
   c. Visit the land-application site only upon initiation of the project; there is no need for regular inspections.
   d. Purchase the crops for the proposed land-application site.

Table 9.2 Economics of Disinfection/Stabilization Sludge Processes Along With Potential Products (Adapted from Sandberg & Reimers, 2013).

Facility Size	10 mgd	100 mgd	Biosolids Products
Pyrolysis treatment	NA	$1000	Activated carbon, biochar and fuels
Thermal treatment or drying	$1200	$600	Fuel/fertilizer/turf grass
Modified commercial fertilizer	NA	$300	Commercial fertilizer
Biological anaerobic digestion	$800	$600	Biogas/soil amender/turf grass
Thermal hydrolysis mesophilic anaerobic digestion pretreatment	NA	$510	Biogas/soil amender/turf grass
Aerobic digestion	$700	NA	Soil amender/turf grass
Advanced open alkaline stabilization	$850	$650	Aglime agent/engineered soils
Advanced closed alkaline stabilization	$650	$500	Aglime agent/turf grass/fertilizer/ engineered soils
Acid stabilization/disinfection	$500	NA	Soil amender/turf grass

Table 9.3 Disposal Type, Distribution, and Average Costs of Biosolids (Adapted from Peccia & Westerhoff, 2015).

Disposal Type	Quantity of Biosolids Produced Using the Technology	Average Cost
Land application (dewatering, storage, stabilization, hauling, land application)	30% of all sludge produced	\$300–\$800 per dry ton
Landfill (tipping fees and hauling)	55% of all sludge produced. Of land-applied sludge, 60% is class B, 40% is class A	\$100–\$650 per dry ton
Incineration (hauling, incineration costs)	15% of all sludge produced	\$300–\$500 per dry ton

3. Which of the following is a beneficial use option for ash?
   a. Landfill cover
   b. Plastic food containers
   c. Tire additive
   d. Vitamin casing
4. The National Institute of Occupational Safety and Health recommends which of the following to provide a more comprehensive set of precautions for use by employers and employees?
   a. Employees purchase their own PPE
   b. Hygiene stations
   c. Liability waivers
   d. Optional training programs
5. What is the difference between land reclamation and land application?
   a. Biosolids application rates are higher
   b. Biosolids application rates are lower
   c. No mixing with amendments of any type
   d. Cannot use alkaline-stabilized biosolids
6. Which of the following product options is the most feasible when using thermal treatment or drying?
   a. Engineered soils
   b. Turf grass
   c. Aglime agent
   d. Biochar
7. Which environmental management system is the most appropriate for WRRFs to use?
   a. ISO 14001
   b. ISO 9000
   c. ISO 17021
   d. EMAS Article 4
8. Which type of PPE is appropriate for a job in which there is the potential for exposure to spray?
   a. Respirator
   b. Face shield
   c. High-visibility jacket
   d. Fall harness

CHAPTER SUMMARY

TYPES OF SOLIDS	■ Any treatment that generates sludge and performs land application and/or incineration, as well as the owners and operators of surface disposal sites, must comply with the requirements under 40 CFR Part 503. ■ 40 CFR Part 503 is a self-implementing rule, meaning, the requirements of Part 503 must be met even in the absence of a permit. ■ In addition to federal rules, the WRRF must also confirm compliance with the state and local regulations for disposing of biosolids. ■ These state regulations may be more stringent than the federal Part 503 rule and may regulate certain types of biosolids more strictly than the federal regulations.
CHARACTERIZATION	■ The physical, chemical, and biological characteristics of biosolids affect both their use and the public's perception of them. ■ The textures and particle sizes of biosolids are as varied as the processes used to create them. ■ Pathogen content of wastewater treatment solids depends on the level of treatment they have received. Raw primary solids contain the most pathogens due to lack of treatment. ■ Biosolids contain various concentrations of macro and micronutrients, including nitrogen and phosphorus.
HANDLING	■ While all wastewater solids and septage must be treated to meet the relevant federal, state, and local regulations, differences in processing methods can result in biosolids with widely varying consistencies, which will affect how they can be transported and beneficially used.
MANAGEMENT	■ Biosolids that meet the requirements of U.S. EPA's Standards for the Use or Disposal of Sewage Sludge (40 CFR 503), as well as state and local standards, can be beneficially used. ■ Some states have training programs for biosolids generators and land appliers. Biosolids personnel should take such programs (whether mandatory or optional) because they will learn how to interact with regulators and what their expectations are. ■ Water resource recovery facilities typically decide whether to use or dispose of biosolids based on the cost of each evaluated option. Biosolids management should include holistic waste management/reuse, value-added product development, and enhanced product performance. Recommended options should lead to simple and safe management alternatives that focus on reusing resources.
SAFETY	■ The risk of worker exposure to infectious agents in solids is likely greatest before treatment. ■ Because the concentration of pathogens declines through natural processes, the potential for pathogen exposure decreases over time. ■ The International Organization for Standardization recommends the following for use by employers and employees: provide appropriate protective equipment, hygiene stations, and training.

References

Alvarez-Campos, O., & Evanylo, G. (2019). *Biosolids use for land reclamation.* http://www.virginiabiosolids.com/wp-content/uploads/2019/04/VBC_Land-Reclamation.pdf

Basta, N. T. (2000). Examples and case studies of beneficial reuse of municipal byproducts. In J. F. Power (Ed.), *Land application of agricultural, industrial, and municipal by-products* (SSSA Book Ser. 6, pp. 481–504). Soil Science Society of America.

Brown, S., & Chaney, R. L. (2000). Combining by-products to achieve specific soil amendment objectives. In J. F. Power & W. A. Dick (Eds.), *Land application of agricultural industrial, and municipal by-products* (pp. 343–360). Soil Science Society of America.

Brown, S., Mahoney, M., & Sprenger, M. (2014). A comparison of the efficacy and ecosystem impact of residual-based and topsoil-based amendments for restoring historic mine tailings in the tri-state mining district. *Science of the Total Environment,* 485–486, 624–632.

Dentel, S. K., Gucciardi, B. M., Griskowitz, N. J., Chang, L. L., Raudenbush, D. L., & Arican, B. (2000). Chemistry, function, and fate of acrylamide-based polymers. In H. H. Hahn, E. Hoffmann, & H. Ødegaard (Eds.), *Chemical water and wastewater treatment VI.* Springer. https://doi.org/10.1007/978-3-642-59791-6_4

Dominak, R. P. (2005). *Long term residuals management plan for the Northeast Ohio Regional Sewer District.* Northeast Ohio Regional Sewer District.

Fisichelli, A. P. (1992). Operators as environmentalists. *Operators' Forum, 9*(1), 14–15.

Haering, K. C., Daniels, W. L., & Feagly, S. E. (2000). Reclaiming mined lands with biosolids, manures and papermill sludges. In R. Barnhisel (Ed.), *Reclamation of drastically disturbed lands* (pp. 615–644). Soil Science Society of America.

Lu, Q., He, Z. L., & Stoffella, P. J. (2012). Land application of biosolids in the USA: A review. *Applied and Environmental Soil Science*, Article ID 201462, p. 11.

Nicholson, F., Bhogal, A., Taylor, M., McGrath, S., & Withers, P. (2018). Long-term effects of biosolids on soil quality and fertility. *Soil Science, 183*(3), 89–98.

Northwest Biosolids. (2018). Soil improvement/land reclamation: Biosolids provide nutrients and organic matter to damaged soils, enabling renewed plant growth. Fact Sheet of the Northwest Biosolids. https://nwbiosolids.org/sites/default/files/2019-10/1805_9127_nwbio_fact_soilimprove_optimized.pdf

Peccia, J., & Westerhoff, P. (2015). We should expect more out of our sewage sludge. *Environmental Science & Technology, 49*(14), 8271–8276. https://doi.org/10.1021/acs.est.5b01931

Sandberg, M. A., & Reimers, R. S. (2013). *Economic assessment of the competitive application for E-beam process in the treatment of waste residuals.* Paradigm International, Inc. Report to HeadworksBIO, Inc.

Sopper, W. E. (1992). Reclamation of mine land using municipal sludge. *Advances in Soil Science, 7,* 351–431.

U.S. Department of Health and Human Services, Centers for Disease Control and Prevention, & National Institute of Occupational Safety and Health. (2002). *Guidance for controlling potential risks to workers exposed to Class B biosolids* (Publication No. 2002–149).

U.S. Environmental Protection Agency. (1995). *A guide to the biosolids risk assessment for the EPA Part 503 Rule* (EPA 832/B-93-005). U.S. Environmental Protection Agency. (2020a). Basic information about biosolids. www.epa.gov/biosolids/basic-information-about-biosolids

U.S. Environmental Protection Agency. (2003). *Environmental regulations and technology: Control of pathogens and vector attraction in sewage sludge* (EPA 625/R-92-013).

U.S. Environmental Protection Agency. (2007). *The use of soil amendments for remediation, revitalization and reuse* (EPA 542/R-07-013). U.S. Environmental Protection Agency, Office of Solid Waste and Emergency Response.

U.S. Environmental Protection Agency. (2009). *Targeted national sewage sludge survey sampling and analysis technical report* (EPA 822/R-08-016).

U.S. Environmental Protection Agency. (2018). A plain English guide to the EPA Part 503 Biosolids Rule. www.epa.gov/biosolids/plain-english-guide-epa-part-503-biosolids-rule

U.S. Environmental Protection Agency. (2020a). Basic information about biosolids. www.epa.gov/biosolids/basic-information-about-biosolids

U.S. Environmental Protection Agency. (2020b). Biosolids laws and regulations. www.epa.gov/biosolids/biosolids-laws-and-regulations

U.S. Environmental Protection Agency. (2021a). Regulatory determinations for pollutants in biosolids. www.epa.gov/biosolids/regulatory-determinations-pollutants-biosolids

U.S. Environmental Protection Agency. (2021b). EPA regional and state contacts for biosolids. https://www.epa.gov/biosolids/epa-regional-and-state-contacts-biosolids

CHAPTER 10

Additional Stabilization Methods

Introduction

The amount of sludge or solids generated increases as a higher quality effluent is produced at a water resource recovery facility (WRRF). This occurs with advances in treatment and more stringent discharge requirements. Effective sludge processing is crucial to the success of the WRRF operation. Sludge stabilization processes further treat the sludge to reduce odors or nuisances, to reduce the level of pathogens, and to facilitate efficient disposal or use of the product. This chapter discusses the following stabilization methods:

- Composting
- Alkaline stabilization
- Heat drying
- Incineration

The resultant stabilized solids from composting, alkaline stabilization, and heat drying are referred to as *biosolids* in this text. These stabilization methods are considered acceptable processes and are widely used for treating sludge to Class A biosolids or Class B biosolids levels for beneficial use and disposal, as referenced in the Code of Federal Regulations (CFR) 40 CFR Part 503. The Part 503 regulations require documentation that one of these processes is used when meeting one of the alternatives listed for pathogen reduction in 503.32(a) for Class A biosolids and 503.32(b) for Class B biosolids. In addition, the U.S. Environmental Protection Agency (U.S. EPA) guidance manual, *Control of Pathogens and Vector Attraction in Sewage Sludge* (U.S. EPA, 1992), includes detailed instructions for operators, biosolids marketers, and engineers on compliance with both vector attraction and pathogen reduction requirements.

Solids: Wastewater solids consist of both organic and inorganic matter that is either dissolved or suspended. *Sludge*: The settleable suspended solids and dissolved organic solids separated from liquids during treatment. Sludge and solids are commonly used interchangeably. *Biosolids:* Sludge that has been treated or stabilized to the extent that it may be beneficially used without nuisance or give off obnoxious odors.

The combustion of sludge during incineration generates stabilized solids in the form of ash. The ash contains acceptable amounts of phosphorus and potassium as a soil amendment. The ash also contains trace metals, which could have adverse environmental effects. Therefore, when ash is used as a soil amendment, the fate of the trace metals applied to the soil needs to be understood along with the effect on the crop. The disposal of the incinerator ash must be in accordance with the solid waste disposal regulations in 40 CFR Part 258. When the incinerator ash is applied to land or placed on other than a municipal solid waste landfill, the regulations under 40 CFR Part 257 must be followed.

LEARNING OBJECTIVES

Upon completing this chapter, for the stabilization methods covered, you will be able to:

- Explain the purpose and function of the processes.
- Understand the operating theory, process equipment, process control logic, and process variables.
- Properly start up, operate, shut down, and maintain the processes.
- Identify potential safety hazards and conduct your duties using safe procedures.
- Troubleshoot when a process does not function properly.

Composting

PURPOSE AND FUNCTION

Vector attraction reduction (VAR) is based on maintaining aerobic conditions for 14 days at 40 °C (104 °F)

Composting is the aerobic decomposition of organic material in the solid phase using controlled temperature, moisture, and oxygen levels. The process involves blending raw or digested sludge with bulking agents to absorb moisture, provide a source of carbon, and increase porosity to improve oxygen transfer.

The 40 CFR Part 503, subpart D, rules identify composting as a process for Class B or Class A pathogen reduction, depending on the time and temperature conditions that are used.

Compost acts as (1) a soil amendment by improving the physical properties of the soil, which provides a better environment for the plant roots, and (2) a fertilizer by providing plant nutrients.

Composting results in a stabilized humus-like product that can be used as a soil amendment in gardening (horticultural) and farming (agricultural) applications. Composting encourages the growth of beneficial bacteria that break down organic material and make minerals and nutrients more readily available to plants. Compost is used to improve soil health and fertility.

THEORY OF OPERATION

All composting methods rely on the same basic process as shown in Figure 10.1. Typically, sludge is dewatered to 15% to 30% solids, by weight, before composting to remove excess moisture from the compost mixture. The blend is achieved by mixing the dewatered sludge and bulking agents to achieve an initial blend of about 40% to 50% solids, by weight. This is a critical step in the process by making the blend porous to allow better air circulation, which provides oxygen throughout the compost pile and enhances the biological activity, providing a finished compost at a faster rate.

Bulking agents such as wood chips, bark, sawdust, straw, and other organic substrates that provide carbon and porosity are added to the sludge.

The compost blend must have an appropriate ratio of carbon (C) to nitrogen (N) to encourage biological activity. The C : N weight in the initial mix should be maintained in the range of 20 to 30. Typical wastewater sludge has a C : N of 5 to 8 and, therefore, the bulking agents are supplying the required additional carbon because most feed sludge provides adequate nitrogen. Bulking agents are screened out of the finished compost product and recovered and reused in subsequent batches.

Composting is an aerobic process and the various process modifications are largely based on how oxygen is introduced to the mix and how aerobic conditions are maintained during the composting process. The aerobic decomposition produces heat, which drives the stabilization and pathogen reduction in the composting material. Whether by forced or natural ventilation, the movement of air through the composting mass also controls the temperature and moisture levels as the process proceeds.

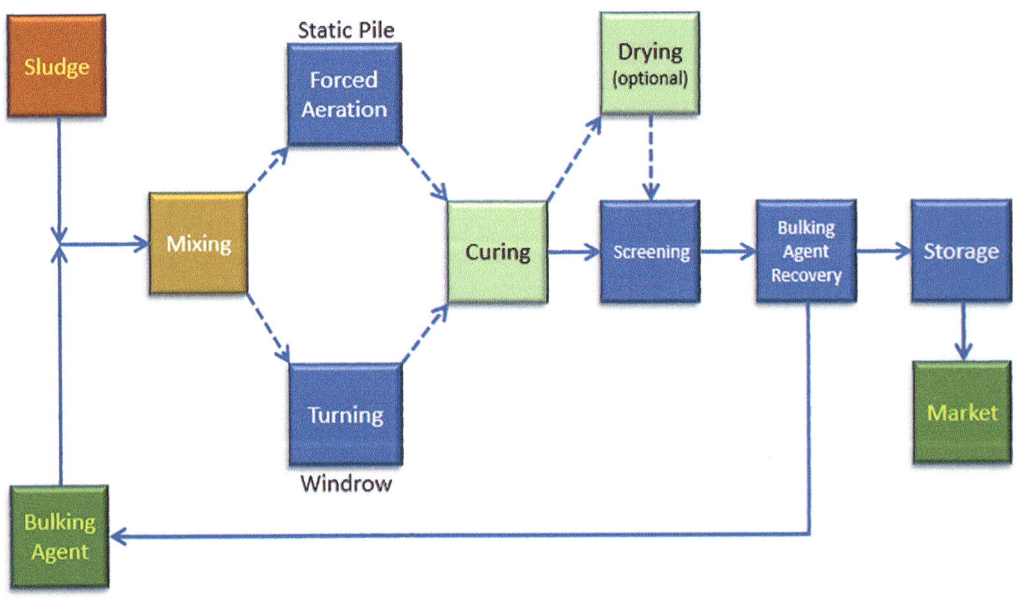

Figure 10.1 Typical Composting Process Flow Diagram

Compost quality is enhanced by aging after active biological composting is complete to eliminate phytotoxins and increase stability. Curing can start after active composting when the biological activity drops to a low level, as indicated by significantly lower temperatures and demand for oxygen and nitrogen, although moisture and aeration needs must still be met. During curing, compost is placed in piles and left undisturbed for several months to provide additional stability to reduce odors and moisture and make the final product more acceptable in all markets. Screening to recover bulking agents may occur before or after curing, but leaving the bulking agent in the compost enhances aeration. Final screening to meet product specifications in certain markets may be required. Generally, all screenings can be recycled with the bulking agents. Mature compost is a dark brown, humic material that will not significantly degrade further.

The composting process includes a curing period after the time and temperature regime required by regulations is met. Curing provides additional stability, reducing odors and moisture.

TYPES OF COMPOSTING SYSTEMS

Composting can be performed by open systems or enclosed systems. Composting with open systems is a labor-intensive process involving the addition of a bulking agent, mixing, windrowing or pile building, screening, and other activities. Some of these operations, however, can be automated. With enclosed systems, most, if not all, operations are automated. The potential for odor generation exists at all composting facilities, especially at poorly designed and operated sites. Also, the potential exists for the generation of significant quantities of dust, which might possibly spread pathogens.

There are two general types of open systems:

- Windrow—Aerobic conditions are maintained through convective airflow and periodic turning.
- Static pile—Blowers attached to pipes located in the base of the piles either blow or pull air through the compost pile to maintain aerobic conditions.

After active composting is complete, the windrow or compost pile is moved to a separate area for curing.

There are two general types of enclosed systems:

- Agitated bin—The composting blend is continuously added and mechanically turned and moved the length of an enclosed concrete bed in a plug flow scenario as blowers provide air through the floor.
- Enclosed vessel—The composting blend and air from blowers are mixed in a batch process.

After active composting, the material is removed mechanically from the enclosed systems and placed in a separate area to cure.

The simplest composting system is an open windrow system. Figure 10.2 shows a picture of a windrow compost pile. The compost blend is stacked in windrows, which are long piles of material between 1.8 m (6 ft) and 4.9 m (16 ft) wide at the base and 0.9 to 1.8 m (3 to 6 ft) high. The rows are turned and mixed periodically, but no mechanical aeration is provided. Compost turners ride over the rows to turn them.

Batch process: Reactants are well mixed in a reactor and left to react for a certain period of time. Baking a cake would be a batch process. *Plug flow process:* Continuously fed reactants travel through the reactor as a plug. As the plug flows, the reaction takes place. The products and unreacted reactants flow out the other end continuously. The small intestine acts as a plug flow reactor by mixing food with digestive juices and pushing the mixture forward for further digestion.

Figure 10.2 Windrow Compost Pile

To meet Class A requirements, the row must be turned five times in 15 days while maintaining 55 °C (131 °F). It may be difficult to meet Class A requirements in cold climates. Raw sludge may also be unsuitable for windrow systems due to high oxygen demand. If the oxygen in the windrow is depleted between when the sludge is turned, significant odors will occur.

AERATED STATIC PILE

In an open aerated static pile system (see Figure 10.3), the pile is constructed over a grid of aeration pipes and the compost can be stacked higher because it is not turned. The height is typically limited by the size of the equipment used to stack the piles. The initial mixing is especially important in the aerated static pile system. Air is forced into or drawn through the pile by aeration blowers.

As shown in Figure 10.4, aerated static pile systems may also be covered by finished compost or different types of geomembranes to retain heat and contain odors. In systems where air is drawn through the pile, the exhaust air can be treated further to minimize odors.

AGITATED BIN–PLUG FLOW

With the agitated bin–plug flow composting system, as shown in Figure 10.5, the compost blend is introduced to a long, narrow concrete bay. A mechanical agitator runs the length of the bay and turns and moves the composting mass along in the bay as fresh blend is added. Air is introduced to the floor of the bay and temperature is continuously monitored along the length of the bay. This type of system is typically enclosed in a building.

ENCLOSED VESSEL

The batch composting takes place in an enclosed reactor where all environmental conditions are controlled. Air is provided and the temperature is continuously monitored. These are generally static bed systems. Figure 10.6 provides a schematic of a typical enclosed composting system.

DESIGN PARAMETERS

The design of the compost blend is a recipe determined based on the type and quantity of sludge and the available bulking agents. Blend design is based on

- Moisture content of the compost blend
- Sludge feedstock solids concentration
- Bulking agents
- Carbon-to-nitrogen ratio of the compost blend

Figure 10.3 Aerated Static Pile

Figure 10.4 Aerated Static Pile with Finished Compost Cover

Figure 10.5 Agitated Bin–Plug Flow

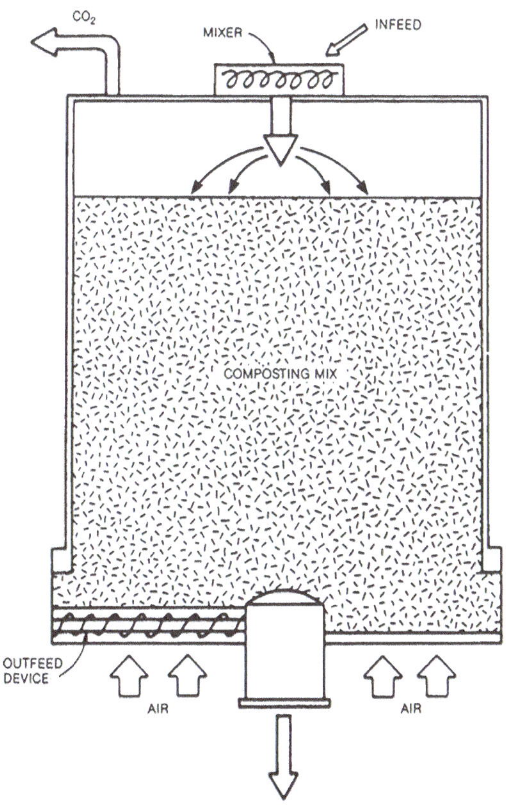

Figure 10.6 Silo-Type Enclosed Composting System

Figure 10.7 presents a flowchart of compost process selection and implementation, which includes the following steps:

1. Determine the end use for finished compost.
2. Determine process requirements for the selected end use.
3. Select the type of composting process (land base and structures required).
4. Determine how the final product will be certified and distributed.

EXPECTED PERFORMANCE

The expected performance of a composting system follows the end use selected in the "Design Parameter" subsection. As a sludge stabilization process, composting renders sludge suitable for beneficial use. Composting has greater potential for public distribution and commercial use than other biosolids products.

EQUIPMENT

The basic facilities for a composting operation include a paved working area; a system for blending sludge and bulking agents; an aeration system or windrow turner; equipment for moving the sludge, bulking agents, and compost; equipment for screening and separating the bulking agent; odor control facilities; areas for bulking agent storage, compost storage, and compost curing; and a system to control runoff (WEF et al., 2018).

Most biosolids composting systems will require dewatered sludge cake in the range of 15% to 30% solids to create a suitable compost blend with the bulking agents. Table 10.1 lists common types of composting equipment.

PROCESS VARIABLES

Three key process variables that must be monitored and controlled in the composting process are (1) moisture, (2) temperature, and (3) nutrient balance (C:N ratio).

PROCESS CONTROL

Moisture, temperature, nutrients, bulking agents, and aeration greatly affect stabilization by composting. Moisture and nutrient balance are established in the initial blend. Moisture and temperature are the most difficult to control in windrow systems because the piles are generally exposed to elements. Windrow systems must have provisions for water sprays and regular turning of the rows.

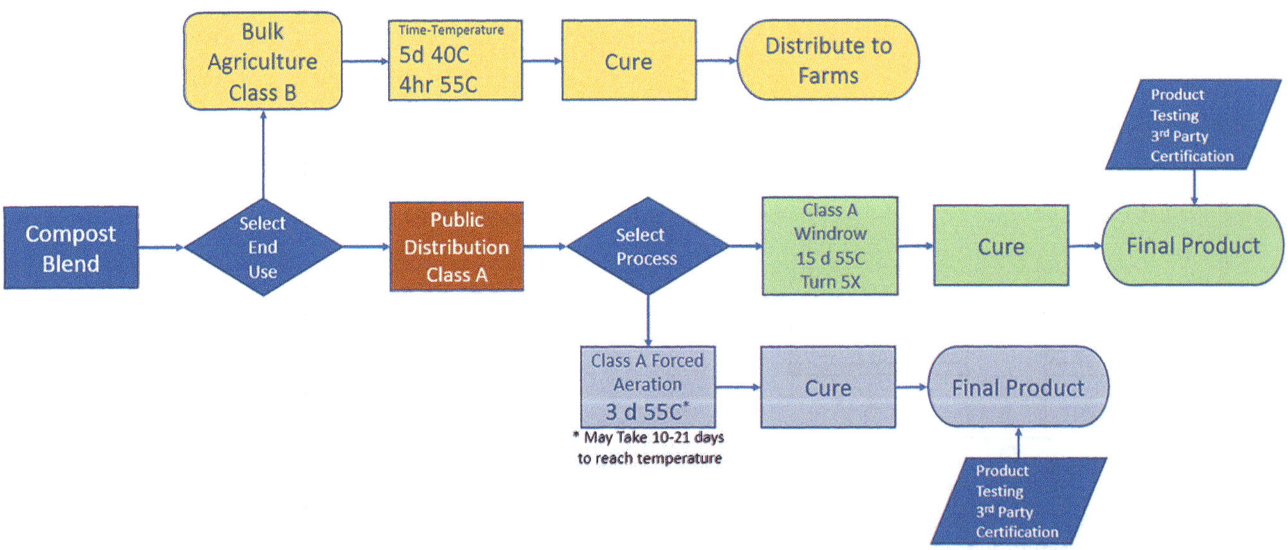

Figure 10.7 Flowchart of the Compost Process Selection and Implementation

Table 10.1 **Composting Equipment**

Function	Equipment	Compost Process
Blending	Pug mill	Any
	Mixing auger	Any
	Mixing truck	Agitated bin (refer to Figure 10.11)
	Front-end loader	Any (refer to Figure 10.8)
Turning	Windrow turner	Open, windrow (refer to Figure 10.10)
	Rotating drum	Closed, drum composter (refer to Figure 10.9)
	Bin agitator	Closed, agitated bin
Aeration	Multistage centrifugal blower	Open, aerated static pile Closed, enclosed vessel Closed, agitated bin
	Rotary lobe blower	Open, aerated static pile Closed, enclosed vessel Closed, agitated bin
Conveyance and material handling	Front-end loader	Any (refer to Figure 10.8)
	Auger	Any
	Skid steer	Any
	Mixing truck	Agitated bin (refer to Figure 10.11)
Screening	Trommel screen	Any (refer to Figure 10.12)
	Vibratory screen	Any
	Bucket screen	Any

Figure 10.8 Front-End Loader (Reprinted with permission from Heather Phillips, City of Olathe, Kansas)

Figure 10.9 Rotating Drum Composter

Figure 10.10 Windrow Turner

Figure 10.11 Mixing/Feeding Truck (Reprinted with permission from Black & Veatch)

Figure 10.12 Trommel Screen (Reprinted with permission from Black & Veatch)

MOISTURE

The rate of biological activity is affected by moisture. At less than approximately 40% moisture, activity begins to decrease; at approximately 60% moisture, the air pore space is substantially blocked, which results in anaerobic areas in the compost piles and odor. The target moisture level of the compost blend is 40% to 50%. Moisture content will also affect materials handling.

TEMPERATURE

As with all biological processes, the rate at which the microorganisms break down the organic matter increases with temperature. The Part 503 regulations also specify the required time and temperature for pathogen reduction. At temperatures higher than 60 °C (140 °F), microbial activity decreases. To meet Class A or Class B requirements, the composting mass must achieve a minimum of 55 °C (131 °F), which is considered the thermophilic temperature threshold. The length of time that 55 °C (131 °F) is maintained is the difference between a Class A and Class B compost. Compost held at this temperature for at least 3 days meets Class A pathogen reduction requirements. To meet Class B requirements, it only needs to be held at this temperature for 4 hours.

CARBON AND NITROGEN RATIO

Carbon and nitrogen are the principal nutrients that affect composting. The C:N affects the microbial activity and the rate of organic matter decomposition. The target C:N is 20 to 30 units of carbon per unit of nitrogen. Carbon values greater than 30 tend to slow the process and result in lower temperatures. Carbon values less than 20 typically affect the compost product rather than the process, as ammonia is released and the nitrogen content of the compost is reduced. The optimal compost blend will retain nitrogen, an essential plant nutrient, in the finished product by minimizing ammonia release.

BULKING AGENTS

When bulking agents are added to the sludge, the amount of their addition creates a way to control moisture, increases the air voids for proper aeration by giving porosity, provides structural support for the compost mass, and acts as an organic amendment for C:N adjustment. Bulking agents can also be used for product modifications. Their addition can change the physical and chemical characteristics of the compost, thereby affecting its range of uses and marketability.

The physical texture of the compost varies depending on the characteristics of the bulking agent and whether it is completely screened out, partially screened out, or left in the final product. The bulking agent can dramatically affect the chemical characteristics of the compost. Sawdust and other slow-degrading, high-carbon materials left in the compost result in an unfavorably high C:N that consumes the nitrogen in the soil and limits the amount available to support plant growth (40 CFR Part 503).

Important bulking agent properties include moisture content, particle size, and absorbency. Bulking agent characteristics affect materials handling, processing time, and product quality. Some bulking agents may require shredding or screening. High-carbon or high-cellulose materials that are not recovered and reused may require long curing times. Pilot testing of new bulking agents should be performed to ensure compatibility with process operations.

AERATION

This section pertains to forced aeration systems. Windrows are aerated by turning and wind action. Aeration provides oxygen for the decomposition process and air movement for temperature control and moisture removal. Higher temperatures are achieved under aerobic conditions than under anaerobic conditions. Under-aeration may result in low temperatures because of anaerobic conditions and excessive aeration may result in cooling of the compost piles.

Studies suggest that 20 to 50 m^3/h (700 to 1800 cu ft/hr) of airflow per dry ton of sludge results in oxygen levels from 5% to 15% throughout the pile (WEF et al., 2018). Oxygen levels can be analyzed with a portable oxygen monitor and should be measured simultaneously with each temperature reading. During the early stages of production, higher aeration rates may be necessary to control temperatures and increase microbial activity.

As air is drawn through the pile, it is gradually heated by the elevated interior temperatures. This temperature increase is accompanied by an increase in specific humidity that allows the air to increase its water-holding capacity. It is this temperature and specific humidity relationship that allows successful air drying even in regions of high relative humidity.

ODOR CONTROL

In composting operations, odor control measures may be necessary whenever incompletely stabilized sludge or compost is exposed to the air. Processes within the composting operation that have a potential for odor generation include mixing, aeration, tearing down of the compost and curing piles, and screening. Odors during composting are best controlled by maintaining an aerobic environment.

Several methods are used to reduce odors, depending on the location of the composting facility in relation to developed areas (especially residential areas). For example, certain facilities can be enclosed such as mixing and screening. Also, exhaust from aeration of the compost can be discharged through biofilters or air scrubbers.

OPERATION

Although composting can be accomplished by a variety of methods, the basic process flow remains unchanged (see Figure 10.1). Even though composting systems may differ in appearance, they all include the following basic steps:

- Mixing of the sludge and bulking agent,
- Composting or microbial decomposition of the organic matter,
- Curing,
- Recovery of the bulking agent or product recycling,
- Storage, and
- Final distribution of the composted material (which may be applied to agricultural fields or marketed as a humus material).

Homogeneous mixing is an extremely important aspect of composting. Partially dewatered sludge cake, typically 15% to 30% solids, is mixed with recycled compost or other bulking agents to provide proper porosity, initial moisture content, and nutrients. As the sludge and bulking material are thoroughly mixed, air distribution flow becomes more uniform, resulting in greater microbial activity and efficiency, odor control, and ease of material handling.

Bulking agent recycle depends on the composting method used and the moisture content of the sludge. Some methods include recycling a portion of the finished compost product.

Curing is an extension of the composting process that results in further organic matter stabilization and volatile solids reduction. The length of the curing period depends on the product's end use and the process efficiency in pathogen reduction, odor control, and volatile solids reduction.

Macronutrients are divided into two classes: primary and secondary. The three primary macronutrients are required in relatively large quantities by plants and include nitrogen (N), phosphorus (P), and potassium (K). The secondary macronutrients are required in lesser quantities relative to the primary category and include calcium (Ca), magnesium (Mg), and sulfur (S). *Micronutrients* are required by plants in far smaller quantities than any macronutrient. Micronutrients required by plants include zinc (Zn), iron (Fe), manganese (Mn), copper (Cu), boron (B), molybdenum (Mo), and chlorine (Cl). Relative to the macronutrients, micronutrients are required in trace quantities.

DATA COLLECTION, SAMPLING, AND ANALYSIS

Information management for composting systems includes monitoring, data collection, analysis, and reporting. Gathered information is needed for process and product quality control, regulatory compliance, good neighbor relationships, and market education. Monitoring product quality is needed both for regulatory compliance and in support of compost marketing and sales efforts. Regulatory compliance parameters include measuring biological content (fecal coliform, salmonella, and Ascaris ova), heavy metal concentrations (typically the 40 CFR Part 503 list of metals), and, in some states, stability and/or percentage of human-made inerts (glass, metal, plastic, etc.). Market-driven monitoring parameters are nutrients (both macro and micro), soluble salts (electrical conductivity), and horticultural characteristics that support healthy plant growth such as water holding capacity, cation exchange capacity, and so on. It is also possible to monitor final product for nonpathogenic biological content (essentially the abundance and diversity of various micro- and higher-order organisms), but the relationships between these microorganism counts, compost quality, and the market's willingness to pay are poorly defined at present.

Monitoring the composting process is required for effective process quality control, and, in the case of temperature data, to verify the compost has met obligatory time-temperature standards (i.e., process to further reduce pathogens [PFRP] and/or VAR). The main process control elements to monitor are temperature, moisture, and either oxygen or carbon dioxide levels (one is the inverse of the other). Measuring these data is traditionally done with manual methods and specialized monitoring equipment (moisture, of course, is often monitored with the rather nontechnical "squeeze" test).

Sample compost piles daily to verify compliance with pathogen reduction requirements. Monitor temperature using a sample probe that penetrates to the center of the pile, as shown in Figure 10.13. A quick measure of moisture content is to use the squeeze ball test. Dig into the upper half of the compost pile at least 76 mm (3 in.). With a gloved hand, grab a handful of material and form it into a ball about the size of a golf ball using normal single-hand pressure. Be careful not to squeeze it so much that moisture is squeezed out of the material. After opening your hand, if the ball falls apart there is insufficient moisture, and more water must be added. If the ball stays together, there is sufficient moisture. If more than a few drops of water can be squeezed out, it may be too wet and dry bulking agent may need to be added to avoid anaerobic conditions.

MAINTENANCE

Composting relies on material handling equipment that is subjected to harsh and abrasive conditions. Equipment should be cleaned, lubricated, and maintained according to manufacturer's recommendations. Composting sites are often dusty, so it is especially important to clean or replace intake filters for aeration equipment and air filters on engine-driven mobile equipment such as front-end loaders and compost turners. Table 10.2 provides a typical maintenance schedule for an aerated static pile composting operation.

TROUBLESHOOTING

Table 10.3 presents a troubleshooting guide for an aerated static pile composting system that will help the operator identify problems and develop solutions. If you operate a reactor system, refer to the troubleshooting guide in the operations and maintenance (O&M) manual provided by the reactor manufacturer.

SAFETY CONSIDERATIONS

Employees working at composting operations should take precautions to reduce exposure to pathogenic organisms. Composting where temperatures reach the thermophilic range can eliminate practically all viral, bacterial, and parasitic pathogens. However, some fungi (e.g., *Aspergillus fumigatus*) are thermotolerant and survive the composting process. Large numbers of spores are released into the atmosphere during certain composting operations (compost screening, dumping, and mixing of compost, and wood chip dumping). To reduce exposure to airborne pathogens, enclose the cabs of heavy equipment, ventilate all enclosed areas properly, and provide dust masks to employees working in dusty areas.

Figure 10.13 Compost Temperature Monitoring (Reprinted with permission from Heather Phillips, City of Olathe, Kansas)

Table 10.2 Aerated Static Pile Maintenance Schedule

Daily	Weekly	Monthly	Quarterly	Every 6 Months	Annually	As Needed	Tasks for Aerated Static Pile Composting
					x		Re-torque foundation nuts and bolts and any rotating equipment.
		x					Screens—Clear dirt and debris to maintain airflow. Note: Replace belts with matched set only.
			x				V-Belt Drives—Check for belt wear, alignment of sheaves, and belt tension.
				x			Clean and inspect the blades for wear, rotational balance, and signs of damage.
					x		Perform vibration analysis on the blowers, screens, mixing equipment, or conveyors.
					x		Check that all shafts are properly aligned.
	x						Check for excessive temperature or chatter in all bearings.
				x			Check surface coatings or paints.
						x	Check alignment of shafts to protect bearings, flexible couplings, wheel, or inlet cones.
					x		Check damper and linkage connections for tightness. Check all automatic dampers for freedom of movement.
						x	Lubricate motors as required by the manufacturer.
						x	Perform voltage and amperage checks. Confirm the motor insulation and power wire analysis.
x							Check screen driver for proper operation.
x							Brush and clean off accumulation of solid particles, wood chips, and plastic from the screen.
						x	Clean hoppers and conveyor(s) for debris and obstructions.
				x			Inspect screen drive for lubrication, oil level, and alignment.
				x			Lubricate and grease all moving parts.
		x					Perform voltage and amperage checks. Confirm the motor drive insulation and power wire analysis.

Table 10.3 Aerated Static Pile Composting Process Control Troubleshooting

Problem	Possible Cause(s)	Remedial Measures
Low compost temperature—compost pile does not reach temperature within a few days after the pile is constructed	Aeration frequency too high	Decrease aeration frequency. If the pile does not come up to temperature within a few days, the pile should be torn down, remixed, and reconstructed. Make sure the bulking agent is not too wet. If it is, dry the material or find drier bulking agent deeper in the bulking agent pile.
	High moisture content	Check mixture ratios and blend in additional compost or bulking agents. Increase aeration by piping hot exhaust from an adjacent pile.
High compost temperature	Aeration frequency too low	Increase aeration frequency.
Compost bay area creating foul odors	Anaerobic conditions	Aeration frequency too low; consider increasing frequency.
	High moisture content	Check mixture ratios and blend in addition compost or bulking agents.
	Balling or clumping	Dewatered sludge cake wet; increase cake dryness.
Large amounts of fine dust created	Compost too dry	Spray pile with water or increase spray frequency.
Compost balling	Dewatered sludge cake too wet	Increase cake dryness.
	Polymer overdosing	Decrease polymer dosage.
	Ineffective blending	Verify mixing operation in truck; improve or increase mixing. Consider changing mix ratios. Verify agitator operation.
Compost pile is emitting odors	Pile is anaerobic due to poor mixing	Increase airflow rate or run blowers continuously.
	Poor air distribution	Check blower operation. Air distribution piping may contain water, which plugs the lines. Check blower timers.

TEST YOUR KNOWLEDGE

1. Before composting, sludge is typically dewatered to
   a. 5% to 10%
   b. 15% to 30%
   c. 75% to 90%
   d. 50% to 65%

2. When performing the squeeze ball test on compost to determine if sufficient moisture is present, you open your hand and the material in your hand retains the shape of a ball. The compost mixture has sufficient moisture.
   ☐ True
   ☐ False

3. The compost blend must have an appropriate ratio of carbon to nitrogen to encourage biological activity and the initial mix should be maintained with a C:N in the range of
   a. 5 to 10
   b. 20 to 30
   c. 10 to 40
   d. 50 to 50

4. The purpose of bulking agents is to provide porosity, structural support, and the source of carbon.
   ☐ True
   ☐ False

5. **The operator should clean hoppers and conveyor(s) from debris and obstructions**
   a. Weekly
   b. Monthly
   c. Annually
   d. As needed

6. **A squeeze ball test is a quick measure of**
   a. Temperature
   b. Moisture content
   c. Compost blend
   d. pH

7. **If the compost pile is emitting odors, the operator should**
   a. Decrease aeration frequency.
   b. Spray the pile with water or increase spray frequency.
   c. Increase airflow rate or run blowers continuously.
   d. Decrease cake dryness.

8. **If anaerobic conditions occur during composting, the following will result:**
   a. High compost temperature
   b. Large amounts of fine dust generated
   c. Generation of foul odors
   d. Reduced moisture content

9. **The target moisture level of the compost blend is**
   a. 45%
   b. 60%
   c. 35%
   d. 65%

Alkaline Stabilization

PURPOSE AND FUNCTION

Sludge that has not been biologically digested or thermally stabilized can be made stable by raising the pH between 11.5 to 12.0 by adding lime and destroying pathogenic and nonpathogenic organisms. The purpose of alkaline stabilization includes the following:

- To create stable biosolids for storage by reducing the number of pathogenic and odor-producing organisms and preventing their regrowth.
- Reduce the mobility (or leaching) of metals from the biosolids to the land on which it is applied, thereby resulting in reduced metal uptake by the crop and the leaching of metals into the soil. This is because the alkaline additives chemically bind up the metals.
- The biosolids also may be used to supplement or replace liming agents on acid soils because they elevate the soil pH. However, alkaline-stabilized biosolids may not be appropriate in areas where the soils are naturally alkaline (e.g., many parts of the western United States).

The alkaline stabilization process involves elevating the pH of the sludge by the addition of chemical for a defined duration. When the sludge is thoroughly mixed with the chemical and held at a pH 12.0 or higher for an adequate amount of time, the microorganisms and pathogens in the sludge are either inactivated or destroyed. Of the alkaline chemicals available in the wastewater industry, lime is the most used and it is the cheapest. Therefore, lime stabilization typically is more cost-effective compared to other alkaline stabilization options.

Lime is an inorganic mineral containing calcium that is produced from limestone. Most of the chemical composition of limestone is calcium carbonate, which, when processed through a kiln, will produce calcium oxide (CaO), also known as *quicklime.*

Although both small and large WRRFs have used lime stabilization, this process is more common at small facilities. At the larger facilities, lime stabilization is used as an interim process when either the primary stabilization process (e.g., anaerobic or aerobic digestion) was temporarily out of service or as a supplemental process for the excess solids produced during peak solids-production events.

Depending on the end use, the alkaline-treated sludge must be (a) stabilized, (b) nonhazardous, (c) non-infectious, (d) reasonably odorless, and (e) physically and chemically suitable for the end use. Major beneficial use options for the alkaline-treated sludge are

- Organic fertilizer or soil amendment
- Agricultural liming agent or conditioner
- Depending on the state and local regulations, both landfill daily and final cover
- Erosion control/slope stabilization

THEORY OF OPERATION

Lime addition to sludge reduces odors and pathogen levels by creating a high pH environment that prevents biological activity from taking place. Gases containing nitrogen and sulfur that are produced during anaerobic decomposition of organic matter are the principal source of odors in sludge. Lime stabilization depends on maintaining the pH of the sludge at a high-enough level for an adequate amount of time to inactivate the microorganism population. This substantially reduces the microbial reactions that can otherwise lead to odor production and vector attraction. The process can also inactivate viruses, bacteria, and other microorganisms that are present.

The lime stabilization process involves a variety of chemical reactions that alter the chemical composition of the sludge. Although the complex chemical reactions between lime and sludge are not well understood, it is likely that mild reactions, such as the splitting of complex molecules by hydrolysis, saponification, and acid neutralization, occur in the high pH environments created during lime stabilization (WEF et al., 2018). The following equations, simplified for illustrative purposes, show the types of reactions that occur:

Hydrolysis is the chemical breakdown of a compound through the reaction with water; with lime, this process is known as *lime slaking*. Saponification is a process that converts fats, oils, or lipids into soap by reaction with alkali salts (e.g., calcium hydroxide).

Reactions with inorganic constituents include

Water:	$CaO + H_2O = Ca(OH)_2$	hydration process or lime slaking
Calcium:	$Ca^{2+} + 2HCO_3^- + CaO \rightarrow 2CaCO_3 + H_2O$	
Phosphorus:	$2PO_4^{-3} + 6H^+ + 3CaO \rightarrow Ca_3(PO_4)_2 + 3H_2O$	
Carbon dioxide:	$CO_2 + CaO \rightarrow CaCO_3$	

Reactions with organic constituents include

Acids:	$RCOOH + CaO \rightarrow RCOOCaOH$
Fats:	Fat + CaO $\rightarrow$ Fatty acids

After lime is added to the sludge and the pH of the sludge increases, the reactions described above occur. During these reactions, the pH decreases; therefore, excess lime is required. The lime reacts with compounds produced by the biological activity, such as carbon dioxide and organic acids. The sludge stabilization will be inadequate if the biological activity of the sludge continues to react with the lime, reducing the pH of the sludge. It would be possible to calculate the lime requirement to raise the pH to a given value if enough is known about a WRRF's sludge.

Quicklime reacts with water to form hydrated lime when added to sludge. This is an exothermic reaction and releases approximately 15 300 calorie/gram-mole (cal/gmol) (2.75 × 10 British thermal unit/pound/mole [Btu/lb/mol]) (WEF et al., 2018). The reaction between quicklime and carbon dioxide is also exothermic, releasing approximately 43 300 cal/gmol (7.8 × 10 Btu/lb/mol) (WEF et al., 2018). These reactions may result in significant temperature rise, particularly when the sludge is low in moisture content. For example, adding 45 g (0.1 lb) of quicklime per gram of solids to a cake containing 15% solids concentration can result in a temperature increase of more than 30 °C (50 °F) (in U.S. customary units):

The hydration of lime or lime slaking is an exothermic reaction. Exothermic reactions are chemical reactions that produce heat.

(0.1 lb CaO) (1 lb mol/56 lb CaO) (27 500 Btu/lb mol) = 49 Btu (52 kJ)

(49 Btu) (1/0.85 lb H_2O) (1 °F/lb H_2O/Btu) = 58 °F (14 °C)

Temperature increases in actual practice will be less, although they can be substantial. In some cases, they can be enough to contribute to pathogen destruction during lime stabilization.

DESIGN PARAMETERS

Biosolids stabilization process selection is dependent on the intended end use of the product. Varying the design parameters of the alkaline stabilization process provides the WRRF with the ability to modify the physical and chemical characteristics of the biosolids product. These design parameters include the sludge type and characteristics, type and dosage of the alkaline material, contact time, pH, temperature, and mixing configurations, thus affording the WRRF diversification and flexibility in their biosolids program. There are typically two types of lime stabilization processes, pre- and post-lime stabilization, which depend on the location of WRRF and the solids treatment. "Pre" and "post" refer to where the lime is applied in the process with respect to dewatering.

PRE-LIME STABILIZATION

In pre-lime stabilization, lime is added to thickened WRRF sludge to meet stabilization requirements, generating a beneficial use product. Pre-lime stabilization typically produces liquid product in compliance with Class B standards. The estimated dose to achieve pH 12.0 is generally 120 g (240 lb) of lime per kilogram (ton) for primary sludge solids at 3% to 6% total solids. Waste activated sludge (WAS) will require a dosage of approximately 300 g (600 lb) of lime per kilogram (ton) of sludge solids. It has been common practice to land apply the Class B liquid biosolids for smaller WRRFs (less than 18.9 ML/d [5 mgd]) with nearby land application sites. Under these circumstances, this practice is quite an economical way of beneficially using municipal biosolids.

POST-LIME STABILIZATION

In post-lime stabilization, lime is added to dewatered WRRF sludge. The purpose of post-lime stabilization is to reduce (Class B) or to eliminate (Class A) pathogenic organisms, to comply with VAR requirements, minimize odors and nuisance conditions, and, finally, to produce a marketable beneficial use product with minimum cost. The majority of the post-lime stabilized biosolids are beneficially land applied or used as a substitution for landfill cover material. Post-lime stabilization typically produces the following beneficial use products:

- Class B process to significantly reduce pathogens (PSRP) material to use as a soil fertilizer for specifically permitted sites.
- Class A PFRP material to use as an agricultural liming agent.
- Landfill cover material in compliance with physical properties required by landfill operation; it can be either Class A or B material.

Sludge that has not been biologically digested or thermally stabilized can be made stable by raising the pH between 11.5 to 12.0 with the addition of lime, destroying pathogenic and nonpathogenic organisms. Estimated dosages to achieve pH 11.5 to 12.0 are generally 100 to 110 kg (200 to 220 lb) of lime per metric ton (ton) for primary sludge solids (100 to 110 g of lime per kilogram of solids) (WEF et al., 2018). Waste activated sludge will require doses from 200 to 400 kg (400 to 800 lb) of lime per metric ton (ton) of sludge solids (WEF et al., 2018). The following example provides an estimate of the amount of lime required to stabilize sludge at an 18.9 ML/d (5 mgd) WRRF:

Example

Given: A 18.9 ML/d (5 mgd) WRRF produces 3026 kg/d (6670 lb/d) of primary sludge solids and 1511 kg/d (3330 lb/d) of secondary solids. Lime stabilization requires the addition of 105 kg/metric ton (210 lb/ton) to raise the pH to 12.0

Find: The kg (lb) of lime required per day.

Solution:

Lime used (kg/d) = Dose (kg lime/metric ton sludge) × sludge (kg/d ÷ 1000 kg/metric ton)

= (105 kg lime/metric ton sludge) × [(3026 kg/d + 1511 kg/d) ÷ 1000 kg/metric ton]

= **476 kg lime/d**

lime used (lb/d) = Dose (lb lime/ton sludge) × sludge (lb/d ÷ 2000 lb/ton)

= (210 lb lime/ton sludge) × [(6670 lb/d + 3330 lb/d) ÷ 2000 lb/ton]

= **1050 lb lime/d**

Lime Stabilization Advantages and Disadvantages

Major advantages of pre- and post-lime stabilization are

- Pathogen and vector reduction requirements of the 503 Regulations are met; such compliance can be achieved without biological treatment (digestion), which is often economically beneficial for small WRRFs.
- Odors are reduced.

The disadvantages of pre- and post-lime stabilization are

- There are limited markets for alkaline-stabilized biosolids, especially for the pre-lime stabilization of liquid biosolids.
- Substantial storage is required due to seasonal and market variations.
- Pre-lime stabilization (i.e., alkaline stabilization of liquid biosolids) can meet only Class B (PSRP) requirements.

EXPECTED PERFORMANCE

Lime-stabilized biosolids have lower concentrations of available nutrients (e.g., nitrogen and phosphorus) because of the inert solids and reactions involved than a comparable mixture of biologically stabilized (anaerobic or aerobic digestion) primary and WAS. Therefore, the land application of lime-stabilized biosolids hold more benefit for agricultural areas with acidic soils.

Alkaline stabilization can be used to meet Class B or Class A pathogen reduction requirements.

To meet Class B stabilization requirements, the pH of the feed dewatered sludge cake must be elevated to more than pH 12.0 for 2 hours, and then maintained above pH 11.5 for another 22 hours to meet VAR requirements. To meet Class A stabilization requirements, the elevated pH is combined with elevated temperatures (70 °C for 30 minutes or other U.S. EPA-approved time and temperature combinations listed in U.S. EPA's *Control of Pathogens and Vector Attraction in Sewage Sludge* [2003]). If the pH remains above 10 to 10.5, microbial activity and the associated odorous gases are greatly reduced or eliminated (U.S. EPA, 1979). However, other odorous gases (e.g., ammonia and trimethyl amine) may be produced under high-pH and high-temperature conditions.

Odorous gases such as ammonia and trimethylamine are produced because of the high pH and temperature conditions.

EQUIPMENT

Large facilities use pebble quicklime, whereas small ones use hydrated lime. For lime-application rates more than 1800 to 2700 kg/d (2 to 3 tons/d), bulk quicklime is typically more economical than hydrated lime. Bagged lime requires a waterproof, well-ventilated storage building; bulk lime requires watertight and airtight storage bins. Bagged lime should be stored on pallets in a dry place for no longer than 60 days. Bulk lime can be pneumatically transferred in bins or conveyed to the bins via conventional bucket elevators or screw conveyors. Figure 10.14 shows a schematic of a typical lime storage and feed system.

Hydrated lime is: "A dry powder obtained by treating quicklime with enough water to satisfy its chemical affinity for water under the conditions of its hydration." (ASTM C5 1-81)

LIME SILOS

Quicklime bins typically have a 55- to 60-deg slope to the bin outlet; hydrated lime bins have a 60- to 66-deg slope. Tall, slender structures with a height-to-diameter ratio of 4 : 2.5 are preferred. The design volume should be based on the average bulk density of the chemical, with an allowance for 50% to 100% extra capacity beyond that required to accommodate a typical delivery. Quicklime and hydrated limes are abrasive, but not corrosive, so steel or concrete bins can be used.

It is very important that the quicklime storage bins be airtight and watertight. Moisture in quicklime storage bins will hydrate the quicklime, which is referred to as *air slaking*. The hydrated quicklime will lump together or cake, preventing it from flowing freely. Figure 10.15 shows a typical lime silo.

Air slaking is caused by the hydration of quicklime with moisture in the air at ambient temperatures. Air slaking in quicklime bins will result in caking issues.

Hydrated lime bins should be equipped with bin agitation and a non-flooding rotary feeder at the bin outlet. Other required appurtenances include air-relief valves, access hatches, and a dust-collector mechanism.

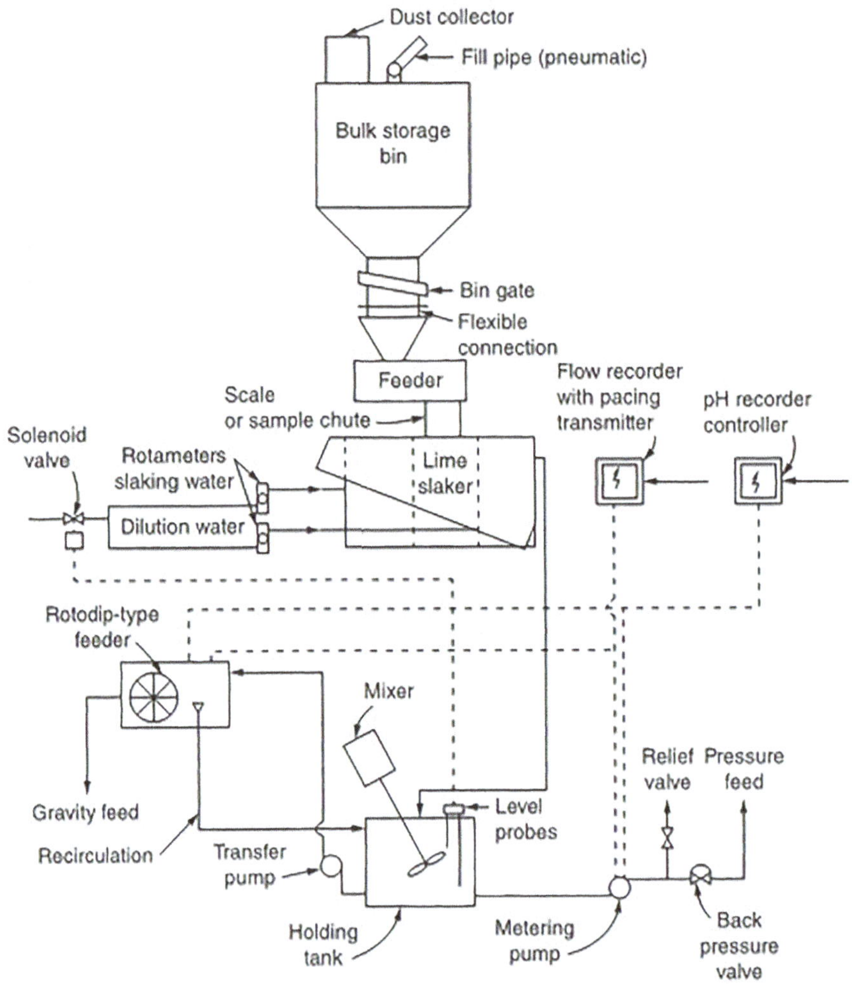

Figure 10.14 Typical Lime Feed System

Figure 10.15 Lime Storage Silos

LIME FEED SYSTEM

Bulk quicklime typically is fed to a slaking device, where oxides are converted to hydroxides, producing a paste or slurry that is further diluted before being piped or pumped to the application points (see Figure 10.16). There are several different designs of suitable dry feeders. The choice depends on facility capacity and the degree of accuracy desired. For example, if a gravimetric-type feeder and slaker combination is indicated, larger facilities can use pebble lime because gravimetric feeders are the most accurate (0.5% to 1% of set rate). This translates into cost reductions in large operations. Large, medium, and small facilities also will find volumetric feeders satisfactory, with their accuracy range being 1% to 5%.

After all chemical reactions between water and quicklime are completed, the lime slurry is stabilized, making it relatively easy to keep in suspension and noncorrosive. However, it is still abrasive and the suggested method for transferring slurry is via gravity and open trough, as long as the slurry is stabilized. If piping and transfer pumping cannot be avoided, the feed loop should be designed with a minimum velocity of 0.9 to 1.5 m/s (3 to 5 ft/sec). On the slurry lines, pinch valves are preferable to ball valves and plug valves. A flexible hose with a velocity less than 0.9 m/s (3 ft/sec) can be used to transfer lime slurry a short distance. In general, feed piping should be at least 50 mm (2 in.) in diameter and have minimal turns and bends.

Slurry pumps typically fall within two categories: centrifugal and positive displacement. Centrifugal pumps are typically used for low-head transfer or recirculation. Replaceable pump liners and semi-open impellers are desired. The pump layout should provide for easy dismantling for cleanout and repairs and should not include water-flushed seals because they tend to scale.

Positive-displacement pumps should be used when the slurry flow must be metered or positively controlled. However, because of the abrasive nature of lime slurry, these pumps are subject to excessive wear and replacement (e.g., pistons and tubing). Turbine pumps and eductors should be avoided because of scaling problems that occur in the pipelines. The lime feed rate can be either controlled by pH or paced with the incoming solids flow.

PROCESS VARIABLES

Unslaked lime is classified by two measurements: available lime content and reactivity. These are important factors in the lime stabilization processes. Reactivity of quicklime varies from different suppliers. The quicklime used in the alkaline stabilization process should slake easily, readily dissolve into a solution, and produce the required temperature rise of 40 °C (72 °F) in a predetermined time.

Available lime content is a measure of the acid neutralization capacity of the unslaked lime. Lime reactivity is a measure of how quickly the exothermic slaking reaction raises the temperature of the water.

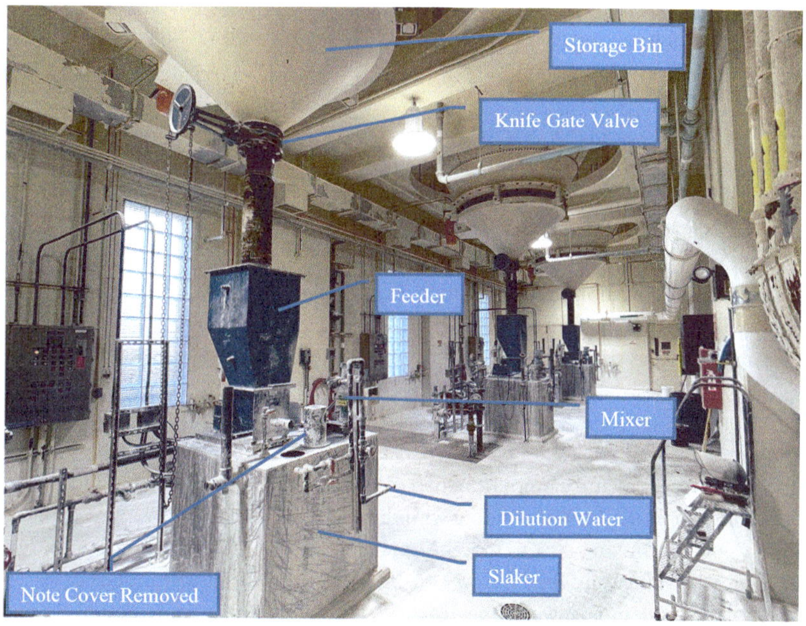

Figure 10.16 Lime Feeder

The three levels of lime reactivity are defined as follows (ANSI/AWWA B202-93 "Standard for Quicklime and Hydrated Lime"):

1. A high reactive lime will show a temperature rise of 40 °C (72 °F) in 3 minutes or less and the reaction will be complete within 10 minutes.
2. A medium reactive lime will show a temperature rise of 40 °C (72 °F) in 3 to 6 minutes and the reaction will be complete within 10 to 20 minutes.
3. A low reactive lime will require more than 6 minutes to raise the temperature to 40 °C (72 °F) and more than 20 minutes to complete the reaction.

Quicklime will deteriorate in storage at a much more rapid rate than hydrated lime. Under good storage conditions, with multiwall moisture-proofed bags, quicklime may be stored up to 6 months; in general, however, it should not be stored longer than 3 months.

Quicklime is produced from limestone (calcium carbonate [$CaCO_3$]). Dolomite is a specific type of limestone that includes magnesium as well (calcium magnesium carbonate [$CaCO_3 + MgCO_3$]).

The chemical composition of hydrated lime generally reflects the composition of the quicklime from which it is derived. A high calcium quicklime will produce a high calcium hydrated lime containing 72% to 74% calcium oxide and 23% to 24% water in chemical combination with the calcium oxide. A dolomitic quicklime will produce a dolomitic hydrate. Under normal hydrating conditions, the calcium oxide fraction of the dolomitic quicklime completely hydrates, but only a small portion of the magnesium oxide hydrates (about 5% to 20%). The composition of a normal dolomitic hydrate will be 46% to 48% calcium oxide, 33% to 34% magnesium oxide, and 15% to 17% water in chemical combination with the calcium oxide. (With some soft-burned dolomitic quicklimes, 20% to 50% of the MgO will hydrate.) A "special" or pressure-hydrated dolomitic lime has almost all (more than 92%) of the oxides hydrated.

Sieving is a method to separate particles of different sizes. The sieve retains particles larger than the sieve openings. The sieve # is a number used to designate the size of the opening.

Hydrated lime is air-classified to produce the fineness necessary to meet the requirements of the user. The normal grades of hydrate used for chemical purposes will have 75% to 95% or more passing a #200 sieve, whereas for special uses, the hydrate may be classified as fine as 99.5% passing #325 sieve. Hydrated lime is packed in paper bags weighing 23 kg (50 lb) net; it is also shipped in bulk.

PROCESS CONTROL

Process control is critical to achieve adequate lime stabilization. Proper control techniques are important because the effects of incomplete stabilization are not readily apparent and may not be seen at the WRRF. Figure 10.17 summarizes the variables that influence lime stabilization, including design, sludge characteristics, lime characteristics, and product considerations.

Sensors, particularly pH electrodes, need to be properly cleaned, calibrated, and maintained. Special pH electrodes are necessary for routine measurements more than pH 10. Sensitive narrow-range pH paper is acceptable for process monitoring.

The absence of odors at the WRRF and acceptable dewatering characteristics are not good indicators of adequate stabilization. Monitoring the pH will ensure that an adequate pH is being maintained long enough to achieve adequate stabilization. Microbiological examinations for indicator organisms should be performed and recorded in accordance with your WRRF permit requirements.

Compared to composting, the lime stabilization process is simpler to operate. Some of the principal disadvantages of lime stabilization are that the sludge can become unstable if the pH drops after treatment and a regrowth of biological organisms occurs, added lime increases the mass to be disposed of, and lime is expensive.

OPERATION

Lime arrives from the supplier in dry form and cannot be added directly to sludge. The dry lime must be made into slurry with the addition of water before blending with the sludge. A lime stabilization system must incorporate the use of a slurry or mixing tank to mix the lime with water, a slurry transfer pump, and a sludge mixing tank to mix the sludge and lime slurry. The process may be either continuous or batch and the slurry–sludge mixing tank must have enough capacity to allow a minimum retention time of 30 minutes of the mixture at a pH of 11.5 to 12.0. The pH of the slurry–sludge mixture should be

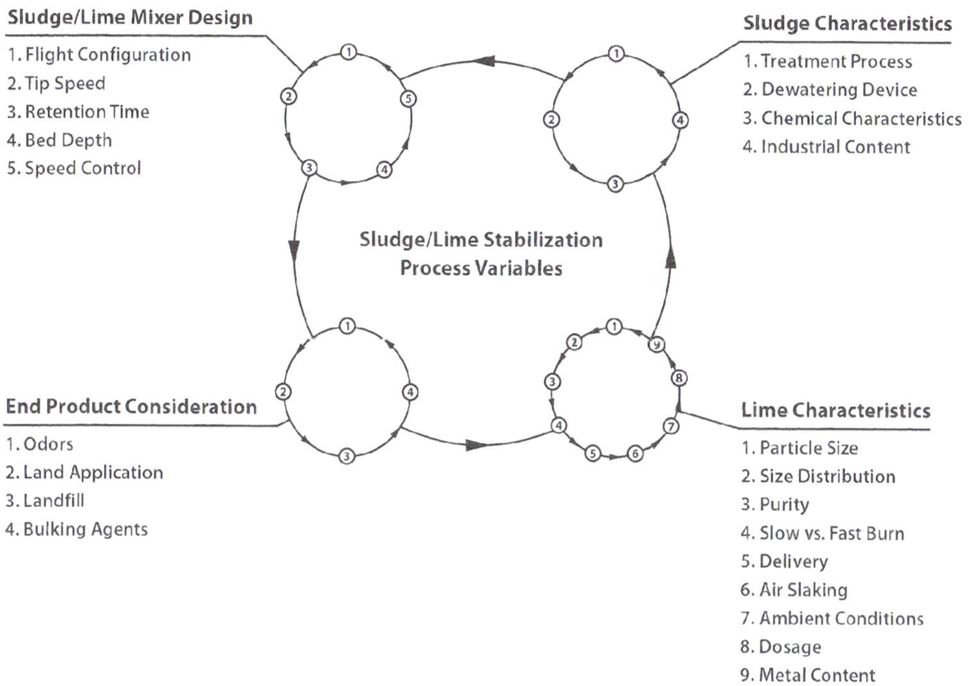

Figure 10.17 Lime Stabilization Process Control Variables

continuously measured and monitored to ensure proper pH adjustment. Greater than 99% of the fecal coliforms and fecal streptococci can be destroyed if the pH is maintained within a 11.5 to 12 pH range, achieving the goals of stabilization.

Dose accuracy and mixing are the most common issues during startup associated with alkaline stabilization. To ensure that the dose measurement is accurate and mixing is homogeneous, it is important that, during startup, (a) sludge is representative of the WRRF's operational conditions and (2) all process components, including storage and metering systems, are tested over the range of operational conditions. If several types of alkaline materials are to be used, each should be tested with the complete process to verify acceptable operation.

DATA COLLECTION, SAMPLING, AND ANALYSIS

The pH of the process must be monitored and recorded to document that permit requirements are met. In addition, the operator must take samples for biological analysis quarterly or as required by the operation National Pollutant Discharge Elimination System (NPDES) permit.

MAINTENANCE

Equipment used in lime stabilization processes requires high maintenance because of the severe conditions caused by handling caustic lime. Proper design and operation of the equipment is necessary to ensure a quality product. Recommended maintenance should be completed by operators and maintenance staff to preserve warranties and performance guarantees. Refer to the equipment manufacturer's operation manual and operate solids treatment equipment in accordance with the manufacturer's recommendations and all NPDES permits. Table 10.4 summarizes the major problems with lime handling systems and suggested improvements.

TROUBLESHOOTING

Table 10.5 presents a troubleshooting guide for lime stabilization.

SAFETY CONSIDERATIONS

Employees should take precautions to reduce exposure to lime when working in a lime stabilization area. The high pH of the lime is very caustic and will damage skin, eyes, and lungs if exposed through contact or inhalation. Use of personal protective equipment should be mandatory for all employees when working

Table 10.4 Major Maintenance Issues of Lime Handling Systems (U.S. EPA, 1984)

Problem	Remedy
Lime dust generation	Install baghouse on lime silos.
Abrasion of dry lime transport lines	Use long radius bends and heavy duty elbows.
Arching or bridging of dry lime in feed hopper	Install live bin systems.
Caking of lime in feed hopper	Operate slaker under negative pressure and install rotary valve under the feeder.
Poor quality slaked lime	Maintain proper temperature and water content in slake.
Malfunctioning of instrumentation located near slaker	Locate instrumentation away from slaker.
Abrasion of lime slurry tanks	Use corrosion-resistant metal tanks (not FRP).
Encrustation of probes used for slurry level control	Use sonic-type (non-contact) level sensor systems.
Clogging of lime slurry transport lines	Install lime slurry recirculation loops, flush transport lines with water, use flexible hoses, avoid sharp bends, and use minimum 51-mm- (2-in.-) diameter piping.
Clogging in lime slurry feed system	Minimize distance between recirculation loop and feeder; use open/close type valves; use rotary cup type feeder.
Encrustation of pH probes	Rotate the use of alternate pH probes.

Table 10.5 Troubleshooting Guide for Lime Stabilization Processes (U.S. EPA, 1979)

Problem	Probable Cause	Solution
Air slaking occurring during storage of quicklime	Adsorption of moisture from atmosphere when humidity is high	Make storage facilities airtight, and do not convey pneumatically.
Feed pump discharge line clogged	Chemical deposits	Provide additional dilution water.
Grit conveyor or slaker inoperable	Foreign material in the conveyor	Check and replace shear pin if necessary; remove foreign material from grit conveyor.
Paddle drive on slaker is overloaded	Lime paste too thick	Adjust water control valve to alter the consistency of the paste.
	Grit or foreign matter interfering with paddle action	Remove grit or foreign materials, try to obtain lime with a lower grit content, or install grit removal facilities in slaker or slurry line.
Lime deposits in lime slurry feeder	Velocity too low	Maintain continuous high velocity by adding a return line to the slurry holding tank.
"Downing" or incomplete slaking of quicklime	Too much water is being added	Reduce quantity of water added to quicklime (detention slaker/water to lime ratio = 3.5:1; paste slaker ratio = 2:1).
"Burning" during quicklime slaking	Insufficient water being added, resulting in excessive reaction temperature	Add sufficient water for slaking (see ratios above).
Sludge retains definite offensive odor after addition of lime	Lime dose too low	Check pH in sludge–lime mixing tank and increase lime dose, if needed; check pH for possible malfunction.

in or around the lime stabilization areas, including, at a minimum, gloves, safety goggles, and a ventilator. Proper housekeeping and storage procedures, as recommended by the chemical manufacturer, should always be followed because the lime materials can be very corrosive if not handled properly.

TEST YOUR KNOWLEDGE

1. Lime addition to sludge reduces odors and pathogen levels by creating a high pH environment that
   a. Prevents chemical reaction from occurring
   b. Physically separates water from the sludge
   c. Prevents biological activity from taking place
   d. Promotes biological activity to take place
2. The reactivity level of lime is defined as high when a temperature rise of 40 °C (72 °F) is achieved in
   a. 30 minutes or less
   b. 3 minutes or less
   c. 60 minutes or less
   d. 6 minutes or less
3. The lime stabilization process is an exothermic reaction (produces heat).
   - ☐ True
   - ☐ False
4. If lime deposits in the lime slurry feeder, the operator should
   a. Reduce the slaker dilution water.
   b. Increase the velocity through the feeder by increasing the recirculation rate back to the to the slurry holding tank.
   c. Reduce the velocity through the feeder.
   d. Increase the dry lime feed rate to the lime slaker.
5. To avoid clogging of lime slurry transport lines, the system should have recirculation loops, flushing connections, avoid sharp bends, and use minimum 51-mm (2-in.) -diameter lines.
   - ☐ True
   - ☐ False
6. The alkaline stabilization process involves the addition of chemical to sludge to
   a. Reduce the pH below 7
   b. Increase the pH to above 12
   c. Neutralize the sludge
   d. Reduce the pH below 4
7. To meet Class B stabilization requirements, the pH of the feed dewatered sludge cake must be elevated to more than pH 12.0 for 2 hours and then maintained above pH 11.5 for
   a. 22 hours
   b. 24 hours
   c. 12 hours
   d. 16 hours
8. Equipment used in lime stabilization processes requires high maintenance because of the severe conditions caused by handling
   a. Acidic chemicals
   b. Caustic lime
   c. Hydrogen sulfide
   d. Explosive gases
9. Pre-lime stabilization and post-lime stabilization refer to where the lime is applied in the treatment process with respect to
   a. Dewatering
   b. Thickening
   c. Secondary treatment
   d. Anaerobic digestion

Heat Drying

PURPOSE AND FUNCTION

Heat drying or thermal drying is a technology that was originally developed for material and chemical processing applications. Heat drying has been successfully applied to WRRF solids and is a proven processing technology that produces a marketable product regulated under the U.S. EPA 40 CFR Part 503 regulations.

The purpose of heat drying is to reduce the mass and volume of material, reduce pathogens, and enhance the material handling properties of the biosolids. Heat drying transfers thermal energy into the solids, raising the temperature to evaporate water and stabilize the product.

> The dried product is referred to as *pellets* or *granules* in the industry. In this section, the term *pellets* will be used.

Heat drying can be used as pretreatment step ahead of incineration. In this case, the purpose of drying is only to remove water and the dried product characteristics are not subject to regulation. This application is referred to as a *scalping dryer* and will not be discussed in this section because the end use process is incineration.

THEORY OF OPERATION

Solids generated in the liquid treatment processes are concentrated in thickening and/or dewatering and dried at high temperatures for stabilization. Table 10.6 shows typical solids concentrations. Heat drying systems generally receive dewatered sludge cake as a feedstock.

Heat drying is a thermal process that removes water or moisture from sludge that has been dewatered to produce a dried product. When the wet dewatered sludge is subjected to heat drying, two processes occur simultaneously, including the transfer of heat from the surrounding environment to evaporate the surface moisture and the transfer of internal moisture to the surface of the solids where it is evaporated. The rate of drying is dependent on the rate at which these two processes proceed.

In the heat drying system, the temperature of the dewatered sludge cake is raised to drive off the water content as vapor by evaporation. The water vapor is carried away by a moist gas (typically air). Through this process, most of the water is removed from the dewatered sludge cake, which results in a significant reduction in both volume and mass. In addition, the temperatures are too low to destroy organic matter, allowing the dried product to retain a significant amount of the nutrients (nitrogen and phosphorus) for beneficial use as fertilizer.

DESIGN PARAMETERS

Heat drying is classified as a U.S. EPA PFRP. Application of a PFRP in the United States will produce a biosolids product meeting the Class A pathogen reduction requirement. To meet Class A standards, the dried product must be 90% solids or greater and have exceeded a temperature of 80 °C (176 °F).

The drying rate depends on both the solids flow and evaporation rates. During drying, a temperature difference between the surface and the inside of the solids develops. This difference in temperature causes moisture to migrate from inside the solids to its surface where it evaporates. Other conditions that affect the process include the temperature, humidity, gas flowrate, direction of gas flow, exposed surface area, and detention time.

Heat drying systems are designed to meet the product characteristics needed for the selected end use. These requirements typically include the solids concentration, size, shape, and uniformity of the product. The system is sized based on the evaporative capacity required to dry the feed sludge to at least 90% solids. The thermal energy needed to meet the system's evaporative capacity is directly proportional to the moisture contained in the dryer feed sludge. The thermal energy consumption is based on the amount of water to be evaporated and the thermal efficiency of the drying system. The thermal efficiency of drying systems may range from approximately 1400 Btu per pound of water evaporated to 1700 Btu per pound of water evaporated. Generally, thermal dryers are rated on evaporation capacity. Cost of fuel is one of the largest operating costs for any thermal drying system. If digester gas or another waste heat source is available, considerable savings in fuel costs can be realized. Therefore, it is important to monitor the sludge feed conditions to the dryer, which include the solids content of the sludge feed and the mass flow to the dryer. These data are used to determine the mass of both the solids and water in the sludge feed. The wetter the feed sludge, the greater the energy and the dryer system size required. To reduce the energy required, the sludge is dewatered before being fed to the dryer.

Another key parameter in sizing thermal dryers is the acceptable hours of operation. Dryers have a fixed evaporative capacity; so, the longer they operate each day, the more solids they can process. A dryer that operates 24 hours per day only needs one-third of the evaporative capacity to process the same quantity of solids as a dryer that operates 8 hours per day. Because heat dryers are expensive, many are sized to run almost continuously (e.g., 24 hours per day for 5 days each week [about 100 to 105 hours/week]) to fit with typical work schedules while leaving some downtime for preventive maintenance. Also, running nearly continuously reduces equipment wear. Frequent heating and cooling cycles increase metal fatigue and, therefore, the rate of component failure.

Table 10.6 Typical Solids Concentrations

Process	Typical Percent Solids
Thickening	2%–10% solids
Dewatering	10%–45% solids
Drying	90+% solids

Product that will be distributed to the public or marketed as fertilizer will be expected to be in the form of a pellet that will not degrade during handling and application. The ability of a dryer to form pellets is a key design parameter. Product with other characteristics may be suitable for bulk agricultural application or non-beneficial use alternatives.

Pelletization is the process of forming the dried solids into pellets and is typically used to control the quality of the product size and shape.

EXPECTED PERFORMANCE

When provided with a homogenous feed, the performance of municipal sludge heat drying is expected to meet or exceed 90% solids in the dried product and meet or exceed the minimum temperature requirement for the U.S. EPA PFRP. The characteristics of product size and shape are based on end user needs and can be defined by the performance requirements of the dryer specifications. The performance of the dryer to meet these requirements depends on the type and consistency of the sludge feed to the dryer, the type of system used, and the intended end use.

Homogenous feed refers to the consistency of the mass loading and the characteristics of the solids. By providing storage upstream of the dryer, this provides a wide spot in the system that dampens out variations, allowing the feed to the dryer to be constant.

EQUIPMENT

Dryer technology is classified by the heat transfer and evaporation method. All dryers must have continuous airflow to remove water vapor and a means to directly exhaust water vapor or condense water from the vapor flow. Hot gases from a furnace or waste heat from co-generation are the most common sources of heat for drying systems at WRRFs.

Heat transfer is the flow of heat (thermal energy) due to a temperature difference. Heat flows from hotter to colder.

Typically, the drying system consists of the dryer, materials handling and storage equipment, heat generation and transfer equipment, air movement and distribution equipment, emissions and odor control equipment, and ancillary systems. Drying systems use different methods for heat transfer, including convection, conduction, and radiation heating. Individual systems may use multiple methods of heat transfer, but they are generally categorized by their primary method of heat transfer.

DIRECT DRYERS

Direct dryers heat the sludge by convection, and heat is transferred directly from hot gases to the sludge. This section will cover two of the more common types of convection dryers—rotary drum and belt dryers. Both are proven processes that produce a Class A biosolids and are best suited for homogenous feed. Compared to belt dryers, rotary drum dryers are typically applied at larger facilities, have a higher capital cost, and require more complicated operation and maintenance. However, rotary drum dryers produce a higher quality product that offers more diverse market opportunities and revenue potential.

Rotary Drum Dryer

Refer to Figure 10.18 for a typical rotary drum process flow schematic. In a rotary drum dryer, the incoming feed (dewatered sludge cake) comes into contact with hot combustion gases as the solids pass through the rotating drum with the aid of internal flights to convey and suspend the solids in the hot gases. The dry solids retention time is controlled by the dewatered sludge cake feed rate and the drum rotational speed. Often, the feed to the drum is a mixture of recycled dry solids, including dust and fines with the incoming wet material. This recycling process is known as *back-mixing*, and allows the operator to

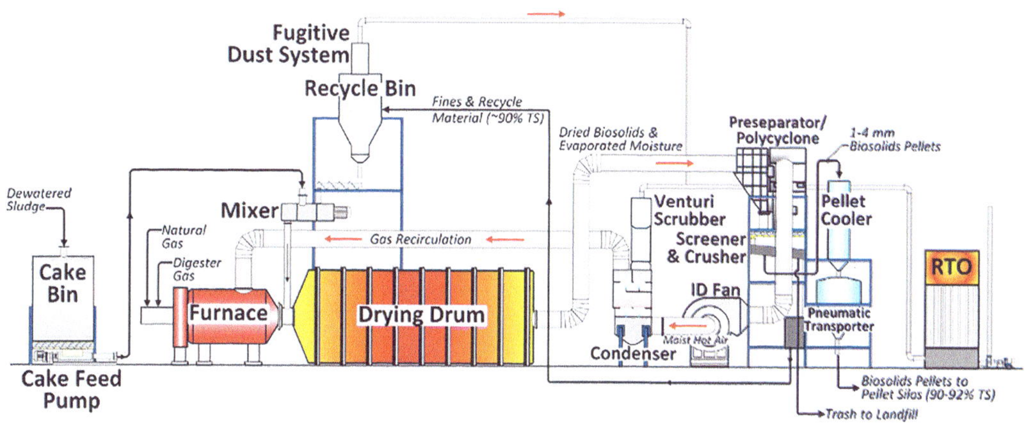

Figure 10.18 Typical Rotary Drum Dryer Process Flow Schematic

adjust the moisture content of the feed material and promotes the formation of a stable pellet in the drum. In most systems, about 70% to 90% of the furnace gases are recycled to enhance the system's thermal efficiency and help maintain an inert atmosphere. Maintaining an inert atmosphere is a safety feature that limits the oxygen concentration in the system, which reduces the risk of explosions. Recycling gas also reduces the volume of exhaust gases to be treated.

Rotary Drum Dryer

- Continuous feed
- Solids heated to approximately 93 °C (200 °F)
- Hot gases heated to approximately 538 °C (1000 °F)
- Creates high-quality product
- More complicated O&M than other dryers
- More complex odor control required
- Typically applied at larger facilities

Referring to Figure 10.18, the system process flow is as follows: The dewatered sludge cake from the cake bin is fed continuously to the mixer by a positive-displacement pump(s) or a screw conveyor, where it is combined and mixed with the recycled dry solids from the system via the recycle bin. The mixture is then fed to the drying drum, where it is combined with the hot gases from the furnace. The dried material is separated from the warm exhaust gas by the polycyclone. The dried material is screened, crushed, and cooled before being transported to storage silos for loadout and hauling to its end use. The exhaust gas separated from the dried material is moved from the polycyclone through a condenser by the induce draft fan to be cooled. Part of the cooled exhaust gas is recycled back to the dryer through the furnace. The remainder of the exhaust gas is treated in air pollution control (APC) equipment and then vented to the atmosphere. Air pollution control systems for drum dryers typically consist of additional particulate removal followed by regenerative thermal oxidation to destroy odors and volatile organic compounds (VOCs).

Figures 10.19 to 10.23 show equipment that are typically part of a rotary drum dryer.

Belt Dryer

Refer to Figure 10.24 for a typical belt dryer process flow schematic.

In belt dryers, dewatered sludge cake is fed on a porous conveyor belt and large volumes of heated gas are blown or drawn through them. The dry solids retention time is controlled by the belt speed. The belts are made of either steel mesh or synthetic material (like the filter media used in a belt filter press). The unit may have one or two belts, and its configuration depends on the manufacturer. The dried product or pellets typically are less uniform in size than those from more complex systems because of the limited materials handling. Screens and conditioners can be used to improve uniformity.

The gas management system is also supplier-specific. To remove moisture from the process, a fan typically removes a portion of the gas from the process air loop and sends it to a condenser. One supplier returns most of the scrubbed gas to the dryer, but diverts a portion to mix with combustion air. Other suppliers recommend that the scrubbed gas be treated by a biofilter or chemical scrubber to control odors before discharge. Recycling most of the process air improves the system's thermal efficiency and reduces the exhaust needing treatment.

The odor-control method is supplier-specific. Biofilters traditionally have been used in Europe to control odors from these systems, but chemical scrubbers also can be used. One supplier uses the furnace to combust exhaust gases.

Belt Dryer

- Continuous feed
- Lower operating gas temperatures than other dryers approximately 127 to 177 °C (260 to 350 °F)
- Larger dryer footprint due to cooler gas temperatures
- Product quality varies
- Potential use of waste heat as an energy source
- Relative ease of operation compared to a rotary drum dryer
- Typically applied at smaller facilities

Belt dryers use a relatively low-temperature gas (about 127 to 177 °C [260 to 350 °F]), so low-grade waste heat can be used as an energy source. This allows for a wide range of energy recovery options. The lower drying temperature is claimed to produce a less odorous air stream.

Referring to Figure 10.24, the process flow of the system is as follows: The dewatered sludge cake is distributed onto a slowly moving horizontal belt enclosed in a housing. As the dewatered sludge cake moves through the belt dryer, the moisture is released into the circulating air. After passing through the drying chamber(s), the dried cake falls off from the belt onto a hopper and is conveyed to a loading or storage facility. The process flow schematic in Figure 10.24 only shows one drying chamber for clarification. However, there may be more than one drying chamber, with each having its own circulating fans and air temperature control. Excess moisture is removed from the air stream in a saturator or vented directly to the atmosphere. Heat for the air circulation loop in each drying chamber is provided with a heat exchanger by indirect contact with steam, hot water, thermal oil, or hot air serving as the heat source.

Figure 10.19 Rotary Drum Mixer

Figure 10.20 Rotary Drum Dryers

Figure 10.21 Rotary Drum Process Induce Draft Fan

Figure 10.22 Rotary Drum Dryers—Wet Bin and Recycle Bin

Figure 10.23 Pellet Cooler and Pellet Loadout Truck

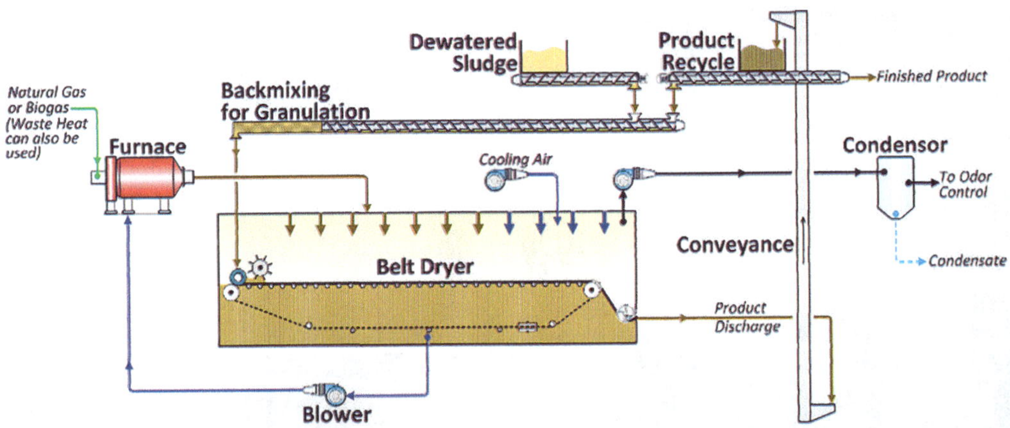

Figure 10.24 Typical Belt Dryer Process Flow Schematic

Indirect Dryers

Indirect dryers heat the sludge by conduction and heat is transferred from a hot surface. This section will cover two of the more common types of indirect dryers: paddle dryers and disc dryers. Both paddle and disc dryers consist of a stationary horizontal vessel and a series of agitators mounted on a rotating shaft. For these dryers, the stationary horizontal vessel is referred to as the *trough*; the agitators are either *paddles*, *discs*, or *flights*; and the rotating shaft is called a *rotor*. The trough may have a jacketed shell through which a heat-transfer medium circulates. Typically, the heat-transfer medium is hot oil or steam. Rotary chamber dryers are similar, except that the trough's outer shell rotates and is heated by gas-fired burners. All these systems have a hollow rotor and agitators through which the heat-transfer medium circulates. The temperature of the heat-transfer medium varies, but ranges from about 150 to 205 °C (302 to 401 °F) and is typically recirculated through the heating system. The surfaces of the trough, agitators, and rotors conduct heat to the dewatered sludge cake.

Dewatered sludge cake is fed directly to the dryer by a positive-displacement pump(s) or a screw conveyor. The dewatered sludge cake is conveyed through the dryer by the rotating agitators. Some systems use a weir at the discharge end of the dryer to submerge the heat transfer surface. The mixing of the feed with recycled product is uncommon with paddle and disc dryers. The dried product exiting the dryer is irregularly shaped and varies in size. This can be attributed to a high concentration of fines in the product caused by the agitation of the dried product by the rotating paddles or discs. Screens or conditioners can be used to produce a more uniformly sized product.

Rotor and agitator assemblies are fabricated of carbon steel, stainless steel, or a combination of both. They also can be made of various special alloys if highly corrosive elements are present. In some systems (primarily the paddle type), the agitators are arranged on the shaft to intermesh, which enhances the relative contact between heated surfaces and solids, thereby increasing the heat-transfer coefficient. They also are self-cleaning. In other systems, stationary agitator ploughs or breaker bars are placed between rotating agitators to improve mixing and prevent cake buildup on the agitator surface.

The water vapor concentration in the vessel creates an inert environment to prevent the potential for explosions or fire. However, enough gas (air) is drawn through the dryer to remove the evaporated water before it can condensate. The dryer is exhausted through a condenser before odor control treatment. Some facilities use chemical scrubbers for odor control treatment, whereas others convey the exhaust to their activated sludge aeration system. The volume of exhaust gas involved is less than that produced by convection systems or direct dryers.

Paddle Dryer

Refer to Figure 10.25 for a paddle dryer schematic.

A paddle dryer conveys the solids through a horizonal vessel that uses heated paddles or augers to move the material. Steam or hot oil is used to heat the paddles. The paddle dryer is a continuous flow device with no back-mixing. The product is not uniform and may be dusty. The paddle dryer may be able to use waste heat from engine generators and has low exhaust volume.

Disc Dryer

The disc dryer is similar in design to a paddle dryer. The surface area available with the disc makes for a compact system. The discs are heated by steam or hot oil.

COMBINATION DRYERS

Fluidized Bed Dryer

The technology was first applied to drying sludge in Europe in the early 1990s and in the United States in the late 1990s. A fluidized bed dryer incorporates both conductive and convection heat transfer. As shown in Figure 10.26, heated tubes at the bottom of the drying chamber provide conductive heating of the sludge. Fluidizing gas circulated continuously over the sludge and tubes, such as when the sludge dries out the particles, are suspended in the fluidizing medium and captured. Fluidized bed dryers incorporate back-mixing to return captured fines to the feed. The main component of a fluidized bed dryer is a stationary, vertical chamber segregated into three zones. The first zone is the wind box (gas plenum),

Paddle Dryer and Disc Dryer

- Continuous feed
- Uses steam or hot oil to heat augers (150 to 205 °C [302 to 401 °F])
- Irregular product with fines making it dusty
- Long startup/shutdown cycle compared to other dryers—3 to 12 hours depending on the heat system used
- Potential use of waste heat as an energy source
- Relative ease of operation compared to a rotary drum dryer
- Typically applied at smaller facilities

Fluidized Bed Dryer

- Continuous feed
- Uses both conductive and convection heat transfer
- Uses steam or hot oil for conductive heat transfer to the sludge
- Creates high-quality product
- More complicated O&M than other dryers
- Long startup/shutdown cycle compared to other dryers
- Potential use of waste heat as an energy source
- Typically applied at larger facilities

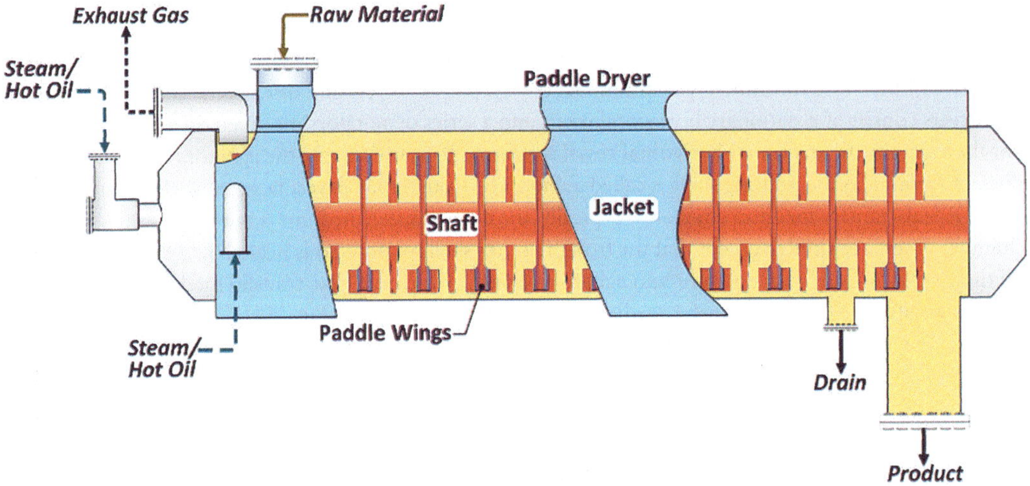

Figure 10.25 Paddle Dryer

which distributes hot fluidizing gases (typically air at about 85 °C [185 °F]) evenly through a bed of solids. The gases are either forced by a blower or induced by a fan through the dryer in a closed-loop system. The second zone is a heat exchanger. A heat-transfer medium (typically steam or hot oil) circulates through the heat exchanger tubes while the fluidizing gas keeps the solids in suspension and removes moisture from the system. The transfer of heat from the tubes to the solids is more efficient when the solids are in suspension. The third zone is a hood, where fluidizing gases exit the dryer and most of the solids are separated from the gas. Then, the exhaust gas is sent to emissions-control equipment (e.g., cyclone separator and condenser), which removes particulates and moisture.

The method for introducing dewatered solids to the dryer is manufacturer specific. Some systems pump solids into the dryer, where a rotating device cuts dewatered cake into small, irregular pieces that fall into the fluid bed. The fluidizing gas contacts and mixes the solids, resulting in a high rate of heat transfer from the gas to the solids. Agitation in the fluidized bed removes rough edges, resulting in solids with a relatively uniform shape. The dried product exits the dryer by overflowing from the chamber via an adjustable weir with a rotary airlock. As with rotary drum systems, fluid-bed dryers will reduce the solids moisture content to less than 10%. These systems are not designed to operate as scalping (i.e., partial drying) systems. Because of the low temperatures involved, fluid-bed systems are well suited for applications where energy recovery is possible.

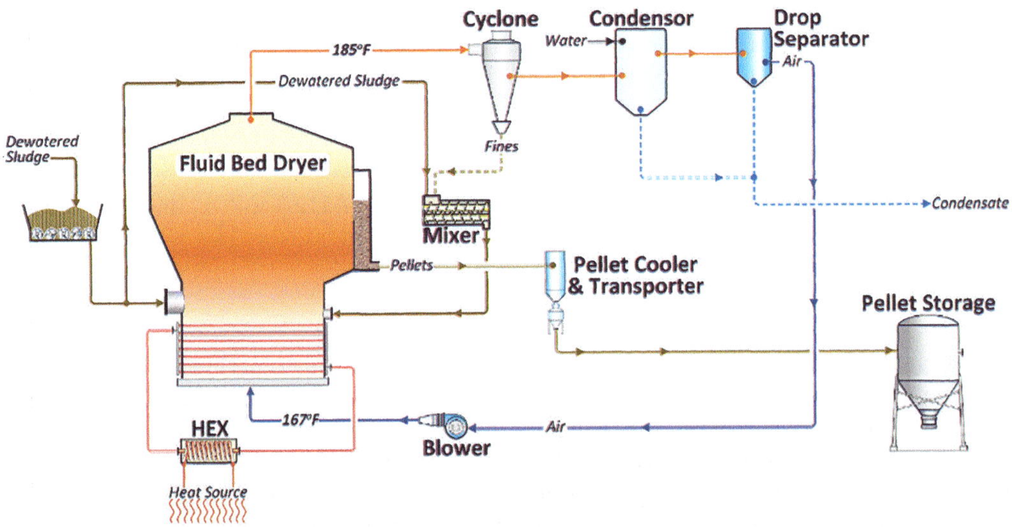

Figure 10.26 Fluidized Bed Dryer

Solar Dryers

Solids have been dewatered naturally (e.g., sand drying beds) for decades, but such systems were susceptible to weather and typically produced solids containing about 70% moisture (for ease of disposal). In the 1990s, German researchers developed systems based on radiant (solar) energy and convective (air) drying theory that could produce solids containing no more than 10% moisture. The U.S. systems often have been designed to produce solids containing less than 25% moisture so they will comply with 40 CFR 503's vector attraction-reduction criteria. However, the Part 503 regulations require that Class A Alternative 3 requirements be met to deem the solar dryer biosolids Class A, which is not required for the previously discussed heat drying technologies. The reason for this requirement is because solar drying systems rely on radiant energy from the sun to meet the Class A temperature and time requirements, making the process sensitive to local weather conditions. These weather conditions include solar radiation (considering typical cloud cover), relative humidity, and temperature. Solar dryers typically consist of a greenhouse structure with a concrete pad and low concrete walls to contain the solids. The dewatered sludge cake is distributed into the greenhouse uniformly onto the pad, either by automated mechanical means or by a manually operated tractor or truck. The sun's radiant energy passes through the greenhouse enclosure (i.e., walls and roof) to heat and evaporate moisture from the sludge. The greenhouse enclosure prevents rain from adding water to the sludge and allows for a semi-controlled greenhouse environment, including air convection to help accelerate evaporation and enclosure of odors that can be processed through an odor control system. Solar drying systems typically provide a mechanical means for mixing and aerating the sludge periodically to promote the drying process. Solar drying is most viable in the southerly latitudes, for municipalities that have land available, and where snow cover is minimal. Radiant heat flooring can also be incorporated to the concrete the slab to reduce required greenhouse space and reduce the effects of weather variations to process performance. Figure 10.27 shows the schematic of a Huber solar dryer along with a photo from inside of the solar dryer.

Solar Dryer

- Uses solar energy to evaporate moisture from solids
- Dewatered sludge cake spread in 152- to-203-mm (6- to-8-in.) layer, left for approximately 20 days
- Product is dirt-like and irregular in size, >70% solids
- Not well suited for primary sludge unless digested
- Space intensive and storage required for periods of low demand for the product
- Odor control typically needed due to the odor potential
- Simple process to operate
- Low operations cost
- Site-specific permitting for Class A biosolids Alternative 3
- Typically applied at smaller facilities

PROCESS VARIABLES

The major variable that affects the heating system operation and size is the evaporative capacity needed. The evaporative capacity is dependent on the following key parameters (a) percent solids of both the feed sludge and product, (b) hours of operation, and (c) rate of heat transfer. For systems with back-mixing, the recycle rate is also an important variable.

Typically, the product percent solids is known and based on the end use. Therefore, mechanically dewatering the feed sludge will reduce the amount of water needed to be evaporated, reducing the size of the dryer and its energy requirements. Also, because dryers have a fixed evaporative capacity, the longer they operate each day, the more sludge they can process. A dryer that operates 24 hours per day only needs one-third of the evaporative capacity to process the same amount of sludge as a dryer that operates 8 hours per day.

When two similar systems are operating at different temperatures, the one with hotter temperatures typically will dry the sludge faster. Although operating temperature may affect the size of the unit, the heat-transfer rate is affected by several factors (e.g., area of exposed, wetted surface, and the unit's heat-transfer coefficient), so temperature will not directly indicate the size.

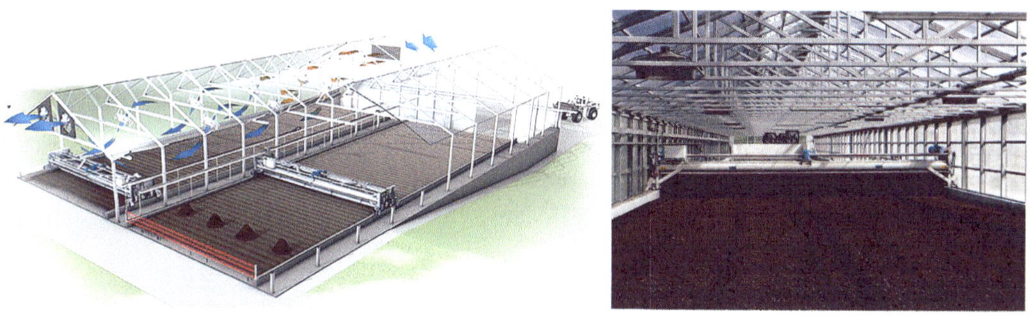

Figure 10.27 Solar Dryer With Sludge Turner (Reprinted with permission from HUBER SE)

PROCESS CONTROL

Dryer systems are highly automated and critical. Temperatures and airflow will be automatically controlled as part of the drying system. The sludge feed rate to the dryer needs to be controlled to keep pace with sludge production from upstream processes, such as sludge dewatering. In many facilities, the dewatering and drying systems may both operate for a limited number of hours per week due to staffing constraints. Most dryer systems include sludge cake storage to provide some operational flexibility for dryer downtime.

There are many types of dryer technology and configurations. Before a dryer system can be successfully operated, sufficient training must be given, and a careful review of the manufacturer's O&M manual should be conducted to learn the nuances of the system.

OPERATION

Efficient and consistent operation of heat drying equipment depends on frequent monitoring, both sensory and analytical. By regularly inspecting the drying equipment during operation, any irregularities in the operating conditions can be detected, such as changes in temperature, pressure, or flowrates. After gaining some operating experience, it is possible to recognize any strange sounds or changes in pitch that may indicate a problem. Early detection of issues can significantly reduce corrective maintenance and downtime.

DATA COLLECTION, SAMPLING, AND ANALYSIS

The temperature of the process must be monitored and recorded, and grab samples of the product should be collected for testing moisture content and nutrients for the fertilizer market. Sampling material in the dryer provides direct insight to dryer performance. Routine sampling analysis and continuous dryer instrument observation are critical to accurately assess, understand, and predict dryer operation.

Collect hourly samples to determine the percent solids concentration of the wet feed mix entering the drum dryer and the percent solids concentration of the dry pellets exiting the pellet cooler. Comparing the results over time can identify trends in the dryer operation. These trends can be very helpful when used to operate, optimize, and troubleshoot the dryer system.

Dryer solids detention time is variable, and it is not monitored. If the wet feed mix and dry pellets are sampled at roughly the same time, the resulting data may not be directly related (the dryer detention time is unaccounted for). Comparing and trending the sample results over time can alleviate this issue and avoid routine dryer detention time calculations. The basis of this sampling methodology relies on insight gained by comparing sample results and the related trends over time.

The wet feed mix and dry pellet samples are collected at approximately the same time. The samples' moisture content changes over time because they are warm and contain water. Therefore, it is important that the time between collecting and analyzing the samples is minimized. The wet feed mix contains the greatest amount of moisture and, therefore, should be analyzed first. Collect samples for each dryer in service, but sample and analyze only one dryer at a time.

MAINTENANCE

Much of the heat dryer equipment is common throughout the rest of the facility. As such, refer to the equipment manufacturer's O&M manual and maintain the dryer equipment in accordance with the manufacturer's recommendations. Preventive maintenance is a necessary part of day-to-day operations. Routine cleaning helps to avoid corrosion caused by the properties of the solids. Multiple units are often used to avoid disruption to treatment works operation when units are out of service. Wear and scale are common maintenance issues associated with the heat drying of sludge. The abrasive nature of the solids causes wear on the heat dryer conveyance equipment. Scale formation on the dryer and piping can be removed by washing with acid or with a hot water pressure washer. Mixing food-grade oil with the solids has been said to help prevent scale formation. However, it is important that the heat drying systems be maintained in accordance with the manufacturer's recommendations to ensure proper operation.

TROUBLESHOOTING

The troubleshooting of heat dryers that are highly mechanized to maintain proper operation are discussed herein. However, it is important that the troubleshooting of heat drying systems should be in accordance with the manufacturer's recommendations to ensure proper operation. Table 10.7 provides general

Table 10.7 Mechanical Dryer Troubleshooting

Problem	Probable Cause	Solution
Final Product Moisture Content High	Outlet temperature too low	Check set point. Check evaporation rate.
	Wet and dry material ratio incorrect (too much water)	Check and change mixture ratio.
	Not enough recycle material (direct dryers only)	Check recycle system (direct dryers only).
Final Product Moisture Content Low	Outlet temperature too high	Check set point.
	Wet and dry material ratio incorrect (not enough wet material)	Check and change mixer ratio.
	Too much recycle material (direct dryers only)	Check recycle system (direct dryers only).
Burner System Problems	Inlet temperature high (burner shuts down automatically)	Wait for burner temperature to drop below set point and reduce evaporation rate.
	Valves to gas train closed	Open valves.
	Control valves not operating properly	Repair valves.
	Gas pressure too low	Increase pressure, adjust regulator.
	No pilot light available	Clean or replace.
	Flame detection system deactivates gas flow	Clean or replace detector.
Equipment Failure	Motor overload at starter	Check and adjust starter amps. Check for equipment clog or jam.
	Motor short circuited	Check and replace motor: Too much material Broken bearing
	Motor stops	Check zero speed switch alarm and correct as needed.
Drum, Belt, or Shaft (paddle) Stops Rotating	Too much material in the dryer	Reduce sludge feed when restarting.
	Motor overload	Reset thermal overloads.
	Drive failure	Inspect drive unit for damage.
Zero Speed Alarm	Drive belt slips	Tighten belt.
	Too much material in dryer	Reduce sludge feed.
Dryer Makes Unusual Noise	Foreign object in the dryer	Shut down, remove object, and inspect for damage.
	Support roller bearing failure	Replace bearings.
	Guide roller bearing failure	Replace bearings.
	Inlet seal failed	Replace seal.
	Outlet seal failed	Replace seal.
	Insulation loose	Repair or replace insulation.
High Inlet or Outlet Temperature Alarm	Burner controller broken	Check and replace.
	Outlet thermocouple broken	Check and replace.
	Wet sludge flow stopped	Check wet sludge slide gate. Check wet material bin level.
	Mixture is too wet and falls onto furnace bottom (rotary drum dryer)	Adjust mixture.
Saturator Outlet Water Too High	Water evaporation capacity too high	Reduce quantity of sludge.
	Water pressure too low	Check system pressure.
	Cooling water nozzles clogged	Check and clean nozzles.
Outlet Set Point Hunting	Feed mixture too wet	Reduce wet feed.
	Evaporation rate too high	Reduce evaporation rate.
	Burner system problem	Check burner system.
Final Product Storage Silo Hot Spot Detected	Stored material hot	Empty silo or begin nitrogen purge.
	Condensation in silo	Check fugitive dust system.
Truck Loading Chute Fault (Limit Switch Not Made)	Chute blocked	Clean fixture.
	Loose/damaged air hose	Reconnect/replace air hose.
	Cylinder faulty	Repair/replace pneumatic cylinder.
	Limit switch faulty	Repair/replace limit switch.
	Loss of air pressure	Check air compressor system.

troubleshooting guidelines for dryers with common mechanical systems including direct dryers, indirect dryers, and combine dryers that have been previously discussed.

SAFETY CONSIDERATIONS

The equipment used for these processes is like other equipment in the solids-handling area, including pumps, valves, centrifuges, conveyors, fans, and exhaust systems. Safety considerations specific to heat dryers are identified as follows:

- Heat drying uses high temperatures. The equipment, valves, and fans should be protected from casual contact using guardrails, covers, and signs to prevent possible burn injuries.
- Fire risk—Keep dryer oxygen levels low by recirculating the dryer exhaust gas and limiting the amount of infiltration air such that the oxygen level in the dryer is held at 3% to 9%. Typically, the oxygen level must be greater than 10% to support combustion. Additionally, dryers are equipped with quench sprays to extinguish a fire or burning ambers in the dryer.
- Dust explosions—Prevent dust from accumulating by keeping the area clean, maintaining positive ventilations pressures, and properly operating the exhaust air system.

TEST YOUR KNOWLEDGE

1. **During heat drying, two processes occur simultaneously, including the transfer of heat from the surrounding environment to evaporate the surface moisture and the transfer of internal moisture to the surface of the solid where it is evaporated.**
   - ☐ True
   - ☐ False
2. **To meet Class A standards with heat drying, the dried product must be 90% solids or greater and a temperature of**
   a. 80 °C (176 °F) is exceeded
   b. 55 °C (131 °F) is maintained for 15 days or longer
   c. 180 °C (356 °F) or higher is achieved for 30 minutes
   d. 55 °C (131 °F) is maintained for 3 days or longer
3. **Heat drying systems are designed to meet the product characteristics needed for the selected end use.**
   - ☐ True
   - ☐ False
4. **The following dryers are typically applied at smaller facilities:**
   a. Rotary drum dryer, fluidized bed dryer, and paddle dryer
   b. Solar dryer, paddle dryer, and belt dryer
   c. Solar dryer, fluidized bed dryer, and belt dryer
   d. Rotary drum dryer, fluidized bed dryer, and belt dryer
5. **Typical solids concentrations of 2% to 10% best describes what process?**
   a. Incineration
   b. Dewatering
   c. Thickening
   d. Drying
6. **Drying systems that heat the sludge by convection are referred to as**
   a. Indirect dryers
   b. Solar dryers
   c. Combination dryers
   d. Direct dryers
7. **The purpose of back-mixing is to**
   a. Increase the amount of exhaust gases for treatment
   b. Increase the feed moisture content
   c. Adjust the moisture content of the feed and promote stable pellet formation
   d. Reduce the system's thermal efficiency
8. **Which dryer has more complicated O&M than other dryers?**
   a. Solar dryer
   b. Belt dryer
   c. Paddle dryer
   d. Rotary drum dryer
9. **Belt dryers operate at gas temperatures of**
   a. 93 °C (200 °F)
   b. 538 °C (1000 °F)
   c. 127 to 177 °C (260 to 350 °F)
   d. 871 °C (1600 °F)

Incineration

Incineration removes all moisture from the dewatered sludge cake, thereby reducing the total volume. It destroys all organic matter by complete combustion, destroys pathogenic organisms, and produces a non-odorous product. Multiple-hearth furnaces and fluidized bed incinerators are the most typical types used for sludge combustion in the United States. Air emissions from incinerators should be monitored on a specific frequency to determine that no toxic or objectionable substances escape destruction. The U.S. EPA regulations for sludge disposal (40 CFR Part 503) limit the pollutant in the feed sludge, as well as in any air emissions from burning sludge for disposal. Specific pollutants limited by the 40 CFR Part 503 regulations include lead, cadmium, chromium, and nickel in the feed sludge and beryllium, mercury, and hydrocarbons in the air emissions from burned sludge. The firing of sludge in an incinerator cannot violate the requirements of the National Emissions Standards for these pollutants, outlined in 40 CFR Part 61, Subparts C and E.

PURPOSE AND FUNCTION

The purpose of incineration is to eliminate pathogens in the dewatered sludge cake and achieve volume reduction. Figure 10.28 shows the volume reduction by incineration. Incineration of dewatered sludge cake completely stabilizes the solids with the benefit of thermal energy recovery and beneficial use of ash. Many of the concerns in sludge disposal, such as the presence of pathogenic organisms and the sludge becoming a solid product with no value that must be disposed of, are eliminated with incineration.

THEORY OF OPERATION

Incineration is a high-temperature oxidation process in the presence of excess air that involves the combustion of waste. Combustion (burning) is the rapid combination of oxygen with fuel, resulting in the release of heat. The typically combustible elements of wastewater solids (carbon, hydrogen, and sulfur) chemically combine in organic solids as grease, carbohydrates, and protein. Although the combustible portion of the wastewater solids has a heat content about equal to that of lignite coal, it only will sustain combustion if enough water has been removed. Therefore, dewatering should be performed before incineration. Dewatered sludge cake with 20% to 35% total solids can be burned with or without auxiliary fuels, depending on the temperature of the combustion air. Adding air provides oxygen to support the combustion. The products of complete combustion are carbon dioxide, water vapor, sulfur dioxide, and inert ash. Good combustion requires proper proportioning and thorough mixing of the fuel and the air. In addition, the initial and sustained ignition of the mixture is required.

The additional air available for combustion over and above the stoichiometric amount required for complete combustion is called excess air.

Heat content is the amount of stored energy that can be converted into actual heating ability. It indicates the amount of heat that is released during combustion. It is expressed in two different ways depending on whether the available heat energy includes or excludes the energy used to vaporize water (contained in the original energy form or created during the combustion process). The gross calorific value, or higher heating value, includes the energy used to vaporize the water. Net calorific value, or lower heating value, is determined by subtracting the heat of vaporization of the water from the higher heating value. Therefore, the energy required to vaporize the water is not realized as heat.

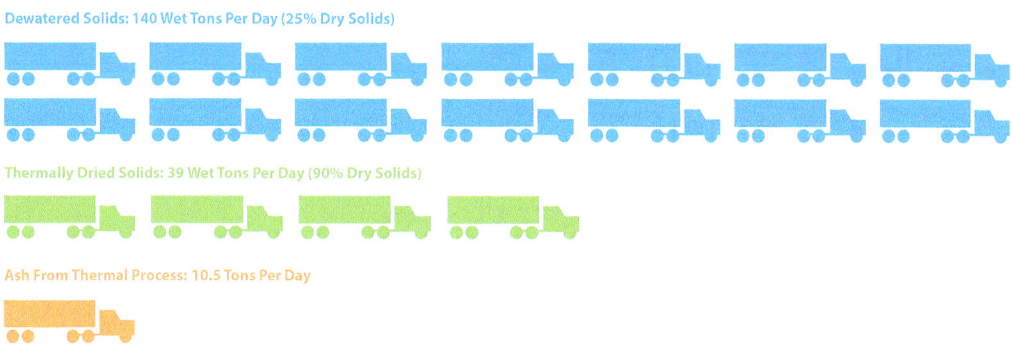

Figure 10.28 Volume Reduction of Sludge by Incineration

Combustion seeks a final equilibrium temperature at which the heat input from the fuel equals the system heat losses. To achieve this, the amount of heat released from combustion must be sufficient to overcome the system heat losses and raise the temperature of all the input streams from ambient levels to those of the system output streams. If the heat released from burning the dewatered sludge cake is sufficient, then this equilibrium temperature is called the *autogenous burning temperature*. When this is achieved, the combustion of the dewatering sludge cake is self-sustaining. If it is not sufficient, supplemental or auxiliary fuel such as natural gas, No. 2 fuel oil, landfill gas, or digester gas must be burned to make up for the heat deficit. The most used auxiliary fuels are natural gas and No. 2 fuel oil. When combustion is sustained by the dewatered sludge cake without the need for auxiliary fuel, this is referred to as *autogenous combustion*.

MULTIPLE-HEARTH INCINERATORS

A multiple-hearth incinerator (MHI) (also referred to as a *multiple-hearth furnace*) has three zones. The top hearths consist of the drying zone where water is evaporated, the combustion zone occurs in the middle hearths, and the bottom hearths comprise the cooling zone where the ash is cooled by incoming combustion air. If auxiliary fuel is required, gas or oil burners are provided in the middle or combustion hearths.

Refractory refers to material that is resistant to heat. It is used as a lining in furnaces to resist high heat temperature (above 538 °C [1000 °F]), thermal shocks, erosion, physical attack, and chemical attack.

An MHI consists of a vertical, refractory-lined steel cylinder containing a series of stacked horizontal refractory shelves (hearths). Refer to Figures 10.29 and 10.30 for a typical MHI cross-section and process schematic, respectively. A hollow rotating shaft runs through the center of the hearths. Rabble arms are attached to this shaft above each hearth, and cooling air may flow through them. Metal blades (rabble teeth) on each arm may be angled in the direction of rotation (forward rabbling) or may be reversed (back rabbling). Back rabbling increases the dewatered sludge cake's detention time and improves drying.

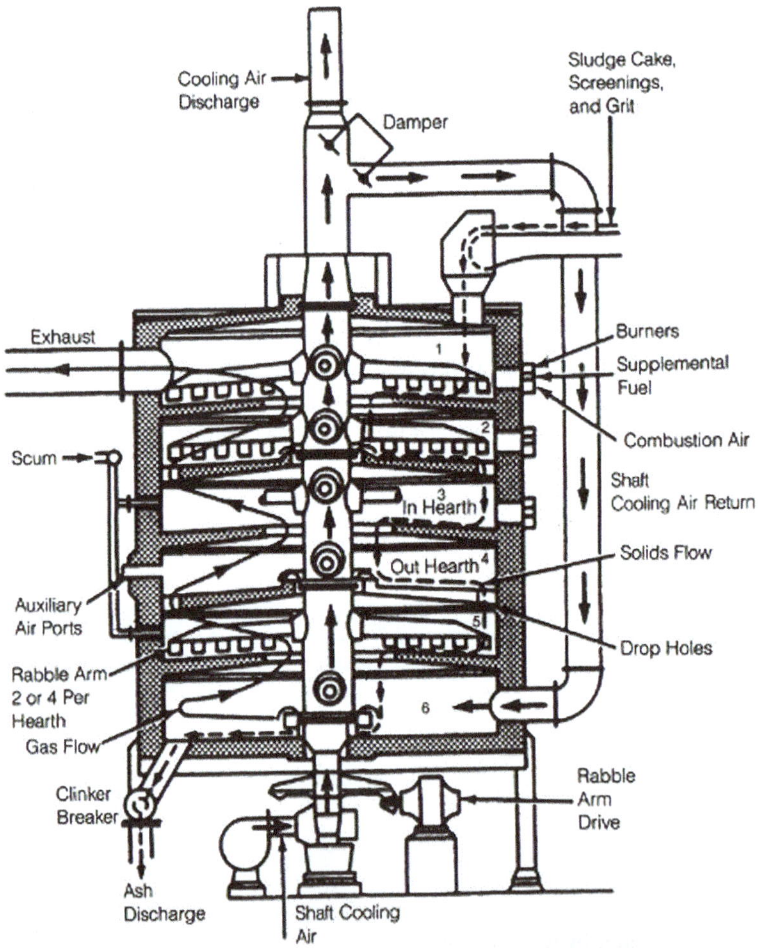

Figure 10.29 Typical Cross-Section of a Multiple Hearth (WEF, 2009)

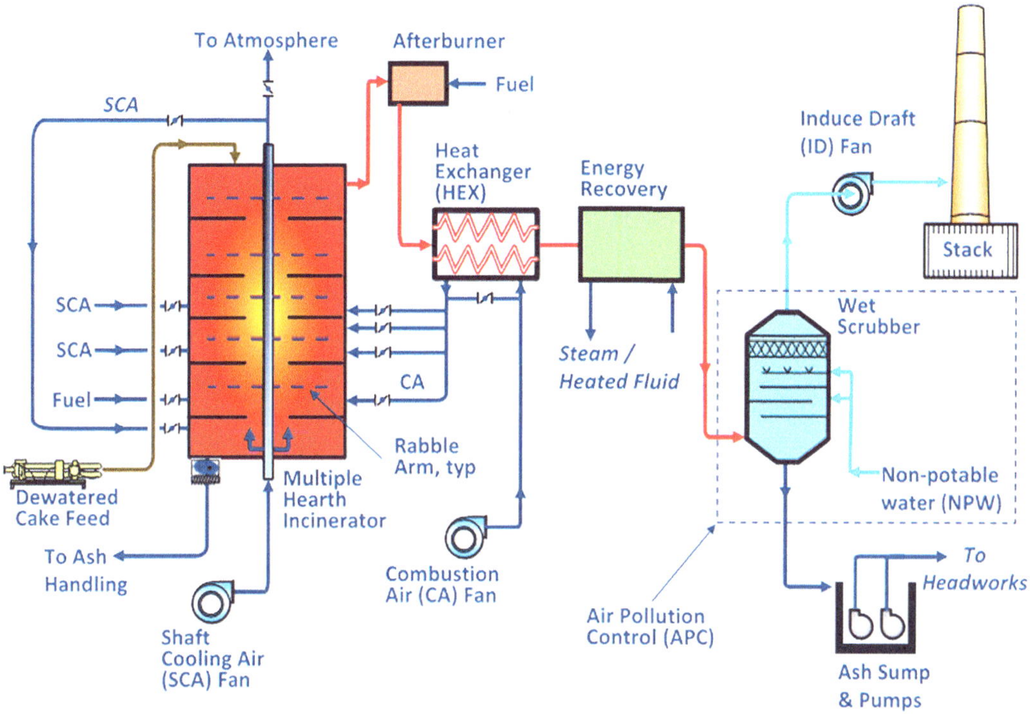

Figure 10.30 Multiple-Hearth Process Schematic

Combustion air enters at the bottom of the hearth and circulates upward through drop holes in the hearths, countercurrent to dewatered sludge cake flow. This countercurrent flow improves performance by maximizing the contact between the dewatered sludge cake and the hot gases. The exhaust exits through the top hearth to the waste heat exchanger (if present) and then to the APC equipment.

Dewatered sludge cake typically flow down through the MHI. This is not a simple vertical drop, however; they move outward and inward on successive hearths, dropping through either peripheral openings in a hearth ("out-hearth") or an annular opening at the center ("in-hearth"). The rabble arms and teeth sweep the dewatered sludge cake forward and radially several times on each hearth, resting after each plowing action. In addition to moving the dewatered sludge cake, the rabble teeth cut, furrow, turn, and open it, thereby exposing new surfaces for drying.

In operation, the hottest hearth in the combustion zone operates from 760 to 871 °C (1400 to 1600 °F) and its location may move up or down from hearth to hearth within the furnace in response to changes in feed rate or characteristics. The shaft and rabble arms are cooled by air blown through the central shaft by the shaft cooling air fan. An air tube runs up the center of the central shaft and air lances extend out to the ends of each rabble arm. The air is blown up through the air tube and a portion, or all, of the air is then recirculated from the top of the central shaft back into the MHI as preheated combustion air.

In addition, ambient air is also injected directly to the middle hearths (combustion zone) by the combustion air fan. The combustion airflow is adjusted by the combustion air fan to keep the hottest hearth below 871 °C (1600 °F) and typically falls in the range of 75% to 125% excess air. A heat exchanger is sometimes used to preheat the combustion air. However, the level that the combustion air can be preheated is limited. This is because as the temperature of the preheated air combustion increases, so do the airflow rates required to keep the hottest hearth below 871 °C (1600 °F). As the products of combustion move up through the drying zone, heat is transferred to the dewatered sludge cake and the evaporation of water from the feed causes these gases to cool, resulting in exhaust gas temperatures at the top of the furnace, often in the range of 482 to 649 °C (900 to 1200 °F), depending on feed solids content and the amount of auxiliary fuel added via the burners in the drying zone. These temperatures are high enough to volatilize some of the organic content in the feed, but not high enough to complete combustion of the volatilized compounds, which translates to higher emission levels of carbon monoxide (CO) and total hydrocarbons (THCs) and the potential for odors in the exhaust stack. As a result, MHIs are often equipped with afterburners, either as a "zero" hearth (above the feed hearth) or an external afterburner chamber located downstream of the furnace exhaust duct. Ideally, afterburner temperatures would be at least 1400 °F to minimize these emissions; however, many systems operate as low as 649 °C (1200 °F) and achieve acceptable performance with lower auxiliary fuel usage. The MHI solids residence time is 40 to 60 minutes. Energy recovery systems have sometimes been used with MHI systems, most commonly to produce steam for power generation, facility heating, or to meet other facility thermal loads.

FLUID BED INCINERATORS

In a fluid bed incinerator (FBI), a bed of sand is set in a fluid motion by passing air through the bed so that all particles are in a homogeneous boiling motion. Enough air is used to keep the sand in suspension, but not enough to carry it out of the reactor. In this state, particles are separated from each other by an envelope of fluidizing gas (combustion air), thereby presenting more surface for a gas-to-solid reaction (e.g., air to carbon). The maximized surface area is what gives most fluidized-bed combustors their high thermal efficiency. Refer to Figures 10.31 and 10.32 for an FBI typical cross-section and schematic, respectively.

The reactor is a closed cylindrical vessel with refractory walls consisting of a windbox section at the bottom of the furnace. Above the windbox is the bed section where the solids are fluidized along with the bed material (sand), and combustion begins. The freeboard section is above the bed zone and is where combustion is completed. The windbox and fluidized bed section are separated by a refractory arch, metal plate, or metal air distribution piping. The windbox-type designation is determined by the temperature of the combustion air as it is introduced to the bed zone, with less than 93 °C (200 °F) being classified as a cold windbox, 316 to 482 °C (600 to 900 °F) being classified as a warm windbox, and above 538 °C (1000 °F) being considered a hot windbox. The nozzles used to distribute air into the bed from the windbox below are called *tuyeres*.

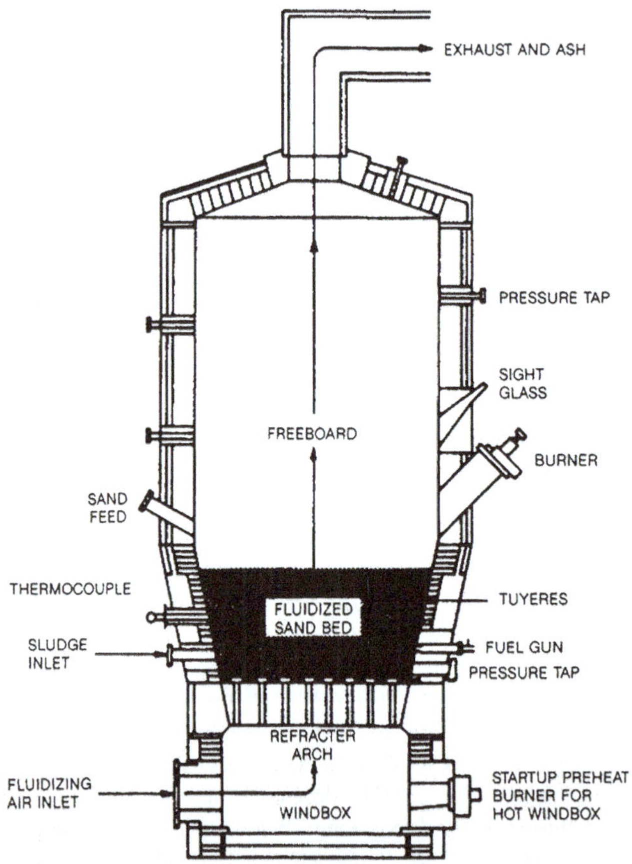

Figure 10.31 Fluid Bed Incinerator

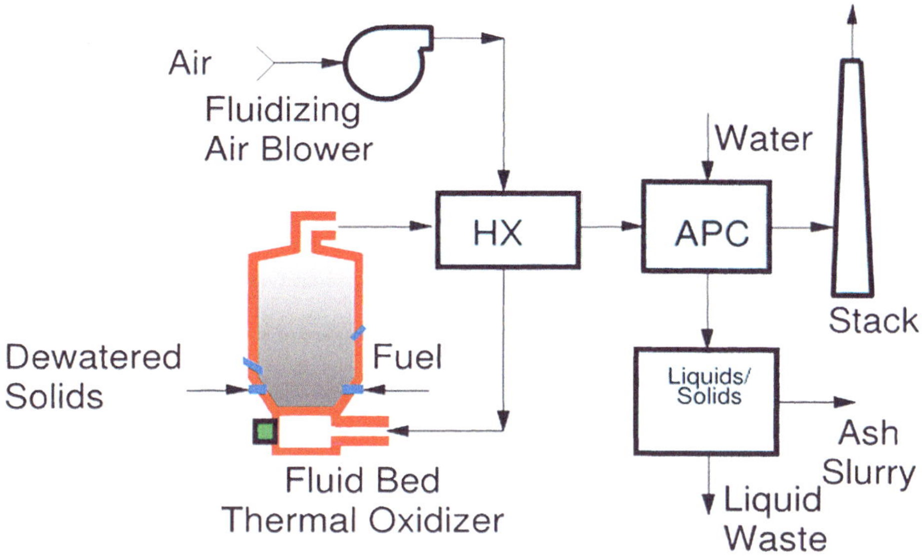

Figure 10.32 Fluid Bed Incinerator Schematic (HX = Heat Exchanger; APC = Air Pollution Control)

The dewatered sludge cake and auxiliary fuel are injected to the fluidized sand bed. The boiling action of the bed results in a rapid dispersion of the sludge solids and optimum contact with the combustion air. The sand bed retains the organic particles until they are reduced to ash. The burning solids and exhaust gas move upward through the furnace in the same direction (co-currently). Exhaust gas with the ash exits the FBI at the top of the freeboard zone and then passes through energy recovery facilities, if included, before it is treated by the APC equipment.

There typically is a temperature difference between the fluidized bed and the freeboard because a portion of VOCs will be combusted in the area above the bed, causing a temperature increase of 65 to 150 °C (150 to 300 °F). In other words, while the bed temperature typically ranges from 730 to 760 °C (1350 to 1400 °F), the freeboard temperature will be significantly hotter (816 to 900 °C [1500 to 1650 °F]). The exhaust gas exposure to the higher freeboard temperatures helps with the complete combustion or burning of the solids, which is responsible for the very low levels of CO and THCs in FBI exhaust.

Ash is commonly removed in a slurry form via the wet scrubbing system, although some designs have a dry ash removal system upstream of the wet scrubbers, such as a cyclone, multiclone, or baghouse. Dry ash systems facilitate beneficial reuse of ash, for example, as an admixture in construction material applications or as a fertilizer amendment based on increased phosphorus and potassium content in the ash.

The most common form of energy recovery in an FBI system is a combustion air preheater, which is typically a shell and tube heat exchanger used to preheat fluidizing air/combustion air to the desired process temperature. Downstream of the air preheater, other supplemental energy recovery devices have sometimes been installed, such as waste heat recovery boilers, thermal fluid heaters, exhaust gas reheaters, or water heaters.

The auxiliary fuel is used, as needed, to maintain the combustion temperature. Gas, fuel oil, or other fuel sources are used as auxiliary fuels. When the dewatered sludge cake is typically greater than 26% solids, the amount of auxiliary fuel needed with FBI systems may be limited or not needed. This is because of the relatively low excess air levels needed and the ability to use very high (up to 649 °C [1200 °F]) air preheat temperatures in FBI systems. As previously discussed, when combustion can occur without auxiliary fuel, this is referred to as *autogenous combustion*.

Fluid bed incinerator systems all have a fluidizing air blower and some systems also have an induced draft fan. The latter are referred to as *push-pull* systems and typically operate at a negative pressure from the freeboard zone of the furnace through to the inlet of the induced draft fan. In the absence of an induced draft fan, the system is referred to as a *push* system and operates under positive pressure through to the final exhaust stack.

The combustion and fluidizing air quantities supplied are determined by the requirements to

- Fluidize the bed to the proper density without blowing sand and incomplete combustion products out of the bed;
- Provide minimum oxygen requirements for complete combustion of all volatiles in the sludge; and
- Provide temperature control to ensure complete deodorization and protection of the refractory, heat exchanger, and flue piping.

Periodic replenishment of the sand is necessary because of scouring in the bed. A port is provided on the side of the reactor into which sand is added, as needed.

Nearly all new thermal oxidation installations in the last 40 years have used fluid bed technology. Many agencies with multiple hearths have required retrofits and operational changes to comply with more stringent emissions. These modifications have included afterburner chambers, exhaust gas recirculation, add-on thermal oxidizers, and improved scrubbers. Other facilities have increased afterburner operating temperatures to reduce "yellow plume" emissions or cyanide concentrations in the scrubber water. Both emissions are signs of incomplete combustion.

DESIGN PARAMETERS

Table 10.8 provides design parameters for incineration.

EXPECTED PERFORMANCE

Incineration of dewatered sludge cake completely stabilizes the solids and achieves volume reduction with the benefit of thermal energy recovery and beneficial use of ash. The best way to operate an incinerator for effective and efficient performance is to consider the following key components of the process:

- **Sludge Fuel Value.** The fuel value in the sludge is important to reduce auxiliary fuel consumption. The combustible elements in sludge are carbon, sulfur, and hydrogen. They primarily exist in organic sludges as grease, carbohydrates, and protein. These materials combine with oxygen to release heat and form carbon dioxide, sulfur dioxide, and water. As the percentage of volatiles increases, the auxiliary fuel consumption decreases. The volatile content of a sludge is maximized by removing inorganics such as grit, by avoiding the use of inorganic chemicals such as ferric chloride and lime in dewatering processes, and by avoiding biological processes such as digestion before incineration.

Table 10.8 Incineration Design Parameters

Parameter	Fluid Bed Incinerator	Multiple-Hearth Incinerator
Flow	Intense back mixing	Counterflow
Heat Transfer	High	Poor
Solids Detention Time	1 to 5 minutes	40 to 60 minutes
Gas Detention Time	5 to 8 seconds	1 to 2 seconds
Combustion Temperature	Bed: 730 to 760 °C (1350 to 1400 °F) Freeboard: 816 to 900 °C (1500 to 1650 °F)	760 to 871 °C (1400 to 1600 °F)
Exit Temperatures	816 to 900 °C (1500 to 1650 °F)	482 to 649 °C (900 to 1200 °F)—No afterburner 760 to 871 °C (1200 to 1400 °F)—w/ afterburner
Excess Air	40% to 50%	75% to 125%

- **Dewatering.** The percent solids of the dewatered sludge cake is important in determining whether the system will require auxiliary fuel or if excess energy is available. As the percent solids of the dewatered sludge increases, the amount of auxiliary fuel reduces. The level of dewatering required to achieve autogenous combustion is dependent on the characteristics of the sludge at your WRRF.
- **Air Requirements.** The air requirements for combustion include the stoichiometric amount of excess air that is necessary to ensure complete combustion. The amount of excess air will vary with the type of incineration, the characteristics of the dewatered sludge cake, and the auxiliary fuel used. Because excess air requires fuel to heat it, excess air is provided in the minimum amount that will ensure complete combustion.
- **Multiple-Hearth Incinerator.** Multiple-hearth incinerators may be operated continuously or intermittently. When not in operation, however, auxiliary fuel is fed to the MHI to maintain temperatures to protect refractory materials. While the temperatures vary in different facilities, the temperature on any given hearth is kept constant to ensure uniform operations. Temperatures typically are interlocked either with the sludge feed and air feed or the auxiliary fuel feed and air feed systems. If the temperature drops, more sludge or fuel and air are added, and vice versa. An analysis of stack conditions is used to determine typical operations. Oxygen, carbon dioxide, and carbon monoxide in the stack are monitored automatically by a continuous emissions monitoring system (CEMS) and compared with preset values. An increase of the carbon monoxide level indicates that incomplete combustion is occurring. If during the same period the oxygen level compares well with the preset value, then either the mixing of sludge and air combustion is inefficient or the temperature has been reduced by the addition of cake that is wetter than normal. When introducing air above the hot (combustion) zone, false high readings can occur. In any event, an analysis of stack gases may be used to determine normal operation and indicate problems. Appropriate steps can then be taken to correct the problem and return to normal operation (refer to the "Troubleshooting" section for information on incineration).
- **Fluid Bed Incinerator.** The basic principles that apply to the operation of MHIs also apply to the operation of FBIs. The fluidized bed does have the advantage of operating intermittently without auxiliary fuel wastage because the sand bed acts as an enormous heat reservoir that holds the heat in the furnace longer. Also, temperature control is simpler in the fluidized bed. During startup, auxiliary fuel is used to raise the temperature of the sand bed to approximately 650 °C (1200 °F). As soon as sludge feed to the furnace begins, the auxiliary fuel rate is adjusted downward to achieve the maximum heat release from the sludge to avoid wasting fuel. This is done by gradually reducing the fuel feed rate to the minimum needed to maintain bed temperatures in the range of 680 to 700 °C (1250 to 1300 °F). During typical operations, the fluidized bed without an air preheater requires more fuel than a multiple-hearth unit; however, if deodorization is required on the multiple-hearth unit, both will use approximately the same amount of fuel.

EQUIPMENT

The following subsections present typical equipment used in MHI incineration.

GENERAL

Figure 10.33 Incinerator Feed Strainer

Figure 10.34 Incinerator Stack

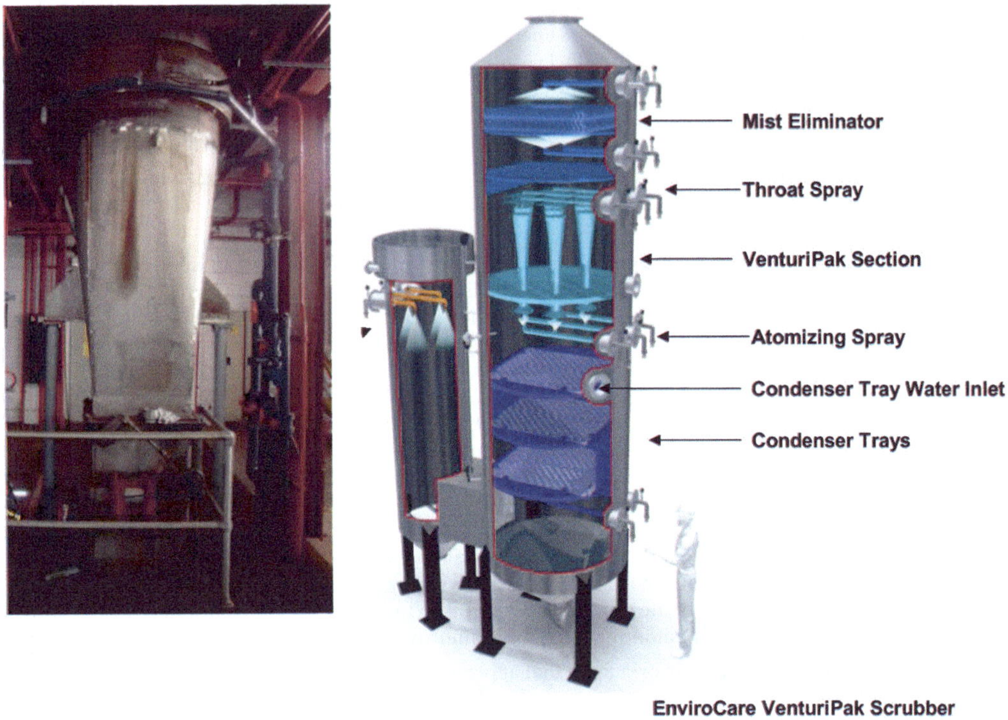

Figure 10.35 Air Pollution Control Equipment—Wet Scrubber (Reprinted with permission from EnviroCare International)

Figure 10.36 Continuous Emissions Monitoring System

TEMPERATURE CONTROL

Figure 10.37 Ambient Air Damper

Figure 10.38 Incinerator Interior—Burner

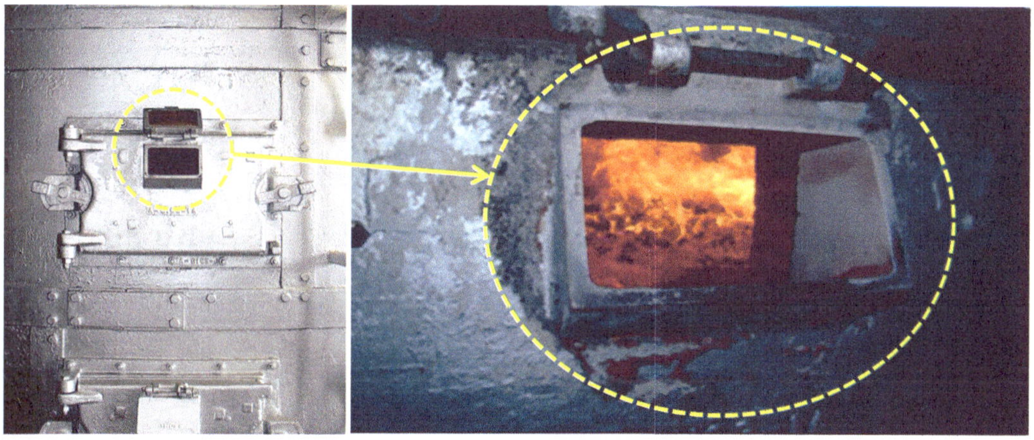

Figure 10.39 Hearth View Port

DEWATERED SLUDGE CAKE FEED CONVEYANCE EQUIPMENT

Figure 10.40 Cake Feed Pumps (Reprinted with permission by Passaro Engineering, LLC)

Figure 10.41 Cake Feed Conveyor (Reprinted with permission by Indigo Water Group)

SLUDGE RABBLING/CENTER SHAFT

Figure 10.42 Incinerator Refractory

Figure 10.43 Incinerator Refractory Bricks

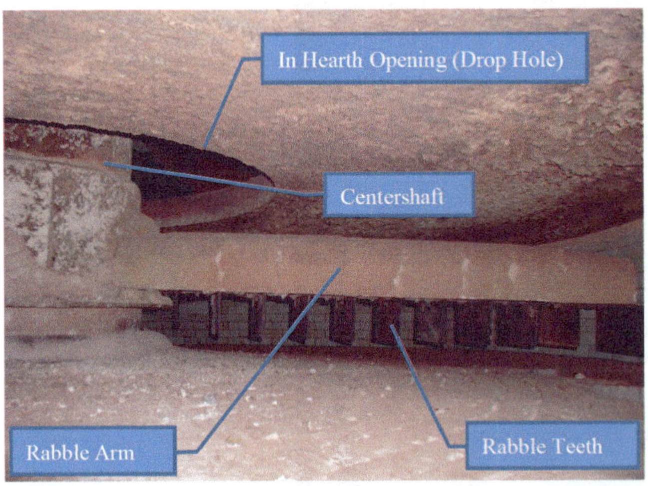

Figure 10.44 Incinerator Hearth

BLOWERS AND FANS

Figure 10.45 Reheat Combustion Fan

Figure 10.46 Center Shaft Cooling Fan

Figure 10.47 Induced Draft Fan

ASH STORAGE AND LOADOUT SYSTEM FOR HAULING AND DISPOSAL

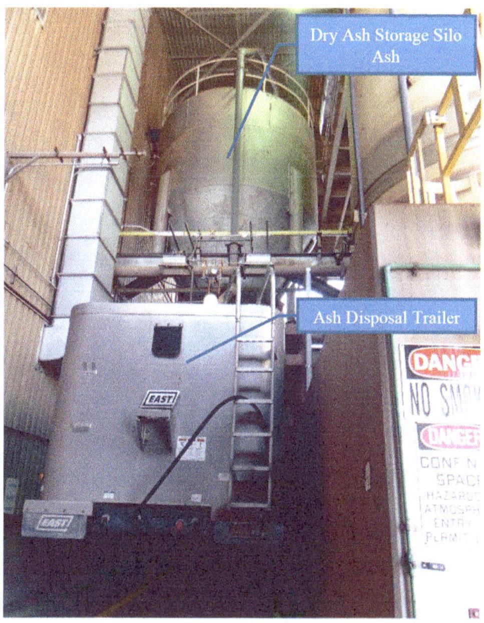

Figure 10.48 Dry Ash Storage and Loadout for Hauling and Disposal

PROCESS CONTROL

Similar guiding principles govern the efficiency of incinerator design and operation:

- Provide sufficient oxygen for complete combustion.
- Regulate time, temperature, and turbulence.
- Minimize sludge feed water content.
- Add auxiliary fuel if necessary.
- Maintain stable sludge feed rate to the incinerator to minimize temperature fluctuations.
- Minimize sludge constituents that act like impurities. Some examples are industrial wastes, fats and greases, polymers, ferric chloride, and lime.
- Create an economically viable operation through heat regulation or recovery or both.

Several guiding principles govern the efficiency of MHI design and operation:

- Feed rate control
- Burner fuel feed control
- Shaft cooling air dampers
- Central shaft speed control

OPERATION

The efficient, safe, and cost-effective operation of an incineration system is dependent on having well-qualified, trained personnel working within a well-managed system with adequate budget support process. Additionally, specialized equipment of MHI and FBI systems require that special attention be given to the management, training, and staffing for these systems to ensure safe and efficient operation.

DATA COLLECTION, SAMPLING, AND ANALYSIS

Operators should monitor for (a) parameters, (b) air emissions permit parameters, and (c) maximum achievable control technology (MACT) and operating parameters. System parameters to be monitored are those during normal operation including temperatures, pressures, draft, percent speed, damper open/close, airflow, and dewater sludge feed rates. Permit and operating parameter monitoring are set during emissions testing as part of the permit and include operating parameters that indicate that the air pollution control system is functioning. In the United States, the MACT and operating parameters are those (a) required by MACT 129 rule (Part 62, Subpart LLL); (b) set during official emissions performance testing and used as surrogate for compliance; and (c) include minimum combustion chamber temperature, minimum scrubber total pressure drop, minimum scrubber total water flowrate, and minimum scrubber liquid pH.

The public perceives that air emissions are greater with thermal oxidation than other biosolids options. This perception generally focuses on stack emissions and does not consider off-site emissions associated with hauling for other options. The basis of this perception stems from older incinerators that emitted much higher air emissions 20 to 30 years ago. In practice, most utilities have upgraded their emission control equipment and are not experiencing chronic air pollution problems. A modern thermal facility should have emission controls and process controls that leave no significant pollutant emissions and no visible stack emissions.

Commonly regulated pollutants include particulate matter; THCs; CO; nitrogen oxides; sulfur oxides; and metals including arsenic, cadmium, chromium, lead, nickel, beryllium, and mercury.

The CEMS required by the U.S. EPA Part 503 regulations provide operators with something they were lacking: immediate feedback. In most cases, this has allowed the operators to better control the thermal oxidation system and comply with emission requirements.

Another factor that greatly reduces emissions is the industrial pretreatment program. The high concentrations of pollutants that could contaminate wastewater solids have been controlled at the source and largely eliminated from sewers. If the pollutants are not in the wastewater solids, they will not be in the emissions.

MAINTENANCE

Incinerators are complex systems and should be maintained in accordance with the manufacturer's recommendations. It is recommended that the incinerator be shut down once a year for inspection and preventive maintenance. The incinerator shutdown should be at a rate less than 38 °C (100 °F) per hour to prevent the refractory from becoming overstressed.

During the inspection of an MHI, the following should be checked and performed:

- Remove slag that forms in the hearths (Figure 10.49), including those that obstruct burners (Figure 10.50) and plug hearth drop holes (Figure 10.51). These slag obstructions are referred to as *impingements*.
- Sagging hearths (Figure 10.52) may cause the rabble arm not to contact the hearth and move ash to the drop holes. Figure 10.52 is a zoomed-in view of Figure 10.53. Notice the gap between the hearth and the center shaft and rabble arm teeth; also notice the cracks in the refractory (bricks) shown in Figure 10.53. These are indications of a sagging hearth.
- Inspect the refractory for defects (Figure 10.53).
- Perform thermocouple calibration, repair, or replacement, as needed.
- Inspect the exterior shell for hot spots or corrosion and evaluate the system elements outside the incinerator per manufacturer recommendations.

Fluid bed incinerators require less maintenance than MHIs because there are fewer moving parts. When the FBI is shut down for annual cleaning and inspection, check the following:

- Slag formation
  - Similar to MHI as previously described
  - In the fluidized sand bed below the auxiliary fuel lines
  - At the burners in the freeboard
  - At thermocouples
  - Within the exhaust ducting
- Thermocouple calibration, repair, or replacement, as needed
- Tuyeres and replace as needed
- The exterior shell for hot spots or corrosion and evaluate the system elements outside the incinerator per the manufacturer recommendations

Figure 10.49 Slag Formation (Courtesy of Kansas City Water)

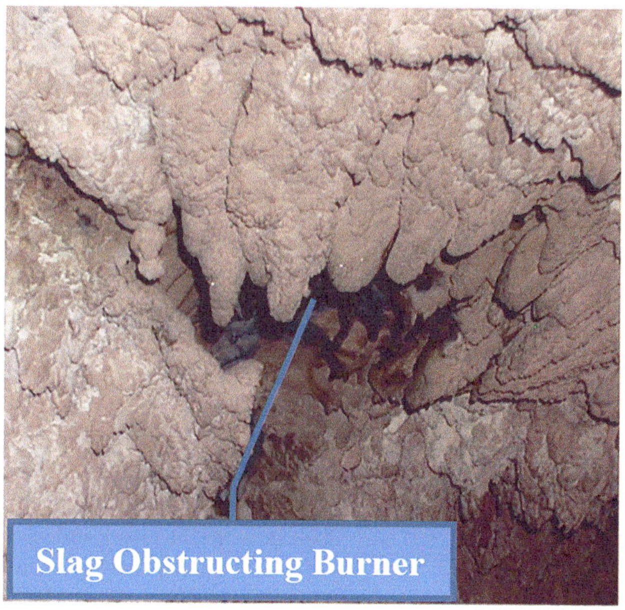

Figure 10.50 Burner Impingement (Courtesy of Kansas City Water)

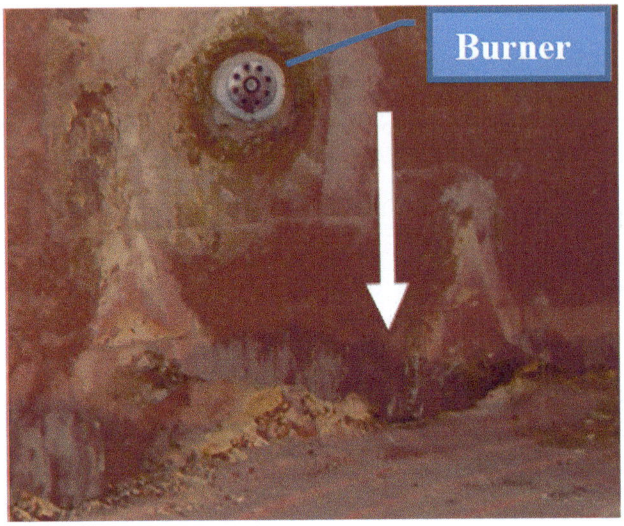

Figure 10.51 Plugged Drop Holes (Courtesy of Kansas City Water)

Figure 10.52 Sagging Hearth (Courtesy of Kansas City Water)

Figure 10.53 Inspect Refractory (Courtesy of Kansas City Water)

TROUBLESHOOTING

Changes in a process variable (sludge feed rate, % dry solids, % volatile solids, etc.) can cause a deviation from operating targets or ranges. Other process variables may need to be adjusted. The operator must take some control actions to return the process back to its desired state. Table 10.9 presents a troubleshooting guide for incineration that will help the operator identify problems and develop solutions.

SAFETY CONSIDERATIONS

As in the entire facility, safe equipment operation and personnel safety must be ensured with incineration. Heat rises to the top of the building and must be adequately vented there and elsewhere to prevent unsafe high temperatures in the building. Because of the large concentration of equipment associated with incineration, noise may be higher than in other areas of the facility. When working around incinerators, personnel should wear hearing protection.

Hydrocarbons can be generated by the incineration process and deposit on the inside of duct work and other places downstream of the air pollution equipment. These deposits can ignite and burn, causing damage to the duct work and related equipment. Ignition could be caused by unanticipated high temperature in the duct work caused by loss of scrubber water or by welding or cutting of the duct to remove or install equipment. Stacks and ducts should be inspected and cleaned periodically to guard against formation of these deposits.

Table 10.9 Troubleshooting Guide for Incineration (U.S. EPA, 1978)

Problem	Probable Cause	Solution
Multiple-Hearth Incineration		
Furnace temperature too high	Excessive fuel feed rate	Decrease fuel feed rate.
	Greasy solids	If fuel is off and temperature is rising, this may be the cause; raise air feed rate or reduce sludge feed rate.
	Thermocouple burned out	If temperature indicator is off scale, this is the likely cause; replace thermocouple.
Furnace temperature too low	Moisture content of sludge has increased	Increase fuel feed rate until dewatering system operation is improved.
	Fuel system malfunction	Check fuel system; establish proper fuel feed rate.
	Excessive air feed rate	If oxygen content of stack gas is high, this is likely the cause; reduce air feed rate or increase feed rate.
Oxygen content of stack gas is too low	Volatile or grease content of sludge has increased	Increase air feed rate or decrease sludge feed rate.
	Air feed rate too low	Check for malfunction of air supply and increase air feed rate, if necessary.
Furnace refractories have deteriorated	Furnace has been started up and shut down too quickly	Replace refractories and observe proper heating and cooling procedures in the future.
Unusually high cooling effect from one hearth to another	Air leak	Check hearth doors, discharge pipe, center shaft seal, air butterfly valves in inactive burners, and stop leak.
Short hearth life	Uneven firing	Check all burners in hearth; fire hearths equally on both sides.
Center shaft drive shear pin fails	Rabble arm is dragging on hearth or foreign object is caught beneath arm	Correct cause of problem and replace shear pin.
Furnace scrubber temperature too high	Low water flow to scrubber	Establish adequate scrubber water flow.
Stack gas temperatures too low (260 to 320 °C [500 to 600 °F]); odors noted	Inadequate fuel feed rate or excessive sludge feed rate	Increase fuel or decrease sludge feed rates.
Stack gas temperatures too high (650 to 870 °C [1200 to 1600 °F])	Excess heat value in sludge or excessive sludge feed rate	Add more excess air or decrease fuel rate.
Furnace burners slagging up	Burner design	Consult manufacturer and replace burners with newer designs that minimize slagging.
	Air–fuel mixture is off	Consult manufacturer.
Fluidized Bed Incineration		
Bed temperature is falling	Inadequate fuel supply	Increase fuel feed rate or repair any fuel system malfunctions.
	Excessive rate of sludge feed	Decrease sludge feed rate.
	Excessive sludge moisture	Improve dewatering system operation.
	Excessive air flow	Reduce air rate if oxygen content of exhaust gas exceeds 6%.
Low (<4%) oxygen in exhaust gas	Low airflow	Increase air blower rate.
	Fuel rate too high	Decrease fuel rate.

Table 10.9 Troubleshooting Guide for Incineration (U.S. EPA, 1978) (*Continued*)

Problem	Probable Cause	Solution
Excessive (>6%) oxygen in exhaust gas	Sludge feed rate too low	Increase sludge feed rate and adjust fuel rate to maintain steady bed temperature.
Erratic bed depth readings on control panel	Bed pressure taps plugged with solids	Tap a metal rod into the pressure tap pipe when reactor is not in operation.
		Apply compressed air to pressure tap while the reactor is in operation after reviewing manufacturer's safety instructions.
Preheat burner fails and alarm sounds	Pilot flame not receiving fuel	Open appropriate valves and establish fuel supply.
	Pilot flame not receiving spark	Remove spark plug and check for spark; check transformer, replace defective part.
	Pressure regulators defective	Disassemble and thoroughly clean regulators.
	Pilot flame ignites but flame scanner malfunctions	Clean sight glass on scanner, replace defective scanner.
Bed temperatures too high	Fuel feed rate too high through bed fuel injectors (fuel guns)	Decrease fuel flowrate through bed fuel guns.
	Bed fuel guns have been turned off but temperature still too high because of greasy solids or increased heat value of sludge	Raise air flowrate or decrease sludge feed rate.
Bed temperature reads off scale	Thermocouple burned out or controller malfunction	Check the entire control system; repair as necessary.
High scrubber temperature	No water flowing in scrubber	Open valves.
	Spray nozzles plugged	Clean nozzles and strainers.
	Water not recirculating	Return pump to service or remove scrubber blockage.
Reactor sludge feed pump fails	Bed temperature interlocks may have shut down the pump	Check bed temperature.
	Pump is blocked	Dilute feed sludge with water if sludge is too concentrated.
Poor bed fluidization	During shutdowns, sand has leaked through support plate	Once per month, clean windbox.

Incinerators are a confined space. Because the sand bed in an FBI is typically silica, the air inside these incinerators should also be tested for silica. The appropriate protective clothing and face masks must be worn. No work should be done on hot surfaces without special protective clothing and heat-resistant gear.

Periodic checks must be made to ensure that systems are functioning properly to prevent any buildup of positive pressure inside an operating incinerator. Positive pressure can cause flames to come out through open doors and can also create combustion gases that are unsafe for personnel.

TEST YOUR KNOWLEDGE

1. The typical excess air range in an FBI is
   a. 50% to 150%
   b. 0% to 100%
   c. 75% to 100%
   d. 40% to 50%

2. The MHI heat transfer is
   a. High
   b. Poor
   c. Excellent
   d. Average

3. The MHI combustion temperature range is between
   a. 730 to 760 °C (1350 to 1400 °F)
   b. 482 to 649 °C (900 to 1200 °F)
   c. 760 to 871 °C (1200 to 1400 °F)
   d. 760 to 871 °C (1400 to 1600 °F)

4. The FBI exit temperature range is between
   a. 816 to 900 °C (1500 to 1650 °F)
   b. 730 to 760 °C (1350 to 1400 °F)
   c. 482 to 649 °C (900 to 1200 °F)
   d. 127 to 177 °C (260 to 350 °F)

5. There typically is a temperature difference between the fluidized bed and the freeboard because a portion of VOCs will be combusted in the area above the bed, causing a temperature increase of 65 to 150 °C (150 to 300 °F).
   ☐ True
   ☐ False

6. The solids detention time in an FBI is
   a. 1 to 5 seconds
   b. 1 to 5 minutes
   c. 1 to 5 hours
   d. 40 to 60 minutes

7. The gas detention time in an MHI is
   a. 1 to 2 minutes
   b. 5 to 8 seconds
   c. 1 to 2 seconds
   d. 2 to 4 minutes

8. The typical excess air range in an MHI is
   a. 50% to 150%
   b. 0% to 100%
   c. 75% to 125%
   d. 40% to 50%

9. Refractory is a material that is resistant to heat. It is used as a lining in furnaces to resist high heat temperature (above 538 °C [1000 °F]), thermal shocks, erosion, physical attack, and chemical attack.
   ☐ True
   ☐ False

10. Incineration of 25% dewatered sludge cake will reduce the volume by
    a. 2 times
    b. 13 times
    c. 4 times
    d. 3 times

CHAPTER SUMMARY

Table 10.10 provides a summary of the four additional stabilization processes discussed in this chapter.

Table 10.10 Summary of the Chapter Additional Stabilization Processes

Composting	
Purpose and Function	Composting encourages beneficial bacteria and results in a stabilized humus-like product that makes minerals and nutrients available to plants. Compost is used to improve soil health and fertility in horticultural and agricultural applications.
Theory of Operation	■ The aerobic decomposition process involves blending dewatered raw or digested sludge with bulking agents to absorb moisture, provide a source of carbon, and increase porosity to improve oxygen transfer, enhancing the biological activity and providing a finished compost at a faster rate. ■ Bulking agents such as wood chips, bark, sawdust, straw, and other organic substrates are added to the sludge. ■ The composting process includes a curing period after the temperature declines following the time and temperature regime required by regulations. During curing, compost is placed in piles and left undisturbed for several months to provide additional stability to reduce odors and moisture and make the final product more acceptable in all markets. Oxygen levels and temperatures are still occasionally monitored.
Design Parameters	■ Composting is the aerobic decomposition of organic material in the solid phase using controlled temperature, moisture, and oxygen levels. ■ Blend design is based on • Moisture content of the compost blend • Sludge feedstock solids concentration • Bulking agents • C:N of the compost blend
Expected Performance	■ Mature compost is a dark brown humic material that will not significantly degrade further. ■ As a sludge stabilization process, composting renders sludge suitable for beneficial use. ■ Composting has greater potential for public distribution and commercial use than other biosolids products.
Lime stabilization	
Purpose and Function	■ To create a stable biosolids for storage by reducing the number of pathogenic and odor producing organisms and preventing their regrowth. ■ Reduce the mobility (or leaching) of metals from the biosolids to the land on which it is applied. ■ Biosolids also may be used to supplement or replace liming agents on acid soils because they elevate the soil pH.
Theory of Operation	■ Lime addition to sludge reduces odors and pathogen levels by creating a high pH environment that prevents biological activity from taking place. Lime stabilization depends on maintaining the pH of the sludge at a high enough level for an adequate amount of time to inactivate the microorganism population. This substantially reduces the microbial reactions that can otherwise lead to odor production and vector attraction. ■ The standard process involves adding enough lime to sludge to raise the pH of the mixture to 12 or more and to maintain a pH of 12 for a minimum of 2 hours. ■ There are typically two types of lime stabilization processes including pre- and post-lime stabilization. • In pre-lime stabilization, lime is added to thickened sludge to meet stabilization requirements, generating a Class B liquid biosolids for beneficial use. It has been common practice to land apply the Class B liquid biosolids for WRRFs with a capacity less than 18.9 ML/d (5 mgd). • The purpose of the post-lime stabilization is to reduce (Class B) or to eliminate (Class A) pathogenic organisms, to comply with VAR requirements, minimize odors and nuisance conditions and, finally, to produce a marketable beneficial use product. Majority of the post-lime stabilized biosolids are beneficially land applied or used as landfill cover material.

Design Parameters	The design parameters of the alkaline stabilization process include the sludge type and characteristics, type and dosage of the alkaline material, contact time, pH, temperature, and mixing configurations.
Expected Performance	Lime-stabilized biosolids have lower concentrations of available nutrients (e.g., nitrogen and phosphorus) because of the inert solids and reactions involved as compared to biologically stabilized (anaerobic or aerobic digestion) primary sludge and/or WAS. Therefore, the land application of lime-stabilized biosolids only benefit agricultural areas needing to neutralize acidic soils.
Heat drying	
Purpose and Function	■ To reduce the mass and volume of material, reduce pathogens, and enhance the material handling properties of the biosolids.
Theory of Operation	■ Heat drying transfers thermal energy into the solids, raising the temperature to evaporate water and pasteurize the product. ■ Dryer technology is classified by the heat transfer and evaporation method. • Direct dryers heat the sludge by convection; heat is transferred directly from hot gases to the sludge, promoting evaporation and removal of water vapor. • Indirect dryers heat the sludge by conduction, or direct contact with a hot surface. ■ All dryers must have continuous airflow to remove water vapor and a means to directly exhaust water vapor or condense water from the vapor flow. Some dryers use combinations of these methods to evaporate and remove moisture. Dryers may also include odor control for the exhaust gases from the dryer.
Design Parameters	Design parameters of heat drying include: ■ Percent solids in the wet sludge feed; ■ Ratio of dried sludge to wet sludge; ■ Quantity of hot combustion gases used for drying; ■ System temperatures; and ■ For systems with back-mixing, the recycle rate is an important variable.
Expected Performance	The expected performance of thermal drying is to meet or exceed 90% solids in the dried product and meet or exceed the minimum temperature requirement for the U.S. EPA PFRP. The product quality in terms of pelletization, size, and uniformity can be included in the dryer performance requirements.
Incineration	
Purpose and Function	Incineration removes all moisture from the sludge, thereby reducing the total volume. It destroys all organic matter by complete combustion, destroys pathogenic organisms, and produces a non-odorous product.
Theory of Operation	Incineration is the combustion of organic substances in the presence of air. The process occurs in two steps. The first step increases temperature to evaporate moisture from the sludge. The second step increases the temperature to the point that the volatile solids fraction of the sludge combusts. Incineration is a volume-reducing step to decrease the cost of landfilling.
Design Parameters	Similar guiding principles govern the efficiency of FBI design and operation: ■ Provide sufficient oxygen for complete combustion; ■ Regulate time, temperature, and turbulence; ■ Minimize sludge feed water content; ■ Add auxiliary fuel, if necessary; ■ Maintain stable sludge feed rate to the incinerator to minimize temperature fluctuations; ■ Minimize sludge constituents that act like impurities. Some examples are industrial wastes, fats and greases, polymers, ferric chloride, and lime; and ■ Create an economically viable operation through heat regulation or recovery or both.
Expected Performance	Combustion of sludge completely stabilizes it with the benefit of thermal energy recovery and beneficial use of ash. Many of the concerns in sludge disposal are eliminated, such as the prevalence of pathogenic organisms, and the sludge becomes a solid product with no value that must be disposed of. The success or failure of sludge combustion is often dependent on the ability to dewater sludge to its autothermal combustion point, reducing or eliminating the need for auxiliary fuel.

Table 10.11 lists the advantages and disadvantages of the four additional stabilization processes discussed in this chapter.

Table 10.11 Comparison of Additional Stabilization Processes

Advantages	Disadvantages
Composting	
High-quality, potentially saleable product suitable for agricultural use	Requires an initial blend of dewatered sludge and bulking agents of about 40% to 50% solids, by weight
	Requires bulking agent and carbon source
Can be combined with other processes	Requires either forced air or turning (labor)
Low capital cost (static pile and windrow)	Potential for pathogen spread through dust
	High operational cost; can be power and labor intensive
	May require significant land area, potential for odors
Lime stabilization	
Low capital cost	Increases volume of sludge for disposal
Easy operation	Land application may not be appropriate where soils are alkaline
Good as interim or emergency stabilization method	Chemical intensive
	Overall cost is site specific
	pH drop after treatment can lead to odors and biological regrowth
Heat drying or incineration	
Significant reduction in sludge volume for disposal	Complex mechanical and control equipment
Total destruction of pathogens	Higher capital cost
	Requires auxiliary fuel source
	Requires time to raise the operating temperature
	Requires relatively uniform dewatered sludge feed
	May generate odors

References

U.S. Environmental Protection Agency. (1978). *Operations manual: Sludge handling and conditioning* (EPA-430/9-78-002). U.S. Environmental Protection Agency, Office Water Program Operations.

U. S. Environmental Protection Agency. (1979). *Process design manual for sludge treatment and disposal* (EPA-625/4-78-012). U.S. Environmental Protection Agency, Office of Technology Transfer.

U.S. Environmental Protection Agency. (1984). *Lime handling systems problems and remedies* (EPA-905/R-84-121). U.S. Environmental Protection Agency: Office of Technology Transfer.

U.S. Environmental Protection Agency. (1992). *Control of pathogens and vector attraction in sewage sludge* (EPA-625/R-92-013). U.S. Environmental Protection Agency, Office of Research and Development, Office of Science, Planning, and Regulatory Evaluation.

U.S. Environmental Protection Agency. (2003). *Control of pathogens and vector attraction in sewage sludge* (EPA-625/R-92-013). U.S. Environmental Protection Agency, Office of Research and Development, Office of Science, Planning, and Regulatory Evaluation.

Water Environment Federation. (2009). *Wastewater solids incineration systems* (Manual of Practice No. 30). Water Environment Federation.

Water Environment Federation, American Society of Civil Engineers, & Environmental & Water Resources Institute. (2018). *Design of water resource recovery facilities* (6th ed., WEF Manual of Practice No. 8; ASCE Manuals and Reports on Engineering Practice No. 76). Water Environment Federation.

Suggested Readings

U.S. Code of Federal Regulations (1993) Title 40, 40 CFR Part 503.

Water Environment Federation, Water Environment Research Foundation, & U.S. Environmental Protection Agency. (2012). *Solids process design and management.* Water Environment Federation.

CHAPTER 11

Odor Control

Introduction

As residential, commercial, and industrial land uses are being developed closer to water resource recovery facilities (WRRFs), there has been a growing concern about the effects of odors from WRRF operations on surrounding communities. Federal, state, and local regulations have been revised or developed to limit odors and other air contaminants from WRRFs and other industrial facilities. Decreases in allowable metals discharge limits at WRRFs have resulted in increased pretreatment for heavy metals. The removal of heavy metals through pretreatment results in increased odors in the collection systems because the heavy metals inhibit biological activity that generates hydrogen sulfide (H_2S) and also binds up hydrogen sulfide as insoluble metal salts (WEF, 2017).

Water resource recovery facility operators must provide adequate odor control to prevent the release of nuisance odors in the surrounding community. Successful control of odor emissions requires proper planning, invested management, integrated design, attentive operation and maintenance, and proactive public communications.

LEARNING OBJECTIVES

Upon completing this chapter, you will be able to understand

- Chemical compounds that cause odors
- Odor generation and concerns in wastewater treatment infrastructure
- Safety concerns with odorous compounds
- Parameters to measure odors
- Monitoring, sampling, and analysis of odorous compounds
- Applicable odor regulations
- The effects of odor on the community
- Odor control technologies, including performance, operation, and maintenance

Purpose and Function

COMPOUNDS THAT CAUSE ODORS

General classes of odorous compounds that are associated with untreated wastewater and WRRFs are listed in Table 11.1. These compounds are typically formed during wastewater transport but can also be formed within treatment processes (WERF, 2007). The formation of odorous compounds is most frequently caused by anaerobic conditions that can develop in sections of the wastewater sewer system and in localized areas of wastewater treatment systems. The two pathways that typically produce the most odor-causing compounds are fermentation and sulfate reduction (Hvitved-Jacobsen et al., 2000).

The decay of organic material under anaerobic conditions can take place throughout the sewer system and in treatment processes, including the sediment in the sewers, biofilms, and in the wastewater itself. This process is referred to as *fermentation*. The fermentation process produces volatile organic compounds (VOCs), including volatile fatty acids (VFAs); nitrogenous compounds such as scatole; aldehydes and ketones like formaldehyde and acetone; and total reduced sulfur (TRS) compounds.

Table 11.2 provides a list of VOCs that are associated with odor in wastewater and are produced by fermentation (Hvitved-Jacobsen, 2002). Table 11.2 also contains the odor threshold of the compounds.

The formation of odorous compounds is most frequently caused by anaerobic conditions that can develop in sections of the wastewater sewer system or in poorly mixed or aerated areas of wastewater treatment systems. The two pathways that typically produce the most odor are fermentation and sulfate reduction.

Anaerobic conditions occur in areas of WRRFs that are essentially free of dissolved oxygen.

Odor threshold is the lowest concentration at which a compound can be detected by human smell.

Table 11.1 Odorous Compounds Associated With Untreated Wastewater and Water Resource Recovery Facilities (Adapted from WERF, 2007)

Class of Compound	Relevant Examples
Inorganic gases	Ammonia, hydrogen sulfide*
Mercaptans*	(Allyl, amyl, benzyl, crotyl, ethyl, methyl, propyl) mercaptans
Other organic sulfur*	Dimethyl sulfide, diphenyl sulfide, thiocresol, thiophenol
Amines	(Dibutyl, diisopropyl, dimethyl, ethyl, methyl, triethyl) amines
Diamines	Cadaverine (1,5-pentanediamine), putrescine (butanediamine)
Other organic nitrogen	Indole, pyridine, skatole
Volatile fatty acids	(Acetic, propionic, butyric, isovaleric) acids

*Total reduced sulfur compound. Total reduced sulfur compounds include hydrogen sulfide, mercaptans, carbonyl sulfide, carbon disulfide, dimethyl sulfide, dimethyl disulfide, and dimethyl trisulfide.

Table 11.2 Examples of Volatile Organic Compounds Associated With Odors and Produced by Fermentation Process Wastewater (Adapted from Hvitved-Jacobsen, 2002)

Substance	Compound	Formula	Threshold Odor (ppb)	Odor Descriptor
Volatile sulfur compounds	Methyl mercaptan	CH_3SH	1	Sweet, pleasant, chloroform-like
	Ethyl mercaptan	C_2H_5SH	0.2	Garlic/onion, sewer, decayed cabbage, earthy
	Allyl mercaptan	$CH_2{=}CHCH_2SH$	0.05	Garlic
	Benzyl mercaptan	$C_6H_5CH_2SH$	0.2	Repulsive, garlic-like odor
	Dimethyl sulfide	$(CH_3)_2S$	1	Decayed cabbage
	Dimethyl disulfide	CH_3SSCH_3	0.3–10	Repulsive
	Thiocresol	$CH_3C_6H_4SH$	0.1	Rancid, skunk-like
Nitrogenous compounds	Methylamine	CH_3NH_2	1–50	Pungent fishy odor resembling odor of ammonia
	Ethylamine	$C_2H_5NH_2$	2400	Strong ammonia-like
	Dimethylamine	$(CH_3)_2NH$	20–80	Floral
	Pyridine	C_5H_5N	4	Sour, putrid, fishy
	Indole	C_8H_7N	1.5	Excreta
	Scatole	C_9H_9N	0.002–1	Strong fecal odor
Acids (VFAs)	Acetic	CH_3COOH	15	Sour vinegar
	Butyric	C_2H_5COOH	0.1–20	Sweet body odor
	Valeric	C_3H_7COOH	2–2600	Sweat, body odor, cheese
Aldehydes and ketones	Formaldehyde	$HCHO$	370	Pungent, suffocating odor
	Acetaldehyde	CH_3CHO	0.005–2	Penetrating, pungent, suffocating odor
	Butyraldehyde	C_2H_5CHO	5	Pungent
	Acetone	CH_3COCH_3	4600	Pungent
	Butanone	$C_2H_5COCH_3$	270	Sharp, sweet odor

Odorous compounds become a nuisance when they are released to the atmosphere and reach surrounding community areas at concentrations above their odor thresholds. The chemical speciation of certain odorous compounds, such as mercaptans, H_2S, VFAs, ammonia, and amines, is pH dependent. Odors are associated with the gas phase of these compounds.

Chemical speciation refers to the distribution of an element among chemical species in a system.

Sulfide exists in solution as H_2S gas and the dissolved ions, hydrosulfide (HS^-) and sulfide (S^{-2}). The amount of each ion is pH and temperature dependent. Figure 11.1 shows the distribution (%) of H_2S speciation as a function of pH. At pH 7.0, there is a 50/50 distribution of H_2S and HS^- present, with a very small amount of S^{-2}. At lower pH (pH < 6) values, the H_2S gas predominates.

Whether ammonia exists in solution as ammonia (NH_3) gas and dissolved ammonium ion (NH_4^+) is also pH and temperature dependent. Figure 11.2 shows the distribution (%) of ammonia speciation as a function of pH. Only NH_4^+ ions are present in neutral solution (pH = 7) at ambient temperatures. As the pH increases, more and more of the ammonia is converted to ammonia gas. At pH 11 or greater, essentially all the ammonia appears as NH_3 gas. At lower temperatures, the percentage of dissolved ammonia relative to ammonium ion decreases.

Ambient temperature is the ordinary temperature of the immediate surroundings.

ODOR SOURCES IN WATER RESOURCE RECOVERY FACILITIES

Any place or process in which wastewater is collected, conveyed, or treated has the potential to generate and release odors to the surrounding area. However, most odor problems occur in the collection system, in primary treatment facilities, and in solids handling facilities. Odors are generated under anaerobic or "septic" conditions that occur when oxygen transfer to the wastewater is limited, such as in a force main or siphon. Under anaerobic conditions, sulfate reducing bacteria reduce sulfate (SO_4^{2-}) that is abundant in most wastewaters to H_2S, which has a low solubility in the wastewater and a strong, offensive, rotten-egg odor (Hvitved-Jacobsen et al., 2000). Sulfate reducing bacteria are typically found in biofilms within sewers and in some WRRF processes as well as in solids that settle out in collection systems (WERF, 2007). Hydrogen sulfide is the most common odor-causing compound in wastewater systems. Due to its low solubility in the wastewater, H_2S is released into the air in areas where there is turbulence such as wet wells, headworks, grit chambers, and primary clarifiers. There are typically other "organic" odorous compounds, such as mercaptans and amines, present in these areas, but H_2S is the most prevalent compound.

Septic conditions occur when bacteria use all the available oxygen while decomposing organic matter in wastewater.

Solubility is defined as the maximum amount of a substance that will dissolve in water at a specified temperature.

Solids handling facilities are another significant odor problem area. In biosolids dewatering and treatment processes, the biosolids commonly undergo extreme turbulence, pH adjustment, and/or thermal treatment. Depending on the nature of the biosolids stream and the treatment used, the odor compounds released can consist of any combination of the following compounds in a wide range of concentrations:

Turbulent flow is the type of flow in which the wastewater undergoes irregular fluctuations, or mixing, in contrast to laminar flow, in which the fluid moves in smooth paths or layers. In turbulent flow, the speed of the fluid at a point is continuously undergoing changes in both magnitude and direction.

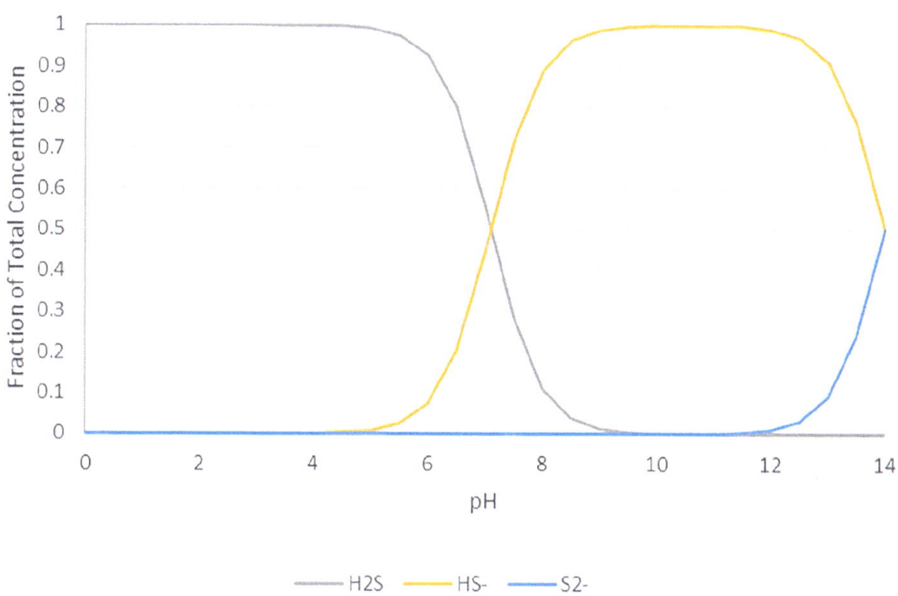

Figure 11.1 Fraction of Total Concentration of Hydrogen Sulfide Speciation as a Function of pH

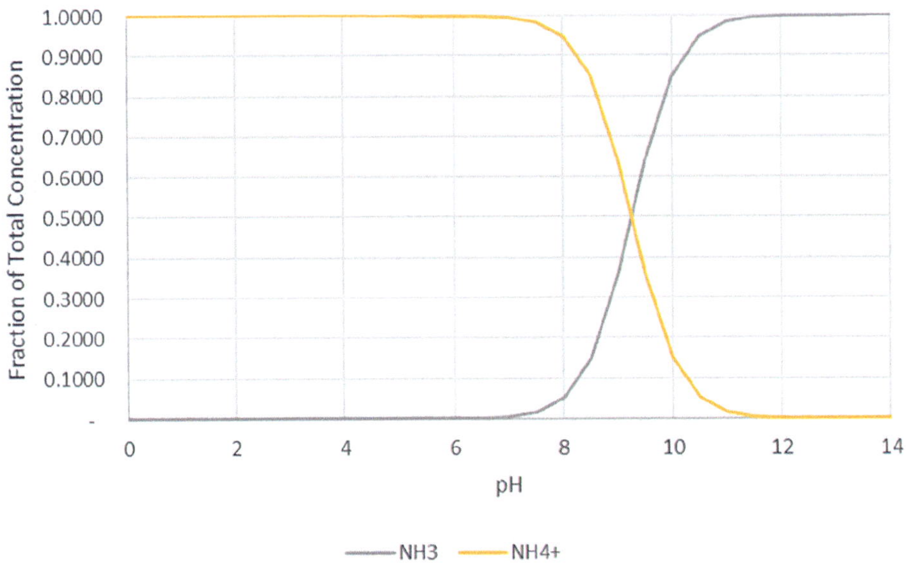

Figure 11.2 Fraction of Total Concentration of Ammonia Speciation as a Function of pH

NH_3, amines, H_2S, organic sulfides, and mercaptans. Additionally, anaerobic sludge digestion creates the conditions in which sulfate reducing bacteria thrive, causing the formation of H_2S that will be a portion of the digester "biogas."

The following sections discuss the mechanisms by which odors can be released to the air at WRRFs.

HEADWORKS (SCREENS/GRIT CHAMBERS)

The following are odor sources at the headworks of WRRFs (U.S. EPA, 1985):

- Dissolved VOCs and H_2S formed in the sewer collection systems can be released due to turbulence encountered from water flowing over weirs or out of pipes and aeration in the case of aerated grit chambers.
- Debris collected on screens can also become a location for biofilm formation where additional fermentation and sulfate reduction can take place if localized anaerobic conditions are present. Sewer collection systems can also be the cause of odors when sediments that contain organic material and bacteria settle out and anaerobic conditions form.
- Organic material trapped in the grit chambers and in the grit conveyance systems will also produce odors.

Grit conveyance system is the equipment used to transfer grit collected at the beginning of a WRRF to a truck, dumpster, or storage hopper. From there, the grit is taken to a landfill or other disposal facility.

PRIMARY CLARIFIERS

Odor sources from primary clarifiers of WRRFs include the following (U.S. EPA, 1985):

- Anaerobic conditions in settled sludge cause odorous compounds to develop and be released. This can occur when sludge removal is incomplete or inefficient. Primary sludge can develop anaerobic conditions or become septic in as little as 1.5 to 2 hours.
- Turbulence caused from effluent falling over the weirs and into the effluent launder can release odorous gases that were previously formed in the system but remained dissolved in the wastewater.
- Scum accumulation, like the sludge collection in the bottom of the clarifier, can quickly become anaerobic and create odor-causing compounds.

SLUDGE THICKENING/DEWATERING

Common sources of odors from sludge thickening/dewatering operations of WRRFs are as follows (U.S. EPA, 2000):

- Sludge, and in particular primary sludge, contains high concentrations of food sources for microorganisms, and odorous compounds can be developed as microorganisms break down these food sources. Sludge handling systems can be a significant cause of odors in a WRRF.

- Anaerobic sludges are more odorous than aerobic sludges.
- Any open component in a sludge handling system is a potential area for odor release.

PROCESS SIDESTREAMS

These streams, such as centrate or filtrare from dewatering sludge, can be high in organic material, including 5-day biochemical oxygen demand (BOD_5)/chemical oxygen demand (COD), and ammonia-nitrogen (NH_3-N). High BOD_5/COD can cause oxygen depletion resulting in anoxic or anaerobic conditions. This can lead to malodorous compound formation (U.S. EPA, 1985).

COLLECTION SYSTEMS

Odors in collection systems are associated with H_2S, also known as *sewer* gas, and form from the anaerobic breakdown of sulfate. This occurs whenever wastewater or sediment is allowed to sit undisturbed for too long. The microorganisms quickly use all of the oxygen and anaerobic conditions occur. The formation of H_2S is primarily a function of the sewer system's design. Two key factors that affect the extent of biological activity and the H_2S production in sewers are biofilm area-to-liquid volume ratio (A/V) and the hydraulic retention time (HRT). A higher A/V would give higher sewer gas (H_2S) production rates and vice versa. A longer HRT means longer time for biological activity, and thus results in higher H_2S levels. Areas subject to generation of H_2S in collection systems include (ASCE & WEF, 2007):

- Sewers with low-velocity conditions and/or long detention times;
- Sewers subject to solids deposition;
- Pumping stations, especially those with long retention times in the wet wells;
- Turbulent areas, such as drop manholes or force main discharge points; and
- Inverted siphon discharges.

To help minimize odor production, minimum full-pipe velocity should not be less than 0.6 m/s (2 ft/sec) to prevent deposition of solids and the formation of H_2S. A velocity of 0.9 m/s (3 ft/sec) is preferred to scour heavier solids. A well-planned line cleaning maintenance program should be used. Wet well detention times should be minimized. Wet wells should be cleaned regularly.

ODOR SENSORY EFFECT

Odor can be defined as sensations resulting from the interaction of volatile chemical compounds inhaled through the nose, making contact with the olfactory area and registering in the brain. The detectability of an odor is the primary characteristic. Additional measurable, but subjective, parameters of perceived odor are

- **Hedonic Tone**—A property of an odor relating to its pleasantness or unpleasantness;
- **Annoyance**—Interference with comfortable enjoyment of life and property;
- **Objectionable**—Causes a person to avoid an area or experience or causes physiological effects; and
- **Strength**—Defined by word scales like "faint to strong."

These odor parameters are subjective because individuals have different interpretations of word scales and they rely on their personal feelings, beliefs, memories, experiences, and prejudices to report them. Written guidelines for subjective odor parameter scales assist individuals (citizens and air pollution inspectors) in reporting observed odor; however, the nature of these parameters remains subjective.

Subjective is relating to the way a person experiences things, including odor, based in his or her own perspectives.

Odors cause an emotional reaction and are subjective. Individuals experience the same odor differently. The following factors affect how individuals experience an odor:

- Stack height or emission area
- Concentration and type of odorous substance (s) and synergistic effects
- Exposure frequency and duration
- Past experience with an odor
- Population type (age, gender)—for example, sense of smell diminishes (gets weaker) with age
- Sensitivity

The synergistic effect is the result of two or more chemicals interacting together to produce an odor effect that is different than the odor they produce when used individually.

- Annoyance
- Sensitization and adaptation
- Weather
- Wind direction

Odor intensity is a logarithmic function of the concentration of the odorous chemical species. Threshold is defined as the degree of stimulation of a nerve or nerve center that produces a response. Odor threshold is the minimum identifiable odor or recognition threshold (Leonardos et al., 1969).

Individuals report a variety of symptoms of odor exposure. These symptoms include the following:

- Headaches
- Nasal congestion
- Eye, nose, and throat irritation
- Hoarseness, sore throat
- Cough
- Chest tightness
- Shortness of breath
- Wheezing
- Heart palpitations
- Nausea
- Drowsiness
- Mental depression

If the concentration of an odor in air is below levels of irritation (levels known to cause eye, nose, or throat irritation in people), the symptoms will pass when you move out of the exposure area. If the concentration of an odor in air is at or above levels of irritation and the exposure duration is longer, the symptoms may last after moving out of the exposure area.

Unpleasant odors can be a warning sign of potential risks to human health. Sometimes, people can smell certain chemicals in the air before they are at harmful levels. However, not all environmental odors are harmful or toxic.

ODOR MEASUREMENT AND CHARACTERIZATION PARAMETERS

There are four measurement and characterization parameters of perceived odor, which are described below (Buonicore & Davis, 1992):

- **Odor concentration**—Odor concentration is measured as dilution ratios and reported as detection threshold or recognition thresholds or as the dilution-to-threshold ratio (D/T) and is sometimes assigned the dimension of odor units per cubic meter. The detectability or odor threshold is a sensory property referring to the minimum concentration that produces an olfactory response or sensation.
- **Odor intensity**—Odor intensity is the strength of the perceived odor sensation and is related to the odorant concentration, which is an entirely different category of measurement. The intensity of an odor is perceived directly without the knowledge of the odorant concentration or of the degree of air dilution of the odorous sample needed to eliminate the odor. Odor intensity is measured by the butanol scale method. In this method, odor intensities are recorded in terms of concentration (in parts per million by volume [ppmv]) of 1-butanol vapor that has an odor intensity that matches the perceived intensity of the odorous sample. Odor intensity is reported as equivalent parts per million butanol, using a referencing scale of discrete butanol concentrations.
- **Odor character descriptors**—Odor character or quality is that property that identifies an odor and differentiates it from another odor of equal intensity. The character describes what the odor smells like using categorical scales and real examples (e.g., fruity, citrus, lemon). The odor character is described by methods known as multidimensional scaling or profiling. In these methods, the odor is characterized by either the degree of its similarity or dissimilarity from a set of reference odors or the degree of applying a scale of various descriptor terms to it. The result is the odor profile.
- **Adaptation**—Adaptation or olfactory fatigue is a phenomenon that occurs when people with a normal sense of smell experience a decrease in perceived intensity of an odor if the stimulus is received

continuously. Depending on the intensity of the stimulus, this self-adaptation and sensory recovery after the stimulus both take place during a relatively short time. Adaptation to a specific odorant generally does not interfere with the ability of a person to detect other odors. Another phenomenon known as *habituation* or *occupational anosmia* occurs when a worker in an industrial situation experiences a long-term exposure and develops a higher threshold tolerance to the odor.

These odor parameters are objective because they are measured using techniques or referencing scales dealing with facts without distortion by personal feelings or prejudices.

Objective parameters are those parameters that can be measured through field monitoring or laboratory testing.

MONITORING, SAMPLING, AND ANALYSIS OF ODOROUS COMPOUNDS

Odor monitoring involves the development of a Sampling and Analysis Plan, which typically includes: (a) odor characterization using standard odor descriptors and intensity, (b) odor field measurements, and (c) ambient air sampling and analysis of odor-causing compounds.

When facility odors affect air quality and cause citizen complaints, an investigation that measures specific odorants using standardized scientific methods may be required. Point emission sources, area emission sources, and volume emission sources can be sampled, and the samples sent to an odor laboratory to be tested for odor parameters, such as odor concentration, odor intensity, odor persistence, and odor characterization. Odor can also be measured and quantified directly in the ambient air, at the property line, and in the community, using standard field olfactometry practices (e.g., odor-intensity referencing scales and field olfactometers). Quantifying ambient odor is often needed for the following purposes:

- Monitoring daily operations (i.e., management performance evaluations)
- Comparing operating practices (i.e., evaluating alternatives)
- Documenting specific events or episodes (i.e., defensible, credible evidence)
- Monitoring compliance (i.e., compliance assurance for permits)
- Determining compliance (i.e., permit renewal)
- Determining status (i.e., baseline data for expansion planning)
- Investigating odor control effectiveness (i.e., scientific testing)
- Verifying odor dispersion modeling (i.e., model calibration)
- Determining specific odor sources (i.e., investigation of complaints)
- Verification of complaints (i.e., notice of violation).

Field measurements can be collected using an olfactometer that is a portable device that determines the D/T, which is a measure of the number of dilutions needed to make the odorous ambient air "nondetectable."

Ambient air samples can also be collected and sent to a laboratory for the following analyses:

- Volatile organic compounds
- Hydrogen sulfide and other TRS (mercaptans, carbonyl sulfide, carbon disulfide, dimethyl sulfide, dimethyl disulfide, dimethyl trisulfide)
- Semi-volatile organic compounds
- Ammonia
- Volatile fatty acids
- Amines

The emitted concentration of odor-causing compounds may be used to perform air dispersion modeling to assess the off-site odor-causing compound levels. The resulting off-site concentrations can be compared to odor detection thresholds and mitigation techniques identified. AERMOD is a steady-state plume model developed by the U.S. Environmental Protection Agency (U.S. EPA) (U.S. EPA, 2004). Where appropriate, the odor-causing chemical compound plume is modeled as either affecting and/or following the terrain. This approach has been designed to be physically realistic and simple to implement while avoiding the need to distinguish among simple, intermediate, and complex terrain, as required by other regulatory models. As a result, AERMOD removes the need for defining complex terrain regimes. AERMOD constructs vertical profiles of required meteorological variables based on measurements and extrapolations of those measurements using similarity (scaling) relationships. AERMOD is designed to run with a minimum of observed meteorological parameters. Vertical profiles of wind speed, wind direction, turbulence, temperature, and temperature gradient are estimated using all available meteorological observations (U.S. EPA, 2004).

ODOR REGULATIONS

U.S. EPA regulates pollutants in outdoor air through the National Ambient Air Quality Standards. However, environmental odors are not regulated by U.S. EPA. Although environmental odors are not nationally regulated in the United States, many cities, counties, and local governments have established nuisance odor regulations.

The following ambient odor limits may be incorporated to odor regulations or into facility permits as compliance-determining criteria when the standard odor testing methods are applied:

- Ambient Odor Limits:
  - Odor concentration as D/T or
  - Odor intensity as part per million butanol
- Source Odor Limits:
  - Odor concentration as odor units per cubic meter or
  - Odor rates as odor units per second

HYDROGEN SULFIDE SAFETY CONSIDERATIONS AND EXPLOSION HAZARDS

Hydrogen sulfide is the most common cause of odors released from collection systems and WRRFs. Hydrogen sulfide is a colorless gas, heavier than air, and has a density of 1.363 g/m^3. Hydrogen sulfide has a flashpoint of –82.4 °C, a lower explosive limit of 4.3%, and an upper explosive limit of 46%. Hydrogen sulfide has a reported odor threshold of 0.0001 to 0.002 ppmv (Hvitved-Jacobsen et al., 2000) and has a characteristic rotten egg smell at concentrations of 3 to 5 ppmv. At concentrations above 100 ppmv, H_2S numbs the olfactory system and prevents odor detection. Hydrogen sulfide in ambient air is not classified as a criteria air pollutant or a hazardous air pollutant under the federal Clean Air Act and, consequently, is not regulated by U.S. EPA. The following is a summary of applicable H_2S regulatory limits:

- There are no international standards limiting H_2S concentrations in ambient air.
- Other countries have short-term standards of H_2S concentrations ranging from 6 to 210 ppbv.
- The World Health Organization recommends a limit of ambient H_2S concentration of 5 ppbv for a 30-minute average time to avoid odor annoyance.
- The Bay Area Air Quality Management District (San Francisco, California) specifies ground concentration less than (<) 60 ppbv for 3 consecutive minutes.

The only enforceable H_2S air quality standards in effect currently are the Occupational Safety and Health Administration (OSHA) regulations. Hydrogen sulfide intoxication is a concern in confined spaces. Hydrogen sulfide is both an irritant and a chemical asphyxiant that affects both oxygen use and the central nervous system. Its health effects can vary depending on the level and duration of exposure. Repeated exposure can result in health effects occurring at levels that were previously tolerated without any effect. Low concentrations irritate the eyes, nose, throat, and respiratory system (e.g., burning/ tearing of eyes, cough, shortness of breath). People with asthma may experience breathing difficulties. The effects can be delayed for several hours, or sometimes several days, when working in areas with low-level H_2S concentrations. Repeated or prolonged exposures may cause eye inflammation, headache, fatigue, irritability, insomnia, digestive disturbances, and weight loss. Moderate concentrations can cause more severe eye and respiratory irritation (including coughing, difficulty breathing, accumulation of fluid in the lungs); headaches; dizziness; nausea; vomiting; staggering; and excitability. High concentrations can cause shock, convulsions, inability to breathe, extremely rapid unconsciousness, coma, and even death. Effects can occur within a few breaths, and possibly a single breath (OSHA, 2005).

The OSHA limits are applicable to the workplace only, and are summarized below (Occupational Health & Safety, 2007):

- Eight-hour permissible exposure limit of 10 ppmv in the workplace
  - Five parts per million by volume exposure; healthy adults can work without a problem
  - Ten parts per million by volume, 8-hr exposure; healthy adults can work without a problem
- Acceptable ceiling concentration of 20 ppmv
- Maximum level of 50 ppmv allowed for 10 minutes if no other measurable exposure occurs
- At H_2S concentrations >50 ppmv, olfactory fatigue prevents odor detection
- Exposure to 100 to 150 ppmv for several hours causes local irritation
- A level of H_2S gas at or above 100 ppmv is immediately dangerous to life and health
- Life threatening (300 to 500 ppmv)

EFFECTS OF ODOR ON THE COMMUNITY

Community members have reported that smelling odors in the air decreases their quality of life and sense of well-being. Odors from environmental sources might also cause health symptoms, depending on individual and environmental factors. Those factors can include age, sex, medical condition, and the level and type of the substance in the environment. The community can get involved in odor monitoring by keeping personal odor diaries to track the odors in the community. Odor diaries can help distinguish odor types and times of the day or night when odors are observed. Results of the odor monitoring events along with citizens' complaints can be assessed within a geographic information systems (GIS) database application. Along with developing site-specific odor descriptors, odor observations can be placed with a GIS-based database that allows the user to find correlations between weather conditions, geographic locations, and time of day to develop the community's odor fingerprints. The evaluation can identify several odor fingerprints within the community and guide the development of mitigation measures.

TEST YOUR KNOWLEDGE

1. **Odor concentration, intensity, persistence, and character descriptors are measurable objective parameters of perceived odor.**
   - ☐ True
   - ☐ False

2. **The only enforceable H_2S air quality standards currently in effect are the OSHA occupational regulations.**
   - ☐ True
   - ☐ False

3. **Hydrogen sulfide intoxication is a concern in**
   a. Open spaces
   b. Dark areas
   c. Cities
   d. Confined spaces

4. **Hydrogen sulfide is both an irritant and a chemical asphyxiant, with effects on both oxygen use and the central nervous system.**
   - ☐ True
   - ☐ False

5. **At concentrations around __________, H_2S numbs the olfactory system and prevents odor detection.**
   a. 10 ppmv
   b. 50 ppmv
   c. 100 ppmv
   d. 200 ppmv

6. **Odor threshold of a compound is**
   a. The highest concentration at which that compound can be detected by human smell
   b. The lowest concentration at which that compound causes severe headaches
   c. The lowest concentration at which that compound causes shortness of breath
   d. The lowest concentration at which that compound can be detected by human smell

7. **Hydrogen sulfide is a predominant odor-causing constituent emitted at WRRFs.**
   - ☐ True
   - ☐ False

8. **__________ have the potential to generate and release odors to the surrounding area.**
   a. Places or processes in which wastewater is collected, conveyed, or treated
   b. Offices
   c. School buildings
   d. Gardens

9. **Hydrogen sulfide is __________ than air and, therefore, often collects in low-lying areas and working sites near the ground.**
   a. Heavier
   b. Lighter
   c. Darker
   d. More volatile

10. To help minimize odor production, minimum full-pipe velocity should greater than __________ to prevent deposition of solids and the formation of H_2S.
   a. 3 m/s (10 ft/sec)
   b. 0.6 m/s (2 ft/sec)
   c. 1.8 m/s (6 ft/sec)
   d. 0.2 m/s (0.6 ft/sec)

11. Adaptation or olfactory fatigue is a phenomenon that occurs when people with a normal sense of smell experience
   a. An aromatic odor
   b. A citrus odor
   c. A rotten vegetable odor
   d. A decrease in perceived intensity of an odor

12. Odors from environmental sources can cause health symptoms depending on
   a. Source of odor
   b. Unpleasantness of odor
   c. Familiarity of odor
   d. Age, sex, medical condition, and the level and type of the odorous substance in the environment

Odor control collection system: A proper odor collection system will be designed and maintained to prevent fugitive odor emissions.

Odor Control Collection Equipment

The basic concept of an odor control collection system is to cover the odor-emitting WRRF areas to contain the odorous offgas, and then remove the odorous air and treat the odors. An odor control cover system is considered to have the following components: a fresh air inlet point, a gastight cover, a foul-air withdrawal point, a fan with ductwork, and an odor treatment system. The fan will create a slightly lower pressure through the air space in a covered tank, pulling fresh air in. This prevents the odorous air from escaping, preventing fugitive odor emissions. Figure 11.3 presents a schematic showing these components.

FRESH AIR INLET POINT

The importance of the fresh air inlet point is often overlooked. The general design concept is that fresh air is drawn under the cover and mixes with foul air, and this mixture flows along under the cover until it is withdrawn. The fresh air serves two purposes: dilution of the odorous air and replacement of volume within the covered area. The fresh air inlet is typically located at one end of a cover system and the foul air withdrawal point is located at the other end.

COVERS AND CONTAINMENT

Covers and odor containment methods vary based on the source of the odor. For process equipment such as tanks and/or reactors, covers are typically flat covers or domed covers. Some process equipment, such as solids handling equipment, is not located at the process equipment itself; rather, it is contained by being within its own room or building. Other systems, such as solids conveyors, may be designed as completely enclosed systems by the manufacturer and may not require an additional cover system.

From the designer's point of view, the goal is to have a cover that is sufficiently gastight so that:

- Foul air does not escape through leak points and lead to odor complaints.
- Excessive fresh air does not infiltrate through leak points along the cover, creating relative dead spots where odorous gas could collect in concentrations that could potentially damage concrete or equipment under the cover.

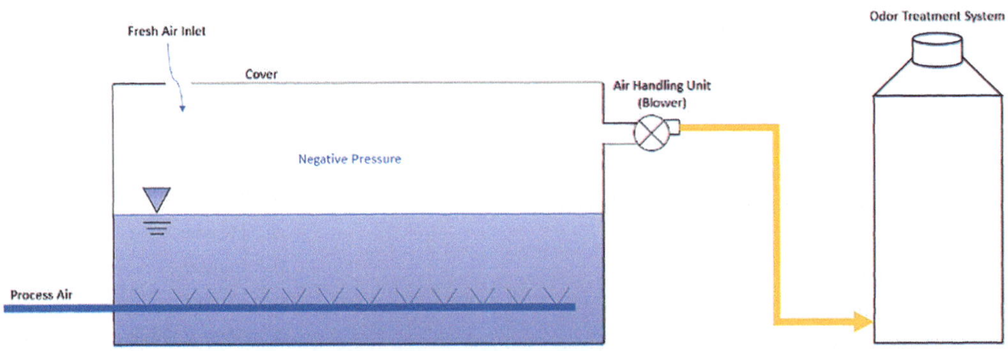

Figure 11.3 Odor Control System Components

- The gas tightness of a cover is important because less leakage means foul air can be removed using a smaller blower and ductwork, reducing capital and operating costs.

From the operations and maintenance personnel perspective, the covers should not restrict normal operations and maintenance activities. If covers are left open to create easy access for inspections, sampling, or equipment access, the original goals of odor containment cannot be achieved. Where needed, the covers should be designed to be removed safely and easily.

CONVEYANCE SYSTEM AND AIR HANDLING UNITS

The foul air withdrawal point is the point where ductwork connects to the cover system. The connection detail could be a duct stub, flanged duct, or flexible coupler. A fan and ductwork move odorous air to the dilution point or treatment process. Once foul air is captured and moved through a ductwork system, it can be treated by various technologies such as a biofilter, wet scrubber, carbon filter, proprietary media, and so on.

The fans and ductwork need to be designed to move the appropriate amount of air through the system. For enclosed or covered process equipment, the conveyance system should be sized to account for the maximum process area of the equipment, such as the air used in an aeration basin, as well as any fresh air entering the system. The system should maintain a slight negative pressure to prevent odors from short-circuiting around the treatment process.

For systems designed for rooms or buildings, the size should be based on air changes per hour (ACH). The use of the room and its contents will dictate the number of ACH required. For example, the National Fire Protection Association (NFPA) Code 820 specifies 12 ACH for areas classified as Class I, Division 1. For occupied areas with low odor, 6 ACH may be suitable; in other occupied areas with strong odors, ACH as high as 20 may be required. Another consideration for room or building systems is to maintain the containment by keeping doors closed.

TEST YOUR KNOWLEDGE

1. **Odor collection systems should maintain negative pressure to ___________.**
   a. Prevent any air from entering the system
   b. Prevent fugitive emissions
   c. Create a vacuum chamber
   d. Stop odors from forming

2. **An odor collection system should prevent air from entering or leaving the system.**
   ☐ True
   ☐ False

Odor Treatment Technologies

There are many different technologies that can be applied to control odors from wastewater collection and treatment systems. These technologies can be categorized into two main groups: vapor-phase technologies, used to control odorous compounds in the air or gas, and liquid-phase technologies, used to control odorous compounds in the liquid wastewater itself. Vapor-phase technologies typically are used in point-source applications such as WRRFs and pumping stations, or for the treatment of biogas. Liquid-phase technologies typically are used in collection systems where control of both odors and corrosion are concerns and/or where multiple-point odor control is an objective.

Vapor-phase technologies ventilate the point sources of odor emissions (wet wells, headworks, etc.). For air treatment, the ventilation system is designed to maintain a negative pressure on the source area at all times. This prevents the "leakage" of odorous air from vents, , access hatches, or other areas. The discharge from the ventilation system is the only route for the odorous air to escape; therefore, it must be treated. Vapor-phase systems are very effective at preventing both fugitive emissions from the ventilated sources and odor problems associated with those sources.

If large amounts of air are ventilated, vapor-phase systems also can be effective at providing adequate ventilation for occupancy and/or reducing the contaminant concentration to levels that do not pose safety or corrosion hazards. Biogas treatment systems primarily involve the removal of hydrogen sulfide to prevent the formation of sulfur dioxide (SO_2), a product of the combustion of gas containing hydrogen sulfide. Most WRRFs that have anaerobic treatment burn the methane-rich biogas produced from the anaerobic sludge digestion (either as waste gas in a flare or in engines, turbines, or boilers to produce power or steam). Because of this practice, the removal of H_2S is often required to meet emissions standards or to prevent corrosion in the combustion equipment. The technologies discussed in this document for treating ventilation air or biogas are carbon adsorption, wet air scrubbing, liquid redox technology, biofiltration, and solid scavengers.

VAPOR-PHASE PHYSICAL-CHEMICAL TREATMENT

There are several different types of adsorption media used for the removal of odor compounds. Activated carbon is the most common due to its large surface area, making it very effective at adsorption of odor compounds. Carbon adsorption is an effective physical–chemical treatment technology for odor control because of the ease of operation, low initial capital cost, and wide range of odor compounds that can be treated (SCORe, 2014). Carbon adsorption is the simplest vapor-phase control technology. There is no ongoing chemical supply to the system and there are no biological processes involved. Adsorption is applicable to a wide range of compounds that are typically in wastewater. Hydrogen sulfide and TRS compounds are removed effectively by carbon adsorption systems, but ammonia and other nitrogen-based compounds are not effectively treated. Various carbon media systems, including activated and impregnated carbon, can be used independently or in combinations to remove many different contaminants.

Van der Waals molecular forces: Include attraction and repulsions between atoms, molecules, and surfaces as well as other intermolecular forces. They differ from covalent and ionic bonding in that they are caused by correlations in the fluctuating polarization of nearby particles (i.e., change in the electric charge of the particles).

In an adsorption system, the gas stream is passed over a bed of adsorbent (carbon or other media) and the odor-causing compounds are attracted to and adhere to the surface of the adsorbent. There are three main mechanisms for adsorption: (a) physical adsorption, (b) chemical adsorption, and (c) reactive adsorption (i.e., combination of both physical and chemical). Each will be explained further in the following subsections.

PHYSICAL ADSORPTION

Generally, physical adsorption is the dominant mechanism for gas-phase adsorption of odor chemicals (Le-Minh et al., 2018). The transport of the odor compounds, in gas phase, allows the odor compounds to attach, via Van der Waals molecular forces, to the macro, meso, and micro pores of the activated carbon, as shown in Figure 11.4. Factors that control the extent and rate of adsorption include size of media pore structure, molecular size, and polarity of the media (i.e., adsorbate) and water trapped in the activated carbon.

Covalent bonding: A chemical bond in which two atoms share electrons, thereby creating a force that hold atoms together. An example of covalent bonds includes gases such as methane, carbon dioxide, or solids such as diamond and quartz.

Ionic bonding: A chemical bond in which two atoms are held together through electrostatic attraction from oppositely charged ions. Examples of ionic bonding include many salts such as sodium chloride or potassium chloride or oxides such as iron oxide, which is commonly referred to as *rust*.

Polarity: Is a molecule or its chemical group having a separate positive and negative electric charge, with a negative charged end and an opposite positively charged end.

As discussed above, polarity is a factor that influences the extent of odor adsorption and consists of three different interactions: (a) orientation effects between polar molecules (i.e., electric charge molecules); (b) dispersion effects between nonpolar molecules and nonpolar surfaces, which distributes the molecules along the activated carbon surface; and (c) induction effects between polar and nonpolar molecules that result in a transmission of electronic charge and potential bonding of the molecules. Activated carbon

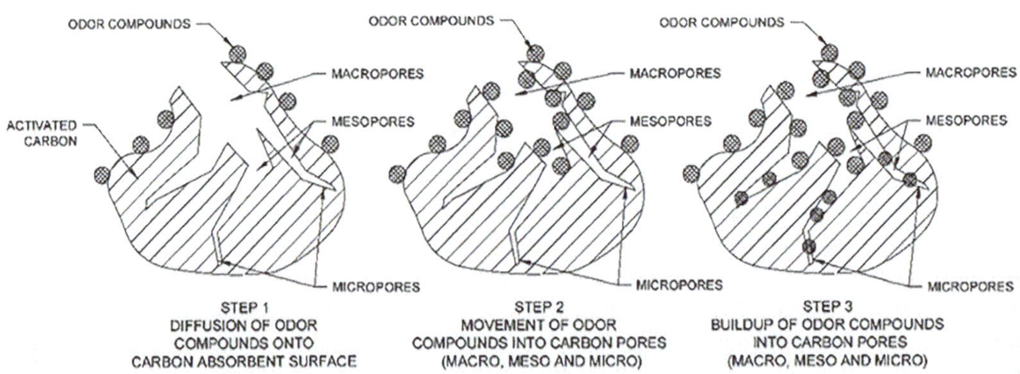

Figure 11.4 Adsorption Process

is very effective as an adsorption media because it includes multiple polar functional groups, thereby allowing multiple polarity interactions.

CHEMICAL ADSORPTION

Chemical adsorption or chemisorption consists of chemical bonding between the odor compounds (i.e., adsorbate) and the media (i.e., adsorbent) catalysts. Media activation (i.e., when activated carbon becomes chemically active) or chemical impregnation (i.e., activated carbon media that has been chemically treated with additional chemical compounds) changes the functional groups and surface chemistry of the media and promotes chemical adsorption.

REACTIVE ADSORPTION

The combination of physical and chemical adsorption is called *reactive adsorption* and is the process that occurs naturally. Under reactive adsorption, odor compounds are physically adsorbed (i.e., on the media and then are chemically transformed [i.e., bonded] in the presence of the media catalysts). The oxidation of chemical compounds can occur through either a catalytic reactive process on the media surface area or the action of impregnated oxidants, chemically adding oxidants to the activated carbon, in the media.

Catalytic reactive process: A process in which the rate of a chemical reaction is increased by adding a catalyst.

ADSORPTION MEDIA

Media Characteristics

Activated carbon media are typically supplied as a granular or pelletized product. Granular media will typically have more surface area but will settle more and create more dust as compared to pelletized media. Additionally, due to the particle size (i.e., increased surface area), granular media will create more pressure drop through the system and may require a large blower and more power consumption. Lastly, pelletized media have higher mechanical strength due to the binders added for production, which may reduce clogging or caking concerns. Table 11.3 summarizes properties of different types of adsorption media, which are discussed below.

Surface area: A 4.0-mm-diameter activated carbon granule has as much internal surface area as a football field (Raina, 2004).

Virgin Activated Carbon

Virgin activated carbon is typically defined as activated carbon that has been chemically treated to attract odor compounds through physical adsorption. Virgin activated carbon is highly effective at adsorbing organic, high molecular compounds that are neutral or nonpolar in nature. The removal performance of virgin activated carbon for small odor compounds, such as hydrogen sulfide, methyl mercaptan, or dimethyl sulfide, is limited. Virgin activated carbon is typically used as a polishing stage or for the removal of low odor compound concentrations. The two most common forms of virgin activated carbon for odor removal are coal or coconut shells. Coal-derived activated carbon is commonly generated from anthracite or bituminous coal grades. It can be supplied in either pellet or granular form. Coconut shell activated carbon has a structure similar to coal and, therefore, has a similar ability to attract odor compounds through physical adsorption.

Impregnated Carbon Media

Impregnated carbon media are activated carbon media that have been chemically treated or "impregnated" with additional chemical compounds. In impregnated carbon media, physical adsorption initially removes the odor compounds and then the impregnated chemicals react with the odor compounds to form a new compound that chemically adsorbs onto the carbon molecules. Because both physical and chemical adsorption are occurring, impregnated carbon media typically have higher removal efficiencies for specific odor compounds such as hydrogen sulfide.

The common impregnated chemicals are caustic compounds such as sodium hydroxide or potassium hydroxide. These chemicals oxidize H_2S to form sulfites, sulfide, or elemental sulfur, which are more readily adsorbed to the impregnated carbon molecules and, therefore, increase removal efficiency.

Other oxidation chemicals can be impregnated, such as metallic oxides or noble metals, to enhance the overall adsorption capacity of the activated carbon. For example, sulfur-reducing compounds react with impregnated salt or metal oxide to form metal sulfide or salts, which are then fixed on the activated carbon. Acid-based chemicals can be impregnated to increase the chemical adsorption of other odor compounds such as ammonia.

Table 11.3 Comparison of Various Dry Adsorption Media (Adapted from WEF, 2020, and Graham, 2006)

Media Type	H_2S Capacity (% of wt.) (ASTM Standard D-6646)	Ignition Temperature	pH of Spent Media (unwashed before disposal)	Advantages	Disadvantages
Virgin activated carbon	47	380 °C/720 °F	Acidic	■ Cost-effective ■ Effective for larger nonpolar molecules such as gasoline or alcohols	■ Lowest removal capacity for reduced sulfur compounds
Chemically impregnated carbon	20–25	150 °C/ 300 °F	Acidic	■ Increased H_2S removal ■ Can modify carbon composition to specific odor compounds	■ Higher potential for bed smoldering ■ Less removal capacity for VOCs such as benzene, toluene, xylenes ■ On-site regeneration requires handling and management of chemicals
Metallic oxide	10–15	NA	NA	■ Oxidizes H_2S to elemental sulfur	■ Less common, may be difficult to obtain ■ Sulfur may transform to dimethyl disulfide
Water regenerable	1–18	420 °C/790 °F	Acidic	■ Regenerate on-site; therefore, reduces media change outs ■ More cost-effective for long-term application ■ Can oxidize sulfur	■ More costly ■ Main chemical removal for H_2S ■ Deteriorates over time and will require media changeout ■ Regeneration liquids are acidic that may require special disposal practices
High-capacity carbon	65–75	540 °C/1000 °F	Neutral	■ High removal capacity for H_2S, can handle high concentrations of H_2S ■ Oxidizes H_2S to form elemental sulfur	■ Sulfur may transform to dimethyl disulfide ■ More costly

High-Capacity Carbon

High-capacity carbon products are typically proprietary products that have been specially designed and manufactured to provide greater capacity for removal of specific odor compounds. Advantages of these systems are longer media life, reduced carbon changeouts, and reduced disposal volumes and related costs. High-capacity carbon is typically more expensive than virgin activated carbon.

Water-Regenerable Carbon

Coal-based activated carbon can be modified so that captured contaminants can be removed by washing, restoring its adsorptive capacity (i.e., regenerable). The regenerable property is developed from the addition of nitrogen-based gas (e.g., ammonia or urea) to the carbon surface during the last stages of activation to form nitrogen functional groups on the carbon molecule surfaces.

The established nitrogen function groups have catalytic properties and, therefore, will oxidize hydrogen sulfide to sulfuric acid. This may create disposal challenges for any collected condensate due to its highly acidic nature. To add the nitrogen functional groups, the activation process does not go to completion and, therefore, the activated carbon has lower surface area due to the underdeveloped pore structure. As compared to standard activated carbon, water regenerable carbon has a lower surface area and, therefore, a lower adsorption capacity for VOCs.

Washing regenerable carbon is not 100% effective and, therefore, the carbon only achieves between 75% to 85% of its original adsorption capacity and will eventually need to be replaced.

Other Non-Carbon Media

Other non-carbon media can be used alone or in combination with activated carbon to optimize odor removal. Typically, these non-carbon media are metal compounds such as permanganate-impregnated zeolite or iron sponge. Permanganate acts as a strong oxidizer that will react with most sulfur compounds. Therefore, it is common to mix permanganate-impregnated media with activated carbon media to extend the life of the media or to treat specific odor compounds (WEF, 2020).

Iron sponge uses hydrated iron oxide in combination with a large surface-area-to-volume material, such as wood chips, to provide both physical and chemical adsorption. This dual adsorption process allows for the removal of high concentrations of hydrogen sulfide, but may require longer residence times and, therefore, larger vessels.

Adsorption Properties

Although many factors affect the adsorption properties of activated carbon (i.e., particle size, porosity, pH), surface area typically has the greatest influence. Most activated carbons have surface areas that range from 600 to 1600 m^2/g. Adsorption properties are determined using standard test methods such as iodine number (ASTM D4607) and butane activity (ASTM D5742), which are used to estimate the micropore volume (e.g., surface area) of activated carbon for comparison and quality control.

Moisture and Humidity

The presence of moisture (i.e., water vapor) may have either a positive or negative effect on the removal of odor compounds in activated carbon systems. If the primary mechanism for removal is physical adsorption, the water vapor molecules will compete with the odor compounds for the surface area of the activated carbon. For chemical and reactive absorption, water vapor is required to complete the necessary chemical reactions for removal. However, even in these systems, too much moisture will fill up the pore spaces of the activated carbon and shorten the life of the carbon. Optimal moisture/humidity levels are highly dependent on the odor compounds and concentrations present (Le-Minh et al., 2018).

REMOVAL EFFICIENCY AND EXPECTED PERFORMANCE

Odor removal and control may require treatment of a wide range of odor compounds and concentrations. Therefore, the long-term performance of any given adsorption media system may be difficult to predict. Additionally, each odor compound will have a different adsorption mechanism (physical, chemical, or reactive) based on media type, concentration, presence of other odor compounds, and operating conditions. The following subsections describe removal efficiency and expected performance for different odor compounds or groups of compounds in more detail.

Reduced Sulfur Compounds

Reduced sulfur compounds include hydrogen sulfide, dimethyl sulfide, carbonyl sulfide, and various mercaptans compounds. Chemical adsorption is the dominant mechanism for removal of reduced sulfur compounds (SCORe, 2014). Impregnated or high-capacity activated carbon that has adsorbed oxygen or other catalysis creates oxidation reactions with reduced sulfur compounds that form sulfur dioxide or elemental sulfur. Virgin activated carbon is the least effective at removal of reduced sulfur compounds because physical adsorption is its primary removal mechanism.

Volatile Nitrogen Compounds

Ammonia, tri-methyl amine, and heterocyclic compounds such as idole, skatole, and pyridines are examples of volatile nitrogen compounds. These compounds have low molecular weight and, therefore, physical adsorption is not effective for removal. Volatile nitrogen compounds are effectively removed with reactive adsorption mechanisms through the addition of metal oxides or acid compounds impregnated in activated carbon (SCORe, 2014). Additionally, water vapor will enhance the reactive adsorption of ammonia and other amine compounds through the reaction of ammonium ions with the activated carbon surface areas that are acidic in nature.

Volatile Organic Compounds

The dominant mechanism for removal of VOCs is physical adsorption; therefore, VOCs with higher molecular weight and/or higher boiling points will have higher adsorptive capacity. As previously discussed, dry conditions are optimal for physical adsorption because there is more surface area available (SCORe, 2014). Virgin activated carbon is typically effective at VOC removal because its primary removal mechanism is physical adsorption.

ADSORPTION EQUIPMENT

Physical/chemical adsorption systems are easy to operate, and they have minimal equipment requirements. At a minimum, the following equipment is needed: a vessel to hold the treatment media, incoming piping for untreated airflow, and discharge piping of treated air. Additional equipment that may be needed includes a blower for active collection of untreated airflow, pressure gauges to measure differential pressure across the media bed, valving and piping manifold if multiple vessels are used, condensate piping and drains, and infrastructure to support media changeouts.

Media vessels are typically constructed of plastics such as: high-density polyethylene (HDPE), fiberglass-reinforced plastic (FRP), or other corrosive resistant materials. For high temperature applications, Type 316 stainless steel is also available but will require ground rods to prevent the accumulation of static electricity.

Additional equipment needed may include a blower for active collection of untreated airflow, pressure gauges to measure differential pressure across the media bed, valving and piping manifold if multiple vessels are used, condensate piping and drains, and infrastructure to support media changeouts. Figure 11.5 presents a typical configuration for an activated carbon system.

PROCESS VARIABLES

Configurations

Activated carbon bed systems can be configured with either single or multiple vessels. Single beds require less effort for media changeouts, but generally require larger vessels. Multiple vessel configurations are typically used for large-capacity systems or where continuous odor treatment is required.

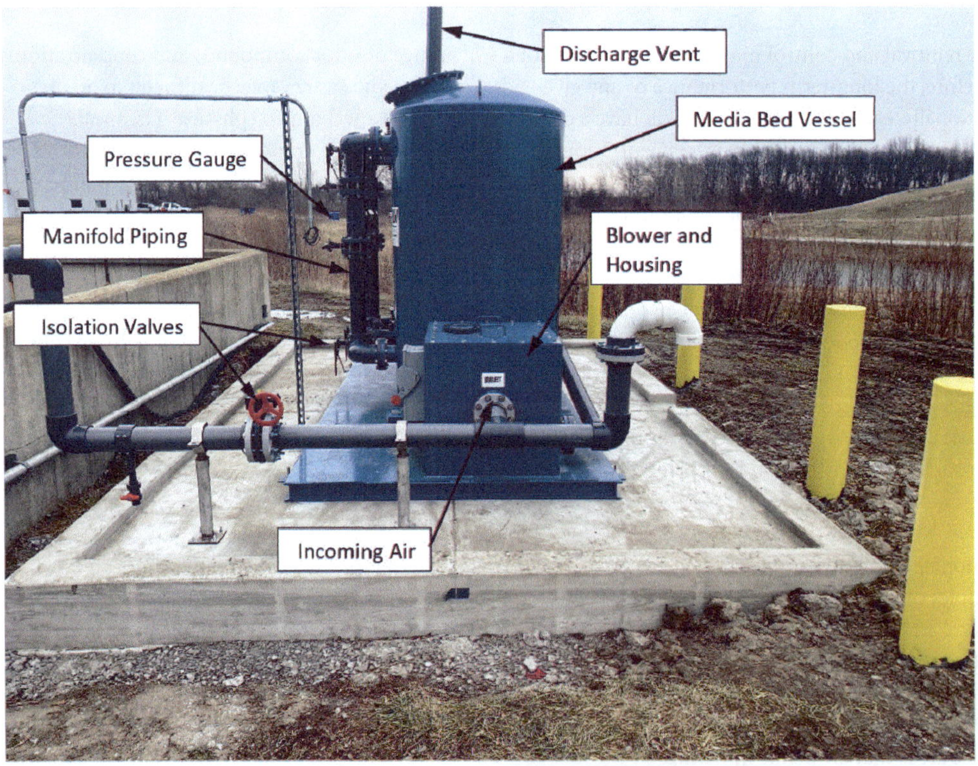

Figure 11.5 Typical Components of Activated Carbon Treatment System

There are several different system configurations for different carbon activated treatment system vessels:

- Vertical—Airflow flows in a vertical orientation direction through the vessel(s), with the inlet typically at the bottom of the vessel and the outlet at the top. This is the most common configuration for small-to-medium-sized capacity treatment systems that allows for the use of both single and multiple beds. Media changeouts for vertical system configurations may be challenging due to the potential design height of the vessel and the limited number of potential cleanout ports/access points. Figure 11.6 shows a vertical flow configuration.
- Horizontal—Airflow flows in a horizontal direction with the inlet on the opposite side of the vessel from the outlet. This configuration is typically used for the large capacity treatment systems where multiple bed configurations can be used. Advantages of a horizontal treatment system are the ease for media changeouts because multiple cleanout ports/access points can be installed in the vessel. Disadvantages include media settling that may create layers of different bed densities in the vessel. The airflow may preferentially move through lower density media layers, thereby short-circuiting through a vessel.
- Radial—Radial flow configurations provide an outside-to-inside airflow pattern that results in a lower pressure drop through the media and can be designed in freestanding, vertical, or single-bed tower configurations. Disadvantages of the radial configuration include shortened media life, rapid breakthrough across the media bed, and short-circuiting.

Figure 11.7 illustrates horizontal and radial flow configurations.

Bed Depth

For dry media (e.g., activated carbon) adsorption systems, chemical compounds are adsorbed until the media bed is saturated. The media bed can be divided into the three zones (ARC Sewer Corrosion & Odour Research Project, 2011):

Zone 1: Initial zone where media (i.e., active carbon) pores are empty and have significant capacity to absorb odor compounds.

Zone 2: Active zone, which is the zone where adsorption is actually occurring.

Zone 3: The media (i.e., activated carbon) pores are completely saturated and, therefore, cannot absorb any more odor compounds.

As odor compounds enter the media zone, the saturated or "mass transfer front" moves through the media unit, which is defined as breakthrough (i.e., inlet chemical concentration equals outlet chemical design concentration). When breakthrough occurs, the activated carbon media have reached their operational life and need to be replaced or regenerated.

Figure 11.6 Vertical Flow Configuration for Activated Carbon Bed System With Standby Treatment Vessel

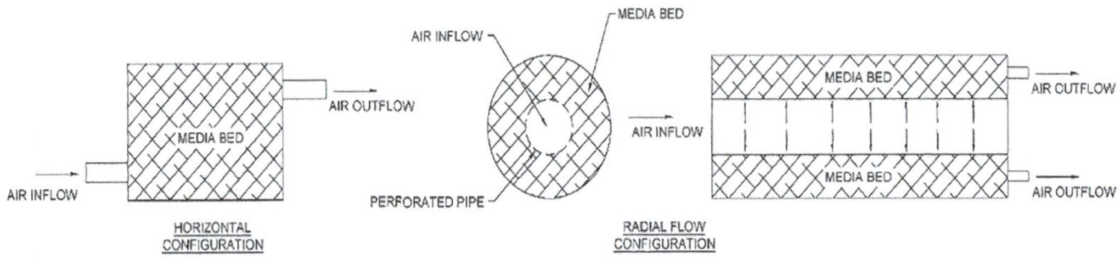

Figure 11.7 Horizontal and Radial Flow Schematics

A breakthrough curve plots concentration versus bed depth, time, or flowrate; they are an important design consideration for dry media treatment systems. Figure 11.8 provides an example of a breakthrough curve. The shape of the breakthrough curve is dependent on several factors including odor compound, odor concentration, bed velocity, bed depth, and relative humidity.

Most activated carbon treatment vessels have bed depths around 1 m (3 ft) because this equals reasonable breakthrough time (i.e., days to months) for most dry media breakthrough curves. Increasing bed depth would reduce media changeouts but increasing bed depth increases pressure drop across the system and, therefore, may require a larger increase and increased electricity requirements.

Bed Velocity

Bed velocities typically range between 0.2 and 0.3 m/s (40 to 60 fpm) to balance higher treatment flowrates versus adequate contact time for adsorption (WEF, 2020). If there is not enough contact time for adsorption, short-circuiting and reduced performance can occur. Higher bed velocities may be possible based on media type and odor concentration levels.

ACTIVE AND PASSIVE PROCESS CONTROL

Active control systems deliver air to the active carbon treatment system through a blower. The sizing of the blower on the inlet side is a function of flowrate and pressure loss through the transmission piping. The blower outlet sizing is a function of the design bed velocity.

Passive system relies on in-place air pressure and chemical gradients to move odor chemicals through the treatment system. The advantage of passive systems is the lack of mechanical equipment, such as blowers, and, consequently, the need for power. Disadvantages include minimal bed velocity, potential for fugitive emissions from source, and a greater risk of bed smoldering, as discussed below, due to low airflow through the treatment system.

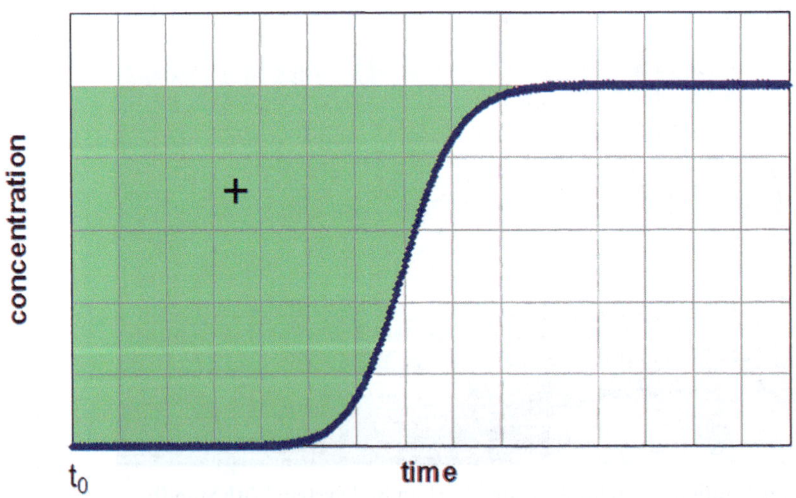

Figure 11.8 Example Breakthrough Curve

SECONDARY OR POLISHING TREATMENT

Activated carbon systems can be used for secondary or polishing odor treatment because these systems can treat a wide range of odor compounds at low concentrations levels.

Water vapor can affect the removal efficiency of activated carbon systems because water molecules compete with odor compounds for absorption sites; thus, excessive water from primary treatment needs to be addressed. Effective methods include the use of preheaters, water traps, or mist eliminators between the primary and secondary treatments to increase temperature and reduce relative humidity.

BED SMOLDERING

Smoldering or fires have occurred in activated carbon treatment system vessels. Melted vessels or vessels on fire have been observed during operation of the systems, especially when chemically impregnated activated carbon is used for the treatment media. Chemically impregnated active carbon has a significantly lower ignition temperature (200 to 225 °C) compared to virgin or catalytic activated carbon (380 to 425 °C) and, therefore, is more susceptible to bed smoldering. Figure 11.9 shows a photograph of smoke coming off of spent chemically impregnated activated carbon.

During active operations, the heat of reaction is removed from the vessel through the discharge of hot gas. During standby operations, especially if the chemical treatment system is designed with two vessels in standby/duty operations, the heat of reaction generates heat, but heat is not being removed and, therefore, the temperature of ignition for chemically impregnated activated carbon can be reached.

Heat of reaction is the heat that is generated when exothermic chemical reactions take place.

OPERATION OF ADSORPTION SYSTEMS

Monitoring Parameters

Various parameters are used to evaluate the performance of the activated carbon treatment system and determine when media changeout should occur to avoid breakthrough. Typically monitoring parameters include:

- Inlet and outlet gas pressure,
- Differential pressure (inlet minus outlet pressure),
- Inlet and outlet gas temperature, and
- Gas volume flowrate.

Figure 11.9 Field Documentation of Exothermic Reactions of Spent Chemically Impregnated Activated Carbon

If odor treatment is tied to a performance criterion (e.g., H_2S concentration less than 15 ppbv at property boundary), a continuous monitoring system for inlet and outlet concentrations can be designed into the treatment system.

The activated carbon media can also be sampled throughout the treatment process to provide better estimates for media changeouts. Multiple sample points can be installed along the media bed to allow for media sample collection and testing. Media can be sampled for percent moisture, mass of hydrogen sulfide, and remaining hydrogen sulfide adsorption capacity.

Media Changeout

Spent media will require regular removal and replacement (i.e., changeout) to provide efficient odor treatment and removal performance.

For small- to medium-sized treatment systems, manual changeout of spent media is the most common method. Media are typically removed manually through vessel hatches using a vacuum truck and replaced with new media. Another method for manual changeouts is replacement of the entire vessel; under this method, the entire vessel, including the spent media, is replaced with a new vessel containing new media.

Advantages of this approach are the cost savings of using existing staff. Laboratory testing may be required to assess the proper disposal (i.e., hazardous versus nonhazardous) of the spent carbon.

It is also possible to hire a contractor to come on-site to remove the existing spent media and replace the new media. This may be as simple as switching out the media vessel such as removing a chemical drum that contains the used media and replacing it with a new chemical drum or it may be more complex such as jetting and vacuuming out spent media from a large vessel.

VAPOR-PHASE CHEMICAL TREATMENT

WET AIR SCRUBBING ABSORPTION

Wet air scrubbing absorption is the most flexible and reliable technology for vapor-phase (gas) wastewater odor control. This technology can be used to treat virtually any water-soluble contaminant. Absorption may be either a physical or chemical process. Physical absorption involves such factors as solubility and vapor-pressure relationships. Chemical absorption involves chemical reactions between the absorbed substance and the absorbing medium. Multistage scrubber systems are often used to treat process emissions, whereas single-stage scrubbers are used to treat facility-ventilating air. Entrainment separators, venturi scrubbers, and spray-type scrubbers are typically used as the first stage in the multistage system to remove particulate and aerosols from gaseous streams so that the subsequent scrubber stages will be more effective in achieving odor removal. Typically, chemical oxidants, such as sodium hypochlorite and chlorine dioxide, are effective scrubbing agents. However, common bases and acids, such as caustic soda and sulfuric acid, are effective at solubilizing and neutralizing many compounds at a lower cost than oxidants. The selection of chemicals to use for scrubbing depends on the odorous compounds present and the emission source (Buonicore & Davis, 1992).

Chemical scrubbers have long been one of the most common odor control technologies used in municipal WRRFs, and their performance for removing odorants is well documented. In a wet air scrubber, the odor contaminants are solubilized from the vapor phase into an aqueous chemical solution. Chemical scrubbing has been shown to be highly effective at controlling H_2S and ammonia odors, which are the most predominant odorous compounds found in municipal wastewater treatment systems.

The removal mechanism is purely chemical and is not subject to upsets like biological processes. The chemical balance in the system is automatically and continuously maintained, even under changing loading conditions, minimizing the chance for odor breakthrough. The application of a multistage scrubber allows for the use of a different chemical solution in each of the stages to efficiently use chemicals and target a wide range of contaminants for treatment.

An exhaust fan moves odorous air through corrosion-resistant ductwork and into the wet scrubber vessel, where it encounters liquid-to-gas contact from the scrubbing solution and packing media with high surface area. The contaminants in the odorous air are chemically absorbed into the liquid scrubbing solution through mass transfer. The deodorized airstream then passes through a mist eliminator, which separates entrained mist droplets from the airstream through inertial impaction and mechanical impingement. The deodorized air is then exhausted to the atmosphere. Figure 11.10 shows a typical wet scrubber. Some gases, such as HCl, can form an aerosol or "acid fog" inside the wet scrubber vessel as it cools.

Mass transfer describes the transport of mass from the gas phase (odorous air) to a liquid phase (scrubbing liquid).

Inertial impaction is the removal of aerosol particles from an airstream when it strikes a collection surface, which changes the momentum of the particle.

Chemical scrubbers achieve odor removal by mass transfer absorption via contact of the airstream with aqueous solution on packing material in a scrubbing chamber (see Figure 11.10). The liquid is typically water, adjusted to the proper pH and oxidation potential by chemicals. Two parameters define the performance of any absorption scrubbing system. These are the efficiency of gas/liquid contact and the amount of chemical within the aqueous phase to promote reaction and absorption. The packing media must have a high active surface area for gas/liquid contact, promoting an even liquid and gas distribution for high mass transfer rates. The wastewater treatment industry has standardized the use of sodium hydroxide and sodium hypochlorite to control H_2S and sulfuric acid to control ammonia odors. Maintaining the chemical conditions favorable for ultimate performance is critical. The desired oxidation–reduction potential (ORP) and pH chemistry is typically adjusted at the site based on the actual gas components present. When necessary to remove ammonia with sulfuric acid, an additional chamber is added. The following chemical reactions describe the H_2S removal processes:

Wet impingement occurs where the air-and-droplet mixture collides with a solid surface. Collision with a surface atomizes the droplets, reducing droplet size and thereby increasing total surface contact area.

- Absorption process (caustic soda only)

$$NaOH + H_2S \leftrightarrow Na^+ + HS^- + H_2O \quad (11.1)$$

- Absorption with oxidation (caustic soda and bleach)

$$H_2S + 2NaOH + 4NaOCl \leftrightarrow Na_2SO_4 + 4NaCl + 2H_2O \quad (11.2)$$

- Absorption of NH_3 with sulfuric acid

$$2NH_3 + H_2SO_4 \leftrightarrow (NH_4)_2\,SO_4 \quad (11.3)$$

Chemical scrubbers are one of the most common odor control technologies used in municipal WRRFs to achieve odor removal by mass transfer absorption via contact of the airstream with aqueous solution on packing material in a scrubbing chamber. The liquid is typically water, adjusted to the proper pH and oxidation potential by chemicals.

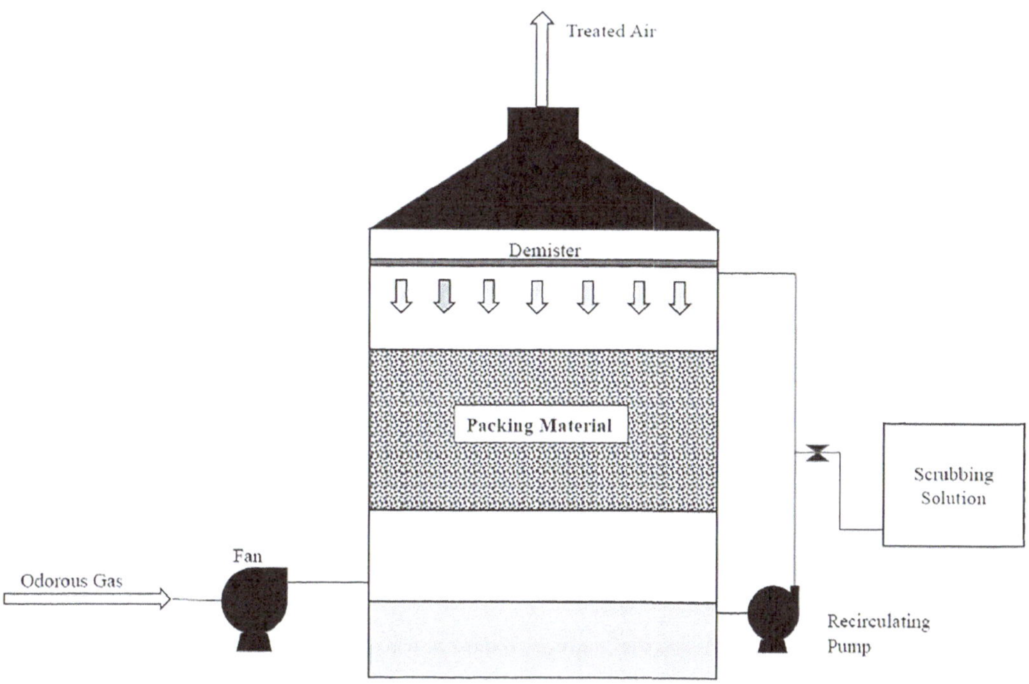

Figure 11.10 Typical Wet Scrubber Unit

The typical treatment scrubbing settings with caustic soda are as follows:

pH = 11.0 to 11.5 standard units

The typical treatment scrubbing settings with caustic soda and bleach are as follows:

pH = 9.5 to 10 standard units; ORP = 600–700 mV

The typical treatment setting for NH_3 scrubbing settings is as follows:

pH = 4.0 to 6.0 standard units

When configured as a two-stage caustic/bleach system, chemical wet scrubbers can achieve 99% removal of H_2S. When a third stage (typically the first in line) is added with sulfuric acid for ammonia removal, ammonia can also be removed in the 99% efficiency range. Caustic wet scrubbers are generally much less effective on organic sulfur compounds, with removal efficiencies between 20% and 70%.

One significant advantage of wet scrubbing systems is the reliability and flexibility provided using chemicals and chemical reactions. A major challenge in the design and operation of wet air scrubbers is minimizing chemical use and cost while maintaining complete, flexible, and reliable treatment. One design used to reduce chemical use is the multistage scrubbing system. As mentioned above, in raw wastewater applications, the most prevalent contaminant is hydrogen sulfide, but other organic odorous compounds typically are present as well. Hydrogen sulfide can be solubilized with a solution of sodium hydroxide.

However, the other odor-causing compounds are best treated by sodium hypochlorite. In a single-stage scrubber system, sodium hydroxide and sodium hypochlorite are used in a recirculating chemical solution. Because sodium hypochlorite is a strong oxidizer, it readily reacts with the sulfide that is solubilized by the sodium hydroxide. Therefore, in a single-stage system, sodium hypochlorite must be added in sufficient quantities to oxidize the hydrogen sulfide and maintain a residual to treat the other odor compounds. In a multistage system, sodium hydroxide is used alone in the first stage to solubilize hydrogen sulfide. Sodium hypochlorite is added to the last stage only. By eliminating the sodium hypochlorite from the sulfide removal process, the consumption rate for this chemical is greatly reduced.

Single-Stage Packed Bed Chemical Scrubbers

Single-stage packed bed chemical scrubbing units may use a single chemical or a combination of two compatible chemicals to remove odors from the air, generally when one odor-causing compound is predominant and adequate reaction time can be provided in a single stage. The scrubbing solution is sprayed over a bed of packing material and the air is passed through the packing bed in a counter-flow pattern.

Multistage Packed Bed Chemical Scrubbers

Multistage packed bed chemical scrubbers are used when concentrations of various compounds present in the air require different types of chemicals to be used in combination, or sequentially, and require reaction times that are longer than can readily be provided in a single-stage scrubber. Single-stage chemical scrubbers are arranged in a series flow pattern to provide the multistage chemical scrubber.

Mist-Type Chemical Scrubbers

Mist-type chemical scrubbers are used when concentrations of various compounds present in the air require different types of chemicals to be used in combination or sequentially. The mist-type chemical scrubbers use atomizing nozzles to produce a fine mist (droplets < 10 microns) to increase the surface area for contact with the scrubbing solution and odorous air.

Air-atomizing nozzles atomize fluids by creating a very fine spray. Liquid and compressed air (gas) are used to produce a mist of atomized liquid at low pressures.

LIQUID REDOX TECHNOLOGY

Liquid redox technology has a relatively long history but is not widely used. Liquid redox units use a chelated metal dissolved in a water solution to remove hydrogen sulfide from a gas stream and convert it catalytically to solid, elemental sulfur. The metal ion in the solution removes electrons (negative charges) from a sulfide ion (S^{2-}) to form elemental sulfur and, in turn, can transfer the electrons to oxygen (O_2) in the regeneration process. Although there are many metals that can perform these functions, iron (Fe) is the most commercially used because it is inexpensive and nontoxic.

Liquid redox processes have limited applications in WRRFs compared to wet scrubbers because they generally have higher capital costs. Where these processes have realized greater demand is in the treatment of anaerobic biogas, where higher levels of hydrogen sulfide and the presence of carbon dioxide cause the lower capital costs of wet scrubbers to be offset by their much higher operating chemical costs. Generally, liquid redox units have operating costs that can be less than 10% of other vapor-phase treatment options because of the regeneration of the active catalyst, the iron solution. However, the lower operating costs are enough to justify the use of liquid redox systems only when the requirement for the removal of hydrogen sulfide reaches 90 to 135 kg/d (200 to 300 lb/d). Above this level, the low operational cost of a liquid redox system easily can achieve less than a 2- to 3-year payback on the initial capital investment.

Liquid redox units focus on digester gas treatment and ventilation air applications with high loadings of hydrogen sulfide. Additionally, because they remove only hydrogen sulfide, liquid redox units used for ventilation air in wastewater facilities often will require polishing with hypochlorite or carbon to remove other odor compounds present.

Ozone contact chambers have been used primarily to control odors from large volumes of air with relatively low concentrations of odor-causing compounds, which can be oxidized to form non-odorous compounds. Ozone can be produced in quantities required for odor control service with on-site ozone generators. The ozone generators pass dry, filtered air through a high-voltage field, which converts a portion of the oxygen to ozone. The air containing ozone is sent through a sparger to a contact chamber.

Spargers are devices that ensure the distribution of gases in liquids by passing the gas through thousands of pores on the surface of the device, causing large amounts of gas to pass a particular place in the liquid. As a result, there is more gas/liquid contact surface area for reactions to take place.

VAPOR-PHASE BIOLOGICAL TREATMENT

BIOFILTRATION

Biofiltration technology can be used to treat a variety of biodegradable, water-soluble contaminants. Figure 11.11 shows a diagram of a biofilter. In a biofilter, the odor air moves upward through an organic media, such as compost, mulch, or peat, while a freshwater source is sprayed from the top and trickles downward. The odorous contaminants are solubilized and move from the air into the fresh water on the

Biofilters are a common odor treatment technology that uses bacteria to remove soluble, sulfur-based compounds.

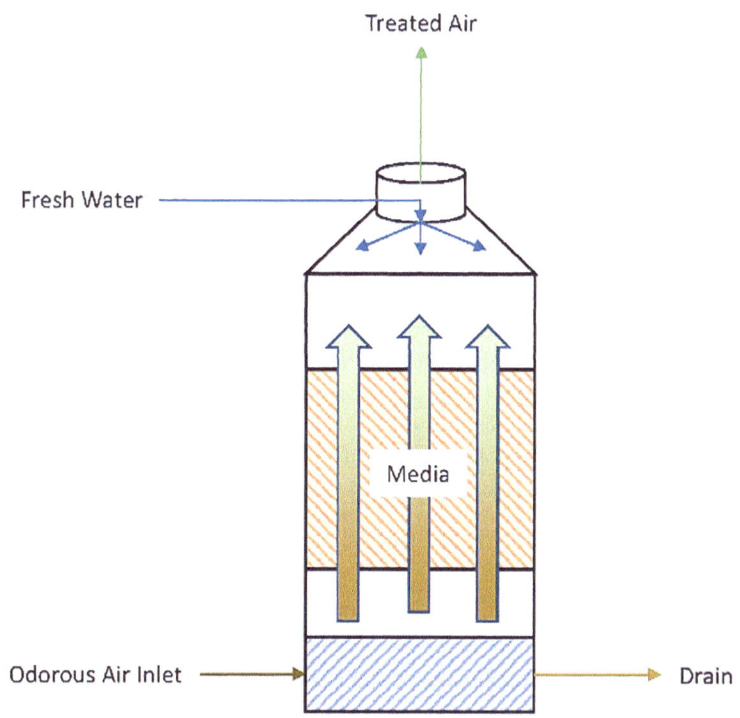

Figure 11.11 Diagram of a Biofilter

surface of the organic medium. The compounds are then degraded by the bacteriological population on this media. Biofilters are very effective at removing sulfur-based odor compounds such as hydrogen sulfide, organic sulfides, and mercaptans. Biofilters generally are not effective at removing nitrogen-based compounds such as ammonia and amines.

Two main challenges in biofiltration systems are the lifespan of the media and control of the biofiltration process. The media used in biofilters can be prone to breakdown. When this occurs, the bed settles and compacts, and increases the headloss through the filter. This causes a decrease in airflow, and fugitive odor emissions typically escape from the air source. Control of the biofiltration process is also an important concern. The biological population within the filter is subject to upsets, as are any organisms. If the media are exposed to wide swings in environmental conditions, the bacteria population can be upset and breakthrough odors may occur.

These challenges can be addressed by using an engineered or "manufactured" biofilter. Engineered biofilters typically address media stability and control issues and are provided with process guarantees. The media stability is addressed through the use of an "engineered media," as contrasted with a "simple media." Simple media typically consist of materials locally available near a specific application (e.g., compost, mulch, wood chips, etc.). Engineered media typically are designed to meet a particular composition and performance specification with a performance guarantee. The engineered media provide high surface area for biological growth to adhere to while also providing structural stability.

Engineered media have a specific combination of organic materials that is followed to ensure consistent and repeatable performance. Process control is addressed through the design of the filter. Most engineered systems use an enclosed design, such as a vessel, to protect the media from temperature and humidity swings. In many cases, the enclosed design is also modular. The media are contained in modules that are arranged in an array to provide a complete biofiltration system. This modular design provides quick and easy construction of the system and extremely easy media replacement. For example, a system that requires treatment of more air than a single module can handle, may have multiple modules arranged in series to achieve the desired air treatment capacity.

BIOTRICKLING FILTERS

As with biofilters, biological trickling filters (or biotrickling filters [BTFs], bioscrubbers) use biochemical processes to degrade odor compounds. Figure 11.12 presents a diagram of a biological trickling filter. A biological trickling filter functions similar to a conventional chemical packed-bed scrubber, with the main difference being that it uses a bioactive solution instead of chemical solution. The bioactive solution is distributed over the top of a synthetic packing media in a vessel, while the odorous air is forced upward. Although most of the biomass in biofilters is attached to media (fixed film), some of the biomass in BTFs is suspended in the bioactive solution. With BTFs, odorous compounds are absorbed or dissolved into the bioactive solution, where the microorganisms can partially oxidize them. Nutrient supplement is commonly required to promote heterotroph biomass growth.

LIQUID-PHASE CHEMICAL TREATMENT

Liquid-phase technologies involve the treatment of a wastewater stream to control the release of odor and corrosion-causing compounds from the stream. Most liquid-phase technologies involve the addition of a chemical to the wastewater to either control the formation of odorous compounds or react with those compounds once they are formed. Liquid-phase treatment is most commonly applied in wastewater collection systems, not WRRFs. It often is applied in a collection system for downstream control at the headworks of a WRRF. Because liquid-phase treatment controls the odor compounds in the wastewater itself, it also provides corrosion control.

With liquid-phase treatment, hydrogen sulfide is prevented from escaping from the liquid into the vapor; therefore, it is not present to cause corrosion on process structures, pipe crowns, and so on. Through proper application of a liquid-phase treatment in a collection system, multiple odor release points such as manholes, air relief valves, and pumping stations can be controlled through one chemical application point. The liquid-phase technologies include iron salts, bioxide process, oxidizers, and anthraquinone (Hvitved-Jacobsen et al., 2000). Each of these is explained in the following subsections.

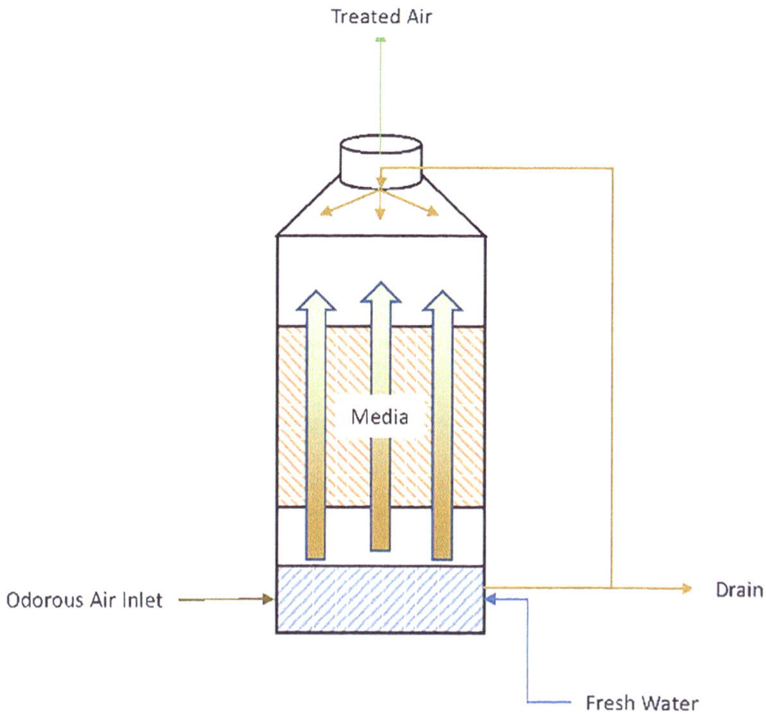

Figure 11.12 Biological Trickling Filter Diagram

IRON SALTS

Iron salts are applied to wastewater to oxidize and/or precipitate dissolved sulfide. Ferrous salts such as ferrous sulfate precipitate sulfide as ferrous sulfide. Ferric salts such as ferric chloride oxidize some of the sulfide to sulfur while reducing the ferric iron to ferrous iron. The remaining ferrous iron then reacts with dissolved sulfide to form the ferrous sulfide precipitate. Ferrous sulfide is a light precipitate, black in color, that is carried downstream to the WRRF. Within the aeration process of the WRRF, the ferrous sulfide is quickly oxidized to sulfate and to ferric iron. The ferric iron can then be used for chemical precipitation of phosphorous. Iron salts can only treat dissolved sulfide. Other odorous compounds are not treated by iron salts. Iron salt solutions are classified as hazardous compounds and often require double-wall tankage and piping systems (Hvitved-Jacobsen et al., 2000).

BIOXIDE PROCESS

The bioxide process adds a nitrate solution (bioxide) to wastewater. The process uses naturally occurring bacteria to biochemically oxidize dissolved sulfide in the presence of nitrate. The sulfide typically is either generated upstream of the injection point where no nitrate exists in the wastewater or is contributed downstream via a septic lateral flow. Nitrate is injected in sufficient quantities to biochemically oxidize the sulfide via the following reaction:

$$8NO_3^- + 5H_2S \rightarrow 5SO_4^{2-} + 4N_2 + 4H_2O + 2H^+ \tag{11.4}$$

This reaction takes place in the bulk flow and in the upper zones of the slime layer. Nitrate is not added in sufficient quantities to fully saturate the slime layer; therefore, sulfide production continues to occur in the lower zones of the slime and is removed in the upper zones or in the bulk flow. The bioxide solution is not classified as a hazardous compound. Nitrate can be added as different salts (e.g., [$Ca(NO_3)_2$]). It is relatively easy and safe to store, handle, and apply (Hvitved-Jacobsen et al., 2000).

OXIDIZING AGENTS

Oxidation for wastewater odor control involves the application of a strong chemical oxidizing agent to the wastewater to prevent sulfur-reducing bacteria from thriving in the sewer. The oxidizing agent is used to chemically react with dissolved sulfide, converting it to sulfate or sulfur. In addition to dissolved sulfide, many other odorous compounds are treated by the oxidizing agent. Many non-odorous compounds also are oxidized, causing application rates to be higher than predicted by just the stoichiometry of the sulfide reaction. Injection of air, pure oxygen, and other oxidizers such as chlorine, ozone, and hydrogen peroxide are used. Chlorine is not typically recommended in collection systems. Hydrogen peroxide is the most commonly used oxidizing agent, but ozone is also used (Hvitved-Jacobsen et al., 2000).

BIOLOGICAL INHIBITORS–ANTHRAQUINONE

Anthraquinone is a chemical compound that interrupts the sulfate-reduction process carried out by sulfate-reducing bacteria in wastewater under anaerobic conditions. Anthraquinone is a nonhazardous compound. When used alone, anthraquinone is specific to the control of sulfide. It can also be used in conjunction with the bioxide process.

TEST YOUR KNOWLEDGE

1. **Activated carbon removes odorous chemicals through the following surface-based process?**
   a. Diffusion
   b. Adsorption
   c. Oxidization
   d. Absorption

2. **Virgin carbon media are activated carbon media that have been chemically treated with additional chemical compounds.**
   ☐ True
   ☐ False

3. **Biofilters are very effective at removing sulfur-based odor compounds such as hydrogen sulfide, organic sulfides, and mercaptans, but generally are not effective at removing nitrogen-based compounds such as ammonia and amines.**
   ☐ True
   ☐ False

4. **Where is liquid-phase treatment typically used for odor control?**
   a. Water resource recovery facility
   b. Wastewater collections system
   c. Emission stacks
   d. Anaerobic digesters

CHAPTER SUMMARY

ODOR CONTROL

- Odorous compounds are formed in wastewater collection, conveyance, and treatment areas.
- Odors in WRRFs are associated with H_2S, also known as *sewer gas*, formed from the anaerobic degradation of sulfate. Other odorous compounds include NH_3; VFAs; nitrogenous compounds such as scatole, aldehydes, and ketones like formaldehyde and acetone; and TRS compounds (mercaptans, carbonyl sulfide, carbon disulfide, dimethyl sulfide, dimethyl disulfide, dimethyl trisulfide).
- The only enforceable H_2S air quality standards in effect currently are the OSHA occupational regulations.
- Technologies that can be applied to control odors include vapor-phase technologies used to control odorous compounds in the ventilation air or gas and liquid-phase technologies used to control odorous compounds in the liquid wastewater itself.
- Technologies used in treating the ventilation air or biogas include carbon adsorption, wet air scrubbing, liquid redox technology, biofiltration, and solid scavengers.
    - In an adsorption system, the gas stream is passed over a bed of adsorbent (carbon or other media) and the odor-causing compounds are attracted to and adhere to the surface of the adsorbent.
    - Activated carbon adsorption is an effective physical–chemical treatment technology for odor control because of the ease of operation, low initial capital cost, and wide range of odor compounds that can be treated. Hydrogen sulfide and TRS compounds are removed effectively by carbon adsorption systems, but ammonia and other nitrogen-based compounds are not effectively treated. Various carbon media systems, including activated and impregnated carbon, can be used independently or in combinations to remove many different contaminants.
    - Other non-carbon media can be used alone or in combination with activated carbon to optimize odor removal. Typically, these non-carbon media are metal compounds such as permanganate-impregnated zeolite or iron sponge.
    - Wet air scrubbing absorption is the most flexible and reliable technology for vapor-phase (gas) wastewater odor control. Absorption may be either a physical or a chemical process. The selection of chemicals to use for scrubbing depends on the odorous compounds present and the emission source.
    - In a biofilter, the odor air moves upward through an organic media, such as compost, mulch, or peat, while a freshwater source is sprayed from the top and trickles downward. The odorous contaminants are solubilized and move from the air into the fresh water on the surface of the organic medium where they are degraded by the bacteriological population on this media.
- Liquid-phase odor treatment is applied in wastewater collection systems, not WRRFs.
- The liquid-phase odor control technologies include iron salts, bioxide process, oxidizers, and anthraquinone.
    - Iron salts are applied to wastewater to oxidize and/or precipitate dissolved sulfide. Ferrous salts such as ferrous sulfate precipitate sulfide as ferrous sulfide.
    - The bioxide process involves the addition of a nitrate solution (bioxide) to wastewater; naturally occurring bacteria biochemically oxidize dissolved sulfide to sulfate in the presence of nitrate.
    - Oxidation for wastewater odor control involves the application of a strong chemical oxidizing agent (air, pure oxygen, and other oxidizers such as chlorine, ozone, and hydrogen peroxide) to the wastewater to prevent sulfur-reducing bacteria from thriving in the sewer.
    - Anthraquinone is a chemical compound that interrupts the sulfate reduction process carried out by sulfate-reducing bacteria in wastewater under anaerobic conditions.

References

American Society of Civil Engineers & Water Environment Federation. (2007). *Gravity sanitary sewer design and construction* (ASCE Manual of Practice No. 60/WEF Manual of Practice No. FD-5). American Society of Civil Engineers.

ARC Sewer Corrosion & Odour Research Project. (2011). Detailed technical subproject research report on ARC corrosion & odour linkage project LP0882016.

Buonicore, A. J., & Davis, W. T. (1992). *Air pollution engineering manual.* Air and Waste Management Association. Van Nostrand.

Graham, J. (2006). Elemental odor control—New trends in activated carbons. *Water Environment & Technology, 18*(6), 46–48.

Hvitved-Jacobsen, T. (2002). *Sewer processes: Microbial and chemical process engineering of sewer networks.* CRC Press LLC.

Hvitved-Jacobsen, T., Vollertsen, J., & Tanaka, N. (2000). An integrated aerobic/anaerobic approach for prediction of sulfide formation in sewers. *Water Science and Technology, 41*(6), 107–115.

Le-Minh, N., Sivret, E., Shammay, A., & Stuetz, R. (2018). Factors affecting the adsorption of gaseous environmental odors by activated carbon: A critical review. *Critical Reviews in Environmental Science and Technology, 48*(4), 1–35.

Leonardos, G. , Kendall, D., & Barnard, N. (1969). Odor threshold determinations of 53 odorant chemicals. *Journal of the Air Pollution Control Association, 19*(2), 91–95, DOI: 10.1080/00022470.1969.10466465.

Occupational Health & Safety. (2007). Human health effects from exposure to low-level concentrations of hydrogen sulfide. http://ohsonline.com/Articles/2007/10/Human-Health-Effects-from-Exposure-to-LowLevel-Concentrations-of-Hydrogen-Sulfide.aspx?Page=2

Occupational Safety and Health Administration. (2005). Hydrogen sulfide fact sheet.

Raina, D. (2004). Activated carbons for wastewater odor control. http://sawea.org/pdf/2004/March-2004/ActivatedCarbonsForWastewaterOdorControl.pdf

SCORe. (2014). SP3: Odour measurement and assessment evaluation of odour treatment technologies. Review of the application of activated carbon for sewer odour abatement sub-project, Sewer Corrosion and Odor Research Project.

U.S. Environmental Protection Agency. (1985). *Odor and corrosion control in sanitary sewerage systems and treatment plants* (EPA-625/1-85/018).

U.S. Environmental Protection Agency. (2000). Biosolids and residuals management fact sheet: Odor control in biosolids management (EPA-832/F/00/067). U.S. Environmental Protection Agency, Office of Water.

U.S. Environmental Protection Agency. (2004). *AERMOD: Description of model formulation* (EPA-454/R-03-004).

Water Environmental Federation. (2017). *Operation of water resource recovery facilities* (7th ed., Manual of Practice 11).

Water Environment Federation. (2020). *Odor emissions and control for collection systems and water resource recovery facilities* (2nd ed., Manual of Practice No. 25).

Water Environment Research Foundation. (2007). *Minimization of odors and corrosion in collection systems.*

Suggested Readings

American Industrial Hygiene Association. (1989). *Odor thresholds for chemicals with established occupational health standards.*

ASTM International. (2004). *Standard practice for determination of odor and taste thresholds by a forced-choice ascending concentration series method of limits* (ASTM E679-04).

National Research Council. (2010). *Acute exposure guideline levels for selected airborne chemicals* (Volume 9). Committee on Acute Exposure Guideline Levels; Committee on Toxicology; National Research Council.

St. Croix Sensory. (2003). Importance of field olfactometry, odor monitoring, and odor regulations. http://www.nasalranger.com/importance.cfm#overview

U.S. Environmental Protection Agency. (1993). *Report to Congress on hydrogen sulfide air emissions associated with the extraction of oil and natural gas* (EPA-453/R-93-045).

CHAPTER 12

Instrumentation

Introduction

Instrumentation provides "eyes and ears" on hundreds or even thousands of different parameters located in every different process at a water resource recovery facility (WRRF). We use instruments to sense and measure critical parameters and send signals back to the main control system. Control systems are programmed to check that these measurements are within the required ranges to keep the facility and processes operating properly. If a parameter shifts outside of the desired range, then the control system will alarm and notify the operator that a system or piece of equipment needs to be checked. Today's WRRFs are so complicated that it would quickly become impossible to run them if we did not have instrumentation constantly checking, measuring, and analyzing for us and sending this information to the control systems.

LEARNING OBJECTIVES

Upon completing this chapter, you will be able to:

- Explain the two categories of parameters that are measured by instruments.
- Describe the instruments that can be used to measure various parameters in a WRRF.
- Explain the operating principles of commonly used instruments.
- Define the components of common instrumentation.
- Explain the importance of properly setting spans, calibrating, and cleaning instruments.
- Identify and correct installation factors and operating conditions that might reduce the usable life of an instrument or analyzer.

Purpose and Function

Instruments reduce the number of labor hours that are required to properly monitor all aspects of a WRRF's operation and performance. They also provide critical safety monitoring that can provide warnings that help save lives or avoid serious injuries. Instruments provide continuous data that allow for quick reactions to changing conditions by making process adjustments before process performance starts to be negatively affected. Process adjustments allow us to minimize chemical usage and run equipment at speeds that will improve efficiency. Instrumentation allows us to meet more stringent permit conditions that provide better protection for the environment and public health.

Applications in Water Resource Recovery Facilities

Instruments are used to measure different categories of parameters:

- Physical properties: These parameters include location, speed, pressure, temperature, level, and flow.
- Analytical parameters: dissolved oxygen, free chlorine, pH, oxidation–reduction potential (ORP), suspended solids, sludge density, ammonia, nitrite/nitrate, and phosphorus

These measurements are turned into electrical signals that can then be used in a control system. We use them in different ways throughout the WRRF to do all sorts of different monitoring and control tasks for us every day. Figure 12.1 shows several examples of typical locations where instruments and analyzers are commonly used. In reality, this visual represents only a small percentage of the number of instruments commonly found at WRRFs today. As discharge permits get more stringent and WRRFs get more complex, plant operations and maintenance (O&M) staff must rely on instrumentation more and more to help properly manage the processes and equipment.

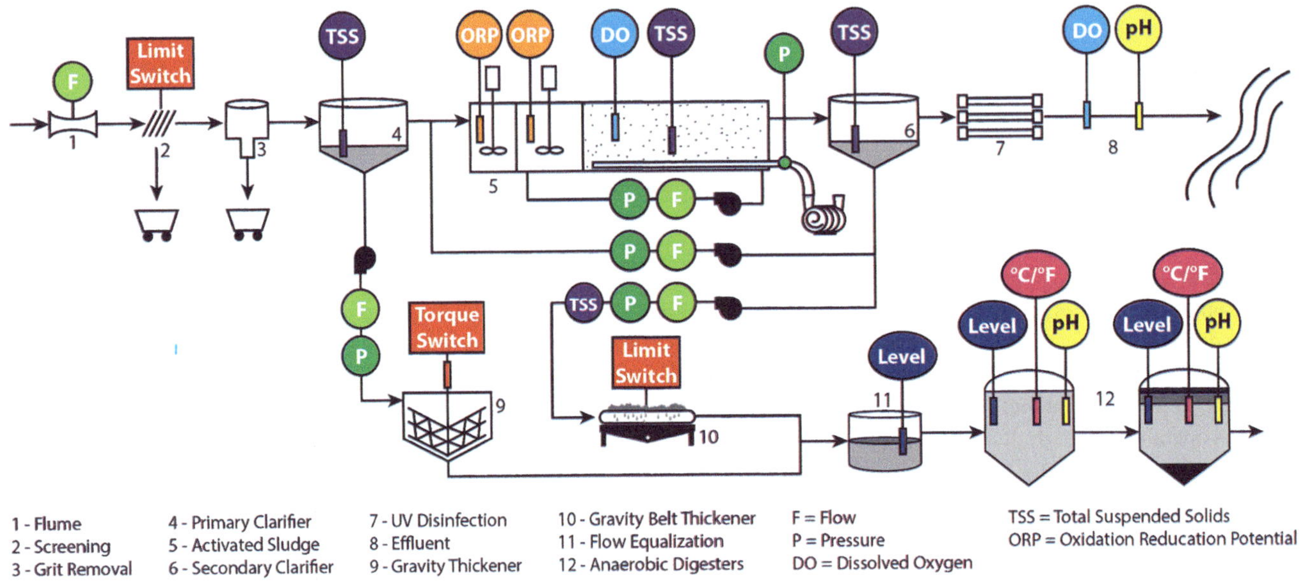

Figure 12.1 **Examples of Instrument and Analyzer Applications at WRRFs**

In this chapter, we will cover how instruments measure different parameters and send the data to the control system. Details on how the controllers use the data will not be covered here. Chapter 13 of this manual discusses supervisory control and data acquisition (SCADA) systems.

Life Cycle of an Instrument

Figure 12.2 presents a schematic of the overall life cycle of an instrument.

Adding a new instrument to a system should always start with clearly defining the reason why the data it will produce are needed (Stage 1: Purpose). Although instruments can provide many benefits, installing unnecessary instruments or producing excess data is not helpful. In fact, having too much data can distract operators from being able to focus on important information.

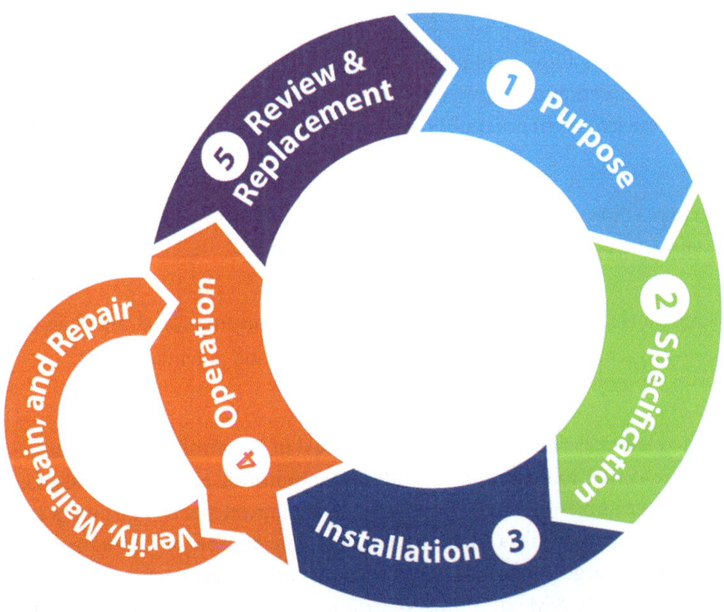

Figure 12.2 **Instrument Life Cycle Schematic**

The medium we are measuring (air, water, chemical stream, etc.) and the area in which the instrument or analyzer will be installed are important factors that must be considered in Stage 2: Specification. The criticality of the data from an instrument will determine the reliability and accuracy required. The communication standard that the control system uses is also included in this stage.

We should never look at the upfront purchase and installation cost alone when weighing the total life cycle cost of an instrument. Many times, the labor and other resources required to operate and maintain (Stage 4: Operation) an instrument over the course of its usable life is considerably more than the original capital and installation cost. Minimizing long-term O&M costs must be considered during upfront planning.

Introduction

In the field of instrumentation, there is some terminology that can be quite confusing. For instance, some terms are used as synonyms that do not really mean the same things. To try to clarify some of this, we are including this section to establish some standardized definitions for frequently used terms.

WHAT IS AN INSTRUMENT?

In simplest terms, an instrument is a measuring device that takes information from a process and provides it in a usable form. As the technology that we use to measure, process, communicate, and use process data evolves, the types and complexity of instruments continues to grow and evolve as well. The words that we use to describe instrumentation have evolved over time. Sometimes, this creates some messy situations in which terminology is not used consistently. Some words are used interchangeably that do not really mean the same things. We will define some common terms in this section to attempt to create consistent definitions that will be used throughout this chapter.

The simplest type of instrument is made up of a sensor and an indicator. Think of a temperature gauge or a pressure gauge. The sensor detects the temperature or pressure and displays that information on a dial or scale. There are no electrical connections or electronics involved. There is no data collected or conveyed. To use this data, a person must be present to read the display and then take any required action.

Instruments that are used to provide data to automated control systems all consist of at least two components: the sensing element and the transmitter (see Figure 12.3). These components may or may not be in one casing; in many instances, manufacturers provide an option for purchasing either an instrument with an integral transmitter (both packaged together in one unit) or a sensor that is connected by cable to a remote-mounted transmitter.

An instrument is sometimes referred to as a *meter*. For example, a flow instrument that is made up of a magnetic flow element and a transmitter is often referred to as a *magnetic flow meter* or a *magmeter*.

Keep in mind that not all instruments are designed to feed data to control systems. There are handheld instruments that are portable and can be carried from process to process to take individual readings. There

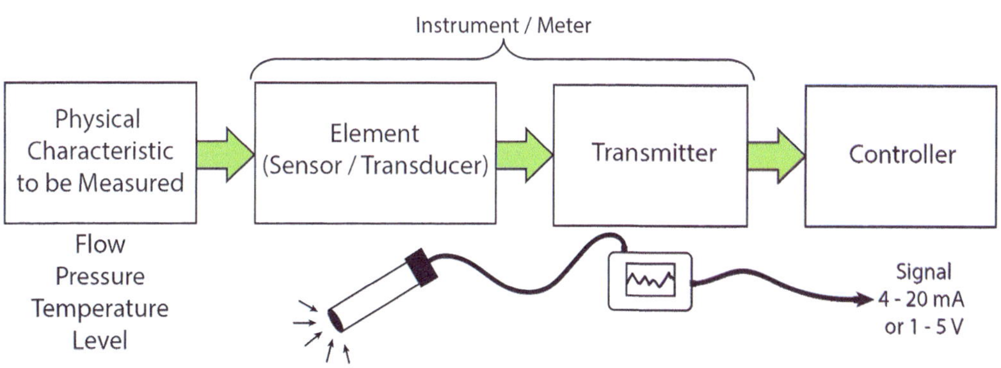

Figure 12.3 Instrument: Element Plus Transmitter

are also benchtop instruments that are used in the lab to measure specific parameters. This chapter is focused more on instruments that are permanently installed to support WRRF control systems. For more information on instruments used in the lab, refer to Chapter 9, Laboratory Procedures, in Wastewater Treatment Fundamentals Book 2.

A complex type of instrument often used in WRRFs to monitor various water quality parameters is an analyzer. These instruments may use several different types of sensors to collect data, and then the transmitter uses these various measurements to make calculations to determine a specific parameter. For example, dissolved oxygen (DO) analyzers typically measure temperature and salinity in addition to percent DO saturation. All three parameters plus barometric pressure are needed to calculate the DO concentration from percent DO saturation. Water quality monitoring instruments that measure parameters such as DO, pH, ORP, suspended solids, turbidity, ammonia, nitrite/nitrate, and phosphorus are all referred to as *analyzers*.

SENSOR/ELEMENT

A sensor detects the physical, electrical, or chemical property and then produces a response that can be converted into a measurement. Keep in mind that the terms *element, sensor, transducer,* and *probe* are often used interchangeably.

There are many different types of sensors, with new technologies being developed all the time. In WRRFs, there are several general sensor categories that are commonly used. Electromechanical sensors take a mechanical stimulus and turn it into an electrical signal. For example, a pressure diaphragm will flex in response to different pressures acting on it. This movement in the diaphragm can be measured as a change in the electrical resistance in the diaphragm material and then converted into a pressure reading. An electrochemical sensor converts a chemical reaction into an electrical signal. An optical sensor converts a specific wavelength measurement or light scatter into an electrical signal. Sonic sensors convert sound wave measurements into electrical signals.

Some sensors must come into direct contact with the process stream to take a measurement, whereas others must be adjacent to, but protected from, the process to take reliable wastewater measurements. Sensors immersed in the process must be constructed out of materials that are compatible with the process environment to minimize fouling and corrosion. The sensor must be able to detect the measured parameter while ignoring or reducing the effect of process interferences. As with any device immersed in direct contact with wastewater or sludge processes, consideration should be given to providing a self-cleaning system for installed sensors. Regardless of whether or not a self-cleaning system is provided, all sensors must be installed so that they are accessible for regular activities such as inspection, cleaning, calibration, and repair.

Other sensor designs are typically installed adjacent to the process stream and do not actually need to touch the process flow to make an accurate measurement. They might be encapsulated by a protective coating or isolated from the process by a diaphragm. Some sensors require an air space above the process. Examples of this are a thermocouple installed in a thermowell and an ultrasonic sensor installed above a process tank. Specific installation details are factored into the instrument's initial setup and calibration. Later in this chapter, we discuss specific installation requirements for each type of instrument.

To minimize signal noise and to protect the sensors, be sure to follow all manufacturer recommendations for proper sensor grounding and surge protection. Cables need to be installed as directed by the manufacturer.

TRANSMITTER

The transmitter typically includes an amplifier and signal converter that changes the signal into a usable form. The transmitter displays the data directly and/or sends the signal to another piece of equipment or the facility's SCADA system via various communication methods. Many transmitters today also include self-diagnostic programming in which internal tests run to verify calibration and proper operation and create an alarm if a problem is detected. It is important to read the manufacturer's O&M manual to understand how the instrument operates and exactly what regular preventive maintenance is required to keep the instrument operating accurately and reliably.

If a transmitter must be installed in a location that would make it difficult to read, the transmitter may be provided with a remote indicator unit that can be installed at eye level in an area that is easy to access.

All transmitters should be installed in areas that provide easy access, safe conditions, a practical way to perform field calibration, and an environment that is correct for the specified transmitter. Minimize the vibration that the transmitter is exposed to and protect against the possibility that the transmitter will be accidentally hit, bumped, or otherwise affected by other equipment or personnel working in the area. Protect transmitters from large temperature variations and other extreme conditions such as direct sunshine. Lastly, provide proper electrical safety provisions and protections.

CONNECTIVITY TO CONTROLLERS

As mentioned earlier, we are focusing on instruments and analyzers in this chapter. Chapter 13 contains more details on control systems, such as SCADA. In this chapter, we will talk about how an instrument sends information to a controller. There are a few different aspects to consider.

First, instruments can be classified as passive or active. A passive instrument does not require its own power source. Its measurement is generated directly by the input from the variable. An example of a passive instrument is a pressure gauge. The pressure being measured directly alters the diaphragm or the bourdon tube and the pressure is shown on the dial face of the gauge. There are no power cables connected to a simple pressure gauge. The reading on the dial shows the current pressure condition that is being measured. No signal is sent to the controller. The data are not recorded or stored for later use. An active instrument, alternatively, must be connected to an external power source such as a battery or utility power to function. An example of an active instrument would be a pressure transmitter. Both power and signal cable are connected to the transmitter. The external power source allows the pressure measurement to be converted and amplified before it is sent as a standardized signal to the controller.

So how, exactly, are instrument data sent to a controller? Different instruments produce different types of signals depending on what type of information they are measuring. Digital signals are binary, which means they are either a 0 or a 1. They are effective at transmitting information such as "ON/OFF" or "OPENED/CLOSED" or "HIGH/LOW." An instrument such as a switch sends digital inputs to the controller, indicating whether the switch is open or closed.

Other instruments are continuously taking measurements to monitor rapidly changing process parameters. They produce analog signals, which are continuous signals. The most common analog standard signal is

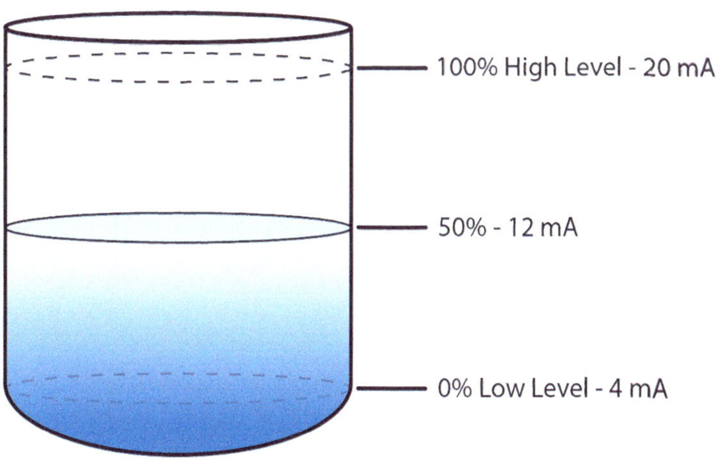

Figure 12.4 Example of Assigning an Analog Scale to Tank Level Measurements

Why does an analog signal run from 4 to 20 mA instead of running from 0 to 20 mA? First, a minimum amount of current is needed to overcome the resistance of the capacitor. Three milliamps are generally needed for this, so 4 mA is a practical minimum to use. The other reason to use a positive number as a minimum is that it provides proof that the system is operating correctly. For instance, if we used 0 mA as the signal for a no flow condition from a flow transmitter, we could not be sure if it were a true zero flow condition or if the instrument was not operating. Using 4 mA (or any other non-zero number) for the minimum reading on our instrument gives us proof that the instrument is operating and actively sending signals. Any time a 0-mA signal is received, it indicates that the instrument is not functioning properly (loss of power, damaged cable, faulty sensor, etc.).

a 4 to 20 milliamp (mA) signal, but other standards such as 1 to 5 volt (V) or 0 to 10 V are also used. They are effective at sending continuous, changing data. For example, a level transmitter measures the level in a tank and sends an analog input to the controller. Different levels in the tank are scaled to match up with a specific milliamp signal. The 4-mA signal is typically considered to be the zero value or the lowest level to be measured. The 20-mA signal typically correlates to the highest level in the tank. Figure 12.4 presents an example of how the signal is scaled to tank level.

When new instruments are installed, they must be set up properly and verified to make sure that the transmitter output signal will be scaled correctly to read across the required span.

TEST YOUR KNOWLEDGE

1. **The labor and resources required to operate and maintain an instrument can exceed the original purchase price and installation cost.**
   - ☐ True
   - ☐ False
2. **Transmitters are always installed in the same case (integral) with their sensors.**
   - ☐ True
   - ☐ False
3. **This is an example of a physical property.**
   a. pH
   b. Pressure
   c. Ammonia
   d. Nitrate
4. **This term is used to describe a measuring device that takes information from a process and converts it into a standardized signal.**
   a. Instrument
   b. Sensor
   c. Solenoid
   d. Probe
5. **Instruments that provide information to automated control systems all consist of at least these two components:**
   a. Sensor and controller
   b. Controller and transmitter
   c. Transducer and controller
   d. Sensing element and transmitter
6. **Optical sensors convert light into a/an**
   a. Pressure signal
   b. Chemical signal
   c. Electrical signal
   d. Wavelength signal
7. **Passive instruments differ from active instruments in this way.**
   a. Active instruments use pressure to generate power.
   b. Passive instruments do not require a power source.
   c. Active instruments only display the current condition.
   d. Passive instruments send signals to controllers.
8. **When using an analog signal, this reading indicates the zero value or the lowest possible measurement.**
   a. 0 mA
   b. 4 mA
   c. 16 mA
   d. 20 mA

INSTRUMENT PERFORMANCE

There are many terms that are used to describe the expected performance of an instrument. It is important to understand these terms so that we select the right instrument for different applications. This section includes an alphabetized list of terms and definitions to be referenced as needed.

ACCURACY

The accuracy of an instrument is the difference between the value that is indicated (reported) by the instrument and the actual value. If the difference is small, the accuracy is high. If the difference between the two values is large, the accuracy of the instrument is low.

BIAS

A bias occurs in an instrument reading when a constant amount is added to or subtracted from a signal. For example, a positive bias of a sensor output would result in all measurements being overreported by the same amount.

DAMPING

Damping is physically or electronically averaging, or damping, field signals to eliminate signal noise and make an instrument's outputs more useful. Damping may vary from fractions of an oscillation to several oscillations. Some damping always occurs in the different components of an instrument such as the electronics, conversions, and membrane. Additional damping may be required to obtain a usable signal for control. See Figure 12.5.

Damping can also refer to the adjustment of control-loop response time. Critical damping is the smallest amount of damping necessary for a controller to respond to a step adjustment without overshooting the control setpoint. Another way to say this is that a critically damped system converges to zero as fast as possible without oscillating. A control system that is underdamped experiences overshoot. An overdamped control system has a response time that is too slow.

DRIFT

Drift is the sum of two error components, the zero shift and the span shift, over some period of time. Drift should be evaluated for the entire instrument (sensor and transmitter), not just the transmitter electronics.

LIMIT OF DETECTION

The detection limit is the lowest concentration of something that can be reliably measured as different from zero. If the detection limit is 5 mg/L and the true sample concentration is 3 mg/L, then the sample result would be reported as <5 mg/L. This is not the same as zero.

LIMIT OF QUANTIFICATION

The limit of quantification is the lowest value of a parameter that can be determined with an acceptable level of accuracy and precision.

OFFSET

The offset (also sometimes referred to as *droop*) is the reading that an instrument gives when there is zero input. It stays constant over the entire range of the instrument. The offset can be corrected for a specific installation by adding or subtracting it from the instrument reading. For example, a pressure sensor is mounted near the bottom of a tank to read the tank level. Because of the structure of the tank and the position of the flanged connection needed to mount the instrument, the sensor ends up being located 0.3 m (1 ft) from the bottom of the tank. When there is 0.3 m (1 ft) of liquid in the tank, the meter will read zero. Therefore, an offset of 0.3 m (1 ft) must be added to the depth reading from the instrument to show the actual depth of the liquid in the tank.

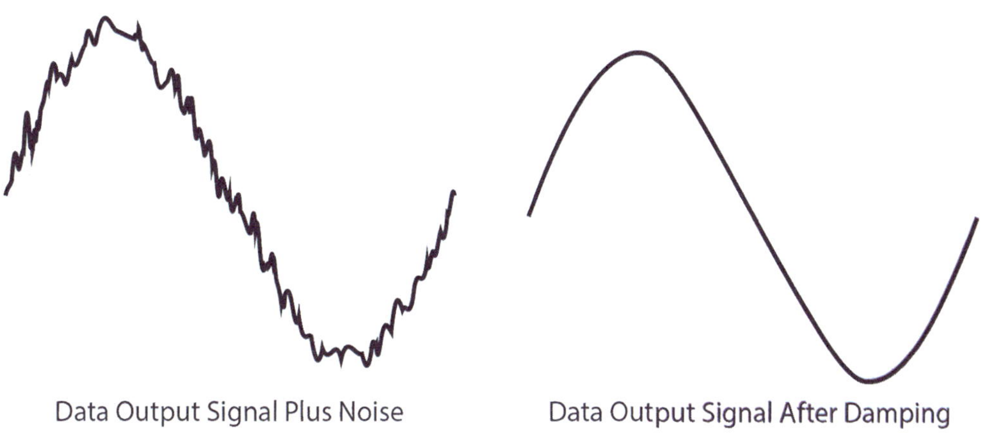

Figure 12.5 Signal Damping

RANGE

An instrument's range is defined as the highest and lowest values of a parameter that an instrument can accurately and/or reliably measure. Many instruments are zero based (i.e., the lowest limit of their range is zero); others start at higher values. Some instruments can function in more than one range depending on how they are calibrated or set up.

An over-range value for an instrument is a measure of its ability to handle conditions outside its range without damage. This is important to know what the upper limit is when extreme conditions occur. For example, the range of a pressure sensor is 0 to 690 kPa (0 to 100 psi). Its over-range pressure is listed as 1379 kPa (200 psi). This means that the instrument will accurately read up to 690 kPa (100 psi) and it can handle pressures up to 1379 kPa (200 psi), but the accuracy of the readings between 690 kPa (100 psi) and 1379 kPa (200 psi) cannot be relied upon. Permanent damage to the instrument will not occur in this range. Once the system's pressures drop back down into the 0 to 690 kPa (0 to 100 psi) range, the readings should be accurate again. Exceeding the maximum over-range pressure will cause permanent damage to the sensor and can result in a structural instrument failure.

REPEATABILITY (REPRODUCIBILITY)

Repeatability is a measure of how consistently a specific instrument can measure and report a value for a certain condition. For example, the repeatability of a specific temperature sensor would be determined by collecting many readings from a circulating constant temperature water bath and then calculating the variation between those readings. Repeatability is reported as a range of values above and below the arithmetic mean (average) of all the readings taken during the test.

SIGNAL NOISE

Signal noise refers to any unwanted and meaningless variations in signal output that can be introduced by interferences somewhere in the system. All instruments introduce some amount of signal noise into a measurement. When instrument signals are used for precision control, damping (i.e., removing the signal noise) becomes more important. Signal conditioners are often included in the transmitter electronics to remove signal noise and amplify the instrument's output signal.

SPAN

The span is the difference between the upper and lower values of an instrument's range. For example, a level transmitter with a range of 3 to 6 m (11 to 21 ft) would have a span of 3 m (10 ft).

SPAN ADJUSTMENT

The span adjustment for an instrument is when we reset the high-end output, which is the value where the span ends or the maximum reading that can be made with the instrument.

STABILITY

Stability of an instrument is a general term that includes errors introduced over time (drift) from many different sources including aging, vibration, temperature extremes, humidity, corrosion, cycling, hysteresis, signal noise, and so on.

ZERO ADJUSTMENT

The zero adjustment for an instrument is when we make changes to the low-end output. This is basically the point where the instrument's span begins or the minimum reading that can be made with the instrument.

ZERO TRIM

A zero trim is performed on a newly installed sensor to adjust the input sensor to compensate for offsets caused by installation conditions and sensor orientation.

ZERO SUPPRESSION

In pressure-level measurement, zero suppression is the ability to set the lower limit to a positive value. This would be useful, for example, when a pressure transmitter is at the base of an elevated storage tank, but is only measuring the water level in the tank. The transmitter is physically mounted a certain distance below the level that we want the instrument to read as a zero tank level so a positive number is added to the pressure reading.

ZERO OFFSET

Zero offset is an instrument's ability to emit a "zero" signal when measuring a low level and to emit actual measurements over the rest of its range. This feature is useful under zero-flow conditions, when a minor drift in calibration could cause totalizers to indicate a flow that does not exist.

TEST YOUR KNOWLEDGE

1. **When an instrument reads consistently high or consistently low, the resulting data have**
   a. Accuracy
   b. Precision
   c. Bias
   d. Damping

2. **The lowest concentration that can be reliably measured as different from zero is the**
   a. Detection limit
   b. Quantitation limit
   c. Accuracy
   d. Precision

3. **A pressure instrument is capable of reading values between 70 and 1720 kPa (10 and 250 psi). What is the span?**
   a. 70 kPa (10 psi)
   b. 1650 kPa (240 psi)
   c. 1720 kPa (250 psi)
   d. 1790 kPa (260 psi)

4. **__________ can be included in a transmitter's electronics to remove signal noise and amplify an instrument's output.**
   a. Bias cancellation
   b. Horns and antennae
   c. Sensors and probes
   d. Signal conditioners

5. **__________ is the ability to set a lower limit to a positive number that is created by instrument installation position, for example, when an instrument is located below the low level of a tank.**
   a. Span adjustment
   b. Instrument accuracy
   c. Zero suppression
   d. Signal damping

Parameters Measured

LOCATION/POSITION

LIMIT SWITCH

Purpose and Function

A limit switch is a simple electromechanical device that is made up of two parts: an actuator that is mechanically linked to an electrical switch (see Figure 12.6). When the actuator is physically moved, the switch is opened or closed to energize or de-energize an electrical circuit. A limit switch directly makes contact with the object it is sensing.

Applications in Water Resource Recovery Facilities

Limit switches are fairly inexpensive and robust and can be used for many different simple tasks. A limit switch is used to stop the travel of an automatic garage door opener when the door reaches the fully open position. Limit switches are often used on motor-operated valves and gates to stop the travel when the gate/valve is fully open or fully closed. They can be used to turn on a ventilation fan when a remote pump station door or a wet well hatch is opened. As a safety precaution, they can shut down a piece of moving equipment if a safety guard is removed or an access door is opened while the equipment is running.

A General Note About Switches

When dealing with switches, you will run into the terms *normally open* and *normally closed*. These terms can be a little confusing. How do we determine what is normal? The "normal" state of a switch is defined as the state of the switch when it is not activated or when it is below its setpoint. Another way to say this is that the "normal" status is where the switch would be with no power connected to it. For example, a

Figure 12.6 Limit Switch Installed on Wet Well Access Hatch (Reprinted with permission by Passaro Engineering, LLC)

normally closed pressure switch is installed on a pump motor circuit and has a setpoint of 172 kPa (25 psi). It is sensing the pump discharge pressure. The pump typically operates between 63 kPa (10 psi) and 103 kPa (15 psi). Most of the time, the switch is closed. If there is a spike in pressure, the switch will be activated at 172 kPa (25 psi) and it will open, break the electrical circuit, and shut off the pump motor. That seems straightforward, right? Keep in mind that the "normal" position of a switch is not always where it spends most of its time. Let's look at another example. A normally open flow switch is installed on an electrical circuit to activate a "low flow" alarm light. If the system is operating correctly, the flow switch is sensing flow; therefore, the switch is activated and is closed. If flow drops to zero, the switch will go to its "normal" position and close, completing the circuit and lighting the low flow alarm light.

How They Function

Limit switches come in several different configurations. The portion that senses the movement can be a lever, a roller, a plunger, a whisker, and so on. Regardless of the shape of the piece that moves, they all function the same way. See Figure 12.7.

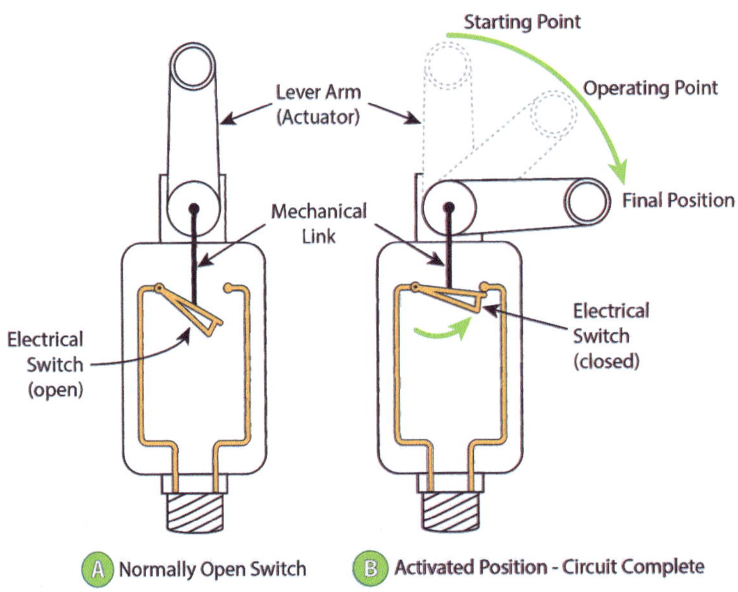

Figure 12.7 Limit Switch

The piece of the switch that moves is called the *actuator*. In this figure, it is a lever arm with a roller on the end. The starting position is the resting position or the "normal" position of the limit switch. Because this example is a "NO" switch, the electrical switch is open when the lever arm is in its starting or resting position. The lever arm is connected to the electrical switch by a mechanical link or pin. When the lever arm is moved past its operating point, the mechanical link pulls the electrical switch closed and completes the circuit.

Installation and Operation

Limit switches must be installed in the right position to allow the actuator to move as needed. Typically, the actuator can be installed in several different positions. Some switches can also be configured to actuate in either a clockwise or a counter clockwise rotational direction. Consult thc installation instructions for your switch.

When installing the switch and the electrical conduit, take care not to position the conduit where liquid can travel through the conduit and run into the body of the switch. Properly install the ground wire.

After installation is complete, actuate the switch to make sure that there are no interferences that will prevent it from moving freely.

Maintenance and Troubleshooting

Keep the moving parts of a limit switch clean and free from debris. The moving mechanical parts will wear out over time. Generally speaking, limit switches are fairly inexpensive and simple to replace, so it is usually more cost effective to purchase a new limit switch rather than replace individual worn components on an older switch. However, replacement parts can typically be purchased from the original manufacturer if you know the model number of the original switch.

SPEED/MOVEMENT

To protect staff and equipment, and avoid potentially huge messes, it is sometimes important to have verification that equipment is operating properly. This can be done with simple sensors that will verify that a piece of equipment is moving as expected when the control system turns it on. They can be called several different names—*zero speed switches, motion sensing switches, proximity switches, speed actuating sensing switches,* and so on. Many of these operate under the same basic principles. There are lots of long names used for something that is quite simple. In this chapter, we are going to cover how proximity switches are used to verify equipment speed and movement.

PROXIMITY SWITCH

Purpose and Function

A proximity switch detects an object without making direct physical contact with it. There are no mechanical moving parts so they can generally work in dusty, dirty areas without issue, and they do not wear out.

Applications in Water Resource Recovery Facilities

Proximity switches can be used to sense movement on rotating, reciprocating, or conveying equipment. They can also be used to sense the presence of material in a tank, hopper, or bin.

How They Work

Proximity switches use several different methods of detecting that a specific object is nearby. They do not come into direct contact with the material they are sensing. We will look at three different kinds of proximity switches: inductive, capacitive, and magnetic.

Inductive proximity switches can be used to detect a metallic object (basically anything made out of iron, aluminum, brass, copper, etc.). Electric current is sent through a coil at the end of the sensor and a magnetic field is generated. When a metallic object gets close to the sensor, it causes a disturbance in the magnetic field that can be detected by the sensor. The change causes the switch to open or close. To sense motion and make sure a piece of machinery is operating at its normal speed, metallic tabs are added to the piece of moving machinery. See Figure 12.8, which shows an example of the tabs being added to the end

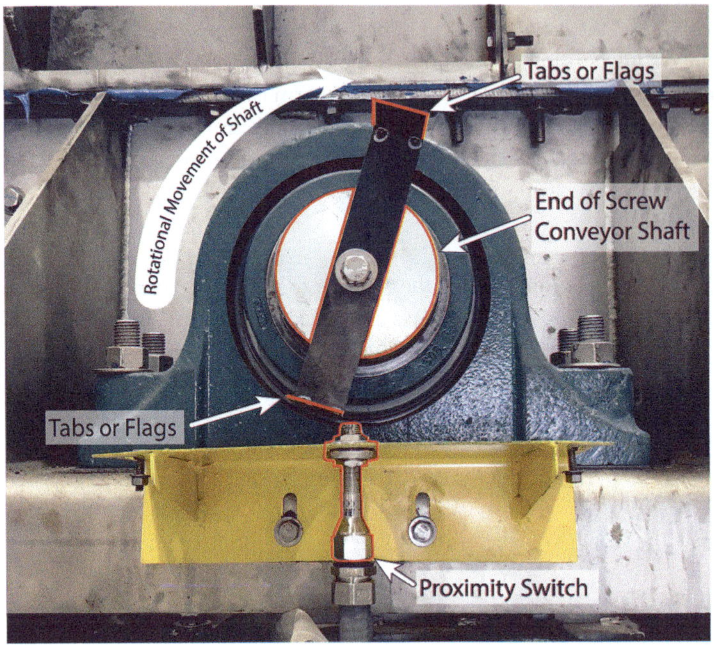

Figure 12.8 Inductive Proximity Switch Used to Sense Shaft Rotation (Reprinted with permission by Passaro Engineering, LLC)

of a rotating shaft. Every time a tab passes near the proximity switch, it momentarily activates the switch. Because the normal operating rotational speed of the shaft is known, the expected timing between switch activations can be calculated. A maximum time between switch activations is programmed as a setpoint in the controller. If the time between switch activations is longer than the setpoint, an alarm can be activated and/or the equipment can be shut down to avoid damage or unsafe conditions from occurring.

Capacitive proximity switches can be used to detect a much wider range of materials than an inductive proximity switch. They can detect any material that can conduct electricity. This includes metallic objects as well as many types of liquids, gels, powders, and so on. They can even sense materials that are inside a container or packaging. For example, they can be used to sense the presence of a material through a tank wall. Their sensitivity can typically be adjusted in the field. The sensor is basically half of a capacitor (one plate). When a conducting material gets close enough to it, it acts like the other plate of a capacitor. The air gap in between is the insulator. The conditions are right for a "temporary" capacitor to form between the sensor and the conductive material through the air gap. The sensor measures the impedance (think of this as just a sudden change in the electrical properties at the sensor). The change in the impedance opens or closes the switch.

What is nuisance tripping? This occurs when a switch is activated (i.e., changes status from open to closed or closed to open) when it is not intended to. For example, a proximity switch is being used to verify rotation on a V-belt-driven shaft. The switch is configured to shut down the equipment if the shaft is rotating too slowly (or not moving at all) because this might indicate an overloaded condition or a broken belt. However, when the shaft initially begins moving when the equipment is first started, it will take a certain amount of time to reach its normal operating rotational speed. If there is not a sufficient amount of time delay provided at startup, there could be nuisance tripping every time there is an attempt to start the equipment.

The third type is the magnetic proximity sensor, which has a permanent magnet mounted on its sensing face. When a ferrous metal object passes through the sensor's permanent magnetic field, the distortion of the field is sensed, which causes the switch to change states.

Installation and Operation

Proximity switches need to be close enough to the object or material they are sensing for the switch to be activated. However, we want to keep the proximity switch far enough away from the target to avoid any accidental contact. This is particularly true when we are sensing motion on a moving piece of equipment. Typically, there will need to be some set up and adjustment to find the exact distance between the sensor and the target that will provide the best performance of the switch. Adjustments to the sensitivity of the sensor may also need to be made. Time delays can be adjusted in the controls to avoid nuisance tripping.

Proximity switches work best when they are installed on a vibration-free support. Locate the switch as far away from other moving equipment as possible to minimize the possibility of sensor interference from nearby unintended targets.

As with all instruments, conduits should be installed with a bottom entry to prevent moisture from entering the instrument through the conduits.

Maintenance and Troubleshooting

Proximity sensors require very little maintenance. They should be wiped clean with a damp cloth occasionally and the instrument and cables should be inspected for any signs of damage.

TEST YOUR KNOWLEDGE

1. A __________ is an electromechanical device made up of an actuator that is mechanically linked to an electrical switch.
   a. Proximity switch
   b. Limit switch
   c. Level switch
   d. Flow switch
2. A __________ detects the presence of an object without actually making contact with it.
   a. Proximity switch
   b. Limit switch
   c. Level switch
   d. Flow switch
3. __________ is the term used when a switch is activated when it was not intended to be.
   a. Bias adjustment
   b. Zero adjustment
   c. Signal damping
   d. Nuisance tripping
4. The "normal" status of a switch is where it would be positioned
   a. After activation
   b. Most of the time
   c. With no power connected
   d. Under typical process conditions
5. In a "motion sensor" configuration, a proximity switch will sense the presence (and, indirectly, the speed) of metallic tabs connected to the __________ of a piece of rotating equipment.
   a. Starter
   b. Shaft
   c. Breaker
   d. Magnet

PRESSURE

Pressure is a measure of force distributed over an area and it is represented by the following formula:

$$\text{Pressure} = \frac{\text{Force}}{\text{Area}} \qquad (12.1)$$

In International Standard Units, it has the units of newtons per square meter, which is also referred to as a pascal (Pa). In U.S. Customary Units, pressure has the units of pounds per square inch (psi).

From the formula, if the force stays the same, the pressure can be increased by decreasing the area over which it is applied. If we have a 10 kg (22 lb) mass sitting on the two styles of table shown below, which one would exert more pressure on the floor? See Figure 12.9.

The mass on the table with the four small legs would spread the force over a much smaller area, creating more pressure at the points where the four legs meet the floor.

Gases and liquids also exert pressure but in slightly different ways. See Figure 12.10.

Look at the pressurized container on the left that is holding only gas. Gases exert pressure evenly in all directions so the pressure gauges located at the three locations shown will all read the same pressure. The pressurized container on the right contains both liquid and gas. The gas pressure will still read the same

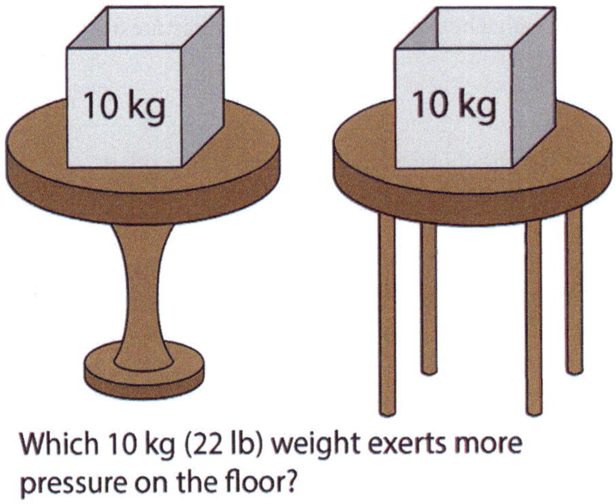

Figure 12.9 Pressure Exerted on a Surface

at any point above the liquid. However, the pressure readings below the liquid level will vary and will be higher, depending on the height of the liquid column above the gauge location. This is due to the additional force or weight of the water column pushing down on the gauge sensors. We can calculate the pressure exerted by the column of liquid using the following formula:

$$\text{Pressure} = \rho g h \quad (12.2)$$

Where

ρ = the density of the liquid
g = acceleration due to gravity (9.8 m/s^2)
h = height of the liquid above the point where the pressure is being calculated

Therefore, in the right-hand figure, the highest pressure will be measured by the gauge at the bottom because this includes the pressure from the gas at the top of the tank plus the pressure from the full water depth in the tank.

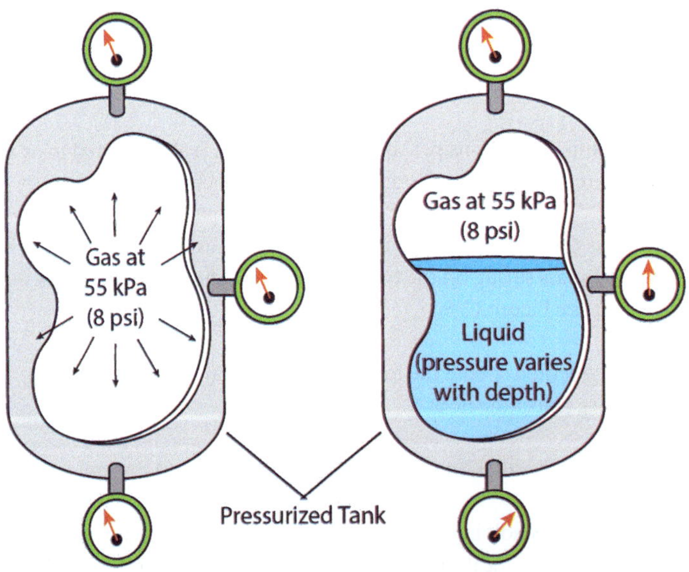

Figure 12.10 Gas Pressure Versus Liquid Pressure

Let's look at an example in which the tank is not pressurized. It has an open top. Let's also say that the tank contains clean water at room temperature. How do we calculate the pressure here? See Figure 12.11. At the top of the water, the gauge pressure will be 0 kPa (0 psi). For each increment of depth that we move our pressure gauge deeper, the pressure will increase by a constant amount. Each meter of water depth will increase the pressure by 9.79 kPa. Each foot of water depth will increase the pressure by 0.433 psi.

All pressure readings are essentially differential pressure measurements because the readings are relative to a reference pressure. The three primary pressure measurements are (see Figure 12.12):

- Gauge pressure—pressure is referenced to atmospheric pressure,
- Absolute pressure—pressure is referenced to a perfect vacuum, and
- Differential pressure—pressure is referenced to second process pressure.

So, how do these pressures relate to each other? See Figure 12.13. When we talk about the operating pressure of a pump, we are referring to gauge pressure. We start by saying that atmospheric pressure at sea level is equal to 0 kPaG (0 psig). The amount of pressure that the pump creates when operating is above and beyond this standard atmospheric pressure and is referred to as *gauge pressure*. Vacuum pressure is a negative number that indicates a pressure below 0 kPaG (0 psig). Looking back at Figure 12.11, all of the pressures listed are gauge pressures because we are using atmospheric pressure as our zero pressure point. If a pressure is just listed as kilopascals or pounds per square inch, it is understood that they are gauge pressures.

PURPOSE AND FUNCTION

Pressure instruments are used to measure the pressure of liquids or gases in a process at the point where the sensor is located.

APPLICATIONS IN WATER RESOURCE RECOVERY FACILITIES

In wastewater collection and treatment systems, pressure instruments are frequently used to monitor pump suction and discharge pressures as well as compressed-air system and aeration system operating conditions. See Figure 12.14. Pressure is used to control pneumatic systems and create proper ventilation in heating, ventilating, and air-conditioning (HVAC) systems. Pressure measurements can also be used to calculate levels and flows, which will be discussed later in this chapter.

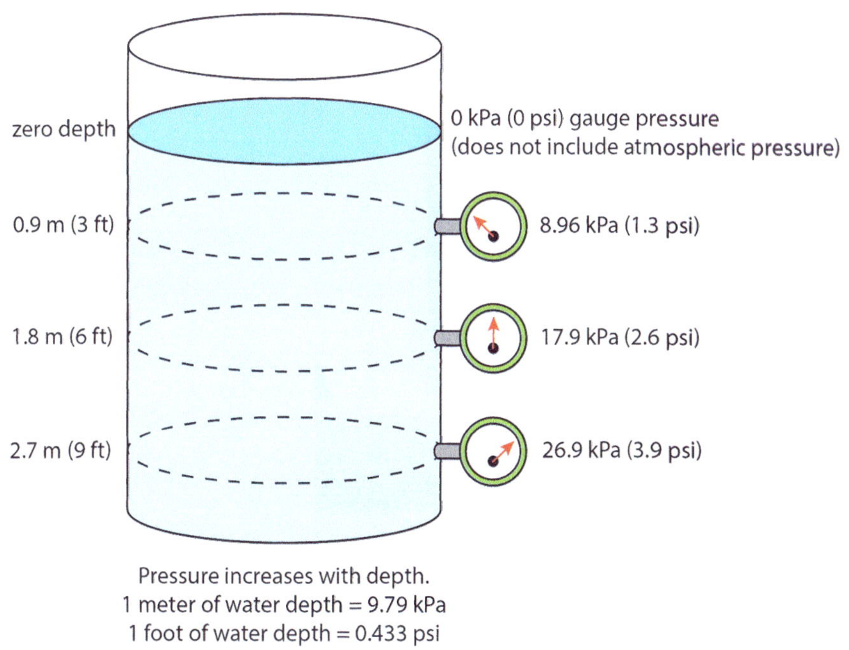

Figure 12.11 Pressure at Different Water Depths

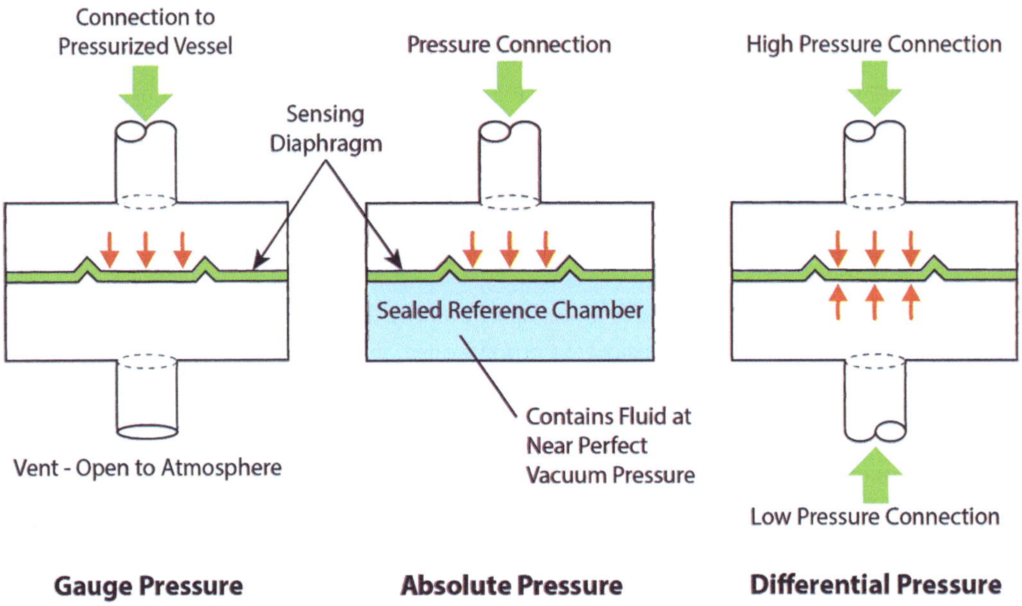

Figure 12.12 Methods of Pressure Measurement

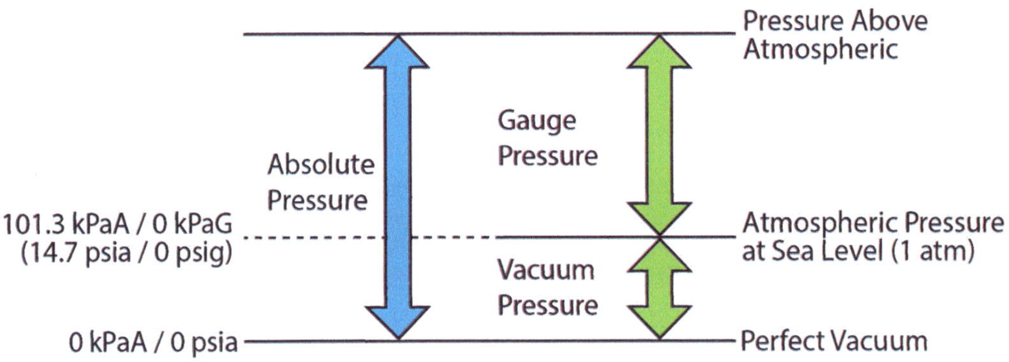

Figure 12.13 Types of Pressure

Figure 12.14 Differential Pressure Transmitter on an Aeration Blower (Reprinted with permission by Indigo Water Group)

HOW THEY WORK

Pressure gauges are located on many systems throughout a WRRF. They are passive instruments that are used for local monitoring and troubleshooting.

There are many different mechanical pressure gauge designs. The three most common designs are C-type bourdon tube, bellows, and diaphragm.

Pressure gauges can be dry or fluid filled. See Figure 12.15. Glycerin is the most commonly used fill fluid, but it can become too viscous (thick) at lower temperatures. Silicone oil is a better choice for very cold weather conditions. Although dry gauges are typically less expensive to purchase than liquid-filled, they also tend to have a shorter useful life. In general, the sealed fluid-fill helps to dampen pressure spikes, reduce vibration and pulsation, and keep moisture and humidity out of the gauge. Silicone oil can also be used in pressure sensors that contain electronic components.

A pressure switch is a pressure sensor that opens or closes an electrical contact when its pressure setpoint has been reached. Some pressure switches have field-adjustable setpoints. They can be configured to activate when their setpoint is reached with rising pressure or when a setpoint is reached with falling pressure.

A pressure transmitter is used when it is important to be able to monitor continuous pressure in a system. A pressure sensor (transducer) senses the continuous pressure changes and sends the measurements to a transmitter, which amplifies and converts the signal into a 4 to 20 mA, which can be used by a control system.

Pressure sensors using diaphragms have replaced traditional bellows sensing elements for many applications in the wastewater industry. The diaphragms are flat or concentrically corrugated metal or ceramic disks that deflect in response to pressure changes. See Figure 12.16.

As the diaphragm deflects, the deflection is measured using various methods such as capacitive, inductive, piezoelectric, piezoresistive, or optical sensors. For example, the deflection changes an associated electrical property (e.g., the distance between the plates in a capacitor, piezoelectric response, loop reluctance, or optical transmittance), which is converted to a pressure reading by the instrument.

Figure 12.15 Fluid-Filled Pressure Gauge

Figure 12.16 Stainless Steel Pressure Diaphragm (Reprinted with permission by Keller America, Inc.)

The instrument assembly shown in Figure 12.17 uses a piezoresistive sensor in the transducer. The vent tube allows atmospheric pressure to reach the reference side of the diaphragm for gauge pressure readings. It is important to follow the manufacturer's recommendations for keeping this vent tube open and free from water.

When a pressure instrument such as a gauge, switch, or transmitter is installed in a system in which rapid pressure changes might occur, a snubber should also be installed to protect the sensing element from damage. A snubber works like a shock absorber. It operates by driving a piston up against an orifice during rapid pressure fluctuations, choking the flow, and reducing the pressure spike reaching the sensing element. Porous-disc-style pressure snubbers are not recommended for most wastewater because they clog easily.

Pressure-sensing elements are typically made of stainless steel, Kynar, Hastelloy, or ceramic. When the process being measured is chemically incompatible with typically available materials, then a diaphragm seal should be considered.

Kynar® (brand name for polyvinylidene fluoride) is a material that has a nonstick surface similar to Teflon™ (brand name for tetrafluoroethylene) and also has high strength properties that give it good resistance to denting, rupture, or other damage from impacts. The nonstick surface helps to prevent oils and grease from sticking to it. Hastelloy® is a nickel-molybdenum alloy that is resistant to corrosion.

Diaphragm seals are helpful to protect pressure instruments installed on wastewater systems. The diaphragm seal system provides a physical barrier to protect the sensing diaphragm from solids and/or corrosive material while still transferring pressure to the sensing element via a fill fluid. Typical fill fluids

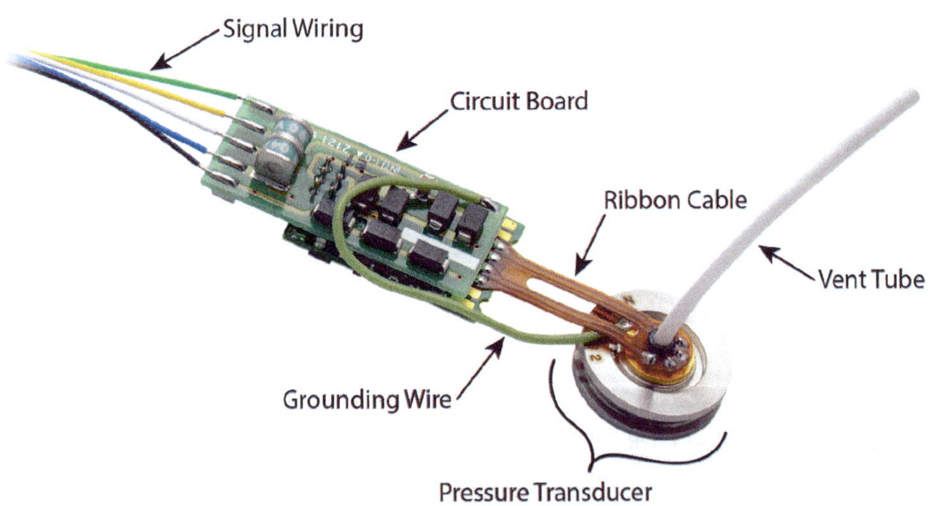

Figure 12.17 Components of a Pressure Instrument Assembly (Reprinted with permission by Keller America, Inc.)

include glycerin, silicone, and Halocarbon. Glycerin is most cost-effective. Silicone performs better in extreme conditions, especially very low temperatures. Halocarbon is best for applications that involve oxidants such as chlorine, oxygen, and hydrogen peroxide.

A flushing connection/drain ring can be installed on the process side of the diaphragm to clear solids away from the sensing face of the diaphragm and provide additional protection to keep critical pressure instruments operating properly.

In general, there are a couple of rules to follow to protect a pressure sensor:

- Protect the sensor from corrosive chemicals;
- Avoid extreme vibration or sharp impacts; and
- Do not directly contact the pressure-sensing membrane with anything harsher than warm, soapy water when you are trying to clean it.

Each sensor is designed to have a normal operating range. They will be able to withstand short periods of higher pressure without causing permanent damage. The instrument's "overpressure" rating is the highest pressure spike the instrument can safely withstand. The "burst pressure" for a pressure sensor is the pressure at which components will fail and permanent damage will occur.

Pressure sensors must be specially designed for the following:

- Very high or very low temperatures
- Exposure to high-pressure water jets
- Corrosive or incompatible materials
- High humidity
- Submergence

If a pressure sensor is mounted higher or lower than the process connection, then an offset will need to be applied to the measurement to compensate for the elevation difference.

Sensors may include a vent tube that allows atmospheric pressure to act on the reference side of the sensor. It is important that moisture is not allowed to enter and/or build up in the vent tube. This will cause unstable instrument readings and will eventually cause instrument failure. Read the manufacturer's O&M instructions to be sure that you are performing tasks required to keep the vent tube open and dry, if applicable.

Electrically, instruments must be properly grounded. Sensors installed in wastewater applications may need to be intrinsically safe or explosion-proof. In hazardous areas, an intrinsically safe barrier should also be added to the instrument circuit. Keep in mind that this will increase the required supply voltage. Watch for signs of signal interference such as unstable readings. Be sure to install all cabling as recommended by the manufacturer.

Pressure instruments will experience hysteresis. This is a phenomenon where an instrument will measure a specific condition differently if the parameter being measured is increasing versus when it is decreasing. See Figure 12.18. The different readings are due to stresses created within the sensor as it reacts to pressure moving from a zero reading up to its midpoint reading versus the stresses created when the sensor reacts to changes from maximum pressure down to the midpoint reading. The difference between the reading at the same pressure condition is called the *hysteresis*. This effect can be minimized by selecting the right materials and design for your specific application. Hysteresis also occurs in temperature instruments.

The word *hysteresis* comes from a Greek word meaning "lagging behind." It is used to describe situations in which the state of a system is affected by its history. In instrumentation, it is used when an instrument's readings are affected by the readings taken immediately before.

A directly connected pressure instrument can be accurate within 0.04% of span; however, typical errors are in the range of 0.1% of span.

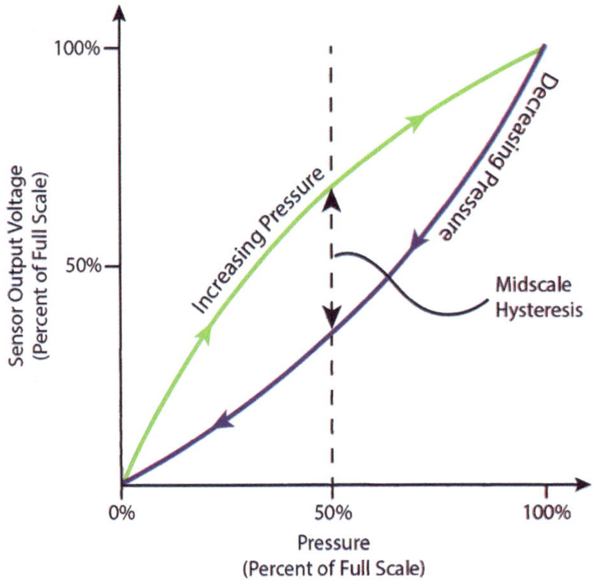

Figure 12.18 Hysteresis

MAINTENANCE AND TROUBLESHOOTING

Refer to the manufacturer's instructions for cleaning. In general, never attempt to clean a sensor by scraping it or sticking an object in to unclog it. In general, warm, soapy water and gentle agitation are recommended to clean a fouled pressure sensor. In some cases, it may be necessary to use a solvent to clean the sensor, but this can be done only after you verify that all sensor materials such as the diaphragm and O-rings are compatible with the solvent. Make sure you carefully follow the manufacturer's recommendations for cleaning a sensor—doing this incorrectly can permanently damage your sensor.

The sensing element should be accessible for maintenance, using diaphragm seals with capillary tubing as required to mount the sensor at a height that is easy to access. Pressure sensors should be installed with a two-valve manifold consisting of a block valve that can be used for process isolation and a bleed valve that can be used to vent, drain, and verify/calibrate the sensor. See Figure 12.19.

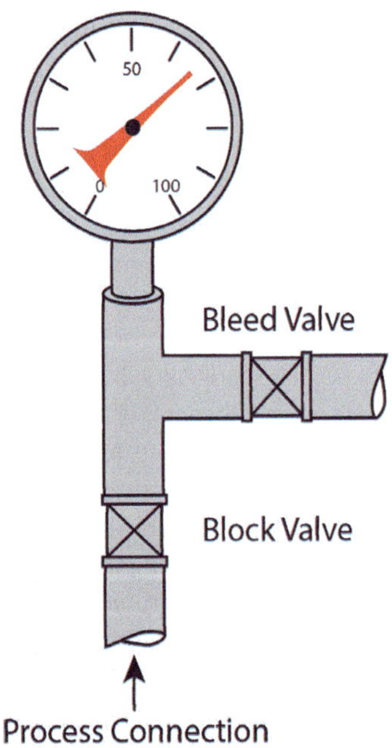

Figure 12.19 Two-Valve Block and Bleed Manifold

Differential-pressure sensors should be installed with three- or five-valve manifolds. Three-valve manifolds consist of high pressure, low pressure, and equalizing valves. Three-valve manifolds still require the sensor to be removed from the process to perform calibrations; as such, they should not be used if calibration of the differential-pressure sensor is anticipated. Five-valve manifolds are preferred for differential-pressure sensors because they allow the sensor to be calibrated in place. Five-valve manifolds consist of high pressure, low pressure, low-pressure calibration, high-pressure calibration, and equalizing valves. For a step-by-step procedure on using manifold valves with pressure instruments, refer to Ashcroft's "Advantages of Manifold Valves Used with Pressure Instruments" white paper listed in the Suggested Readings section at the end of this chapter.

Flanged sensors should be used for processes that are prone to fouling. They are quicker and easier to remove from the piping, will not require the wiring to be disconnected to screw the fitting back together, and are less likely to leak.

For applications involving solids-bearing liquids, flushing provisions and/or diaphragm isolation may be needed. Connections should be at least 25 mm (1 in.) in diameter. Diaphragm flanges are available with integral flushing connections or diaphragm extensions to extend the sensing diaphragm into the process.

TEST YOUR KNOWLEDGE

1. __________ pressure is a pressure that is referenced to atmospheric pressure.
   a. Absolute
   b. Differential
   c. Gauge
   d. Switch

2. __________ is a force distributed over an area.
   a. Proximity
   b. Speed
   c. Flow
   d. Pressure

3. The pressure exerted by a column of liquid is calculated by multiplying the gravitational acceleration times the height of the liquid column times the __________.
   a. Dielectric constant
   b. Density of the liquid
   c. Electrical current
   d. Capacitive potential

4. Gas inside a sealed container will exert pressure
   a. Varying by location in container
   b. The same in all directions
   c. The same as a liquid
   d. Varying by density gradient

5. At sea level, atmospheric pressure would be equal to
   a. 0 kPaG (0 psig)
   b. 0 psia
   c. 0 atm
   d. Perfect vacuum

6. Pressure __________ are flat ceramic or corrugated metal disks that deflect when subjected to changing pressures.
   a. Switches
   b. Transmitters
   c. Diaphragms
   d. Seals

7. __________ on gauge pressure sensors must be kept open and dry so that atmospheric pressure can reach the reference side of the sensor.
   a. Vent tubes
   b. Electrical switches
   c. Actuators
   d. Transmitters

8. Which fluid is used in fluid-filled pressure instruments that operate in extreme cold conditions?
   a. Glycerin
   b. Salt water
   c. Halocarbon
   d. Silicone

9. The best thing to clean pressure sensors with is
   a. A dishwashing scrub pad
   b. Steel wool
   c. Soapy water
   d. A bottle brush

10. __________ occurs in pressure instruments when an instrument reads different values for the same pressure condition when measuring increasing pressures versus when reading decreasing pressures.
    a. Span adjustment
    b. Hysteresis
    c. Temperature correction
    d. Zero suppression

11. Block and bleed valves are used when a pressure instrument is
    a. Isolated and drained
    b. Used with corrosive chemicals
    c. Subjected to pressure spikes
    d. Disassembled and rebuilt

TEMPERATURE

The following three types of temperature sensors are typically used at WRRFs: thermocouples, resistance temperature detectors (RTDs), and thermistors. Thermocouples are rugged and inexpensive. Resistance temperature detectors are used when accuracy is important. Thermistors are often used where temperature compensation of flow and level meters are needed because they have a large response to temperature changes and they do not need a reference point.

PURPOSE AND FUNCTION

Temperature instruments monitor and/or transmit temperature data from the point where the temperature sensor is in a piece of equipment or in a process.

APPLICATIONS IN WATER RESOURCE RECOVERY FACILITIES

Temperature is monitored and controlled as an important process control parameter in anaerobic digesters and some other types of biological reactors such as deammonification reactors. Boiler temperatures are monitored and regulated. Heat exchangers will have inlet and outlet temperature monitoring. We also monitor temperature on individual pieces of equipment to protect them from damage. The temperature at the rotor/stator interface on a progressing cavity pump will be monitored to prevent overheating and damage if the pump is run dry by mistake. See Figure 12.20. Motors winding temperatures are monitored so that they can be shut down if they overheat. Changes in bearing operating temperatures can warn us about a bearing issue before a failure occurs. This is particularly important in larger, high-speed equipment such as dewatering centrifuges.

Figure 12.20 Temperature Sensor (Thermocouple) Installed on a Progressing Cavity Pump (Reprinted with permission by Passaro Engineering, LLC)

HOW THEY WORK

Thermocouples are made by welding two different metal wires together at the sensing point of the instrument (see Figure 12.21). The wires are two dissimilar metals (metal alloys that have very different thermoelectric properties [i.e., they react differently when they are heated]). The electrons flow away from the heated tip at different rates, which sets up a self-powered electrical current that is called the *Seebeck effect*. It is measured in millivolts across the cold junction of the instrument. Thermocouples are relatively inexpensive and moderately accurate. They are also durable and work well over a broad temperature range.

Resistance temperature detectors are simple instruments that require an external current or voltage source to function. They can be made with 2-, 3-, or 4-wire arrangements; the added wires increase the precision of the instrument. Platinum RTDs are the standard because they sense a wider temperature range and have better repeatability, but other materials such as copper and nickel are also available. Resistance temperature detectors are highly accurate and can operate over a very wide temperature range of –196 to 420 °C (–321 to 788 °F). The drawback to using RTDs is that they are very sensitive to vibration and any impacts or jostling.

A thermistor is a solid-state device, the resistance of which changes as its temperature varies. It is a rugged sensor that is not susceptible to damage from vibration and minor impacts such as being dropped or bumped. Thermistors have good sensitivity and accuracy characteristics. Their main limitation is that their temperature ranges are typically limited to 100 °C (180 °F).

INSTALLATION AND OPERATION

All temperature sensors inserted into process streams need protection. Thermocouples, RTDs, and thermistors typically are inserted to a thermowell mounted in contact with the process fluid. A thermowell is a thin-walled metal tube capped at the process end. Temperature sensors can be spring-loaded to ensure positive contact with the thermowell wall to minimize sensing lag time. Conductive paste can be used to coat the temperature sensor in the thermowell to make heat transfer more efficient. The conductive paste must be selected for the instrument's temperature range. Thermowells are typically constructed of carbon steel or stainless steel; however, manufacturers offer a wide variety of materials depending on the application's temperature, corrosiveness, abrasiveness, or any other special process environment.

In some applications, it is not possible to mount the thermal sensor through the tank or pipe wall for direct contact with the process material being measured. The temperature probe is mounted externally without a thermowell; however, higher measurement errors and slower response need to be factored into using these measurements to control a process temperature.

Temperature sensors must be protected from corrosive or erosive environments because physical changes in the sensor will result in permanent damage. Thermocouple sensors can also lose calibration if wires become annealed at high temperatures. Annealing happens when a temperature shock occurs (e.g., removing the sensor from a hot process directly into a cold ambient temperature). It can also be caused by excessive vibration or wires installed under tension.

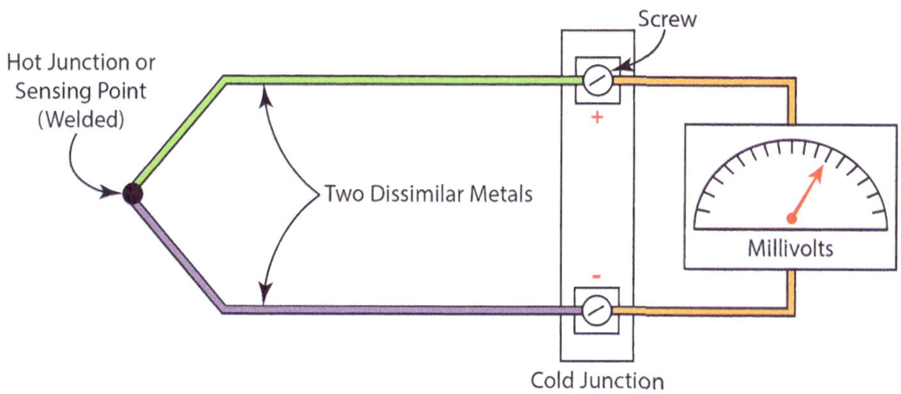

Figure 12.21 Diagram of a Thermocouple

Table 12.1 Accuracy and Repeatability of Temperature-Measurement Devices (WEF, 2013)

Device	Accuracy	Relative Repeatability
Thermocouple	1% (0.1 °C [0.18 °F])	Poor
Resistance Temperature Detector (4-wire)	0.5% (0.05 °C [0.09 °F])	Good
Thermistor	0.1 °C (0.18 °F)	Medium

Care must be taken to not stress or twist the temperature sensor during installation. Do not bend the element. If welding is required, make sure to carefully remove the temperature sensor from the thermowell before welding. The accuracy and precision of temperature-measurement devices depend on proper installation. Table 12.1 shows typical accuracies for properly installed equipment. Thermocouple accuracy also depends on the type and size of element used and the fit of the curve used to convert millivolts to temperature.

MAINTENANCE AND TROUBLESHOOTING

Temperature sensors and thermowells should be regularly wiped clean and inspected. Temperature readings should be verified periodically and outputs calibrated as needed. A resistance decade box or similar resistance standards are typically used to simulate temperature sensor inputs to temperature instruments.

TEST YOUR KNOWLEDGE

1. **The most accurate type of temperature sensor is**
   a. Thermistor
   b. Thermowell
   c. Resistance temperature detector
   d. Thermocouple

2. **__________ is the temperature sensor that is the most rugged (can handle more vibration and minor impacts).**
   a. Thermistor
   b. Thermowell
   c. Resistance temperature detector
   d. Thermocouple

3. **Thermowells can be filled with __________ to increase the sensor's response time.**
   a. Salt water
   b. Conductive paste
   c. Electrolyte solution
   d. Spray foam

4. **Resistance temperature detectors are often made out of __________ because this provides the broadest temperature range and the best repeatability.**
   a. Gold
   b. Platinum
   c. Copper
   d. Nickel

LEVEL

Like the other types of instruments discussed earlier in this chapter, there are two general categories for level measurements: level switches that take single point measurements and level transmitters that take continuous level measurements.

LEVEL SWITCHES

In this section, we will cover two types of level switches that are often used in WRRFs: float switches and conductivity switches.

Float Switch

Purpose and Function

Float switches are used to open or close a circuit when a specific liquid level in a tank, vessel, or structure is reached.

Applications in Water Resource Recovery Facilities

Float switches are often used to turn pumps on and off, for example, in a sump. A float switch would turn the pump on when the sump is full and then another float switch would turn the pump off when the sump level is low, before the pump runs dry and is potentially damaged. Float switches are also often used to trigger flooding alarms or high-level alarms when water starts to accumulate in areas that should be dry.

How They Work

Floats are simple devices that use microswitches or reed switches to open or close circuits to provide controls or alarms based on a certain liquid level being reached. Mercury switches used to be common but are not used anymore to avoid the risk of a mercury leak and contamination. A float switch is in its normal position when it is hanging straight down. As liquid level increases around it, the float (which is buoyant) begins to float upward. When the float reaches a horizontal position, it is activated and the switch changes status. Often, weights are added to the float cable to ensure that it is pulled straight down while the float moves upward with the liquid level and the float activates at the same elevation every time. See Figure 12.22.

Installation and Operation

Overall, float switches are easy to install and adjust and they are cost-effective. Installation involves hanging the float at the right elevation so that when it is pushed up by the liquid level, it reaches a horizontal position, which activates the switch at the desired elevation. Wet wells or tanks with considerable turbulence and mixing can cause floats to bounce around, sometimes causing the switch to activate by mistake. Adding weights to the cable can help avoid this situation. In some applications, the floats are installed within a stilling well where the level still rises and falls and there is plenty of room for the float to move to its horizontal position, but the turbulence is blocked.

Floating debris or heavy foam or scum can prevent a float from being activated as designed. Floating debris will need to be removed from the tank. Sprayer systems can sometimes help to keep scum and foam from building up excessively on the surface of a process tank. In some installations, stilling wells can help isolate debris, foam, and scum from the floats as well.

Maintenance and Troubleshooting

Rags and stringy material can become wound around the cable and the float and prevent the float from moving properly as the water level goes up and down. It is important to remove this debris as soon as it begins to collect on the floats. Cleaning will also reduce the amount of odor that is generated in the tank or wet well.

When inspecting a float, check that there is no visible damage to the cable or to the float itself.

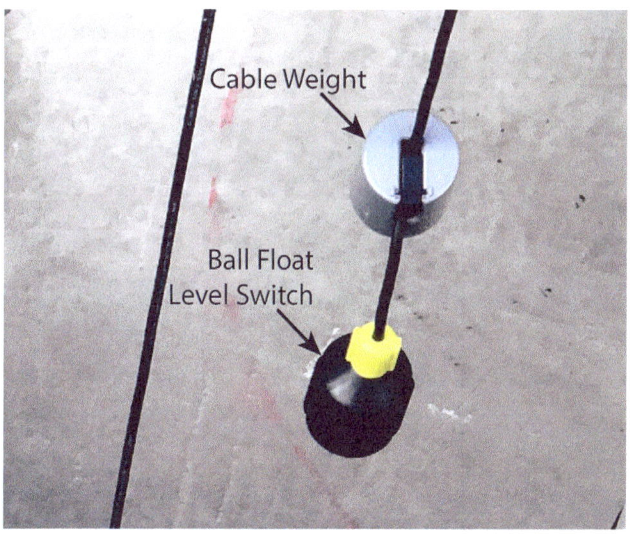

Figure 12.22 Float Switch with Cable Weight (Reprinted with permission by Indigo Water Group)

Conductivity Switches

Purpose and Function

Every existing or created substance is conductive to some degree. For instance, glass has low conductivity and copper has relatively high conductivity. Conductivity switches are used to detect a change in conductivity in the area where it is installed.

Applications in Water Resource Recovery Facilities

Conductivity is used in many ways in wastewater instruments. In this section, we are focusing on how a change in conductivity can be used to detect the level of a liquid in a tank or channel.

How They Work

Conductivity switches are simple instruments with no moving parts. Rods are used to measure depths up to approximately 4 m (13 ft), and ropes with conductance probes on the ends can be used to measure up to 15 m (50 ft). The sensors are typically made of stainless steel. For highly corrosive liquids, the rods or probes can be made from corrosion-resistant materials such as Monel® (nickel-copper alloy), titanium, and Hastelloy® C.

The sensitivity of the instrument is adjustable for use with different process liquids. The minimum conductivity of the liquid to be sensed is 10 µS/cm.

While the liquid is below the end of a rod, the air acts as a dielectric (i.e., an insulator), preventing any flow of electricity. However, when the liquid reaches the end of the rod, the liquid is then able to conduct electricity and complete the circuit. In single-rod or rope configurations, the circuit is completed through the conductive liquid to the grounded tank. See Figure 12.23. If the tank cannot be grounded, two rods or ropes of the same length are used to create a completed circuit when the conductive liquid reaches the bottom of the rods/ropes.

Conductivity switches can be configured to provide up to five-point switching with a single instrument. Tasks such as pump on-and-off control, dry run protection for pumps, and chemical fill tank overfill protection can be handled by a conductivity switch. They are simple to install and cost-effective. All wetted parts must be compatible with the liquid being sensed. Over time, the sensors will erode.

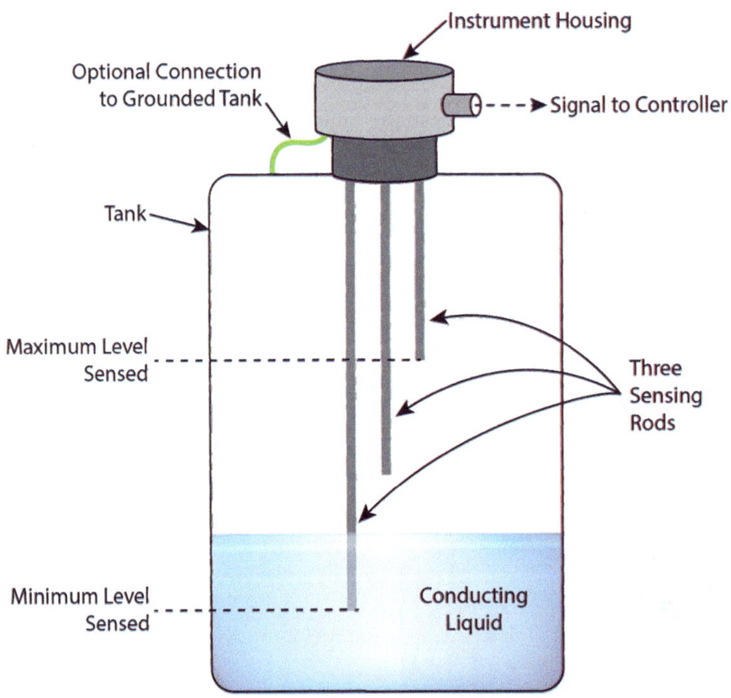

Figure 12.23 Conductivity Switches

Installation and Operation

Conductivity sensors can be installed in any orientation within a process using manufacturer-provided insulators to maintain spacing for bare electrodes. It is important to maintain at least a 3-cm (1-in.) clearance between bare sensors and conductive process pipes or tank walls.

Conductivity sensors typically have errors of less than 1.0% and precision of less than 0.2% over their rated temperature range, provided they have automatic temperature compensation.

Maintenance and Troubleshooting

Conductivity switches typically require little maintenance other than occasional cleaning to remove buildup on the sensors. If deposits and buildup on the sensor do not easily wash away, it may be necessary to soak it in a dilute acid solution before cleaning. Always verify the recommended approach with the manufacturer's O&M manual before beginning the work.

LEVEL TRANSMITTERS

Level transmitters are used where continuous level signals are required. Bubbler systems used to be a very common way to measure liquid levels in wet wells or tanks. However, due to the need to maintain the compressor system, monitor and repair leaks in the air lines, and regularly replace the bubbler tubing, these systems have fallen out of favor. In this section, we will cover three types of level transmitters that are often being installed in WRRFs today: ultrasonic, radar, and pressure-sensing level instruments.

Ultrasonic Level Transmitters

Purpose and Function

Ultrasonic level instruments provide noncontact measurement of a liquid surface level.

Applications in Water Resource Recovery Facilities

Ultrasonic level transmitters are used in a variety of ways in a WRRF. They can be used to sense the level inside a chemical tank. See Figure 12.25. They are often used with Parshall flumes or weirs to detect the upstream liquid level that is used to calculate the flow through the flume or over the weir. They are used to monitor filter levels to trigger backwash cycles when a setpoint level is reached. Ultrasonic sensors can be located upstream and downstream of an influent screen to monitor the headloss across the screen.

How They Work

Ultrasonic level instruments are referred to as "time-of-flight" instruments. The element (or sensor) contains a horn that creates pulses of very high frequency sound waves, much higher than the human ear can hear. It releases the sound waves toward the liquid surface and then records how long it takes for the

Figure 12.24 Ultrasonic Sensor Measuring Channel Level (Reprinted with permission by Richard Weigand, CET)

waves to reflect off the surface and return to the sensor. See Figure 12.25. Because sound waves travel at a constant speed, the transmitter can convert the time measurement into a distance travelled. Because the dimensions of the tank and the position of the sensor were entered when the instrument was originally put into service, the transmitter has all of the information needed to output a continuous signal showing the depth of water in the tank.

Sensors are typically provided with polyvinylidene fluoride (PVDF) or Teflon® (polytetrafluoroethylene [PTFE]) facings to resist corrosion.

Ultrasonic waves can be absorbed if they pass through airspace that contains vapors. Foam will also absorb ultrasonic waves. Applications with these conditions, such as inside anaerobic digesters, are not ideal for this technology.

Installation and Operation

During initial setup, the distance from the sensor to the low water level (0% full) and the distance from the sensor to the highest tank level (100%) of the tank are entered. The output signal for the transmitter is calibrated to these levels.

Ultrasonic sensors must be installed so that the sound waves hit the target surface perpendicular (90° angle). If the mounting bracket is not square and level, sensor readings will not be accurate.

Ultrasonic sensors have deadbands or blanking distances. They cannot sense a surface that is too close to the sensor because they cannot send the pulse and receive the pulse at the same time. Ultrasonic elements typically have a deadband (or blocking distance) of 0.07 to 0.3 m (0.2 to 1.0 ft). Always verify the deadband for a specific instrument with the instrument's manufacturer.

The sensor should not be mounted close to the filling point of the tank.

There should be no internal tank structures located within the path of the ultrasonic sound waves (signal beam). This will disrupt the wave path and reflection and cause performance issues.

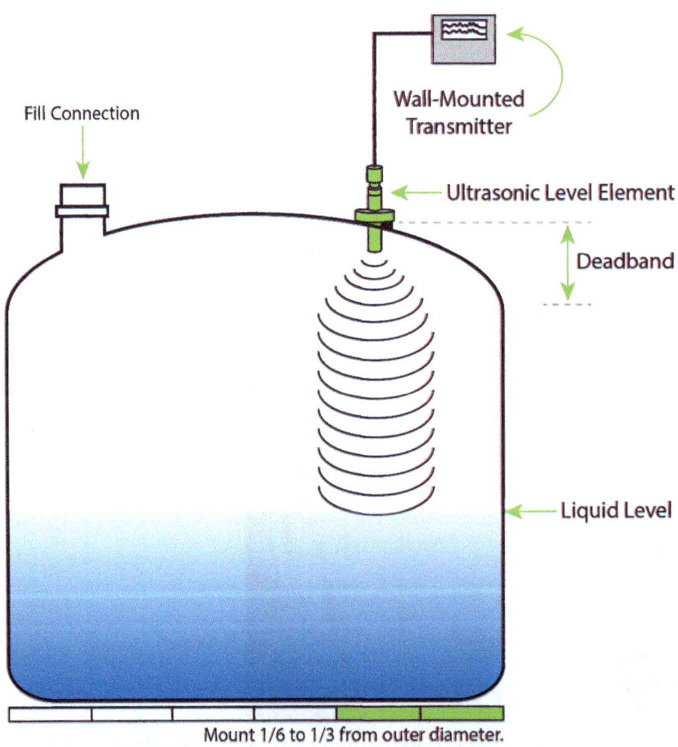

Figure 12.25 Ultrasonic Level Measurement in a Tank

The sensor should not be installed in the middle of a domed top tank. This position can cause secondary sound wave echoes that bounce off the top and create sensor performance issues. Depending on the total diameter and depth of the tank, it is recommended to install the sensor one-sixth to one-third of the diameter of the tank in from the side. Follow the installation recommendations of the instrument manufacturer.

Ultrasonic sensors require a smooth, calm surface to properly reflect the waves back to the sensor. In turbulent areas, stilling wells may need to be provided.

As with all instruments, make sure ultrasonic level instruments are properly grounded. For installations in hazardous locations, additional precautions must be taken.

Ultrasonic level instruments that are installed correctly in the appropriate applications typically have errors less than 0.25% and are repeatable within 0.1% of span.

Maintenance and Troubleshooting

Brackets and mounting hardware can loosen over time and slip. Periodically check to make sure that the sensor head is mounted securely and the ultrasonic waves will travel perpendicular to the liquid surface being measured.

There must be a clear path from the sensor to the liquid surface. Sensors should be periodically cleaned and inspected.

Noncontact Radar (Microwave) Level Transmitters

Purpose and Function

Noncontact radar level instruments also use "time-of-flight" calculations to determine the level of a liquid in a tank. They use two different sensor designs: noncontact sensors or guided wave level sensors. We will cover the use of noncontact radar instruments in this chapter.

Applications in Water Resource Recovery Facilities

Radar level transmitters can be used in many of the same installations in a WRRF as ultrasonic level instruments are used. They do tend to be more expensive than ultrasonic level instruments. Radar sensors can be configured to overcome some vapor and foam issues that cause problems for ultrasonic sensors. For example, low frequency radar level sensors have been successful at measuring level inside an anaerobic digester.

How They Work

Noncontact radar sensors send out radar wave signals. The antenna portion of the sensor detects the radar waves that have reflected off the liquid surface and returned back to the sensor. The waves travel at the speed of light. The time that is measured from when the signal is released until it returns to the antenna is proportional to the distance of the reflective surface. See Figure 12.26.

Because reflectivity of the liquid surface is a key parameter in the sensor's performance, it is better to measure liquids with high dielectric constants because they tend to reflect better and enable the sensor to measure longer ranges.

What is a dielectric constant? It is a unitless number that compares the electric permeability of a material (gas, liquid, solid) with the electric permeability of a vacuum. A dielectric is an electrical insulator. A material's dielectric constant is a measure of how well a material insulates, or a measure of how well it stores electrical energy. A material that is a good electrical insulator will have a high dielectric constant.

Larger antenna cone diameters can detect weaker return signals, so the largest antenna size that will work for a specific installation should always be used. Larger antennae also produce smaller signal cone (beam) angles, which reduces the amount of interference picked up from other internal structures in the tank.

In processes where vapors and condensation might be present, low-frequency radar sensors are recommended. Foam can cause some issues for noncontact radar instruments. With light, airy foam, the radar will "look" through it and measure off the more reflective water surface. Thick, dense, frothy foams can cause inconsistent water surface measurements. Lower frequency radar instruments have more reliable performance when foam is present.

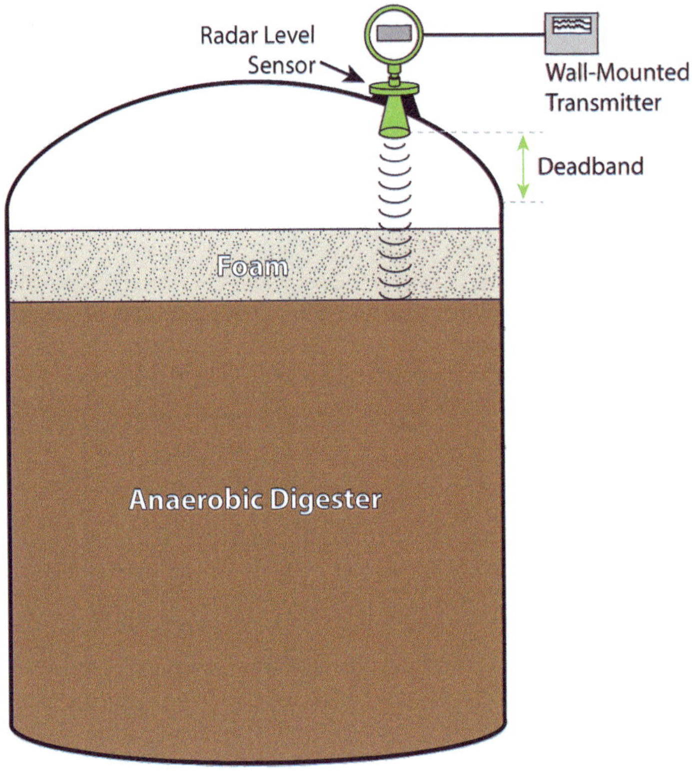

Figure 12.26 Noncontact Radar Level Instrument

Installation and Operation

During initial set up, the distance from the sensor to the low water level (0% full) and the distance from the sensor to the highest tank level (100%) of the tank are entered and the output signal for the transmitter are calibrated to these levels.

Radar sensors must be installed so that the radar waves hit the target surface perpendicular (90° angle). If the mounting bracket is not square and level, sensor readings can be affected. The antenna should be installed to extend into the tank below the mounting pipe/nozzle.

Radar sensors have deadbands or blanking distances immediately below the sensor where they cannot sense a surface because they cannot send the pulse and receive the pulse at the same time. Radar elements tend to have larger deadbands (or blocking distances) than ultrasonic sensors that can range from 0.15 m (0.5 ft) to more than 0.6 m (2.0 ft). Always verify the deadband for a specific instrument with the manufacturer.

The sensor should not be mounted close to the filling point of the tank.

There should be no internal tank structures located within the path of the radar waves (signal beam). Be aware that if the instrument is installed in a non-metallic tank, the walls will not be detected, but objects outside of the tank may be.

The sensor should not be installed in the middle of a domed top tank. This position can cause secondary echoes of radar waves that bounce off the cover of the tank and create sensor performance issues. Depending on the total diameter and depth of the tank, it is recommended to install the sensor one-quarter of the diameter of the tank in from the side. Follow the recommendations of the instrument manufacturer.

Radar sensors require a smooth, calm surface to properly reflect the waves back to the sensor. In turbulent areas, stilling wells may need to be provided.

Generally, condensation does not affect radar signals. If the sensor is operating in a cold area inside a warm process tank, insulation can be added to the sensor to reduce the temperature difference at the sensor and the amount of condensation that forms.

As with all instruments, make sure radar level instruments are properly grounded. For installations in hazardous locations, additional precautions must be taken.

Radar level instruments that are installed correctly in the appropriate applications typically have an accuracy of ±2 to 10 mm (0.08 to 0.4 in.) and are repeatable within ±1 mm (0.04 in.).

Maintenance and Troubleshooting

Sensors should be periodically cleaned and inspected.

Sensors must be installed perpendicular to the liquid surface they are sensing. Periodically check to make sure that the sensor is properly secured, level, and square.

There must be a clear path from the sensor to the liquid surface. Keep the content of the tank as free from debris, scum, and foam as possible.

Pressure-Sensing Level Transmitters

Purpose and Function

Pressure-sensing instruments can be used to calculate the height of a liquid column above the sensor. If a tank is enclosed and the top portion of the tank contains pressurized gas, a differential pressure instrument must be used to calculate the liquid level.

Applications in Water Resource Recovery Facilities

Submersible transducer-style pressure sensors are often used to sense wet well levels for pump control. Pressure-sensing level transmitters mounted through a pipe fitting in the tank wall are commonly used in WRRFs to sense the level inside a tank, such as a chemical storage tank or an anaerobic digester. See Figure 12.27.

Figure 12.27 Pressure-Sensing Level Transmitter

How They Work

In the pressure section of this chapter, we reviewed the basics of how pressure instruments work. The equation for pressure is as follows:

$$\text{Pressure} = \rho g h \tag{12.3}$$

Where

ρ = the density of the liquid
g = acceleration due to gravity (9.8 m/s^2)
h = height of the liquid above the point where the pressure is being calculated

Did you know that water is at its highest density at 4 °C (39.2 °F)? Ice is less dense that liquid water—that is why it floats. Warm water is less dense than cold water. For most calculations, we use the standard water density at room temperature of approximately 1.0 kg/L (8.34 lb/U.S. gal). When more precise numbers are needed, this will need to be adjusted for temperature.

When we have clean water in a tank, we know the density is 1.0 kg/L (8.34 lb/U.S. gal). Therefore, we can calculate that each meter of water height above a gauge will add 9.79 kPa (each foot of water height above a gauge will add 0.433 psi) of pressure to a gauge reading. See Figure 12.28.

When the tank is holding something other than clean water, the pressure increase must take into account the density of the material in the tank. Also, temperature effects on the density need to be taken into consideration. Considering how important density is to the pressure readings, it is obvious that measuring tank levels with a pressure-sensing instrument is a lot simpler if the liquid contents of a tank have consistent density characteristics. However, anaerobic digester content characteristics can change significantly. The more accurate the density numbers entered for the liquid are, the higher the accuracy of the level calculation.

Pressure-sensing level transmitters are very effective at measuring liquid levels in a tank, even if foam is present. This is because foam is mostly made up of air. It adds a negligible amount of weight to the liquid column so it does not affect the water level calculations based on the pressure readings.

Installation and Operation

There are two basic kinds of pressure-sensing level instruments that we will cover: the pressure transducer that is immersed within a tank and the pressure transmitter that is mounted on the outside of the tank with a pressure-sensing diaphragm measuring the pressure.

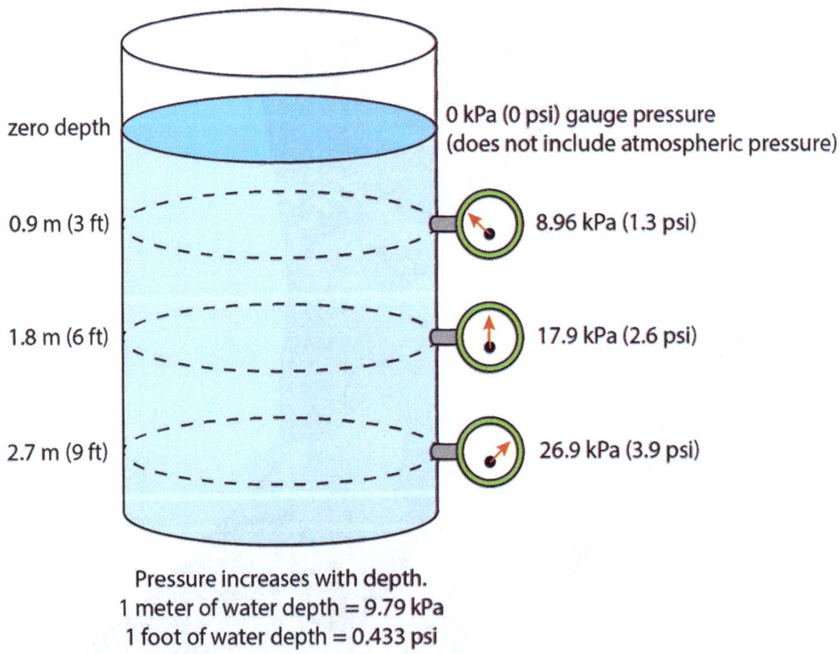

Figure 12.28 Pressure Increase With Water Depth

The submersible pressure transducer is the pressure-sensing element at the end of a cable. It is lowered down into the liquid that is being measured. When the transducer is at the proper location, the cable is carefully secured at the top to maintain the proper location. For examples of different styles of submersible pressure transducers, see Figure 12.29 and Figure 12.30.

The submersible pressure transducer can only be used in tanks or wet wells where no mixing equipment is present. If needed, the transducer can be installed within a stilling well to help reduce the amount of turbulence it is subjected to. The transducer might also be weighted down to help prevent turbulence from affecting the sensor's performance. Transducers should not be permanently attached to a ladder or to a wall because it will be necessary to remove them from the process for cleaning and inspection.

The pressure transducer must be set at or below the lowest water level that must be measured and then calibrated for the required range. If the transducer is removed for inspection, the sensor should be recalibrated after it is reinstalled.

It is important that the transducer and cables be made of materials that are compatible with the process liquids being measured. Common materials for the transducer are stainless steel and Kynar®.

Pressure transducers that are installed correctly in the appropriate applications typically have a static accuracy of ±0.5% full scale.

The other type of pressure-sensing level transmitter is a unit installed outside of the tank where only the sensing diaphragm is in contact with the media being measured. If the tank is open at the top (atmospheric pressure), only one pressure sensor needs to be mounted at the bottom of the wall of the tank to read a gauge pressure. The process line is connected to the high-pressure side of the transmitter and the low-pressure side is vented. If the tank is closed and there is any gas pressure above the liquid, then an additional pressure sensor above the liquid level is needed so that the differential pressure can be used to calculate the water column height.

For most wastewater applications and especially for anaerobic digesters, it is recommended to use extended diaphragms on each sensor that allows a flush mount at the inside wall of the tank. Then, fill the sensing lines that are connected to the transmitter with a fill fluid that will transfer the pressure from the wall diaphragm to the transmitter while protecting the sensing lines from debris, blockages,

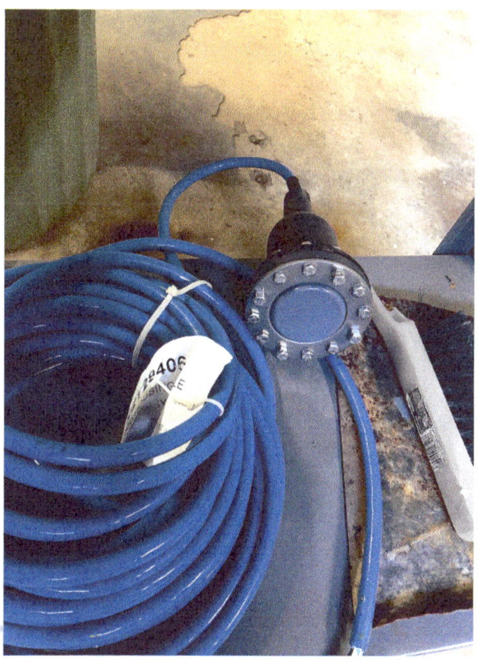

Figure 12.29 Pressure Transducer Style 1 (Reprinted with permission by Passaro Engineering, LLC)

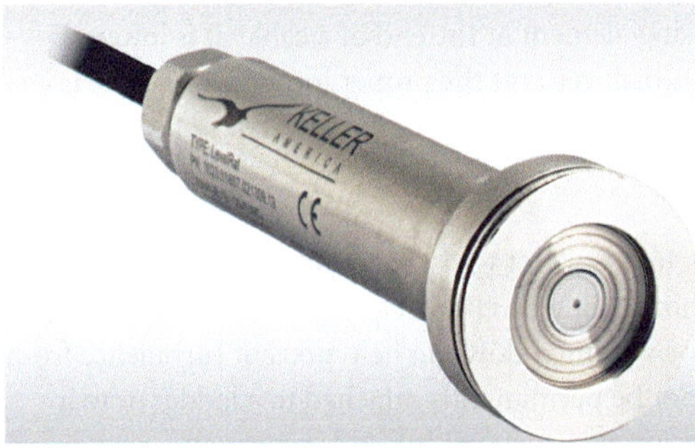

Figure 12.30 Pressure Transducer Style 2 (Reprinted with permission by Keller America, Inc.)

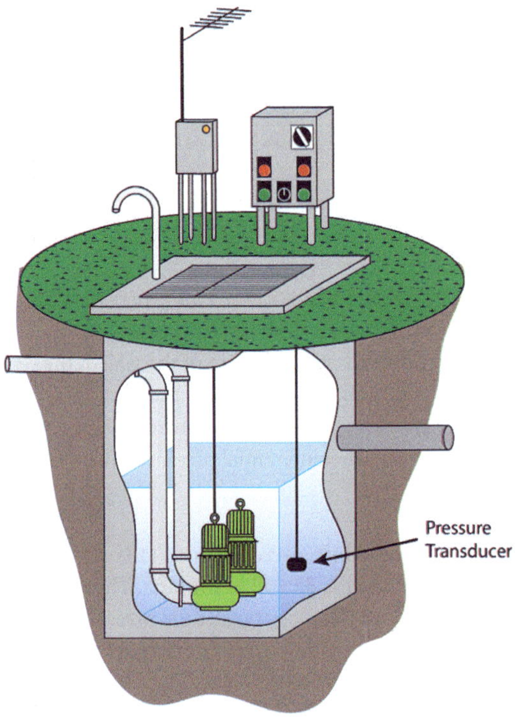

Figure 12.31 Pressure Transducer Submerged in Wet Well

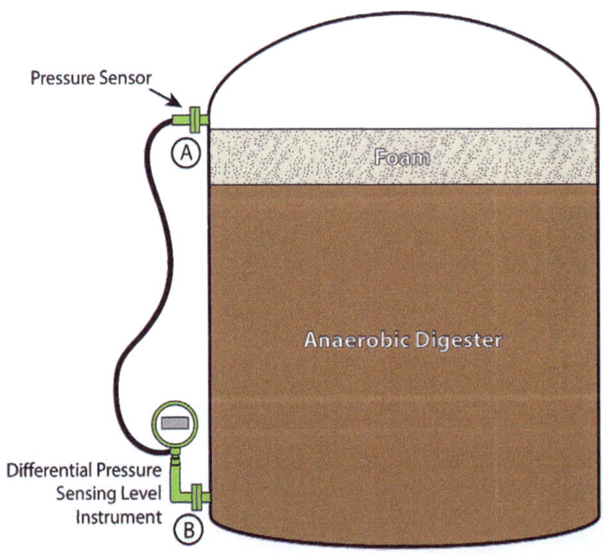

Figure 12.32 Differential Pressure-Sensing Level Instrument

condensation, and possible corrosion. The fill fluid should be selected based on the temperature range that the instrument will experience. Glycerin and water are commonly used as the fill fluids because they are most cost-effective. Silicone is effective for systems that experience a broader range of temperature extremes. Halocarbon is best when dealing with chlorine, oxygen, or other oxidizing process chemicals. All air bubbles must be removed from the fluid and the sensing lines. Follow the manufacturer's installation instructions.

During initial set up, the distance from the sensor to the low water level (0% full) and the distance from the sensor to the highest tank level (100%) of the tank are entered, and the output signal for the transmitter is calibrated to these levels.

Pressure-sensing level instruments that are installed correctly in the appropriate applications and have accurate liquid density inputs have accuracies of ±0.075% of span.

Maintenance and Troubleshooting

Perform regular calibrations at a frequency that will maintain the required performance of the instrument.

Brackets and mounting hardware can loosen over time and slip. Periodically check to make sure that the sensor and transmitter mounting hardware are tight and all components are square and level.

When the process tank is taken out of service for cleaning and maintenance, the sensing diaphragms should be very gently cleaned and inspected. Never scrape or scrub the diaphragm to remove deposits. Consult the manufacturer for recommendations on cleaning options if the diaphragms are not easily cleaned.

See the section covering pressure instruments earlier in this chapter for more information.

TEST YOUR KNOWLEDGE

1. **A __________ is activated when water level rise causes it to move from a vertical position to a horizontal position.**
   a. Pressure transmitter
   b. Flow switch
   c. Float switch
   d. Conductance transmitter

2. **__________ are often added to float switch installations to reduce the effect of turbulence in the wet well/tank and to help ensure the switch is consistently activated at the same elevation.**
   a. Buoyancy correction factors
   b. Cable weights
   c. Automatic flushing devices
   d. Cable retractors

3. **__________ can prevent a float from being activated at the proper liquid level.**
   a. pH fluctuations
   b. Temperature extremes
   c. Inconsistent liquid column density
   d. Floating debris and stringy material

4. **__________ both provide point level control.**
   a. Ultrasonic level instruments and float switches
   b. Noncontact radar and pressure-sensing level transmitters
   c. Conductivity and float switches
   d. Conductivity and impedance level transmitters

5. **Conductivity switches provide a noncontact level sensing.**
   ☐ True
   ☐ False

6. **Bubbler systems are a new technology that is becoming popular for continuous level measurement.**
   ☐ True
   ☐ False

7. **__________ uses high frequency sounds waves and "time of flight" to calculate a tank liquid level.**
   a. Ultrasonic level sensor
   b. Noncontact radar level sensor
   c. Conductivity switch
   d. Proximity switch

8. **The __________ on a "time-of-flight" sensor is created by the fact that the signal cannot be sent and received at the same time.**
   a. Bias
   b. Zero suppression
   c. Deadband
   d. Hysteresis

9. **The sensor on a "time-of-flight" instrument should not be installed in the middle of a domed cover because that placement can create __________.**
   a. Instrument grounding issues
   b. Secondary echo issues
   c. Interference with cover supports
   d. Magnetic field generation

10. **Radar and ultrasonic sensors must send signals perpendicular to the liquid surface.**
   ☐ True
   ☐ False

11. **__________ on a noncontact radar instrument increase(s) the range of the instrument.**
   a. Higher voltage power
   b. Heavier cable
   c. Larger antennae
   d. Stronger magnets

12. **When setting up the 4 to 20 mA output from a radar level sensor, the 50% tank level should have what mA output signal?**
   a. 4 mA
   b. 8 mA
   c. 12 mA
   d. 16 mA

13. **A tank full of clean water has an open top. A pressure-sensing level instrument is used to measure the water level. If the water level in the tank is 1.22 m (4 ft), what pressure will the transmitter be reading?**
   a. 11.93 kPa (1.73 psi)
   b. 14.82 kPa (2.15 psi)
   c. 19.34 kPa (2.81 psi)
   d. 23.48 kPa (3.41 psi)

14. **It is necessary to use a differential pressure transmitter when you are measuring level in**
   a. A tank that is open to atmospheric pressure
   b. A tank that is covered and vented
   c. A tank that is covered and pressurized
   d. A tank that is open and in a cold climate

15. **To measure continuous tank level, the submersible pressure transducer must be installed**
   a. Below the lowest tank level
   b. At the midpoint of the tank depth
   c. At the maximum tank level
   d. In a thermowell next to the tank

16. **Pressure-sensing systems in wastewater systems with diaphragm seals and sealed sensing fluid lines benefit from protecting sensing lines against blockages and preventing possible corrosion.**
   ☐ True
   ☐ False

FLOW

There are many types of flow meters, including electromagnetic, ultrasonic, flumes and weirs, differential-pressure, mechanical, and mass-flow. Although most require full pipe, a few technologies can be used in open pipes and channels. Table 12.2 provides an overview of typical flow meter characteristics.

Although some flow meters directly measure volumetric or mass flow, most compute the flowrate from velocity measurements. This requires accurate measurement of the flow area to determine a volumetric flowrate. For meters that are factory-installed in a spool piece, the diameter is under the control of the instrument manufacturer. For other types of meters, the area must be carefully measured. This is the pipe internal diameter for full pipes. Converting velocity to flowrate also requires a known flow profile, which requires adequate upstream and downstream lengths of straight pipe to allow stable performance of these types of meters. The effect of obstructions varies depending on the type of obstruction and the characteristics of the meter. Straight-pipe requirements can be reduced by inserting flow-conditioning elements in the pipe upstream of the meter.

For open-channel or partially filled pipes, a level measurement must be used to compute the flow area. Open channel flow measurement is covered in the beginning of Chapter 9, Laboratory Procedures, in Wastewater Treatment Fundamentals Book 2. In this chapter, we will focus on two of the more popular closed-pipe flow meter designs used in WRRFs: full pipe flow measurement with electromagnetic flow meters and gas flow measurement with thermal mass flow meters.

Table 12.2 Typical Flow Meter Characteristics (WEF, 2013)

Method	Type[a]	Fluid[b]	Range	Rev.[c]	Head Loss	Accuracy	Constraints
Magnetic	V	L	0.09–9 m/s (0.3–30 ft/sec)	Y	None	Excellent	Conductivity >5 microsiemens/cm
Ultrasonic transit	V	L	0.1–12 m/s (±0.5–40 ft/sec)	Y	None	Good	TSS < 5%
Ultrasonic doppler	V	L/G	±0.1–12 m/s (±0.5–40 ft/sec)	Y	None	Good	TSS < 0.2%
Pitot tube	V	L/G		N	Medium	Good	
Coriolis	M	L/G		Y	Low	Excellent	
Turbine	Vol	L/G		N	High	Good	
Thermal	M	L/G		N	Low	Good	More accurate with gases
Orifice plate	V	L/G		Y	High	Good	
Vortex	V	G		N	Low	Good	
Variable area	Vol	L/G		N	Medium	Good	

[a]V = velocity, M = mass flow, Vol = volumetric flow
[b]L = liquid, G = gas
[c]Rev. = reverse flow measurement, Y = yes, suitable for bi-directional flow

ELECTROMAGNETIC FLOW METERS

Purpose and Function

Electromagnetic flow meters (often referred to as *magmeters*) are currently the most popular method of measuring flow through a closed pipe. They have a long life, are cost-effective, and offer accurate and precise measurements when installed correctly. They also do not have any components that protrude into the flow path so they create no headloss. They also do not catch any rags and stringy material in the waste stream that would require extra cleaning.

Applications in Water Resource Recovery Facilities

Magmeters can be found on many systems in WRRFs. They often measure pumped discharge flow, especially where the pumps or variable frequency drives are adjusted by the WRRF's control system. Return sludge and waste sludge flows are often monitored and/or controlled with magmeters. See Figure 12.33. Feed pumps are another common application. If plant influent or effluent flows are pumped, they are often monitored and reported using magmeters.

How They Work

Electromagnetic flow meters are based on Faraday's principle of electromagnetic induction, in which the voltage induced by an electrical conductor moving through a magnetic field is proportional to the conductor's velocity.

$$E = k \times V \times B \times D \tag{12.4}$$

Where

E = electrical voltage
k = proportionality constant
V = velocity of fluid
B = magnetic field strength
D = distance between electrodes

Figure 12.33 Magnetic Flow Elements Installed on Sludge Lines (Reprinted with permission by Indigo Water Group)

What is an electrode? It is a point of transition. It conducts electricity, allowing it to enter or leave an object or substance. In magmeters, electrodes make electrical contact with the flowing fluid. They carry the electrical current that is generated by the water flowing through the magnetized flow tube.

The transmitter power supply energizes the magnetic coils that are installed around the flow tube. See Figure 12.34. This creates a magnetic field across the flow tube. If the process liquid has enough conductivity (a minimum of 5 microsiemens/cm), it will act as an electrical conductor and induce a voltage as it flows through the tube. As shown in Equation 12.4, the voltage produced is equal to the product of the magnetic field strength, the distance between the electrodes, flow velocity, and a proportionality constant. The induced voltage is measured by the electrodes and converted, using Faraday's equation above, into the velocity of the liquid moving through the flow tube. Flowrates can then be calculated based on the known tube size (cross-sectional area).

Electromagnetic flow meters work well in wastewater applications because there are no components that protrude into the process liquid flow. They do not create any headloss as the fluid moves through them. Flow tube liner and electrode materials must be selected correctly for best performance from a magmeter. Liners generally are available in PTFE, polyurethane, or neoprene. The electrodes are made of stainless steel or a nickel alloy. Temperature is one of the most important criteria for choosing these materials. They also must be compatible with the process fluid that is being measured. This is important during the original design of a system but should also be checked when process changes are considered, such as moving to a different type of chemical feed.

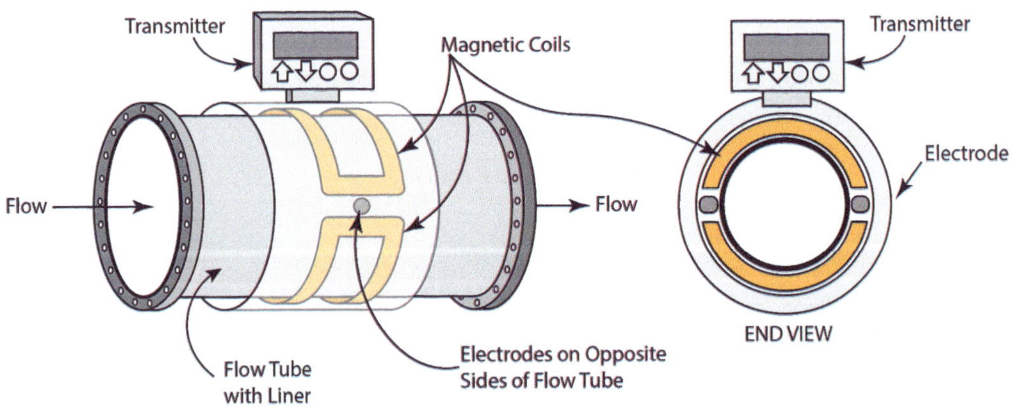

Figure 12.34 Electromagnetic Flow Meter Diagram

Installation and Operation

To provide stable and accurate measurements, almost all flow meter technology requires a specified distance of straight run before the meter and after the meter. Flow must be laminar through the flow element. If these guidelines are not followed, the meter's accuracy and repeatability could be negatively affected. Magmeters are less sensitive to turbulence through the meter than other flow meter designs. Most magmeter manufacturers recommend a minimum of 5 pipe diameters of straight pipe before the meter and 2 pipe diameters after the meter to ensure the most accurate readings.

What is laminar flow? It is a flow condition in a fluid where all of the particles are moving in parallel lines. In other words, there is no turbulence in the flow.

If the required straight run distances cannot be provided upstream and downstream of the flow meter, a flow conditioner may be available to help create laminar flow conditions. It is important to work with the flow meter manufacturer to determine if a flow conditioner might improve a magmeter's performance under specific installation conditions.

Magnetic flow sensors should be installed with the flow travelling in a specific direction. There is typically a flow direction arrow permanently marked on the sensor body. For meters that may experience reverse flow, the sensor should be installed with the flow arrow pointing in the primary flow direction. Some additional configuration may be necessary to allow the flow transmitter to properly function during reverse flow conditions and to have totalizer flows show correctly. See Figure 12.35. Refer to your O&M manual for details.

Magnetic flow elements need to be installed on a section of pipe that is always full of process liquid, with no trapped air bubbles. Flow tubes should also be free of any deposited sludge or solids particles. For these two reasons, it is best to install a magnetic flow meter on a vertical section of pipe where the flow is travelling in the upward direction. Horizontal installations also work well, but considerations should be made to make sure that the pipe is always running full, no air is being trapped in the flow tube, and that solids deposits are regularly cleared out of the flow element. Manufacturers typically recommend a minimum velocity of 0.9 m/s (3 fps) through the flow element to help prevent solids from building up on the bottom. Another issue that can occur is the buildup of coatings or deposits in the flow tube. For example, a fats, oils, and grease coating can build up in some process pipes. Special or more frequent cleaning may be required in certain situations. Self-diagnostic tests in some magnetic flow meter transmitters can identify and send error messages when solids or coatings are negatively affecting the

Figure 12.35 Flow Transmitter With Totalized Flow (Reprinted with permission by Passaro Engineering, LLC)

accuracy of the flow measurements. Although these diagnostics keep improving, it is still important for flow elements to be cleaned and inspected regularly, especially for flow meters that are relied upon for sensitive process control or permit reporting.

A gasket must be installed at each end of the flow meter where it attaches to the process piping. A gasket should also be installed on either side of a grounding ring. Magnetic flow meters are very sensitive to stray currents that might be present in the process flow. It is important to follow the manufacturer's recommendations on properly grounding the flow element.

All flow meters need enough space for maintenance, or meter body removal, either for calibration or for maintenance. Isolation valves located several pipe diameters away from the meter can be very helpful in limiting the amount of pipe that must be drained to remove a flow element from a process line. Bypass lines should be provided around critical flow meters where the process cannot be shut down. Flow elements should be easy to access.

Critical meters should be required to have full factory calibration documentation and verification of the meter's accuracy. Electronic verification units that check the performance of the electronic meter components are available from some manufacturers. They verify that the flow meter is performing the same as when the unit left the factory. This does not verify that the field flow reading is accurate.

Most magnetic flow meters have an accuracy of 0.5%.

If the meter is being used for control, repeatability is often a more important factor to consider than error. Repeatability in instrumentation is the maximum error for repeated measurements of the same sample, with all other conditions unchanged between measurements except time. In some control applications where meters are installed in less-than-ideal locations, an offset can be added in the control system to give the correct rate. A meter with good repeatability is often acceptable for control because of the consistent readings.

If the flow element is to be installed in an area that may be flooded, it is important to purchase a flow element with submergence protection. In these conditions, the transmitter must be a remote mount that is positioned above possible high-water level.

Maintenance and Troubleshooting

Magnetic flow meters have no moving parts and, therefore, require very little maintenance. There is nothing to lubricate and there are no moving parts that will wear over time. Make sure that the flow element is clean and clear of solids and air bubbles. Inspect the flow element transmitter and cables for any sign of wear or damage. If a flow meter is performing erratically, verify that the transmitter is receiving a stable power supply and that the meter is properly grounded. Refer to the troubleshooting section included in the manufacturer's O&M manual.

THERMAL MASS FLOW METERS

Purpose and Function

Thermal mass flow meters are insertion-type meters in which the sensor tip is made up of two parts: a heated element and a heat sensing element. Knowing the properties of the material flowing through the pipe allows the meter to calculate the mass flowing past the sensing element based on the rate of the dispersion of the heat. In WRRFs, this type of meter is used primarily to measure the flowrates of gases.

Applications in Water Resource Recovery Facilities

Thermal mass flow meters are often used to measure flowrates through aeration headers, digester gas systems, and natural gas lines.

How They Work

Thermal mass flow meters have quick response times and are effective at measuring fairly broad ranges of gas flows. They can also safely operate under highly corrosive conditions with combustible gases. Because the thermal properties of the gas being measured must be known and used to calculate the mass flowing through the pipe, it is important that the gas characteristics are consistent. This is true for the air in the aeration header and the natural gas coming from the utility. However, digester gas, also known as *biogas*, can be more variable and cause some accuracy issues that must be watched out for.

The flow element is inserted so that the sensor tip is located at the centerline of the pipe. See Figure 12.36. The sensor includes two RTDs—one of them is heated and the other is the sensing element that is used as a reference. There are two different approaches for using these elements to compute mass flow.

The constant temperature differential method keeps the temperature sensed at the gas temperature sensing RTD constant. The amount of heat dispersed will increase at higher flowrates and decrease at lower flowrates. The amount of electricity required to power the heated sensor and keep the temperature constant is used to calculate the mass flowrate of gas moving past the two sensors.

Figure 12.37 shows a more detailed closeup of the RTDs in the sensor tip.

The constant current method sends the same amount of electricity to the heated sensor and then monitors the change in the temperature detected at the gas temperature sensor RTD. The temperature will decrease at higher flowrates and increase at lower flowrates. The change in temperature and the thermal properties of the gas being measured are used to calculate the flowrate of the gas through the pipe at any given time.

Installation and Operation

The flow element must be positioned properly. Manufacturers typically stamp a flow direction arrow on the stem of the element. It is important that this arrow is properly aligned with the direction that the gas is flowing through the pipe. Also, the element must be installed properly so that the sensor tip is positioned in the center of the pipe. This positioning has a larger effect on performance for thermal mass flow meters installed in smaller pipes.

Flow transmitters can be mounted integral with the flow element or they can be remote-mounted nearby for better access and viewing the readouts on the indicator. See Figure 12.38.

Eddy currents and turbulence in the flow will negatively affect the performance of the flow meter. Manufacturers generally recommend a minimum straight run of 20 pipe diameters before the flow meter and 10 pipe diameters after the flow meter to minimize this effect. If this layout is not possible, a flow conditioner may be used to create the needed pipe flow profile at the meter.

Thermal mass flow meters typically have an accuracy of ±2% reading and a repeatability of ±0.5% reading.

Maintenance and Troubleshooting

Periodically, verify the output calibration and recalibrate as needed. This can be done using a resistance decade box. Follow the manufacturer's recommended procedure.

What is a Resistance Decade Box? It is a passive testing device that contains groups of switches arranged in "decades" that allow the user to select a combination of resistance values to add up to a wide range of different test values. For example, the first decade might have a 5-ohm resistor, a 2-ohm resistor, and two 1-ohm resistors. The second decade might have a 50-ohm resistor, a 20-ohm resistor, and two 10-ohm resistors. The third decade might have a 500-ohm resistor, a 200-ohm resistor, and two 100-ohm resistors. Depending on what needs to be tested, some decade boxes have six or more decades to allow a large range of test values to be selected. Users can "build" the manufacturer recommended total test resistance by selecting the right combination of individual resistors to use for the test.

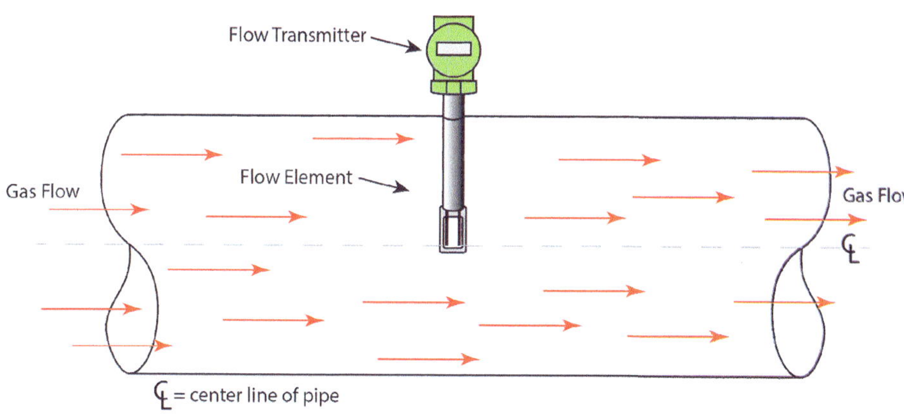

Figure 12.36 Thermal Mass Flow Meter Installation

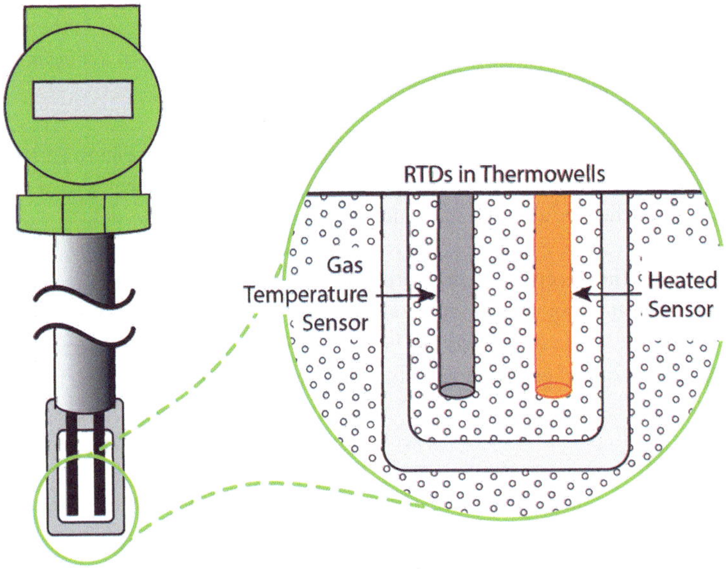

Figure 12.37 Thermal Flow Sensor RTDs

Periodically inspect the electrical connections; make sure the terminal connections are tight and there are no signs of corrosion. Inspect the moisture barriers and seals on the enclosure protecting the electronics and make sure that no moisture is entering the enclosure. Inspect electrical wiring for any sings of damage.

Check to make sure that there is no process gas leaking from the seals at the location that the sensor enters the pipe.

Periodically remove the flow element from the pipe and inspect it for any debris or scale buildup. Clean the flow element as necessary with a soft brush and, if needed, with a solvent that is compatible with sensor materials.

Figure 12.38 Digester Gas Flow Meter With Remote Indicating Transmitter (Reprinted with permission by Indigo Water Group)

TEST YOUR KNOWLEDGE

1. For proper operation, most closed-pipe flow meters required __________ flow through/past the flow element.
   a. Turbulent
   b. Laminar
   c. High-velocity
   d. Low-velocity

2. Where installation conditions do not allow for the minimum straight run of pipe requirements for a closed-pipe flow meter to be met, a __________ can be installed to improve the flow characteristics.
   a. Flow conditioner
   b. Flow classifier
   c. Velocity exciter
   d. Venturi port

3. In an electromagnetic flow meter, the liquid flow through the pipe acts as a(n) __________.
   a. Permanent magnet
   b. Grounding rod
   c. Electrical capacitor
   d. Electrical conductor

4. In an electromagnetic flow meter, the __________ of the liquid moving through the pipe is calculated from the measured induced voltage, the flow tube diameter, and the magnetic field strength.
   a. Temperature
   b. pH
   c. Velocity
   d. Turbulence

5. In an electromagnetic flow meter, the induced voltage is measured across the __________.
   a. Junctions
   b. Terminals
   c. Diaphragms
   d. Electrodes

6. Magmeters are capable of measuring reverse flows.
   - ☐ True
   - ☐ False

7. The flow element of a thermal mass flow meter contains a heated sensor RTD and a __________.
   a. Gas temperature sensing RTD
   b. Magnetic coil
   c. Electrode
   d. Density compensating element

8. __________ should be installed for critical flow meters that are installed on pipes that cannot be taken out of service.
   a. Isolation valves
   b. Quick disconnect couplings
   c. Backup generators
   d. Bypass lines

9. A __________ is a passive testing device that is set up with flexibility to select combinations of resistors to create required testing values.
   a. Block and bleed valves
   b. Resistance decade box
   c. Handheld calibration unit
   d. Magnetic resonance indicator

10. Thermal mass flow meters are more effective at measuring flowrates in gases than in liquids.
    - ☐ True
    - ☐ False

11. The flow element for a thermal mass flow meter must be installed in a specific orientation to the flow direction for the meter to operate properly.
    - ☐ True
    - ☐ False

WATER QUALITY PARAMETERS

The instruments in the following subsections measure water quality parameters including DO, chlorine residual, pH, ORP, conductivity, total suspended solids (TSS), interface/sludge blanket level, and turbidity. These instruments are typically available as bench-top, portable handheld, and in-process units. In each case, the instrument consists of a sensor, sometimes called a *probe,* and a processor with display. Bench-top and handheld units typically stand alone and do not report to SCADA or other automated data collection and management systems. The processor of a bench-top or handheld unit is commonly referred to as the meter. Figure 12.39 presents an example bench-top pH analyzer (meter). The probes (B) generally have shorter life spans than the meters (A) and may be purchased separately. In-process units, like the one shown in Figure 12.40, typically display instantaneous readings locally on a panel or other display that may also be capable of displaying historic data in tables or charts. In-process units often report to SCADA systems where data are archived for long-term storage and trending. Some systems are capable of managing multiple probes and sensors from one control panel, as shown in Figure 12.41.

The terms *sensor*, *probe*, and *meter* are often used interchangeably, especially when talking about bench-top and portable units. A probe is a small device, especially an electrode, that is inserted to something to test or measure a condition. In this section, examples of conditions are pH, ORP, and ammonia concentration.

Figure 12.39 Bench-Top pH Analyzer (Reprinted with permission by Indigo Water Group)

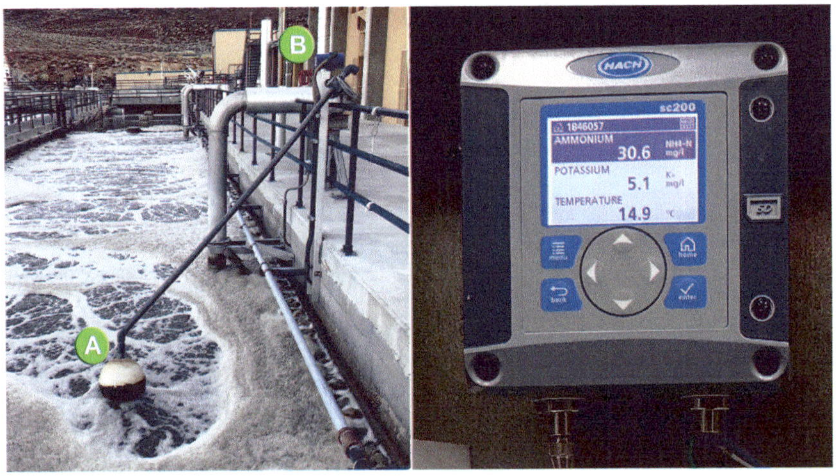

Figure 12.40 In-Basin Probe and Local Display (Reprinted with permission by Indigo Water Group)

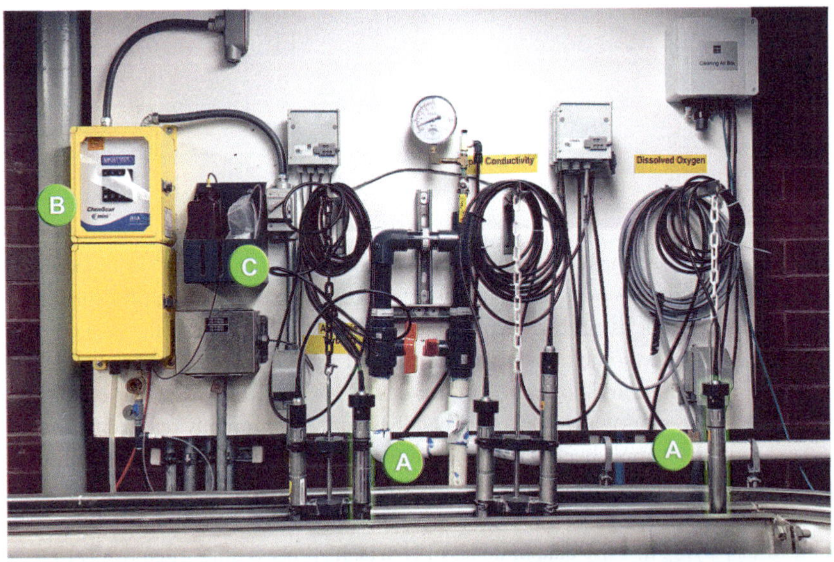

Figure 12.41 Multiple Probes With Local Display (Reprinted with permission by Indigo Water Group) (A) Individual probes, (B) Local control panel, (C) Reagents

DISSOLVED OXYGEN

Purpose and Function

Dissolved oxygen meters (also referred to as *DO analyzers*) quantify the concentration of DO in a liquid indirectly by comparing the amount of DO in the liquid to the maximum amount of DO that it can theoretically hold. This measurement is shown as percent of saturation (more on this later). For samples and processes that are open to the atmosphere, the concentration of DO in a liquid is related to three parameters: barometric pressure, temperature, and salinity. The DO sensor (also referred to as the *DO probe*) measures the percent saturation and then the processor (meter) calculates the DO concentration in milligrams per liter using the barometric pressure, temperature, and salinity of the water. That means that these parameters must either also be measured by the DO sensor or be measured separately and manually entered into the meter by the operator.

Two types of DO sensors (probes) are used in water and wastewater treatment: electrochemical and optical. Electrochemical probes include polarographic and galvanic types. The polarographic probe is more commonly called a "Clark electrode" after its inventor. Both polarographic and galvanic probes have a membrane that DO must travel through to be measured. Optical probes use a membrane coated with an oxygen-sensitive dye to measure DO. Oxygen molecules don't pass through the membrane. Instead, they interact with the dye molecules and light to produce a signal. Clark electrodes were once the most common type of DO probe used in wastewater treatment. Optical probes have grown in popularity since their introduction and are now the most common due to their low maintenance requirements.

Applications in Water Resource Recovery Facilities

Dissolved oxygen is typically monitored in biological wastewater treatment processes (e.g., aeration basins and aerobic digesters). These processes use aerobic bacteria to treat wastewater, break down organic compounds, and convert ammonia into nitrite and nitrate. Maintaining the proper DO concentration in the aeration basin is necessary to keep aerobic microorganisms alive. Too much aeration is wasteful and inefficient and can interfere with a facility's ability to remove nitrate. Too little oxygen can allow troublesome microorganisms to proliferate, resulting in poor process performance, foaming, and sludge settling problems.

Dissolved oxygen sensors may be used to control blower output and run time to maintain a desired DO concentration. They may also be used to monitor DO in the final effluent for permit compliance.

Salinity is the amount of salt present. Most people think of salt as the sodium chloride that they use in cooking. Technically, a salt is a chemical compound made up of positive and negative ions. Salinity includes sodium chloride plus other salts such as calcium sulfate and potassium hydroxide.

How They Function

Dissolved Oxygen Concentration

Dissolved oxygen probes measure the relative concentration of DO in a liquid. To understand what that means, we need to understand the difference between air bubbles and DO as well as how DO concentration is affected by other parameters. Air bubbles are pockets of air inside a liquid. Dissolved oxygen is free oxygen molecules dissolved into the liquid. The oxygen molecules fit in between water molecules, as shown in Figure 12.42. Oxygen contained within air bubbles is not yet dissolved. It must move out of the air bubble and into the liquid to be dissolved.

Barometric pressure is the pressure created by the atmosphere. At sea level, barometric pressure is 1 atmosphere or 101.3 kPa (14.7 psi). In Denver, Colorado, the mile-high city, barometric pressure is only 83.4 kPa (12.1 psi).

Three variables affect how much oxygen a liquid can hold: temperature, salinity, and barometric pressure. Figure 12.43 summarizes these relationships. As conditions change, the quantity of DO also changes.

When two variables are inversely proportional, they change opposite one another. When one increases, the other decreases.

Temperature affects oxygen solubility the most. Cold water can hold more dissolved gases—including oxygen—than warm water. If a warm can of soda is opened, the undissolved carbon dioxide will be released, causing the drink to foam up and out of the can. When a cold can of soda is opened, that doesn't happen to the same extent because the gas remains dissolved in the soda. As water temperature increases, the amount of gas the water can hold decreases. These two variables are inversely proportional. Table 12.3 presents some examples of DO solubility at different temperatures.

Salinity is measured in parts-per-thousand. Trace metals and some organic compounds are reported in parts per trillion, which is also shown as ppt.

As salinity increases, the amount of oxygen the water can hold decreases. Salinity is inversely proportional to oxygen solubility. One way to think about it is that the dissolved salts are taking up some of the space between water molecules that could have held oxygen. Seawater holds about 20% less oxygen than fresh water at the same temperature and pressure. Most modern DO sensors automatically compensate for

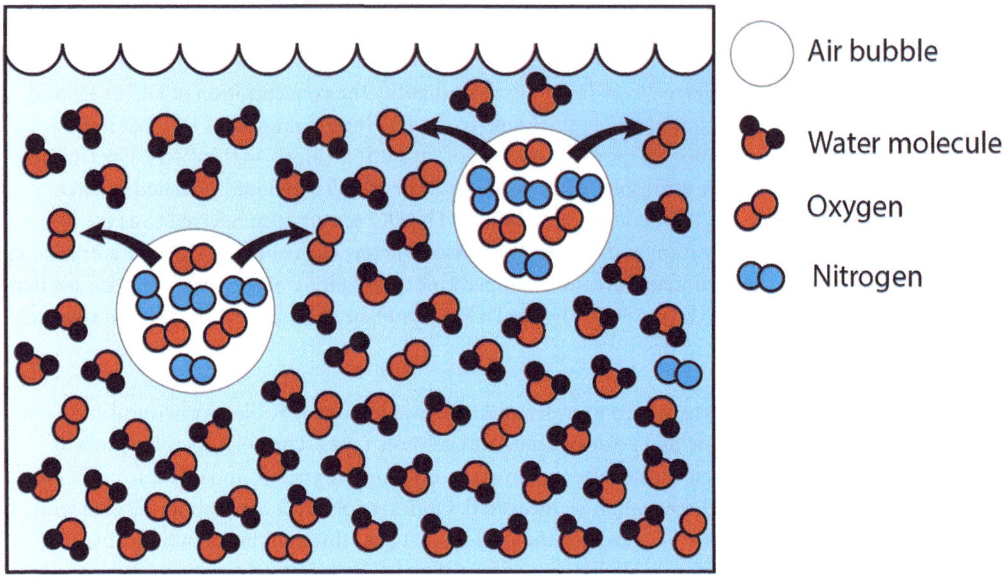

Figure 12.42 Oxygen Passing From Air Bubbles Into the Water

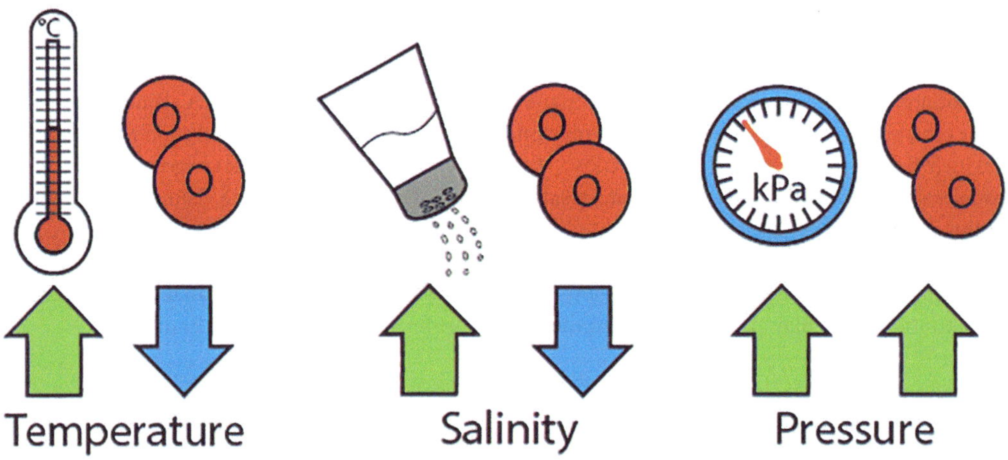

Figure 12.43 Solubility of Oxygen in Water (Reprinted with permission by Indigo Water Group)

Table 12.3 Oxygen Solubility Versus Water Temperature

Temperature	Oxygen Solubility at Sea Level and Zero Salinity
0 °C (32 °F)	14.62 mg/L
10 °C (50 °F)	11.29 mg/L
20 °C (68 °F)	9.09 mg/L
40 °C (104 °F)	6.41 mg/L

Note. Selections from YSI Dissolved Oxygen Tables (https://www.ysi.com/File%20Library/Documents/Technical%20Notes/DO-Oxygen-Solubility-Table.pdf)

salinity by simultaneously measuring the conductivity of the sample. If the sensor does not measure conductivity, the salinity will need to be entered into the meter manually. Salinity is entered in parts-per-thousand (ppt).

Unlike temperature and salinity, barometric pressure and oxygen solubility are directly proportional. As pressure increases, the amount of oxygen the water can hold also increases. As atmospheric pressure increases, the pressure of the oxygen in the atmosphere also increases. More oxygen molecules are able to push into the water. Some modern DO instruments have a built-in barometric pressure sensor that will automatically compensate for changes in barometric pressure. However, many modern and older instruments lack this feature. For these instruments, barometric pressure must be entered manually.

Water at every temperature, salinity, and pressure combination has some maximum amount of oxygen that it can hold. Water holding exactly the maximum amount of oxygen for that temperature, salinity, and pressure is said to be at saturation or is saturated. Water holding less than the maximum amount of oxygen for those conditions is undersaturated. Under some circumstances, water can become supersaturated. That is, it holds more oxygen than it should for a given temperature, salinity, and pressure. This can happen when DO that should have escaped to the atmosphere remains trapped in the water. This can occur when cold water is warmed without being disturbed. It can also occur in facultative ponds when algae produce an excess of oxygen in still water. Shaking or stirring the water will release the excess oxygen and the DO concentration will go down.

Dissolved oxygen is typically shown as milligrams per liter when operators are setting target DO concentrations for treatment processes and when recording measurements from DO sensors. Dissolved oxygen can also be expressed as "% saturation." We define 100% saturation as the amount of oxygen the water would hold at sea level and a particular temperature and salinity. To find the % saturation for other elevations, divide the barometric pressure at the instrument's elevation by the barometric pressure at sea level. For example, a DO sensor located at 1843 m (6047 ft) above sea level is exposed to a barometric pressure of 81.1 kPa (11.8 psi). The barometric pressure at sea level is 101.3 kPa (14.7 psi). The percent saturation for the higher elevation is 81.1 kPa ÷ 101.3 kPa • 100 = 80% (11.8 psi ÷ 14.7 psi • 100 = 80%). Table 12.4 presents examples of percent saturation at other elevations. Units of percent saturation are important when calibrating DO sensors.

Dissolved Oxygen Sensors

There are two categories of DO sensors used in wastewater treatment: electrochemical and optical. Electrochemical sensors include polarographic and galvanic types. Table 12.5 summarizes some of the differences between sensors.

Polarographic and galvanic sensors measure DO in the same way, but are made from different materials. Both sensor types contain an anode, a cathode, a conducting solution, and an oxygen-permeable membrane as shown in Figure 12.44. The anode and cathode may be compared to the negative and positive ends of a battery, respectively. When connected through a conducting solution, a complete

Permeable means the oxygen passes through the membrane.

Table 12.4 Pressure, Elevation, and Percent Saturation

Barometric Pressure		Elevation		
kPa	psi	Meters	Feet	% Saturation
101.3	14.7	0	0	100
91.2	13.2	880	2887	90
81.1	11.8	1843	6047	80
70.9	10.3	2909	9545	70

Note. Selections from YSI Dissolved Oxygen Tables (https://www.ysi.com/File%20Library/Documents/Technical%20Notes/DO-Oxygen-Solubility-Table.pdf)

Table 12.5 Comparison of DO Sensors (Data from GLI International, 2000; Hach Company, 2010; Hach Company, 2012; YSI Inc. /Xylem Inc., 2022)

Parameter	Electrochemical: Polarographic (Clark)	Electrochemical: Galvanic	Optical
Measurement Range	0 to 40 mg/L		0 to 20 mg/L (0 – 200% saturation)
Accuracy	±0.05% of range		Below 5 mg/L: ±0.1 mg/L Above 5 mg/L: ±0.2 mg/L
Interferences	Any substance that can be reduced at the cathode such as free chlorine or that coats the membrane (oils, biofilm). Hydrogen sulfide interferes by reacting with the electrode to form Ag_2S, which poisons the electrode.		Crude oils, Free chlorine residual
Membrane	Y	Y	N
Materials	Silver anode with gold or platinum cathode	Zinc anode with silver cathode	Light source, luminescent dye, detector
Fill Solution (electrolyte)	Y	Y	N
Fill Solution Type	Potassium chloride	Sodium chloride	None
Constant Voltage Applied?	Y	N	N
Consumes Oxygen?	Y	N	N
Requires Flow Across Sensor?	Y	Y	N

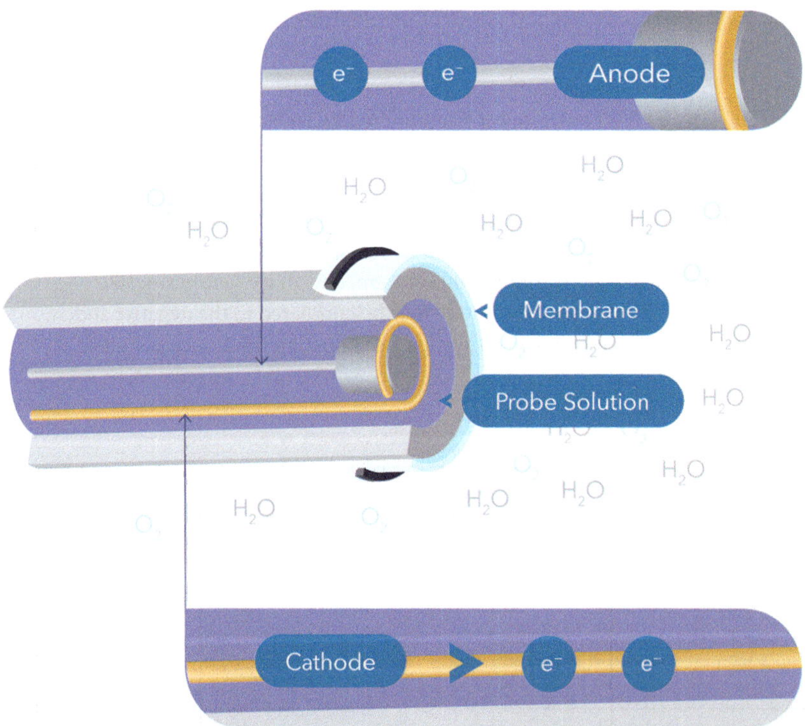

Figure 12.44 Electrochemical Sensor
© 2021, YSI, a Xylem brand
YSI.com/DO

electrical circuit is made and current can flow between the anode and cathode. The conducting solution may be potassium chloride, sodium chloride, or another salt depending on the type of sensor. The conducting solution may also be called *electrolyte, electrolyte solution,* or, simply, *fill solution.*

An electrolyte is a substance that produces ions when dissolved in solution.

Oxygen molecules diffuse through the membrane and react with the cathode. In the process, oxygen molecules consume electrons and hydrogen ions and are converted into water. This chemical reaction generates an electrical signal that is transferred from the cathode to the anode and on to the instrument. The higher the DO concentration, the greater the electrical signal will be. Because oxygen is consumed by the measurement, the liquid being tested must flow continuously over the membrane surface. This can be accomplished by either agitating the liquid or by moving the sensor.

The membrane may be stretched over the end of the probe body and be held in place with an O-ring. This looks very similar to stretching a piece of plastic wrap over then end of cylinder and holding it in place with a rubber band. Membranes can also be an integral part of a threaded membrane cap that screws onto the end of the probe body. Both membranes and membrane caps are consumable items that require regular replacement. The body of the probe is filled with electrolyte solution. Dissolved oxygen will move from one side of the membrane to the other, always moving from the side with the highest DO concentration to the side with the lowest DO concentration. When the membrane is working correctly, the electrolyte's DO concentration is about equal to the DO concentration in the liquid being measured. The movement of oxygen through the membrane is rapid. Most of the time, the oxygen concentration equalizes within a few seconds.

The diffusion process is critical. Any substance that coats the membrane, such as grease or even biofouling, can interfere with the measurement.

Biofouling occurs when bacteria and other microorganisms colonize the membrane surface. A slime layer forms that can prevent oxygen from reaching the membrane.

Figures 12.45 and 12.46 present diagrams of polarographic and galvanic sensors. Polarographic sensors use a gold cathode and silver anode. Gold and silver do not give up electrons easily and are close together on the galvanic series. They don't spontaneously transfer electrons from one to the other. Voltage must be applied to polarize the sensor. As a result, polarographic sensors require a 5- to 15-minute warmup period after they are turned on before they can be used or calibrated. Galvanic sensors use a silver cathode and an anode made of a metal that easily gives up electrons, typically zinc. These materials easily transfer electrons between them and are far apart on the galvanic series. The sensor is continuously polarized and does not require a warmup period. Galvanic sensors have a shorter life span than polarographic sensors.

The galvanic series lists metals in order of how likely they are to corrode when in contact with another metal. Noble metals like gold and silver are less likely to corrode, whereas more reactive metals like zinc and magnesium are more likely to corrode. The farther apart two metals are on the galvanic series, the easier it becomes for electrons to move between them.

Polarization is when electrical charges are separated. Applying voltage to a polarographic electrode creates the negatively charged cathode and positively charged anode.

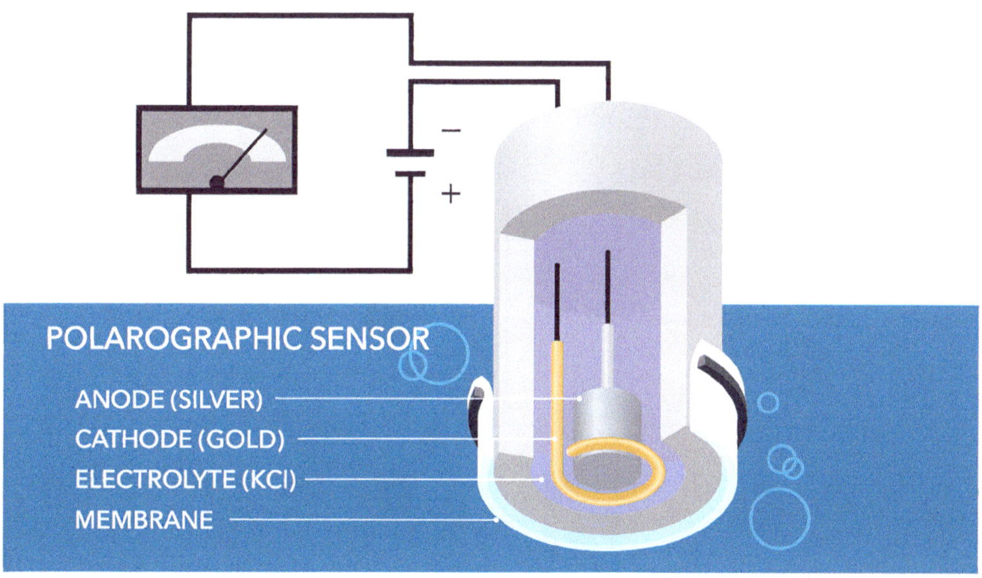

Figure 12.45 Polarographic Dissolved Oxygen Sensor
© 2021, YSI, a Xylem brand
YSI.com/DO

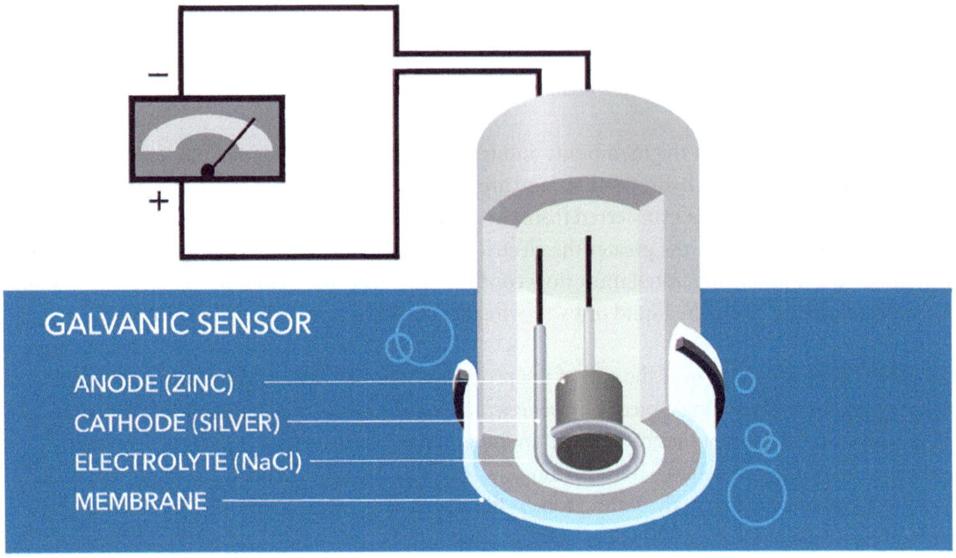

Figure 12.46 Galvanic Dissolved Oxygen Sensor
© 2021, YSI, a Xylem brand
YSI.com/DO

Optical DO sensors consist of at least one light-emitting diode (LED), a photodiode or photodetector, and an endcap (see Figure 12.47). The endcap contains an oxygen-permeable membrane. The inner layer of the membrane is coated with a luminescent dye. A blue or green light from the LED strikes the luminescent dye, which absorbs the light energy. This temporarily "excites" the dye molecules. The dye molecules can't hold onto the light energy and quickly release it as red light. The red light is then detected by a photodiode, as shown in Figure 12.48. If DO is present, the oxygen molecules will react with the dye molecules. That interaction "quenches" the light, so less light is released back for the photodiode to read. As DO concentrations increase, quenching increases and less light is rereleased. Luminescent DO sensors are capable of accurately reading lower concentrations of DO than membrane-type probes.

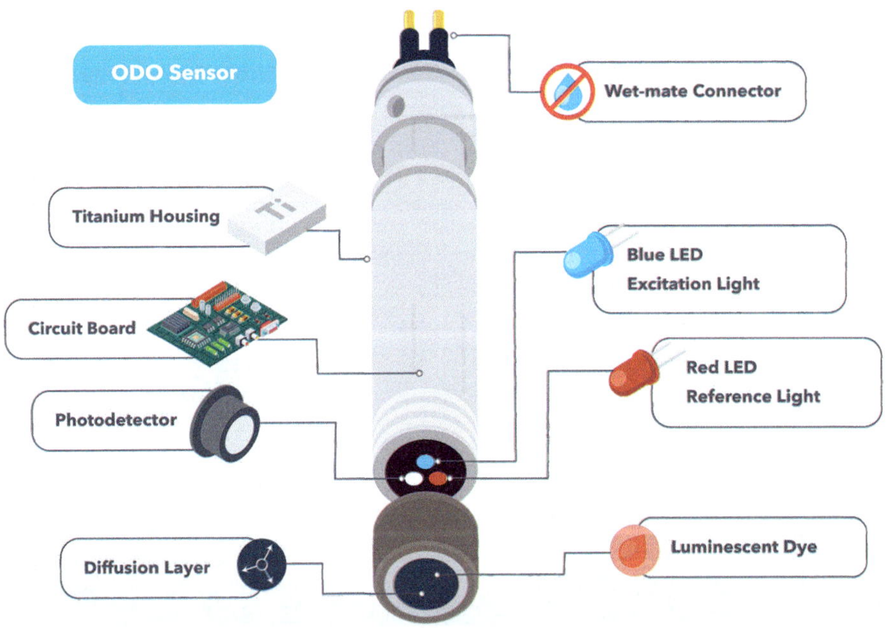

Figure 12.47 Optical DO Sensor
© 2021, YSI, a Xylem brand
YSI.com/DO

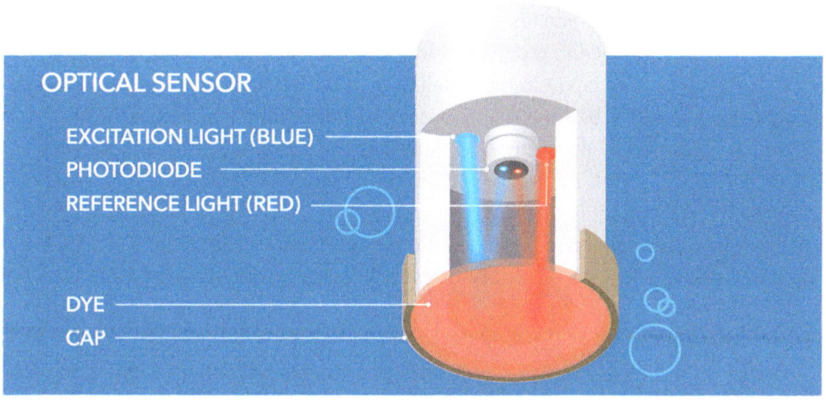

Figure 12.48 Close-Up View of an Optical Dissolved Oxygen Sensor
© 2021, YSI, a Xylem brand
YSI.com/DO

Installation and Operation

Dissolved oxygen sensors are typically placed at strategic points in an aeration basin (e.g., one-quarter, halfway, and three-quarters through the process) to adequately control the process. Some facilities install DO sensors before the aeration tank to monitor anoxic or anaerobic conditions and help combat DO sag. The sensor installed at the three-quarter point (i.e., after aeration) is used to determine whether the process is overaerated and wasting money. As with all analytical instruments, care should be taken to install the sensing element in an area where there is constant flow over the entire range of flows. Dead spots and areas where grease and foam are likely to accumulate should be avoided.

The sensing point is typically several meters (feet) from the basin wall and about 0.46 m (18 in.) below the wastewater surface. The sensors should have convenient mountings and adequate control wire so they can be removed easily and often for maintenance. Handrail-mounted instrument booms can also place a floating buoy several meters (feet) from the wall and retrieve it for regular maintenance, as shown in Figure 12.49. Sensors should not be installed directly over aerators or air diffusors.

Figure 12.49 Dissolved Oxygen Sensor in an Aeration Basin (A = Sensor location and B = Meter) (Reprinted with permission by Indigo Water Group)

Upon initial installation, new electrochemical sensors must be conditioned (polarized) by allowing the sensor to sit, connected to power, undisturbed in water for at least 12 hours before calibration (GLI International, 2000; Hach Company, 2010). Upon initial installation of both electrochemical and optical sensors, the operator must enter a value for barometric pressure or distance above sea level and a salinity correction value if they are not measured by the meter. The initial calibration of all sensor types must be done in water-saturated air.

Dissolved oxygen sensors may be calibrated using one of three different methods (see Table 12.6). Each method may be used with or without a zero DO solution to generate a single-point or two-point calibration, respectively. A zero-solution calibration can help increase accuracy when measuring low DO concentrations; however, a sensor should never just be zeroed. Zeroing is always done while also calibrating using one of the other three methods at the same time. Some manufacturers recommend cleaning the end of the probe with a soft wet cloth and mild, soapy solution before calibration.

Maintenance and Troubleshooting

The sensor must be kept reasonably clean to maintain measurement accuracy. During process operations, a portable meter is typically used to check in situ instruments and ensure that the entire process has adequate DO. The portable meter must be calibrated in the laboratory before using it to verify the calibration of in-basin probes. Because of the diurnal nature of the aeration process, this monitoring should occur at the same time every day or week.

Depending on process conditions, some installed DO sensor units may need to be gently wiped and rinsed each week to control growth on the sensing head (see Figure 12.50). The time between cleanings (days, weeks, etc.) depends on wastewater characteristics and can only be determined by experience. For example, a sensor operating in oily or greasy wastewater may need more frequent cleaning or membrane replacement if fouled. Consult the manufacturer's O&M manual for a cleaning procedure that will not damage the oxygen-permeable membrane. Generally, wiping with a soft, wet cloth or sponge is acceptable (see Figure 12.51).

Electrochemical sensor membranes should be kept wet or they will immediately begin to dry out. After 24 hours of drying, loss of electrolyte through the membrane can cause inaccurate measurements. Membrane dry out and electrolyte evaporation can be prevented by temporarily placing the sensor in a container of clean water until it can be returned to the treatment process.

Electrochemical sensor membranes should be replaced regularly (typically, quarterly or more often), depending on process conditions. Electrolyte is consumed and/or can become contaminated over time. If

Table 12.6 Calibration Methods for DO Sensors

Zero	Zero DO solution may be created from 20 g sodium sulfite dissolved in 1 L of water. Sodium sulfite removes DO. Ensure there are no bubbles on the membrane during calibration. Bubbles can be removed by tapping the side of the probe or by wiping the end of the probe with a soft wet cloth. If zero DO solution is unavailable, the probe can be placed in return activated sludge or waste activated sludge instead. Again, ensure there are no bubbles on the membrane during calibration. Some manufacturers recommend soaking the probe for up to 15 minutes in the zero DO water before calibrating.
Air Calibration	Remove probe from process. Clean membrane as directed by manufacturer. Place probe in a calibration bag with a few drops of water to create a moist environment. The goal is to create an oxygen-saturated environment in air. This defines the 100% saturation point for the probe. Follow manufacturer's directions for calibration. Operator must enter altitude or barometric pressure.
Sample Calibration	Probe remains in sample, calibrated against another calibrated instrument that is also measuring the sample. Operator must enter altitude or barometric pressure.
Saturation Calibration	Probe remains in sample. Sample must be aerated or known to be at saturation. Operator must enter altitude or barometric pressure.

Figure 12.50 Biofouling of an Optical DO Sensor (Reprinted with permission from Cory Moose, City of Midland, Texas)

a sensor drifts, gives erroneous readings, or cannot be calibrated, replace the electrolyte. Never add fresh electrolyte to existing electrolyte. The sensor body must be drained of old electrolyte first.

Over time, a layer of silver chloride will form on the anode of polarographic sensors. This reduces the amount of voltage and, as a result, the amount of oxygen reduced. This will eventually result in an error message, an unstable probe, or an inability to calibrate the probe. A silver reference electrode can help compensate for the accumulation of silver chloride on the anode. The silver chloride can also be removed by cleaning. Always follow the manufacturer's recommendations for your sensor. Some manufacturers provide a cleaning solution for soaking electrodes. Electrodes are typically soaked for 2 hours followed by rinsing with distilled water and gentle brushing to remove deposits (GLI International, 2000). If cleaning solution is not provided, first try mechanical cleaning by removing (unscrewing) the probe cap and gently polishing the silver and gold electrodes with wet, 400-grit sandpaper. Wet with distilled water. Be sure to use two separate pieces of sandpaper—one for each electrode so gold isn't transferred to the silver electrode and vice versa.

Figure 12.51 Cleaning the End of a Sensor (Reprinted with permission by Passaro Engineering, LLC)

The silver and gold coatings are thin. Take care to polish as little as possible to prolong the life of the probe. When finished, rinse well with distilled water. When mechanical cleaning is complete, the silver electrode should be free of silver chloride patina and the gold electrode should have a matte finish without any silver showing. Obtain a new membrane cap and fill it with fresh probe solution (electrolyte). Replace the cap by screwing it onto the end of the probe. The probe caps should not be reused.

Some probes use loose membranes with O-rings instead of membrane caps. In this case, place the membrane as directed by the manufacturer and install a lubricated O-ring to hold the membrane in place. After assembly, check the membrane to be sure it doesn't have any creases, holes, or tears. If it does, start over with a fresh membrane. When using loose membranes instead of membrane caps, the probe solution (electrolyte) is typically added directly to an outer tube or casing (probe body). Then, the sensor is placed inside the outer casing. If mechanical cleaning isn't enough, chemical cleaning of the electrode may be done by soaking the electrode in a 14% ammonium hydroxide solution for 2 to 3 minutes or in a 3% ammonium hydroxide solution overnight. As always, consult the O&M manual of the probe manufacturer before performing any maintenance.

With galvanic sensors, byproducts do not form on the sensor. They flake off to form a milky-white suspension in the electrolyte. Maintaining a galvanic sensor requires more frequent replacement of the electrolyte, but the electrodes do not require polishing.

Besides cleaning, optical DO sensors require little maintenance. However, abrasive environments or excessive cleaning can remove the luminescent material from the sensing cap. External calibration effort is minimal because the instrument internally calibrates itself to a red LED of known intensity; however, periodic air calibrations and a manufacturer's calibration are required when a new cap is installed. New sensors are factory-calibrated and should not require field calibration after installation. Optical sensors should be air-calibrated with protection from direct sunlight. Clean and inspect sensors for damage quarterly. Replace sensor caps every 2 years.

Sensor error and precision depend on the manufacturer and cell type. Electrochemical, oxygen-permeable membrane probes typically have errors of less than 0.10% of span and are repeatable within 0.05% of span.

Typical errors of luminescent DO sensors are less than 0.1 mg/L for solutions with less than 1 mg/L of DO. The error increases to 0.2 mg/L for solutions with more than 1 mg/L of DO. The sensor's precision is typically 0.05 mg/L, with a typical repeatability of 0.01 mg/L or 0.01% of saturation. Sensitivity should be 0.05% of the span.

TEST YOUR KNOWLEDGE

1. **For electrochemical DO probes to work properly, the liquid being tested must flow over the membrane continuously.**
   - ☐ True
   - ☐ False

2. **Dissolved oxygen passes easily through membranes coated with grease.**
   - ☐ True
   - ☐ False

3. **With optical DO sensors, light output increases with increasing DO concentration.**
   - ☐ True
   - ☐ False

4. **Electrochemical DO sensor membranes must be kept wet to prevent loss of electrolyte through the membrane.**
   - ☐ True
   - ☐ False

5. **When water quality parameters are measured, this comes into direct contact with the water**
   a. Instrument
   b. Processor
   c. Relay
   d. Probe

6. Dissolved oxygen sensors read
   a. Percent saturation
   b. Oxygen concentration
   c. Mass of oxygen
   d. Barometric pressure

7. The two types of DO sensors used in wastewater treatment include
   a. Clark and galvanic
   b. Polarographic and galvanic
   c. Optical and electrochemical
   d. Optical and polarographic

8. Which of the following parameters has the greatest effect on DO concentration?
   a. Salinity
   b. Barometric pressure
   c. Temperature
   d. pH

9. Warm water can hold __________ DO than cold water at the same pressure and salinity.
   a. More
   b. Less
   c. Half
   d. Double

10. As barometric pressure increases, the DO concentration at saturation __________.
    a. Increases
    b. Decreases
    c. Remains unchanged
    d. Matches salinity

11. Samples that contain the maximum amount of oxygen for a given temperature, pressure, and salinity are said to be __________ with DO.
    a. Undersaturated
    b. Saturated
    c. Supersaturated
    d. Hyposaturated

12. The purpose of the electrolyte in polarographic and galvanic sensors is to
    a. Equalize the salt concentration in the sample
    b. Transfer corrosion byproducts away from the cathode
    c. Form a connection between the anode and cathode
    d. Provide a pathway for oxygen to move up the probe

13. This type of DO probe requires a 5- to 15-minute warmup period before use.
    a. Galvanic
    b. Optical
    c. Reeding
    d. Polarographic

14. How long must electrochemical DO probes be conditioned before they are calibrated for the first time?
    a. 6 hours
    b. 12 hours
    c. 18 hours
    d. 24 hours

15. This critical parameter must be entered by the operator when calibrating DO probes.
    a. pH
    b. Temperature
    c. Percent salinity
    d. Barometric pressure

16. This substance may accumulate on the anode of a polarographic (Clark) electrode over time.
    a. Silver chloride
    b. Biofilm
    c. Gold bromide
    d. Salt

17. The sensor cap for an optical DO sensor should be replaced:
    a. Every 30 days
    b. Every 3 months
    c. Once a year
    d. Every 2 years

18. Aggressive cleaning of optical DO sensors can
    a. Damage the DO permeable membrane
    b. Remove luminescent material from the cap
    c. Improve sensitivity at low DO concentrations
    d. Deposit silver chloride on the gold anode

CHLORINE RESIDUAL

Purpose and Function

Most WRRFs are required to disinfect their effluent to reduce the number of pathogens before it can be discharged to a stream, river, pond, lake, or ocean. Some facilities are exempt from this requirement because the treatment processes they use reduce pathogens to acceptable levels without disinfection. Water resource recovery facilities may choose from several different disinfection methods including chlorination, ozonation, and ultraviolet light.

Disinfection is not sterilization. Sterilization kills all living organisms. Disinfection reduces pathogens to acceptable levels.

For more information on chlorine disinfection, see Chapter 10 of Book 1 of the Wastewater Treatment Fundamentals series.

Chlorination is the process of adding liquid or gaseous chlorine, sodium hypochlorite (bleach), or calcium hypochlorite to water or wastewater. Facilities that chlorinate are typically required to dechlorinate as well. Dechlorination removes chlorine that remains after disinfection is complete so it cannot harm the environment. Dechlorination is typically accomplished by adding sulfur dioxide, sodium bisulfite, or sodium metabisulfite to the treated water immediately after disinfection is complete.

The forms of chlorine primarily used for wastewater disinfection are chlorine gas (Cl_2), sodium hypochlorite (NaOCl), and calcium hypochlorite ($Ca[OCl]_2$). Each form of chlorine reacts similarly with water to form hypochlorous acid (HOCl) according to the reactions shown in Equations 12.5 through 12.7. Hypochlorous acid (HOCl) breaks down further to form hydrogen ions (H^+) and hypochlorite ions (OCl^-), as shown in Equation 12.8.

$$Cl_2(g) + H_2O \leftrightarrow HOCl + HCl \tag{12.5}$$
Chlorine gas + Water ↔ Hypochlorous acid + hydrochloric acid

$$NaOCl + H_2O \leftrightarrow HOCl + NaOH \tag{12.6}$$
Sodium hypochlorite + Water ↔ Hypochlorous acid + Sodium hydroxide

$$Ca(OCl)_2 + H_2O \leftrightarrow HOCl + Ca(OH)_2 \tag{12.7}$$
Calcium hypochlorite + Water ↔ Hypochlorous acid + Calcium hydroxide

$$HOCl \leftrightarrow H^+ + OCl^- \tag{12.8}$$
Hypochlorous acid ↔ Hydrogen ion + Hypochlorite ion

The chemical equilibrium between hypochlorous acid (HOCl) and hypochlorite ion (OCl^-) is strongly pH dependent, as shown in Figure 12.52. When the pH is near 7.6, the percentages of HOCl and OCl^- are equal. The equilibrium point is where the green and blue lines cross in Figure 12.52. Notice that it

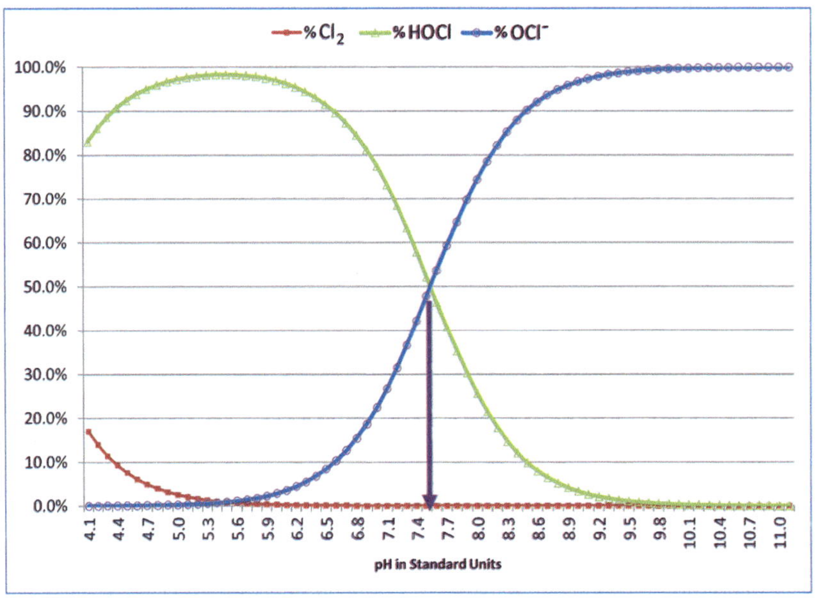

Figure 12.52 Hypochlorous Acid and Hypochlorite Ion Equilibrium at 20 °C (68 °F) (Reprinted with permission by Indigo Water Group)
Equilibrium point at 20 °C is pH 7.58. Equilibrium point at 0 °C is pH 7.82 and the equilibrium point at 30 °C is pH 7.50. Calculated from equations and constants in Morris (1966).

is located at pH 7.6. The equilibrium point moves slightly to the right as the water gets colder (higher pH) and moves slightly to the left (lower pH) as the water gets warmer. This effect is barely noticeable in domestic wastewater because water temperatures tend to remain with a narrow range between about 10 and 20 °C (50 and 68 °F). When the pH is greater than 8.5, HOCl is less than 10% of the total while OCl^- is more than 90% of the total. When the pH is less than 6.5, the equilibrium shifts to more than 90% HOCl and less than 10% OCl^-.

In addition to forming HOCl and OCl^-, chlorine reacts with other compounds present in the wastewater such as hydrogen sulfide (H_2S), ferrous iron (Fe^{+2}), manganese (Mn^{+2}), nitrite (NO_2^-), and organic compounds. These chemical reactions use up the oxidizing power of chlorine, converting it from HOCl into chloride ion (Cl^-). Chlorine can also react with ammonia to form chloramines (see Figure 12.53). Chloramines are disinfectants just like HOCl and OCl^-, but take longer to achieve the same level of disinfection.

The amount of chlorine added to the wastewater is the *dose*. The chlorine that is used up (i.e., consumed) in chemical reactions is the chlorine *demand*. The *residual* is the amount of chlorine remaining after the demand has been satisfied. Dose, demand, and residual are related, according to Equation 12.9. The residual chlorine concentration can be measured using a variety of test methods. The dose can be estimated if the concentration and volume of chemical being added are known. The demand can only be calculated if both the dose and residual are known. Demand cannot be measured.

$$\begin{gathered}\text{Dose} = \text{Demand} + \text{Residual} \\ \text{or} \\ \text{Chlorine Added} = \text{Chlorine Used} + \text{Chlorine Remaining}\end{gathered} \quad (12.9)$$

Chlorine residual may be measured as either the free chlorine residual or the total chlorine residual. Free chlorine residual measures HOCl and OCl^-. It does not measure chloramines. Total chlorine residual includes HOCl, OCl^-, and chloramines (combined residual) (see Equation **12.10)**.

$$\text{Total Chlorine Residual} = \text{Free Residual} + \text{Combined Residual} \quad (12.10)$$

Applications in Water Resource Recovery Facilities

Process analyzers are used to measure free or total chlorine residual. Results may be reported to regulatory agencies as part of permit compliance. They may also be used to control chemical feed rates for both chlorination and dechlorination.

How They Function

Chlorine residual may be measured using either the photometric or amperometric methods. The photometric method uses a chemical that combines with residual chlorine to produce a color change. The intensity of the color change is directly proportional to either the free or total chlorine residual, depending on the reagents used. The amperometric method uses an electrode that reacts with gaseous hypochlorous acid to generate an electric signal.

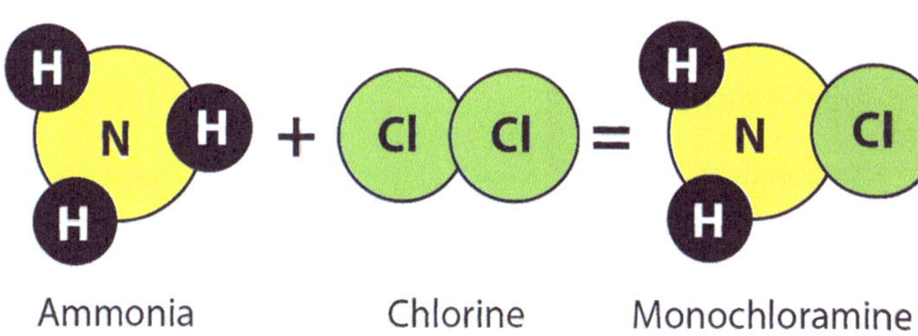

Figure 12.53 Monochloramine Formation

N, N-diethyl-p-phenylenediamine Colorimetric Method Analyzer

The N, N-diethyl-p-phenylenediamine (DPD) method for residual chlorine was first introduced in 1957 and has since become widely used for determining free and total chlorine in water and wastewater. This method can measure free or total residual concentrations between 0.03 mg/L and 10 mg/L as chlorine (Cl_2) (Hach Company, 2020).

The DPD colorimetric method analyzer injects reagents to a sample collected from the treatment process and then measures the color change to determine the chlorine residual. It's almost like having a tiny chemistry lab that collects and analyzes samples on demand. If chlorine residual is present, the sample will turn a shade of pink or magenta (see Figure 12.54). The darker the color, the greater the chlorine residual concentration. This is a batch process that typically analyzes a sample for chlorine concentration every 90 seconds to 10 minutes.

Figure 12.55 presents a depiction of the chemical basis for the DPD–free chlorine reaction. The DPD reagent reacts with chlorine to form two products: a Würster dye and imine. The Würster dye is responsible for the magenta color in the DPD colorimetric test. When DPD reacts with small amounts of chlorine at a near neutral pH, the Würster dye is the primary product. At high residual chlorine concentrations, the unstable colorless imine is favored. This can result in the apparent "fading" of the colored solution. In other words, it is a sample that initially colors magenta, but then fades to colorless. Occasionally, operators performing this test will interpret a colorless sample as having no measurable chlorine, when, actually, an excess amount of chlorine is present. If this is a problem, adding additional reagent will turn the sample magenta; however, doing so will prevent the instrument from accurately reading chlorine residual. For proper measurement of high residuals, the sample must be diluted and reanalyzed.

The DPD test may be used to measure either free or total chlorine residual. Figure 12.56 presents a depiction of the chemical basis for the DPD–free chlorine reaction. The only differences between the two testing methods are (a) the total chlorine residual test adds potassium iodide in addition to the DPD reagent and (b) the reaction time is longer—about 3 minutes—for the total chlorine residual test. Chloramines take longer to react with the DPD reagent and produce full color development.

The DPD–Würster dye color has been measured photometrically at wavelengths ranging from 490 to 555 nm. The absorption spectrum shows double peaks with the highest points at 512 and 553 nm. For maximum sensitivity, absorption measurements should be made between 510 and 515 nm. The measurement wavelength is determined by the manufacturer.

Figure 12.54 N, N-diethyl-p-phenylenediamine Color Change (Reprinted with permission by Indigo Water Group)

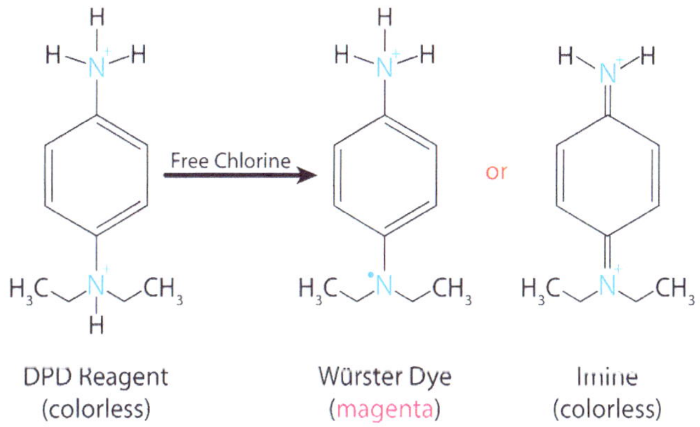

Figure 12.55 Reaction of Free Chlorine With DPD Reagent

A typical analyzer is equipped with a two-reagent system based on the DPD chemistry. The DPD indicator is prepared by adding powdered DPD reagent salt to the acidic indicator solution. The powder readily dissolves in the solution. The mixed solution is stable for at least 2 months. The buffer solution for the total chlorine analyzer is a citrate that also contains iodide. The free chlorine buffer reagent is a maleate buffer. The buffer reagent and its complementary indicator reagent are added in equal volumes to a captured portion of sample. Premixing the two reagents before introduction to the sample is important. This is done by putting a T union in the reagent feed lines before the colorimeter block.

An online DPD instrument typically operates as follows:

1. The sample inlet line of the pump or valve module is opened, allowing pressurized sample to flush the sample tubing and the colorimeter sample cell.
2. The sample inlet is closed, leaving fresh sample in the cell. The cell volume is controlled by an overflow weir.
3. As the sample inlet line closes, the reagent lines open, allowing buffer and indicator solutions to fill the tubing in the pump or valve module.
4. The untreated sample (sample blank) is measured for reference before the reagent is added. This compensates for sample color and turbidity.
5. The reagent outlet block opens, allowing precise volumes of buffer and indicator to blend and enter the colorimeter cell to mix with the sample.
6. After about a 1-minute delay so the color can fully develop, the magenta color is measured at 510 nm (the sample measurement; the time is a function of temperature).
7. The concentration (as "x" mg/L of chlorine) is calculated and then displayed.

Note that the color development time is increased to 3 minutes when measuring total chlorine residual. This sequence may be repeated every 1.5 to 10 minutes as set by the operator.

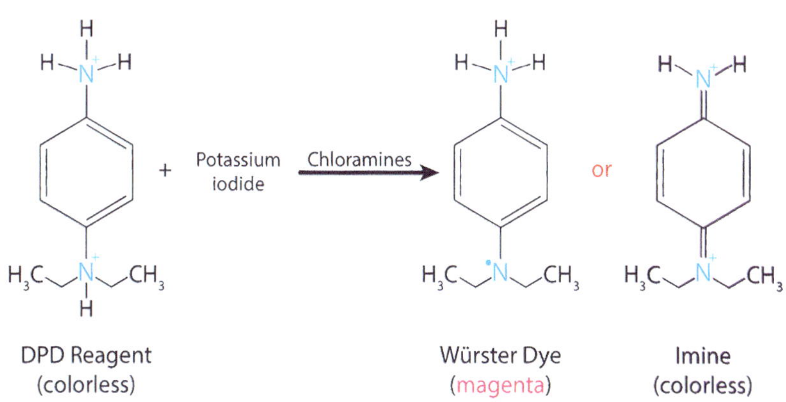

Figure 12.56 Reaction of Chloramines With DPD Reagent

Amperometric Analyzer

Amperometric chlorine analyzers are similar to electrochemical DO probes. They consist of a gold cathode, silver/silver chloride anode, and electrolyte fill solution—all contained within the probe body. Direct current voltage is applied to both the anode and the cathode. Similar to DO probes, chlorine probes also have a semi-permeable membrane that forms a barrier between the electrodes and electrolyte solution from the liquid being measured. The membrane is made from PVDF and is hydrophobic. It allows gaseous HOCl to pass through while rejecting OCl^- and other ions. Because the percentages of HOCl and OCl^- are dependent on pH, some analyzers add a buffer solution to the samples. The buffer solution ensures that the pH stays within a narrow range. A pH measurement can be used instead of a buffer to compensate for the change in dissociation when the pH of the water is higher than 4.5 and less than 9.

Hydrophobic means water loving. Water molecules are attracted to the membrane surface.

Gaseous hypochlorous acid that passes through the membrane reacts with the gold cathode. This chemical reaction consumes electrons (provided by the direct current) and converts hypochlorous acid into water (H_2O) and chloride ion (Cl^-). The chloride ion then reacts with the silver electrode to form silver chloride. That reaction generates an electron. The two reactions together produce an electrical current, measured in milliamps. As chlorine residual concentration increases, so does the electrical current generated.

Typical amperometric chlorine residual analyzing systems consist of a sample chamber, flow regulator, reagent solutions with associated metering pump or mechanism, anode–cathode measurement cell (the probe), and an electronic signal converter.

The measurement cell is sensitive to temperature; the reading can change as much as 3% per 1 °C of temperature change. To compensate, a temperature sensor in the measurement cell provides data to the electronics converter that are used to correct the indicator and output signals.

Chlorine sensors are available in three measurement ranges: trace, standard, and high (Endress+Hauser, 2022). The trace measurement range runs from 0 to 5 mg/L HOCl. Standard and high measurement ranges run from 0 to 20 mg/L HOCl and 0 to 200 mg/L HOCl, respectively. Operators should select the sensor with a range that is closest to the expected process concentrations. For most WRRFs, this will be either the trace or standard measurement ranges.

Installation and Operation

Samples for chlorine residual are typically pumped from the contact tank immediately after the chlorine is added to control chlorine dosing, and then again at the end of the contact tank to control the dechlorination process. Chlorine residual analyzers are typically installed in a conditioned sample, building as close as possible to the sample source to reduce the sample piping length. Limiting the length of sample lines reduces the potential chlorine demand in the sample piping caused by biofouling. For low-residual sampling systems, it is important to provide either an automatic or manual sample line cleaning system (bleach addition). In most instances, it is prudent to prevent the cleaning solution from flowing into the chlorine-sensing element to prevent possible poisoning of the sensing element.

Because of sensitivity to violations of National Pollutant Discharge Elimination System (NPDES) permits, it is common to install redundant chlorine analyzers at compliance-related control points. The sample point should be where the sample is homogenous and satisfies the NPDES permitted monitoring location, yet does not contribute unnecessary lag in the chlorine residual analysis. In some instances, installation of a mechanical mixer may be required to develop complete mixing of the disinfectant. Installing a sample bypass line to drain adjacent to the analyzer is recommended to help reduce sample lag time.

For membrane amperometric residual analyzers installed in situ, the probe should be installed on the end of a polyvinyl chloride (PVC) pipe into a well-mixed sample location in the process. Redundancy should also be considered for these sensors when they are compliance-related measurements. Sufficient cable should be retained to allow the probe and mounting pipe to be removed without disconnecting the conductors.

Maintenance and Troubleshooting

Photometric systems that use the DPD reagent use small-diameter tubing to move both the process sample and reagents. The tubing and all associated fittings must be cleaned and/or replaced regularly to prevent bacteria from growing within it and forming a biofilm. Biofilms are problematic because they react with chlorine residual and create demand, which can result in artificially low chlorine residual results. Biofilms can also plug sample and reagent lines when not removed promptly. Other maintenance for photometric systems includes replenishing reagents and periodically verifying the internal calibration by analyzing a certified standard.

Some amperometric systems include balls, beads, or grit to routinely clean the electrode surfaces that periodically need replacement. The electrolyte must be replaced periodically according to the manufacturer's recommendations. Membranes typically need to be replaced every 3 to 6 months. More cleaning may be required, depending on water conditions and any clogging or coating that might prevent the membrane from operating. The anodes and cathodes also need to be replaced periodically, depending on system conditions.

Chlorine residual analyzer errors are within 3% of full scale. Standard full-scale spans vary from 1 to 20 mg/L. The measuring error also varies from 0.03 to 0.5 mg/L, depending on the range selected. The analyzers' precision typically is within 1% of full scale. With automatic temperature compensation, accuracy and precision should hold over a sample temperature range of 0 to 50 °C (32 to 120 °F).

TEST YOUR KNOWLEDGE

1. This variable has the biggest effect on the percentages of hypochlorous acid and hypochlorite ion in water.
   a. pH
   b. Temperature
   c. DO
   d. Alkalinity
2. This form of chlorine dominates at low pH.
   a. Chloramine
   b. Hypochlorite ion
   c. Hypochlorous acid
   d. Free
3. Which of the following can be measured?
   a. Dose
   b. Demand
   c. Residual
   d. Extra
4. N,N-diethyl-p-phenylenediamine reagent turns __________ in the presence of residual chlorine.
   a. Yellow
   b. Blue
   c. Orange
   d. Magenta
5. At high concentrations of residual chlorine, the DPD test may appear this color
   a. Magenta
   b. Clear
   c. Orange
   d. Blue
6. One difference between the free chlorine and total chlorine test methods is
   a. The free chlorine test measures chloramines.
   b. The total chlorine test has a shorter react period.
   c. The free chlorine test turns magenta, then clear.
   d. The total chlorine test reagent includes potassium iodide.
7. Amperometric chlorine analyzers measure __________ as it passes through the membrane and reacts with the anode.
   a. Free chlorine
   b. Total chlorine
   c. Hypochlorous acid
   d. Hypochlorite ion
8. Amperometric chlorine residual measurements are sensitive to __________.
   a. Temperature
   b. Alkalinity
   c. Dissolved oxygen
   d. Ammonia
9. Small-diameter tubing used for reagents must be cleaned or replaced regularly to prevent
   a. Reagent contamination
   b. Biofilm growth
   c. Quenching
   d. Bypassing

pH

Purpose and Function

The term *pH* is traditionally used as a convenient representation of the concentration of hydrogen ions in a solution. The pH scale ranges from 0 to 14, with a neutral reading of 7 (see Figure 12.57). Readings below 7 are acidic and readings above 7 are basic. A reading of exactly 7 is defined as neutral. Raw wastewater typically has a pH near 7, although industrial and other nondomestic discharges can cause the pH to be much higher or lower.

The term *pH* is a mathematical abbreviation for the inverse log of the hydrogen ion concentration:

$$pH = -\log[H^+] \tag{12.11}$$

A mole is a unit of measurement used with atoms and molecules. Shoes come in a pair, paper comes in a ream, pencils come in a gross, and eggs come in a dozen. Atoms and molecules come in moles. One mole contains 6.022×10^{23} atoms or molecules. One millimole is one thousandth of a mole or 6.022×10^{20} atoms or molecules. Why such a huge number? Because atoms and molecules are so small that we need a large quantity of them in one place to measure them accurately. One mole of hydrogen atoms weighs 1.00794 g.

Where $[H^+]$ is the actual hydrogen ion concentration in moles per liter (mole/L) in a solution. The –log piece of the equation means that pH is expressed on a logarithmic or log scale. With a log scale, each segment of the scale increases by a factor of 10, so a sample with a pH of 4 has 10 times more hydrogen atoms than a sample with a pH of 5.

Because pH is expressed on a log scale, it takes a big change in hydrogen ion concentration to make a small change in the readout on the pH meter. Operators should be aware that pH measurements alone are rarely adequate for process control and that they should be paired with measurements of alkalinity. Often, by the time a pH change is noticeable, the process has degraded substantially and will be more difficult to bring back within normal operating range.

Applications in Water Resource Recovery Facilities

pH sensors are used to monitor conditions in WRRFs, track biological treatment process conditions, and control acid–base additions for pH adjustment. While some biological processes such as activated sludge systems and aerobic digesters can tolerate a pH variance of 6 to 9, others are more pH-sensitive. Anaerobic digestion, for example, requires pH to be between 6.6 and 7.6 at all times. In all biological treatment processes, the microorganisms responsible for treatment can cease to function or even die if the pH falls too low or goes too high. This can result in process failure.

Regulators often require that the pH of treatment facility influent and effluent be measured to indicate overall conditions. Specific industrial discharges may also require pH monitoring to provide advance warning of possible conditions that would be toxic to wastewater treatment processes. An online pH monitor can also provide feedback in other processes requiring pH adjustment. For example, pH adjustment may be required to neutralize low-pH industrial wastes, enhance phosphorus removal by alum addition, or optimize nitrification or denitrification processes.

How They Function

Glass pH Sensor

Glass pH sensors consist of a glass bulb fused to the end of a glass tube (see Figure 12.58). The glass is specially formulated to interact with hydrogen ions. The outer surface of the glass bulb is the gel layer. The glass bulb and tube are filled with an electrolyte solution that is buffered to pH 7. There is a silver/silver chloride electrode in the center of the tube. The electrode is connected back to the meter to form

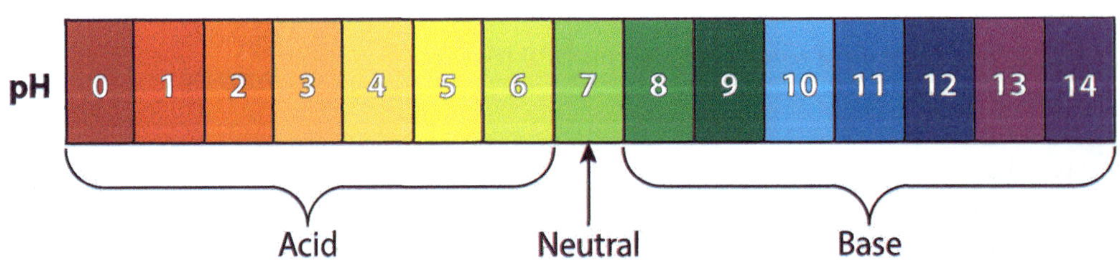

Figure 12.57 pH Scale

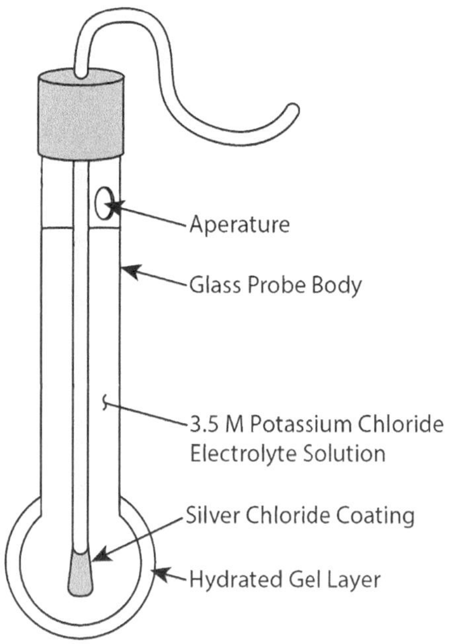

Figure 12.58 pH Electrode Without Reference Electrode

an electrical connection. This entire assembly is contained within an outer tube that is also filled with electrolyte (see Figure 12.59). A second silver/silver chloride electrode—the reference electrode—is housed within the outer tube. The outer tube may be made from glass, but is more commonly made of plastic or polycarbonate. The inner and outer tubes are connected by a junction. The junction prevents the solution being measured from coming into direct contact with either the electrolyte or electrodes. The junction may be ceramic, Teflon, or other material. Small holes in the junction create an electrical path between the two electrodes without the free flow of analyte.

A small voltage is applied to both electrodes. This charges the outer layer of the glass bulb with electrons and the bulb develops a negative charge (see Figure 12.60). Hydrogen ions, being positively charged, are attracted to the gel layer. This occurs inside and outside the glass bulb. Inside the bulb, the electrolyte is at pH 7. That means, we know how many hydrogen ions will be attracted to the inside of the bulb. Outside

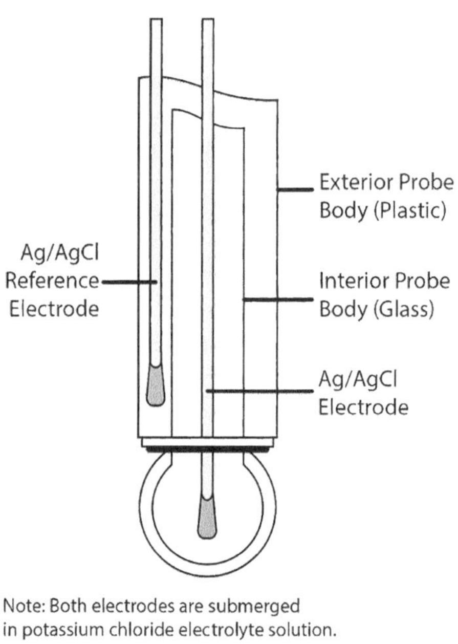

Figure 12.59 Glass pH Electrode

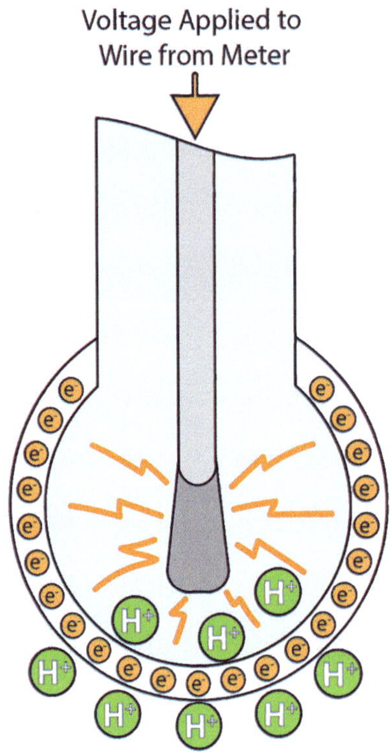

Figure 12.60 Hydrogen Ion Interaction With Gel Layer

the bulb, in the liquid being measured, the number of hydrogen ions can be lower, higher, or equal to the number of hydrogen ions inside the bulb. If there is a difference in hydrogen ion concentration on the two sides of the glass bulb, a potential difference develops. This is measured as voltage by the meter. pH values lower than 7 (more hydrogen ions outside than inside) produce a positive voltage. pH values higher than 7 (fewer hydrogen ions outside than inside) produce a negative voltage.

The potential difference between the primary electrode and reference electrode varies from sensor to sensor according to design preferences among manufacturers. It also changes as the sensor ages. For this reason, pH sensors must be periodically calibrated against buffer solutions of known pH. pH sensors may be calibrated using one to three buffer solutions. Buffers at pH 4, 7, and 10 are typically used.

The electrode gain is a function of temperature. So, most commercial pH sensors include automatic temperature compensation. A temperature sensor in the process fluid adjusts amplifier gain to compensate for changes in electrode gain that are caused by temperature.

Ion-Selective Field Effect Transistor

Ion-selective field effect transistors (ISFETs) are an alternative to glass pH electrodes. They use a transistor with a pH-selective layer instead of a glass bulb, which makes them less prone to breakage (see Figure 12.61). Similar to the gel layer of a glass electrode, hydrogen ions are attracted to the pH-selective layer. The buildup of hydrogen ions on one side of the selective layer attracts negative ions on the other side of the selective layer. The result is a separation of charge, which generates an electric current. As hydrogen ion concentration increases, the current also increases. Similar to glass pH probes, ISFETs also have a reference electrode that is kept separate from the liquid being measured.

Installation and Operation

When pH control is not the goal, a pH analyzer can be installed with other online analytical instruments in a sample system. Flow-through pH sensors are typically used in sample system installations. The flow-through sensor should be designed so electrode tips are flush with the tube wall and do not obstruct flow. To make maintenance easier, bypass and shutoff valves should be installed to provide for instrument

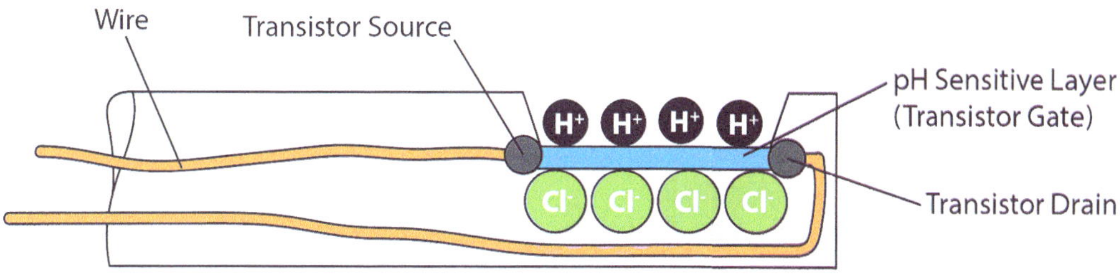

Figure 12.61 Ion-Selective Field Effect Transistor pH Probe

removal and service; in addition, the sensor should be located within view of the local display to facilitate standardization. Figure 12.62 presents an example of a local display. Sensors with electrodes that are easily removed from flow-through housing allow for easier cleaning and replacement. Providing a work surface for buffer solution containers is also helpful.

Submersed electrode assemblies with integral preamplifiers are typically used in tank and open-channel installations. The electrode assembly is typically attached to a PVC pipe with a bracket mounted on a guardrail or through the floor using a pipe cap to hold the probe at the required depth. Regardless of the mounting method chosen, it should be designed so the pipe and electrode assembly can be removed for maintenance without using tools. All fastening devices should be secured to prevent them from being dropped into the tank or channel. Enough spare cable is also needed to allow the sensor and pipe assembly to be lifted clear of the tank without having to unwire conductors.

A submersion probe installed in a well-mixed zone will provide a representative sample of the process. If the probe is installed in an open channel, it should be in a free-flowing zone. The electrode assembly and support pipe installation should be designed to inhibit collection of debris.

Maintenance and Troubleshooting

The sensor must be kept reasonably clean to maintain accurate measurements. The pH sensor shown in Figures 12.63 and 12.64 was installed in an influent channel. Grease, fibers, rags, and debris can accumulate on the sensor. Grease can interfere with the ability of hydrogen ions to reach the gel layer of a glass probe and the pH sensitive layer of ISFET. pH sensors should be cleaned and calibrated at least once every 3 months. Actual cleaning frequency will be determined by the liquid being monitored and operator experience. For example, a sensor placed in wastewater with a high grease content will require more frequent cleaning.

Figure 12.62 Example of a Local Display for pH (Reprinted with permission by Indigo Water Group)

Figure 12.63 End View of a Glass pH Electrode (Reprinted with permission by Indigo Water Group)

Glass pH sensors can be cleaned by rinsing with distilled or tap water. A mild soap that does not contain lanolin may be used to remove greasy deposits. A soft polymer brush may be used to remove more persistent deposits. Some manufacturers recommend soaking in a dilute hydrochloric acid solution (<5% HCl) for no more than 5 minutes to remove mineral deposits followed by thorough rinsing with distilled water. Cleaning procedures for ISFET sensors are similar. Use hot water and a mild soap to remove oily and greasy films. Use a dilute hydrochloric acid solution (<3% HCl) to remove mineral deposits followed by rinsing thoroughly with distilled or tap water (Endress+Hauser, n.d.).

Glass sensors are protected to some extent by an extension of the main probe body. This can be seen in Figures 12.63 and 12.64 as the black "teeth" on opposite sides of the end of the probe. It can still be easy to knock the end of the probe against something when putting it into or taking it out of a treatment process. Handle them carefully to prevent breakage. A broken glass pH probe cannot be repaired and must be replaced.

During process operations, a portable pH meter can typically be used to verify accuracy. Water resource recovery facility pH calibrations are performed using standard buffer solutions (typically, 4.0, 7.0, or 10 using appropriate buffers based on the configured range of the instrument). Glass pH probes may be calibrated with a single buffer; however calibrating with at least two buffers is recommended. Ion-selective field effect transistors sensors require a two-buffer calibration.

Figure 12.64 Side View of a Glass pH Electrode (Reprinted with permission by Indigo Water Group)

Methods of reporting performance specifications vary among manufacturers. When these are adjusted to equal units of measure, the error and stability claimed by different manufacturers varies greatly. Manufacturer claims for pH meter errors range from 0.01 to 0.2 pH units. Without temperature compensation, an additional error of 0.002 pH units per 1 °C difference from the calibration temperature can be expected. The precision of pH meter measurements varies by manufacturer from 0.01 to 0.04 pH units.

Stability (drift) is an important performance parameter that indicates how often meters must be recalibrated. Manufacturer claims for stability vary from 0.002 to 0.2 pH units per week.

TEST YOUR KNOWLEDGE

1. A sample at pH 4 contains __________ times more hydrogen ions than a sample at pH 5.
   a. 1
   b. 10
   c. 100
   d. 1000

2. The bulb of a pH sensor is made from
   a. Plastic
   b. Teflon
   c. Glass
   d. Metal

3. pH measurements use a reference electrode to measure the __________ between the electrolyte and the liquid being tested.
   a. Gravimetric potential
   b. Chloride disparity
   c. Piezometric charge
   d. Potential difference

4. The hydrogen ion concentration is higher on the outside of a glass pH sensor than it is on the inside. The solution being measured is
   a. Acidic
   b. Basic
   c. Neutral
   d. Charged

5. pH sensors must be calibrated periodically using
   a. Process water
   b. Buffer solutions
   c. Zero oxygen solution
   d. Water-saturated air

6. The main component of an ISFET pH sensor is the
   a. Gel layer
   b. Glass bulb
   c. Transistor
   d. Electrolyte

7. This substance can interfere with pH measurements by preventing hydrogen ions from interacting with the sensor.
   a. Alkalinity
   b. Suspended solids
   c. Grease
   d. Detergent

8. Which of the following cleaning methods should be used to remove mineral deposits from a pH sensor?
   a. Dilute hydrochloric acid
   b. Rubbing alcohol
   c. Mild dish detergent
   d. Soft polymer brush

9. Ion-selective field effect transistor sensors require a __________-point calibration.
   a. One
   b. Two
   c. Three
   d. Four

10. pH measurements may be affected by __________.
    a. Flow
    b. Suspended solids
    c. Dissolved oxygen
    d. Temperature

OXIDATION–REDUCTION POTENTIAL

Purpose and Function

Oxidation–reduction potential analyzers measure a water's potential to oxidize (or reduce) organic matter or compounds in water. *Oxidation* means to add oxygen, lose hydrogen, or lose electrons. A good example of oxidation is the conversion of ammonia (NH_3) to nitrate (NO_3^-). *Reduction* means to remove oxygen, add hydrogen, or gain electrons. A good example of reduction is the conversion of nitrate (NO_3^-)

Oxidation–reduction reactions are often short-handed as RedOx.

to nitrogen gas (N_2). Oxidation–reduction reactions are chemical reactions that take place in wastewater all the time. Oxidation and reduction always occur together with one chemical substance being oxidized, while another is reduced. If the two chemical reactions are physically separated, but joined by a wire, the electrons lost from one chemical substance will travel down the wire to the other chemical substance. The difference in electrical charge between the two points is the electrical potential or potential difference, also known as *voltage*. Hence the term, *oxidation–reduction potential.*

Oxidation–reduction potential is measured in millivolts. There are 1000 mV in 1 V. Essentially, ORP measures the difference in potential (charge) between all the different chemical substances in wastewater. Wastewaters with higher concentrations of oxidized substances like nitrate, nitrite, and DO tend to have positive ORPs. Wastewaters with higher concentrations of reduced substances like hydrogen sulfide (or a lack of oxidized substances) tend to have negative ORPs.

The ORP value for a given wastewater will then be dependent on all species in the matrix whose concentrations will be affected by the following: the amount of time spent in the sewer, septic conditions, water temperature and amount of nitrification, denitrification, and orthophosphate removal. The presence of a strong oxidant such as chlorine or reductant such as sulfite will affect the reading of an ORP instrument and reductant NH_4^+ or oxidants NO_2^-, NO_3^-, or PO_4^{3-}.

Oxidation–reduction potential changes when DO is present and when it isn't. It is useful for monitoring conditions in anaerobic, anoxic, and aerobic processes. In general, ORP values lower than –200 mV indicate anaerobic conditions, values between –200 and +200 mV indicate anoxic conditions, and values higher than +200 mV indicate aerobic conditions. Typically, the ORP in an anoxic process must be below +50 mV to trigger denitrification. These ranges are based on an ORP probe with a silver/silver chloride reference electrode and a saturated potassium chloride (KCl) fill solution. The actual ORP reading depends on the specific reference electrode being used and can be affected by water quality. Facilities that use ORP for process control should determine which ORP ranges indicate good operating conditions for their processes rather than relying on typical ranges given in books like this one. Oxidation–reduction potential measurements are typically taken in situ by ORP probes mounted directly in the activated sludge basins.

When measurements are taken directly from the treatment process instead of transporting them to a laboratory, they are "in place" or in situ measurements.

Applications in Water Resource Recovery Facilities

Monitoring and control of chlorination and dechlorination were the first applications for ORP in WRRFs. The addition of chlorine produces an increase in the ORP signal, whereas the addition of sulfite produces a decrease in the signal. The following two limitations have created difficulties for automatic dosing of chlorine and sulfite:

- The ORP is not a concentration measurement and cannot define quantities.
- The base value changes with wastewater conditions, as previously described.

The solution is to always work with differential information, that is, the difference between a measurement before and a measurement after.

Oxidation–reduction potential is also used for monitoring and controlling nutrient removal processes. Oxidation–reduction potential sensors (probes) are typically placed in anaerobic and anoxic zones of activated sludge treatment processes. Oxidation–reduction potential sensors in anaerobic zones are used to monitor enhanced biological phosphorus removal. In anoxic zones, they monitor denitrification. Oxidation–reduction potential is used in sequencing batch reactors to determine when air should be cycled on and off. They may also be placed in aerobic digesters for the same purpose. Anyplace where biological activity needs to be monitored or controlled is a candidate for ORP. Dissolved oxygen probes can only give process control information in aerobic processes. Oxidation–reduction potential probes can also be used in aerobic processes in addition to anaerobic and anoxic processes.

Different types of microbial activity occur within their own ranges of ORP. Table 12.7 lists typical ORP ranges using a silver/saturated potassium chloride reference electrode, which is the most commonly used ORP probe. An ORP probe may be useful in determining the type of biological activity likely to be occurring at any point in the treatment process.

Table 12.7 Oxidation–Reduction Potential Ranges for Various Biochemical Reactions (WEF & ABC, 2018)

Microbial Activity	Approximate ORP (mV)
Carbon oxidation (conversion of 5-day biochemical oxygen demand to biomass, carbon dioxide, and water with oxygen)	+50 to +200
Polyphosphate accumulation	+50 to +250
Nitrification	+150 to +350
Denitrification	–50 to +50
Polyphosphate release	–40 to –175
Volatile fatty acid formation	–40 to –200
Sulfide formation	–50 to –250
Methane formation	–200 to –400

Note. Values are for a silver/silver chloride electrode with saturated potassium chloride fill solution.

How They Function

Oxidation–reduction potential probes look very similar to pH probes; they are also constructed similarly to pH probes. Combined ORP probes contain two measurement devices: a primary electrode and a reference electrode. The primary electrode is submerged in the sample; in this case, wastewater. The reference electrode is submerged inside the probe in a standard solution (fill solution). The measurement from the primary electrode is compared to the reference electrode by the electronics in the meter to produce an ORP reading in millivolts. Different ORP probes produce different ORP readings for the same solution, based on the specific reference electrode and fill solution used. Typical reference electrodes include standard hydrogen, saturated calomel electrode, or silver/silver chloride electrode. A solution showing zero millivolts with a hydrogen electrode would read the approximate values shown in Table 12.8. Most ORP probes use a silver wire in saturated potassium chloride (KCl) solution as the reference electrode and a platinum wire as the primary electrode (Spencer, 2012). The key point to take away from this discussion is that ORP readings can be quite different from one probe to the next. Don't worry if the result from the ORP probe in the lab doesn't exactly match the result from the ORP probe in the treatment process. If the in-process probe is a silver/silver chloride electrode with 4-Molar potassium chloride (KCl) fill solution and the laboratory probe is a saturated calomel electrode, the readings produced by placing each probe in the exact same water will differ (in theory) by –41 mV. In practice, the difference can be much larger.

Saturated means the solution is the highest possible concentration for a particular temperature and pressure. Hot tea can dissolve more sugar than cold tea. The point where no more sugar can be dissolved into the tea is the sugar saturation point.

Oxidation–reduction potential measurements are affected by temperature. Because the effect isn't easy to correct for, ORP probes typically do not incorporate temperature sensors and do not adjust the ORP result. Rather, the operator must make adjustments to how the ORP reading is interpreted. Changes in the hydrogen ion (H^+) concentration (pH) will also change ORP readings. Fortunately, in most well-operating treatment processes, the pH does not change very much from day to day.

Table 12.8 Oxidation–Reduction Potential Reference Electrodes (WEF & ABC, 2018)

Reference Electrode	ORP (mV)
Standard hydrogen electrode	0
Saturated calomel electrode	–241
Silver/silver chloride, 1-Molar KCl*	–236
Silver/silver chloride, 4-Molar KCl*	–200
Silver/silver chloride, saturated KCl*	–197

*A 1-Molar solution contains 1 mole per liter.

Installation and Operation

Mounting requirements are the same as those for pH electrodes. Because the surface area of the platinum electrode has a large influence on the measurement, both in terms of response time and actual value, an installation point where there are turbulences will reduce the need for cleaning.

Maintenance and Troubleshooting

An ORP electrode will need cleaning, some polishing, and verification. Oxidation–reduction potential standards are primarily used for verification of electrode response rather than calibration. This is because standards such as ZoBell's solution or Light's solution provide different readings, but do not match each other. As a result, multipoint calibration is not possible.

The oxidation state and surface chemistry on the platinum sensor can significantly influence the response over the useful life of ORP electrodes. The quality of the finish must be maintained and each manufacturer has their own recommendations for maintenance. The requirements depend on the application.

After verification of electrode response, the sensor will need to readjust to the wastewater and will take some time to reach stability. In addition, the oxide layer must be reformed to get the full response. The time required for these functions varies among manufacturers.

Because ORP is a nonspecific measurement (i.e., the measurement notes the effect of the concentration of all dissolved species in the medium), there is little value to discuss error and interferences. It is not a concentration measurement.

Typically, readings from various instruments (sometimes with exactly the same sensor type and transmitter) differ by a significant margin (50 to 100 mV), even though the sensors are in the same wastewater. To make the problem more perplexing, all of the sensors show identical readings in an ORP standard. Trends in ORP are often more valuable than the absolute ORP reading. Operators should track ORP and learn which values work best for their treatment processes.

In conclusion, an ORP instrument is a trend indicator and best results are obtained using either the difference between two measurements (before and after) in a continuous process or the rate of change in a batch process. The readings must be carefully interpreted. They depend on the condition of the sensor and the makeup of the wastewater being tested.

TEST YOUR KNOWLEDGE

1. **Two ORP probes made from the same materials and with the same fill solution will always measure the same ORP value.**
   - ☐ True
   - ☐ False
2. **Oxidation–reduction potential probes are similar to pH probes with an electrode and reference electrode.**
   - ☐ True
   - ☐ False
3. **Both ORP and pH must be corrected for temperature effects and incorporate temperature sensors.**
   - ☐ True
   - ☐ False
4. **When a substance is oxidized it**
   a. Gains oxygen
   b. Loses oxygen
   c. Gains water
   d. Loses water
5. **What will the ORP most likely be in an aeration basin with a DO concentration of 2 mg/L?**
   a. –300 mV
   b. –150 mV
   c. 0 mV
   d. +150 mV
6. **Maintenance of ORP sensors requires**
   a. Regular calibration
   b. Polishing
   c. Acid soaking
   d. Electrolysis

TOTAL SUSPENDED SOLIDS

Purpose and Function

Total suspended solids meters may be used to estimate the solids concentrations.

Applications in Water Resource Recovery Facilities

Online solids analyzers may be installed in most wastewater treatment processes as long as the nature of the solids (e.g., concentration and variability) and the site (e.g., basin, channel, or pipe) is known before selecting an instrument (see Table 12.9). Currently, most solids instruments use near-infrared technologies, although instruments based on microwave, nuclear, or ultrasonic technologies are also available. Solids analyzers are available as both handheld and in situ versions (see Figure 12.65). Some facilities have solids analyzers in their activated sludge basins and waste activated sludge lines. This allows the sludge age calculation to be done automatically in SCADA.

How They Function

This section describes near-infrared (optical) and microwave analyzers. Nuclear and ultrasonic solids analyzers have rarely been used at wastewater treatment facilities and, therefore, are not discussed here.

Online solids analyzers work by developing an electronic correlation between the sludge's characteristics and actual solids concentrations. This correlation is established during calibration, which compensates for some variation in solids characteristics. However, if solids characteristics vary significantly, so do the correlation constants and recalibration will typically be required. Most optical-based instruments, for example, are sensitive to size distribution and significant color changes. Other instruments may be sensitive to air bubbles, conductivity, or precipitate buildup.

Calibration generally involves collecting a sample from the treatment process being monitored and analyzing it by an approved laboratory testing method, typically gravimetric analysis. At the same time the sample is collected, the solids concentration is measured and stored within the solids meter. The laboratory

Table 12.9 Treatment Processes and Their Typical Solids Concentrations (WEF, 2013)

Process	Solids Concentration	Possible Interference	Purpose
Facility Influent	200 to 1000 mg/L	Grease, rags, color change	Monitor solids load to facility
Primary sludge	10 000 to 100 000 mg/L	Grease, rags, color change	Optimize sludge pumping, monitor digester loading
Mixed liquor suspended solids	400 to 8000 mg/L	Attached growth	Monitor pounds under aeration, calculate sludge age, calculate food-to-microorganism ratio
Return activated sludge/waste activated sludge	400 to 30 000 mg/L	Attached growth	Optimize sludge pumping, automatic control of wasting rate, calculate and/or control sludge age
Clarifier blanket depth	400 to 30 000 mg/L	Attached growth	Optimize sludge pumping
Secondary effluent	1 to 1000 mg/L	Attached growth	Compliance monitoring
Thickening feed	400 to 30 000 mg/L	Grease, rags, color change	Control and optimize solids loading rate or polymer dose
Dewatering feed	10 000 to 100 000 mg/L	Grease, rags, color change	Control and optimize solids loading rate or polymer dose
Thickening/dewatering sidestream (underflow, filtrate, centrate)	50 to 10 000 mg/L	Grease, rags, color change, air	Monitor thickening/dewatering efficiency, calculate percent solids capture, calculate solids loading in recycle streams
Thickening/dewatering solids	3% to 30% mg/L	High solids, abrasives	Calculate percent solids capture, optimize polymer dose, monitor moisture content

Figure 12.65 Handheld TSS Meter (Right) and Handheld Temperature Sensor (Left) (Reprinted with permission by Passaro Engineering, LLC)

result is later entered as the true concentration for the stored sample. The solids meter then extrapolates from the single calibration point to estimate the solids concentrations of future samples. How often the calibration must be done depends on how variable the solids characteristics are. In an activated sludge process, calibration may be needed weekly, monthly, or less often if floc particle size and density are very consistent.

Gravimetric analysis for TSS takes a known sample volume and filters it through a 1.2-µm filter disc. The collected solids are the dried at 104 °C to remove moisture. The filter is weighed before and after analysis to obtain the net increase in weight from the filtered solids. The solids concentration is then calculated in milligrams per liter.

Near-Infrared Analyzers

Near-infrared sensors use light that is between 860 and 880 nm and invisible to the naked eye, but visible to video camera systems. Because the light is outside the visible spectrum, sensors are able to compensate to some degree for the color of the particles, the color of the water, and changes in the color of the particles or water.

Optical suspended solids analyzers typically consist of a controller or interface device that communicates with a sensing head that is either inserted to a pipe or suspended in a process tank. There are two types of optical solids analyzers developed for use in the wastewater treatment industry: transmittance (two-beam and four-beam alternating light) and reflectance (single- and multiple-angle backscatter). Transmittance analyzers have an approximate range of 200 mg/L to 15% solids. The transmittance meter measures the relative loss of light and converts the signal to a solids concentration. The four-beam version uses two emitters and two detectors, creating an array of light paths that produce a matrix of results. A ratiometric comparison of these signals produces a reading that effectively eliminates interferences, compensates for fouling, and allows measurement in harsh environments. In addition, the analyzers may have an integral purge port that allows solenoid-activated air or water pulses to clean deposits from the sensing head at user-programmed intervals.

Reflectance analyzers can accurately measure a wider range, generally between 0.5 mg/L and up to 10% solids concentration depending on sludge type. The multiple-angle version consists of a light source and multiple detectors positioned with their optical axes between 90 deg and 140 deg from each other. The emitted light is backscattered by solids particles and received by the detectors. Analyzers typically use a mechanical device (e.g., a wiper) or an ultrasonic cleaning system to keep the light source and detector windows clean.

Microwave Analyzers

Online microwave analyzers typically consist of wafer-style stainless steel process sections that are fitted between existing American National Standards Institute (ANSI) 150 pipe flanges. They measure total solids based on the measurement of microwave propagation time or phase difference of microwaves traveling through the sludge. Microwaves are a form of electromagnetic radiation, and their propagation velocity, or flying time, is dependent on the medium in which they travel. In water, microwaves travel at much slower speeds than in sludge. The dry solids content can be calculated based on the time it takes the microwaves to travel through the measured mass.

Microwave analyzers using phase difference operate on the principle that when microwaves travel through a substance, the physical density can be determined by measuring the phase lag of the waves. The phase difference correlates to density, which can be related to the total solids concentration after compensation for ambient conditions.

Microwave analyzers are a microprocessor-controlled unit that displays the total solids reading in terms of percent or grams per liter. They can provide both digital and analog outputs.

Advantages of microwave technology include the following: the measurement method is independent of process medium, is insensitive to flowrate, requires only single-point calibration, and is capable of measuring high solids content.

Online microwave analyzers are easy to calibrate. Because there is a linear relationship between microwave time of flight, a single-point calibration is performed using grab-sample percent solids results.

The analyzers are not sensitive to upstream or downstream fittings. Color and color changes do not affect the reading, provided they do not absorb the spectrum used such as, potentially, ferric chloride. However, the analyzers can be sensitive to extreme changes in conductivity, excessive entrained air, and partially filled pipes. The pipe where the sensing element is installed must be full and under pressure at all times. To ensure a successful installation, engineers and operators should advise the manufacturer of process characteristics before installation.

The instrument head of an optical analyzer is typically made of plastic or stainless steel. The lens may be constructed of epoxy or scratch-resistant glass (e.g., quartz or sapphire) and should resist grease accumulation.

Online microwave analyzers are available in various pipe sizes. The transmitter is made of 316 stainless steel and can operate up to 16 bar (232 psig). Units designed to measure higher cake solids are designed with ANSI 600 flanges and can operate up to 100 bar (1440 psig). Teflon (PTFE) coatings and purge cleaning devices are available for applications with potential fouling problems.

Installation and Operation

Optical suspended solids analyzers may be installed as submersible probes. Various devices are available to attach them to concrete walls or handrails. Most manufacturers also have an insertion assembly so the probe can be installed in a pressurized pipe. The insertion assembly allows the meter to be installed and removed from service for maintenance while the process pipe is in service and pressurized. However, each manufacturer has different specifications for the maximum allowable pressure for the insertion sensor. Therefore, it is important to check with the manufacturer before designing an insertion application.

Meters should be installed near a work platform for easy access. Sample taps should be installed near the instrument to collect samples to verify instrument readings. The online microwave analyzer is a flow-through spool piece or an insertion probe. It should be installed so the pipe is full and pressurized at all times. The instrument is not sensitive to upstream or downstream fittings and, because it is fully ported, can be installed next to an electromagnetic flow meter.

Solids analyzers will have a local display showing the current measurement, as shown in Figure 12.66. They are typically mounted as close to the point of measurement as possible to enable easy calibration and troubleshooting.

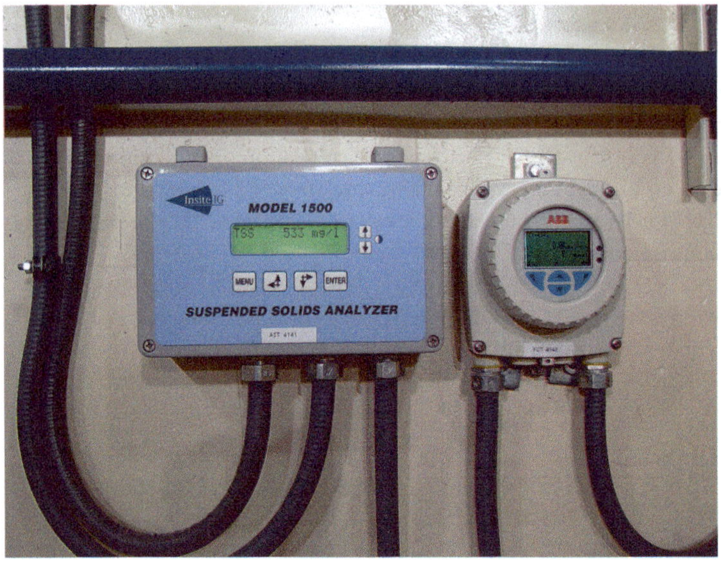

Figure 12.66 Suspended Solids Analyzer Wall-Mounted Display (Reprinted with permission by Indigo Water Group)

Gradual precipitation or other buildup on the analyzer's sensors may degrade accuracy. Therefore, locations with potential buildup problems may want to use Teflon or similar coatings to reduce buildup issues. Some manufacturers provide a purge cleaning system to keep critical sensor parts clean. Another option is to install the analyzer in bypass piping, which will allow for periodic cleanings.

Maintenance and Troubleshooting

All optical instruments will need periodic cleaning and occasional recalibration to maintain an accurate output. The actual cleaning and recalibration interval depends on site-specific conditions. Typically, a sensing head only needs to be wiped with a rag once every 2 to 3 weeks. Cleaning may include removing the probe from the process and washing it with a mild detergent or, possibly, a mild acid or caustic, to remove solids buildup. Daily or weekly samples may be collected to verify instrument output.

Suspended solids analyzers with effective cleaning devices require little operator intervention. Analyzers with air-purging systems effectively remove debris (e.g., rags or plastic wraps) that may block sensing elements. Operators should assess fouling characteristics at their facility and implement appropriate maintenance schedules.

Calibration requirements vary by manufacturer and installation conditions. Analyzers may only need to be calibrated after 6 months of operation to account for varying floc conditions. Analyzers can be dependent on solids color, such as color changes associated with the addition of ferric chloride; for intermittent process changes, recalibration of the instrument should be performed upon each change or it should be acknowledged that the reading may have additional error. Particle size and shape typically vary with seasonal process changes; thus, optical meters should be seasonally recalibrated so the instrument can continue to produce an accurate output. Although a one-point calibration is typically recommended, most sensors can perform multiple-point calibrations to increase accuracy.

Online microwave analyzers require little to no maintenance and rarely require recalibration. In processes with fats, oils, and grease or installations that allow precipitates to deposit on the sensors, internal coatings or purge cleaning systems should be used.

The error of an optical sensor is typically 5% of reading, and several ranges of operation are available. Its precision is typically 1% of reading or 0.1 g/L, whichever is greater. The main interferences with optical sensors are stray light and direct sunlight. To minimize stray light, it is important to avoid installing the sensor in a manner that will allow the emitted light to reflect off a pipe or tank wall back to the sensors. For example, in a closed-PVC-pipe application, the sensor should be inserted at an angle or dark-colored

PVC should be used. To avoid interference from sunlight in clear water applications (e.g., facility effluent), the sensor should be immersed deep enough so that sunlight cannot directly affect the detectors. In higher-concentration applications, submerging the sensor to a depth at which sunlight cannot penetrate is recommended.

Online microwave analyzers can measure up to 16% solids, with some specialty applications used for up to 35% solids. Repeatability is ±0.01% and sensitivity is 0.001% of the current reading.

TEST YOUR KNOWLEDGE

1. Optical solids sensors must be recalibrated when solids characteristics change.
   - ☐ True
   - ☐ False
2. Microwaves travel faster through water than sludge.
   - ☐ True
   - ☐ False
3. This parameter can affect solids measurements by optical sensors.
   a. Color
   b. Conductivity
   c. Air bubbles
   d. Temperature
4. Transmittance-based solids analyzers can measure solids concentrations up to
   a. 1%
   b. 5%
   c. 10%
   d. 15%
5. This type of solids analyzer is capable of accurately reading solids concentrations as low as 0.5 mg/L.
   a. Transmittance
   b. Reflectance
   c. Microwave
   d. Electromagnetic
6. Microwave solids analyzers are sensitive to
   a. Upstream fittings
   b. Dissolved oxygen
   c. Conductivity changes
   d. System pressure
7. Optical solids analyzers should be calibrated at least
   a. Weekly
   b. Monthly
   c. Seasonally
   d. Annually
8. Which of the following solids analyzers should be selected for monitoring dewatered cake?
   a. Transmittance
   b. Reflectance
   c. Optical
   d. Microwave

INTERFACE/SLUDGE BLANKET LEVEL

Purpose and Function

Traditionally, operators measured sludge blanket depths manually by inserting a core sampler into the clarifier blanket. The core sampler consists of several interchangeable 1.5-m (5-ft) sections of clear PVC tube with markings 0.3-m (1-ft) apart and a bottom 1.5-m (5-ft) section with a foot valve (check valve) that is seated by the water's static head. When assembled, the core sampler should be at least 0.3 m (1 ft) longer than the clarifier depth at the measurement point. The core sampler is gently lowered vertically down to the clarifier floor, allowing the displaced water column (including the sludge blanket) to fill the sampler through the unseated valve. The goal is to collect a representative sample of both the water column and sludge blanket. The top of the sampler must not be submerged when its bottom is resting on the clarifier floor or the core sampler will underestimate sludge depth. Once the sampler touches the bottom, the operator lifts it back out vertically with a quick jerk. The movement combined with the static head of water in the sampler (water depth) seats the ball in the valve at the bottom of the sampler. The operator

Core samplers are more commonly known as *sludge judges*.

then "reads" the blanket depth by looking for the point on the tube between the sludge and the clearer water above it. The height of the sludge blanket is typically estimated to the closest 15 cm (6 in.).

An interface/sludge blanket level (ISBL) analyzer, or sludge blanket detector, automatically identifies and reports the depth of the sludge blanket in a clarifier. It can replace the manual core sampler and, more importantly, eliminate the variability of manual readings.

Applications in Water Resource Recovery Facilities

Sludge blanket level is typically monitored in the following locations:

- The primary clarifier, to monitor the primary sludge level or control primary sludge pumps;
- The secondary clarifier, to monitor the secondary sludge level or control secondary sludge pumps; and
- Gravity thickeners, to monitor the sludge level or control thickened sludge pumps.

How They Function

Ultrasonic Interface Sludge Blanket Level Analyzers

Ultrasonic ISBL analyzers are similar to level sensors used in tanks or channels. In other applications, sound waves travel through air and bounce off the top of the liquid. In clarifiers, the sound waves travel through water and reflect off the different densities of sludge in the tank. Some manufacturers use fixed-frequency systems, whereas others scan within a range of frequencies to find the peak reflectance for that particular sludge. Typically, systems that perform a scan are more accurate and perform better in a wider range of applications. Because ultrasonic signals reflect off the different densities of sludge in the tank, the user can select which density is considered the sludge interface so that it agrees with a comparison measurement. Likewise, systems with graphical displays can show the sludge profile in the tank.

Electronics are typically rail-mounted over the point where the blanket needs to be measured. In circular clarifiers with a bottom that slopes toward the center of the clarifier, blanket depth is measured one-third of the distance in from the outer wall. The sensor is mounted on a rigid or hinged mount with its face submerged, and the distance between the sensor face and the tank floor is measured immediately after installation and keyed into the analyzer. The blanket height is calculated as the difference between this distance and the echo depth to the top of the blanket, similar to a level transmitter. Most systems can display the blanket interface as either the distance from the top of the blanket to the bottom of the tank ("sludge height" or "blanket depth") or as the distance from the top of the blanket to the surface of the water ("sludge level").

Ultrasonic ISBL analyzers consist of a probe with an ultrasonic transducer to transmit and receive signals; a probe-mounting arm; and an electronics module to receive, transmit, and process the signal.

Table 12.10 lists construction materials typically available for various components. The choice of construction materials depends on the reactivity of the environment and the monitored substance.

Optical Interface Sludge Blanket Level Analyzers

Optical ISBL analyzers are similar to turbidity-style suspended solids sensors. They function by transmitting two or more beams of light across a gap and calculate how much light was absorbed by media

Table 12.10 Typical Construction Materials for ISBL Analyzers (WEF, 2013)

Component	Material
Interface/sludge blanket level probe body (wetted part)	Chlorinated polyvinylchloride (CPVC), PVC, polyoxymethylene, Ryton, 316L stainless steel
Electronics module enclosure	Aluminum, 316L stainless steel, polycarbonate
Probe mounting arm (ultrasonic)	Aluminum, 316L stainless steel, galvanized steel, CPVC, PVC
Retractable cable (optical)	316L stainless steel

in the gap. A typical optical ISBL system consists of a suspended solids sensor with a long cable wound onto a spool. In a clarifier, the sensor will continuously measure while being slowly lowered deeper by unwinding the spool. The interface is defined by the operator as a certain concentration of suspended solids. Once that concentration is reached, the sensor is retracted from the clarifier to prepare for the next measurement. These systems can show both the interface level and the profile of the solids concentration in the tank.

Optical ISBL analyzers consist of a probe with a light source to transmit and receive signals; a retractable reel and cable; an electronics module to receive, transmit, and process the signal; and interlocks between the reel and clarifier mechanisms. The interlock prevents the reel from lowering the instrument head when skimmer or rake equipment are in the vicinity.

Electronics are typically placed in a rail-mounted cabinet over the point where the blanket needs to be measured. The housing is large enough to accommodate both electronics to analyze the data and a retractable motorized reel to lower and raise the sensor through the sludge.

Installation and Operation

Installation of ISBL analyzers can be divided into the following three classifications: in the path of skimmer arm or surface flights; out of the path of skimmer arm or surface flights; and a fixed submerged point out of the path of skimmer arm, sludge collectors, and surface flights.

An online analyzer should be installed at the point where the core sampler was traditionally used. In a circular clarifier, this would be a point along the bridge that is easily accessible and typically in the path of a surface scum skimmer. In a rectangular clarifier, this would be a point on the bridge typically over the cross collector and in the clarifier flight's path. The probes are mounted on a two-piece mounting pole that is hinged in the middle so the unit can swing out of the way of the flights or skimmers.

Optical units that must be lowered into the water could become snagged in the skimmer arm or flights. To avoid this, manufacturers typically either

- interlock the reel mechanism with the clarifier drive or chain-and-flight drive so the skimmer or flight cannot move while the probe is lowered and retracted or
- if there is enough time to lower and raise the probe after the skimmer or flight has passed, a limit or proximity switch is tripped when the skimmer arm passes underneath the ISBL analyzer, triggering the reel on which the optical probe is mounted.

Offline With Skimmers or Flights

This mounting location is not a physical challenge to ISBL analyzer operations. In a circular clarifier, this location would be the annular ring between an inboard launder that is mounted away from the clarifier wall. In a rectangular clarifier, this location would be over the cross-collector sump. While minimizing the skimmer or flight interference problem, neither location is ideal for measuring sludge blankets.

Fixed Submerged Location

Currently, only optical units can be adapted for fixed submerged installation. The sensor is wall-mounted below the water surface at a fixed elevation and provides continuous feedback on light absorbance. This installation eliminates the reel mechanism, but requires the tank to be drained when personnel need to access the probe for cleaning and maintenance. An alternative is installing the probe at a fixed elevation in an offline area and mounting it on a support that can be retrieved from the top without draining the tank. Fixed submerged systems have the potential drawback of only providing a sludge blanket level measurement at a single point; typically, this will just represent whether the sludge blanket level is rising or falling and works best in blankets up to 0.6-m (2-ft) thick.

Maintenance and Troubleshooting

The sensing element of the ISBL analyzer may need periodic cleaning, depending on the facility and wastewater involved. Rags, plastics, grease, and scum will interfere with sensor functions. Although some sensors come equipped with features to clean the sensor head, periodic manual cleaning will be necessary.

Retractable reels on an optical sensor need to be serviced periodically to ensure smooth operation. In an interlocked system, a jammed reel mechanism will keep the interlocks in place and prevent the drive from operating. In a limit-switch-operated system, reel failure will result in the extended cable being wrapped around the skimmer and the sensor being damaged or lost.

At best, manual core samplers have an error of 75 mm (3 in.); for most facilities, an error of 15 cm (6 in.) is acceptable. Online ISBL analyzers have an error between 2.5 to 38 mm (0.1 to 1.5 in.). Manufacturers report the analyzers' precision to be in the same range as accuracy. In addition, when ultrasonic ISBL analyzers are installed in line with scum skimmers and surface flights, they are hinge-mounted and designed to swing out of the way of the skimmers and flights. This interruption in measurement can be accounted for through signal averaging or through internal tilt sensors, which hold the reading when the angle of the sensor is greater than a certain angle. These tilt sensors also allow for measurement compensation when the sensor is not installed perfectly plumb. For optical systems, an external switch needs to be installed so that the sensor does not become tangled in the skimmer or rake mechanism.

TEST YOUR KNOWLEDGE

1. **Optical interface sludge blanket analyzers are lowered into the clarifier blanket each time a measurement is desired.**
   - ☐ True
   - ☐ False
2. **Rags, plastic, and grease can interfere with ISBL measurements.**
   - ☐ True
   - ☐ False
3. **Clarifier blanket depths may be manually measured with a**
   a. Rake arm
   b. Sludge judge
   c. Interface/sludge blanket level analyzer
   d. Radar
4. **Ultrasonic interface sludge blanket analyzers use __________ to determine sludge blanket depth in clarifiers and gravity settlers.**
   a. Sound waves
   b. Radar waves
   c. Light waves
   d. Proton waves

GAS DETECTORS

PURPOSE AND FUNCTION

At facilities that handle wastewater and sludges, dangerous gasses can be produced or build up within an area and quickly create dangerous/potentially lethal conditions. Combustible and/or toxic gases such as hydrogen sulfide (H_2S), methane (CH_4), carbon monoxide (CO), and others can quickly build up to dangerous levels in certain areas. In other situations, oxygen (O_2) can be displaced by other gases and create a low DO condition, which is also very dangerous. Several types of gas detecting systems are frequently used at wastewater facilities to monitor areas and provide warnings if unsafe gas conditions are present. Gas detecting monitors can also be used on foul air treatment systems such as scrubbers and biofilters to confirm that dangerous gases such as hydrogen sulfide have been sufficiently removed by the systems.

Confined space areas are defined by the U.S. Occupational Safety and Health Administration (OSHA) as having the following characteristics:

- Not designed for continuous occupancy
- Large enough for workers to enter to perform certain tasks
- Limited or restricted means for entry or exit

Refer to OSHA's webpage for resources on safety in and around confined spaces.

APPLICATIONS IN WATER RESOURCE RECOVERY FACILITIES

To keep people safe, we need to have properly functioning alarm systems that continuously monitor a potentially dangerous area and provide alarms if an area is unsafe before personnel enter it. Permanently mounted gas monitoring systems are used to perform this function in areas where dangerous gas conditions may occur and are regularly occupied by people. See Figure 12.67.

When people need to enter areas that are rarely occupied, such as manholes and wet wells, a portable monitor can be lowered into the area to test it before entry. The monitor can then be worn as a part of a person's personal protective equipment during entry to monitor and alarm if conditions change and could soon become unsafe.

Figure 12.67 Permanently Mounted Multiple Gas Monitoring System (Reprinted with permission by Passaro Engineering, LLC)

HOW THEY FUNCTION

Gas detectors use different types of sensors:

- Electrochemical sensors for oxygen or toxic gases
- Catalytic (pellistor or bead) sensors for combustible gases
- Infrared sensors for hydrocarbons (combustible gases)

Gas sensors can operate in diffusion mode or in pumped airflow mode. Diffusion mode is where air circulates around the sensor and the gas is drawn into the sensor by the natural force of diffusion, moving from a higher concentration to a lower concentration. There is no forced airflow so the sensor can only detect gases in the area immediately around it. A sensor operating in pumped mode uses a pump and tubing to pull a sample from a remote location and bring it to the sensor. See Figure 12.68. This allows the sensors, alarms, and other components to be safely located outside of corrosive or explosive areas; additionally, the area can be tested and determined to be safe before a person enters.

Electrochemical Sensors

These are basically fuel cells where the sensing electrode is treated with a catalyst that will react with a specific target gas. The gas is diffused into an electrolyte solution through a membrane or a diffusion barrier on the cell. If a target gas is present, it reacts with the catalyst and produces a weak electrical current that flows between the sensing electrode and the counter electrode. The reference electrode provides a zero-reference point so that the electrical potential between the sensing and counter electrodes can be calculated and then converted into a target gas concentration (see Figure 12.69).

Figure 12.68 Gas Monitoring System with Pumped Gas Monitoring (Reprinted with permission from MSA Safety)

Electrochemical sensor performance can be affected by several different factors:

- Temperature
- Exposure to high gas levels
- Extreme vibration or shock
- Humidity

The ideal operating condition for an electrochemical sensor is 20 °C at 60% relative humidity. However, sensors are designed to compensate for temperature changes over a broad range, typically –30 to 50 °C. Overall sensor life can be shortened if it is repeatedly exposed to temperature extremes. Electrochemical

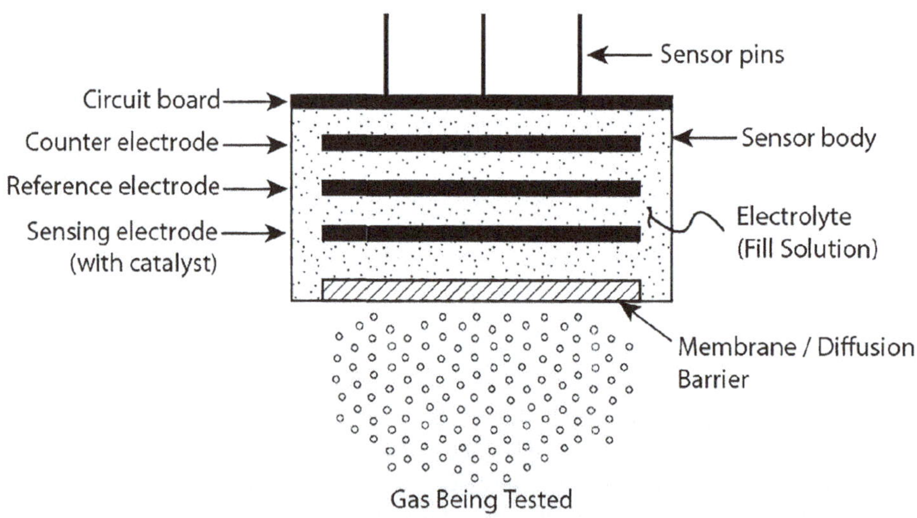

Figure 12.69 Components of an Electrochemical Sensor

sensors may be permanently damaged if they are exposed to very high gas concentrations, although this is fairly rare. Sensors can be damaged if they are dropped or installed in a high vibration area.

Humidity levels may be the most significant factor affecting sensor life. If a sensor is installed in an area with high relative humidity, water can be absorbed into the electrolyte solution, causing dilution and even leakage from the sensor body and pins. If a sensor is installed in an area with low relative humidity, the electrolyte can evaporate over time, causing a slower response time. The electrolyte can be "restored" to its original quantity by placing it in an area with approximately 60% relative humidity.

Relative humidity is the amount of water vapor that is present in air. It is measured as a percentage of the total amount of water vapor that air can hold at a specific temperature. Warm air holds more water vapor than cold air. Low relative humidity conditions often occur in spaces that are heated or cooled with traditional HVAC systems.

The catalyst in an electrochemical sensor begins to be consumed from the day it is produced. Therefore, it is not ideal to store sensors for a long time before they are installed. Most manufacturers recommend a maximum storage time of 6 months. After installation, the sensor will stabilize to the local temperature and relative humidity. It is a good idea to retest sensors several weeks after installation to verify proper performance after stabilization.

The best way to know if your electrochemical sensors are working properly is to perform a bump test and/or a calibration. This is covered in the maintenance section that follows the sensor descriptions.

Catalytic (Pellistor or Bead) Sensors

A catalytic sensor uses small (1-mm long) porous ceramic beads that have platinum-plated wire coils embedded in them (see Figure 12.70). The beads are sometimes referred to as *pellistors*, which is a combination of the words *pellet* and *resistor*.

Two beads (pellistors) are installed in the sensor in a circuit referred to as a *Wheatstone bridge*. See Figure 12.71. The detecting bead or active bead is treated with a catalyst that will oxidize when exposed to a flammable/combustible gas. The compensating bead is treated with a catalyst so that it will not oxidize when exposed to a flammable/combustible gas. When electrical current is sent through the circuit, the resistance in the beads causes them to heat up to approximately 500 °C. When the detecting bead is exposed to a flammable/combustible gas, the gas ignites on the bead and increases its temperature. The other bead's temperature does not change. The difference in temperature can be measured as a difference in resistance between the two beads. This difference is proportional to the amount of flammable/combustible gas that is present.

What is the difference between a flammable and a combustible substance? Both types of materials will ignite (burn) when exposed to an ignition source such as a spark or a flame. However, flammable materials burn at lower temperatures than combustible materials. The temperature at which a material will begin to burn is called its *flash point*. In other words, flammable materials have lower flashpoints than combustible materials.

Because of the high temperature that a pellistor operates at and the ignition that will occur if a flammable/combustible gas is present, the catalytic sensors are sealed behind a flame arrestor or sinter disk.

Pellistor beads are more likely to be poisoned than electrochemical sensors. Exposing the sensor to some types of cleaning solutions or lubricant spills or aerosols can cause sensor failures. Other compounds and chemicals called "inhibitors" can cause a more gradual reduction in sensor response time and the overall ability to measure specific gases. Exposure to compounds such as fire extinguisher chemicals and refrigerants such as freon cause sensor inhibition.

A flame arrestor is a passive safety device typically made out of metal that allows gases to pass through it under normal conditions, but will stop a flame from passing through to prevent an explosion or a back flash fire from spreading. They are typically made up of many small parallel labyrinth-like pathways that cool the flame or the gas to extinguish the flame. A sinter disk is smaller, but operates under the same principle—it is made up of many small metal filings or powder that are pressed together in a disk with many small openings that allow gas to pass through but extinguish a flame.

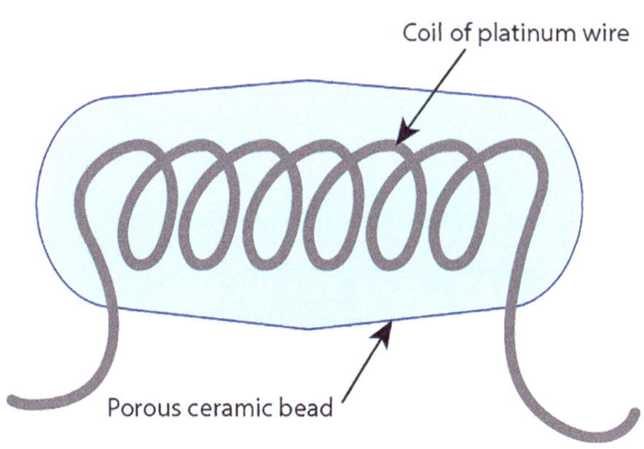

Figure 12.70 Schematic of a Bead or Pellistor Sensor

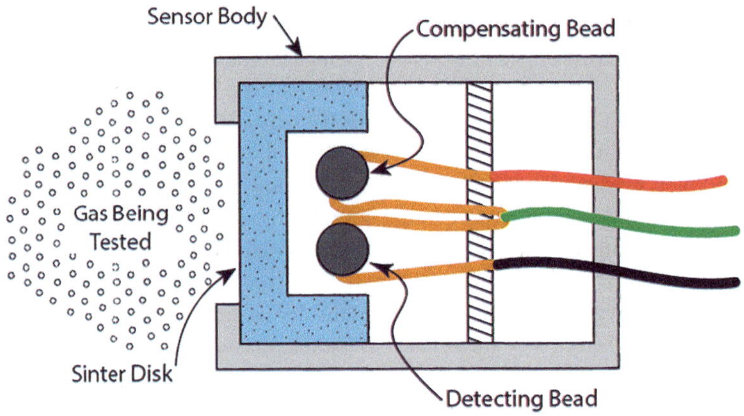

Figure 12.71 Basic Components of a Catalytic Gas Sensor

The phrase *poisoning a sensor* refers to immediate and irreversible damage that occurs when a catalytic sensor is exposed to chemicals that block the pores of the bead or bind to the catalyst and reduce the activity of the sensor. Common catalytic sensor "poisons" include silicon, sulfur, lead, and phosphorus compounds. Carbon soot can also build up on the bead and cause poisoning if a sensor is exposed to high concentrations of flammable/combustible gases.

Pellistors operate under high temperature conditions and are affected by gases and chemicals that they are exposed to. Their sensitivity will degrade over time. Sensor failure may only be detected by regular checks by a bump test and calibration.

Infrared Sensors

Infrared sensors are often used to detect hydrocarbons/combustible gases such as methane. The sensor has a gas inlet and outlet that allow the air that is being tested to flow through a chamber (see Figure 12.72). There is an infrared light source that passes through the chamber. The detector measures the infrared after it passes through the gas. Infrared is changed in specific ways when it passes through different gases. Lenses and optic filters are used to help the detector identify the changes to the infrared sensor that occur when the target gases are present. Most infrared sensors include a second sealed chamber (no target or interfering gas present) that can be used as a reference measurement.

Infrared sensors do not include components that are consumed or physically affected during the measurement. Therefore, they tend to have a longer expected life than the electrochemical or catalytic sensors. Infrared sensors are typically more expensive to purchase initially; however, they require less monitoring and calibration so their related O&M costs are typically lower.

INSTALLATION AND OPERATION

Proper installation is important for gas detectors to operate properly. Here are a few general installation practices to check for:

- Sensors operating in diffusion mode must be installed in areas where gas leaks or collection are most likely to occur. Gases that are less dense than air will tend to collect up near the ceiling in an enclosed area. Gases that are more dense than air will tend to collect down near the floor.
- Electrochemical sensors should have gas inlets installed pointing downward.
- Infrared sensors are installed so the infrared light beams travel horizontally.

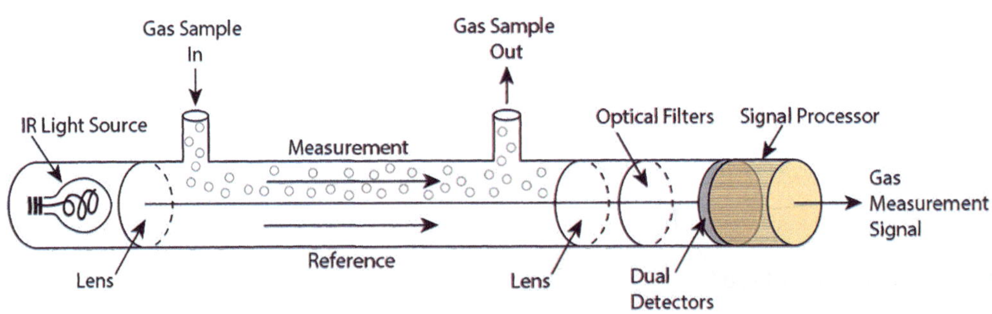

Figure 12.72 Components of an Infrared Gas Sensor

Keep all sensors protected from water, oil, and dust. Many sensors have guards that help to keep them dry and clean. Make sure that all of the guards are properly installed.

Do not allow paint to get on a sensor.

Do not spray sensors with cleaning solutions to clean them. It is always best to protect them from dirt, dust, water, and aerosols. If a sensor is accidentally directly contaminated with a substance, check your O&M manual and/or with your local service representative to determine how to best clean the sensor without causing permanent damage.

MAINTENANCE AND TROUBLESHOOTING

The best way to ensure that a gas detector is functioning properly is to perform regular bump tests. An alarm concentration of a test gas is passed over a sensor to make sure it is detected and that an alarm is properly activated. This test verifies that a sensor is able to detect the target gas and is creating an alarm to warn personnel about potentially unsafe conditions. All sensors will eventually need to be replaced. A bump test may be the only way to determine when a sensor has failed. A bump test tells us the status of the sensor at the moment the test is performed. It does not help to predict when a failure will occur. Some gas detectors have internal checks that will activate alarms to notify users about an error in the sensor or detector. Carefully read the manufacturer's O&M manual to determine the bump test procedure and testing schedule that is appropriate for your application.

Calibration is critical to ensure that gas detectors will provide alarms and warnings as intended. Detector readings must be checked against known concentrations of test gases. Test gases should be certified, traceable test gases and must only be used before their expiration date.

Detector manufacturers can provide bump tests and calibration kits that include the test gases and accessories needed (see Figure 12.73). Always follow the accompanying directions carefully to ensure that the detector will protect personnel as needed. Detailed records of all bump tests and calibration should be maintained.

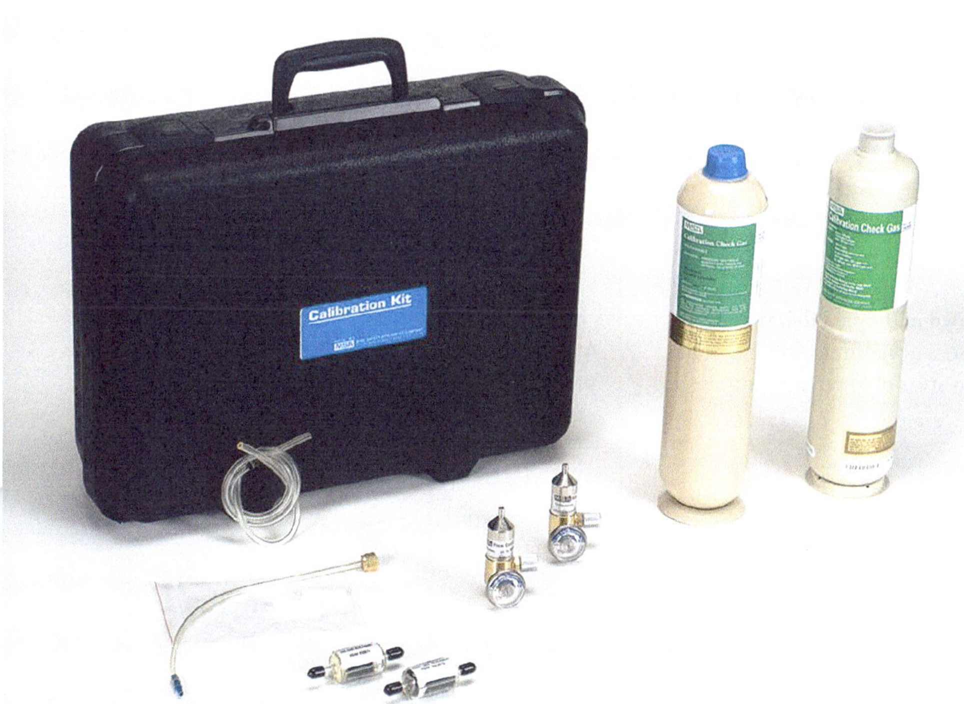

Figure 12.73 Gas Detector Bump Test/Calibration Kit (Reprinted with permission from MSA Safety)

All sensors will eventually fail and need to be replaced. Manufacturers can provide information on expected sensor life spans. Some sensors, such as oxygen sensors, that are continuously exposed to their target gas have predetermined life spans. Other sensors may rarely be exposed to their target gas. This can have a big effect on how quickly a sensor catalyst is consumed.

Environmental conditions can also affect the useful life span of a sensor. The presence of sensor poisons and inhibitors can cause quick sensor failures. All of these factors should be considered. Sensor replacement can be done on a set schedule before the sensor fails. However, this will waste some useable life of each sensor. Sensor replacement can be done only after a bump test confirms sensor failure. However, this creates a period of time between the failure and the next bump test during which there is no gas detection or alarms, increasing risks for facility personnel. Increasing the bump test frequency can minimize the amount of time there could be an unknown sensor failure, but this also increases labor and testing costs. Facility operators need to weigh the risks versus the costs of more frequent testing or sensor replacement and find the right solution for their application. Manufacturers regularly incorporate new technology to improve self-testing and reliability of gas detectors. When replacing parts or detectors, make sure you are checking for products that incorporate the latest improvements.

Sensors do degrade over time and should not be ordered too early and stored for long periods of time. However, a reliable source that can quickly provide replacement sensors should be identified and accounts should be set up before parts are needed.

TEST YOUR KNOWLEDGE

1. __________ at WRRFs monitor for hydrogen sulfide, methane, and low oxygen conditions.
   a. Differential pressure instruments
   b. Dissolved oxygen analyzers
   c. Fuel emissions testing
   d. Gas detectors

2. __________ are areas that are not designed for regular occupancy, are large enough for a person to enter, and have limited means for entry and exit.
   a. Confined space areas
   b. Office space cubicles
   c. Hazardous materials storage areas
   d. Control rooms

3. __________ gas sensors are located in the area that they are testing, and gases enter the sensors by naturally moving from areas of higher concentration to areas of lower concentration.
   a. Forced air
   b. Pumped in
   c. Diffusion mode
   d. Capillary action

4. __________ sensors are basically fuel cells where the sensing electrode is treated with a catalyst that will react with a specific target gas.
   a. Catalytic
   b. Electrochemical
   c. Infrared gas
   d. Optical scatter

5. __________ sensors have an active bead that is treated with a catalyst that will oxidize when exposed to a flammable/combustible gas.
   a. Catalytic
   b. Electrochemical
   c. Infrared gas
   d. Optical scatter

6. __________ sensors monitor for changes to the infrared light beam that occur when target combustible gases/hydrocarbons are present.
   a. Catalytic
   b. Electrochemical
   c. Infrared gas
   d. Optical scatter

7. A __________ is a passive device with many serpentine pathways that break a flame up and cool it down to extinguish it.
   a. Combustion chiller
   b. Flame alarm
   c. Fire extinguisher
   d. Flame arrester

8. Sensor __________ is immediate and irreversible damage that occurs when a catalytic sensor is exposed to chemicals that block the pores of the bead or bind to the catalyst and reduce the activity of the sensor.
   a. Poisoning
   b. Sickness
   c. Death
   d. Destruction

9. Under the right conditions, gas sensors could last forever.
   - ☐ True
   - ☐ False

10. For most sensors, they start to be consumed from the first day they are produced.
    - ☐ True
    - ☐ False

11. A __________ is used to verify that a gas sensor is operating correctly.
    a. Gas conductance test
    b. Bump test
    c. Resistive decade box
    d. Magnetic reactivity monitor

CHAPTER SUMMARY

Device and Parameter Measured	Applications	Key Points
Limit switch (detects position)	■ Detect door and hatch positions ■ Detect safety guards	■ Simple actuator with mechanical linkage to an electrical switch ■ Must contact what it senses ■ Will wear over time
Proximity switch (detects presence)	■ Sense equipment motion and speed	■ No contact with objects sensed ■ No moving parts ■ Uses inductive, capacitive, or magnetic detection methods
Pressure switch (point pressure)	■ Actuate switches at a single pressure setpoint	■ Diaphragm
Pressure transmitter (continuous pressure)	■ Continuously monitor pressure	■ Diaphragm
Temperature sensors	■ Monitor temperature on equipment or in processes	■ Thermocouples ■ RTDs ■ Thermistors
Level switch (point level)	■ Actuate an electrical switch at a single level setpoint	■ Float switches ■ Conductivity switches
Level transmitter (continuous level)	■ Continuously monitor the liquid level in a tank, channel, wet well	■ Ultrasonic level ■ Noncontact radar ■ Pressure-sensing level–transducers and pressure transmitters
Flow meters (closed pipe flow)	■ Monitor pump discharge flow ■ Monitor influent/effluent flow ■ Monitor flow for process control such as return activated sludge and waste activated sludge flows	■ Magnetic flow meters (closed pipe liquid) ■ Thermal mass flow meters (closed pipe gas)
Dissolved oxygen sensor (measures dissolved oxygen)	■ Maintains DO concentrations in aerobic treatment processes such as activated sludge and aerobic digesters. ■ May be used to control blower output and run time. ■ May be used for permit compliance in final effluent.	■ Polarographic and optical sensors ■ Range of 0 to 40 mg/L ■ Measurement dependent on barometric pressure, temperature, and salinity ■ Measures percent saturation and calculates concentration ■ Electrolyte and membranes require regular replacement.
Chlorine residual (measures free or total chlorine residual)	■ Monitors chlorine residual at addition point to maintain desired dose. ■ Monitors chlorine residual in final effluent for permit compliance. ■ May be used to control chlorine and/or dichlorination chemical feed rate.	■ Photometric and amperometric methods ■ Photometric method relies on reaction between DPD reagent and chlorine residual to produce a magenta color. ■ Color intensity is proportional to residual concentration. ■ Excessive chlorine residual may turn samples clear or yellow instead of magenta.

		■ The amperometric method relies on passage of HOCl through a semi-permeable membrane. ■ HOCl reacts with electrodes to produce an electric signal. ■ Electric signal is proportional to residual concentration.
pH (measures hydrogen ion concentration)	■ Monitor biological treatment processes for pH change ■ May be used to control acid-base additions for pH adjustment ■ May be used to enhance chemical phosphorus removal	■ Glass sensors and ISFET sensors ■ Both sensor types measure pH by comparing an external voltage (electrical potential) to a reference electrode. ■ Electrical potential is proportional to hydrogen ion concentration.
ORP (measures water's potential to oxidize or reduce organic matter and nitrogen compounds in water)	■ Monitor chlorination and dichlorination ■ May be used to control chlorine addition and/or dichlorination chemical addition ■ Monitor anaerobic zones for enhanced biological phosphorus removal ■ Monitor anoxic zones for denitrification potential ■ Monitor aerobic processes to determine when maximum oxidation has been achieved ■ Control on/off cycling of aeration in sequencing batch reactors and aerobic digesters	■ Measures how oxidized or reduced wastewater is ■ Expressed in millivolts ■ Depends on all chemical species present ■ ORP range can predict which biological reactions are likely to be taking place. ■ Similar to pH probes, but with a platinum wire anode ■ Cannot be calibrated, but function can be verified
TSS (measures suspended solids)	■ Monitor solids loading to various treatment processes ■ Automated calculation of sludge age	■ Infrared (optical) sensors and microwave technology ■ Optical sensors may be either transmittance or reflectance type ■ Reflectance type work better for low solids concentrations ■ Optical sensor measurements are affected by solids size and color. ■ Microwave analyzers work well for high solids applications like dewatered cake.
Interface/sludge blanket analyzer (measure blanket depth)	■ Monitors sludge blanket depth in • Primary clarifiers • Secondary clarifiers • Gravity thickeners	■ Ultrasonic and optical interface analyzers ■ Ultrasonics use sound waves to measure blanket depth. ■ Optical interfaces determine blanket depth as a change in solids concentration as the sensor is lowered from the clarifier surface into the sludge blanket. ■ Measurements are taken in the same location as a traditional manual grab sample with a sludge judge.
Gas detectors (measure hazardous gases such as methane and hydrogen sulfide and low oxygen conditions)	■ Monitor wet wells ■ Monitor anaerobic digester buildings	■ Electrochemical sensors test for toxic gases. ■ Catalytic sensors test for hydrocarbons, flammable/combustible gases, and oxygen. ■ Infrared sensors monitor for hydrocarbons and combustible gases.

Download U.S. Environmental Protection Agency (EPA) reports and manuals by visiting the U.S. EPA National Service Center for Environmental Publications website at https://nepis.epa.gov and searching for the report number.

References

Endress+Hauser. (n.d.). *Operating Instructions: ISFET sensors, BA01609C/07/A2/01.16, 71324851.*

Endress+Hauser. (2022). Digital free chlorine sensor, Memosens CCS51D. https://www.endress.com/en/field-instruments-overview/liquid-analysis-product-overview/free-chlorine-digital-sensor-ccs51d?t.tabId=product-specification

GLI International. (2000). Operating manual: Membrane dissolved oxygen sensors (Model 5400, 5500, and 5600-series).

Hach Company. (2010). *Dissolved oxygen sensor, DOC023.97.80082, user manual* (1st ed.). Hach Company. https://www.hach.com/clark-submersion-dissolved-oxygen-sensor/product-downloads?id=7640086016&callback=qs

Hach Company. (2012). *LDO Sensor, DOC023.97.80170, user manual* (4th ed.). https://www.hach.com/dissolved-oxygen-sensors/ldo-sc-luminescent-dissolved-oxygen-sensors/family-downloads?productCategoryId=35546331647

Hach Company. (2020). *DOC023.97.80614, CL17sc, user manual* (6th ed.).

Morris, J. (1966). The acid ionization constant of HOCl from 5 °C to 35 °C. *Journal of Physical Chemistry, 70*(12), 3798–3806.

Spencer, M. (2012). *The everyman's guide to the miraculous but misunderstood ORP sensor.* Water Analytics, Inc. http://www.wateranalytics.net/sites/default/files/Aquametrix%20ORP%20White%20Paper.pdf

Water Environment Federation. (2013). *Automation of water resource recovery facilities* (4th ed., WEF Manual of Practice No. 21).

Water Environment Federation & Association of Boards of Certification. (2018). *Wastewater treatment fundamentals I—Liquid treatment.*

YSI Inc./Xylem Inc. (2022). *Dissolved oxygen measurement.* YSI Inc./Xylem Inc. https://www.ysi.com/parameters/dissolved-oxygen

Suggested Readings

American Public Health Association, American Water Works Association, & Water Environment Federation. (2017). *Standard methods for the examination of water and wastewater* (23rd ed.). American Public Health Association.

Ashcroft. (2019). Diaphragm seal installation and maintenance manual. https://www.ashcroft.com/wp-content/uploads/2020/08/manual-100-200-300-threaded-seal.pdf

Certified Balance and Scale. (2014, May 12). *Polographic probe maintenance.* Sycamore, Illinois. https://www.youtube.com/watch?v=d0Sw7znYMDU

Goncalves, T. (2017). *Advantages of manifold valves used with pressure instruments. Ashcroft.* Advantages of Manifold Valves Used with Pressure Instruments (tristatetechnicalsales.com)

International Safety and Equipment Association. (n.d.) *Recommendations for DRPGM calibration and verification.*

Metcalf & Eddy/AECOM. (2014). *Wastewater engineering: Treatment and resource recovery.* McGraw Hill Education.

Palin, A. (1957). The determination of free and combined chlorine in water by the use of Diethyl-p-phenylene Diamine. *Journal of the American Water Works Association, 49*(7), 873–880.

Spitzer, A., & Boyes, W. (2002). *The consumer guide to magnetic flowmeters.* Copperhill and Pointer.

U.S. Occupational Safety and Health Administration. (n.d.-a). Confined spaces. https://www.osha.gov/SLTC/confinedspaces/

U.S. Occupational Safety and Health Administration. (n.d.-b). Hydrogen sulfide. https://www.osha.gov/SLTC/hydrogensulfide/index.html

Water Environment Federation. (2014). *Moving toward resource recovery facilities.*

Water Environment Federation. (2016). *Operation of water resource recovery facilities* (7th ed., WEF Manual of Practice No. 11).

Water Environment Federation. (2018). *Design of water resource recovery facilities* (6th ed., WEF Manual of Practice No. 8; ASCE Manuals and Reports on Engineering Practice No. 76).

CHAPTER 13

Supervisory Control and Data Acquisition (SCADA) Systems

Introduction

Supervisory control and data acquisition (SCADA) systems are an important tool for effective, efficient monitoring and control of water resource recovery facilities (WRRFs). Chapter 12 described the instruments that measure process conditions in a facility. This chapter discusses how SCADA systems collect data from those instruments and equipment and allow operators to conveniently view the data (i.e., monitor the process) and make changes to improve operations (i.e., control the process).

LEARNING OBJECTIVES

Upon completing this chapter, you will be able to:

- Understand the importance of a SCADA system to facility operation.
- Describe how data flow through a SCADA system.
- Understand the roles of major components of a SCADA system including control panels, input/output devices, controllers, network, and operator interfaces.
- Understand how a SCADA system is used for monitoring and control.
- Understand the importance of the data collected by SCADA.
- Be able to recognize problems with the SCADA system's operation.

Purpose and Function

The purpose of a SCADA system is to provide situational awareness to operators and to collect/store data for use throughout the utility. Supervisory control and data acquisition systems, as shown in Figure 13.1, function as the intermediary between the operator and the instrumentation/controls throughout the utility. Supervisory control and data acquisition systems take data in and send commands to field devices. They display the data in a way that easily allows the operator to understand and control the process. As with the eyes, ears, and nose of a human body, the instrumentation senses the world around the facility. The SCADA system, working in conjunction with the WRRF operator, acts like the brain in the body to make sense of the information from the instrumentation and take action to keep the facility under control.

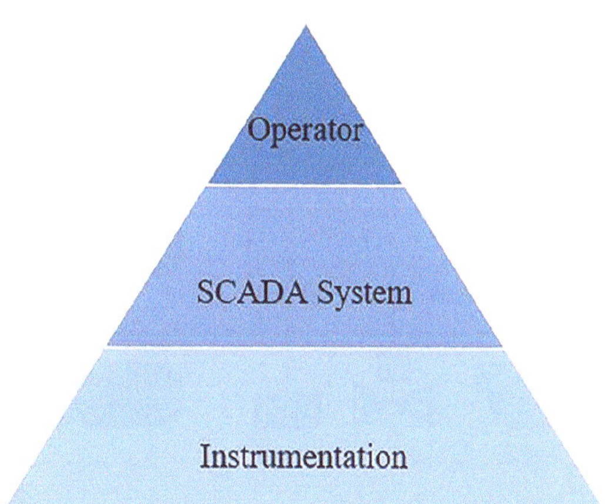

Figure 13.1 Supervisory Control and Data Acquisition System High-Level Overview

A cyber-physical system is a system made of interacting computers, humans, and physical devices engineered for function through physics and logic.

Supervisory control and data acquisition systems are used in many industries, not just wastewater.

Instrumentation provides the foundation for the entire SCADA system. If the data from instrumentation are not reliable or accurate, the process cannot be properly controlled by SCADA. If the system gets "garbage in," it will provide "garbage out."

Theory of Operation

Supervisory control and data acquisition is a system made up of elements that bring data from the physical equipment and processes in the field to the operator and the organization. As a cyber-physical system, the computer-based SCADA system can directly change the operation of WRRF process equipment. For example, SCADA may remotely adjust the opening of a valve or start a mixer. The SCADA system in your facility may also be referred to as a *process control system* (PCS) or a *distributed control system* (DCS). These systems have the same basic core components and functions that are connected to each other as presented in Figure 13.2.

The lowest level of a SCADA system is made up of field devices and instrumentation (discussed in Chapter 12). These are shown in level 1 of Figure 13.2. Instrumentation includes the sensors and other indicating equipment that directly touch the physical equipment being monitored or controlled. The information provided by these devices is referred to as *signals*. Supervisory control and data acquisition systems monitor many signals that are generally divided into two types. The first type is called a *digital signal* because the information provided by the signal can only represent one of two possible values (e.g., off or on, normal or failed, absent or present). An extra set of electrical contacts that act as a simple switch may be mounted directly on a pump motor starter such that the contacts on the switch open when the motor starter is off and close when the motor starter is on. The two possible values of the signal provided by the contacts are off and on. The second type of signal is known as an *analog signal*. In general, analog signals can vary continuously within a range of possible values. For example, a pressure sensor can be placed on pump discharge piping to provide a signal that directly relates to the gauge pressure within the discharge pipe. Here, the pressure value could be anything from zero all the way up to some maximum pressure.

Analog sensors are typically paired with additional equipment and transmitters that convert a sensor reading to a more reliable form. This pairing is so the signal can be sent over a longer distance without degradation or errors from electrical interference. Often, the transmitted signal takes the form of an electrical circuit whose electrical current varies between 4 and 20 milliamps around the loop that forms the circuit. One benefit of using 4 milliamps as the "live zero" reading for a sensor is that the SCADA system can determine if a sensor has failed or wiring has been broken if the amperage drops to 0 milliamps.

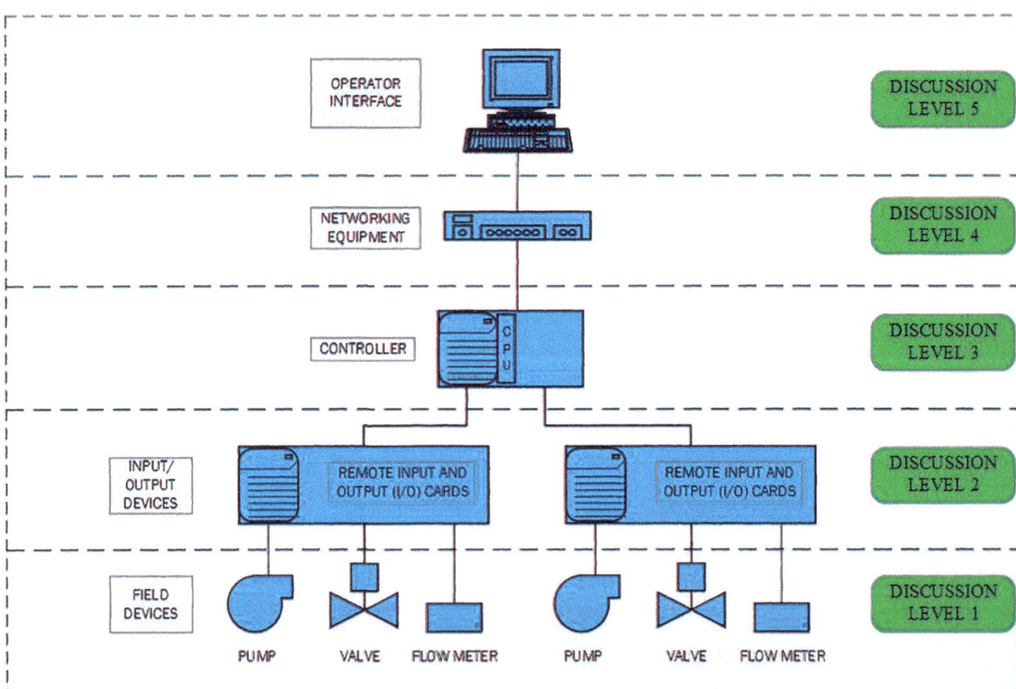

Figure 13.2 Core Components of a Facility Supervisory Control and Data Acquisition System

Wiring connects each sensor/transmitter back to the SCADA system input/output devices. Input/output devices are shown as Level 2 of Figure 13.2. Errors in transmission caused by the industrial environment for analog signals may occur. To avoid these errors, special cabling is used that incorporates twisted pairs of wires covered by additional electrical noise shielding.

Supervisory control and data acquisition system input/output devices convert a measured electrical signal (such as voltage or amperage) into a numerical value inside the SCADA computers.

To allow for flexible combinations of different types of input/output devices to match the local needs of the facility, SCADA systems use modular input/output circuit boards that can be quickly replaced upon failure. The input/output boards receive the analog and digital signals from the field equipment and convert the electrical signal into the computer values that the SCADA controllers can understand. Each individual signal is referred to as an *input/output point.* Supervisory control and data acquisition vendors offer a wealth of different types of input/output boards. Common inputs include analog (4 to 20 milliamps [mA] or 1 to 5 volts direct current [VDC]) and digital inputs (24 VDC or 120 volts alternating current). Common outputs include analog outputs (4 to 20 mA) and digital outputs (relay or solid state). Less common input/output boards might include, for example, those for measuring temperature directly from thermocouples.

Computers store values as tiny electrical switches that are either on (1) or off (0).

While the field wiring from each sensor to an input/output board as described above is very common, there are also many options for digital communication directly from the field equipment to the controller using modern digital communication networks.

The controller is the next level of core components of a SCADA system, as presented in Figure 13.2. Controllers are programmable. Controller programming converts input/output signals into more common units we use to monitor the facility such as degrees Celsius, feet, or percent open. Controller programming can assign labels to signals. The labels are called *tag names.* Tag names are used as references for the signals throughout SCADA system programming. Controllers may also be programmed to perform simple calculations, such as calculating equipment runtime or flow totals. Controllers may also be programmed to perform more advanced sequences of equipment operation or automatic control of facility processes. For example, a controller in the facility may automatically follow a sequence to start and stop multiple pumps to control a wet well level within predefined limits. A controller may automatically vary the speed of recycle pumps to maintain a desired recycle rate.

The next level of core components of a SCADA system, as presented in Figure 13.2, is networking. Controllers communicate to each other and to the highest level of core components of a SCADA system via SCADA networks. Supervisory control and data acquisition networking can take many forms. Supervisory control and data acquisition networks can include wired and/or wireless portions. Wired networks may be via copper and/or fiber optic cabling. Many SCADA networks now use communication networks based on the same technology used for Internet communications known as Ethernet and using Internet protocol. Note that, depending on the design of the PCS, networking may play a part in communication between any of the levels of core components shown in Figure 13.2.

Level 5, the highest level of SCADA system core components as presented in Figure 13.2, is the operator interface level. The primary function of the operator interface level is to allow an operator to monitor and control the process. The operator interface receives information from the controller. The operator can view the information. This is typically done through a computer terminal or touch screen called a *human-machine interface* (HMI). The operator interface may only display information or it may allow the operator to make adjustments to the treatment process. For example, an operator may use the interface to remotely turn a pump on or off or to open or close a valve. Operators may also use the interface to adjust process setpoints. A setpoint is a desired value or an upper or lower limit that is specified by the operator. A good example of using setpoints is to specify minimum and maximum dissolved oxygen concentrations in an activated sludge process. Once the operator has adjusted the setpoints, the HMI passes that information down to the controller. The controller will then send signals through input/output cards to make adjustments to field devices so the treatment process stays within the setpoints. In this example, the HMI would pass on the lower and upper limits for dissolved oxygen concentrations to the controller. The controller programming would then send signals through input/output cards to allow the blower to run longer to increase dissolved oxygen until it reached the upper setpoint concentration before turning the blower off. The controller programming could also call for some other action, like changing which valves are open or closed on different air headers. Those changes should affect the treatment process, changing the field device readings. The field devices then report back to the input/output cards, which report to the controller, which reports to the operator interface. Information flows up and down the SCADA architecture in both directions.

The operator interface programming uses tag names to keep track of different variables.

The operator interface typically shows a stylized schematic of the treatment process or the entire WRRF that may include equipment animations. The operator should be able to see at a glance what equipment is in service. Color coding is often used to show on/off and alarm status. Operators can typically get more information about a particular process or piece of equipment by selecting it through the interface. Selection through touch screens can be as simple as tapping on the piece of equipment or process with a finger. Dropdown menus may also be used. Once selected, operators may be able to access graphs of historic data for that piece of equipment as well as other information.

Operator interfaces typically include additional functionality to alert operators in the event of an alarm, including displaying and organizing alarms in a summary display. The "Operations" section describes operator interface displays in more detail.

TEST YOUR KNOWLEDGE

1. **Label the sections on the following high-level overview of a SCADA system.**

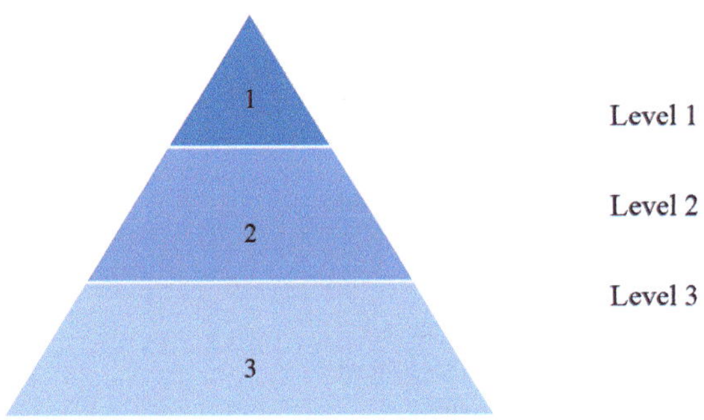

2. **Correctly label the five core components of a SCADA system as shown in the following image.**

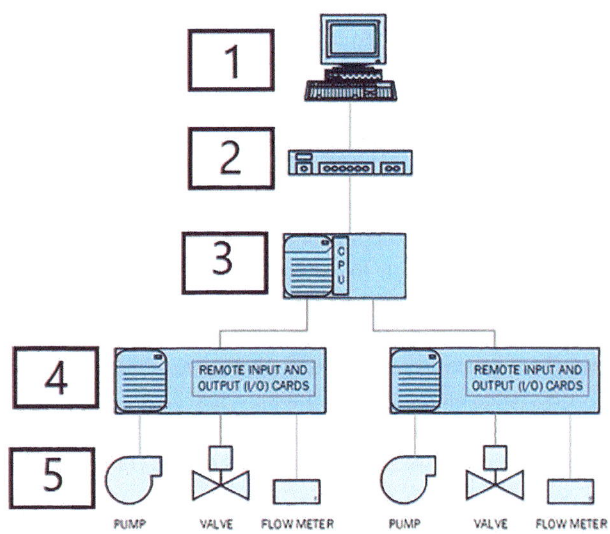

Equipment

Supervisory control and data acquisition systems differ from many traditional office computer systems because they are designed to be highly reliable. They provide accurate and timely data for the operation of critical infrastructure in an industrial environment. In this section, we present examples of the components that make up a typical SCADA system to help you to recognize the components in the field.

The SCADA equipment and software that make up the SCADA system may be from different vendors. This equipment is brought together to form a fully functional system by professionals known as *system integrators*. Often, in larger facilities, the system may be a DCS that is provided by a single vendor. As a result, some of the terms used for components may vary. For example, typical names for controllers provided by system integrators include programmable logic controllers (PLCs) and remote telemetry units (RTUs). Typical names for controllers used by DCS vendors include distributed process controller, distributed control unit (DCU), RTU, or some similar variations.

CONTROL PANELS

For safety and protection of equipment, controllers and/or input/output devices are located in cabinets known as *control panels*. Figure 13.3 shows the inside of a typical control panel located in a facility.

As shown in Figure 13.3, the major components inside a typical control panel include:

- Controller and/or input/output devices
- Equipment for distributing power to panel devices and for input/output signaling, including circuit breakers, fuses, and power supplies
- Control equipment such as relays
- Wiring terminals to make the signal wire connections between input/output devices and equipment in the field, such as transmitters, safe and secure

Some equipment often mounted on the door of the control panel include a power disconnect for the panel, control switches, pilot lights, and a local operator interface (LOI) panel.

As mentioned previously, SCADA control panels may come from a single vendor or be made from components from multiple vendors; thus, the control panels in your facility may vary significantly from the picture presented in Figure 13.3.

INPUT/OUTPUT DEVICES

Figure 13.4 presents a closeup view of typical PLC-based input/output devices.

Typically, input/output circuit boards are grouped together or slid into a chassis that provides structure to hold the boards. Input/output circuit boards of this type are also often referred to as *input/output cards*. The grouping of input/output cards is often referred to as an *input/output rack*. The back of the input/output rack includes a modular connector into which the input/output card mates or slides into. The modular connector connects the input/output card to power and a communication channel. The input/output card can receive configuration information and share the signal values with the rest of the SCADA system through the communication channel. If the input/output rack is located a significant distance from the controller, the rack may also include a communication card to communicate back to the controller. The control panel that holds the input/output rack will also include a power supply for the rack. If the process monitored or controlled by the signals connected to the input/output cards is critical, redundant power supplies may be used.

Figure 13. 3 Typical Control Panel

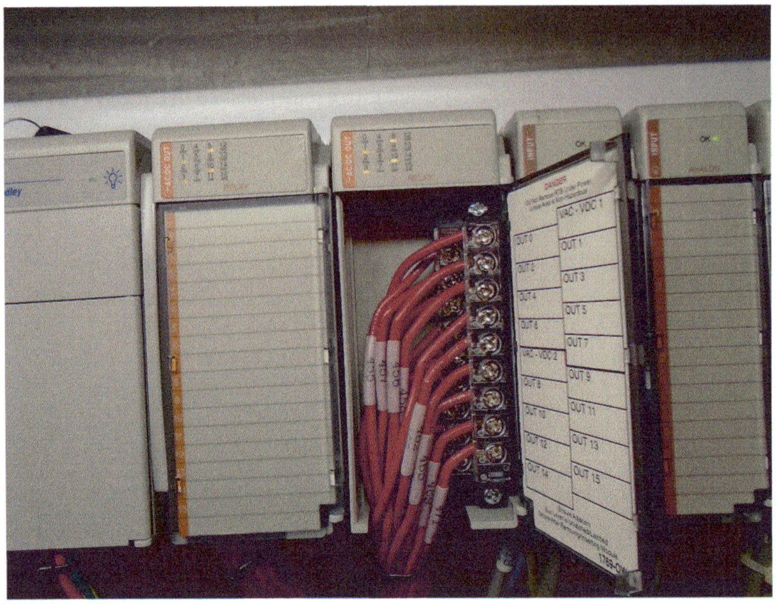

Figure 13.4 Typical Input/Output Rack

Typical input/output cards include cards for receiving analog signals (analog inputs), cards for receiving status information such as on/off or open/close (digital inputs), cards for sending analog commands (analog outputs), and cards for sending commands such as run/stop (digital outputs).

The visible front of input/output devices typically includes status light emitting diodes (LEDs) that provide the first level of troubleshooting. For example, a digital input card has an LED for each signal being monitored. If the status for the signal being monitored is "on," the LED for that input/output point on the card will light. If the point should be "on" and the LED is not lit, then we might suspect a problem in the field or with the wiring. If the LED is lit and the SCADA system does not see the signal, we might suspect an issue with the SCADA system, especially its programming.

Note that input/output devices can take many other forms than the typical input/output rack described and pictured above.

A PLC is like a tiny industrial computer that is responsible for several core SCADA system tasks.

CONTROLLER

Figure 13.5 presents a closeup view of a typical controller. In the figure, the controller is the card with the key switch and green Ethernet cable. In this example, the controller is a PLC. As with input/output devices, controllers can take many forms depending on the required functionality, including processing power, memory, amount and type of input/output devices, the criticality of the process, and cost.

As shown in Figure 13.5, a controller is often installed in the same chassis and/or cabinet with its input/output cards. It is often powered by the same power supply that powers the input/output cards. If the process is critical, two controllers may be installed for redundancy. If one controller fails, the other controller automatically takes over. The controller communicates with local input/output cards through the communication channel in the back of the input/output. The controller may communicate to remote devices either through communication ports located on the controller itself or through communication cards installed in the input/output rack.

The front of the controller often includes diagnostic LEDs that indicate if the controller has power, is running its program, or has experienced a fault. The controller may also have LEDs for diagnosing communications.

NETWORK

Figure 13.6 shows a closeup view of a typical networking device. The networking device is the blue device with green Ethernet cables. The networking device may be located in a control panel to allow

Figure 13.5 Typical Controller

Figure 13.6 Typical Network Switch

communication from a controller to the rest of the SCADA system. The device shown is known as an *Ethernet switch.*

Ethernet is a standards-based method for communicating between computers. The standards define the requirements for the devices, the wiring, and how signals are sent so they can be understood at the other end.

Ethernet communications in a facility are based on the same standards used for the Internet. For security, care must be taken to ensure the facility control system is never connected directly to the Internet. Also, care must be taken to ensure that no unauthorized devices (such as laptops from a visitor) are connected to the Ethernet switch. These security measures will help prevent the possible spread of computer viruses on the SCADA network.

The form of networking equipment you will find in your control panels may differ greatly based on the functions required, including the number of connections, the type of communication cable used, the criticality of the communications, and the complexity of the network.

An Ethernet port is an opening on an Ethernet switch or other piece of computer equipment that accepts and holds a network communication cable.

The networking equipment typically includes LEDs that can be used to diagnose networking problems. For example, for each network connection on an Ethernet switch, there are typically at least two LEDs that indicate the port is connected (linked) to another device and that communication is actively occurring among other diagnostics.

The form of the networking cable connected to the network device may also vary based on functional requirements. Ethernet cabling may take the familiar form of a copper-based cable with a clear plastic connector, or the cabling may use of fiber optics. Fiber optics are often used where electrical noise may be an issue, for long runs of cable, and/or for cabling between buildings. Electrical noise is unwanted amperage or voltage that is induced on a copper-based cable as a result of long runs of cable near equipment or cabling that can produce the noise. Electrical noise can cause errors in understanding the message being sent over the cables. This is similar to how it is difficult for a person to understand a conversation in a noisy room. Older forms of SCADA communication make use of twisted-pair cables and coaxial cables. You may also find wireless communication being used for locations where cabling is not feasible.

OPERATOR INTERFACES

LOCAL OPERATOR INTERFACES

Figure 13.7 shows a LOI mounted on the front of a control panel.

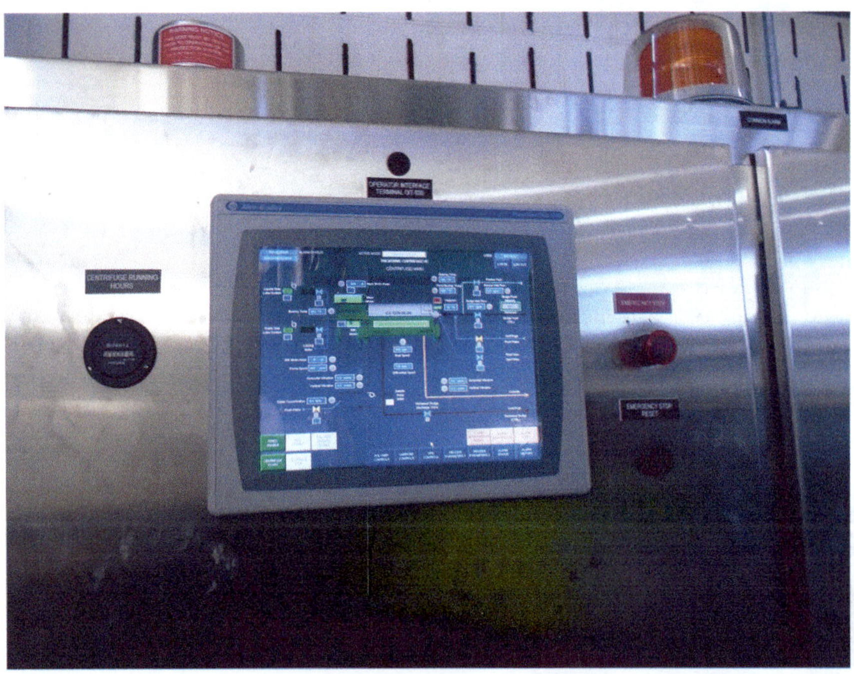

Figure 13.7 Typical Local Operator Interface

Local operator interfaces allow operators to have a view of the process that is close to the equipment being controlled. Because they are near the equipment, they provide a convenient way to check proper monitoring and control during activities such as maintenance or troubleshooting. Local operator interfaces also can provide a valuable means to monitor and control the local process should there be a problem with the larger SCADA system in the facility, especially the failure of communications from the local area to the rest of the SCADA system.

Local operator interfaces may be part of a process system that was packaged by an equipment vendor to provide a more detailed view of the local process than might be available on the facility-wide SCADA system. In such situations, the displays may be unique to that process or vendor. This allows the vendor to supply a standard package across many utilities.

HUMAN-MACHINE INTERFACE

A critical part of a SCADA system is the HMI. It is through the HMI that the operator monitors and controls the facility, often from dedicated control rooms in the facility.

The core functions provided by an HMI include:

- Communication to/from controllers to gain process information and send commands
- Alarm notification including classification, prioritization, organization, management, and acknowledgment of process alarms
- Displaying process information on graphical displays that also allow the operator to send commands or adjust setpoints
- Historical data collection for retrieval and display and reporting

The HMI function may be provided by a single personal computer. In larger facilities, the HMI function is provided by a system of computers that each play a part. Often, the core functions are performed by dedicated computers as follows:

- Supervisory control and data acquisition servers communicate to the controllers and provide process data, including alarming, to the other components of the HMI.
- A SCADA workstation presents graphical displays with process information, alarms, and animations based on information received from the SCADA servers. At the workstation, the operator makes setpoint adjustments and issues commands that are sent through the SCADA servers to the controllers.
- The historian server communicates to the SCADA servers to collect process values and stores the process values for later retrieval. Values are often retrieved by the SCADA workstation to show how a process condition has been trending over time. The historical values are also used for reporting and sharing with other business systems.

The term *server* is used for high-end computers that are designed to provide more reliable and higher levels of storage and processing than standard personal computers. They act as a central source of information that is "served" to other computers.

Often, the various server computers and associated networking equipment may be found in a server rack near the control room.

The term *workstation* is used for computers that are directly used by operators and have good video capabilities. Workstations request and display data from servers.

TEST YOUR KNOWLEDGE

1. Supervisory control and data acquisition equipment is the same as office computer systems.
   - ☐ True
   - ☐ False
2. Supervisory control and data acquisition systems are cyber-physical systems.
   - ☐ True
   - ☐ False
3. Control panels often include circuit breakers, fuses, and power supplies.
   - ☐ True
   - ☐ False
4. It is OK to connect any laptop computer to a SCADA Ethernet switch.
   - ☐ True
   - ☐ False

Operation

PROCESS MONITORING AND CONTROL

INTRODUCTION

This section discusses the general components of a SCADA HMI and how they are used to control and monitor the process data and state. This section includes the topics of displays, navigation, setpoints, alarm management, and trending.

MONITORING THROUGH DISPLAYS

Here we will discuss the basic types of displays that will likely be encountered on any SCADA system and how navigation between them can be performed. We will also touch on the topic of display standardization.

The purpose of display standardization is to provide a logical organization to not just the displays themselves, but also the data displayed on them. This standardization will include things such as display layout, colors, graphical elements, popups/faceplates, fonts, and line weights. Your organization will likely have a standard color set for statuses such as running/stopped and open/closed and for alarms. These colors will be present throughout the SCADA system and have the same meaning regardless of the process being displayed. This standardization of graphical elements is important to reduce operator confusion when moving between process areas, and to increase operator situational awareness to identify and respond to anomalous events more easily.

A *display* will be defined as any graphic that fills an entire monitor or workspace, depending on your organization's SCADA package and configuration.

Supervisory control and data acquisition display layouts are standardized so that different regions of the display provide features such as:

- Navigation to other displays
- A clear title describing the process or data being depicted
- A depiction of the process that conveys all the data necessary for safely operating said process
- A method by which to open any relevant popups or faceplates

Figures 13.8 through 13.11 present examples of various types of standardized displays.

Popups are defined as a graphic that can be opened from a display but does not replace said display or completely fill another monitor or workspace.

Popups will generally contain information or controls that are not immediately relevant for monitoring or controlling the day-to-day operations of the associated process, or, in the case of highly automated systems, will contain configuration and settings for the automated process.

Faceplates are simply popups that correspond to a specific device, such as a pump or valve. These faceplates will generally contain buttons for manual controls. Depending on the level of instrumentation on your specific system, devices may have additional faceplates that provide in-depth status information. For example, detailed diagnostic codes might be presented here for viewing by maintenance personal in the event of the failure of a piece of equipment.

Graphical elements refer not only to the symbols used to depict items such as pumps, valves, gates, tanks, and so on, but also to objects like pushbuttons, radio buttons, and text entry fields. Your organization should have a standard regarding these elements.

There are two navigational concepts commonly used in SCADA systems: drill down and lateral navigation.

> Drill down is a navigational concept in which displays start as very high-level overviews of not just a process, but an entire facility. These overviews incorporate navigational elements allowing operators to "drill down" to more location- or process-specific displays, which can then be drilled down further into device-specific popups or faceplates. For example, you may have a facility that divides itself between a "wet" and "dry" side. The "main" display would show only information that is relevant to the whole facility. It may also show very specific high-level information for each of the sides of the facility. Navigational elements would allow the operator to "drill down" to the "wet" side where they would have a display showing high-level information for all the "wet" processes. From there they could "drill down" again to a specific process display and then once again to device faceplates or maintenance popups. An example of high-level information is a display showing which pieces of

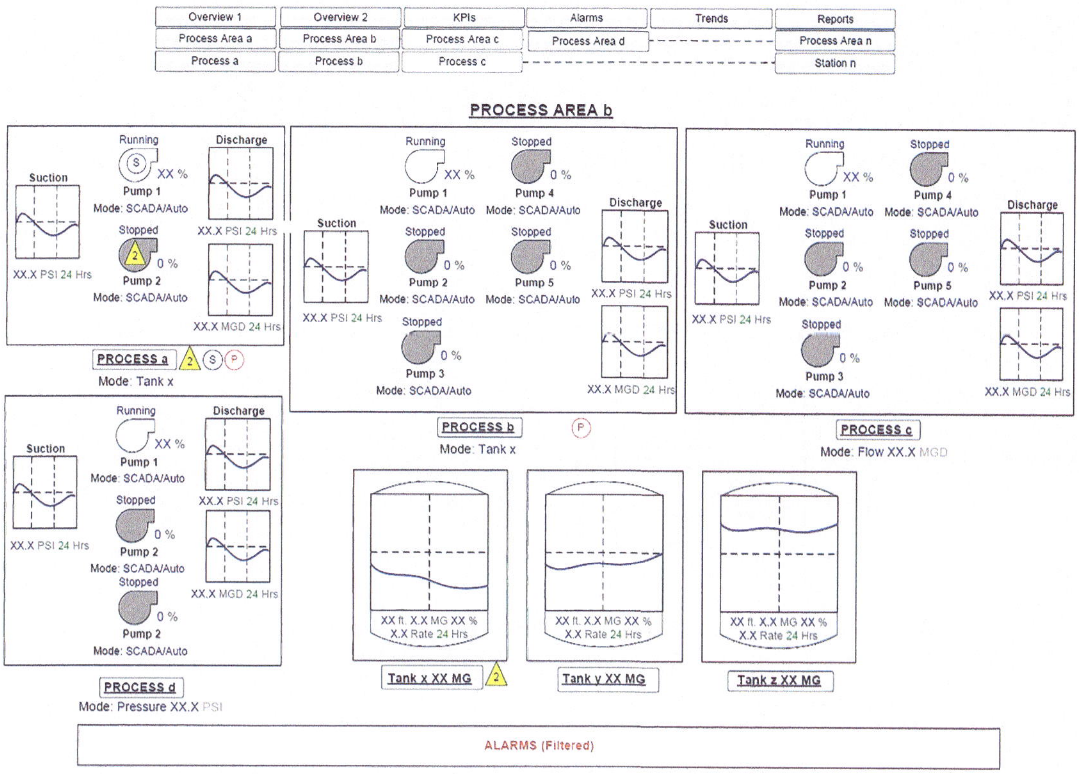

Figure 13.8 Process Area Overview Example

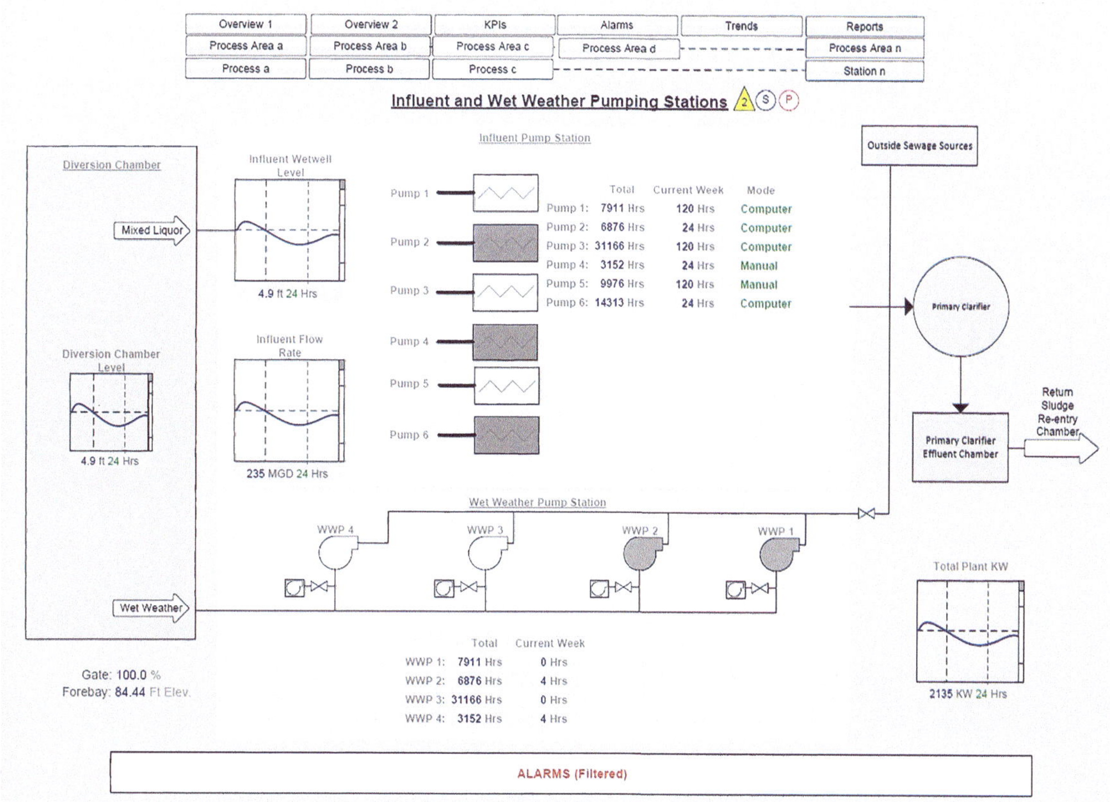

Figure 13.9 Standardized Screen Layout

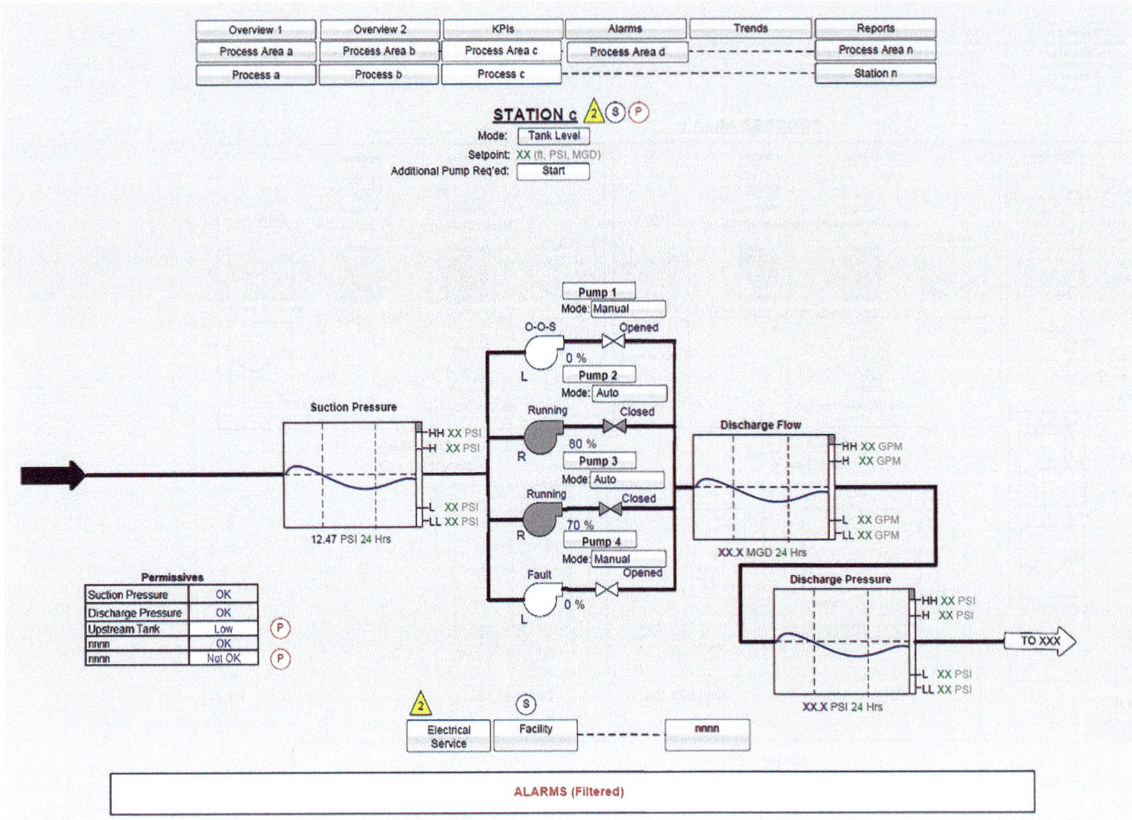

Figure 13.10 Process Monitoring Example

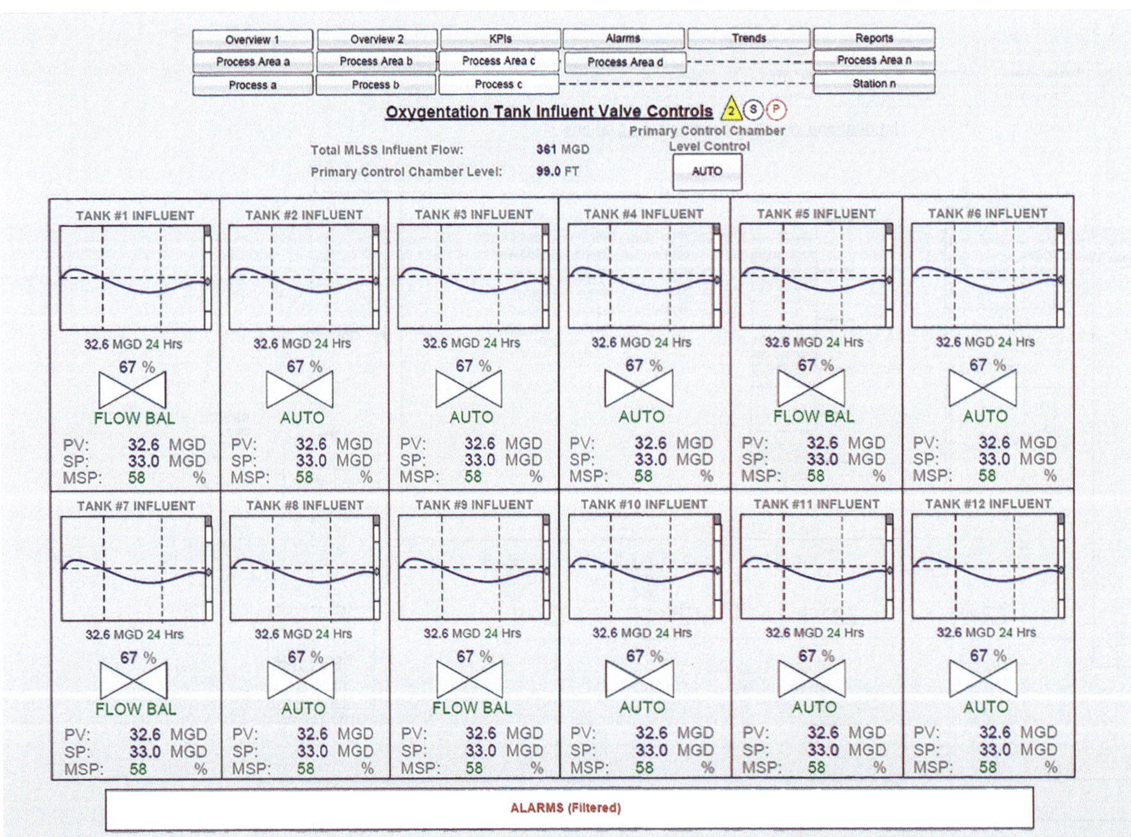

Figure 13.11 System Monitoring Example

equipment are in service. Drilling down will reveal more specific information such as the discharge pressure and flow output for a particular pump. An example of drill-down navigation would be moving between display levels, as shown in Figure 13.12.

Lateral navigation is the concept of moving between displays of the same level without having to go through the preceding level. This navigation goes hand in hand with drill down because, without it, navigation becomes poorly optimized. An example would be moving between Process "X" and Process "Z," as shown in Figure 13.12.

Navigational elements are display and selection conventions used to allow the operator to logically and quickly show a desired display. Examples include pull-down menus, arrows that can be selected to advance forward or backward in a proccss, and buttons on the screen for accessing frequently used displays.

TEST YOUR KNOWLEDGE

. **Moving between displays of the same level is called?**
   a. Drill down
   b. Lateral navigation
   c. Starboard navigation
   d. Clickthrough

ENTERING SETPOINTS AND COMMANDS

Faceplates, like other displays and popups, should have a standardized layout so that controls are as similar as possible between devices of the same type. The following types of elements are commonly found on faceplates:

- Static text. This text indicates the device the faceplate is for and what buttons and dynamic text do. This text does not change.
- Dynamic text. This text calls out device status and state. This text updates to reflect the condition of the process or piece of equipment.

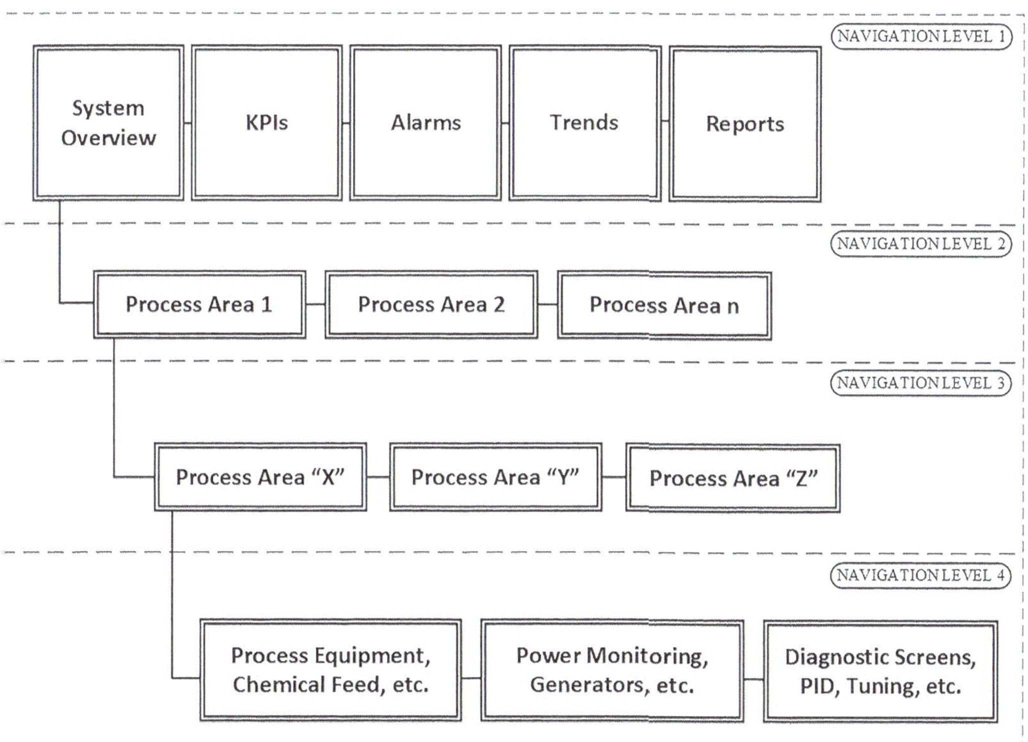

Figure 13.12 Navigation

- Text entry forms. Forms and fields are places where an operator can enter, for example, the desired pump speed.
- Buttons such as "Start," "Stop," "Auto," "Manual," "In service," and "Out of service."
- A method by which to close the faceplate when done.

Regardless of whether the setpoint entry is on a faceplate, popup, or display, the data entry method should be the same, be clearly different from dynamic text that cannot be changed directly by the operator and should include some form of data verification. For example, if the normal range of dissolved oxygen concentration in a process is between 0 and 3 mg/L, then data verification should alert and prevent the operator from entering –1 mg/L or 7 mg/L. Data verification helps avoid issues due to mistyping on a keyboard. Data verification simply means that the SCADA system should know the allowed range for each setpoint, and both notify the operator and disallow any setpoint changes to a value outside of the said range.

Command elements are display and selection conventions used to allow the operator to send information to the controller. Examples include selectable push button symbols for start and stop and selectable entry fields, which allow for changing a setpoint when selected.

Buttons or other command elements should be used for operator-initiated commands such as "Start," "Stop," "Auto," "Manual," and so on, regardless of the element used. The SCADA system should provide feedback to indicate that the element has been successfully interacted with. A button will typically change its color. A toggle, like "Auto/Manual," may remain after the status changes to indicate the current control mode. Figures 13.13 and 13.14 present examples of popups with various setpoint entry and command elements.

TEST YOUR KNOWLEDGE

1. **What is the key difference between a faceplate and a popup?**
   a. Faceplates represent controls for a specific device
   b. Faceplates show video of the operator's face
   c. Faceplates only show controls
   d. Popups can't have controls for individual devices

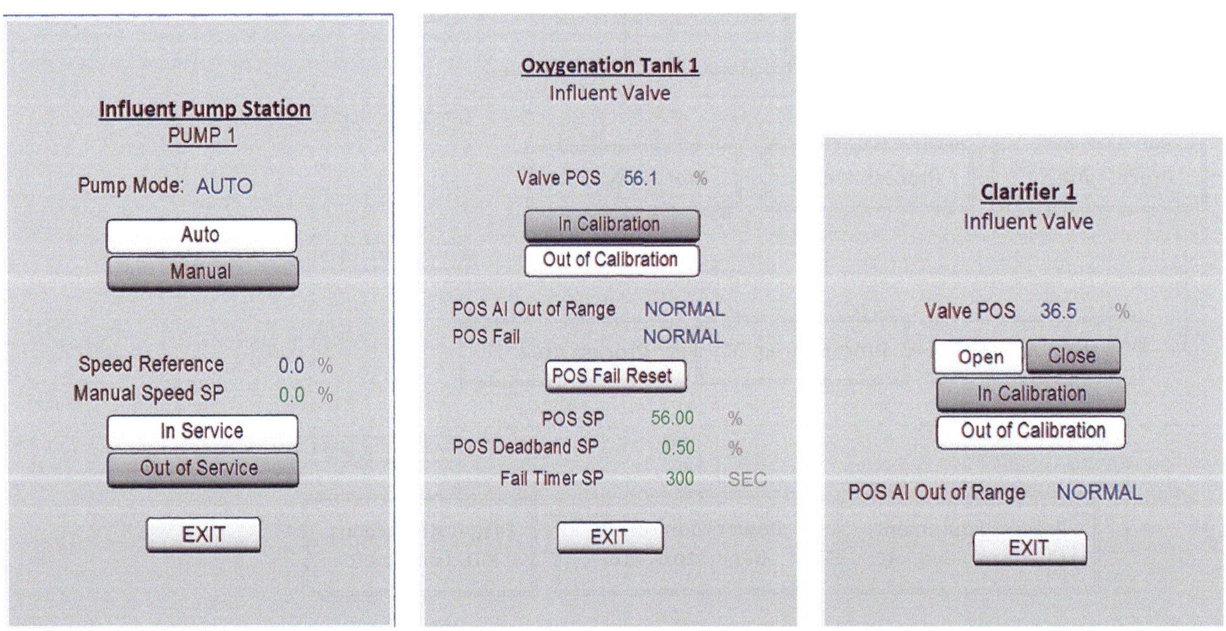

Figure 13.13 Device Control Popups

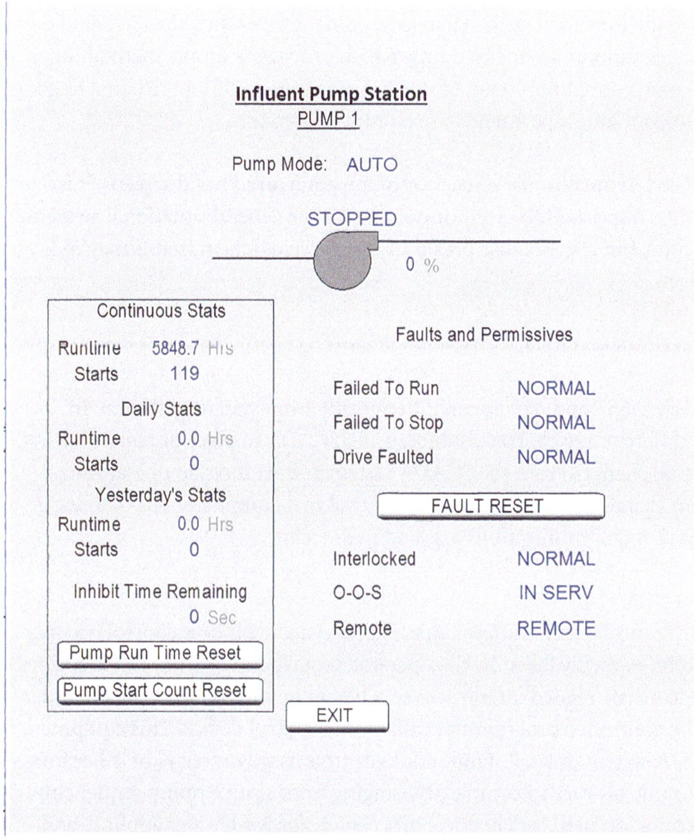

Figure 13.14 Device Status Popup

ALARM MANAGEMENT

Alarm management covers concepts involving alarm generation, monitoring, classification, and acknowledgment. A key fact regarding alarms is that not all alarms are equally critical to operations. For example, an operator's response to a chemical spill alarm should be vastly different than the response to a motor vibration or temperature alarm. A common approach to this is to have four tiers of alarms: critical, high, medium, and low. Each tier should have a documented and agreed upon minimum response time by the operator.

Alarm acknowledgment is a concept shared by most SCADA packages. By acknowledging an alarm, an operator indicates that they have seen it and are in the process of resolving it. Acknowledged and unacknowledged alarms will be displayed differently to draw the operator's attention to the new, unacknowledged alarm. In both cases, alarms are color coded by priority and displayed with the highest priority alarms at the top and the lowest at the bottom.

OPERATION TRENDS

Trends allow an operator to monitor process values as they change over time. This could be as simple as a graph showing the level of a wet well over the past hour. A more complex trend could display how the flow of a pump varies with the pump speed or header pressure. Trends can be embedded directly into displays and popups for process values. This concept is called *embedded trending*. Embedded trending allows operators to evaluate the status of a process and possibly catch and rectify alarm conditions before they occur.

When creating SCADA screens for facility operations, technicians will also configure several standard trends that have been identified as critical to either day-to-day operations or process troubleshooting. These are referred to as *preconfigured trends*. Preconfigured trends often allow the ability to review more history and see more detail than may be presented by an embedded trend. For example, an operator might open a preconfigured trend of facility influent over time to review normal daily variation of facility influent flow and then change the time to view the change in influent flow when the last rain event occurred. Many SCADA packages support ad hoc trending, which provides a simple method for operators to view a trend, similar to a preconfigured trend, of one or more values without needing to go to a supervisor or SCADA technician to configure a permanent, preconfigured trend.

If trends are provided, either embedded in operational screens or on preconfigured trend screens, it is important to monitor these trends. It is important to monitor not just for the general situational awareness of how the facility or process is running, but also because breaks or unexpected data in trends may indicate an issue with historical data collection.

CONTROL MODES

LOCAL/REMOTE

Local control is defined as the equipment that will be operated from a location physically close to the equipment or its control panel. *Remote control* is defined as the equipment that will be controlled from the SCADA system.

In reference to control modes, the terms *local* and *remote* can differ greatly from system to system. In some cases, the exact terms may be different as well. For instance, using SCADA in place of *remote control.* Generally, this control mode cannot be changed from the SCADA system and requires an operator to physically visit the device in question to make a change. It is often critical that equipment has both local and remote control, especially in cases of communications loss and maintenance.

AUTO/MANUAL

Whereas *local/remote* refers to where control occurs, *auto/manual* is concerned with how control occurs. In manual control, all controls must be explicitly taken by the operator. Note that there may be exceptions in which equipment is shut off due to alarm-related interlock even when in manual control. When under automatic control, equipment will be controlled programmatically by the control device. This equipment could be a PLC or DCU in the SCADA system or local stand-alone controls based on relays or other forms of automatic controllers. Automatic controls vary in complexity, ranging from simple pump sequencing based on level to complex batch automation used for chemical processing. For wastewater applications, the most common automatic controls will be sequencing and process and instrumentation diagram (PID) loop controls. Without getting too far into the math governing them, PID loop controls use a configurable equation to vary an output variable, such as pump speed, to maintain an input value, such as flow, at setpoint. Figure 13.15 shows a PID tuning popup, which includes an embedded trend to help the technician in the tuning process. Figure 13.16 shows a typical way PID loop controls will be displayed on an operational screen.

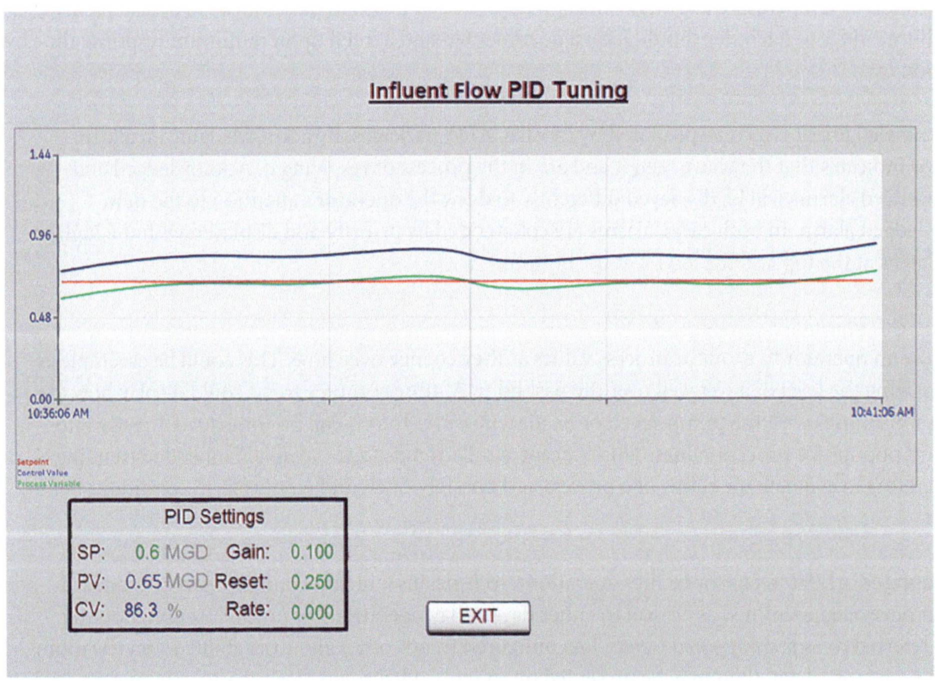

Figure 13.15 Process and Instrumentation Diagram Tuning Popup Example

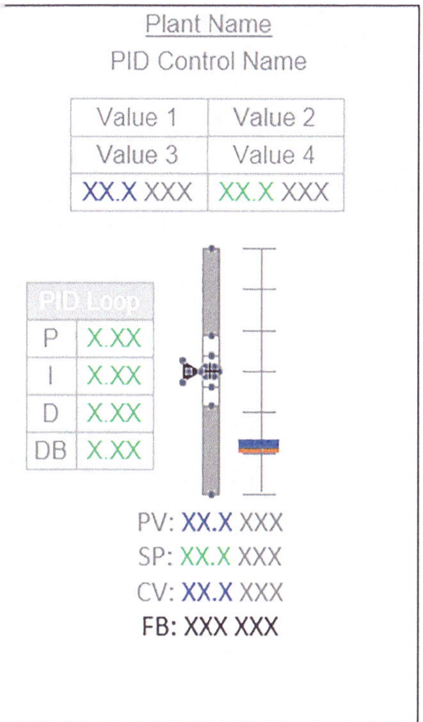

Figure 13.16 Process and Instrumentation Diagram Popup Example

SERVICE MODES

As with local/remote mode, the actual definitions of "in service" and "out of service" can vary greatly from system to system. There are two common configurations, which are as follows:

- In service/out of service is purely internal to the SCADA system and does not indicate anything about the physical status of the device.
- In service/out of service is linked to the power feed for the device and indicates whether the device is receiving power or not.

TEST YOUR KNOWLEDGE

1. **What is the concept of "drill down" when referring to navigation?**
2. **What is lateral navigation and why is it important?**

Data Collection and Analysis

DATA COLLECTION

An important function of SCADA is the collection and storage of data for reporting and analysis. Collection and storage of data may be done in various ways in a SCADA system. These ways range from data collected for displaying trends directly on a local operator interface dedicated to one process in the facility to data being collected by a dedicated historian server for use by the entire utility. Supervisory control and data acquisition systems are configured to store, at a minimum, the most important operational information such as flows, levels, chemical analysis, and so on. The SCADA system stores these process values with a timestamp to permit trending and reporting.

Because the data collected are used to make operational decisions and are also often used for regulatory reporting, attention must be paid to the availability and accuracy of the data that is collected. Frequent reviews of operational trend displays help ensure that the facility is operating properly. Frequent reviews also ensure that accurate data are being collected for use by the rest of the organization.

REPORTING/ANALYSIS

Supervisory control and data acquisition systems are often the primary source of operational data for a utility. These data are used on graphic displays, trends, and dashboards for real-time decision-making by operators and supervisors. The data are also used for less real-time processes such as monthly regulatory reporting. The data may be shared with maintenance management systems to help generate preventive maintenance or repair work orders. The data may also be shared with other groups within the utility, such as engineering, who use the data in models and visualization software to optimize planning and processes.

Maintenance

CONTROL PANELS

The wiring found in control panels may contain power from multiple sources, including sources external to the panel, so it is important that only qualified staff work within control panels.

External inspection of control panels is important to ensure proper operation. Only qualified maintenance staff should inspect the interior of SCADA control panels. Typical external inspections include the following procedures:

- Check that panel doors are properly closed for the safety of staff and to protect the control panel equipment from the environment.
- Check that no liquids are dripping on or from the panel.
- Listen for a change in fan noise that might indicate the failure of a cooling fan.
- Periodically confirm that equipment mounted on the front of the panel is functional. Control panels are often provided with a push button to ensure that lights are not blown. Confirm that any local operator interface is functional.
- Look for evidence of insects and rodents in the vicinity of the panel because they can damage wiring and equipment should they be able to enter the conduit systems and panel.
- If a problem with a control panel is suspected, a work order for further inspection/repair by qualified staff should be created.

OPERATOR INTERFACES

Routine inspection of operator interfaces includes:

- Periodic operation through the operator interface to confirm functionality.
- Periodic cleaning of screens, keyboards, and mice.
- Listening for changes in fan noise that might indicate the impending failure of a fan.

Qualified staff should perform periodic software updates to maintain system functionality and security.

Troubleshooting

A SCADA system is a tool that helps, not replaces, an operator. It is important to maintain an overall awareness of process operation so that failures of SCADA monitoring and control can be quickly recognized and acted upon. Some tips for recognizing that there is an issue with the SCADA system include:

- Review the SCADA alarm log/summary display. Reviewing this summary should be a routine procedure for facility operators.
- Make a habit of frequently reviewing trend displays to confirm that the SCADA system is properly collecting, storing, and displaying data. When reviewing trends, look for:
  - Gaps in data on the trend.
  - Trend lines that are perfectly flat over time when a variation is expected. This flat trend line may indicate that a signal is not being updated periodically and the data you are viewing on the SCADA system may be misleading.
  - Excessive variability from one reading to the next that might indicate that the reading is being affected by electrical noise or that an instrument requires attention.

- Excessive oscillation in a reading, which may indicate that a PID loop may need to be tuned or the associated equipment needs maintenance.
- Review diagnostic graphical displays. These displays may show alarms for individual pieces of control system equipment.
- Continuously do reasonability checks for the data being presented by the SCADA system. For example, if the flow into a tank is constant and the flow out of a tank has increased, you might reasonably expect to see the level of the tank dropping over time.

TEST YOUR KNOWLEDGE

1. **When inspecting a panel, externally, which of the following would cause the panel to fail:**
   a. The panel exterior is dry, with no liquids dripping on it.
   b. There is a loud grinding sound coming from the fan.
   c. The panel-mounted display is in good working order.
   d. All of the panel latches are properly closed and secured.

CHAPTER SUMMARY

	SCADA Systems
PURPOSE AND FUNCTION	■ Supervisory control and data acquisition systems provide situational awareness to operators and collect/store data for use throughout the utility.
THEORY OF OPERATION	■ Core components of a SCADA system include field devices, input/output devices, controllers, networking equipment, and operator interfaces.
EQUIPMENT	■ Supervisory control and data acquisition equipment is different from traditional office computer systems. ■ Supervisory control and data acquisition equipment is mounted in control panels for safety and to protect the equipment from harsh industrial environments. ■ Input/output cards translate the signals from field devices into the 1's and 0's that computers understand. ■ Input/output cards are often organized in input/output racks that communicate data to controllers. ■ Controllers are programmed to perform calculations and execute automatic control routines. ■ Operator interface equipment communicates to controllers via networking equipment. ■ Local operator interfaces provide a means for operators to have a view of the process that is close to the equipment being controlled. ■ The HMI is a critical part of a facility SCADA system because it is through the HMI that operators monitor and control the facility.
OPERATION	■ Operators monitor the process through SCADA displays, popups, and faceplates. ■ Operators can navigate from one display to another. ■ Operators can interact with the SCADA system by entering setpoints and commands. ■ The SCADA system provides a means to manage alarms as they are received, including acknowledgment of an alarm. ■ Trends allow operators to monitor process values over time. ■ Control modes implemented in SCADA systems include "Local/Remote" and "Auto/Manual." ■ "In Service/Out of Service" is also an important status monitored or tracked by a SCADA system.
DATA COLLECTION AND ANALYSIS	■ Collection and storage of facility operational data for reporting and analysis is a key function of a SCADA system. ■ The data collected by SCADA systems are used by operators and for planning and process optimization by others in the utility.
MAINTENANCE	■ Operators should perform a routine external inspection of control panels and periodic cleaning of operator interface devices. ■ Qualified staff should perform any internal control panel maintenance or operator interface software maintenance.

TROUBLESHOOTING

- Situational awareness, including continuous reasonability checks, is a critical part of recognizing an issue with a SCADA system.
- Supervisory control and data acquisition system tools, including trending and alarming, assist in recognizing issues with the SCADA system.

Suggested Readings

Hollifield, B., Oliver, D., Nimmo, I., & Habibi, E. (2008). *The high performance HMI handbook: A comprehensive guide to designing, implementing and maintaining effective HMIs for industrial plant operations* (1st ed.). Plant Automation Services.

Hollifield, B., & Habibi, E. (2010). *Alarm management: A comprehensive guide* (2nd ed.). International Society of Automation.

CHAPTER 14

Leadership and Management

Introduction

Proper operation of a water resource recovery facility (WRRF) begins with good leadership. Leadership comes in many forms. From the WRRF manager to the operators, all have a role in leading. Training and working effectively with coworkers are major operational responsibilities.

A WRRF leader should continue to learn about treatment as well the utility, facility personnel, and ways to communicate and lead more effectively. Leadership positions such as supervisor, superintendent, or manager require additional skills beyond sampling, process control, and other treatment tasks. Your utility may also require additional qualifications, such as a higher license level or more education, for a leadership position.

This chapter covers a wide variety of topics that will help leaders with some of the more common tasks and problems seen throughout the wastewater sector. This is by no means an exhaustive list of what a leader needs to know to effectively provide direction at your WRRF. Leaders are always leaning, and it is our hope that you continue your education on both treatment and leadership throughout your career.

LEARNING OBJECTIVES

Upon completing this chapter, you will be able to

- Read and understand a National Pollution Discharge Elimination System (NPDES) permit for a WRRF;
- Understand the purpose and layout of a typical discharge monitoring report (DMR) and other reports related to WRRF operation;
- Understand the purpose of spill plans and stormwater pollution prevention plans;
- Understand the purpose and background for industrial pretreatment programs;
- Understand requirements for sampling and recordkeeping;
- Plan a sampling schedule;
- Understand types of budgets, how to set and follow a budget, and how to procure goods and services;
- Write and maintain standard operating procedures (SOPs);
- Understand the purpose of asset management as it relates to a WRRF;
- Recognize the various types of maintenance and understand the components of a maintenance plan;
- Understand how capital projects are initiated, planned, funded, and delivered;
- Understand contract methods, bidding, and requests for proposals;
- Learn skills for handling personnel matters;
- Learn traits and responsibilities of a good leader;
- Understand the value of engaging the public and customers about wastewater treatment;
- Plan a tour of a WRRF for the public; and
- Learn the basics of emergency planning and response.

Permits and Reporting

Water resource recovery facilities have a range of permits relating to discharges to the water and air, as well as permits related to solids management. It is important that operators understand not only the permits, but the purpose, testing, and reporting as it relates to each one. Having this knowledge will enable the operator to better understand their role in their WRRF meeting the permit requirements, and to understand the importance of proper sampling requirements and overall operations of the facility.

DISCHARGE PERMITS

In the United States, the NPDES addresses water pollution by regulating discharges to water. Before getting into the definition of the NPDES permit, it is important to understand the difference between point source and nonpoint source pollution. A point source pollutant comes from a single place (or point) and the source of the pollution is easily identified, such as the outfall of a WRRF. The term *point source* has been through a 25-year process of being defined in the Clean Water Act by the U.S. Environmental Protection Agency and, as a result, the definition is very broad. According to the final definition, these single sources include pipes, channels, and ships, to name a few. Inversely, nonpoint source pollution, such as agricultural runoff, can come from many different locations, without a single outfall, making it difficult to locate or manage the source of a pollutant.

To control the pollution from point sources, U.S EPA, under the Clean Water Act, created the requirement for point source discharges to hold an NPDES permit. As defined by U.S. EPA, "the Clean Water Act prohibits anybody from discharging *pollutants* through a *point source* into a *waters of the United States* unless they have an NPDES permit. The permit will contain limits on what you can discharge, monitoring and reporting requirements, and other provisions to ensure that the discharge does not hurt water quality or people's health". U.S EPA delegates authority to state, tribal, and territorial governments to oversee the NPDES program, as shown in Figure 14.1.

To *assimilate* something means to take it in or to incorporate it. With natural waterways such as streams, creeks, rivers, and lakes, studies can be performed to measure how much of a pollutant can be broken down naturally in the waterway before water quality is negatively affected. This natural capacity for breaking down pollutants is called the *waste assimilative capacity* of the waterway.

The NPDES permits protect local waterways by specifying the level of a pollutant concentration to be discharged into a waterway. The acceptable level is based on the amount that can be assimilated in a waterway without degrading the water quality. It is up to the permit holder, the utility, or WRRF to decide

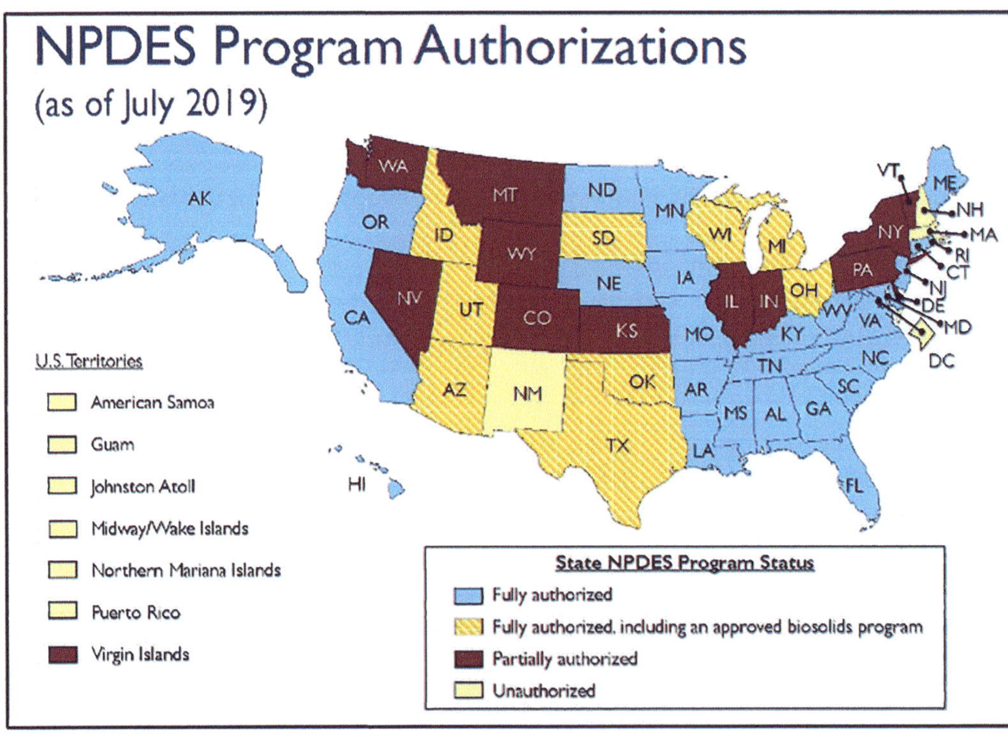

Figure 14.1 National Pollutant Discharge Elimination System Permit Writing Authorization by State (https://www.epa.gov/npdes/npdes-program-authorizations)

on which processes to implement to meet the permit limitations. These pollutants include things like biochemical oxygen demand (BOD), total suspended solids, temperature, nutrients, and pH, to name a few. These permits ensure that the overall goals for the specific waterways are achieved, both at the state and/or federal levels. Figure 14.2 provides an example of a state-controlled **NPDES permit pollution summary.**

In some cases, the water discharged from a WRRF may be cleaner than the receiving water. With the passage of the Clean Water Act in the United States in 1972, the regulation of point source discharges through NPDES permits, and the work of WRRFs, there have been notable improvements in the water quality of many U.S. waterways. Rivers and waterways, which had been extensively polluted, now have fish again and are safe for recreation. The restoration of the Cuyahoga River, located in Ohio, is an example of a success of the Clean Water Act. The river caught fire at least 14 times, including an infamous fire in 1969, which helped to spur the environmental laws requiring the cleanup of the nation's waterways.

OPERATORS AND NATIONAL POLLUTANT DISCHARGE ELIMINATION SYSTEM PERMITS

Operators should have access to their facility's permit(s) and be familiar with all the requirements. In addition to the discharge limits (see Figure 14.2), permits may contain requirements for operator grade, backup power, upsets, notifications, groundwater monitoring, pretreatment, and biosolids handling. An NPDES permit can be long and technical but contains critical information and requirements that an operator must incorporate to daily facility operations. All facility staff should read and understand the entire document. Posting a copy of the discharge limits (Figure 14.2) in the control room or WRRF laboratory allows for quick access to permit limits.

National Pollutant Discharge Elimination System permits are typically issued for 5 years. Renewal applications must be submitted 180 days before the permit expires. The renewal application may require additional sampling such as pollutant scans and toxicity, so be sure to know your permit dates and plan accordingly. If an NPDES permit expires before the state or U.S. EPA issues a renewed permit and the renewal application has been submitted on time, then the WRRF must continue to operate under the expired permit until a new permit is issued.

1. **Treatment Plant Effluent Before Disposal** – During the period beginning on May 1, 2014 and lasting through the expiration date, the permittee is authorized to land apply effluent and is identified as outfall serial 001. Such discharge shall be limited and monitored by the permittee as specified below:

EFFLUENT CHARACTERISTICS	DISCHARGE LIMITATIONS					MONITORING REQUIREMENTS		
	Pounds per Day		Other Units					
	Monthly Average	Weekly Average	Monthly Average	Weekly Average	Daily Maximum	Measurement Frequency	Sample Type	Sample Point
Flow	---	---	0.066 MGD	0.066 MGD	---	Daily	Continuous	Effluent
Carbonaceous Biochemical Oxygen Demand - 5 Day ($CBOD_5$)*	14	22	25 mg/l	40 mg/l	---	2/Month	24 Hour Composite	Effluent
Total Suspended Solids (TSS)	50	74	90 mg/l	135 mg/l	---	2/Month	24 Hour Composite	Effluent
Nitrate (N)	---	---	MR mg/l	MR mg/l	---	2/Month	24 Hour Composite	Effluent
Dissolved Oxygen (DO)	---	---	1.0 mg/l Minimum at all times			5/Week	Grab	Effluent
PH	---	---	6.0 - 8.5 Standard Units			Daily	Grab	Effluent

* $CBOD_5$ is defined as the five-day measure of the pollutant parameter carbonaceous biochemical oxygen demand ($CBOD_5$).

Part III
Page 19 of 34
Permit No.

Figure 14.2 Effluent Limits Table From National Pollutant Discharge Elimination System Permit (Courtesy of Beaufort-Jasper Water & Sewer Authority)

DISCHARGE MONITORING REPORTS

A DMR is required for all WRRFs that discharge into surface waters of the United States. This is the data reporting requirement of the NPDES permit.

In the fall of 2015, U.S. EPA created and published the "NPDES Electronic Reporting Rule," which requires all reporting be submitted electronically. The rule replaces most paper-based NPDES reporting requirements with electronic reporting. Those states that oversee the NPDES permitting (see Figure 14.1) are required to share these electronic data with the U.S. EPA, including inspections, permits, and violations.

When submitting the DMR to either the state, territorial government, or U.S. EPA, a standard format is used. The document contains the permit number, monitoring data, and the signature of the responsible party for the facility. Figure 14.3 shows a blank DMR from U.S. EPA's website. States and territories may have their own version of the federal form, which may include additional reporting requirements. Check with your state regulators or U.S. EPA if you have questions on DMR format or submission.

A designated signatory authority is a person who certifies, under penalty of law, that the information submitted in the report was prepared under their direction and supervision in accordance with the system designed to ensure that qualified personnel properly gather and evaluate the information submitted. The designated signatory authority is typically a leader or officer of the utility or WRRF.

Water resource recovery facilities are required to have a designated signatory authority for DMRs. This person is typically the responsible party for the WRRF and is signing, under penalty of law, that all data are representative, accurate, and complete.

ANNUAL BIOSOLIDS REPORT

Chapter 9 describes some of the options to handle biosolids within a WRRF (see Figure 14.4). Some facilities incinerate their solids, some landfill, and others create a beneficial reuse product such as Class A or Class B biosolids, compost, biochar, and so on. All WRRF facilities are required to report annually to the U.S. EPA and/or state authority, depending on the way in which they handle their solids. Biosolids reporting requirements are outlined in the NPDES permit.

Facilities that prepare or apply biosolids for land application, incinerate biosolids, or own, operate, or use a surface disposal site, are required to adhere to U.S. EPA's Biosolids Permit Program. This program stems from U.S. EPA's 40 CFR, part 503 Rule. Each year, these permitted facilities are required to submit a

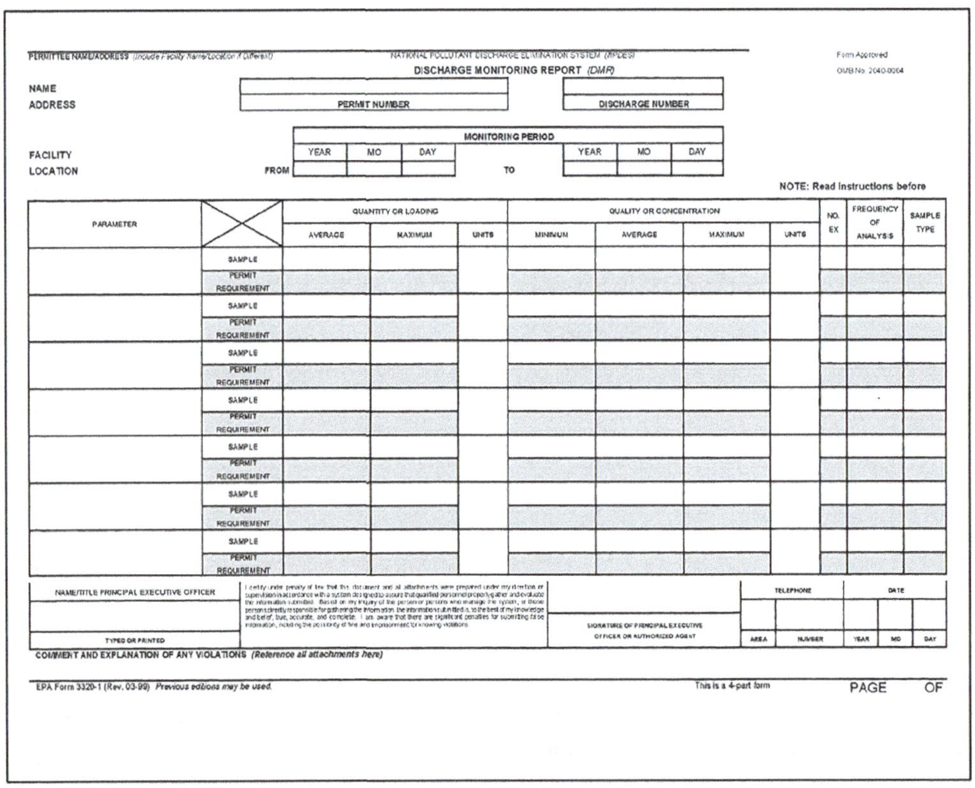

PERMITTEE NAME/ADDRESS *(Include Facility Name/Location if Different)*

NATIONAL POLLUTANT DISCHARGE ELIMINATION SYSTEM *(NPDES)*

DISCHARGE MONITORING REPORT *(DMR)*

Form Approved
OMB No. 2040-0004

NAME
ADDRESS

PERMIT NUMBER | DISCHARGE NUMBER

FACILITY
LOCATION

MONITORING PERIOD

	YEAR	MO	DAY		YEAR	MO	DAY
FROM				TO			

NOTE: Read instructions before

PARAMETER		QUANTITY OR LOADING			QUALITY OR CONCENTRATION				NO. EX	FREQUENCY OF ANALYSIS	SAMPLE TYPE
		AVERAGE	MAXIMUM	UNITS	MINIMUM	AVERAGE	MAXIMUM	UNITS			
	SAMPLE										
	PERMIT REQUIREMENT										
	SAMPLE										
	PERMIT REQUIREMENT										
	SAMPLE										
	PERMIT REQUIREMENT										
	SAMPLE										
	PERMIT REQUIREMENT										
	SAMPLE										
	PERMIT REQUIREMENT										
	SAMPLE										
	PERMIT REQUIREMENT										
	SAMPLE										
	PERMIT REQUIREMENT										

NAME/TITLE PRINCIPAL EXECUTIVE OFFICER

TYPED OR PRINTED

I certify under penalty of law that this document and all attachments were prepared under my direction or supervision in accordance with a system designed to assure that qualified personnel properly gather and evaluate the information submitted. Based on my inquiry of the person or persons who manage the system, or those persons directly responsible for gathering the information, the information submitted is, to the best of my knowledge and belief, true, accurate, and complete. I am aware that there are significant penalties for submitting false information, including the possibility of fine and imprisonment for knowing violations.

SIGNATURE OF PRINCIPAL EXECUTIVE OFFICER OR AUTHORIZED AGENT

TELEPHONE		DATE		
AREA	NUMBER	YEAR	MO	DAY

COMMENT AND EXPLANATION OF ANY VIOLATIONS *(Reference all attachments here)*

EPA Form 3320-1 (Rev. 03-99) *Previous editions may be used.* This is a 4-part form PAGE OF

Figure 14.3 Blank Discharge Monitoring Report Form (U.S. EPA, n.d.-b)

Figure 14.4 Biosolids Being Dewatered on a Belt Filter Press (Courtesy of Beaufort-Jasper Water & Sewer Authority)

biosolids annual report in mid-February. This report covers the previous year and includes information such as the total tons of solids processed. Per the 503 requirements, WRRFs with a design flowrate equal to or greater than 1 mgd that serve 10,000 people or more or are required to have an approved pretreatment program (Class I Sludge Management Facility) all fall under this requirement.

Like the DMR, biosolids reports may be submitted electronically depending on your state. The biosolids report also has a signature requirement like DMRs.

SPILL RESPONSE PLAN AND STORMWATER POLLUTION PREVENTION PLANS

The protection of local waterways does not stop at the point source discharge of the WRRF outfall. It is also critical to protect local waterways as well as the soil on and around the WRRF site by minimizing and controlling process or chemical spills. This is done by creating a stormwater pollution prevention plan (SWPPP) for the facility and developing a spill response plan.

An SWPPP outlines how a facility plans to reduce pollution associated with stormwater, and typically also includes best management practices (BMPs). Not all facilities are required to have an official SWPPP, which is determined by the Standard Industrial Classification system. Spill response plans include how a facility will contain and clean specific spills and should drill down to cover specific chemicals used in treatment processes and maintenance as well as diesel or other fuel for vehicles or generators. Check with your state regulators to determine if your WRRF needs an SWPPP or spill plan and what is required for the plans.

Stormwater pollution prevention plans and spill plans must be reviewed and updated regularly. Operators should be trained annually on the requirements and procedures in the SWPPP and spill plan.

INDUSTRIAL PRETREATMENT

Industrial pretreatment is covered extensively in Chapter 1. U.S. EPA gives the utility or publicly owned treatment works the authority to oversee the pretreatment programs for their WRRF. This program again supports the Clean Water Act by protecting the WRRF from industrial and commercial dischargers that may harm the treatment facility or process, as well as reducing pollutants to concentrations that may be managed and removed by the WRRF, with no effect on the environment. Pretreatment programs enforce three types of pretreatment standards (U.S. EPA, n.d.-c, n.d.-d):

- Uniform general prohibitions;
- Categorical pretreatment standards; and
- Local limits, which are specific to the utility and include numeric discharge limits as well as required BMPs.

Pretreatment standards protect not only the WRRF and its collection system and workers, but also the local receiving waters, biosolids, and, of course, public health.

The standards applied to individual industries depend on the industrial user group applied to the industry. Industrial facilities are split into three types of groups: industrial users, significant industrial users, and categorical industrial users, with categorical industrial users having specific limits that are set by U.S. EPA for the specific industry in which they fall.

U.S. EPA or your state regulators may require an annual or semi-annual report on activities pertaining to industrial users, including inspections, sampling events, violations, significant noncompliance, and status changes. The reporting requirements are outlined in the NPDES permit. For those utilities and WRRFs that have a pretreatment program, the responsibility for various components of the program may be the work of a WRRF operator, laboratory staff, or a separate pretreatment department.

OTHER REPORTS

There are other permits and reports that are required for WRRFs, depending on the processes and technologies in use at the WRRF. For example, backup generators of certain sizes require Title V air permits. Reporting requirements vary by state, so read and comply with your WRRF's Title V air permit requirements.

In April 2016, U.S. EPA developed the final rule for sewage sludge incinerators (SSI) to fall under section 129 (or Maximum Achievable Control Technology standards) of the Clean Air Act. This required U.S. EPA to set numerical emissions limitations for nine pollutants from SSI units. The nine pollutants are cadmium, carbon monoxide, total mass basis dioxins/furans and toxic equivalency basis dioxin/furans, hydrogen chloride, lead, mercury, nitrogen oxides, particulate matter, and sulfur dioxide.

TEST YOUR KNOWLEDGE

1. **Maximum Achievable Control Technology 129 is required by the NPDES permit for stormwater monitoring.**
   - ☐ True
   - ☐ False
2. **True or False: A residential discharger requires an industrial pretreatment permit.**
   - ☐ True
   - ☐ False
3. **A DMR report is required monthly only for WRRFs with permit violations.**
   - ☐ True
   - ☐ False
4. **Local limits are limits set by the utility or WRRF for industrial dischargers.**
   - ☐ True
   - ☐ False
5. **The Safe Drinking Water Act regulates point source pollution.**
   - ☐ True
   - ☐ False
6. **An operator should read and know the requirements of the entire NPDES permit, not just the effluent limits table.**
   - ☐ True
   - ☐ False
7. **A DMR contains important information related to the WRRF. Which of the following is found in the DMR:**
   a. The permit number
   b. Unit process information
   c. Population served
   d. Facility size
8. **In terms of biosolids permitting, WRRFs are required to report on a __________ basis.**
   a. Monthly
   b. Quarterly
   c. Bi-monthly
   d. Annual

;ampling and Records

.ll WRRFs collect and analyze samples for process control and to meet permit requirements. A WRRF or .tility may have its own laboratory for in-house sampling and analysis. Some parameters are not feasible or sending to a commercial or outside laboratory. Chlorine residual, pH, and dissolved oxygen (DO) are arameters measured on-site that cannot be sent to an outside laboratory. Biochemical oxygen demand, .utrients, and toxicity are commonly sent to contract laboratories. Chapter 9 of *Wastewater Treatment 'undamentals*, Volume II, covers laboratory procedures.

J.S. EPA and state regulators have sampling and recordkeeping requirements like 40 CFR Part 136. State nd federal regulations are the first place to start when setting up a new program or updating the current ne. Most of the testing requirements are spelled out in the NPDES permit. Sampling may occur daily, veekly, monthly, or annually. Collection may be a grab sample or a composite sample (see Figure 14.5).

In *Wastewater Treatment Fundamentals*, Volume II, a whole chapter is dedicated to flow measurement, proper sampling techniques, quality assurance and quality control, and analytical methods that operators most commonly perform. Additional information is available in Water Environment Federation's *Basic Laboratory Procedures for the Operator-Analyst*. For any questions on test procedures or sample hold time not answered in these two references, check *Standard Methods for the Examination of Water and Wastewater*, a comprehensive reference book for laboratory methods used at a WRRF. The guide is updated regularly, so look for the latest version.

:ONTRACT LABORATORIES

Jsing a contract laboratory or outside laboratory is very common. There are always tests that are not ost-effective to run at the WRRF because of special equipment, reagents, disposal, or limited staffing. .dditionally, some tests may require an accredited laboratory or special certifications. Having more than ne contract laboratory will give you a backup and allow you to take split samples. Some laboratories pecialize in certain tests and will have better prices and/or better quality.

'rice is one factor when choosing a contract laboratory. The following are other factors to consider:

- Proximity to the WRRF
  - How will the samples get to the laboratory?
  - How long will it take? Will hold time be an issue?
  - Is pickup by the contract laboratory an option? Is overnight delivery feasible?

A grab sample is an individual, discrete sample collected at one time. A grab sample represents the wastewater characteristics at the time of sample collection. A composite sample is collected and combined over a period of time, typically 24 hours, either by mixing discrete samples or by continuous sampling, and analyzed as a single sample. A composite sample may be time- or flow-proportioned. A composite sample represents the wastewater characteristics over a period of time.

igure 14.5 Operator Rodney Kiser Collects an Effluent Sample From the Composite Sampler for lational Pollutant Discharge Elimination System Permit Compliance (Courtesy of Beaufort-Jasper Vater & Sewer Authority)

- Capacity
  - Does the contract laboratory have the capacity to run all your tests?
  - If additional testing is needed, will it be an issue?
- Testing methods
  - Is the laboratory certified for the method?
  - Is the method appropriate for my sample? The U.S. EPA method for wastewater samples may not be the same as the drinking water methods.
- Quality
  - How is the laboratory's quality assurance and quality control (QA/QC)?
  - Are the data reliable?
- Sample bottles
  - Who provides the bottles?
  - How quickly can you get bottles after ordering them?
  - Will the bottles have the preservatives for specific tests?
- Results
  - How quickly will the test results be provided?
  - What format will results be in?
  - How will results be delivered? Mail, email, fax?

A WRRF should have an agreement or contract with all its outside laboratories that lists the scope of services, prices, and contact information.

LABORATORY RECORDKEEPING AND RETENTION

There is no shortage of data and records at a WRRF. How do you decide what is important and what to keep? The place to start is the NPDES permit because all monitored parameters must have a record. In addition to records for permit data, a WRRF also has records on operational data for process control, safet data sheets (SDSs), equipment maintenance, and spare parts inventory.

A laboratory information management system (LIMS) and computerized maintenance management system (CMMS) are valuable for keeping historical information and data in a searchable format. A LIMS keeps records on samples and process data. A CMMS can track equipment and facilities for maintenance and work orders.

Most requirements for recordkeeping come from regulations. Your utility may also have a policy on records. National Pollutant Discharge Elimination System records must be kept for 3 years. Sludge records must be kept for 5 years. In addition to NPDES records, there are regulatory requirements for keeping material safety data sheets/SDSs, employment records, DMRs, laboratory bench sheets, pretreatment information, and injury reports/Occupational Safety and Health Administration (OSHA) records, and other state and federal rules. Requirements vary for record retention so many utilities keep paperwork for many years—even decades.

Besides regulations, there are many advantages to keeping records for a long time. Historical data are useful for evaluations and planning to look at long-range trends (see Figure 14.6) in loadings or treatment efficiencies.

LABORATORY CERTIFICATION AND STANDARDS

All NPDES permit sampling must meet state and federal regulations. Most states have a laboratory certification program to ensure data integrity and reliability. The laboratory certification program includes inspections, proficiency testing, and evaluation of methods and SOPs. These requirements apply to both contract laboratories and WRRF laboratories.

Sampling that is done for process control and is not reported does not require certification. For example, a WRRF may measure ammonia in the aeration basins for process control. Certification is not required for process control sampling.

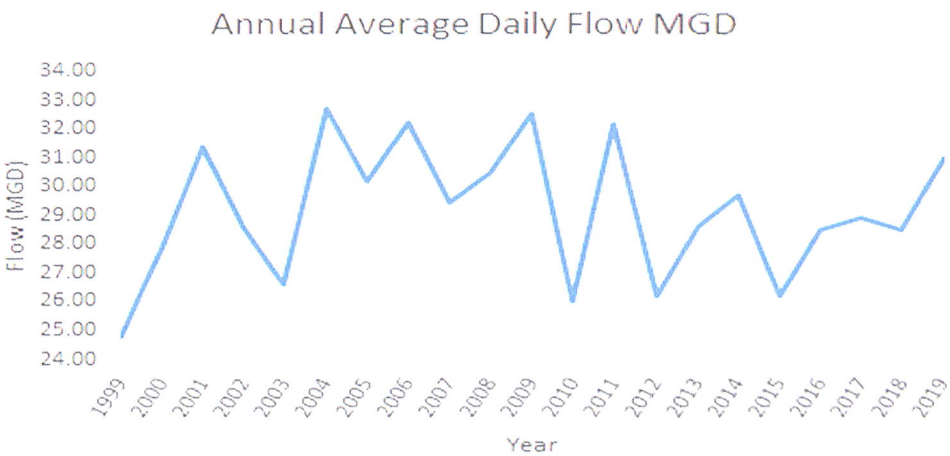

Figure 14.6 Twenty Years of Annual Flow Data (Courtesy of Genesee County Drain Commissioner: Division of Water and Waste Services)

TEST YOUR KNOWLEDGE

1. When setting up a sampling schedule, where is the first place to start?
   a. Outside laboratory
   b. Facility management
   c. NPDES permit or regulatory requirements
   d. Internet search

2. What is the most important information to know about an outside laboratory?
   a. Price
   b. QA/QC
   c. Method
   d. Shipping method

3. The pH, DO, and chlorine residual are generally tested on-site and not sent to an outside laboratory.
   ☐ True
   ☐ False

Budgeting and Financial Management

OVERVIEW

In addition to treatment and environmental regulations, utilities also have financial rules to follow. As a WRRF leader, you may be approving timecards, ordering supplies, signing off on invoices, preparing budget requests, or doing other financial tasks. A basic understanding of utility finance will help you succeed in a management role.

Budgeting and financial management are vital skills for WRRF leaders and utility managers. Studies have shown the need for more funding to address the cost of operation and much needed infrastructure repair and replacement. At the same time, there are limited sources of revenue and budget cuts. Balancing the financial needs of a utility is essential.

DEVELOPING A BUDGET

The development of an annual budget is a collaborative process. Many departments within the utility build a budget together. The budget must maintain levels of service while planning for future growth and addressing aging infrastructure. Budgets must consider the available funds and the needs of the utility. The budget should be available to ratepayers. The budget process may include public input. The budget reflects the goals and objectives of your utility. Developing a budget requires a detailed knowledge of your water and wastewater system. The budget is an important document that projects revenue and expenses for the following year and, once adopted, provides a blueprint for the future (U.S. EPA, 1994).

The *capital* budget is used to build or expand a WRRF and the *annual* budget pays for the operation of the WRRF.

A budget is a comprehensive document for the entire utility. Common practice is to divide the budget, both revenue and expenses, into categories. Broadly, there are two types of budgets:

- Annual or operations budget: This includes salaries and benefits, electricity and gas, chemicals, supplies, software licenses, maintenance (preventative, reactive), contract laboratories, service contracts such as biosolids disposal, and grounds maintenance.
- Capital program: This includes construction or expansion of facilities, vehicles, equipment, and rehabilitation of facilities. Capital budgets are used for things that can be capitalized or depreciated.

TYPES OF BUDGETS

Each utility has its own process for budgets. It is important to understand how your utility develops and itemizes its budgets. Recommended resources for further reading include the Governmental Accounting Standards Board, the Governmental Finance Officers Association, and the International City/County Management Association for detailed information on the types and styles of budget. Table 14.1 outlines the most common types of budget formats.

FOLLOWING A BUDGET

Once a budget has been approved by your utility, it is your responsibility to follow the budget. Funds for the WRRF must be spent according to how they are allocated. A WRRF leader must monitor spending to avoid going over the budget. Each utility has its own methods and tools to track expenses and allocations. Training can help you with the processes and tools to track budgets. Talk to the accountants or clerks at

Table 14.1 Common Forms of Budgeting

Budget Format	Description
Line item	This is the most common budget format used by water and wastewater utilities. It identifies basic operation needs and revenues of the utility by individual line item. For example, salaries, benefits, chemicals, supplies, insurance, and contracts as well as professional services such as consulting, accounting, and so on. From this, spending plans and schedules are established.
Activities-based	This type of budget is structured by specific activities of the organization/utility. This may include separate budgets for primary treatment, solids handling and disposal, collection systems, or other activities determined as important to the successful understanding of the organization. All costs and expenditures for each activity are separated into units such as salaries, utility costs, chemicals, equipment, and asset replacement. Activity-based budgeting is beneficial because it allows for clear development of performance measures (e.g., key performance indicators) to help manage the finances of the utility.
Zero-base	Requires the full justification of all expenditures each year and assumes that each unit starts with a "zero budget." This method allows for the prioritization of all expenditures to a final level of budgeting and for the selection of each set of expenses against all others, which the elected board or council must then evaluate. This method is different than the others because it removes the historical precedent of previous budget years.
Capital	Capital budgets are based on long-term financial plans of multiple years (typically 5 to 10) for the development, study, design, construction, and implementation of programs and projects for infrastructure and large improvements that support the utility. These plans are often developed to support issuance of bonds or impact or development fees. A capital budget and work must be coordinated with the operational budget to ensure the cost of operating the new capital works are accounted for.
Performance	Developed based on specific performance standards for the total unit and subunits of the budget and estimated based on the expected performance results.

our utility about where to find financial information. Get to know the financial tools, which are typically oftware systems and internal spreadsheets and reports. Often, a utility gets budget data from a computer ystem then shares and monitors the budget with spreadsheets.

able 14.2 presents a simplified operational budget. The data are from the midpoint of a fiscal year. This is n example of internal tracking to follow a budget.

A fiscal year is a 12-month budget period that may not be the same as a calendar year. For example, a utility's fiscal year may be July 1 to June 30.

spreadsheet, like that shown in Table 14.2, could be used by staff to track spending and progress. Monitoring the budget regularly helps to make sure the budget is used and not overspent. A more detailed ersion may be broken down on a monthly basis. In Table 14.2, coagulants are currently over budget at this oint in the fiscal year. Management may be concerned that spending will continue at this rate and exceed he budget. As the WRRF leader, you can explain why the chemical expense was so high. For example, extra hemicals were needed to manage high loadings from a seasonal industrial discharger, and later in the year he usage will be reduced. Similarly, grounds maintenance may be high due to a storm event and snow or lebris removal. If a process upset requires extra staff, which will increase over time, inform the accounting lepartment sooner rather than later. Accountants and finance managers are often more accepting of udget overages if they receive an early warning and understand the reasons why.

RATE SETTING AND INCOME

Jser charges or rates are the main source of income for utilities. Utilities typically have separate rates for esidential customers, commercial customers, and wholesale customers. Most utilities have rates based on olume. A customer pays based on the amount of water used such as \$3.50 per 4.5 m^3 (1000 gal). The sewer harge is based on the water volume and the assumption that a portion of the water used ends up in the sewer. here may also be a base charge or facility charge that is the same billing cycle. A wastewater-only utility may se a flat rate for sewer charges because water service is provided by a different utility and wastewater is not ypically metered. Rates are typically calculated by accountants or financial managers. Just like the budget, nput on WRRF operations is essential for projecting expenses to calculate an appropriate rate.

able 14.2 Simplified Operational Budget

	Annual Budget	Spent to Date	Remaining	% of Budget
Chemicals	$175,000.00	$105,000.00	$80,000.00	60%
Coagulants	$100,000.00	$60,000.00	$40,000.00	60%
Polymer	$35,000.00	$20,000.00	$15,000.00	57%
Disinfection	$50,000.00	$25,000.00	$25,000.00	50%
Utilities	$1,300,000.00	$625,000.00	$675,000.00	48%
Electricity	$1,100,000.00	$500,000.00	$600,000.00	45%
Natural Gas	$120,000.00	$85,000.00	$35,000.00	71%
Water	$60,000.00	$30,000.00	$30,000.00	50%
Fleet Fuel	$20,000.00	$10,000.00	$10,000.00	50%
Maintenance	$400,000.00	$181,000.00	$219,000.00	45%
Equipment	$250,000.00	$100,000.00	$150,000.00	40%
Fleet Vehicles	$15,000.00	$7,500.00	$7,500.00	50%
Grounds	$5,000.00	$3,500.00	$1,500.00	70%
Instrumentation	$30,000.00	$15,000.00	$15,000.00	50%
Unit Processes	$100,000.00	$55,000.00	$45,000.00	55%
Total	$1,875,000.00	$911,000.00	$974,000.00	49%

Most state and federal financial assistance programs, such as construction grants or state revolving fund programs, require that a utility have sound fiscal operation. Documented proactive asset management practices are being required in more areas. In Ontario, Canada, for example, all utilities are required to implement full asset management programs.

In addition to user rates, other sources of income for utilities include

- Development charges or impact fees. When a new subdivision or industrial facility is constructed, a utility may charge a fee to the developer to recover the cost of the infrastructure and the additional system capacity needed to serve the new development. These fees, measured in cost per liters per day or gallons per day ($/LPD or $/GPD), are used to build or upgrade facilities required for growth. These fees also reduce the financial effect on existing customers.
- Surcharge fees. These fees are for industries that discharge high-strength wastewater, which costs more to treat. The fees are often part of the pretreatment program. For example, a utility may charge $0.65 per 0.45 kg (1 lb) of BOD for high-strength discharges over 250 mg/L of BOD.
- Government grants and loans. State revolving funds offer low-interest loans for large projects like facility expansions. State and local governments may also offer special grants or low-interest loans for energy efficiency improvements and consolidations.

AFFORDABILITY

A growing challenge for utilities is affordability. How can a utility efficiently and effectively serve all of its customers? This can be answered in a variety of ways, and there is no one-size-fits-all approach. A common model for rates is a decreasing block rate where the more water that is used, the lower the rate. The first X m^3 (gal) may cost \$3.00/m^3, whereas volumes beyond that quantity cost \$2.00/m^3. This rate structure gives the advantage to larger users. This model does not benefit most of your customers.

For a combined utility, a high-income residential user may use a lot more water for lawn irrigation and for a house with more fixtures. A low-income customer may have a small apartment and use very little water. A rate structure that charges less for high volumes favors the large homeowner and disadvantages the low-income user. This effect can be reduced with an increasing block rate structure, where higher volumes cost more.

Businesses and industries sometimes discharge a higher quantity of pollutants. The higher the level of pollutants, the harder the WRRF must work to treat the wastewater. This could include needing larger equipment, more personnel, increased energy costs, more sampling, additional chemicals, or even additional treatment processes. Most WRRFs are designed to treat mostly domestic wastewater. One way to improve affordability and ensure that high-strength dischargers and heavy users are paying the true costs is with surcharge fees.

The rate structure should reflect your region's economic foundation. The services a utility provides are for the public good and a necessity for human health. This should be reflected in rates that are affordable for all and adequate to provide the utility's level of service. Thought should be given to a rate that can be afforded by all members of the community.

Having an affordability or assistance program will help you engage more with your customers. It also helps garner public outreach and establish overall better relationships in the community.

Procurement is the act of finding or obtaining goods or services needed to meet utility goals or business purposes. Procurement includes the overall process and final act of purchasing.

PROCUREMENT

Understanding procurement of goods and services is essential for all staff, and especially WRRF leaders. Procurement includes a range of activities, from buying supplies to selecting a contractor for a facility upgrade. There are state and federal requirements as well as rules local (utility) rules for procurement. Typical elements involved in the procurement process include

- Goods such as equipment, pumps, or work vehicles;
- Supplies such as chemicals or power; and
- Services like consulting, biosolids disposal, and maintenance contracts.

Table 14.3 Levels of Procurement and Approval

Dollar Value	Procurement Method	Authorization for Purchase	Method of Purchase
0–$1,000	User obtains price quote. Only one bid required.	Supervisor	Payment requisition, credit card, petty cash, or purchase order
$1,001–$10,000	A minimum of three written quotes are required.	Superintendent or Manager	Purchase order and/or contract
$10,001–$100,000	A formal tender must be issued and advertised.	Director or General Manager	Purchase order and contract
$100,001 and above	A formal tender must be issued and advertised.	Council or Board approval	Purchase order and contract

Following all the rules for procurement is important. For example, your utility may allow sole-source purchases up to a certain amount, say $10,000. Written quotes may be required for anything between $10,000 and $50,000. Purchases above $50,000 may require a formal bid process with director or council approval. The process is typically similar for obtaining services, like engineering or landscaping. Payment for goods and services may be done with a credit card, purchase orders, invoices, or formal contracts with pay applications. Table 14.3 shows an example of procurement policies for goods at a utility.

In addition to the rules already listed, many organizations have ethics and social guidelines for procurement. These policies are for avoiding conflicts of interest, preferential purchases from minority- or female-owned businesses, or preferential purchases from local suppliers.

SELECTING A CONTRACTOR

Engineering and construction contracts are discussed below in the "Capital Improvements and Projects" section. Service contracts are typically based on the overall price rather than ranking on experience or skills. As the work becomes more complex, so too does the evaluation process. Table 14.4 below provides some examples of work and the typical bid type used.

BIDS AND REQUESTS FOR PROPOSALS

Bids and requests for proposals (RFPs) are procurement methods to ensure cost certainty and fairness. Bids and RFPs are used when hiring constructors and consulting engineers as well as in equipment and vendor selection. Generally, an RFP is used to hire professional services such as engineering or architecture firms. Bids are used for equipment purchases or construction contracts. Requests for proposals allow the utility to select based on qualifications, experience, and price. Bid selection is primarily based on price.

Table 14.4 Types of Service Contracts

Type of Contract	Bid Type
Minor Maintenance	This type of work includes lawn maintenance and housekeeping. Price is typically the main factor in selection.
Major Maintenance	This type of work is a larger project with a costly activity and/or requires specialized staff to perform. Examples include centrifuge maintenance, turbo-blower maintenance, and so on. In some cases, the lowest bid could be used if all contractors meet your requirements. In other cases, there may only be one contractor capable of the work and a sole-source bid is necessary.
Operational	Some utilities own their infrastructure but hire a third party to operate part or all of the system or perform certain tasks like locating utilities. In this case, a competitive bid process is often chosen with evaluation criteria similar to hiring an engineering firm. Price is one component of the selection, along with skills and experience.

CAPITAL BUDGET PLANNING

Capital planning must consider how much a project costs, how much money is coming in, and when the money will be spent. The overall budget for a capital improvement plan (CIP) is based on the utility cash flow. The cash flow is typically a 3- or 5-year forecast of income from developer fees, grants, loans, and user rates with project costs and timing. Budgets for individual capital projects often include design, construction, materials, equipment, land purchase, and a contingency. The role of operators in CIP projects is discussed later in this chapter.

TEST YOUR KNOWLEDGE

1. **How does a capital budget differ from an operational budget?**
   a. It is funded solely from user rates.
   b. Only a manager can develop a capital budget.
   c. It typically has a longer time horizon (5 to 10 years).
   d. It is always higher than an operating budget.

2. **Operating budgets are predominantly funded by user rates.**
   ☐ True
   ☐ False

3. **Which of the following does not typically follow a procurement process:**
   a. Purchase of chemicals
   b. Hiring staff
   c. Hiring a contractor
   d. Disposal of biosolids

4. **A line-item budget is best defined by**
   a. Its total cost line-by-line
   b. Being structured around specific activities at the utility
   c. Key performance indicators and performance standards
   d. Basic operational needs and revenues of the utility by individual line item

Writing and Updating Standard Operating Procedures

An SOP is a set of written instructions that lists all actions or steps related to an event or process. An SOP provides a consistent method to follow. When an SOP is followed, an event or process is handled in the same way, regardless of the person conducting it. Standard operating procedures are written for events ranging from routine daily operations to emergency responses, and everything in between.

An SOP might describe the oil change procedure for an aerator motor, the sampling process for a water test (see Figure 14.7), or the notification process for changes made to a piece of software. Standard operating procedures should be specific to the organization using them to address each organization's unique system and needs. Even when standard methods are being used, SOPs add details and clarify organization-specific steps to a process, such as where an employee should record that an oil change was completed.

BENEFITS

Standard operating procedures provide several benefits: they ensure consistency, help new employees get up to speed quickly, and ensure consistent performance with seasoned staff. Although experienced staff may feel confident with daily tasks, it is common for employees to have different ideas on how to complete a task. Standard operating procedures help ensure that tasks are done correctly and consistently, no matter who is working on them.

Some advantages of using SOPs include

- Promoting quality through consistency, even if there are personnel changes. For example, following an SOP for daily rounds means the WRRF has the same checks and data, whether the operator is a trainee or 20-year veteran.
- Ensuring staff perform tasks consistently and in compliance with the rules. Standard operating procedures can be used as part of staff training programs. For example, laboratory methods for

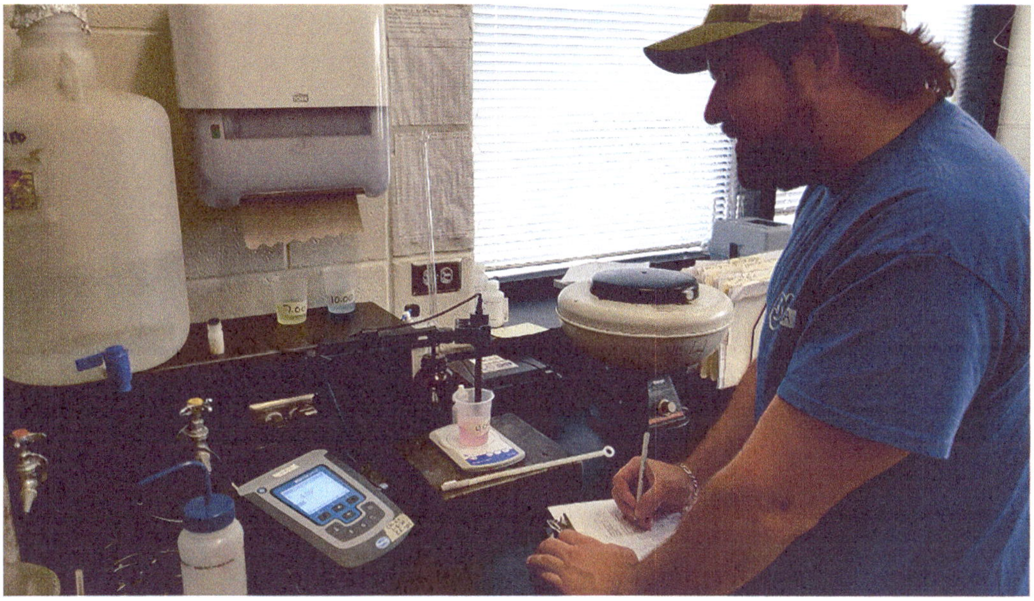

Figure 14.7 Port Royal Water Resource Recovery Facility Operator Tyler McTeer Calibrates the pH Meter According to the Standard Operating Procedure (Courtesy of Beaufort-Jasper Water & Sewer Authority)

calibrating equipment or analyzing samples that follow an approved SOP will meet laboratory certification requirements (see Figure 14.7).

- Minimizing confusion and increasing safety. For example, an SOP for shift change to share the facility's status or lockout/tagout for working on a pump.
- Providing insight when historical data are being evaluated on how the data were collected. Improving the value of historical data and allowing for meaningful comparison with more current data.

COMMON STANDARD OPERATING PROCEDURES AT WATER RESOURCE RECOVERY FACILITIES

At a WRRF, SOPs are commonly used for both regulated and nonregulated events, such as the following:

- Process control;
- Equipment maintenance and asset management;
- Startup and shutdown for equipment, unit processes, and the facility as a whole;
- Sampling and laboratory methods;
- Compliance reporting;
- Human resources tasks such as hiring, promotion, discipline, and performance evaluations;
- Inspections; and
- Financial tasks such as travel reimbursement or purchase order requests.

WRITING A STANDARD OPERATING PROCEDURE

The first step a utility should take when drafting SOPs is to identify all the business processes or procedures that need documentation. These should include commonly repeated processes as well as processes that may not be repeated often but need to be performed in a specific way due to health and safety or regulatory requirements.

A *business process* is a set of related tasks that, in the end, provide a service or product or accomplish an organization goal. A public WRRF is not a business, but still has business processes.

For each procedure, the WRRF or utility should decide who should be involved in creating the SOP, reviewing the final document, distributing the SOP, and training the appropriate staff. Standard operating procedures are created and followed for daily activities, such as facility rounds, and infrequent events like changing the belt on the belt filter press.

Each organization should create an SOP template to use as a starting point. Standard operating procedures can range from simple checklists to much more detailed instructions. In general, an SOP typically includes the following information:

- **Document Metadata Information:** Each page in the SOP should include the document title or identification number, a revision number and date, the author or person responsible for the document, and the page number. The page number format should be "# of #" so the reader can verify that they have the entire document. This information can be placed in the header or footer of each page.
- **Title:** The title block or page of an SOP should include all document metadata as well as approval signatures and version information.
- **Purpose:** A brief paragraph to explain the intended goal in assembling the SOP.
- **Scope:** A description of what activities are covered and who is responsible for them. An organization may also list what is outside the scope of the SOP here.
- **Definitions/Glossary:** Definitions of any words, phrases, acronyms, or abbreviations that might be confusing to someone unfamiliar with the system or process being described.
- **Procedures:** Step-by-step descriptions of how to perform tasks. Information can include methods, equipment and supplies lists, inspection and audit procedures, safety or health warnings, equipment calibration information, specific considerations to prevent damage to the equipment, and records management.
- **Related Documents:** A list of all related documents maintained by the organization that add context to the SOP. This can include references to policies, guidance documents, or other SOPs.
- **Roles and Responsibilities:** Sometimes included as part of the procedure, the SOP should state who is responsible, accountable, informed, or consulted for each step.
- **Appendices:** Supporting documentation that is needed to complete the procedure. Examples include manufacturers' instruction guides, regulations, and process workflow diagrams.
- **Revision History and Approval Signatures:** A record of who authored/approved the SOP and when changes were made. This is important for long-term document management.

All SOPs should be accessible from a common library or central storage location. Individual employees or groups can maintain their own copies. There should always be one set that is maintained as the official record. For example, an SOP for chlorine cylinder exchange may be stored in the chlorine room, but staff should understand that the SOPs in the central library are the official copies and are considered the system of record. When an SOP is updated, the copy in the central library and any additional copies that exist should be updated at the same time.

Because senior staff typically identify needed SOPs, they should also ensure that SOPs are clear to all members within an organization. When writing an SOP, consider all the information and steps needed for the reader to successfully complete the task or process. Even though people who have worked for the organization for a while may be familiar with terms and abbreviations used by your organization, terms should be explained in every SOP so that someone new to the WRRF can follow the SOP. Alternatively, an organization can maintain a central glossary for unique acronyms or terms and each individual SOP can point toward this.

All staff must be made aware of all SOPs that exist. Staff should be trained on the value of SOPs and directed to follow existing SOPs and report discrepancies. For common procedures, print and laminate a copy of the SOP and post it in the appropriate location: for example, consider posting the lockout/tagout SOP in the electrical room. Water resource recovery facility leadership must reinforce the importance of SOPs by following them and addressing any inconsistencies among staff.

MAINTAINING STANDARD OPERATING PROCEDURES

Over time, SOP libraries need to be maintained to ensure that they are current and that they cover all the activities an employee will encounter in their day-to-day job. An SOP library may be on paper in a binder or saved on a computer network. Typically, changes to SOPs will be required due to process change or operational change. Changes may include

- New equipment that has special maintenance needs
- Aging equipment with changing maintenance needs
- New regulations that require different ways of working

By keeping SOPs up-to-date and easy to locate, organizations show their commitment to ensuring that actions are performed consistently and correctly.

TEST YOUR KNOWLEDGE

1. Who is the intended audience of an SOP?
   a. New employees
   b. Experienced employees
   c. Everyone in an organization
   d. People within an organization who will be performing the task described in the SOP

Asset Management and Maintenance

PURPOSE

To maintain WRRF infrastructure for reliable operations while managing funding and meeting customer needs, a WRRF should define, develop, and manage assets through an asset management program. *Asset management* is defined by U.S. EPA (n.d.-a) as

> "a process water and wastewater utilities can use to make sure that planned maintenance can be conducted and capital assets (pumps, motors, pipes, etc.) can be repaired, replaced, or upgraded on time and that there is enough money to pay for it. It is the practice of managing infrastructure capital assets to minimize the total cost of owning and operating these assets while delivering the desired service levels. Many utilities use asset management to pursue and achieve sustainable infrastructure. A high-performing asset management program includes detailed asset inventories, operation and maintenance tasks, and long-range financial planning."

Assets are equipment, piping, infrastructure, buildings, vehicles, or other objects that hold value to a utility in providing a service to ratepayers and carrying out the utility's mission.

Given the broad scope, a functional and useful asset management program takes resources and attention; however, the benefits can greatly outweigh the cost of implementation.

ASSET MANAGEMENT PLANS

To create an asset management program, the utility must first create an asset management plan, which also includes life cycle planning. Life cycle planning looks at the asset from the initial purchasing phase all the way through to replacement with a new asset. Creating an asset management plan will take input from staff and other resources that understand the operational, maintenance, design, and information technology aspects of the equipment and processes. Some WRRFs hire staff to oversee the asset management program and to lead the team through development of the plan. Once developed, the program manager continues to oversee the program and helps to guide decision-making on financial planning. Figure 14.8 shows typical considerations for a successful plan.

Risk matrices are another tool used in asset management plans. *Risk* is defined as the consequence of failure, multiplied by the chance of failure, resulting in a measure of "criticality" (see Eq 14.1). The consequences of the failure of an influent pump are much higher than the failure of a scum pump. The likelihood of failure for a new pump is lower than that for an old pump. By creating a risk matrix, the planning group can see potential risks for the WRRF. The capital planning is directed based on immediate to long-term needs. Figure 14.9 illustrates a typical risk matrix. The assets that fall in the green area have a very low consequence of failure and a low risk if the asset fails. In looking at the circle in this area, the utility in this instance will likely allow that asset to run to failure due to the overall low risk. At the opposite end of the spectrum, if an asset has been tagged as both an extreme risk and has an extreme likelihood of failure, the utility should plan to quickly replace the particular asset. Again, this evaluation helps to prioritize the overall replacement strategy of assets for the utility.

$$\text{Risk} = (\text{Consequence of failure}) \times (\text{chance of failure}) \qquad (14.1)$$

Asset management plans will have the following structure:

Level of Service
- Current and Desired Level of Service

Future Demands
- Assessment of Flows and Loads on Plant

Resilience
- System Resilience Assessment

Current State of Assets
- Asset Register Summary
- Asset Performance
- Asset Condition
- Asset Valuation
- Historical Costs

Risk Assessment
- Asset Risk Assessment

Life Cycle Management Plan
- O&M Plan
- Replacement Plan
- Asset Creation / Acquisition / Augmentation Plan
- Asset Disposal Plan

Financial Summary
- Forecast of O&M Costs
- Forecast of Asset Replacement Costs
- Forecast of Capital Project Costs

Plan Improvement and Monitoring
- Improvement Action Plan

Figure 14.8 Typical Components of an Asset Management Plan (**Courtesy of HRSD, Asset Management Team**)

LIFE CYCLE MANAGEMENT PLANNING

The life cycle approach to managing assets is an attempt to balance asset performance, cost, and risk. Utilities can make the best use of their money by assessing the financial effects across the entire asset life cycle, which includes planning, creation, operation, maintenance, and disposal/replacement. Life cycle management includes analyzing the whole life costs of the asset.

With the life cycle approach and using the risk matrix, utilities may determine it best to run an asset to failure. This might be a sump pump in a containment area or an exhaust fan in a nonhazardous area. The purpose of using the option of running an asset to failure is to reduce the amount of time and effort spent on preventive maintenance to better focus efforts on critical WRRF equipment. The aerator on an oxidation ditch is a critical asset, so an asset management program would track the equipment's condition, likelihood of failure, and the consequence of failure to assess the risk it poses to facility performance. An appropriate maintenance strategy is developed based on this risk assessment.

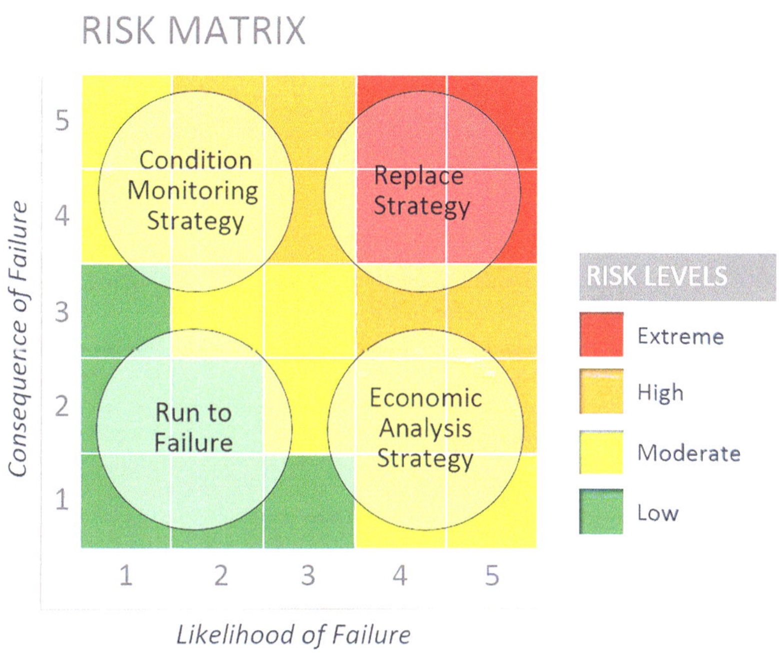

Figure 14.9 **Risk Matrix (Courtesy of HRSD, Asset Management Team)**

TYPES OF MAINTENANCE AND MAINTENANCE PLANNING

Proper and timely maintenance is critical for WRRFs to meet their desired service levels. Level of service includes cost, quality, efficiency, redundancy, reliability, safety, and responsiveness. The WRRF must first define its level of service before proceeding with developing a maintenance plan. The needs of the WRRF can be met by well-thought-out planning and oversight, and financial planning can be best managed to make "the right investment at the right time." Two broad categories of maintenance are preventive and corrective maintenance. Maintaining equipment includes the prescribed preventive maintenance and reactive maintenance, which involves addressing problems noticed while conducting preventive maintenance. Preventive maintenance includes tasks such as oil changes and checks for alignment, vibration, and bearing condition. Equipment manufacturers typically create a schedule of these preventive maintenance tasks based on hours of run time or other service length.

Level of service is defined as a statement of objectives or outputs defined by a utility as to its intents on delivery to customers, while factoring in regulatory requirements and defined performance expectations. A written level of service document defines the goals of how the utility services are provided.

The second category, corrective maintenance, means the equipment has broken down and requires either repair or replacement. Ideally, most maintenance at a WRRF falls under the preventive maintenance category instead of corrective maintenance. Once a utility has established its current mix of corrective versus preventive maintenance, it is good to consider setting incremental goals to shift the balance toward more preventive maintenance and reduce the need for corrective maintenance.

Deciding between maintaining or running to failure is another way to address the level of service. In some cases, it makes the most sense to run the equipment until it fails, never doing any level of preventive maintenance. This directly relates to the WRRF's comfort in risk in running the asset to failure, or where the asset falls in the risk matrix, shown in Figure 14.9.

Preventive maintenance typically incorporates the manufacturers' recommendations on maintenance. This includes maintenance such as adding and replacing oil, replacing gaskets, or just overall inspection of the asset. Many WRRFs use CMMSs or asset management programs to manage preventive maintenance schedules and work orders. Some computer applications are capable of both CMMS and asset management functions.

Computer maintenance management system programs consolidated maintenance information and assist in the overall processes of maintenance operations. Computer maintenance management system software can also be used to track life cycle costs and help WRRF operators in deciding whether to replace an asset or continue to repair the asset.

TEST YOUR KNOWLEDGE

1. Preventive maintenance happens after equipment breaks.
   - ☐ True
   - ☐ False
2. Corrective maintenance happens after the equipment breaks.
   - ☐ True
   - ☐ False
3. By using a risk matrix, WRRFs can prioritize capital spending based on equipment criticality.
   - ☐ True
   - ☐ False
4. All equipment should be used until it dies. Replacing equipment before if fails is a waste of money.
   - ☐ True
   - ☐ False

Capital Improvements and Projects

As a WRRF leader, part of your work may look beyond day-to-day operations into long-term planning. Reliability, capacity, budget, and level of service are all important for a utility, as is maintaining comprehensive budgets to sustain the utility's level of service. This is especially true when it comes to capital projects.

A capital project is a project that involves purchasing, building, replacing, or rehabilitating an asset. Capital projects involve assets that can be depreciated.

PROJECT DEVELOPMENT

CAPITAL PROJECTS

Most utilities have a CIP, which is a series of capital projects funded over several years. Capital projects range from replacing a piece of equipment to constructing new WRRFs. Smaller capital projects, such as major equipment replacement, may be created and carried out by WRRF staff, possibly with the utility's engineering staff. Larger capital projects (see Figure 14.10) are often complex and involve multiple specialties and departments as well as outside consulting services. For example, when a new facility or new process train is needed at an existing WRRF, an engineering company is hired for their skills in process control and electrical, structural, mechanical, civil, and architectural design.

Creating a CIP involves determining which capital projects should be funded and when. This can vary based on a number of factors, such as

- Master plans, which contain growth projections, facility capacities, and life of assets, among other things;
- Asset management plans and CMMS data;
- Organizational policies and goals;

Figure 14.10 Construction of the Hardeeville Water Reclamation Facility Expansion Was a Capital Project (Courtesy of Beaufort-Jasper Water & Sewer Authority)

- Federal and state regulations and meeting NPDES permit limits;
- Internal resources such as funding and staffing to successfully execute a capital project;
- System improvements or capacity increases such as preventing bypasses and overflows, addressing capacity or process limitations, modernizing controls, and so on;
- Reducing energy use; and
- Funding allocated by an external source at the state or federal level, a grant or loan, with defined metrics and/or timelines.

Each utility will have its own process for developing a CIP and its projects. Typically, the engineering department or a consultant of a utility handles the CIP development and project management. An operator with an idea or request for a capital project should talk to the utility engineers or managers for direction on the capital project request process.

OPERATORS AND CAPITAL IMPROVEMENT PLAN PROJECTS

Operations and maintenance (O&M) staff should be involved in all aspects of the CIP development and implementation. They should be the first source of information and are a crucial part of project development and providing feedback on potential projects. Operations and maintenance staff have valuable insight to the current state of the WRRF and can aid engineering staff, consultants, and management with the CIP. They provide data on facility performance, insight on equipment reliability, and feedback on facility and building layouts. They should also have the opportunity to understand project drivers, pros and cons of options, and provide input on final decisions. At the end of the project, they will be responsible for the O&M of the system and ensure it performs well.

An example of WRRF staff informing capital projects is a secondary solids control issue. An operator notices that an excessive amount of time is spent monitoring and maintaining the secondary clarifiers because of excessive algae growth in the weirs and/or solids imbalances between multiple clarifiers. Operators are responsible for identifying these shortcomings in processes and informing management. An operator can also provide some recommendations to improve this process, including

- Brushes attached to the scum scraping mechanism to brush off excessive algae and launder covers to prevent light from reaching the weirs;
- Replacement of aging or failing pumps, which are not distributing flow to the clarifiers equally; or
- An engineering evaluation of the solids distribution between clarifiers as the operator has noticed the solids blanket continues to vary between clarifiers while the pumps appear to operate consistently and similarly across each clarifier.

This is all valuable information that an operator can provide to address a WRRF deficiency with a CIP project.

PROJECT DESIGN

Utilities often contract services such as laboratory, engineering, and other specialties because they may not have the facilities or internal expertise necessary to perform specific or infrequent work. Generally, for a medium or large capital project, a design engineer will be hired to prepare engineering drawings and specifications. A contractor will be selected to build according to the engineer's plans and specifications. This subsection will address project design and engineering selection as engineering firms typically manage the construction contractors.

Key areas for operator input during the design and bid process or an RFP selection include

- Experience with the vendor or consultant: Has this vendor delivered on previous projects or is it a new vendor? Similar questions should be asked regarding consultant services.
- Experience with equipment: Has this pump been reliable in the past? Is the technology familiar or will it require specialized knowledge and support? Is service from this manufacturer reliable or difficult to obtain?
- Potential operations/construction conflicts or effects: Do you anticipate any challenges with this design such as delays, limitations, or roadblocks in construction such as nonfunctional gates, which will prevent taking down a process train, or a tie-in, which requires a facility shutdown?

ENGINEER SELECTION

Requirements vary by state, but generally engineers are not selected solely on price for engineering services. The National Society of Professional Engineers recommends the following steps for selecting an engineer (it is also important to be familiar with your utility's selection process, particularly when it comes to the size and scope of the project):

1. Prepare and issue a request for qualifications, including experience on similar projects.
2. Collect and score qualification statements to identify the qualified firms.
3. Issue an RFP to the qualified firms.
4. Collect and score proposals and, in some cases, interview the top scoring firms.
5. Select the top scoring firm and negotiate the price. Note that, in some states, the price may be part of the proposal.
6. If the negotiation is not successful, a utility may return to step 4.

Operations and maintenance staff may be among the reviewers for engineer selection for a WRRF project.

EQUIPMENT SELECTION

Selection of equipment is one of the most critical areas for O&M staff input on capital projects. Operation and maintenance staff will be working with this equipment for years or decades to come and should have a say. Key questions and considerations during the equipment selection process include

- Does the existing equipment operate well or should a new model be considered? For example, should a raw wastewater pump be replaced with a non-clogging style or a non-clogging impeller if the pump is compatible with it. In some cases, the existing equipment works well but is at the end of its life, and there is no need to replace it with a new technology. Instead, a like-for-like replacement is sufficient.
- Will new technology and equipment improve the process and/or reduce labor? For example, replacing membrane-based DO sensors with optical sensors can significantly reduce maintenance while improving accuracy. New technology is also an area for operators to listen to engineers: An existing process or piece of equipment may be adequate, but there could be better ways of operating that should be considered.
- Will changes lower energy and operational costs? For example, greater turndown of equipment such as turbo-style blowers and peristaltic chemical pumps may reduce energy consumption or chemical consumption.

After the equipment type or process has been selected, the equipment procurement for a capital project can come in many forms, such as

- Multiple vendors submit bids for similar equipment, which is then evaluated by cost and/or selection criteria developed by the utility.
- Sole-sourcing of equipment—A specific manufacturer is specified to provide equipment. This might happen when, for example, no comparable alternative exists or a replacement is required and there are existing pieces of equipment from the same vendor (e.g., blowers) where another vendor cannot work with the existing equipment.
- Pre-purchasing of equipment—The equipment is purchased directly by the utility and then provided to the contractor to be installed. This is most often done when there is a long lead time for equipment (e.g. clarifier mechanisms).
- Contract allowance—This is typically done when the specific model of equipment is not yet defined, but an approximate cost is known. The contractor is given a certain amount of money in the contract to purchase the latest model of the equipment later in the project. This is common when technology is changing rapidly (e.g., supervisory control and data acquisition servers) and the project proceeds over several years.

ENGINEERING DRAWING REVIEW

A utility manager will often select key O&M staff with the experience and detailed knowledge of a facility to review engineering design drawings. Each type of drawing should be reviewed by staff with strengths in that area. For example, process and instrumentation drawings should be reviewed by staff that are intimately familiar with the WRRF and how it operates to provide relevant feedback. Maintenance staff can provide valuable feedback on mechanical and electrical design drawings. Reviewers should be informed of the expectations on level of review and what feedback is needed so that they can use their time wisely and the project receives the best feedback.

PROJECT EXECUTION

Contracts related to procurement of services, except for engineering and construction services, were discussed previously. Contracts, particularly engineering and construction contracts, play a significant role in capital projects and can have serious financial and legal implications. Major elements to a contract as well as delivery methods will be discussed further in the following two sections.

CONTRACT ELEMENTS

The hiring approach described in the "Engineer Selection" section above is commonly referred to as a *competitive bid contract selection* and is also applicable to hiring construction contractors. Hiring of both engineering and non-engineering contractors can also take the form of sole-source contracts. Sole-source contracts are often based on your organization's procurement policies and are typically related to the overall cost of the project, as discussed previously.

Contract between a utility and an engineering firm or construction contractor for a capital project should address the following basic elements:

- Scope of work
- Timing: duration, deadlines, and work schedule
- Payment and payment frequency
- Project deliverables such as plans, specifications, reports, permits, and so on
- Any additional services such as inspection services, materials testing, instrument calibration, and so on
- Legal elements such as contract termination and who has the right to terminate the contract
- Limits to effects on operations

Construction contract documents can be hundreds of pages long, and printed versions may be multiple volumes. Construction contracts typically include the signed agreement, general and supplementary conditions, technical specifications, drawings, bonds, insurance, notice of award, and notice to proceed. Supplemental information such as invitation to bid, instructions to bidders, shop drawing requirements, written interpretations, or clarifications may also be part of the contract documents. Contracts often include requirements for training staff and providing O&M manuals.

Well-written contracts and specifications for construction and engineering services are critical to successful projects. Once signed, a contract can be modified only by issuing a written change order or work change directive. Changes to the contract may also change the cost and timing of the project.

DELIVERY METHODS

Choosing the best delivery method for a capital project is one of the most crucial project decisions and will have significant effect throughout the project and beyond. For all delivery methods, there will be an owner, engineer or designer, and a contractor. As the owner, the utility must understand the various methods and which best suit your project from the beginning. The delivery method influences the ownership, costs, and effect of changes at each stage. Common delivery methods are described below:

- Design-bid-build is the traditional way projects are delivered. The design is produced by a designer, the bid phase determines who will construct the design, and the constructor builds the project once selected from the bid phase.
- Design-build overlaps design with the construction phase. The contractor typically is responsible for design and construction of the project. This can result in shorter schedules and lower cost.
- Construction management at risk is a project with a defined schedule and price. The company responsible for construction provides input during the design phases and becomes the general contractor during the construction phase.
- Multi-prime has three distinct and sequential phases: design, bid, and build. In this delivery model, the owner has control over the entire process and both the designer and the contractor work for the owner

(Design Build Institute of America, 2015). Although it is similar to design-bid-build with three distinct phases, under the multi-prime delivery method the owner contracts directly with separate contractors for specific and designated elements of the work, rather than with a single general or prime contractor. The owner is then responsible for managing the contractors.

- Design-build-operate uses a single contract with a company to design, construct, and operate a new or improved facility. The utility remains the owner.

CONSTRUCTION

Besides being involved in project development and design, operators and maintenance staff are key to the startup of a WRRF project and must be involved in bringing new equipment or processes online or taking equipment out of service. Construction at and around a WRRF can be very disruptive. An operator can help reduce some of the disruption by being involved in the project process from the beginning. Projects at an existing WRRF should have a section in the plans and specifications for maintenance of facility operations. This provides guidance on construction and commissioning new processes and equipment while minimizing facility interruptions. At the end of a capital project, the operators are often the ones to give the final signoff on new equipment and/or processes. Because operators observe the WRRF daily and are in a position to judge performance, operators typically are the ones to discover problems that lead to punch list items or warranty claims.

INNOVATION

Innovation should not be a top-down phenomenon. All WRRF staff have valuable information to share and asking for input and feedback is essential for both staff empowerment and continued improvements at a WRRF. Staff can provide valuable feedback on funding of projects and expected costs, better ways to design a facility based on their experience, as well as optimizations to the facility. State, regional, and national conferences like the Water Environment Federation's Annual Technical Exposition and Conference (WEFTEC) provide great opportunities to learn about innovative technologies that may be suitable for your WRRF or utility.

TEST YOUR KNOWLEDGE

1. **Which of the following do not describe a type of procurement:**
   a. Evaluating multiple bids based on predetermined criteria
   b. Reusing old materials on-site
   c. Pre-purchase of goods or services
   d. Contract allowance

2. **Procurement of engineering services is typically not based on**
   a. The firm's size and ability to perform the work
   b. The lowest bid
   c. The firm's experience on similar projects
   d. Scoring criteria developed by the utility

3. **Which of the following is the most common delivery method for a project:**
   a. Multi-prime
   b. Design-build-operate
   c. Design-bid-build
   d. Design-build

4. **What is not a key element of an engineering contract?**
   a. Project scope
   b. The experience of an engineer
   c. Schedule

5. **Operations and maintenance staff have valuable insight and data to contribute to the design of a CIP project.**
   ☐ True
   ☐ False

Emergency Planning and Response

As a leader at a WRRF, an ongoing and important task is readying the facility and staff to handle different types of emergencies, whether human-caused threats or natural disasters. Having the correct plans in place to deal with these events is key to your ability to both withstand and recover from events, big or small. The purpose of establishing an emergency response plan (ERP) is to have a single guide to help you prevent, minimize, mitigate, and recover when faced with an emergency or incident.

This section will discuss the basics of emergency planning and response, vulnerability assessments, and the roles of people within the ERP, in addition to ERP updates and training.

TYPES OF EMERGENCIES

Geography plays a role in what kind of natural disasters a WRRF may face. Other emergencies may be the result of human activities independent of weather or climate. Some emergencies a WRRF may need to prepare for include

- Blizzard
- Tornado
- Earthquake
- Hurricane (see Figure 14.11)
- Flash floods
- Widespread flooding
- Contamination
- Power failure
- Personal injury
- Labor strikes
- Construction accident
- Terrorism
- Civil unrest
- Cyber attacks

Figure 14.11 Flooding in a Sewer Lift Station After Hurricane Matthew in 2016 (Courtesy of Beaufort-Jasper Water & Sewer Authority)

- Pandemic or local epidemics
- Active shooter
- Toxic spills or dumping
- Protests
- Supply chain disruptions
- Explosions
- Wildfires
- Facility fires

A vulnerability is a weakness, an area of exposure or risk, or a threat.

VULNERABILITY ASSESSMENTS

An important first step in creating an ERP is conducting a vulnerability assessment. Knowing your risk allows better preparation and protection for a WRRF by making security or process-related upgrades. During the vulnerability assessment process, a list is made of potential threats or events that may result in damage to a utility's assets or disruption in a utility's operation. These threats can either be human-caused or natural. Threats can be ranked by results of a specific attack. After a list has been established, it is important to think of immediate actions that can be taken to minimize risk. All this information and data should be taken into consideration when choosing a strategy to reduce risk and increase resiliency through design, operation, and management of a WRRF. Because of its sensitive nature, the vulnerability assessment document should be marked confidential and not shared publicly.

Utilities that provide both drinking water and wastewater services have a regulatory requirement under the American Water Infrastructure Act to prepare and update both the vulnerability assessment (called a *risk and resiliency assessment*) and the ERP. These documents are for the entire utility, not just a WRRF. The standard for conducting a vulnerability assessment can be found in the latest version of *Risk and Resilience Management of Water and Wastewater Systems* (AWWA, 2021).

INCIDENT MANAGEMENT

In 2001, the U.S. government, through the Federal Emergency Management Agency (FEMA), created the National Incident Management System (NIMS). The NIMS provides a "consistent nationwide template to enable partners across the Nation to work together to prevent, protect against, respond to, recover from, and mitigate the effects of incidents, regardless of cause, size, location, or complexity." States, municipalities, agencies, utilities, and stakeholders of all sizes can use the NIMS for command and coordination of incidents, resource management, and information management. The NIMS is not an emergency response plan. Rather, the NIMS is a set of concepts for all threats, hazards, and events that is scalable, flexible, and adaptable for incidents of any size (FEMA, 2017). The advantage of using the NIMS is the flexibility, standardization, and unity of effort with other agencies. State governments, local governments, and first responders use the NIMS, so responding to an incident and working together will be easier. The FEMA offers online training in the NIMS at training.fema.gov. Your state or local emergency management agencies may also offer in-person or online training.

The Federal Emergency Management Agency's National Preparedness Goal identifies what it means for a community to be prepared for all types of disasters and emergencies. This includes WRRFs. The five mission areas, as shown in Figure 14.12, are

1. Prevention: Prevent, avoid, or stop an imminent, threatened, or actual act of terrorism.
2. Protection: Protect our citizens, residents, visitors, and assets against the greatest threats and hazards in a manner that allows our interests, aspirations, and way of life to thrive.
3. Mitigation: Reduce the loss of life and property by lessening the effect of future disasters.
4. Response: Respond quickly to save lives, protect property and the environment, and meet basic human needs in the aftermath of a catastrophic incident.
5. Recovery: Recover through a focus on the timely restoration, strengthening, and revitalization of infrastructure, housing and a sustainable economy, as well as the health, social, cultural, historic, and environmental fabric of communities affected by a catastrophic incident (FEMA, 2015).

For a WRRF, prevention may involve adequate fencing and security to protect against vandalism, theft, or terrorism. Protection at a WRRF includes spill plans. Mitigation at a WRRF could be raising a wet well

Figure 14.12 Incident Management Cycle

level or moving electrical controls so that future flooding is not as damaging (see Figure 14.11). Response happens when WRRF staff stay at the facility overnight or following days after a blizzard or hurricane to maintain operations. Recovery is the cleanup and rebuilding that happens in the weeks or months after an event or disaster. Each incident or event is an opportunity to assess vulnerabilities and improve infrastructure and plans. Response and recovery activities lead to mitigation projects and additional protection measures and make the WRRF more resilient.

An incident action plan (IAP) is a written or oral plan with objectives, tactics, and support activities for an operational period, typically 12 or 24 hours. Every incident should have an action plan, although not all incidents require a written plan. If an incident becomes more complex or lasts more than 1 day, or involves multiple agencies, then a written IAP is recommended.

Some utilities have an emergency operations center (EOC), which is a room or building for coordinating information and resources during an incident. An EOC may be a temporary facility at the scene of an incident or a permanently established facility in a central location.

INCIDENT COMMAND STRUCTURE

Part of the NIMS is the incident command structure (ICS), which is a "standardized approach to the command, control, and coordination of on-scene incident management" (FEMA, 2017). The ICS provides a common structure for personnel so that multiple agencies and organizations can work together effectively. Incident command structure is used by government agencies at all levels as well as private sector organizations and nongovernmental organizations. The NIMS and ICS can be used for natural disasters like a hurricane or blizzard or for planned events, such as a parade or the Super Bowl. A chlorine gas release at a WRRF is an example of an incident where multiple agencies may be working together on scene, including WRRF staff, hazmat personnel, and emergency medical services first responders.

An advantage of ICS is the flexibility to respond with the number of personnel needed for an incident. There is always an incident commander, but the other roles are activated, as needed, depending on the size and nature of the incident. If an incident evolves and changes in scope or scale, positions may be activated and the number of staff may be expanded. The ICS system includes five major functional areas, staffed as needed, for a given incident: command, operations, planning, logistics, and finance/administration. Figure 14.13 presents an ICS diagram.

The incident commander is responsible for overall management of the incident, which includes establishing incident objectives (the IAP) and ensuring that activities accomplish the objectives. In a situation where multiple agencies are working together, there may be a unified command where a group

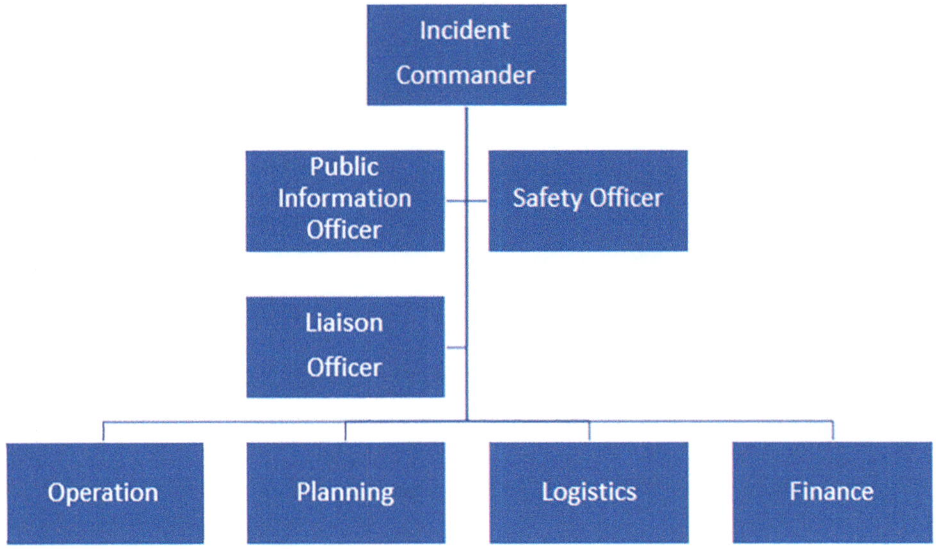

Figure 14.13 Incident Command Structure

of agency leaders share the incident commander responsibilities and jointly determine the objectives and priorities. Depending on the size, the incident commander may activate other roles, as needed.

The incident commander approves the release of information about the incident, and the public information officer (PIO) is responsible for interfacing with media and the public. The PIO collects, verifies, plans, and shares information on the incident and monitors media for relevant information.

Because of the importance of safety, the incident command staff include a safety officer who monitors operations, assesses situations and hazards, advises on health and safety matters, and conducts safety briefings.

For a large event, the incident commander may appoint a liaison officer who is a point of contact for the utility for coordination for other agencies.

The operations section manages tactical operations during an incident to meet the IAP objectives. Operations activities typically "focus on saving lives, reducing the immediate hazard, protecting property and the environment, establishing situational control, and restoring normal operations" (FEMA, 2017). The operations section chief organizes the section based on the scale and type of incident. Most WRRF staff will be part of the operations section, although individuals may have other roles within ICS.

The planning section collects information on the situation and resources status, evaluates it, and processes the information for use in developing the action plans. The planning section chief may disseminate information through the IAP, maps, or a status board. The planning section writes the IAP.

The logistics section is responsible for providing facilities, services, and material support, including transportation, communications, supplies, equipment maintenance and fueling, medical services, and all off-incident resources. The logistics section identifies anticipated and known incident service and support requirements and requests additional resources, as needed.

The finance/administration section handles managing all financial aspects of the incident. This section is typically only for large incidents.

EMERGENCY RESPONSE PLAN MAINTENANCE

Emergency response plans are not static documents and should not be regarded as just another binder on a shelf at a WRRF. Instead, ERPs should be familiar to all WRRF staff through training and regular updates.

RAINING AND EXERCISES

raining and its importance often go undervalued for emergency management. The Federal Emergency Aanagement Agency offers free online training on the NIMS and ICS. All WRRF staff should know and nderstand their roles in an incident. For example, a person that oversees a pretreatment program might e incident commander in a situation of a spill within the collection system. That same individual would ot be involved if the WRRF were under a cyberattack. Incidents vary in severity and in warning levels. weather forecast allows for some preparation for a storm event. An incident with no warning, such as n earthquake, requires response from whomever is at the facility that day. A severe event may require esponse activities outside of typical duties.

n effective way to learn where staff fit in various incident management situations is to do a series of abletop exercises at the WRRF or utility level. Think of some scenarios that WRRFs might face and work hrough them as a team. Consider the preparedness areas (protection, prevention, mitigation, response, nd recovery) and the ICS roles: Who performs what roles and who is available as backup. Make sure to erform multiple exercises with people in different positions so you know what works best for a particular tuation within the WRRF.

Tabletop exercises are discussion sessions where staff meet in an informal setting to discuss their roles during an incident and their responses to a particular situation.

MERGENCY RESPONSE PLAN UPDATES

n ERP should be reviewed and updated annually. Some utilities have a regulatory requirement to pdate their ERPs. A section for updates should be contained within the ERP as well as a section for orrections. Over time, things at a WRRF utility will change (for example, equipment, processes, personnel, nd suppliers). The annual review should include verification that all contact information is correct. he annual review is also a good time to reach out to local support agencies, such as the police or fire epartment, to make sure they are aware the ERP is being updated and to ask them if they have had ny changes within their organization. Make sure the local first responders and emergency managers nderstand what happens at your facility and particular risks that exist (high voltage, chemicals, etc.) so hat if their assistance is needed, they can be prepared and be familiar with who, where, and why they are esponding. This can save valuable time if a situation arises in which help was urgently needed.

EST YOUR KNOWLEDGE

Which of the following are not one of the phases of emergency management:
   a. Mitigate
   b. Recover
   c. Expect
   d. Prepare

How often should you update your ERP?
   a. Every 5 years
   b. When you have the time
   c. Never
   d. Yearly

What is not the responsibility of the incident commander?
   a. Make sure staff know the set proprieties
   b. Approving press releases
   c. Ensure everything is done in a safe manner
   d. Sharing information with the media

What is a natural disaster you might face at a WRRF?
   a. Active shooter
   b. Union strike
   c. Hurricane
   d. Cyberattack

5. **A vulnerability assessment should consider only threats from human action, such as cyberattacks or bombs, because natural disasters cannot be prevented.**
   ☐ True
   ☐ False

6. **National Incident Management System and ICS are for police, fire fighters, and first responders only and are not suitable for a water utility.**
   ☐ True
   ☐ False

7. **One of the most important things learned from a vulnerability assessment is**
   a. Learning how to manage peak flows
   b. Knowing where your facility is at risk
   c. Predicting what event will happen at your facility
   d. To help set guidelines about employee behavior

8. **Match the following terms:**
   a. Prevention
   b. Protection
   c. Mitigation
   d. Response
   e. Recovery
   1. To restore, strengthen, and revitalize infrastructure, housing, and a sustainable economy in a timely manner, as well as the health, social, cultural, historic, and environmental fabric of communities affected by a catastrophic incident.
   2. To avoid or stop an imminent, threatened, or actual act of terrorism.
   3. To reduce the loss of life and property by lessening the effect of future disasters.
   4. To keep safe our citizens, residents, visitors, and assets against the greatest threats and hazards in a manner that allows our interests, aspirations, and way of life to thrive.
   5. To react quickly to save lives, protect property and the environment, and meet basic human needs in the aftermath of a catastrophic incident.

9. **Which member of the ICS should be commenting to the media?**
   a. The incident commander
   b. Employees leaving the scene
   c. The public information officer
   d. The board chairperson

10. **Your vulnerability assessment should be**
    a. Shared with other facilities so you can see if you have common risk
    b. Read by every employee as part of their training
    c. Read by key officials and kept in a safe spot with limited access
    d. Made a public document on your website

Personnel Management

As a WRRF leader, managing people is a central part of your work. Personnel management is one of the tougher tasks of being a leader, but it can also be rewarding. Personnel issues are often more subjective than the treatment problems at a WRRF. There are no formulas in this section. Consistency is important, as is remembering that each situation and employee is different. What works in certain circumstances or for one employee may not be successful with another. This section presents some ideas and suggestions. Finding what works best for you in your context is often found through personal experience. Some of the situations covered in this section are workforce planning, recruitment, promotion, feedback and disciplin[e], conflict management, and working with unions.

WORKFORCE PLANNING

Workforce is a term for workers engaged in a specific activity.

A WRRF's workforce is shown in the organizational chart and defined through job descriptions for all personnel at the facility. The job descriptions should include all the duties of the position and how the position fits into the organization's structure. Job descriptions should be reviewed and updated annually.

Workforce planning means making sure that your utility has the right people in the right jobs with the righ[t] skills to successfully operate a WRRF. The workforce planning process includes these steps:

Institutional knowledge is the combination of experiences, processes, expertise, and information possessed by employees. At a WRRF, institutional knowledge can span decades and include process control insights, infrastructure locations, project history, and perspectives that define the WRRF's history and culture.

1. Analyzing current staffing
2. Identifying future workforce needs
3. Determining gaps between current levels and future needs
4. Developing and implementing plans to close the gap

Workforce planning should be conducted with your utility's human resources department. When analyzing current staffing at your WRRF, consider the overall experience of your staff. Many WRRFs have multigenerational workforces with operators ranging in service time. As the WRRF workforce ages, it is good to know about upcoming retirements to prepare to either promote or hire new people to fill these roles. Remember that more experienced workers have accrued skills and knowledge from many years on the job. If possible, hire trainees or new operators before the impending retirements to make sure the experienced operators have a chance to fully train and mentor incoming employees and share their vast institutional knowledge about various aspects of the WRRF or utility.

Recruitment includes listing or advertising positions, receiving and reviewing applications, conducting interviews, and hiring (i.e., employment testing, background checks, and job offers).

RECRUITMENT

Recruitment is an important part of maintaining or growing a WRRF workforce. Many factors will come into play in setting up a recruitment plan. Public utilities and investor-owned utilities typically have different hiring rules, and you will likely be working with your human resources department on recruitment. Understanding the hiring process for your utility or WRRF and the makeup of the community, both socially and economically, will factor into the recruitment methods as well. When recruiting, a plan is needed that reaches all types of people to create the best staff for the WRRF. Having a clear and precise recruitment strategy will help identify, hire, and retain the best staff possible, which is key to the ongoing success of the WRRF.

A recruitment plan should include options for advertising vacancies such as the local newspaper, Internet jobsites, and professional association job boards. The Water Environment Federation (WEF; Alexandria, Virginia) and most state associations maintain online job listings for recruiting WRRF personnel (see Figure 14.14). When recruiting, it is good practice to list all the criteria for the job, including not just the salary, but also requirements such as written testing, interviews, background checks, drug tests, medical physicals, and shift work. Being clear in the beginning and making sure candidates understand what the job entails is crucial for recruiting employees and retaining them. Be open to questions from candidates as well.

When recruiting for a WRRF, consider posting positions internally and externally at the same time. Candidates should be evaluated for the certification or license level and technical ability as well as nontechnical skills like work ethic, commitment, communication skills, willingness to learn, attendance, and other qualities that make an operator or maintenance technician a good employee.

PROMOTION

As a WRRF leader, you have likely been through your utility's promotion process. Some utilities have a policy to encourage promotion from within. It is best to search both inside and outside your utility to find the best candidate. Licensing and certification requirements and/or civil service exams may also be

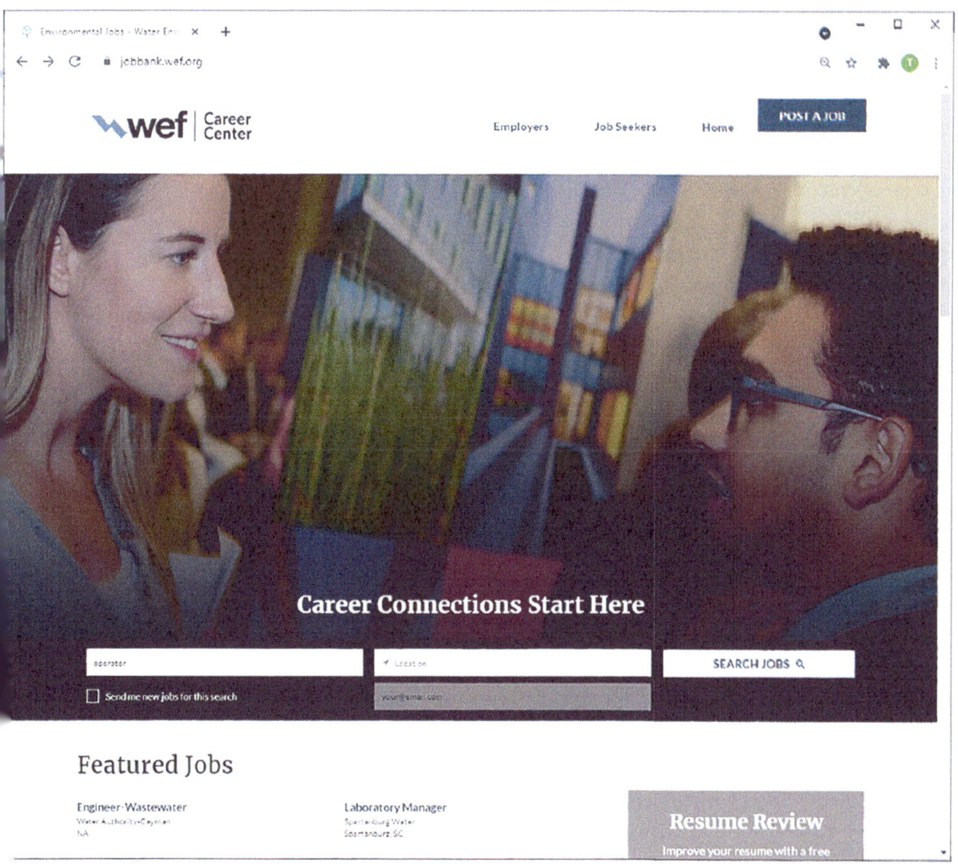

igure 14.14 WEF Career Center Website (http://jobbank.wef.org) Contains Listings for Hundreds of ater and Wastewater Jobs

factors for promotion eligibility. Promotion processes vary. Your utility may post the position then choose from among the applicants, or a promotion may be made from a pool of qualified candidates without an application process.

Making a promotion decision can be tricky. Some employees think promotions should be based solely on seniority or length of service, and that may have been the practice at your utility or WRRF in the past. However, the promotion should go to the best candidate. Keep a clear mind going into the process of promoting to make the choice easier.

When considering a candidate for promotion, review the employee's past performance. (Performance reviews are discussed in the next section.) Has the employee met the performance goals? What is the employee's everyday attitude? Is the employee reliable and responsible? Request insight and recommendations from several of the employee's cohorts: Often, people may have their own personal reasons why someone should or should not get promoted, so gathering opinions from multiple people will help get you a better picture of coworkers' feelings as a whole. Also, remember to be as transparent as possible so all candidates make sure they know why they were or were not selected. Often, in a WRRF, the person being promoted will become the direct supervisor of the people they have worked with for many years and the people they were up against for this promotion. The newly promoted person now has many new tasks and skills to learn. Do not make this process harder by not having a fair and transparent process. Candidates who did not receive a promotion may want feedback, so let them know, as far as your utility rules allow, how they can improve. Anytime an employee can learn from an experience, it will make the workforce stronger.

FEEDBACK AND DISCIPLINE

Feedback between an employee and supervisor can be formal or informal. Formal feedback includes annual performance reviews and documented disciplinary situations. Informal feedback is still intentional and planned, but without the paperwork, and includes coaching and mentoring. Praise, encouragement, guidance, correction, and direction are all forms of feedback that an employee should receive as part of their regular work. The informal feedback is an important precursor to formal feedback. Do not save this feedback for the annual evaluation or let a problem go too long without being addressed. Instead, coach your employees regularly and frequently throughout the year to improve performance.

Feedback is an open conversation between a leader and employees and can be used to express both positive and negative feelings. Leaders should communicate with their staff about all of the positive things that are being done and acknowledge their efforts. Leaders can also discuss areas for improvement. Remember, feedback should be a procedure to help correct performance. For example, if an employee is filling out part of the logbook incorrectly, this is a problem for the WRRF, but not a violation of the employee handbook or a disciplinary situation. The employee should receive feedback that they are not filling out their logbook correctly, which is causing problems with reporting, and they should be instructed on how to correctly complete the logbook. The situation could escalate from informal feedback and coaching to formal discipline if the employee does not improve performance or is found to be falsifying the logbook.

Feedback goes both ways. As a WRRF leader, be willing to ask questions, listen to your employees, and receive feedback. Sometimes, talking things out together is the best way to create a productive work environment and avoid bigger problems in the future.

Discipline is an aspect of personnel management that can weigh the heaviest, particularly if a WRRF leader is in charge of the people that they previously worked with and come up the ranks with. It is common to become friends with coworkers, which can make discipline difficult. Leaders may worry that the people they lead now will think that the person has changed, that they are picking on them because they never liked them, or that friendships that have grown over the years are being ruined. Being fair and firm will help curtail tension between boss and peer. Make sure employees know what is expected of them at all times. All employees should have a copy of the employee handbook, and WRRF leaders must have a clear understanding of the utility's rules. The handbook should highlight what is not acceptable and have a progressive discipline plan that includes informal counseling, oral warning, written warning, probation, suspension, transfer, demotion, and termination based on the nature of the violation. If you are uncertain about handling a disciplinary situation, contact your human resources department for guidance. Situations involving harassment, discrimination, or pay inequity should include the human resources department and may also require a labor attorney.

CONFLICT MANAGEMENT

Conflicts can come in a variety of directions: between two employees, between an employee and supervisor, among upper management, between departments, or within a department. In handling any of these situations, managing conflict correctly is important. Having a structure for clear communication in place is a first step toward minimizing conflict. Also, it is important that everyone inside the utility trust and respect each other and other departments. Some situations, such as a safety situation, require an immediate response. In other situations, particularly with an employee in another department or separate shift, it is better to notify the appropriate supervisor to handle the problem. When conflict is handled in a manner that is swift, precise, and fair, it will lead to fewer and fewer conflicts, which will make for a better workforce culture.

WORKING WITH UNIONS

Many WRRFs may have a union that represents employees of that facility. When dealing with unions, it is particularly important to respect workers' rights. Also important is knowing and following the union contract, so keep a copy on hand for reference. Having a union does not take away from the duties of a WRRF leader who still must enforce rules and guidelines. Rather, the union relationship provides certain ways for enacting and following the rules. State laws on unions and workers' rights vary widely, so it is especially important to know and comply with the local laws that protect union workers. If questions arise on the union relationship, check with human resources staff or reference the contract on file. A mutual respect between the union and management will go a long way toward creating a positive work culture.

A union or labor union is an organization of workers formed to protect the rights and interests of its members. Labor unions give workers power to negotiate for better working conditions and benefits through collective bargaining. Labor unions are less common in right-to-work states where union membership is not required as a condition of employment.

TEST YOUR KNOWLEDGE

1. **When planning your workforce, it is good to know the following:**
   a. The experience level and service time of your staff
   b. If you have enough friends on you staff
   c. Who seems like trouble
   d. Do I have enough staff until I retire
2. **When recruiting for your staff, it is good practice to**
   a. Make sure my friends have a job.
   b. Know the social and economic makeup of the local community.
   c. Make sure I will have the same interest as the people I am recruiting.
   d. Recruit people who are the opposite of me so they challenge my decisions.
3. **When disciplining employees, it is good to**
   a. Be fair and firm.
   b. Punish people based on how my day is going.
   c. Make sure people who like me do not get in trouble.
   d. If I do not like someone, now is the time to take it out on them.
4. **When dealing with a conflict between one of your employees and someone from another department, it is good to**
   a. Yell more at the employee that is not in your department.
   b. Call the manager of the employee in the other department so they can deal with their employee.
   c. Because the employees are from different departments, it is okay for them to have conflict.
   d. Discipline both employees yourself.
5. **When promoting an employee, it is best to**
   a. Choose your friend.
   b. Choose the person that has been there the longest.
   c. Choose the candidate that is most qualified for the job.
   d. Promote everyone that applied so you do not have to make a tough decision.
6. **When receiving feedback on a task from an employee, you should**
   a. Consider the motivation behind the feedback.
   b. Disregard it; the employee needs to follow training and SOPs.
   c. Listen and work with the employee to see if their feedback helps.
   d. Tell the employee to do what they see is fit.
7. **When working with unions, it is important not to**
   a. Understand the employees' rights.
   b. Interpret the union contract to fit your perspective.
   c. Ask the human resources department for help.
   d. Follow proper procedure.

Leadership

This section looks at some of the traits, qualities, and characteristics that help make a person a good leader. It will also look at the roles and responsibilities of being a leader in an organization. There are lots of books available on leadership, and this section is only an introduction. Leadership is not a single trait or quality, rather, a combination of intangible traits such as the characteristics shown below (see Figure 14.15).

Something that is intangible is difficult to define or grasp; it is something that you cannot physically touch. For example, when someone has charisma, they are pleasant to be around, but it may be difficult to pinpoint exactly why.

EMOTIONAL INTELLIGENCE

Many people are familiar with the acronym "IQ," but perhaps fewer people know about "EQ," which stands for emotional quotient or emotional intelligence, that is, the idea that intelligence is more than just technical skills or knowing facts. Emotional intelligence was made popular in an article by Daniel Goleman, who identified five components of emotional intelligence at work (see Figure 14.16) (Goleman, 2004):

- Self-awareness: The ability to recognize and understand one's character, feelings, motives, and desires and their effect on others.
- Self-regulation: To think before acting, the ability to control or direct impulses and to suspend judgment.
- Motivation: The reason for action or work that go beyond money, persistence in pursuing goals.
- Empathy: The ability to understand and share the feelings of someone else.
- Social Skill: A competence in managing relationships, facilitating interaction, building networks, and finding common ground.

Leaders in any organization will deal with people when they are in good moods and when they are not. It is easy to say that people should keep home lives and work lives separate; however, it is much harder to do. People do bring their personal issues to work and leaders must know how to handle that. One's ability to handle people at their highs and at their lows is going to be what sets someone apart as a leader. Earning people's trust and helping them negotiate tough times is what is going to set true leaders apart from just being a boss or manager. A WRRF operator needs extensive technical knowledge to manage the equipment and processes. A WRRF leader also needs high emotional intelligence to manage people and staff successfully.

TRAINING

The role of a leader is to put the right people in the right positions so the WRRF, the utility, and personnel all flourish. Sometimes, building new skills or confidence is needed to prepare personnel for their new—or next—role. A leader shares responsibility for an employee's success and can facilitate success with training.

Figure 14.15 Leadership Characteristics

Figure 14.16 Components of Emotional Intelligence at Work

One of the wonderful benefits of being a member of the WRRF community is that there is a vast amount of training available to everyone. Training is available through state member associations, on the national level in different forms, and perhaps conducted at your own utility. Training may take the form of an in-person class, independent study from a book, or an online course.

One example of leadership training is WEF's Water Leadership Institute. As part of this program, participants engage in management training and leadership development. The program has a blended learning approach that includes the examination of complex challenges and networking with public and private sector practitioners. Participants build skills and knowledge needed to effect change within their organizations.

Training can be on a variety of topics and frequency. Safety is a regular training topic at WRRFs. Training may be technical, where operators increase their wastewater treatment knowledge. Training may also be on soft skills that help individuals grow or on human resources related topics like diversity and harassment. Training should not be limited to new employees. Refresher training is good practice for existing operators. In some cases, annual training may be required on certain topics. Leaders will be tasked with finding programs that work best for their WRRF and members of their team so that everyone will continue to learn and grow within an organization.

It is also important to have a strong in-house training program (Figure 14.17). These programs are where employees will learn the day-to-day operations of a specific facility. It is important that employees have a clear understanding of the safety and skills required for them to succeed at their required job tasks. The amount of training an employee receives can affect what the employee will become in the future. Employee training should be monitored closely to make sure they are competent at the tasks they must perform for their position.

Water resource recovery facility operators will also have to become certified within the state where work. Make sure you check with your state licensing board or member association for certification requirements and examination information. Training for WRRF operator certification and training for daily operations at your WRRF will partially overlap, but not completely, because certification exams will have content or processes that may not be a part of your WRRF. For example, an operator at a WRRF with aerobic digesters will need to learn about anaerobic digestion from a book or course because both will be on the exam.

Figure 14.17 Water Resource Recovery Facility Operator Reviewing Training Documents (Courtesy of Buffalo Sewer)

STRATEGIC PLANNING

As mentioned previously, WRRF leaders will oversee the long-term success of the WRRF. They will have to plan for many different scenarios including workforce, financial stability, and innovation. Long-term and overall planning and goal setting should set a 1-year goal, 3-year goal, 5-year goal, and 10-year goal. These goals should all correlate with the long-term success of the WRRF and its staff. A strategic plan should also have the following:

- A mission statement;
- A vision statement;
- The utility's core values;
- Measurable goals; and
- Performance standards.

It is important to note that, inside a WRRF, there are many different departments and that a WRRF is typically part of a larger utility. A strategic plan may be developed for the entire utility or a single facility. The WRRF leader will need to make sure that each department has its own individual plans and that these plans are in agreement with overall utility plans and its mission and vision.

SAFETY

An important goal for every utility and WRRF leader is making sure everyone goes home at the end of the day safely. As a WRRF leader, it is your goal to make sure that your organization is up to date with all federal and state safety regulations. A leader should model safety by wearing appropriate personal protective equipment (PPE; see Figure 14.18) and following all safety procedures. For a more in-depth look at safety, refer to Chapter 2 of this book.

Figure 14.18 Personal Protective Equipment (Courtesy of Buffalo Sewer)

EST YOUR KNOWLEDGE

In terms of WRRF leadership, what does "EQ" stand for?

a. Equalization tank
b. Emotional intelligence
c. Intelligent quotient
d. Equilibrium

When planning for your organization, you should set a short-term goal of (X) year(s) and a long-term goal of (X) year(s).

a. 1, 10
b. 5, 3
c. 1, 3
d. 5, 10

Which of the following is not a WRRF leader EQ trait:

a. Self-regulation
b. Motivation
c. Empathy
d. Strictness

4. What are the goals of a WRRF training program?
   a. To gain credit hours
   b. To get a free day off
   c. To learn skills to help both with leadership and job performance
   d. To advance toward a college degree

5. As a WRRF leader, you are not in charge of what?
   a. Facility design
   b. Safety
   c. Promoting
   d. Discipline

ustomer Service and Public Engagement

ustomer service is critical for a successful utility. Today's customer is not just ratepayers, but the entire ommunity (AWWA, 2003). For a WRRF operator, customer service and public affairs may seem to be ther people's jobs. Indeed, most utilities have a customer service department to handle billing, set up ccounts, and answer questions. Larger utilities may also have public affairs or communications staff to ngage with the public. However, there is still a customer service and public engagement role for operators.

Vater resource recovery facility operators are representatives of the utility and may receive questions nd complaints from customers while they are out in the community. Water resource recovery facilities ave some special customers like septage haulers, industrial users, and effluent or biosolids customers. ublic education, facility tours, and science, technology, engineering, and math (STEM) events promote nderstanding and appreciation of the vital roles WRRFs play in the community. An operator may be called n to present at a board or council session or represent the utility in a public meeting or with the media.

ommunicating with the public is a chance to nurture community support and create a positive image of e WRRF as a good neighbor that provides a vital service and protects the environment.

OMMUNICATING WITH CUSTOMERS

Vhen talking with a customer of any kind, always be polite, factual, and professional, and always listen. Do ot ignore a customer question or complaint. Keep good records by logging the time, date, name, nature of e interaction, and the resolution. Figure 14.19 presents an example of a complaint documentation form.

n operator in uniform may be stopped in a store or other area out in public to be asked a question or hear complaint. The topic, such as a water bill, may not even be relevant to a WRRF operator. Still, you should sten politely and refer the person to the right contact or the main utility phone number and/or website.

he most common complaints about a WRRF are odor complaints. Chapter 11 of this book covers odors a WRRF. Each odor complaint should be investigated promptly and addressed. Odor can be a nuisance the public and a sign of a problem or process upset. A WRRF that was built in a rural area may now be rrounded by developed or suburban areas as the service area grows. Existing processes or structures may ave to be changed because of the new neighbors. For example, a primary clarifier may be retrofitted with vers and a scrubber to reduce odors around the WRRF. Greenspace with plenty of trees will help disperse e smell as well as make the facility look better.

File Use Only: ______

Genesee County Water and Waste Complaint Form

Date: ______ Time: ______

Taken By: ______

Complaintant Information

Name: ______

Address: ______

Home Phone: ______ Work Phone: ______

Fax Number: ______ Pager / Cell: ______

Complaint/ Issue

______

Resolution

______

Other Information

______

Given to: ______ Date / Time: ______

Figure 14.19 Example of a Complaint Form (Courtesy of Genesee County Drain Commissioner: Division of Water & Waste Services)

INTERACTING WITH WASTEWATER CUSTOMERS

Water resource recovery facility operators often interact with special customers. These may include industrial users as part of a pretreatment program, sludge and septage haulers for facilities that accept special wastes, and resource customers for effluent, biosolids, and/or energy.

Customers who use effluent for irrigation are concerned about water quantity and quality. The manager of a golf course may need to talk to a WRRF operator to coordinate effluent supply for irrigation in the summer. Irrigation customers' water quality needs are not the same as the WRRF permit limits. There may not be permit limits for salinity or total dissolved solids for a WRRF, but it will harm or kill turf grass and other plant life if levels of them are high. If an irrigation customer has a problem with effluent quality, work together to find the cause. Also, if an operator knows that there is a problem with the effluent that will damage turf, the operator should notify the irrigation customer immediately before any turf damage occurs.

A WRRF may have a program and facilities for receiving special wastes like fats, oils, and grease, sludge, and septage. These haulers may have questions about operating hours and access to the WRRF. Special wastes should be tracked by volume, hauler, and source. Taking a sample from each delivery is a good idea. There should be a process for paying for deliveries that may or may not involve an operator. For example, a waste hauler may purchase tickets from the customer service department, which is equipped for handling payments, then hand over the tickets at the WRRF when dumping the waste.

Many WRRFs designed to treat flows larger than 3.8 ML/d (1 mgd) have a pretreatment program for controlling industrial discharges. Depending on the size of the pretreatment program, the work of coordinating samples, scheduling inspections, and responding to events may belong to a wastewater operator. Maintaining an open relationship with these industrial users is beneficial for managing the waste streams, but is also a regulator relationship, which may require confrontation. Make sure you know your state's and utility's pretreatment rules.

A WRRF that land applies biosolids has another path for public interaction. There is often interaction with a contractor for hauling or with a farmer or landowner for scheduling and lining up fields. Sometimes, it can be as simple as meeting a farmer in the field or as complex as going to a public meeting or a regulator for approval to apply in certain areas. However, the program is set up, make sure you have all the information, data, and documentation to land apply biosolids.

PUBLIC ENGAGEMENT

Most customers have no idea what happens after they flush. Many people never think about what happens to wastewater once it goes down a drain. Some people may be uncomfortable talking about wastewater. Water resource recovery facility operators have a special opportunity to promote the value of WRRFs in making our communities cleaner and safer. The Water Environment Federation and state member associations have resources to help with public outreach.

Service disruptions, rate increases, and emergencies should not be the only times that the public hears from their utility. Do not be afraid or unwilling to share the positive stories about the vital work being done in a collection system and WRRF. Operators are essential workers who provide a critical community service. Engaging with the public about your work is a chance to increase their knowledge, improve their opinion, and gain their support. An informed public will be more trusting and accepting. Effective public engagement can strengthen the utility's image as a good neighbor and vital part of the community.

GIVING FACILITY TOURS

Facility tours are a fantastic opportunity to engage the public of all ages (see Figure 14.20). A wastewater operator who gives a tour should plan a safe route around the WRRF that avoids chemical storage, electrical rooms, and basins and walkways without handrails. Every facility layout is different. If possible, create a route that follows the flow through the WRRF.

Start with the headworks or preliminary treatment, even if it smells. Seeing the screenings, rags, and trash that must be removed from the influent is a chance to explain why wipes and trash should never be flushed. Walk through the treatment process. Finish with the effluent and show the clean water leaving the WRRF. Point out how much cleaner is the effluent than was the influent. Discuss the samples and testing done by operators to ensure the WRRF runs properly.

Remember, some people may be embarrassed talking about wastewater. School children (and some adults) may express their discomfort loudly. If a visitor is bothered, acknowledge that wastewater might be gross, then challenge the visitor to imagine what would happen if the WRRF was not there and wastewater went straight into the river.

Most people do not know what happens after the toilet is flushed, so use terms that are clear to non-operators. Instead of talking about "SRT," "F:M," or "mixed liquor," use a metaphor of people at a party to explain the microorganisms in the activated sludge process. For example, *process control* can be described as keeping the bugs happy by getting the right amount of food (organics) for the right amount and right kind of guests (bugs). Explain the biological processes as natural. What the receiving stream would do in days, the WRRF does in hours by providing the proper conditions.

Figure 14.20 Middle School Students on a Tour of the Port Royal Water Reclamation Facility, Beaufort, South Carolina. The Tour Was Part of "Trick or Treatment," a Halloween-Themed Science, Technology, Engineering, and Math Event for Fifth and Sixth Graders (Courtesy of Beaufort-Jasper Water & Sewer Authority)

A favorite activity of students and adults is to see the activated sludge microorganisms under a microscope. Put a drop of mixed liquor on a microscope slide and look for flagellates, ciliates, and rotifers. A digital camera in place of an eyepiece can be connected to a monitor for group viewing and capturing images and video (see Figure 14.21), although kids often like to look in the microscope.

Figure 14.21 Port Royal Water Reclamation Facility Operator Ryan Yoder With Microscope, Camera, and Monitor Setup Used to Show Activated Sludge Microorganisms on Facility Tours (Courtesy of Beaufort-Jasper Water & Sewer Authority)

PUBLIC SPEAKING

A WRRF operator may be required to attend or address a utility board or municipal council meeting. Some may find this intimidating, but public speaking is a skill that can be learned. The best way to become a good public speaker is practice and preparation. When preparing a presentation, think about your audience and the message or information to convey. Before giving a presentation, rehearse it in front of a friendly audience or video. Utility meetings, conferences, and public meetings are opportunities to practice and enhance public speaking skills. Toastmasters meetings or a public speaking course are another way to practice and improve your skills.

TEST YOUR KNOWLEDGE

1. **Odor complaints should be ignored if the WRRF was built before the complainer's house was built.**
   - ☐ True
   - ☐ False

2. **Which parts of the facility should be part of a tour with middle school kids?**
   a. Electrical room
   b. Chemical storage areas
   c. Basins without handrails
   d. Clarifiers with handrails

3. **When giving a facility tour, an operator should use big words and complex explanations of the treatment process.**
   - ☐ True
   - ☐ False

4. **If a ratepayer approaches you in the grocery store and complains about their bill, what should you do?**
   a. Politely refer them to customer service.
   b. Explain your utility's budgeting process and how you estimate power and chemicals for the fiscal year.
   c. Ignore them.
   d. Make a speech about regulatory burdens, workforce needs, and critical infrastructure.

CHAPTER SUMMARY

PERMITS AND REPORTING	■ National Pollutant Discharge Elimination System permits protect waterways by specifying an acceptable level of pollutant allowed to be discharged. ■ Discharge monitoring reports are required for all WRRFs that discharge into surface waters in the United States. ■ Operators should know their permits as well as the testing and reporting requirements.
SAMPLING AND RECORDS	■ Most testing requirements are spelled out in the NPDES permit. ■ There are often sampling requirements beyond the NPDES permit. ■ An operator must be familiar with the various laboratory requirements, including hold times, as well as appropriate record retention.
BUDGETING AND FINANCIAL MANAGEMENT	■ Developing and following a budget are key skills for operators, especially in a management role. ■ Understanding how rates are set and affordability helps ensure customers are being treated fairly. ■ Procurement of goods and services and following the utility policies are essential.
WRITING AND UPDATING SOPS	■ Standard operating procedures are a set of written instructions that document all actions associated with a defined event. ■ Standard operating procedures are helpful for training new employees and ensuring experienced employees follow all the required steps. ■ Each utility has a structure that SOPs should follow and a maintenance procedure to keep SOPs current. ■ There can be legal consequences when an SOP is not followed.
ASSET MANAGEMENT AND MAINTENANCE	■ Asset management is a process to reduce a utility's risk by prioritizing and budgeting for the renewal of equipment, tanks, buildings, and so on. ■ Asset management plans are part of an asset management program and include life cycle planning. ■ Asset management plans use risk matrices to quantify the consequence of failure and the probability of failure to identify the criticality of assets. ■ Computerized maintenance management system programs are essential for an asset management program and allow for quantification and appropriate levels of service (e.g., maintenance).
CAPITAL IMPROVEMENTS AND PROJECTS	■ Capital improvements are often guided by a utility's CIP and require significant funding. ■ Capital projects often rely on contractors for construction and engineers for design; it is important to understand the various hiring methods as well as project delivery models (e.g., design-build versus design-bid). ■ Operator input is essential, and operators should be providing feedback throughout capital projects.

EMERGENCY PLANNING AND RESPONSE	■ It is important to have plans in place at your WRRF to handle human-caused threats and natural disasters (e.g., cyberattacks, earthquakes, flooding, etc.). ■ Conducting vulnerability assessments is key to identifying risks and planning for emergencies. ■ The incident management cycle is key to emergency planning and response. ■ Emergency response plans should be developed for all potential emergencies based on your vulnerability assessments. ■ Emergency response plans should be updated annually.
PERSONNEL MANAGEMENT	■ Managing staff is a central part of your work and one of the more difficult aspects as a leader. Knowing how to approach each situation under varying conditions is a learned skill. ■ Workforce planning ensures your utility has the right people in the right roles in the present while also anticipating future staffing needs and progression. ■ Feedback can be both formal and informal and it is highly dependent on the leader as well as the staff. It is important to meet all formal feedback requirements provided by the human resources department, but informal feedback is a powerful tool for leading staff. ■ Disciplining staff should always be objective and fair. It is important to understand if there are requirements for disciplining unionized staff because that can result in management issues.
LEADERSHIP	■ Leadership is not a single trait or quality, but a combination of certain behaviors and expertise. Most people can be leaders and it is important to understand which leadership skills are your strongest to develop them further. ■ Leadership training is important for your utility to identify leaders and provide them with knowledge and experience to grow as leaders. ■ Emotional intelligence is a key skill for leaders to have and develop.
CUSTOMER SERVICE AND PUBLIC ENGAGEMENT	■ Communicating with the public is an important part of a WRRF leader's job. ■ It is important to always listen, to be kind and professional, and never ignore a public question or complaint. ■ The most common complaint is often an odor complaint, but other types of complaints include billing issues, traffic concerns, and noise.

References

American Water Works Association. (2003). *The evolving water utility: Pathways to higher performance.*

American Water Works Association. (2021, May 1). *Risk and resilience management of water and wastewater systems* (AWWA J100-21).

Design Build Institute of America. (2015). Choosing a project delivery method. dbia.org/wp-content/uploads/2018/05/Primers-Choosing-Delivery-Method.pdf

Federal Emergency Management Agency. (2015, September). National Preparedness Goal (2nd ed.). www.fema.gov/sites/default/files/2020-06/national_preparedness_goal_2nd_edition.pdf

Federal Emergency Management Agency. (2017, October). *National incident management system* (3rd ed.). www.fema.gov/sites/default/files/2020-07/fema_nims_doctrine-2017.pdf

Goleman, D. (2004). What makes a leader? *Harvard Business Review.*

U.S Environmental Protection Agency. (n.d.-a). *Asset management for water and wastewater utilities.* www.epa.gov/sustainable-water-infrastructure/asset-management-water-and-wastewater-utilities

U.S Environmental Protection Agency. (n.d.-b). NPDES permit basics. www.epa.gov/npdes/npdes-permit-basics

U.S. Environmental Protection Agency. (n.d.-c). *Pretreatment standards and requirements—Categorical pretreatment standards.* www.epa.gov/npdes/pretreatment-standards-and-requirements-categorical-pretreatment-standards

U.S. Environmental Protection Agency. (n.d.-d). *Pretreatment standards and requirements—General and specific prohibitions.* www.epa.gov/npdes/pretreatment-standards-and-requirements-general-and-specific-prohibitions

U.S. Environmental Protection Agency. (1994). *Utility manager's guide to water and wastewater budgeting.* nepis.epa.gov/Exe/ZyPURL.cgi?Dockey=200041RD.txt/

Suggested Readings

Bradberry, T., & Greaves, J. (2009). *Emotional intelligence 2.0.* TalentSmart®.

Smith, R. K. (2019). *Water and wastewater laboratory techniques* (2nd ed.). Water Environment Federation

U.S. Environmental Protection Agency. (n.d.). NPDES pretreatment local limits: Additional information. www.epa.gov/npdes/npdes-pretreatment-local-limits-additional-information

Water Environment Federation. (2017). *Operation of water resource recovery facilities* (7th ed., Manual of Practice No. 11).

Water Environment Federation. (2022). *Basic laboratory procedures for the operator-analyst* (6th ed.).

List of Acronyms

% = percent
$(NH_4)_2SO_4$ = ammonium sulfate
°C degrees = Celsius
°F degrees = Fahrenheit
A/V = area-to-liquid volume ratio
ABC = Associated Boards of Certification (now Water Professionals International, or WPI)
ac = acre
ACH = air changes per hour
AERMOD = steady-state air dispersion model used to assess air pollutant effects
ANSI = American National Standards Institute
AOP = advanced oxidation process
APC = air pollution control
API = American Petroleum Institute
BACT = best available control technology
BMP = best management practice
BNR = biological nutrient removal
BOD = biochemical oxygen demand
BOD_5 = 5-day biochemical oxygen demand
BTFs = biotrickling filters
Btu = British thermal unit
BWRO = brackish water reverse osmosis
C = carbon
$C_6H_8O_7$ = citric acid
Ca^{2+} = calcium
$CaCO_3$ = calcium carbonate
cal/gmol = calorie/gram-mole
cCOD = colloidal chemical oxygen demand
CDC = Centers for Disease Control
CEB = chemically enhanced backwash
CECs = compounds of emerging concern
CEMS = continuous emissions monitoring system
CFR = Code of Federal Regulations
CFU = colony-forming unit
CH_3COOH = acetic acid
CH_4 = methane
CIP = capital improvement plan and clean-in-place
CIU = categorical industrial user
Cl_2 = chlorine gas
cm = centimeter
CMMS = computerized maintenance management system
CO = carbon monoxide
CO_3 = carbonate
COD = chemical oxygen demand
COS = carbonyl sulfide
CPI = corrugated plate interceptor
CPVC = chlorinated polyvinyl chloride
CS_2 = carbon disulfide
CT = concentration and time
cu ft = cubic feet
cu ft/h = cubic feet per hour
CWA = Clean Water Act
d = day
D/T = dilution-to-threshold ratio
DAF = dissolved air flotation
DALYs = disability adjusted life years
DCS = distributed control system
DCU = distributed control unit
deg = degree
DMR = discharge monitoring report
DO = dissolved oxygen
DPC = distributed process controller
DPD = N,N-diethyl-p-phenylenediamine
EBPR = enhanced biological phosphorus removal
EDC = endocrine-disrupting compound
EFM = enhanced flux maintenance
EMS = environmental management systems
EOC = emergency operations center
EPS = extracellular polymeric substance
ERD = energy recovery device
ERG enforcement response guide
ERP = emergency response plan and Enforcement Response Plan
F:M = food-to-microorganism ratio
FC = fecal coliforms
Fe = iron
FEMA = Federal Emergency Management Agency
FOG = fats, oils, and grease
FRP = fiber-glass-reinforced plastic
ft = feet
ft/s = feet per second
G = Camp-Stein velocity gradient
g = gram
GAC = granular activated carbon
gal = gallon
gfd = gallons per square foot per day
GIS = geographic information system
GLUMRB = Great Lakes Upper Mississippi River Board
GOX = gaseous oxygen
gpd/sq ft = gallons per square foot per day
gpm = gallons per minute
h = hour
H^+ = hydrogen ion
H_2S = hydrogen sulfide
H_2SO_4 = sulfuric acid
HAZCOM = hazard communication program
HAZMAT = hazardous materials
HAZWOPER = hazardous waste operations and emergency response

HCl = hydrochloric acid
HDPE = high-density polyethylene
HMI = human–machine interface
HOCl = hypochlorous acid
HOD UV = hydro-optic disinfection ultraviolet light
h = hour
HRT = hydraulic retention time
HS^- = hydrosulfide
HVAC = heating, ventilating, and air conditioning
I&I = infiltration and inflow
IAP = incident action plan
IC = internal circulation
ICS = incident command structure
IFAS = integrated fixed-film activated sludge
in. = inch
ISBL = interface/sludge blanket level
ISFET = ion-selective field effect transistor
ISO = International Organization for Standardization
IU = industrial user
JHA = job hazard analysis
K = selectivity coefficient
kg = kilogram
kgal = kilogallon
KI = potassium iodide
kPa = kilopascals
kPaG = kilopascals gauge
kW = kilowatt
L = liter
L:W = length-to-width ratio
lb = pound
LED = light emitting diode
LEL = lower explosive limit
LIMS = laboratory information management system
LMH = liters per square meter per hour
LOI = local operator interface
LOTO = lockout/tagout
LOX = liquid oxygen
LRV = log removal value
m = meter
m/s = meters per second
m/s^2 = meters per second squared
m^2 = square meters
m^3 = cubic meter
m^3/d = cubic meters per day
m^3/h = cubic meter per hour
mA = milliamp
MACT = maximum achievable control technology
MAHL = maximum allowable headworks loading
MAIL = maximum allowable industrial loading
MBBR = moving bed biofilm bioreactor
MBR = membrane bioreactor
MCLs = maximum contaminant levels
MCRT = mean cell residence time
MEWP = mobile elevated work platform
mg = milligram
mg/L = milligrams per liter
Mg^{2+} = magnesium
mgd = million gallons per day
MHI = multiple-hearth incinerator
min = minute
mL = milliliter
ML/d = megaliters per day
MLR = mixed liquor recycle
MLSS = mixed liquor suspended solids
MLVSS = mixed liquor volatile suspended solids
mm = millimeter
mol = mole
MOPO = maintenance of plant operations
MPN = most probable number
MSDS = material safety data sheet
mV = millivolt
N = nitrogen
Na^+ = sodium
$NaHSO_3$ = sodium bisulfite
NaOCl = sodium hypochlorite
NaOH = sodium hydroxide
NBRC = National Blue Ribbon Commission for On-Site Non-Potable Water Systems
NDP = net driving pressure
NFPA = National Fire Protection Association
NH_3 = ammonia
NH_3-N = ammonia-nitrogen
NH_4^+ = ammonium ion
NH_4Cl = ammonium chloride
NH_4-N = ammonium as nitrogen
NH_4OH = ammonium hydroxide
NIMS = National Incident Management System
NIOSH = National Institute for Occupational Safety and Health
NLs = notification levels
NO_2^- = nitrite ion
NO_2-N = nitrite as nitrogen
NO_3^- = nitrate ion
NO_3-N = nitrate as nitrogen
NOM = natural organic matter
NPDES = National Pollutant Discharge Elimination System
NSPS = New Source Performance Standards
NTU = nephelometric turbidity unit
O&M = operations and maintenance
O_2 = oxygen
OCl^- = hypochlorite ion
OH^- = hydroxide ion
ONWS = on-site non-potable water reuse system
ORP = oxidation–reduction potential
OSHA = U.S. Occupational Safety and Health Administration
OWS = oil-water separator
P = phosphorus
Pa = pascal
PAA = peracetic acid
PAC = powder activated carbon
PCBs = polychlorinated biphenyls
pCOD = particulate chemical oxygen demand
PCS = process control system

PDR = pressure decay rate
PDT = pressure decay test
PEL = permissible exposure limit
PES = polyethersulfone
PFA = performic acid
PFAS = per- and poly-fluoroalkyl substances
PFOA = perfluorooctanoic acid
PFOS = perfluorosulfanoic acid
PFRP = process to further reduce pathogens
PID = process and instrumentation diagram
PIV = powered industrial vehicle
PLC = programmable logic controller
PO_4 = orthophosphate
PO_4^{-3} = phosphate ion
PO_4-P = orthophosphate as phosphorus
POTW = publicly owned treatment works
ppbv = parts per billion by volume
PPCPs = pharmaceuticals and personal care products
PPE = personal protective equipment
ppm = parts per million
ppmv = parts per million by volume
ppt = part per trillion
ppt = parts per thousand
PRCS = permit-required confined space
PRV = pressure-reducing valve
psi = pounds per square inch
psig = pounds per square inch gauge
PSRP = process to significantly reduce pathogens
PTFE = polytetrafluoroethylene
PVC = polyvinyl chloride
PVDF = polyvinylidene fluoride
Q = flow
QA = Quality Assurance
QA/QC = quality assurance/quality control
QACs = quaternary ammonium compounds
QAPP = Quality Assurance Project Plans
QC = Quality Control
RAS = return activated sludge
RFP = request for proposal
RFQ = request for qualifications
RTD = resistance temperature detector
RTU = remote telemetry unit
s = second
S^{-2} = sulfide
SAP = sampling and analysis plans
SBR = sequencing batch reactor
SBS = sodium bisulfite
SCADA = supervisory control and data acquisition
scfm = standard cubic feet per minute
sCOD = soluble chemical oxygen demand
SDI = silt density index
SDS = safety data sheet
SFPUC = San Francisco Public Utilities Commission
SIU = significant industrial user
SMP = soluble microbial products
SO_2 = sulfur dioxide
SOP = standard operating procedure
SOUR = specific oxygen uptake rate
sq ft = square feet
SRT = solids retention time
SSI = sewage sludge indicator
SUR = surface underflow rate
SVI = sludge volume index
SVOC = semivolatile organic compound
SWPPP = stormwater pollution prevention plan
SWRO = seawater reverse osmosis
TCLP = toxicity characteristic leaching procedure
TDS = total dissolved solids
THC = total hydrocarbon
TIR = total internal reflection
TKN = total Kjeldahl nitrogen
TMDLs total maximum daily loads
TMP = transmembrane pressure
TOC = total organic carbon
TPHs = total petroleum hydrocarbons
TRS = total reduced sulfur
TSS = total suspended solids
TWL = top water level
U.S. EPA = United States Environmental Protection Agency
U.S. = United States
UASB = upflow anaerobic sludge blanket
UCL = upper control limit
UCMR = Unregulated Contaminant Monitoring Rule
UIC = underground injection control
UV = ultraviolet
UVC LED = ultraviolet C light-emitting diode
UVT = ultraviolet transmittance
V = volt
V/G/C = viruses, *Giardia*, and *Cryptosporidium*
VAR = vector attraction reduction
VDC = volts direct current
VFA = volatile fatty acid
VFD = variable-frequency drive
VOC = volatile organic compound
VSR = volatile solids reduction
VSS = volatile suspended solids
W:D = width-to-depth ratio
WAS = waste activated sludge
WET = whole effluent toxicity
WEV = water expulsion
WLAs = waste load allocations
WRRF = water resource recovery facility

Answer Keys

CHAPTER 1

TEST YOUR KNOWLEDGE, PAGE 9

1. False. The Pretreatment Act does not distinguish between commercial and industrial users. It only defines domestic and industrial wastewater.
2. b.) BOD and TSS are often used to set surcharges. These are additional fees for an industrial discharge if these parameters are higher than a given threshold.
3. True. "Pass-through" refers to pollutants that are not removed through treatment at the WRRF and are discharged to surface water.
4. True. The national pretreatment program is a component of the NPDES program.
5. False. Pretreatment programs, including permitting and administrative and enforcement tasks, can also be administered by U.S. EPA and NPDES state authorities.
6. b.) Source control measures. The other three are specifically called out in pretreatment standards and requirements.
7. False. All WRRFs treating more than 19 000 m^3/d (5 mgd) or smaller WRRFs with significant industrial discharges must have pretreatment programs. Smaller WRRFs without significant industrial discharges are not required to.
8. True. Though, sometimes, the U.S. EPA can also act as control authority for small WRRFs without an established pretreatment program.
9. a.) Procedures to investigate and respond to industrial user noncompliance. The ERP can help guide the WRRF in how to respond to noncompliance events.
10. d.) Hazardous waste production is handled separately from pretreatment requirements.
11. True. This enables WRRFs to set more stringent limits based on site-specific conditions, treatment processes, or NPDES permit requirements.
12. True. This is one of the most commonly used tools to establish local limits.
13. True. Although industrial discharge permits may include parameters such as TDS, ORP, heavy metals, TPH, TOC, WET testing, and a TCLP procedure (for sludge).
14. False. Flow proportional sampling should be implemented wherever possible, except for those parameters that cannot be stored for the sampling duration and must be collected as grab samples. Also, to reflect the average water quality of the total volume of water discharged over the sampling period.
15. True. Different reports are required for the two types of users, and are outlined in Table 1.2.
16. True. A local limit typically applies at the point of discharge to the sewer system, whereas a categorical pretreatment standard generally applies after treatment.
17. Water resource recovery facilities must inspect each industrial discharger at least once per year, per 40 CFR 403.8(f)(2)(v). Although U.S. EPA regulations establish a minimum frequency for inspections, additional inspections by the WRRF might be necessary depending on issues such as the variability of an SIU's effluent, the effect of the SIU's discharge on the WRRFs, and the facility's compliance history.

TEST YOUR KNOWLEDGE, PAGE 11

1. False. Regulation of PFAS continues to evolve as toxicity and treatability information develop.
2. False. Industrial facilities can implement their own wastewater treatment and discharge under an NPDES permit provided by a state or the U.S. EPA.
3. True. "Emerging concern" are the key words here. For example, PFAS have been in heavy use for decades, but have only recently been considered a significant water contaminant.
4. False. Treatment for these contaminants is most often through oxidation or sorption but depends on the specific chemistry of each contaminant.

TEST YOUR KNOWLEDGE, PAGE 16

1. False. Activated sludge treatment removes many degradable constituents associated with industrial wastewater and can, to some degree, absorb toxicity or polymer effects. However, polymers used at industrial facilities often pass through the treatment system and adversely affect polymer use at the WRRF for settleability, phosphorus precipitation, or solids dewatering processes.
2. False. Some food processors have high phosphorus and nitrogen compared to BOD and TSS. This is true in wastewaters high in protein, for example. Other food processes, such as sugar beet processing, result in wastewater with very high COD but very low nitrogen and phosphorus.
3. c.) Cooling water takes heat from a process and transfers it to cooling water, thus a high-volume cooling water discharge might have a thermal load that may affect the biological treatment performance or effluent temperature.
4. True. Quaternary ammonium compounds are strong disinfectants that can inhibit activated sludge bacteria, especially nitrifying bacteria that convert ammonia to nitrate.
5. False. The effect of polymer flocculants on WRRF processes is most often interference with flocculants or coagulants used at the WRRF to enhance settling in secondary clarifiers or chemical phosphorus removal processes.

6. True. Landfill leachate discharges have high nutrient concentrations, which can overload the WRRF's capacity to remove specific nutrients like ammonia or phosphorus.
7. True. The example provided in the text saw negative effects when leachate was only 2% of total WRRF influent flow.
8. False. Equalization at an industrial site refers to using a large tank to store industrial wastewater before discharge to balance variability in discharge flow and water quality and minimize effects on downstream wastewater treatment operations.
9. True. These chemicals can provide slug loads of high-pH or low-pH water to the WRRF, if not pH-adjusted before discharge.
10. b.) High heavy metals concentrations. Food and beverage operations typically do not use significant amounts of heavy metals in their operations to limit human exposure through food products.
11. True. Bench testing is often needed to evaluate and verify the efficacy of different chemical treatment options.

TEST YOUR KNOWLEDGE, PAGE 20

1. True. These can be used to remove fine solids and oily material from wastewater.
2. False. "High-strength" refers to high amounts of organics, frequently reflected by high concentrations of COD and/or BOD.
3. False. Heavy metal wastewater should be source separated as much as possible for separate treatment. Source pretreatment processes like chemical precipitation, ion exchange, and membrane separation are more effective with concentrated heavy metal wastewaters.
4. True. Most heavy metals prefer to stick to organic solids rather than staying with the water, thus increasing biosolids metal concentrations. In addition, as solids are digested and reduced in volume, metals concentration will increase.
5. d.) The incentives for high-strength BOD pretreatment for the food industry are to reduce BOD and TSS surcharge costs given the levels in Table 1.3.
6. c.) The primary negative effects of heavy metals on WRRFs are toxicity and biosolids quality. Metals can inhibit activated sludge treatment, and metals tend to accumulate in WRRF sludge solids.
7. False. Chemical precipitation is usually the first step considered for metals pretreatment. Ion exchange is considered a secondary polishing step for metals removal due to its higher cost compared to chemical precipitation.

CHAPTER 2

TEST YOUR KNOWLEDGE, PAGE 28

1. False. The main goal is to prevent workers from being injured.
2. True
3. A
4. C

TEST YOUR KNOWLEDGE, PAGE 29

1. True
2. B
3. True

TEST YOUR KNOWLEDGE, PAGE 35

1. D
2. D

TEST YOUR KNOWLEDGE, PAGE 38

1. True
2. D
3. True
4. C
5. True
6. C

TEST YOUR KNOWLEDGE, PAGE 41

1. B
2. B
3. False. Lockout/tagout is a legal requirement per the OSHA standard for The Control of Hazardous Energy (Lockout/Tagout), Title 29 Code of Federal Regulations (CFR) Part 1910.147 and addresses the practices and procedures necessary to disable machinery or equipment, thereby preventing the release of hazardous energy while employees perform servicing and maintenance activities.

TEST YOUR KNOWLEDGE, PAGE 46

1. True
2. True
3. True
4. C
5. B

TEST YOUR KNOWLEDGE, PAGE 49

1. C
2. A
3. C

TEST YOUR KNOWLEDGE, PAGE 52

1. A
2. C
3. B
4. B

TEST YOUR KNOWLEDGE, PAGE 62

1. False. Equipment and machines do not always come from the manufacturer with the proper guards. These may need to be purchased in addition, or fabricated.
2. C
3. True
4. C
5. True
6. True

7. False. Cranes and hoists are required to be inspected frequently and periodically. Various items will be inspected anywhere from daily to monthly to annually based on OSHA and the manufacturer's instructions.
8. Label the parts of the PIV with the following terms: Fork Mast Overhead Guard Counterweight Tilt Cylinder Load Backrest

9. False. Modified cranes can only be used after the modifications and associated structure are thoroughly checked for the new rated load by a qualified engineer or the equipment manufacturer.
10. False. Powered industrial vehicle training can be a combination of lecture, discussion, interactive computer learning, video, written material, and practical training. An evaluation must also be performed to evaluate the operator's competency to operate.
11. False. Operators/occupants should never increase their height by standing on rails or using a ladder or other means of increasing height. A larger or different type of MEWP should be used instead.

CHAPTER 3

TEST YOUR KNOWLEDGE, PAGE 76

1. A
2. A
3. A
4. D
5. D

TEST YOUR KNOWLEDGE, PAGE 78

1. D
2. C
3. D
4. C

TEST YOUR KNOWLEDGE, PAGE 87

1. D
2. False
3. C
4. A
5. A
6. C
7. C
8. C
9. C
10. B

TEST YOUR KNOWLEDGE, PAGE 96

1. C
2. A
3. C
4. A
5. False. Short-circuiting occurs when water does not mix properly, creating dead zones.
6. True
7. A
8. D
9. B
10. C

TEST YOUR KNOWLEDGE, PAGE 110

1. A
2. D
3. B
4. D
5. A
6. Filter ripening, effective filtration, breakthrough
7. D
8. D
9. D

TEST YOUR KNOWLEDGE, PAGE 117

1. A
2. C
3. A
4. C
5. A
6. B
7. A
8. A
9. A
10. B

TEST YOUR KNOWLEDGE, PAGE 123

1. B
2. A
3. B
4. C
5. B
6. B
7. A
8. A
9. A
10. B

CHAPTER 4

TEST YOUR KNOWLEDGE, PAGE 142

1. A
2. True
3. False
4. False
5. False
6. Floc, filaments, higher life forms, and bulk water
7. False
8. False
9. Protozoa are single celled and metazoa are multicellular.
10. True
11. True
12. Free swimmers have cilia covering the entire organism, whereas the crawler ciliate has fused-together cilia called cirri.
13. Metazoa
14. False
15. True
16. Very large surface-to-volume ratio
17. True
18. False
19. False
20. True
21. True
22. True
23. False
24. True
25. True

TEST YOUR KNOWLEDGE, PAGE 161

1. False
2. False
3. True
4. False
5. False
6. True
7. False
8. False
9. False
10. True
11. True
12. False
13. False
14. False
15. True
16. True

TEST YOUR KNOWLEDGE, PAGE 180

1. MBBR
2. Dissolved oxygen
3. True
4. 2 to 5 days
5. True
6. True
7. False
8. True
9. True
10. True
11. True
12. False
13. True
14. False
15. True
16. False
17. True
18. False
19. False
20. False
21. False
22. False
23. False

CHAPTER 5

TEST YOUR KNOWLEDGE, PAGE 186

1. D
2. D
3. D

TEST YOUR KNOWLEDGE, PAGE 189

1. A
2. False. The factors that affect the ability of ozone to disinfect include target pathogen, the pathogen's exposure to ozone, particulate shielding, and temperature.
3. B

TEST YOUR KNOWLEDGE, PAGE 196

1. A

TEST YOUR KNOWLEDGE, PAGE 199

1. A
2. B
3. D

TEST YOUR KNOWLEDGE, PAGE 200

1. C
2. A
3. A

CHAPTER 6

TEST YOUR KNOWLEDGE, PAGE 205

1. False
2. B
3. D
4. True
5. C
6. B
7. A

TEST YOUR KNOWLEDGE, PAGE 210

1. B
2. False
3. True
4. C
5. True
6. A
7. B

TEST YOUR KNOWLEDGE, PAGE 211

1. True
2. False
3. D
4. A

TEST YOUR KNOWLEDGE, PAGE 215

1. C
2. B
3. C
4. C
5. B
6. D
7. D
8. A
9. B

TEST YOUR KNOWLEDGE, PAGE 222

1. False
2. True
3. C
4. A
5. True
6. D

TEST YOUR KNOWLEDGE, PAGE 224

1. A
2. False
3. B
4. A

TEST YOUR KNOWLEDGE, PAGE 226

1. D
2. False
3. False

TEST YOUR KNOWLEDGE, PAGE 229

1. B
2. B
3. B
4. C
5. D

TEST YOUR KNOWLEDGE, PAGE 231

1. B
2. A
3. B
4. A

TEST YOUR KNOWLEDGE, PAGE 233

1. False
2. A
3. A
4. B

TEST YOUR KNOWLEDGE, PAGE 234

1. True
2. D
3. B

TEST YOUR KNOWLEDGE, PAGE 237

1. C
2. C
3. C
4. B

TEST YOUR KNOWLEDGE, PAGE 241

1. True
2. A
3. B

TEST YOUR KNOWLEDGE, PAGE 245

1. False
2. D
3. A
4. A

TEST YOUR KNOWLEDGE, PAGE 248

1. False. Polyamide reverse osmosis membranes may not be cleaned with sodium hypochlorite due to chemical damage to the membranes.
2. D
3. B
4. B

TEST YOUR KNOWLEDGE, PAGE 250

1. C
2. A
3. B

TEST YOUR KNOWLEDGE, PAGE 252

1. C
2. C
3. B
4. C
5. B

CHAPTER 7

TEST YOUR KNOWLEDGE, PAGE 258

1. True
2. C
3. C

TEST YOUR KNOWLEDGE, PAGE 260

1. False
2. True

TEST YOUR KNOWLEDGE, PAGE 265

1. False
2. True
3. C
4. True

TEST YOUR KNOWLEDGE, PAGE 270

1. True
2. True
3. B

TEST YOUR KNOWLEDGE, PAGE 278

1. True
2. True
3. B
4. A

TEST YOUR KNOWLEDGE, PAGE 284

1. False
2. True
3. True

TEST YOUR KNOWLEDGE, PAGE 286

1. False
2. True

CHAPTER 8

TEST YOUR KNOWLEDGE, PAGE 294

1. D
2. A
3. C
4. A

TEST YOUR KNOWLEDGE, PAGE 298

1. C
2. A
3. C
4. D
5. A

TEST YOUR KNOWLEDGE, PAGE 301

1. B
2. D
3. C
4. C

CHAPTER 9

TEST YOUR KNOWLEDGE, PAGE 310

1. d.) U.S. EPA has mandated requirements to help ensure that biosolids are processed, handled, and treated in a manner that minimizes potential risk to human health.
2. c.) The treatment processes proposed for Class A biosolids eliminate pathogens, including viruses, whereas treatment requirements for Class B biosolids do not completely eliminate pathogens in the biosolids. Class EQ is above Class A in that it also considers regulated pollutants in addition to meeting the pathogen and VAR requirements.
3. a.) The primary federal regulation pertaining to biosolids is 40 CFR Part 503.
4. b.) These state regulations may be more stringent than the federal Part 503 rule and may regulate certain types of biosolids more strictly than the federal regulations.
5. c.) Selenium
6. a.) To minimize unwanted pollutants in the influent, which helps to reduce contaminants throughout the plant
7. c.) The crop needs typically determine which biosolids characteristics are most important, and this is typically the nutrient content (nitrogen, phosphorus, etc.).
8. b.). Land-applied biosolids also must meet VAR requirements via thermal drying, volatile solids reduction (VSR), oxygen uptake rate, time and temperature, or pH increases.

TEST YOUR KNOWLEDGE, PAGE 320

1. d.) An approved permit is the only one of these that is required. The rest in the list are recommended but not required by law at the federal level.
2. a.) All other options should not be part of the program development. The public should be involved throughout, and the site should be visited on a regular basis.
3. b.) Ash is not currently approved for food-grade uses.
4. b.) Hygiene stations provide locations for handwashing. All other selections are not recommended.
5. a.) The application rates are higher in land reclamation because the application is not just to provide nutrients.
6. b.) The production of turf grass has the greatest potential for economic return with thermal treatment. However, fertilizers are a more common option.
7. a.) ISO 14001. ISO 14001 sets out the criteria for an environmental management system and can be certified. It maps out a framework that a company or organization can follow to set up an effective environmental management system.
8. b.) Management and employee representatives should work together to determine which job duties are likely to result in this type of exposure, to conduct appropriate on-site monitoring, and to determine which type of PPE is needed in conjunction with a qualified safety and health professional.

CHAPTER 10

TEST YOUR KNOWLEDGE, PAGE 337

1. B
2. True
3. B
4. True

5. D
6. B
7. C
8. C
9. A

TEST YOUR KNOWLEDGE, PAGE 347

1. C
2. B
3. True
4. B
5. True
6. B
7. A
8. B
9. A

TEST YOUR KNOWLEDGE, PAGE 358

1. True
2. A
3. True
4. B
5. C
6. D
7. C
8. D
9. C

TEST YOUR KNOWLEDGE, PAGE 375

1. D
2. B
3. D
4. A
5. True
6. B
7. C
8. C
9. True
10. B

CHAPTER 11

TEST YOUR KNOWLEDGE, PAGE 389

1. True
2. True
3. D
4. True
5. C
6. D
7. True
8. A
9. A
10. B
11. D
12. D

TEST YOUR KNOWLEDGE, PAGE 391

1. B
2. False. A collection system needs to allow air to enter the system to replace the odorous air that is removed by the conveyance system.

TEST YOUR KNOWLEDGE, PAGE 406

1. B
2. False. Virgin carbon media are activated carbon media that have not been used for treatment or regenerated previously.
3. True
4. B

CHAPTER 12

TEST YOUR KNOWLEDGE, PAGE 416

1. True
2. False. They may also be wall-mounted nearby. Additional remote indicators may be installed in other areas.
3. B
4. A
5. D
6. C
7. B
8. B

TEST YOUR KNOWLEDGE, PAGE 419

1. C
2. A
3. B
4. D
5. C

TEST YOUR KNOWLEDGE, PAGE 423

1. B
2. A
3. D
4. C
5. B

TEST YOUR KNOWLEDGE, PAGE 431

1. C
2. D
3. B
4. B
5. A
6. C
7. A
8. D
9. C
10. B
11. A

TEST YOUR KNOWLEDGE, PAGE 434

1. C
2. A
3. B
4. B

TEST YOUR KNOWLEDGE, PAGE 445

1. C
2. B
3. D
4. C
5. False. When the liquid contacts the bottom of the conductance rod or sensor, it activates the switch.
6. False. Bubbler systems are an older technology that has fallen out of favor due to high maintenance costs.
7. A
8. C
9. B
10. True
11. C
12. C
13. A
14. C
15. A
16. True

TEST YOUR KNOWLEDGE, PAGE 453

1. B
2. A
3. D
4. C
5. D
6. True
7. A
8. D
9. B
10. True
11. True

TEST YOUR KNOWLEDGE, PAGE 464

1. True
2. False. Dissolved oxygen does not pass easily through coated membranes.
3. False. Light output decreases with increasing light output.
4. True
5. D
6. A
7. C
8. C
9. B
10. A
11. B
12. C
13. D
14. B
15. D
16. A
17. D
18. B

TEST YOUR KNOWLEDGE, PAGE 471

1. A
2. C
3. C
4. D
5. B
6. D
7. C
8. A
9. B

TEST YOUR KNOWLEDGE, PAGE 477

1. B
2. C
3. D
4. A
5. B
6. C
7. C
8. A
9. B
10. D

TEST YOUR KNOWLEDGE, PAGE 480

1. False. Oxidation–reduction potential readings depend on many factors. Similar probes can give different results.
2. True
3. False. Oxidation–reduction potential readings are difficult to correct for temperature. They do not incorporate temperature sensors.
4. A
5. D
6. B

TEST YOUR KNOWLEDGE, PAGE 485

1. True
2. True
3. A
4. D
5. B
6. C
7. C
8. D

TEST YOUR KNOWLEDGE, PAGE 488

1. True
2. True
3. B
4. A

TEST YOUR KNOWLEDGE, PAGE 494

1. D
2. A
3. C
4. B

5. A
6. C
7. D
8. A
9. False. All sensors will fail. Bump tests help us determine exactly when they need to be replaced.
10. True
11. B

CHAPTER 13

TEST YOUR KNOWLEDGE, PAGE 504

1. Level 1: Operator
   Level 2: SCADA System
   Level 3: Instrumentation Level
2. Component 1 is a computer.
   Component 2 is networking equipment.
   Component 3 is a controller or PLC.
   Component 4 is an input/output device.

TEST YOUR KNOWLEDGE, PAGE 509

1. False. Supervisory control and data acquisition equipment differs because it is designed to be highly reliable and provide accurate and timely data in an industrial environment.
2. True
3. True
4. False. Only authorized devices should be connected to the facility SCADA network.

TEST YOUR KNOWLEDGE, PAGE 513

1. B

TEST YOUR KNOWLEDGE, PAGE 514

1. A

TEST YOUR KNOWLEDGE, PAGE 517

1. In navigation, "drill down" is the concept of moving between layers of displays (i.e., more general displays at the top and more specific displays at the bottom).
2. Lateral navigation is the idea of moving between displays of the same level without first having to move to a higher-level display. Lateral navigation is important because it reduces the number of clicks needed to navigate between processes.

TEST YOUR KNOWLEDGE, PAGE 519

1. B

CHAPTER 14

TEST YOUR KNOWLEDGE, PAGE 528

1. False. Maximum Achievable Control Technology 129 relates to the Section 129 requirement for SSI emissions.
2. False. Industrial pretreatment permits are required for nonresidential dischargers who are either a categorical industry or a significant user.
3. False. All WRRFs are required to submit a monthly DMR summary report regardless of any violations.
4. True
5. False. The Clean Water Act regulates point source pollution through the NDPES program. The Safe Drinking Water Act regulates drinking water, not wastewater.
6. True
7. A
8. D

TEST YOUR KNOWLEDGE, PAGE 531

1. C
2. B
3. True

TEST YOUR KNOWLEDGE, PAGE 536

1. C
2. True
3. B
4. A

TEST YOUR KNOWLEDGE, PAGE 539

1. d.) Standard operating procedures should be used by both new and experienced employees. Not every employee needs to be aware of every SOP. Rather, employees associated with the action described in the SOP should be considered the audience.

TEST YOUR KNOWLEDGE, PAGE 542

1. False. Preventive maintenance is done while the equipment is operating normally.
2. True
3. True
4. False. The failure of critical equipment can be extremely costly. An asset management program can help determine which critical equipment should be replaced before failure and which equipment can be run to failure.

TEST YOUR KNOWLEDGE, PAGE 546

1. B
2. B
3. C
4. B
5. True

TEST YOUR KNOWLEDGE, PAGE 551

1. C
2. D
3. D
4. C
5. False. Vulnerability assessments consider threats from both human sources and natural disasters—anything that may damage an asset or disrupt operation.
6. False. National Incident Management System and ICS are valuable tools for managing any kind of incident, especially when working with other first responders like the police.

7. D
8. a = 2, b = 4, c = 3, d = 5, e = 1
9. C
10. C

TEST YOUR KNOWLEDGE, PAGE 555

1. A
2. B
3. A
4. B
5. C
6. C
7. B

TEST YOUR KNOWLEDGE, PAGE 559

1. B
2. A
3. D
4. C
5. A

TEST YOUR KNOWLEDGE, PAGE 563

1. False. All complaints should be handled courteously and promptly.
2. d.) Do not take visitors to electrical rooms, chemical storage areas, or on structures without handrails. Take tour groups to unit processes that can be accessed safely.
3. False. The operator should describe the facility in understandable terms.
4. A

Periodic Table

PERIODIC TABLE

group

element name → HYDROGEN
atomic number → 1
chemical Symbol → H
atomic weight (u) → 1,008

Alkali metal
Alkaline earth metal
Lanthanide
Actinide
Transition metal
Post-transition metal
Metalloid
Nonmetal
Halogen
Noble gas
● Solid
● Liquid
● Gas
● Unknown

period	1	2	3	4	5	6	7	8	9	10	11	12	13	14	15	16	17	18
	IA	IIA	IIIB	IVB	VB	VIB	VIIB	VIIIB	VIIIB	VIIIB	IB	IIB	IIIA	IVA	VA	VIA	VIIA	VIIIA
1	HYDROGEN 1 H 1,008																	HELIUM 2 He 4,003
2	LITHIUM 3 Li 6,941	BERYLLIUM 4 Be 9,012											BORON 5 B 10,81	CARBON 6 C 12,01	NITROGEN 7 N 14,01	OXYGEN 8 O 16,00	FLUORINE 9 F 19,00	NEON 10 Ne 20,18
3	SODIUM 11 Na 22,99	MAGNESIUM 12 Mg 24,31											ALUMINIUM 13 Al 26,98	SILICON 14 Si 28,09	PHOSPHORUS 15 P 30,97	SULFUR 16 S 32,07	CHLORINE 17 Cl 35,45	ARGON 18 Ar 39,95
4	POTASSIUM 19 K 39,10	CALCIUM 20 Ca 40,08	SCANDIUM 21 Sc 44,96	TITANIUM 22 Ti 47,87	VANADIUM 23 V 50,94	CHROMIUM 24 Cr 52,00	MANGANESE 25 Mn 54,94	IRON 26 Fe 55,85	COBALT 27 Co 58,93	NICKEL 28 Ni 58,69	COPPER 29 Cu 63,55	ZINC 30 Zn 65,39	GALLIUM 31 Ga 69,72	GERMANIUM 32 Ge 72,59	ARSENIC 33 As 74,92	SELENIUM 34 Se 78,96	BROMINE 35 Br 79,90	KRYPTON 36 Kr 83,80
5	RUBIDIUM 37 Rb 85,47	STRONTIUM 38 Sr 87,62	YTTRIUM 39 Y 88,91	ZIRCONIUM 40 Zr 91,22	NIOBIUM 41 Nb 92,91	MOLYBDENUM 42 Mo 95,94	TECHNETIUM 43 Tc (98,91)	RUTHENIUM 44 Ru 101,1	RHODIUM 45 Rh 102,9	PALLADIUM 46 Pd 106,4	SILVER 47 Ag 107,9	CADMIUM 48 Cd 112,4	INDIUM 49 In 114,8	TIN 50 Sn 118,7	ANTIMONY 51 Sb 121,8	TELLURIUM 52 Te 127,6	IODINE 53 I 126,9	XENON 54 Xe 131,3
6	CAESIUM 55 Cs 132,9	BARIUM 56 Ba 137,3	LANTHANUM 57 La 138,9	HAFNIUM 72 Hf 178,5	TANTALUM 73 Ta 180,9	TUNGSTEN 74 W 183,9	RHENIUM 75 Re 186,2	OSMIUM 76 Os 190,2	IRIDIUM 77 Ir 192,2	PLATINUM 78 Pt 195,1	GOLD 79 Au 197,0	MERCURY 80 Hg 200,6	THALLIUM 81 Tl 204,4	LEAD 82 Pb 207,2	BISMUTH 83 Bi 209,0	POLONIUM 84 Po (210,0)	ASTATINE 85 At (210,0)	RADON 86 Rn (222,0)
7	FRANCIUM 87 Fr (223,0)	RADIUM 88 Ra (226,0)	ACTINIUM 89 Ac (227,0)	RUTHERFORDIUM 104 Rf ---	DUBNIUM 105 Db ---	SEABORGIUM 106 Sg ---	BOHRIUM 107 Bh ---	HASSIUM 108 Hs ---				COPERNICIUM 112 Cn ---						
Lanthanides					CERIUM 58 Ce 140,1	PRASEODYMIUM 59 Pr 140,9	NEODYMIUM 60 Nd 144,2	PROMETHIUM 61 Pm (144,9)	SAMARIUM 62 Sm 150,4	EUROPIUM 63 Eu 152,0	GADOLINIUM 64 Gd 157,3	TERBIUM 65 Tb 158,9	DYSPROSIUM 66 Dy 162,5	HOLMIUM 67 Ho 164,9	ERBIUM 68 Er 167,3	THULIUM 69 Tm 168,9	YTTERBIUM 70 Yb 173,0	LUTETIUM 71 Lu 175,0
Actinides					THORIUM 90 Th (232,0)	PROTACTINIUM 91 Pa (231,0)	URANIUM 92 U (238,0)	NEPTUNIUM 93 Np (237,0)	PLUTONIUM 94 Pu (239,1)	AMERICIUM 95 Am (243,1)	CURIUM 96 Cm (247,1)	BERKELIUM 97 Bk (247,1)	CALIFORNIUM 98 Cf (252,1)	EINSTEINIUM 99 Es (252,1)	FERMIUM 100 Fm (257,1)	MENDELEVIUM 101 Md (256,1)	NOBELIUM 102 No (259,1)	LAWRENCIUM 103 Lr (260,1)

Index